physica status solidi c
conferences and critical reviews
www.pss-c.com

Editor-in-Chief

Martin Stutzmann, Garching

Regional Editors

Martin S. Brandt, Garching
Peter Deák, Budapest
José Roberto Leite, Saõ Paulo
John I. B. Wilson, Edinburgh

Managing Editor

Stefan Hildebrandt, Berlin

Proceedings

Third International Conference on
Magnetic and Superconducting Materials (MSM '03)

Monastir, Tunisia
1–4 September 2003

Guest Editors

M. Akhavan, E. V. Antipov, B. Barbara, A. Barone, J. Bass, A. Cheikhrouhou,
H.-U. Habermeier, and S. Mahmood

1 · 7 · 2004

physica status solidi (c) – conferences and critical reviews

Editor-in-Chief:	Martin Stutzmann
Managing Editor:	Stefan Hildebrandt
Production Editors:	André Danelius, Heike Höpcke, Irina Juschak
Editorial Assistance:	Katharina Fröhlich, Margit Schütz
Editorial Office:	physica status solidi Bühringstr. 10, 13086 Berlin, Germany Telephone: +49 (0) 30/47 03 13 31, Fax +49 (0) 30/47 03 13 34 e-mail: pss@wiley-vch.de
Publishers:	WILEY-VCH Verlag GmbH & Co. KGaA
Postal Address:	Bühringstr. 10, 13086 Berlin, Germany
Publishing Director:	Alexander Grossmann
Ordering:	Subscription Service, WILEY-VCH Verlag GmbH & Co. KGaA Postfach 10 11 61, 69451 Weinheim, Germany Telephone +49 (0) 62 01/60 64 00, Fax +49 (0) 62 01/60 61 84 e-mail: subservice@wiley-vch.de or through a bookseller
Printing House:	Druckhaus Thomas Müntzer GmbH, Bad Langensalza, Germany Printed on chlorine- and acid free paper.

physica status solidi (c) – conferences and critical reviews is published twelve times per year by WILEY-VCH Verlag GmbH & Co. KGaA.

Annual subscription rates 2004 for pss (a) and (c) *or* pss (b) and pss (c):

		Institutional*	Personal
Europe	Euro	4244/3858	298
Switzerland	SFr	7269/6608	444
All other areas	US$	5406/4914	394

Annual subscription rates 2004 for pss (a), pss (b) and pss (c):

		Institutional*	Personal
Europe	Euro	8488/7716	596
Switzerland	SFr	14538/13216	888
All other areas	US$	10812/9828	788

* print **and** electronic/print only **or** electronic only
pss (a) and/or pss (b) subscriptions: Now including pss (c) – conferences and critical reviews

Postage and handling charges included. All WILEY-VCH prices are exclusive of VAT.
Prices are subject to change.

Single print issues may be ordered by ISBN at www.wiley-vch.de or through your local bookseller.

ISBN 3-527-40522-4

© 2004 WILEY-VCH Verlag GmbH & Co. KGaA, Weinheim

All rights reserved (including those of translations into foreign language). No part of this issue may be reproduced in any form, by photoprint, microfilm or any other means, nor transmitted into a machine language, without written permission from the publisher.

For our American customers:
physica status solidi (c) – conferences and critical reviews (ISSN 1610-1634) is published twelve times per year by WILEY-VCH Verlag GmbH & Co. KGaA, Boschstr. 12, 69469 Weinheim, Germany. Periodicals postage paid at Jamaica, NY 11431. Air freight and mailing in the USA by Publications Expediting Service Inc., 200 Meacham Ave., Elmont, NY 11003. US Postmaster: send address changes to: physica status solidi (c), c/o WILEY-VCH, 111 River Street, Hoboken, NJ 07030.

Valid for users in the USA:
The appearance of the code at the bottom of the first page of an article (serial) indicates the copyright owner's consent that copies of the article may be made for personal or internal use, or for the personal or internal use of specific clients. This consent is given on the condition, however, that the copier pay the stated per-copy fee through the Copyright Clearance Center, Inc. (CCC) for copying beyond that permitted by Sections 107 or 108 of the U.S. Copyright Law. This consent does not extend to other kinds of copying, such as copying for general distribution, for advertising or promotional purposes, for creating new collective works, or for resale. For copying from back volumes of this journal see 'Permissions to Photo Copy: Publishers Fee List' of the CCC.

Contents

Full text on our homepage at: http://www.pss-c.com

This Table of Contents is organized according to the topics presented at the conference. Articles with page numbers marked (a) and (b) are reprinted from phys. stat. sol. (a) **201**, No. 7, 1379–1442 (2004) and (b) **241**, No. 6, 1167–1250 (2004), respectively. You may find papers with phys. stat. sol. (a), phys. stat. sol. (b), and phys. stat. sol. (c) citations in the three sections of this volume, separated by coloured sheets for easy orientation.
Note from the Publisher: This issue of physica status solidi (c) has been produced from publication-ready manuscript files, written by the authors using the provided Word or LaTeX templates.

Preface	1575
Committees	1576

Theoretical magnetism

Quantum dynamics of atomic magnets *(invited)*
    B. Barbara, R. Giraud, and A. M. Tkachuk . . . . (b) 1167–1173

Persistent current and Hall effect due to spin chirality *(invited)*
    G. Tatara . . . . (b) 1174–1179

Energy-level diagrams of high-spin and low-spin molecules *(invited)*
    H. De Raedt, S. Miyashita, and K. Michielsen . . . . (b) 1180–1185

Current-driven excitations in magnetic multilayers: a brief review *(invited)*
    J. Bass, S. Urazhdin, Norman O. Birge, and W. P. Pratt Jr. . . . . (a) 1379–1385

The Fe_8 molecular magnet: A proving ground for the semiclassical theory of spin *(invited)*
    Anupam Garg . . . . (c) 1581–1586

Spin-waves in ferromagnetic double layers: effect of a lateral patterning
    N. Sergeeva, S.-M. Chérif, A. Stachkevitch, M. Kostylev, and Y. Roussigné . . . . (c) 1587–1590

Brillouin light scattering study of magnetic dots
    S.-M. Chérif, S.-Y. Roussigné, and P. Moch . . . . (c) 1591–1594

Field-induced phase transitions (FIPT) in molecular magnets
    B. Barbara, V. V. Kostyuchenko, A. S. Mischenko, and A. K. Zvezdin . . . . (c) 1595–1599

Spin-accumulation contribution to the magnetic nanobridge magnetoresistance
    A. K. Zvezdin, K. A. Zvezdin, and D. Pullini . . . . (c) 1600–1603

Tunable refractive index of magnetic fluids and its applications
    Chin-Yih Hong, H. E. Horng, and S. Y. Yang . . . . (c) 1604–1609

Relationship between the giant magnetoimpedance effect and the relaxation of magnetic permeability
    J. Íñiguez, V. Raposo, D. García, O. Montero, P. Hernández-Gómez, and C. de Francisco . (c) 1610–1613

Magnetic materials: fabrication, characterization, properties

Substrate surface engineering for tailoring properties of functional ceramic thin films *(invited)*
    H.-U. Habermeier . . . . (c) 1614–1619

Optical and magneto-optical study of the Au/Co/Au/Cu multilayer grown on vicinal Si(111) surfaces *(invited)*
    W. Cheikh-Rouhou Koubaa, B. Bartenlian, P. Beauvillain, A. Brun, P. Georges,
    T. Maroutian, and V. Mathet . . . . (c) 1620–1624

Magnetisation reversal dynamics in an ultrathin magnetic film and the creep phenomenon *(invited)*
    J. Ferré, V. Repain, J.-P. Jamet, A. Mougin, V. Mathet, C. Chappert, and H. Bernas . . . . (a) 1386–1391

Spintronics: perspectives for the half-metallic oxides *(invited)*
    A. M. Haghiri-Gosnet, T. Arnal, R. Soulimane, M. Koubaa, and J. P. Renard. . . . . . . (a) 1392–1397

Crystal chemistry of non-perovskite manganese oxides – implications for magnetic properties *(invited)*
    P. Strobel, A. Ibarra-Palos, M. Pernet, S. Zouari, W. Cheikh-Rouhou,
    and A. Cheikh-Rouhou . . . . (c) 1625–1630

Liquid phase epitaxy (LPE) grown Bi, Ga, Al substituted iron garnets with huge Faraday rotation for magneto-optic applications *(invited)*
    P. Görnert, T. Aichele, A. Lorenz, R. Hergt, and J. Taubert . . . . (a) 1398–1402

Oxygen and fluorine doping in Sr_2MnGaO_5 brownmillerite *(invited)*
    E. V. Antipov, A. M. Abakumov, A. M. Alekseeva, M. G. Rozova, J. Hadermann,
    O. I. Lebedev, and G. Van Tendeloo . . . . (a) 1403–1409

Effect of Fe doping on the structural and magneto transport properties in $Pr_{0.67}Sr_{0.33}MnO_3$ perovskite manganese *(invited)*
    W. Boujelben, M. Ellouze, A. Cheikh-Rouhou, R. Madar, and H. Fuess . . . . (a) 1410–1415

Deficiency effects on the physical properties of the lacunar $La_{0.5}Ca_{0.5-x}MnO_3$ manganese oxides *(invited)*
    I. Walha, W. Boujelben, M. Koubaa, A. Cheikh-Rouhou, and A. M. Haghiri-Gosnet . . . (a) 1416–1420

Magnetic and electrical properties of the lacunar $La_{0.7}Ca_{0.3-x}\square_xMnO_3$ and $La_{0.7-x}\square_xCa_{0.3}MnO_3$ oxides *(invited)*
    I. Kamoun, W. Boujelben, A. Cheikh-Rouhou, H. Roussel, and R. Madar . . . . (c) 1631–1636

^{55}Mn NMR study in magnetically ordered state of perovskite manganites *(invited)*
    K. Shimizu, W. Boujelben, and A. Cheikh-Rouhou . . . . (a) 1421–1427

Neutron diffraction studies of magnetic and superconducting compounds *(invited)*
    W. B. Yelon, Q. Cai, W. J. James, H. U. Anderson, J. B. Yang, X. D. Zhou,
    and H. A. Blackstead . . . . (a) 1428–1435

Observations of magnetic domain structures and phase segregation in single-crystal $Nd_{1/2}Sr_{1/2}MnO_3$ using X-ray scattering
    Mohammad E. Ghazi and P. D. Hatton . . . . (c) 1637–1640

Jahn-Teller distortion ordering in single-crystal $Nd_{1/2}Sr_{1/2}MnO_3$
    Mohammad E. Ghazi and P. D. Hatton . . . . (c) 1641–1644

Fe doping effects on the structural and magnetic properties in $Pr_{0.5}Sr_{0.5}Mn_{1-x}Fe_xO_3$ with $0 \le x \le 0.3$
    A. Ammar, S. Zouari, and A. Cheikh-Rouhou . . . . (c) 1645–1648

Magnetism and giant magnetoresistance in $La_{0.7}Sr_{0.3}Mn_{1-x}M_xO_3$ (M = Cr, Ti) systems
    N. Kallel, K. Fröhlich, M. Oumezzine, M. Ghedira, H. Vincent, and S. Pignard . . . . . (c) 1649–1654

Effect of Fe doping on the physical properties of LaKMn$_{1-x}$Fe$_x$MoO$_6$ double perovskite with $0 \leq x \leq 0.2$
    S. Megdiche, M. Ellouze, A. Cheikh-Rouhou, and R. Madar . . . . (c) 1655–1659

Ferromagnetic behaviour in PrKMnMoO$_6$ double perovskite oxide
    S. Megdiche, M. Ellouze, A. Cheikh-Rouhou, and R. Madar . . . . (c) 1660–1663

Study of the temperature dependence of magnetic permeability, selectivity and D.C. resistivity of LaKMnMoO$_6$
    S. A. Saafan, S. Megdiche, M. A. Elkestawy, and A. Cheikh-Rouhou . . . . (c) 1664–1668

Structural and magnetic study of the double-perovskites Ba$_2$(Fe, B′)$_2$O$_6$ (B′ = Mo, W and Re)
    N. Rammeh, K. G. Bramnik, H. Ehrenberg, C. Ritter, H. Fuess, and A. Cheikh-Rouhou . . (c) 1669–1674

Effect of praseodymium doping on the structural and magnetic properties of La$_{1.2-x}$Pr$_x$Sr$_{1.8}$Mn$_2$O$_7$ bilayer manganese oxides
    M. Triki, S. Zouari, N. Chniba, and A. Cheikhrouhou . . . . (c) 1675–1678

Synthesis and characterization of SmNiO$_3$ thin films
    N. Ihzaz, S. Pignard, J. Kreisel, H. Vincent, J. Marcus, J. Dhahri, and M. Oumezzine . . . (c) 1679–1682

High field induced spin reorientations in Ho$_{0.24}$Y$_{2.76}$Fe$_5$O$_{12}$
    A. Bouguerra, S. Khène, and G. Fillion . . . . (c) 1683–1686

Optimized lithography and etching processes for a magnetic oxide micro-device
    R. Soulimane, M. Koubaa, A. M. Haghiri-Gosnet, B. Mercey, W. Prellier, Ph. Lecoeur,
    G. Poullain, and R. Bouregba . . . . (c) 1687–1690

Magnetic and structural properties of intermetallic compounds Nd$_{2-x}$R$_x$Fe$_{17}$ (R = Sm, Gd)
    M. S. Ben Kraiem, M. Ellouze, A. Cheikh-Rouhou, and Ph. L'Héritier . . . . (c) 1691–1696

Chemical hydrogenation effects on R$_2$Fe$_{14}$B compounds with (R = Ce, Nd and Gd)
    M. S. Ben Kraiem, M. Ellouze, A. Cheikh-Rouhou, and Ph. L'Héritier . . . . (c) 1697–1700

Magnetic and structural properties of Sm$_{1.5}$Gd$_{0.5}$Fe$_{17-x}$Co$_x$N$_y$ with $0 \leq x \leq 3$ and $y \approx 3$
    M. S. Ben Kraiem, M. Ellouze, A. Cheikh-Rouhou, and Ph. L'Héritier . . . . (c) 1701–1705

The crystallographic and magnetic properties of Nd$_{2-x}$Gd$_x$Fe$_{16}$Co solid solution and its nitrides
    M. S. Ben Kraiem, M. Ellouze, A. Cheikh-Rouhou, and Ph. L'Héritier . . . . (c) 1706–1710

Structural and magnetic studies of 1% Ho:Gd$_{0.99-x}$Lu$_x$ alloys
    I. A. Al-Omari, A. Rais, M. S. Lataifeh, and A. A. Yousif . . . . (c) 1711–1715

Structure and hyperfine parameters of nanocrystalline R_{1-s}(Fe, M)$_{5+2s}$
    A. Nandra, S. Sab, E. Dorolti, L. Bessais, and C. Djéga-Mariadassou . . . . (c) 1716–1718

Magnetostriction and thermal expansion of polymer-bonded Nd$_4$Fe$_{77.5}$B$_{18.5}$ nano-composite
    M. R. Alinejad, A. Amirabadizadeh, N. Tajabor, F. Pourarian, and H. Kanekiyo . . . . . (c) 1719–1723

A comparative study on magnetostrictive strain in GdCo$_5$ and Gd$_{0.9}$Pr$_{0.1}$Co$_5$
    A. Amirabadizadeh, N. Tajabor, M. R. Alinejad, and F. Pourarian . . . . (c) 1724–1727

Structure and electrical resistivity of Gd$_{1-x}$Pr$_x$Co$_5$ compounds
    A. Amirabadizadeh, N. Tajabor, M. R. Alinejad, and F. Pourarian . . . . (c) 1728–1731

Grain size effect on magnetic properties of Fe–28Cr–15Co permanent magnets as a function of Mo content
    Z. Ahmed and A. Ul Haq . . . . (c) 1732–1735

Giant magnetoresistance and microstructure of FeCo–Al_2O_3 nanogranular films
    Changzheng Wang, Zhenghong Guo, Yonghua Rong, and T. Y. Hsu (Xu Zuyao) . . . . . (c) 1736–1739

Structural, magnetic and magneto-transport properties of thermally evaporated Fe/Cu multilayers
    K. Bouziane, M. Al-Busaidi, A. Gismelseed, and A. Al-Rawas . . . . (c) 1740–1743

Co surface modification by bias sputtering in $Cu/Co(V_b)/NiO/Si(100)$ magnetic multilayer structures
    A. Z. Moshfegh and P. Sangpour . . . . (c) 1744–1747

Growth process and characterization of magnetic semiconductors based on GeMn alloy films
    N. Pinto, L. Morresi, R. Murri, F. D'Orazio, F. Lucari, M. Passacantando, and P. Picozzi . (c) 1748–1751

Magnetoresistance effect and magnetoanisotropy of Co/Cu multilayered films prepared by electron beam evaporation
    Y. Ueda, H. Adachi, W. Takakura, C. L. S. Rizal, and S. Chikazawa . . . . (c) 1752–1755

Magnetic properties and magnetoresistance effect in Co/Au, Ag nano-structure films produced by pulse electrodeposition
    C. L. S. Rizal, A. Yamada, Y. Hori, S. Ishida, M. Matsuda, and Y. Ueda . . . . (c) 1756–1759

Magnetism of Ni overlayers on Fe(111)
    Naseem T. Shawagfeh and Jamil M. Khalifeh . . . . (c) 1760–1764

Stability and magnetism of Ni–Fe alloyed overlayers on Fe(001)
    Naseem T. Shawagfeh and Jamil M. Khalifeh . . . . (c) 1765–1768

The effect of Bi mole ratio on phase formation in $Bi_xY_{3-x}Fe_5O_{12}$ nanoparticles
    J. Amighian, A. Hasanpour, and M. Mozaffari . . . . (c) 1769–1771

The coupling between antiferromagnetic transition and martensitic transformation in γ-MnFe based alloys
    Jihua Zhang, Wenyi Peng, Ping Lu, and T.Y. Hsu (Xu Zuyao) . . . . (c) 1772–1775

The longitudinally driven giant magneto-impedance effect of a Co-based amorphous ribbon
    Jianchao Zheng, Chengyuan Dong, Shipu Chen, and T. Y. Hsu (Xu Zuyao) . . . . . . (c) 1776–1779

Influence of Nb-addition on LDGMI effect in CoFeSiB amorphous ribbons
    Chengyuan Dong, Shipu Chen, and T.Y. Hsu (Xu Zuyao) . . . . (c) 1780–1783

Influence of field direction on magnetization measurement for NbTi wire
    Dali Mao, Ling Jiang, and Chengkang Chang . . . . (c) 1784–1787

Anomalous behavior of electrical resistivity in $NdFe_{11}Ti$
    N. Tajabor, A. Amirabadizadeh, M. R. Alinejad, and F. Pourarian . . . . (c) 1788–1791

Investigation of the garnet–perovskite transition in Nd doped YIG by means of magnetic disaccommodation
    P. Hernández-Gómez, C. De Francisco, C. Torres, J. Iñiguez, V. Raposo,
    J. M. Perdigao, and A. R. Ferreira . . . . (c) 1792–1795

Magnetism of iron in face-centered cubic 4d metals
    M. Elzain, A. Al Rawas, A. Yousif, A. Gismelseed, A. Rais, I. Al Omari, and H. Widatallah (c) 1796–1799

Coercive properties of epitaxial magnetic garnet films after heat treatment in reducing atmosphere
    G. Vértesy and I. Tomáš . . . . (c) 1800–1804

Mössbauer spectroscopy

On the hyperfine and magnetic properties of the alloy system $Fe_{3-x}Mn_xSi$ *(invited)*
  S. H. Mahmood, A.-F. Lehlooh, A. S. Saleh, and F. E. Wagner . . . . (b) 1186–1191

Mössbauer studies of the mechanically alloyed Cu–30 at% Fe
  I. A. Al-Omari. . . . . (c) 1805–1808

Mössbauer and structural studies of $Fe_{0.7-x}V_xSi_{0.3}$ alloy system
  I. A. Al-Omari and H. H. Hamdeh . . . . (c) 1809–1812

Theoretical superconductivity

Effects of d-wave symmetry in high-T_C grain boundary Josephson junctions *(invited)*
  Antonio Barone, Filomena Lombardi, Antonia Monaco, Ettore Sarnelli, Francesco Tafuri,
  and Gianluca Testa. . . . . (b) 1192–1198

First-principles study on the creation of holes in high T_c cuprates *(invited)*
  C. Ambrosch-Draxl, E. Ya. Sherman, H. Auer, and T. Thonhauser . . . . (b) 1199–1203

Magnetic resonant excitations in high-T_c superconductors *(invited)*
  Y. Sidis, S. Pailhés, B. Keimer, P. Bourges, C. Ulrich, and L. P. Regnault . . . . . . (b) 1204–1210

Coexistence of superconductivity and magnetism in low dimensional conductors *(invited)*
  A. BenAli, S. Charfi-Kaddour, C. Pasquier, R. Bennaceur, and M. Héritier. . . . . . (b) 1211–1215

Quasi-one dimensional organic conductors: interplay between a magnetic field and the dimensionality *(invited)*
  S. Haddad, S. Charfi-Kaddour, C. Nickel, M. Héritier, and R. Bennaceur . . . . . . (b) 1216–1222

Effects of magnetic field on the cuprate high-T_c superconductor $La_{2-x}Sr_xCuO_4$ *(invited)*
  B. Lake, G. Aeppli, N. B. Christensen, K. Lefmann, D. F. McMorrow, K. N. Clausen,
  H. M. Rønnow, P. Vordewisch, P. Smeibidl, M. Mankorntong, T. Sasagawa, M. Nohara,
  H. Takagi, and N. E. Hussey. . . . . (b) 1223–1228

Normal state understanding within a Fermi liquid approach *(invited)*
  I. Sfar, S. Charfi-Kaddour, M. Héritier, and R. Bennaceur . . . . (b) 1229–1235

Normal state properties of BEDT compounds *(invited)*
  N. Joo, D. Meddeb, S. Charfi-Kaddour, R. Bennaceur, and M. Héritier . . . . (b) 1236–1241

Superconducting phase in κ-$(BEDT-TTF)_2X$ compounds
  R. Charguia, A. Ben Ali, S. Charfi-Kaddour, R. Bennaceur, and M. Héritier . . . . . (c) 1813–1816

Superconducting phase of fullerite family
  D. Meddeb, S. Charfi-Kaddour, R. Bennaceur, and M. Héritier . . . . (c) 1817–1820

Temperature dependence of vortex flux pinning in melt-textured superconductors
  I. A. Al-Omari, M. K. Hasan (Qaseer), A. Rais, and K. A. Azez . . . . (c) 1821–1823

Possible coexistence of antiferromagnetism, spin–glass, and superconductivity in $ScFe_4Al_8$ and YFe_4Al_8 single crystals
  V. M. Dmitriev, J. Stępień-Damm, W. Suski, E. Talik, and N. N. Prentslau . . . . . (c) 1824–1827

High temperature superconductors as a two-dimensional electron gas
  M. R. Mohammadizadeh and M. Akhavan. . . . . (c) 1828–1831

Advances in doping MgB_2: tuning the Fermi level to the "shape resonance" by Sc substitution
S. Agrestini, C. Metallo, M. Filippi, G. Campi, C. Sanipoli, A. Saccone, S. De Negri,
M. Giovannini, A. Latini, and A. Bianconi . . . . (c) 1832–1835

The role of spin diffusion quasiparticle in CMR/HTSC heterostructures
S. Soltan, J. Albrecht, G. Cristani, and H.-U. Habermeier . . . . (c) 1836–1839

Electrical resistivity of magnetic fluids
B. A. Al Shalabi and H. M. El-Ghanem . . . . (c) 1840–1845

Superconducting materials: fabrication, characterization, properties

$YBa_2Cu_3O_7/La_{2/3}Ca_{1/3}MnO_3$ superlattices showing simultaneously ferromagnetic and superconducting order *(invited)*
H.-U. Habermeier and G. Cristiani . . . . (a) 1436–1440

Ba@Pr or Pr@Ba in R123 HTSC – To be or not to be SC *(invited)*
M. Akhavan. . . . . (b) 1242–1250

The BM_5Se_9 phases (B = Al, Ga, Ge, Sb, Sn; M = V, Nb, Ta): superconductors or ferromagnets?
A. Leblanc-Soreau, P. Molinié, and J. C. Jumas . . . . (c) 1846–1850

Conduction mechanism in Pr-doped $GdBa_2Cu_3O_7$
M. R. Mohammadizadeh and M. Akhavan. . . . . (c) 1851–1854

Role of Pr doping in Gd1113
M. Kariminezhad, H. Khosroabadi, and M. Akhavan . . . . (c) 1855–1858

High pressure effects in YBCO and YSCO
H. Khosroabadi, B. Mossalla, and M. Akhavan. . . . . (c) 1859–1862

Hole carrier transfer by apical oxygen in YBCO
H. Khosroabadi and M. Akhavan. . . . . (c) 1863–1866

Role of Pr/Ba disorder in Pr123 superconductor
H. Khosroabadi, M. Modarreszadeh, P. Taheri, and M. Akhavan . . . . (c) 1867–1870

Electrical and magnetic properties of Pr-doped Nd123
P. Maleki, H. Khosroabadi, and M. Akhavan . . . . (c) 1871–1874

Normal state conduction and TAFC in Gd(BaLn)123 (Ln = La, Nd)
M. Mirzadeh, H. Khosroabadi, and M. Akhavan . . . . (c) 1875–1878

Appearance of a new superconducting phase in $Gd(Ba_{2-x}Pr_x)Cu_3O_{7+\delta}$
M. R. Mohammadizadeh and M. Akhavan. . . . . (c) 1879–1882

Effects of Pr doping and magnetic field on vortex pinning in Gd-123 based HTSC
M. R. Mohammadizadeh and M. Akhavan. . . . . (c) 1883–1886

Magnetic field effects on electrical resistivity of $(Gd_{1-x}Pr_x)Ba_2Cu_3O_{7-\delta}$ and $Gd(Ba_{2-x}Pr_x)Cu_3O_{7+\delta}$
M. R. Mohammadizadeh and M. Akhavan. . . . . (c) 1887–1890

Thermally activated phase slip and variable range hopping in $Tm(Ba_{2-x}Pr_x)Cu_3O_{7+\delta}$
Z. Mokhtari, H. Khosroabadi, and M. Akhavan . . . . (c) 1891–1894

Fabrication of BSCCO thin films using sputtering technique
Hadi Salamati, Parviz Kameli, and Mohammad Akhavan . . . . (c) 1895–1898

Power law behavior of $Tl_1Ba_2Ca_2Cu_3O_9$ superconducting tapes
  B. A. Albiss, A. El-Ali, and K. A. Azez . . . . (c) 1899–1903

The noise power spectral density in thin epitaxial $YBa_2Cu_3O_{7-\delta}$ films
  A. Labrag, A. Taoufik, S. Senoussi, and A. Ramzi . . . . (c) 1904–1907

Glass temperature and critical current density in $YBa_2Cu_3O_{7-\delta}$ thin films
  A. Ramzi, A. Taoufik, S. Senoussi, and A. Labrag . . . . (c) 1908–1911

The $YBa_2Cu_3O_{7-\delta}$ anomalous second peak and irreversible magnetic field in the magnetization hysteresis cycles
  A. Taoufik, A. Ramzi, S. Senoussi, and A. Labrag . . . . (c) 1912–1915

Corrosion process of MgB_2 superconductor in a moisture atmosphere
  M. Annabi, A. M'Chirgui, F. Ben Azzouz, M. Zouaoui, and M. Ben Salem . . . . (c) 1916–1919

Effects of nano-Al_2O_3 particles on the superconductivity of Pb-doped BSCCO
  M. Annabi, A. M'Chirgui, F. Ben Azzouz, M. Zouaoui, and M. Ben Salem . . . . (c) 1920–1923

Non-magnetic anion substitution in $(TMTSF)_2ClO_4$: consequences on superconductivity
  N. Joo, C. Pasquier, P. Auban Senzier, D. Jérome, and K. Bechgaard . . . . (c) 1924–1927

Effects of gamma irradiation on the superconducting properties of $TlBa_2Ca_2Cu_3O_x$ tapes
  A. El-Ali, B. A. Albiss, and K. Khasawinah . . . . (c) 1928–1934

Identification of pinning centres in high T_c superconducting thin films by AC susceptibility
  D.-G. Crété, R. Bernard, J.-H. Pommereau, C. Gadois, S. Berger, J. Briatico, J.-P. Contour, O. Durand, J.-L. Maurice, J. Grollier, and K. Bouzehouane . . . . (c) 1935–1939

Magnetic properties of superconducting ceramics $Bi_{1.6}Pb_{0.4}Sr_2Ca_2Cu_3O_{10+d}$ prepared by different methods
  A. Aït-Kaki, O. Belkhen, A. Amira, and M.-F. Mosbah . . . . (c) 1940–1943

Structure and transport properties of (calcium, fluorine) co-doped Y-based superconducting ceramics, effect of heat treatment
  A. Amira, M.-F. Mosbah, A. Leblanc, P. Molinié, and B. Corraze . . . . (c) 1944–1947

Praseodymium and oxygen role on magnetic properties of $PrBa_2Cu_3O_{6+x}$
  A. Harat, G. Fillion, P. Haen, J. Hejtmanek, M. F. Mosbah, and M. Guerioune . . . . (c) 1948–1951

Effect of doping on properties of Bi-based superconductors
  L. D. Sýkorová, O. Smrčková, and V. Jakeš . . . . (c) 1952–1956

Electron backscattered Kikuchi diffraction technique: for a better understanding of epitaxial superconducting film growth on buffered Ni (RABiTS) tapes
  S. Donet, P. Chaudouet, F. Weiss, C. Jimenez, H. P. Ng, C. E. Bruzek, and J. M. Saugrain . (c) 1957–1960

Superconductivity in high-quality $(Hg_{0.9}Re_{0.1})Ba_2CaCu_2O_{6+\delta}$ HTSC thin films
  A. Salem, G. Jakob, and H. Adrian . . . . (c) 1961–1964

Other topics

Neutron powder thermo-diffraction: a very useful tool for the study of crystallisation kinetics and phase segregation in metastable materials *(invited)*
  P. Gorria, D. Martínez-Blanco, J. A. Blanco, J. S. Garitaonandia, J. Campo, and R. I. Smith (c) 1965–1970

Theoretical study of diluted magnetic semiconductor trilayers
  H. Dakhlaoui and S. Jaziri . . . . (c) 1971–1975

Zeeman coupling and Swap action in spin-based diluted magnetic semiconductor quantum dot quantum computer
  A. Hichri and S. Jaziri . . . . (c) 1976–1980

Microstructure evolution after thermal treatments of nanocrystalline Ni_3Al and Ni_3Al+B produced by filling
  M. Khitouni and N. Njah . . . . (c) 1981–1984

Synthesis of potassium chloroapatites, IR, X-ray and Raman studies
  H. El Feki, M. Amami, A. Ben Salah, and M. Jemal . . . . (c) 1985–1988

DOI: The fastest way to find an article online is the *Digital Object Identifier* (DOI).
Starting in Vol. 0, No. 2 (2003), DOIs have been printed in the header of the first page of every article. On the WWW, one can find an article for example with a DOI of 10.1002/pssc.200306190 at **http://dx.doi.org/**10.1002/pssc.200306190.

Please use the DOI of the article to link from your home page to the articles in Wiley Interscience.

The DOI is a result of a cross-publisher initiative to create a system for the persistent identification of documents on digital networks. More information is available from **www.doi.org**.

Preface

The Third International Conference on Magnetic and Superconducting Materials (MSM03) belongs to a series of conferences, held biannually, aiming at providing a forum to the scientists in the magnetic and superconducting materials areas over the world.

The first conference of the series (MSM99) was held in Iran with the proceedings published by World Scientific in 2000, and the second conference (MSM01) was held in Jordan with the proceedings published in Physica B **321** (2002).

MSM03 was organized by the Materials Physics Laboratory, Sfax University (Laboratoire de Physique des Matériaux de la Faculté des Sciences de Sfax), with many domestic and international supporting institutions. It was held in Monastir (Tunisia), 1–4 September 2003, with over 150 participants and keynote lecturers attending from the following countries: Algeria, Austria, China, Czech Republic, Egypt, France, Germany, Hungary, Iran, Italy, Japan, Jordan, Morocco, Netherlands, Pakistan, Poland, Russia, Spain, Sudan, Sultanate of Oman, Taiwan, Tunisia, United Kingdom and United States of America.

Altogether, about 170 papers on a variety of subjects relevant to the topic of the conference were presented, out of which 42 were keynote lectures.

The submissions were peer-reviewed, and ultimately 115 articles were selected for publication in this journal. However, it must be noted that 13 of 39 keynote speakers did not submit their manuscripts for publication.

Invited and other speakers were distinguished members of the international scientific community who are interested in pure sciences and materials research, and involved in the fabrication, characterization and investigation of the physical properties of magnetic and superconducting materials. High-caliber scientists attended the conference contributing to its success and the event resulted in new international relationships in research and cooperation.

The Chairman of the Organizing Committee was Professor Abdelwaheb Cheikhrouhou, Materials Physics Laboratory, Sciences Faculty of Sfax (Tunisia) and the Co-Chairmen were Professor Sami Mahmood, Dean of Sciences at Yarmouk University (Jordan) and Professor Mohamed Akhavan from the Sharif University of Technology (Iran).

The four-day conference consisted of several oral and poster sessions, followed by social programs in the evenings. The success of the event could be measured during the closing session on the last day, when several delegates emphasized the high-quality science that had been evident at the conference.

A post conference three-day tour to the south of Tunisia (Matmata, Douz City: the gate of desert and the mountains oasis: Tamerza, Mides and Chebika) was also arranged.

The conference was generously sponsored by:

– The Tunisian Ministry of High Education, Scientific Research and Technology
– The Tunisian Secretary of State for Scientific Research and Technology
– The Tunisian National Office of Tourism
– The Abdus Salam International Centre for Theoretical Physics (ICTP)
– French Institute for Cooperation in Tunisia
– Tunisian-Italian Scientific Partnership
– British Gas
– Tunisian Society for Electricity and Gas
– Imex Olive Oil
– Confiserie TRIKI "Le Moulin"

The next MSM conference in 2005 will be held in Morocco.

Abdelwaheb Cheikhrouhou
March 2004

Committees

Organizing Committee

Prof. A. Cheikhrouhou	University of Sfax, Tunisia (Chairman)
Prof. M. Akhavan	Sharif University of Technology, Iran (Co-Chairman)
Prof. S. Mahmood	Yarmouk University, Jordan (Co-Chairman)
Prof. R. Bennaceur	University El Manar, Tunisia
Prof. M. Ben Salem	University El Manar, Tunisia
Dr. W. Boujelben	University of Sfax, Tunisia
Dr. W. Cheikhrouhou	University of Sfax, Tunisia
Dr. M. Ellouze	University of Sfax, Tunisia
Prof. M. Ghedira	University of Centre, Tunisia
Prof. S. Kaddour	University El Manar, Tunisia
Dr. S. Zouari	University of Sfax, Tunisia

International Advisory Committee

Prof. E. V. Antipov	Moscow State University, Russia
Prof. B. Barbara	Laboratoire Louis Néel, Grenoble, France
Prof. A. Barone	Naples University, Italy
Prof. J. Bass	Michigan State University, USA
Prof. J. R. Clem	Ames Laboratory, Iowa State University, USA
Prof. J. Ferré	Laboratoire de Physique des Solides, Orsay, France
Prof. A. Fert	CNRS-Thomson CSF, Orsay, France
Prof. D. Fiorani	ISM, CNR, Roma, Italy
Prof. H. Fuess	Darmstadt University of Technology, Darmstadt, Germany
Prof. H.-U. Habermeier	Max-Planck-Institute, Stuttgart, Germany
Prof. M. N. Khan	GIK Institute of Engineering Science and Technology, Pakistan
Prof. K. Kitazawa	Tokyo University, Japan
Prof. C. Lacroix	Laboratoire Louis Néel, Grenoble, France
Prof. M. B. Maple	IPAPS, University of California, San Diego, USA
Prof. M. Murakami	ISTEC, Tokyo, Japan
Prof. W. Nolting	Humboldt-University of Berlin, Germany
Prof. H. R. Ott	LFET, Zurich, Switzerland
Prof. F. Pourarian	Carnegie Mellon University, USA
Prof. F. Rachdi	Université Montpellier 2, Montpellier, France
Prof. C. N. R. Rao	J. Nehru Centre for Advanced Scientific Research, Banglor, India
Prof. B. Raveau	Laboratoire CRISMAT, Université de Caen, France
Prof. F. Weiss	LMGP, Institut National Polytechnique de Grenoble, France
Prof. W. B. Yelon	University of Missouri, Rolla, USA
Prof. L. Yu	Beijing, China, ICTP, Italy

Scientific Committee

Prof. M. Akhavan	Sharif University of Technology, Iran
Prof. H. Al Ghanem	Jordan University of Science and Technology, Jordan
Prof. H. Alloul	Laboratoire de Physique des Solides, Orsay, France
Prof. B. Barbara	Laboratoire Louis Néel, Grenoble, France
Prof. A. Barone	Naples University, Italy
Prof. R. Bennaceur	University El Manar, Tunisia
Prof. M. Ben Salem	University El Manar, Tunisia
Prof. C. Dgega	LCMTR, CNRS, Thiais, France
Dr. C. Dubourdieu	LMGP, Institut National Polytechnique de Grenoble, France
Prof. A. El Ali	Yarmouk University, Jordan
Prof. J. Ferré	Laboratoire de Physique des Solides, Orsay, France
Dr. O. Fruchart	Laboratoire Louis Néel, Grenoble, France
Prof. P. Gorria	Universidad de Oviedo, Oviedo, Spain
Prof. M. Ghedira	University of Centre, Tunisia
Prof. M. Heritier	Laboratoire de Physique des Solides, Orsay, France
Prof. S. Kaddour	University El Manar, Tunisia
Prof. M. N. Khan	GIK Institute of Engineering Science and Technology, Pakistan
Prof. M. Lataifeh	Mu'tah University, Jordan
Prof. Ph. L'Heritier	Laboratoire de Physique des Solides, Orsay, France
Prof. S. Maegawa	Kyoto University, Kyoto, Japan
Prof. S. Mahmood	Yarmouk University, Jordan
Dr. L. Ranno	Laboratoire Louis Néel, Grenoble, France
Prof. B. Raveau	Laboratoire CRISMAT, Université de Caen, France
Dr. H. Salamati	Isfahan University of Technology, Iran
Prof. M. A. Shahzamanian	University of Isfahan, Isfahan, Iran
Prof. N. Tajabor	Ferdowsi University of Mashad, Mashad, Iran
Prof. G. Tatara	Osaka University, Japan
Prof. A. Ul-Haq	Pakistan Council of Scientific and Industrial Research, Pakistan
Prof. F. Weiss	LMGP, Institut National Polytechnique de Grenoble, France
Prof. W. B. Yelon	University of Missouri, Rolla, USA
Prof. M. Zargar-Shooshtari	Iran

Invited Speakers

1. M. Akhavan — Sharif University of Technology (Iran)
2. H. Alloul — Laboratoire de Physique des Solides, Orsay (France)
3. C. Ambrosch-Draxl — Institute for Theoretical Physics, University Graz (Austria)
4. M. Angst — Solid State Physics Laboratory, ETH Zürich (Switzerland)
5. E. V. Antipov — Moscow State University (Russia)
6. B. Barbara — Laboratoire Louis Néel, Grenoble (France)
7. A. Barone — Naples University (Italy)
8. J. Bass — Michigan State University (USA)
9. K. Behnia — LPQ (CNRS), ESPCI, Paris (France)
10. R. Bennaceur — University El Manar (Tunisia)
11. Ph. Bourges — Laboratoire Leon Brillouin, Gif/Yvette (France)
12. A. Cheikhrouhou — University of Sfax (Tunisia)
13. W. Cheikhrouhou-Koubaa — University of Sfax (Tunisia)
14. H. De Raedt — University of Groningen (Netherlands)
15. C. Dubourdieu — LMGP, Institut National Polytechnique de Grenoble (France)
16. M. Farle — University of Duisburg-Essen (Germany)
17. J. Ferré — Laboratoire de Physique des Solides, Orsay (France)
18. G. Fillion — Laboratoire Louis Néel, Grenoble (France)
19. A. Garg — Northwestern University (USA)
20. P. Görnert — Innovent e.V. Jena (Germany)
21. P. Gorria — Universidad de Oviedo (Spain)
22. H.-U. Habermeier — Max-Planck-Institut, Stuttgart (Germany)
23. S. Haddad — University El Manar (Tunisia)
24. A.-M. Haghiri — IEF, Université Paris-sud, Orsay (France)
25. M. Heritier — Laboratoire de Physique des Solides, Orsay (France)
26. S. Kaddour-Charfi — University El Manar (Tunisia)
27. B. Lake — Oxford University (U. K.)
28. S. Maegawa — Kyoto University (Japan)
29. S. Mahmood — Yarmouk University, Jordan (Jordan)
30. L. Ouahab — Université de Rennes 1 (France)
31. F. Rachdi — Université Montpellier 2 (France)
32. H. Raffy — Laboratoire de Physique des Solides, Orsay (France)
33. A. Rinkevich — Institute of Metal Physics, UD of RAS, Ekaterinburg (Russia)
34. K. Shimizu — Toyama University (Japan)
35. P. Strobel — Laboratoire de Cristallographie, Grenoble (France)
36. G. Tatara — Osaka University (Japan)
37. O. Thomas — Université Aix-Marseille III (France)
38. P. Weinberger — Center for Computational Materials Science, TU-Vienna (Austria)
39. W. B. Yelon — University of Missouri, Rolla (USA)

The Best of Physics

Plasma Processes and Polymers New
ISSN print: 1612-8850
ISSN online: 1612-8869
Volume 1 2004 2 issues
Volume 2 2005 9 issues

www.plasma-polymers.org

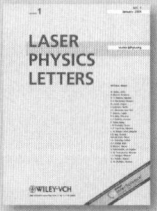
Laser Physics Letters New
ISSN print: 1612-2011
ISSN online: 1612-202X
12 issues per year

www.lphys.org

Annalen der Physik
ISSN print: 0003-3804
ISSN online: 1521-3889
12 issues per year

www.ann-phys.org

Astronomical Notes
ISSN print: 0004-6337
ISSN electronic: 1521-3994
10 issues per year

www.an-journal.org

ChemPhysChem
ISSN print: 1439-4235
ISSN electronic: 1439-7641
12 issues per year

www.chemphyschem.org

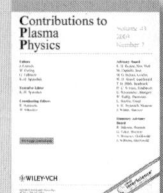
Contributions to Plasma Physics
ISSN print: 0863-1042
ISSN electronic: 1521-3986
8 issues per year

www.cpp-journal.org

Macromolecular Chemistry and Physics
ISSN print: 1022-1352
ISSN online: 1521-3935
18 issues per year

www.mcp-journal.de

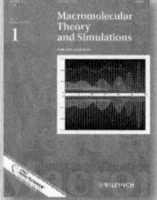
Macromolecular Theory and Simulations
ISSN print: 1022-1344
ISSN online: 1521-3919
9 issues per year

www.mts-journal.de

Progress of Physics
ISSN print: 0015-8208
ISSN online: 1521-3978
12 issues per year

www.fp-journal.org

physica status solidi (a)
applied research
ISSN print: 0031-8965
ISSN electronic: 1521-396X
15 issues per year

www.pss-a.com

physica status solidi (b)
basic research
ISSN print: 0370-1972
ISSN online: 1521-3951
15 issues per year

www.pss-b.com

physica status solidi (c)
conferences
ISSN print: 1610-1634
ISSN online: 1610-1642
12 issues per year

www.pss-c.com

Stay on top of the latest developments in your field

Sign up for WileyInterScience© profiled e-mail alerts based on keywords, author names, and other parameters of your choice.

Direct links from your article to the references. **CrossRef** is the innovative multi-publisher reference linking system enabling readers to move seamlessly from a reference in a journal article to the cited publication, typically located on a different server and published by a different publisher.

Get your free sample copy! At
www.interscience.wiley.com

Wiley-VCH
E-Mail: service@wiley-vch.de,
Phone: +49 6201 606 400, FAX: +49 6201 606 184

John Wiley & Sons, Ltd.
E-Mail: cs-journals@wiley.co.uk
Phone: +44 (0) 1243 779777, FAX: +44 (0) 1243 843232

physica pss status solidi a

www.pss-a.com

applied research

The following pages have been reprinted from
phys. stat. sol. (a) **201**, No. 7, 1379–1442 (2004) as part of the
Proceedings of the 3rd International Conference on
Magnetic and Superconducting Materials (MSM '03)
held in Monastir, Tunisia, 1–4 September 2003.

Current-Driven Excitations in Magnetic Multilayers: A Brief Review

J. Bass[*], S. Urazhdin, Norman O. Birge, and W. P. Pratt Jr.

Department of Physics and Astronomy, Center for Sensor Materials,
and Center for Fundamental Materials Research, Michigan State University, East Lansing, MI 48824

Received 1 September 2003, accepted 3 March 2004
Published online 20 April 2004

PACS 73.40.Jn, 75.60.Jk, 75.70.Cn

In 1996, Berger and Slonczewski independently predicted that a large enough spin–polarized dc current density sent perpendicularly through a ferromagnetic layer could produce magnetic excitations (spin–waves) or reversal of magnetization (switching). In the past few years, both current-driven switching and current-driven excitation of spin–waves have been observed. The switching is of potential technological interest for direct 'writing' of magnetic random access memory (MRAM) or magnetic media. The spin–wave generation could provide a new source of dc generated microwave frequency radiation. We will describe what has been learned experimentally about these two related phenomena, and some models being tested to explain these observations.

© 2004 WILEY-VCH Verlag GmbH & Co. KGaA, Weinheim

1 Introduction

Giant Magnetoresistance (GMR) in ferromagnetic/non-magnetic (F/N) metallic multilayers is now a major subject in studies of metallic magnetic materials, in part because of its importance for devices such as the read-heads of hard discs. GMR can be described as the change in current passing through the multilayer due to a change in magnetic order–specifically the change of the magnetizations of closest F-layers from parallel (P) (usually low resistance state) to anti-parallel (AP) (usually high resistance state) as an applied magnetic field H is reduced from above the saturation field of the F-metal to beyond its negative coercive field.

Since its prediction in 1996 [1, 2] and supporting evidence of its presence starting in 1998 [3, 4], there has arisen great theoretical [5–14] and experimental [15–32] interest in the inverse phenomenon, current-driven excitations in magnetic multilayers: either reversal of layer magnetization, or generation of spin-waves. This interest lies both in trying to understand the physics underlying the new phenomenon and its potential for device use: magnetization reversal for both magnetic memories and magnetic media, and spin-wave generation for production of high frequency radiation. In present magnetic devices or media, moments are reversed via externally generated magnetic fields. It would be much simpler to reverse a moment simply by applying a current pulse perpendicularly through the magnetic layer itself. This possibility is now under study.

In this paper, we first describe the most widely used model of current-induced excitations, briefly note others and some issues still to be resolved, and then review what has been learned from experiment.

---

[*] Corresponding author: e-mail: bass@pa.msu.edu, Phone: 1-(517) 355-9200 (Ext. 2201), Fax: 1-(517) 353-4500

2 Theory

The most widely used model is the semi-classical spin-torque model of Slonczewski [1], in which a spin-polarized current exerts a torque on an *F*-layer if that layer's moment is not collinear with the direction of current polarization. The current's spin–polarization might be produced, e.g., by passing through another (polarizing) *F*-layer. If the current density is large enough, current in one direction causes the moment of the affected *F*-layer to rotate in one sense, and current in the other direction causes it to rotate in the opposite sense. If the moment of the polarizing layer is fixed in direction, for example by making that layer much thicker or much wider, then the moment of the affected layer can be switched by a large enough current (in a low magnetic field), or set into rotation (in a field high enough to inhibit reversal). The threshold for excitation is set by competition between the torque and magnetic damping. Models involving a current-induced effective magnetic field [18] or a quantum threshold for excitations [2, 3] have also been proposed. Still to be determined by experiment are the relative importances in different circumstances of classical moment rotation versus quantum phenomena, such as incoherent generation of spin–waves. Differences between the two should appear, for example, in details of thermal activation of magnetization switching at finite temperature. Here, the topic of debate is whether the appropriate temperature is just the phonon temperature, T_{ph}, [15, 24, 28] or an effective current-dependent temperature, T_m, [19, 30, 31]. Experiments must also determine the length scales in the problem [11, 12].

3 Experiments

The first experimental evidence for a current-driven excitation was obtained at 4.2 K using a Ag point contact to a [Co(1.5)/Cu(2.0–2.2)] multilayer (all thicknesses in this paper are given in nm) [3]. Since data from that study were included in an MSM02 paper two years ago [33], they won't be shown again here, but we just note that peaks in dynamic resistance d*V*/d*I* were seen for currents only in one direction, and the current needed to produce a peak grew linearly with the applied field *H* (in this case applied perpendicular to the layers). These observations were taken as evidence of excitation of spin-waves. A subsequent experiment [3] showed that microwave radiation applied to such a point contact (by placing the sample and contact within a microwave cavity), gave rise to an additional dc response that was attributed to resonant excitation of spin-waves. Further studies with point contacts at 4.2 K showed similar peaks for single *F*-layers [4, 16], occasional more complex peak structures [3, 4], hysteretic switching [4], current-driven excitations at $H = 0$ for ferromagnetically coupled multilayers [18], and evidence [26] that the critical current at a particular perpendicular field is proportional to the exchange energy density of the *F*-metal (Fig. 1).

The first evidence of hysteretic current-driven magnetic switching in a lithographically patterned magnetic nanopillar was reported for Co/Cu/Co at room temperature (295 K) in [20], and reproduced by others listed above. Switching has also been seen in single and multilayer nanowires electrodeposited into nanochannels [19]. Figure 2 [17] shows an example of a patterned Co(40)/Cu(6)/Co(2.5) nanopillar with dimensions ~60 × 130 nm. The elongated shape with sharpened ends is to facilitate formation of only a sin-

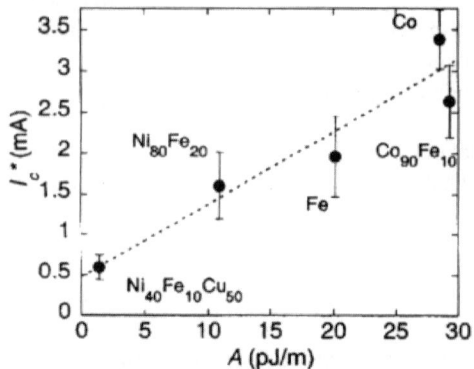

Fig. 1 I_c vs exchange energy density *A* for a variety of different ferromagnetic metals and alloys. Here I_c is the critical current for excitation at the applied field, H_{app}, where the net field on the magnetization is zero: $H_{app} = M_{eff} - H_{ex}$, where M_{eff} is the effective saturation magnetization density and H_{ex} is the effective interlayer exchange field. (From Pufall et al., [26]).

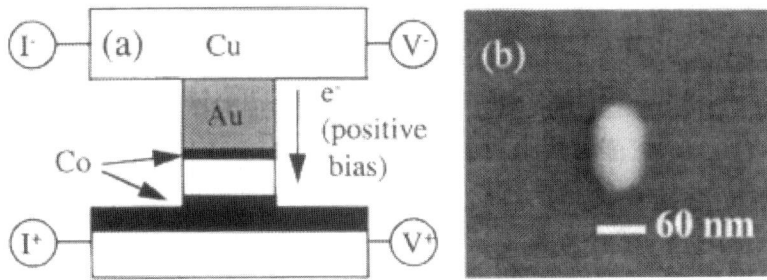

Fig. 2 a) Schematic cross-section of a nanopillar that has been patterned partly into the bottom (thicker) Co layer. (b) A scanning electron micrograph of the Co nanopillar before the top Au electrode was deposited. (From Albert et al. [17]).

gle magnetic domain. The dipolar coupling between the Co layers can be varied by modifying the amount of patterning of the thick bottom Co layer. Dipolar coupling is minimized by stopping in the middle of the Cu layer, leaving the entire bottom Co layer extended. Partial patterning of the bottom layer as in Fig. 2 gives some antiferromagnetic dipolar coupling. Because Py/Cu/Py (Py = Permalloy = $Ni_{84}Fe_{16}$) often gives stable switching also at 4.2 K, it allowed the first comparative studies of switching at 4.2 K and 295 K [30].

Figure 3 shows data for a Py(30)/Cu(10)/Py(6) trilayer with dimensions about 70 nm × 130 nm and patterned to minimize dipolar coupling between the Py layers. The close agreement between the maximum changes in dV/dI with both H and current I, evidences complete switching. A signature of direct current-driven switching is the asymmetry of the switching, where positive switching current, I_s^+, increases dV/dI as the sample switches from a parallel (P) orientation of the two Co layer magnetizations to an anti-parallel (AP) orientation, but negative switching current, I_s^-, decreases dV/dI as the sample

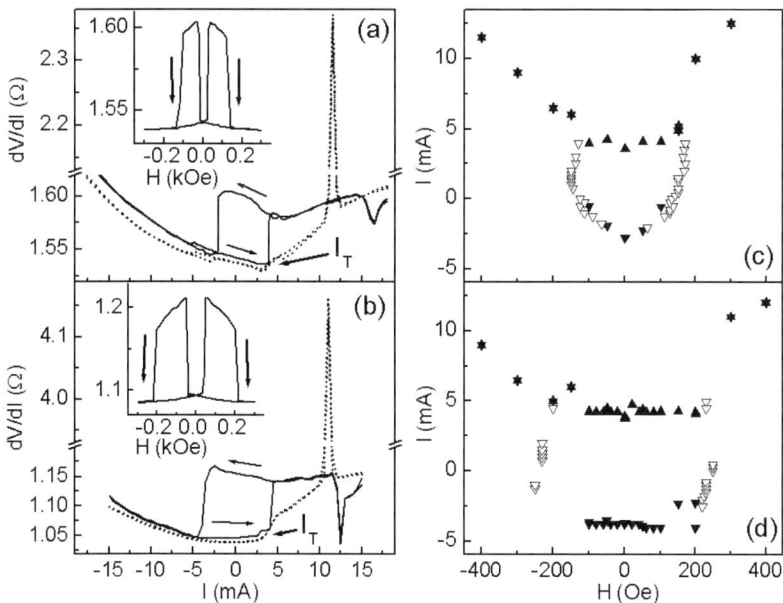

Fig. 3 a) 295 K, (b) 4.2 K: dV/dI vs I for a Py(20)/Cu(10)/Py(6) trilayer nanopillar at 295 K for $H = 50$ Oe (solid curves) and $H = -500$ Oe (dashed curves). Arrows show scan directions; I_t is the threshold current for a linear rise in dV/dI. Insets show dV/dI vs H at $I = 0$. (c) 295 K, (d) 4.2 K: I vs H switching diagrams for the patterned Py layer, from current switching at fixed H (solid symbols) or field switching at fixed I (open symbols). Downward triangles show AP to P switching, upward P to AP. Coinciding up and down triangles mark reversible switching peaks. (From Urazhdin et al. [30]).

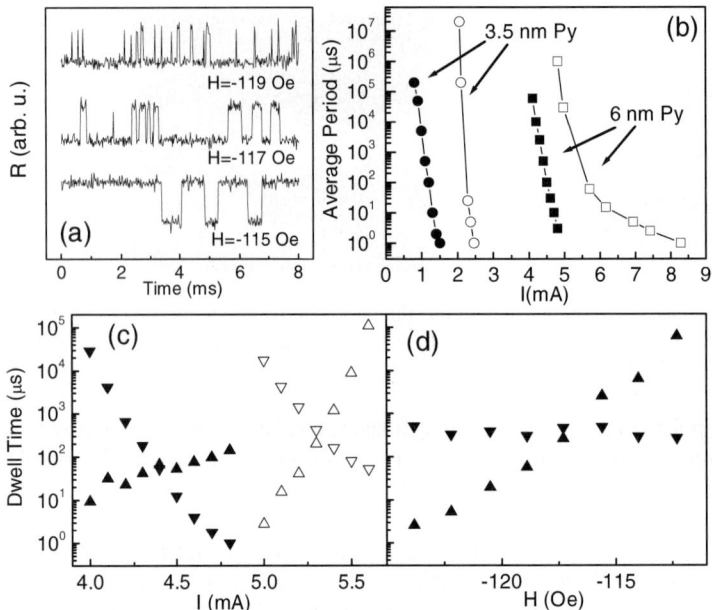

Fig. 4 a) Telegraph noise time traces (offset for clarity) of the Py(20)/Cu(10)/Py(6) static resistance $R = V/I$ at $I = 4.4$ mA, $T = 295$ K for various H. The jumps are closely the same. b) Average telegraph noise period vs I with H adjusted closely linearly with I to make the average dwell times in the AP and P states equal. Solid circles: Py(20)/Cu(10)/Py(3.5) at 295 K; open circles same sample at 4.2 K; solid squares: Py(20)/Cu(10)/Py(6) at 295 K; open squares at 4.2 K. c) P (downward triangles) and AP (upward triangles) dwell times vs I for a Py(20)/Cu(20)/Py(6) sample. Filled symbols: 295 K; open symbols: 4.2 K. d) Similar dwell times vs H for $I = 4.4$ mA at 295 K. (From Urazhdin et al., [30]).

reverses from AP to P. In contrast, switching induced by the self-field of the current gives a symmetric pattern vs I similar to those of the MR curves vs H in the insets to Fig. 3 i.e. dV/dI is in the lower P-state at both large negative and positive I and in the higher AP-state at intermediate I. We now summarize some results of further studies. We start with the H–I 'phase-diagram' for switching in samples with weak coupling between the two F-layers. Figure 3c, d shows that the switching diagrams for Py/Cu/Py at 4.2 K and 295 K are qualitatively similar, but much squarer at 4.2 K. As first shown completely in [23], the diagrams are nearly symmetric in H but highly asymmetric in I. Beyond a certain value of negative I^-, switching disappears. Beyond a certain value of positive I^+, switching becomes non-hysteretic (reversible), as shown by a change from a non-reversible step to a reversible peak. Figure 3a, b shows that, in the reversible regime, dV/dI has a threshold current, labeled I_t, above which the data increase almost linearly up to where the sharp peak occurs. As illustrated in Fig. 3a, b, further studies show [30] that I_t is closely equal to the switching current I_s at zero field.

Switching just beyond the hysteretic to non-hysteretic transition was first shown in Co/Cu/Co [24] to be associated with telegraph noise switching (see, e.g., Fig. 4a for Py/Cu/Py) between the P and AP

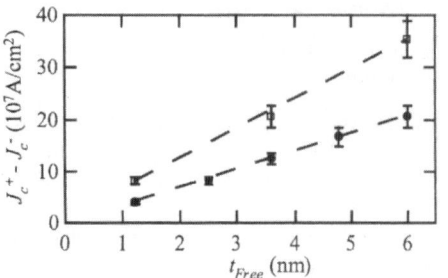

Fig. 5 Critical current density difference, $\Delta J_c = J_c^+ - J_c^-$, at $H = 0$ vs the thickness of the switching Co layer, t_{free}, in a Co/Cu/Co(t_{free}) nanopillar for current ramp rates of 300 mA/sec (open squares) and 0.1 mA/sec (closed circles). The dashed lines are guides to the eyes. (From Albert et al., [25]).

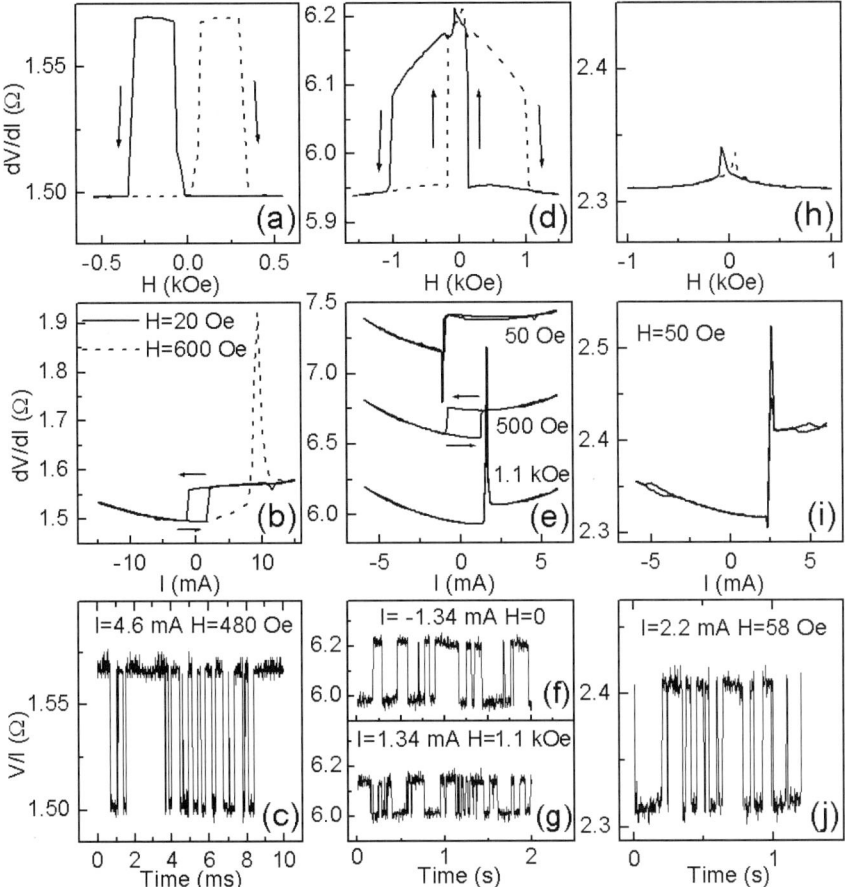

Fig. 6 Magnetoresistance, Current Switching, and Telegraph Noise in uncoupled (a–c), antiferromagnetically coupled (d–g), and ferromagnetically coupled (h–j) trilayer nanopillars. In hysteretic plots, arrows show scan direction. (a, d, h) dV/dI vs H at $I = 0$. (b, e, i) dV/dI vs I at the listed values of H. In (e) the curves are offset for clarity. (c, f, g, j) Telegraph noise traces, $R = V/I$ vs time, at the listed values of H and I. (From Urazhdin et al., [31]).

states, and the variation of the switching period τ with I and H was examined at 295 K. More recently [30], the switching peaks in Figs. 3a, b were shown to occur where the dwell times in these two states are approximately equal, $\tau_P \approx \tau_{AP}$. Figure 4b shows that, when they are equal, this common dwell time (or period) decreases exponentially with I, with slopes that are similar at 4.2 K and 295 K. The similarity of these slopes, as well as the detailed variations of τ with I and H shown in Figs. 4c, d were taken as evidence that the effective temperature for switching is not simply T_{ph} (which would differ by a factor of 70 between 4.2 K and 295 K), but rather a current-driven temperature, T_m, that can differ substantially from T_{ph} when the current I exceeds I_s.

An observation [25] that the switching current, I_s, is proportional to the thickness of the free F-layer (Fig. 5), was interpreted as evidence of an interfacial source for the current-driven excitations.

Antiferromagnetic coupling between the two F-layers can be produced magnetostatically by at least partly patterning the lower F-layer as shown in Fig. 2. Since exchange coupling between two metallic F-layers oscillates with the thickness of the spacer layer separating the F-layers, ferromagnetic coupling can be induced by leaving the bottom layer unpatterned and choosing a spacer thickness corresponding to ferromagnetic coupling. Figure 6 [31] shows the different switching behaviors associated with different couplings. Figure 6a–c shows the magnetoresistance (MR) at $H = 0$ and the low field hysteretic and

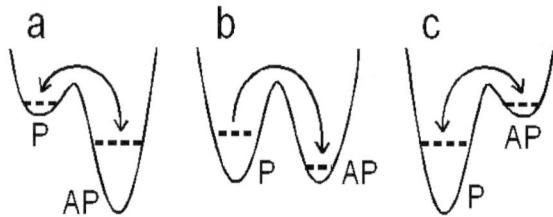

Fig. 7 Schematics of effective two-level energy diagrams for current-driven switching. Dashed lines indicate the effective magnetic temperatures, T_m. (From Urazhdin et al. [31]).

high field reversible switching with telegraph noise expected for an uncoupled sample. Figure 6d–g shows the MR at $H = 0$ for an antiferromagnetically coupled sample; such a sample can display reversible switching and telegraph noise at low fields. Figure 6h–j shows that the MR for a ferromagnetically coupled sample can be almost zero; such a sample can display reversible switching and telegraph noise at all fields. These different behaviors can be explained in a simple (but not unique) way using the two-level diagrams of Fig. 7. We start at $H = 0$. In uncoupled samples, the P and AP wells have the same depth (Fig. 7b) and a large applied current I (or $-I$) is needed to increase T_m^P (T_m^{AP}) enough to excite the system from either initial state. Antiferromagnetic coupling facilitates the AP state, leading to Fig. 7a. Now P to AP transitions are excited by the ambient temperature, and a large reverse current is needed to produce large enough T_m^{AP} to generate telegraph noise. Conversely, ferromagnetic coupling facilitates the P state, leading to Fig. 7c, where a large enough forward current will excite telegraph noise. Turning now to the effect of changing H, increasing H from $H = 0$ favors the P-state. For uncoupled samples, increasing H causes Fig. 7b to gradually change to Fig. 7c, shifting the switching from hysteretic to reversible with telegraph noise. For antiferromagnetic coupling, Fig. 7a changes to Fig. 7b and then to Fig. 7c, shifting the switching from reversible at negative I to hysteretic to reversible at positive I. For ferromagnetic coupling, Fig. 7c stays dominant for all H, and the switching is reversible at all positive I.

As noted above, in a high magnetic field, current-driven excitations in magnetic multilayers excited by a point contact have been seen as peaks in dV/dI and attributed to excitation of spin-waves, i.e. less than full reversal of the moment. In contrast, recent data [27] for nanopillars evidence complete moment reversal. Recent high speed [28] spin–transfer studies seemed generally consistent with a simple spin–torque model. Detailed high frequency studies are just beginning. One study of nanopillars [29] revealed complex spectra above a minimum current, with both broad backgrounds and peaks that vary in location and relative sizes with current and field. The authors conclude that several different types of magnetic excitations must be present. Another, using a lithographed point contact [32] to a multilayer, reported results generally consistent with spin–torque driven excitation of a single domain nanoparticle.

4 Summary and conclusions

Point contact and nanopillar studies have revealed a fair amount about current-driven excitations in magnetic multilayers. In small fields, the excitation structure of uncoupled layers seems simple – just switching of the magnetization of one layer. Both coupling and higher fields lead to complications, coupling causes reversible switching even at low fields, and higher fields cause more complex excitations, including telegraph noise, spin-waves, and perhaps even chaotic magnetization dynamics. The sources of peaks in point contact spectra are not yet completely clear. Many observed phenomena can be described at least qualitatively (and sometimes quantitatively – see, e.g., [26, 28]) by a simple semi-classical spin-torque model. However, evidence of complications from several experiments suggests that a full understanding of all observations is not yet achieved.

Acknowledgements Please insert the following as in the original draft: The authors acknowledge support from the MSU CFMR, CSM, the Keck microfabrication facility, the US NSF through grants DMR 02-02476, 98-09688, NSF-EU 00-98803, and Seagate Technology.

References

[1] J. Slonczewski, J. Magn. Magn. Mater. **159**, L1 (1996); J. Magn. Magn. Mat. **195**, L261 (1999); J. Magn. Magn. Mater. **247**, 324 (2002).
[2] L. Berger, Phys. Rev. B **54**, 9353 (1996); IEEE Trans. Magn. **34**, 3837 (1998); J. Appl. Phys. **89**, 5521 (2001).
[3] M. Tsoi et al., Phys. Rev. Lett. **80**, 4281 (1998); ibid, **81**, 493 (E) (1998); Nature (London) **406**, 46 (2000); Phys. Rev. Lett. B **89**, 246803 (2002).
[4] E. B. Myers et al., Science **285**, 867 (1999).
[5] Ya. B. Bazaliy et al., Phys. Rev. B **57**, R3213 (1998).
[6] A. Brataas et al., Phys. Rev. Lett. **84**, 2481 (2000).
[7] X. Waintal et al., Phys. Rev. B **62**, 12,317 (2000).
[8] C. Heide et al., Phys. Rev. B **63**, 064424 (2001).
C. Heide, Phys. Rev. Lett. B **87**, 197201 (2001).
[9] A. A. Kovalev et al., Phys. Rev. B **66**, 224424 (2002).
[10] Y. Tserkovnyak et al., Phys. Rev. B **66**, 224403 (2002).
[11] M. D. Stiles and A. Zangwill, J. Appl. Phys. **91**, 6812 (2002); Phys. Rev. B **66**, 0114407 (2002).
[12] S. Zhang et al., Phys. Rev. Lett. **88**, 236601 (2002).
[13] K. Xia et al., Phys. Rev. B **65**, 220401(R) (2002).
[14] A. Shapiro et al., Phys. Rev. B **67**, 104430 (2003).
[15] Z. Li and S. Zhang, cond-mat/0302339 (2003).
[16] Y. Ji, C. L. Chien, and M. D. Stiles, Phys. Rev. Lett. **90**, 106601 (2003).
[17] F. J. Albert et al., Appl. Phys. Lett. **77**, 3809 (2000).
[18] W. H. Rippard, M. R. Pufall, and T. J. Silva, Appl. Phys. Lett. **82**, 1260 (2003).
[19] J.-E. Wegrowe et al., Europhys. Lett. **45**, 626 (1999); J. Appl. Phys. **91**, 6806 (2002).
[20] J. A. Katine et al., Phys. Rev. Lett. **84**, 3149 (2000).
[21] J. Grollier et al., Appl. Phys. Lett **78**, 3663 (2001); Phys. Rev. B **67**, 174402 (2003).
[22] F. B. Mancoff and S. E. Russek, IEEE Trans. Magn. **38**, 2853 (2002).
[23] J. Z. Sun et al., Appl. Phys. Lett. **81**, 2202 (2002).
[24] E. B. Myers et al., Phys. Rev. Lett. **89**, 196801 (2002).
[25] F. J. Albert et al., Phys. Rev. Lett. **89**, 226802 (2002).
[26] M. R. Pufall et al., Appl. Phys. Lett. **83**, 323 (2003).
[27] B. Oezyilmaz et al., Phys. Rev. Lett. **91**, 067203 (2003).
[28] R. H. Koch et al., Phys. Rev. Lett. **92**, 088302 (2004).
[29] S. I. Kiselev et al., Nature **425**, 380 (2003).
[30] S. Urazhdin et al., Phys. Rev. Lett. **91**, 146803 (2003).
[31] S. Urazhdin et al., cond-mat/0304299 (2003); Appl. Phys. Lett. **83**, 114 (2003).
[32] W. H. Rippard et al., Phys. Rev. Lett. **92**, 027201 (2004).
[33] J. Bass and W. P. Pratt Jr., Physica B **321**, 1 (2002).

Magnetisation reversal dynamics in an ultrathin magnetic film and the creep phenomenon

J. Ferré[*,1], **V. Repain**[1], **J.-P. Jamet**[1], **A. Mougin**[1], **V. Mathet**[2], **C. Chappert**[2], and **H. Bernas**[3]

[1] Laboratoire de Physique des Solides, UMR CNRS 8502, Université Paris-Sud, 91405 Orsay, France
[2] Institut d'Electronique Fondamentale, UMR CNRS 8622, Université Paris-Sud, 91405 Orsay, France
[3] Centre de Spectrométrie de Masse et de Spectrométrie Nucléaire, UMR CNRS 8609, Université Paris-Sud, 91405 Orsay, France

Received 1 September 2003, revised 27 February 2004, accepted 27 February 2004
Published online 26 April 2004

PACS 61.80.Jh, 75.60.Jk, 75.70.Ak, 78.20.Ls

The domain structure and magnetisation reversal phenomenon in ultrathin magnetic film structures with perpendicular anisotropy are reviewed. The domain wall motion is studied in virgin and He ion irradiated Pt/Co(0.5nm)/Pt films. The field (H) dependence of the mean wall velocity v does not follow a trivial Arrhenius law, where Ln (v) is proportional to H. At low field, it is perfectly interpreted within the so-called creep theory as soon as the domain wall moves in a weak pinning media. The analogy between the motion of vortices in a type-II superconductor and of walls in a magnetic ultrathin film is emphasised.

© 2004 WILEY-VCH Verlag GmbH & Co. KGaA, Weinheim

1 Introduction

From a fundamental or applied point of view, it is important to understand the field-induced magnetisation reversal process in ultrathin magnetic film structures. Only too crude phenomenological models were proposed so far. As in thicker magnetic layers, the reversal is still controlled by domain nucleation and wall propagation [1, 2]. Precise studies require time dependent observations of the evolution of the magnetic domain structure in constant applied fields. In homogeneous thin films with only few extrinsic defects, and under rather small fields, nucleation occurs at a restricted number of sites followed by a quasi-uniform domain wall motion. We focus here essentially on the field-induced motion of a domain wall in a weakly disordered ultrathin magnetic layer. The problem of a moving one-dimensional (1d) interface in a random medium with weak disorder is not only relevant in magnetism. We point out, for example, the analogy between slow domain wall motion in quasi-perfect magnetic layers and the movement of vortices in type-II superconductors. From this point of view, ultrathin magnetic films can be considered as canonical for solving more general problems in physics.

This paper reviews recent results obtained in ultrathin magnetic films with perpendicular anisotropy, essentially virgin or He-irradiated Pt/Co(0.5 nm)/Pt samples [3–7]. Large magnetic domains are present in their demagnetised state; this is discussed first. At low field, dynamics can be well explained within the so-called "creep" theory that is used to explain slow motion of interfaces or lattices in various fields such as the charge density waves, the vortices in superconductors, the wetting phenomenon, the fluid invasion in a porous material, etc. Slow dynamics allows also to test the stability of written information bits in magnetic recording media. At higher field, the so-called pinning field, H_{crit}, sets the boundary between the thermally assisted regime for wall motion and the viscous regime. In the last case, nano-defects are less efficient and the study of this regime has a practical importance for ultra-fast switching

---

[*] Corresponding author: e-mail: ferre@lps.u-psud.fr, Phone: 0169156063, Fax: 0169156086

of the magnetisation that determines the writing time in memories. Similar behaviour occurs also in ultrathin films with in-plane anisotropy [2], but the involved parameters differ markedly.

2 Samples and experimental methods

The virgin Pt/Co(0.5 nm)/Pt films have been prepared by sputtering on a sapphire substrate in a good vacuum [4]. The Pt and Co crystallites have typical size of 10 nm, and atomically flat terraces are even smaller. An rms roughness of less than 0.4 nm for the upper Pt layer has been measured by atomic force microscopy. Such films exhibit a highly square room temperature hysteresis loop with large perpendicular anisotropy.

We have previously demonstrated that the magnetic properties of Pt/Co/Pt films can be strongly modified under an uniform irradiation by light He ions at low fluence [6–8]. The coercivity, the anisotropy and the Curie temperature are significantly reduced when increasing the fluence. For an uniform irradiation fluence of 8×10^{15} He ions/cm^2, the room temperature perpendicular coercivity of the Pt/Co(0.5 nm)/Pt film is reduced to 50 Oe [6]. Magnetic domains in the virgin film have been visualised and their field-induced wall velocity mesured from high spatial resolution magneto-optical microscopy [2].

3 Domains in demagnetised samples

In the past, the existence of domains in ultrathin magnetic films was a subject of controversies [9–11]. It is well known that domain formation results from the competition between exchange and dipolar interactions. For vanishingly small dipolar effects, as in ultrathin magnetic films with perpendicular anisotropy, one expects large size domains in the equilibrium state. However, extrinsic defects can pin walls and therefore limit the domain size. For ultrathin Co layers ($e < 2$ nm), the magnetisation is uniform in the depth of each domain since "e" is far smaller than the exchange length (7 nm). In counterpart, when anisotropy is reduced, for example by increasing the thickness of the Co layer, smaller size up-down domains or even dendritic structures prevail in the demagnetised state. The dependence of the domain size with Co thickness has been determined for Au/Co/Au films and interpreted within a simple model [12]. As expected, magnetic domains in demagnetised or in-field states expand over hundreds of microns in our ultrathin thin Co layers as soon as $e < 0.6$ nm.

4 Field-induced magnetisation reversal

For Co layers thinner than 1.2 nm, the magnetic hysteresis loop is highly square [2, 4]. Magnetisation dynamics can be probed by performing magnetic aftereffect measurements, i.e. investigating the relaxation of the magnetisation under a constant field. However, a full description of the magnetic reversal process requires a perfect knowledge of the spatial and time dependence of the nucleation and domain wall motion process; this can be successfully studied by magneto-optical microscopy [2, 13]. At a given applied field, a wall propagates quite uniformly at a constant velocity. This propagation is limited by a collective wall pinning associated with a large assembly of nano-defects (grain boundaries or steps between atomically flat terraces).

Below the depinning field H_{crit}, the domain wall motion is thermally activated and its velocity increases with field as a consequence of the reduction of the associated energy barriers by the Zeeman energy (Fig. 1). This is confirmed from our low temperature magneto-optical domain imaging experiments showing that the wall velocity is finite only above H_{crit} [2, 14]. Note that H_{crit} increases when lowering the temperature, since it depends on the anisotropy [2]. For $H > H_{crit}$, domain wall motion is better described by a viscous process, the wall velocity being proportional to H. One believes that, at large enough field, the wall velocity is limited by the collective precessional mode of switching of the magnetisation.

From magneto-optical image snapshots of the wall motion under successive square pulses of field with different amplitude and duration [2, 3], we deduced the variation of the mean wall velocity v with H in a Pt/Co(0.5nm)/Pt film over a rather large dynamic range (Fig. 2). This result is consistent with the

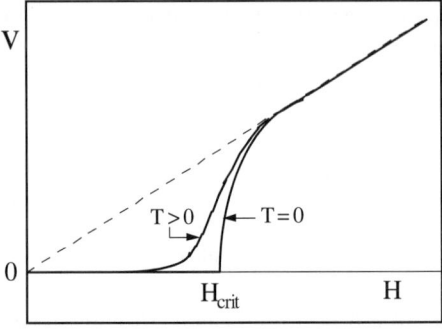
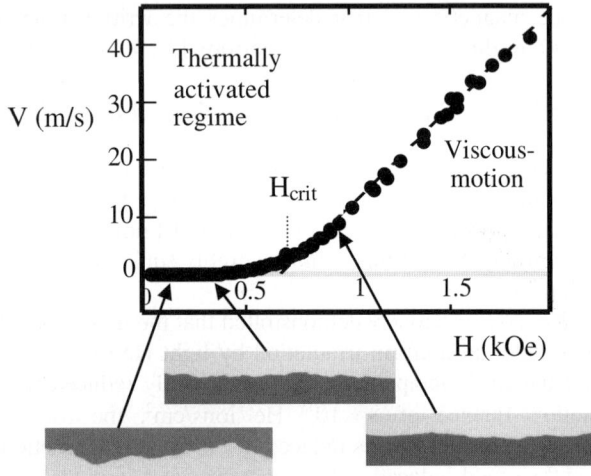

Fig. 1 Schematic variation of the field-induced domain wall velocity v as a function of the applied field. H_{crit} is the depinning field.

Fig. 2 Variation of the domain wall velocity as a function of the applied field H for the virgin Pt/Co(0.5 nm)/Pt film, at room temperature. Shnapshot images of the wall are shown for three values of H.

predicted scheme (Fig. 1). As expected and demonstrated by simulations [15], the domain wall roughness resulting from the pinning of the wall by nano-defects, does not vary much in the thermally activated regime, but vanishes in the viscous regime at high field [2] (Fig. 2). Unfortunately, it is often difficult to investigate domain wall motion at high field (for $v > 50$ m/s in our films) since nucleation of reversed domains takes place everywhere and hinder a precise determination of the wall velocity.

But what is the mechanism explaining the viscous domain wall mobility above H_{crit}? Is it related to the usual damping effect in the high frequency limit? The involved mechanism is not phonon-assisted since, as we found previously for a Au/Co/Au film, the slope of the $v(H)$ curve, i.e. the wall mobility, does not vary when the temperature is changed between 320 K to 150 K [16]. As figured out on Fig. 1, we believe that only the asymptotic mobility relates to the usual damping coefficient. As already found for Au/Co/Au, some field-velocity anomalies appear at high enough field [16]. We interpreted them as a Walker breakdown in the wall velocity, the associated field being dependent of the film thickness [17].

Under a low uniform He ion irradiation, the magnetic anisotropy of Pt/Co/Pt films is reduced, but also the structural disorder [7], so that one has a tool to reduce the wall pinning energies. As a consequence, H_{crit} is reduced (Fig. 4), and a drastic slowing down of the wall motion is observed at very low field. This allowed us to check predictions on the creep theory [18], as briefly reported below.

5 Domain wall motion in very low field

5.1 Theoretical background

The creep theory, relative to the slow motion of a moving elastic interface in a weakly disordered medium has been proposed first to explain the dynamics of vortices in type II superconductors [18]. It is based on scaling arguments and has been extended more recently to magnetic thin film media [3].

Let us consider a thin magnetic film where domain wall motion is limited by a large assembly of weak pinning defects giving wall jaggedness (Fig. 3). Here $u(x)$ represents the amplitude of the displacement of the wall, and L precises the position along Ox. The wall jaggedness is then only determined by the dimensionality of the wall (a 1d-interface) and that of the randomly disordered medium [19]. This means that the domain wall thickness and its detailed internal spin structure are neglected. This is the case of our ultrathin Co-layer with large perpendicular anisotropy and thickness far smaller than the exchange length for cobalt. The wall jaggedness can be described by the roughness exponent ζ appearing in the

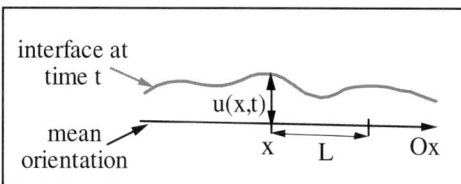

Fig. 3 Schematic representation of the domain wall.

following time and thermal averaged correlation function:

$$\langle\langle [u(x+L) - u(x)]^2 \rangle\rangle = u_c^2 (L/L_c)^{2\zeta} . \quad (1)$$

The theoretical value of the universal roughness exponent ζ is 2/3 for a $d = 1$ interface. The collective pinning length, L_c, is the magnetic counterpart of the Larkin length for superconductors [18]. In the magnetic case, $L_c = [(\varepsilon_{el}\xi^2)/\Delta]^{1/3}$ can be estimated from the comparison between the elastic and pinning energies, where ε_{el} is the lineic wall tension, ξ the mean distance between defects, and Δ the pinning force. Neglecting the dipolar energy term, which is justified in our ultrathin film, the total magnetic energy of the system writes as the sum of the wall energy, the pinning energy and the Zeeman magnetic energy. The integral form of the Edwards-Wilkinson equation for the energy then writes:

$$E(u) = \int dx[(\gamma_0/2)(du/dx)^2 + \Delta\gamma(u,x) - F_{mag}u] . \quad (2)$$

The first term of expression (2) represents the wall elastic energy, the second term its fluctuations due to local pinning, and the third term the Zeeman magnetic energy proportional to H. This perturbative approach assumes small wall curvature and weak pinning. Weak pinning ($\Delta\gamma/\gamma_0 = 0.07$) is realised in our sample since, as discussed above, it is linked to small perturbations. For a portion of interface of length $L > L_C$, the characteristic distance $u(L)$ between two metastable positions, and the energy barrier scale as:

$$u(L) = u_C (L/L_C)^\zeta \quad \text{and} \quad E(L) \sim u^2/2 \sim U_C (L/L_C)^{2\zeta + d - 2} \quad (3)$$

with $U_C = (\varepsilon_{el}\xi^2\Delta)^{1/3}$. $E(L)$ gives a typical value of the pinning energy since, in the quasi-static limit, it is assumed that the elastic energy compensates the pinning energy. These scaling relations are general and do not depend on the disorder as soon as it is weak enough. One only assumes that the interaction range associated to one pinning defect is far smaller than L, as fulfilled in our case. After minimising the expression of the total energy with respect to L, one obtains the expression of the involved energy barrier E_H for a field H far smaller than the depinning field H_{crit}:

$$E_H = U_C (H_{crit}/H)^\mu \quad \text{with} \quad \mu = (2\zeta + d - 2)/(2 - \zeta) . \quad (4)$$

A value $\mu = 1/4$ is calculated for a thin 1d magnetic wall. The depinning field can be estimated from the comparison between the Zeeman and pinning energies:

$$H_{crit} = [(\varepsilon_{el}\xi)/M_s e L_C^2] \quad (5)$$

and finally, from expression (4), the domain wall velocity expresses as:

$$v(H) = v_0 \exp[-(U_C/kT)(H_{crit}/H)^\mu] . \quad (6)$$

5.2 Experimental proof of a creep regime

The roughness exponent, deduced experimentally, is $\zeta = 0.69 \pm 0.07$ [3], i.e. consistent with the theoretical value 2/3. In order to verify the creep theory, the $v(H)$ law (expression (6)) has been tested experimentally for our Pt/Co(0.5 nm)/Pt film. For this purpose, a plot of Ln $[v(H)]$ is more appropriate (Fig. 4). The Arrhenius thermally activated law [20]:

$$v(H) = v_0 \exp[\Delta E/kT] = v_0 \exp[V_p M_S (H - H_{crit})/kT] \quad (7)$$

is usually considered to interpret domain wall motion in thin magnetic films. Here V_p is the propagation volume. Such a simple law predicts a linear variation in a Ln $[v(H)]$ plot, which strongly disagrees with

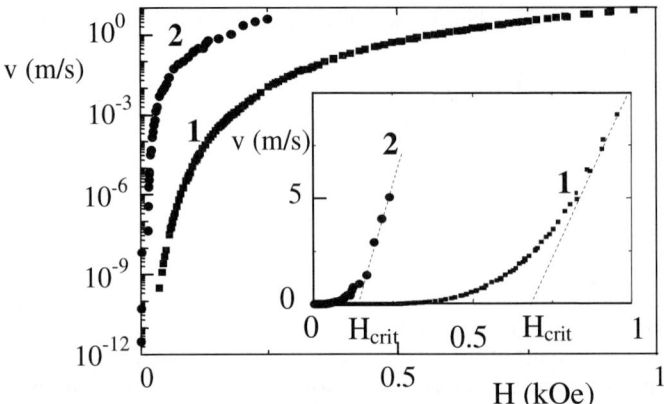

Fig. 4 Semi-logarithmic plot of the domain wall velocity v versus the applied field H, (1) for the virgin and (2) He ion irradiated Pt/Co(0.5 nm)/Pt film. In inset: linear $v(H)$ plot for the same samples.

our experimental result (Fig. 4). For the virgin Pt/Co(0.5 nm)/Pt film, an abrupt slowing down is clearly evidenced when H tends to zero. This behaviour is perfectly consistent with the creep velocity expression (6), predicting a critical slowing down at $H = 0$. The validity of the creep theory with $\mu = 1/4$ is then unambiguously stated since the experimental exponent $\mu = 0.24 \pm 0.04$ [3].

For the same Pt/Co(0.5 nm)/Pt sample, uniformly irradiated at a fluence of 8×10^{15} He ions/cm^2, H_{crit} is highly reduced, and the $H = 0$ slowing down of the wall velocity is even much more pronounced than for the non-irradiated film (Fig. 4). The weakening of the pinning force can be tentatively explained considering that the Co and Pt ions intermixing at Co/Pt interfaces, induced by He irradiation, tends to smooth most of the nano-irregularities that pin the wall [6, 7]. Moreover, the anisotropy is reduced under He irradiation. As in the virgin film, one finds again an exponent μ close to 1/4, which means that the creep law is still valid at weaker disorder. This supports the universality of this regime in a large range of weak disorder. In agreement with the creep theory which predicts that the size of field-induced reversed magnetised areas diverges at $H = 0$, we have evidenced that the domain wall moves by larger and larger jumps when H goes to zero. Jumps in magnetisation then become rare events with typical size of a few micron square for $H = 2$ Oe in the irradiated sample, while they are not detectable optically at much higher field. Weak pinning is not fulfilled in many systems, in particular for virgin Co layers thicker than 1.2 nm, i.e. when the anisotropy becomes too small and the dipolar field too efficient; this leads to rougher or even dendritic domain structures [2].

5.3 Analogy between superconductors and magnetic thin films

As discussed above, we extended to magnetism the creep theory proposed first for superconductors [18]. In this analogy, the Lorentz force related to the current density is replaced by the Zeeman force, proportional to H. Thus, the critical depinning field plays the same role as the critical current. The Larkin-Ovchinikov length is replaced by the collective pinning length L_C over which the wall can be considered as rigid. L_C is a new characteristic nanometer scale length, which becomes relevant in laterally nano-patterned magnetic films.

6 Conclusion

This contribution reports on magnetisation reversal in ultrathin films with perpendicular anisotropy. In films with planar magnetic anisotropy, the relatively strong dipolar contribution will modify the above predicted behaviour, leading to a smoothing of the domain walls whose contours are less sensitive to disorder. Theory is mainly focused here on the creep phenomenon present at low field, and experimentally justified for the Pt/Co(0.5 nm)/Pt model system. In spite of the reduction of the disorder in the He ion beam irradiated sample, the v(H) variation remains universal, in agreement with the creep theory. It will be interesting now to extend this model in the presence of well known correlated defects, and to perform experiments in systems that can be easily modelled. The only published results so far concern the interaction of a domain wall with a linear defect through the magnetoelastic interaction [22].

Acknowledgement We wish to thank S. Lemerle for his pionnering work in this field of research.

References

[1] J. Ferré, J. P. Jamet, and P. Meyer, phys. stat. sol. (a) **175**, 213 (1999).
[2] J. Ferré, in Spin dynamics in confined magnetic structures, edited by B. Hillebrands and K. Ounadjela, Springer Heidelberg, p. 127 (2001).
[3] S. Lemerle, J. Ferré, C. Chappert, V. Mathet, T. Giamarchi, and P. Le Doussal, Phys. Rev. Lett. **80**, 849 (1998).
[4] V. Mathet, T. Devolder, C. Chappert, J. Ferré, S. Lemerle, L. Belliard, and G. Guentherodt, J. Magn. Magn. Mater. **260**, 295 (2003).
[5] M. Kisielewski, A. Maziewski, M. Tekielak, J. Ferré, S. Lemerle, V. Mathet, and C. Chappert, J. Magn. Magn. Mater. **260**, 231 (2003).
[6] J. Ferré, C. Chappert, H. Bernas, J.-P. Jamet, P. Meyer, O. Kaitasov, S. Lemerle, V. Mathet, F. Rousseaux, and H. Launois, J. Magn. Magn. Mater. **198–199**, 191 (1999).
[7] T. Devolder, J. Ferré, C. Chappert, H. Bernas, J. P. Jamet, and V. Mathet, Phys. Rev. B **64**, 064415 (2001).
[8] C. Chappert, H. Bernas, J. Ferré, V. Kottler, J. P. Jamet, Y. Chen, E. Cambril, T. Devolder, F. Rousseaux, V. Mathet, and H. Launois, Science **280**, 1919 (1998).
[9] Y. Yafet and E. M. Gyorgy, Phys. Rev. B **38**, 9145 (1988).
[10] B. Kaplan and G. A. Gehring, J. Magn. Magn. Mater. **128**, 11 (1993).
[11] R. Allenspach, J. Magn. Magn. Mater. **129**, 160 (1994).
[12] M. Speckmann, H. P. Oepen, and H. Ibach, Phys. Rev. Lett. **75**, 2035 (1995).
[13] J. Ferré, in Magnetism and Synchrotron Radiation, Lecture Notes in Physics, edited by E. Beaurepaire et al., Springer, Heidelberg, p. 316 (2001).
[14] A. Lyberatos and J. Ferré, J. Phys. D: Appl. Phys. **33**, 1060 (2000).
[15] A. Lyberatos, J. Magn. Magn. Mater. **186**, 248 (1998).
[16] A. Kirilyuk, J. Ferré, and D. Renard, IEEE Trans. Magn. **29**, 2518 (1993).
[17] S. V. Tarasenko, A. Stankiewicz, V. V. Tarasenko, and J. Ferré, J. Magn. Magn. Mater. **189**, 19 (1998).
[18] G. Blatter, M. V. Feigelman, V. B. Geshkenbein, A. I. Larkin and V. M. Vinokur, Rev. Mod. Phys. **66**, 1125 (1994).
[19] D. S Fisher, Phys. Rev. B **43**, 10728 (1991).
[20] M. Labrune, S. Andrieu, F. Rio, and P. Bernstein, J. Magn. Magn. Mater. **80**, 211 (1989).
[21] T. Shibauchi, L. Krusin-Elbaum, V. M. Vinokur, B. Argyle, D. Weller, and B. Terris, Phys. Rev. Lett. **87**, 267201 (2001).

Spintronics: perspectives for the half-metallic oxides

A. M. Haghiri-Gosnet[*], **T. Arnal, R. Soulimane, M. Koubaa,** and **J. P. Renard**

Institut d'Electronique Fondamentale – IEF, UMR CNRS 8622, Université Paris Sud, Bâtiment 220, 91405 Orsay Cedex, France

Received 1 September 2003, accepted 5 March 2004
Published online 20 April 2004

PACS 72.25.–b, 75.47.Lx, 85.40.Hp, 85.75.–d

Due to their total spin polarization (~100%), the half-metallic oxides, such as $La_{0.7}Sr_{0.3}MnO_3$ (LSMO), CrO_2, Fe_3O_4 and Sr_2FeMoO_6, are promising materials for producing very high magneto-resistive (MR) responses in magnetic devices. Extremely large MR values have been obtained by tunneling through vertical tunnel junctions, as well as through nano-interfaces or domain walls in planar devices. The state of the art of the spin-polarized half-metallic devices in a two-electrode configuration is described. Moreover, some recent results, that confirm the high spin polarization of these magnetic oxides, are detailed. In case of planar devices, nanofabrication is shown to allow transferring of nanopatterns in a sub-50 nm scale. The technical challenges to integrate these half-metallic oxides devices in the industrial context of magnetic random access memories (MRAMs) are finally discussed.

© 2004 WILEY-VCH Verlag GmbH & Co. KGaA, Weinheim

1 Operational principles

Spintronics is a new approach to electronics, where the information is carried out by the spin (up or down) of the carrier, in addition to the charge (electron or hole) [1]. Spin-polarized transport occurs naturally in any material presenting an imbalance of the spin populations at the Fermi level. This shift in energy of the two densities of states (spin up or spin down) is the source of the magnetic moment of any ferromagnetic metal, associated to a spin polarization P (see Fig. 1). In fact, the spin polarization P traduces the inequality in mobility and number of the spin-up and spin-down carriers. The basic two-terminal device is a magnetic tunnel junction (MTJ), which principles are depicted in the Fig. 2. Two electrodes of the magnetic material are separated by a very thin insulating barrier layer, through which the spin-polarized carriers tunnel. Because the spin-up electrons can only tunnel into spin-up empty states, no tunneling occurs when the magnetic moment is in an anti-parallel configuration in both electrodes (see Fig. 1b) and the resistance of the device becomes very high. This device is able to sense the direction of the external magnetic field and its response, defined as the difference in resistance between the two configurations – parallel and anti-parallel –, is called the magneto-resistance (MR). The two major applications of this basic MTJ device concerns the read-heads for magnetic hard disk drives and the non-volatile magnetic memories (MRAMs) [1, 2]. Because such devices are non-volatile, the market for these applications is great and estimated around $1 billion year in USA. The most drastic MR effects should be observed for the most highly polarized currents. Therefore, several efforts are continuously made to find a fully polarized ferromagnetic material (with $P = 100\%$). Half-metallic ferromagnetic materials appear as potential candidates, and a lot of work is under progress to synthesize magnetic oxides, such as $La_{0.7}Sr_{0.3}MnO_3$ (LSMO), CrO_2, Fe_3O_4 and Sr_2FeMoO_6. The band-diagram of half-metals has the peculiarity to exhibit a large shift in energy of the two densities of states (Fig. 1a). The Fermi level is thus cutting only one spin-band producing a 100% spin polarization. The MR response of MTJ based on

---
[*] Corresponding author: e-mail: anne-marie.haghiri@ief.u-psud.fr, Phone: +33 1 69 15 78 37, Fax: +33 1 69 15 41 11

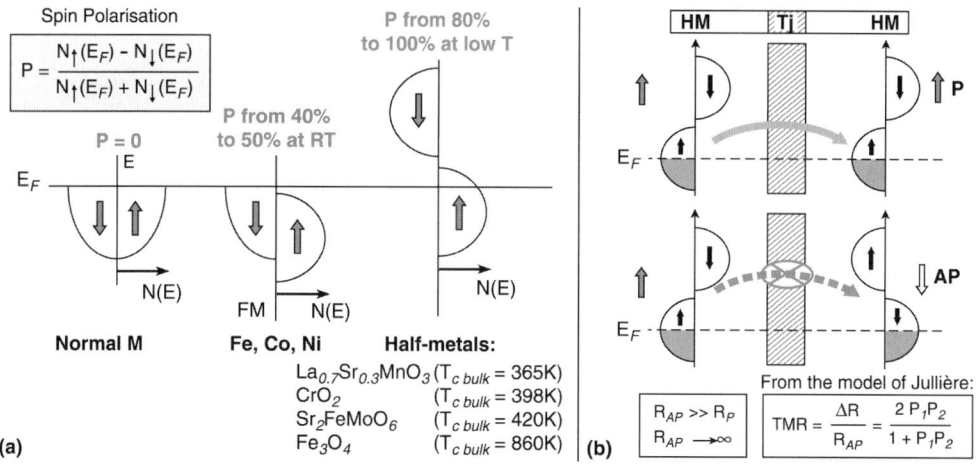

Fig. 1 (online colour at: www.interscience.wiley.com) Schematic representations of a) the densities of states of (i) a normal metal, (ii) a ferromagnetic metal and (iii) a half-metal, b) a tunnel junction with two half-metallic electrodes (the direction of the external field is either up (green arrow) or down (yellow arrow)).

such half-metals should be in theory infinite, which explains the large challenge and the fascination of researchers towards these oxides.

The research on high-temperature cuprates superconductors has strongly renewed the interest in magnetic oxides and a large part of the recent studies has been devoted to the mixed-valence manganese oxides exhibiting a metal-insulator transition accompanied by so-called colossal magneto-resistance (CMR) effects [3]. These oxides have a rich and complex physics related to the large importance of electron-lattice and electron–electron interactions. Their structural, magnetic and transport properties are intricately related. With its Curie temperature (T_C = 360 K) above room temperature, $La_{0.7}Sr_{0.3}MnO_3$ (LSMO) has been the most extensively studied manganite [4]. In order to attain larger TMR values at room temperatures, new half-metallic oxides with higher T_C, such as CrO_2, Sr_2MoFeO_6 [5] and Fe_3O_4 have been tentatively introduced in junctions.

2 Growth of thin films

For applications, the growth of high quality single-crystal epitaxial films has to be first controlled. Pulsed laser deposition (PLD) technique is the most straightforward method for complex oxides due to its simplicity (a dense ceramic target is ablated by a pulsed laser under oxygen creating a dynamic plasma that condensates on the heated substrate) and also because the stoichiometry of the target material is more often kept during deposition. Moreover, low pressure PLD under a strong oxidizing gas (ozone, atomic oxygen, ...) allows a "cell-by-cell" step-flow growth, which can be monitored *in-situ* using reflective high-energy electron diffraction (RHEED).

Today, the growth of high quality LSMO manganite thin films is mastered [6] and opens the way to all oxide or oxide-metal devices [7]. On $SrTiO_3$ (STO), the LSMO film appears single crystalline in a perfect "cube-on-cube" epitaxy. The film is free of any defects, dislocations and secondary phases inclusions as observed on the high resolution electron beam (HREM) cross-section image presented in the Fig. 2b. Moreover, at 300 K, M (H) loops along both the [110] and [100] in-plane easy axes are perfectly square traducing an easy domain wall propagation during reversal of magnetization [6].

The growth of half-metallic oxides with higher T_C, such as CrO_2, Sr_2MoFeO_6 and Fe_3O_4, is still under progress. More particularly, Sr_2MoFeO_6 has to be deposited at very temperatures to avoid the presence of Fe inclusions or secondary phases, such as $SrMoO_4$ [8]. Similarly, the growth of the spinel Fe_3O_4 should be further optimized to prevent the formation of anti-boundaries phases, that are reducing the magnetic properties of the thin epitaxial film [9].

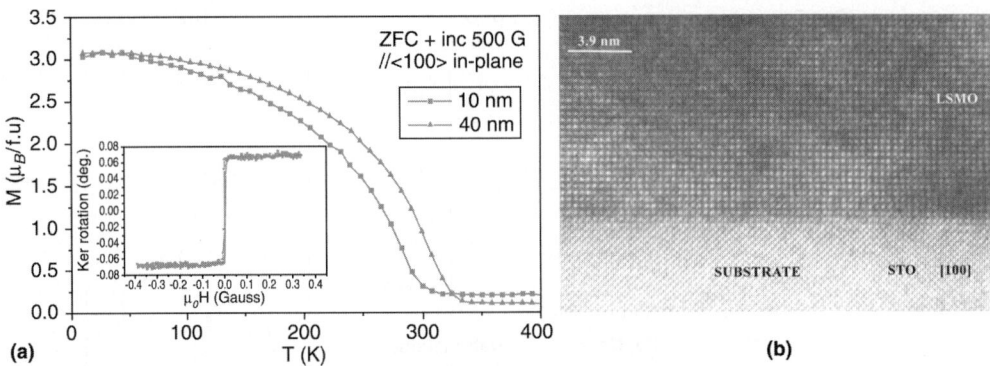

Fig. 2 (online colour at: www.interscience.wiley.com) a) M(T) and M(H) curves of a 40 nm-thick LSMO film, b) cross-section HREM image.

3 Vertical tunnel junctions

A TMR LSMO/STO/LSMO junction was realized for the first time by researchers at IBM Watson research Center in 1996. The MR response at low temperature of one 2.5×12.5 μm^2 rectangular was about 86% at 4.6 K [10]. Because the MR of such a device decreases more rapidly with increasing temperature than expected from the spin-polarization dependence of LSMO [11], researchers have attempted to improve the device fabrication process. Recently, extremely large TMR responses of up to 1800% were obtained by M. Bowen et al. in LSMO/STO ($t_{STO} = 2.8$ nm)/LSMO/Co/Au devices, for which the Ar ion etching IBE process has been optimized [12]. During the etching process with a neutralized ion beam, the sample-holder was water-cooled for reducing structural damages of LSMO. The extremely large TMR response (1800%) at 4 K leads to a spin polarization of the LSMO of at least 95% at the interface with STO. Moreover, the temperature dependence of the TMR in this optimally etched junction is shown to vanish only at $T \sim 280$ K ($T/T_C = 0.7$). Note that the vanishing temperature of the IBM junctions was $T \sim 200$ K ($T/T_C = 0.55$). In the Fig. 3, the temperature dependence of the TMR response obtained by these research groups is plotted for comparison with the polarized charge carrier density (PCCD). PCCD is the difference in the charge carriers of the different spins and is defined as DOS (E_r) × M (DOS density of states and M magnetization). The PCCD temperature dependence at the surface boundary of a LSMO film was deduced from SEPS measurements by Park et al. [11]. In the case of an optimized etching process, one should note the comparable behavior of PCCD and MR for $T/T_C > 0.4$ with a similar vanishing $T/T_C = 0.75$ (for MR = 0). In conclusion, it appears that the strain should not be a limiting factor towards obtaining large TMR response in fully polarized manganite vertical junctions exhibiting high epitaxial quality interfaces. Moreover, the mixed valence of Mn ions should be kept through the junction since the termination of LSMO at the (top-LSMO) interface LSMO/STO should be La$_{2/3}$Sr$_{1/3}$O, as deduced from EELS measurements coupled with HREM observations [13].

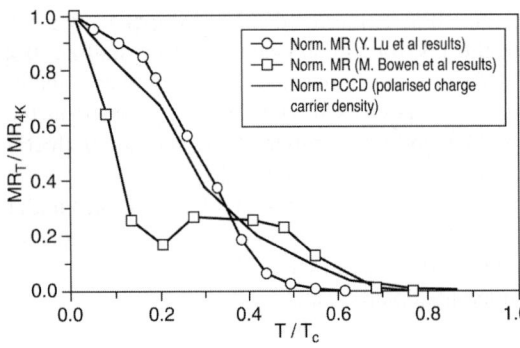

Fig. 3 Temperature dependence of the TMR response of different LSMO/STO/LSMO junctions after [10] and [12] The variation of the polarized charge carrier density PCCD is plotted for comparison (after [11]).

In order to attain larger TMR values at room temperatures, new half-metallic oxides with higher T_C, such as Fe_3O_4, have tentatively been introduced in junctions. Large negative TMR values have been measured by Suzuki et al in Fe_3O_4/$CoCr_2O_4$ (CCO)/LSMO [14]. The CCO insulating barrier is a better barrier than the commonly used STO and MgO, because it crystallizes in the same spinel structure than Fe_3O_4, producing a better epitaxy at the bottom Fe_3O_4/CCO interface. Under a field of 0.4 Tesla, the TMR was –25% at 50 K and –0.5% at room temperature. The inverse MR confirms the predicted negative spin polarization of Fe_3O_4 (P_{theor} = –80% from photoemission studies). From the MR value, the authors have deduced the polarization to be –39% in the Fe_3O_4/CCO/LSMO junctions.

4 Artificial planar junctions

To obtain large MR effects, the low field magneto-resistance (LFMR), across a defect, which can be either a grain boundary, or a domain wall, has been also used in planar devices. The LFMR is clearly related to the alignment of ferromagnetic domains from both sides of the defect. It is attributed to spin-dependent transmission of electrons across the grain boundary, which was evidenced in nice experiments performed on manganite films deposited on a STO bi-crystal substrate [15]. In these experiments, the contribution to the resistance due to the spin misalignment across the grain boundary can be accurately measured. Both the GB resistance and the magnetic field dependence are shown to vary strongly with the disorientation angle.

The introduction of artificial defects with dimensions as small as 40 nm in one single oxide layer has also been investigated. For such planar thin films devices, the fabrication is simpler than the conventional process of vertical tunnel junctions [12], with only one or two process levels. The first example concerns the growth of epitaxial films on pre-etched substrates using focused laser beam [16], for which a significant low-field tunnel MR is observed to develop across these artificial interfaces that originate in the epitaxial film from the top of the substrate laser-induced crack. Another approach concerns the pinning of magnetic domain walls in planar nanostructured devices. It has been demonstrated that the introduction of a nano-constriction in a thin ferromagnet film favors both the pinning of domain wall (DW) inside the constriction and a lateral size reduction of the constrained-DW [17]. In such a constrained DW, the spin of the electrons cannot rotate for alignment with magnetization, inducing a large increase of the resistance. If a magnetic field is applied or a pulsed current is injected, the DW can move and disappear. A large MR effect can be observed and gigantic values should be recorded in a half-metal. We have recently proposed a new concept of planar magnetic memory element, in which nanotrenches are introduced at a nanometer scale using high-resolution electron beam lithography (HREBL) and ion beam etching (IBE) [18]. These nanotrenches are shown to actually reduce the DW width, leading to strongly enhanced DW resistance. Sharp and large resistance switches (MR ~40% at 77 K), that result from the appearance and annihilation of the DWs, have been observed. Room temperature sharp resistance switches, with a MR of 16%, have been recorded for the first time in a manganite-based device. Today, we propose a simplest single-step process for elaborating simultaneously the magnetic memory bridge and the nano-constrictions. Those nanoconstrictions are nanokinks with sufficiently low lateral dimensions for pinning and constraining purposes. This new concept of planar memory is schematically described in the Fig. 4.

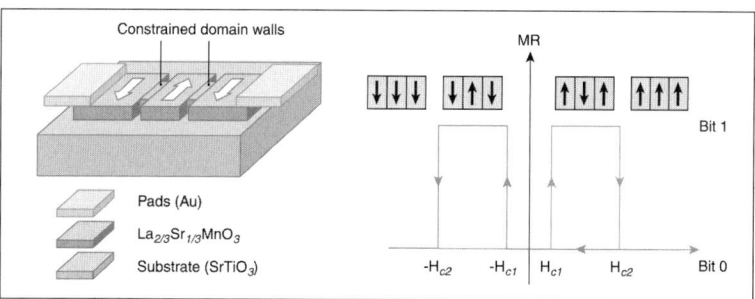

Fig. 4 Principe of the planar nanostructured magnetic memory with nanokinks.

In the last part of this review paper, we report on the optimization of the single-step nanolithography planar process, that allows generating the core-element of a spin-polarized magnetic memory in La$_{0.7}$Sr$_{0.3}$MnO$_3$ (LSMO). Taking benefit of the proximity effects due to backscattered electrons, the conventional electron-beam patterning process at 30 KeV has been optimized to generate sub-50 nm-wide nanokinks in the magnetic microbridge.

Different geometries have been tested for generating the nanokinks. The best layout is decrypted in the Figure 5a): the nanokink has been divided using both rectangular and triangular-shape patterns and the lateral width d of the kink was varied in the range 65 nm – 100 nm. To take benefit of the proximity effects observed between the two adjacent rectangles of the bridge (D_p), electron doses were decreased in both the lateral stripes and the triangles ($D_1 < D_2 < D_p$). This high resolution nanopatterning step has been performed using a 30 KeV SEMFEG microscope equipped with a beam-scanning system and coupled with a computer assisted design software ELPHY (from Raith company). Exposures were carried out on a single 300 nm-thick layer of polymethyl-polymethacrylate (PMMA) resist (molecular weight 950 K). The beam diameter was reduced down to 1 nm producing a minimal beam current of 17 pA. A small contamination dot of several nanometers in diameter was generated by a single-spot exposure of 1 minute for both focus adjustment and astigmatism correction. The current was measured using a small Faraday shaft. The exposure time per pixel was fixed at 4.8 µs and the pixel-to-pixel distance was of 7.6 nm, corresponding to a writing speed of 1.6 nm/s. All the electronic doses given below have been normalized to the value of 140 µAs/cm^2 ($D = 1$). A conventional development of 2.5 minutes in a 1:3 mixture of methyl-isobutyl-ketone (MIBK)/isopropyl alcohol (IPA) was performed. After development, the Al metallic mask for etching has been evaporated and the nanopatterns have been transferred in LSMO using an dry ion beam etching (IBE) under argon. During IBE, the sample holder was rotating and tilted at 45° to produced vertical profiles. A large current density of ions (0.5 mA/cm^2) was associated to a high voltage of 500 V. During the last step, small Cr/Au contacts lines were patterned on top of the nanostructured bridge in a 4-points configuration. The devices were finally inspected using SEM imaging mode at a high magnification (>X50000) for measuring the final size of the nanokinks.

The proximity effects of the two adjacent rectangles of the bridge (D_p) has a direct consequence on the width of the nanokink: after development, the width of the kink, which was 65 nm-wide in the pattern, appears to be always lower than 50 nm (see Fig. 5a). The optimal doses for both rectangles, lateral stripes and triangles, that produce a kink with the exact designed length (0.5 µm), are given in the Fig. 5: $D_p = 2.2$ (308 µC/cm^2), $D_2 = 0.8$ (112 µC/cm^2) and $D_1 = 0.4$ (56 µC/cm^2). For these optimal doses, the kink appears

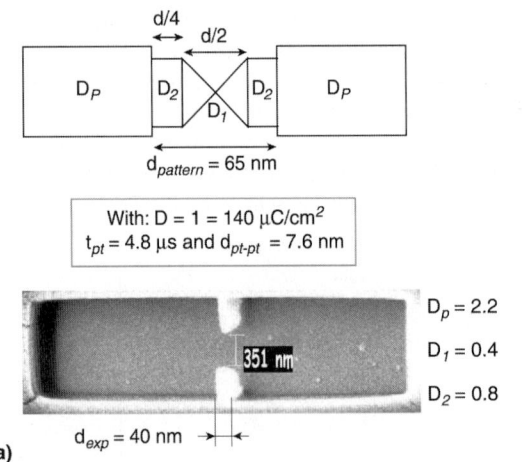
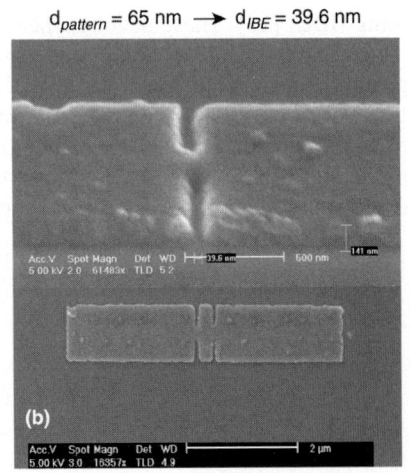

Fig. 5 a) Optimized geometry that produces the strongest PMMA nanobridge and tilted SEM image of this nanobridge after development of PMMA, b) SEM images of the smallest nanokinks (~40 nm) transferred in the LSMO thin film (after the Ar IBE process and before removal of the Al mask). The patterned width of the kink was 65 nm. The bottom view is a top view of the final device before the last level of Cr/Au pads.

© 2004 WILEY-VCH Verlag GmbH & Co. KGaA, Weinheim

to be 45 nm-wide. This corresponds to a 40% lateral reduction in width, that reflects the strong importance of the proximity effects, due to backscattered electrons from LSMO surface under PMMA resist.

Note that, at this nanolithography step, the kink has the appearance of a small resist bridge in suspension over the LSMO surface. The brittle nanobridge will act as a mask during the Al evaporation, allowing the transfer towards a kink during the Al lift-off process. The shape of the brittle nanobridge has been optimized using six different geometries of the device. The geometry presented in Fig. 5 produces the strongest nanobridge. This nanobridge exhibits long feet for a better strength, and the desired 0.5 µm depth is preserved.

The strength of nanobridges has been confirmed during the Al lift-off process, because the lateral width of the transferred in Al kink approaches the value, previously measured after development of PMMA. The transferred in LSMO kinks are presented in Fig. 5b. As a conclusion, no significant lateral reduction is observed between the PMMA development and the IBE process: the kinks are ~40 nm wide after IBE. This important result confirms that both our lift-off and IBE processes are fully optimized for preserving lateral dimensions. A general top view of the magnetic memory element can also be observed in Fig. 5b. This image has been recorded after etching before removal of the Al mask, and before the last level of contact pads. Note than the bubbles and defects observed on the surface are only inherent to Al.

This technological process illustrates the ability to pattern in a sub-50 nm range the manganite thin films. The exciting result proves that the use of half-metallic materials for spintronic devices is entering a new phase: today, one can think to a possible integration of such prototypes in an industrial context.

5 Prospects for future integration of half-metallic oxides

The growth of LSMO is well mastered from all viewpoints (crystallinity, strain effects on magnetic anisotropy, magnetoresistance, etc.) as well as its patterning at low dimensions. The challenge is now to realize devices in an industrial environment, such as sensor elements of non-volatile MRAMs, position sensors, contact-less potentiometers and bolometers. From the industrial viewpoint, two obstacles have to be alleviated first: (i) the high growth temperature (usually ~700 °C), which is not currently used in conventional processes of microelectronics; (ii) the use of a single-crystal substrate; which is expensive and does not allow growth on large surfaces. Moreover, to reach large MR effects at room temperature, the growth of new oxides with higher T_C (SFMO, Fe_3O_4 and CrO_2) should rapidly be mastered.

Acknowledgements We wish to thank A. Anane, W. Prellier, Ph. Lecoeur, B. Mercey, C. Dupas and C. Dubourdieu for fruitful scientific discussions and A. Charrier for graphic illustrations. This work was partly supported by the "Advanced Magnetic Oxides For Responsive Engineering" (GRD1-1999-10502) AMORE project.

References

[1] G. A. Prinz, Science **282**, 1660 (1998).
   J. F. Gregg et al., J. Phys. D, Applied Physics **18**, R121 (2002).
[2] S. A. Wolf et al., Science **294**, 1488 (2001).
[3] R. von Helmolt, J. Wecker, B. Holzapfel, L. Schultz, and K. Samwer, Phys. Rev. Lett. **71**, 2331 (1993).
[4] A. M. Haghiri-Gosnet and J. P. Renard, J. Phys. D, Applied Physics **36**, R127–150 (2003).
[5] K.-I. Kobayashi, T. Kimura, H. Sawada, K. Terakura, and Y. Tokura, Nature **395**, 677 (1998).
[6] M. Koubaa, A. M. Haghiri-Gosnet, R. Desfeux, Ph. Lecoeur et al., J. Appl. Phys. **93**, 5227 (2003).
[7] J. Z. Sun, L. Krusin-Elbaum, A. Gupta et al., IBM J. Res. Develop. **42**, 89 (1998).
[8] M. Besse, F. Pailloux, A. Barthelemy et al., submitted.
[9] J. F. Bobo, D. Basso, E. Snoeck, C. Gatel et al., Eur. Phys. J. B **24**, 43 (2001).
[10] Y. Lu, X. W. Li, G. Q. Gong, G. Xiao, A. Gupta, Ph. Lecoeur et al., Phys. Rev. B **54**, R8357 (1996).
[11] J.-H. Park, E. Vescovo, H.-J. Kim, C. Kwon, R. Ramesh, and T. Venkatesan, Phys. Rev. Lett. **81**, 1953 (1998).
[12] M. Bowen, M. Bibès, A. Barthélémy, J.-P. Contour, A. Anane et al., Appl. Phys. Lett. **82**, 233 (2003).
[13] F. Pailloux, D. Imhoff, T. Sikora, A. Barthélémy et al., Phys. Rev. B **66**, 014417 (2002).
[14] G. Hu and Y. Suzuki Y, Phys. Rev. Lett. **89**, 276601-1 (2002).
[15] N. D. Mathur, G. Burnell, S. P. Isaac et al., Nature **387**, 266 (1997).
[16] J. Fontcuberta, M. Bibès et al., J. Magn. Magn. Mater. **211**, 217 (2000).
[17] P. Bruno, Phys. Rev. Lett. **83**, 2425 (1999).
[18] J. Wolfman, A. M. Haghiri-Gosnet et al., J. Appl. Phys. **89**, 6955 (2001).

Liquid phase epitaxy (LPE) grown Bi, Ga, Al substituted iron garnets with huge Faraday rotation for magneto-optic applications

P. Görnert[*,1], **T. Aichele**[1], **A. Lorenz**[1], **R. Hergt**[2], and **J. Taubert**[3]

[1] INNOVENT Technologieentwicklung, Prüssingstr. 27B, 07745 Jena, Germany
[2] Institut für Physikalische Hochtechnologie, Winzerlaer Str. 10, 07745 Jena, Germany
[3] Carl Zeiss Jena GmbH, Carl-Zeiss-Promenade 10, 07745 Jena, Germany

Received 1 September 2003, revised 15 March 2004, accepted 15 March 2004
Published online 11 May 2004

PACS 78.20.Ls, 81.15.Lm, 81.70.Fy

High quality BiYIG layers can be grown on GGG and above all GGCMZ substrates by standard LPE for magneto-optic applications. The misfit stress in growing BiYIG layers with high Bi content leads to the development of morphological instabilities, which were avoided by use of lattice-matched GGCMZ substrates. Optimisation of flux melt composition and undercooling results in a Bi content up to 1.8 f.u. and a specific magneto-optical sensitivity of 0.085°/mT μm – saturation induction 54 mT; wavelength 633 nm. BiYIG layers are suitable for the imaging of magnetic structures.

© 2004 WILEY-VCH Verlag GmbH & Co. KGaA, Weinheim

1 Introduction

It is well known that Bi substituted iron garnets possess a huge specific Faraday rotation of some degrees per micrometer, where the Faraday rotation increases strongly with the Bi content [1–4, 19, 20]. High quality Bi substituted iron garnet layers are grown best of all by liquid phase epitaxy (LPE) on pure gadolinium gallium garnet (GGG) substrates or Ca, Mg, Zr substituted GGG substrates (GGCMZ) with low defect density [5 and the literature cited there]. Layers with a thickness of some micrometers are excellently suitable for magneto-optic applications to detect the local distribution of magnetic fields and for non-reciprocal optical elements such as isolators. The sensitivity of the magneto-optic sensors depends on the Faraday rotation as a function of the relevant magnetisation component and on the susceptibility. The Faraday rotation can be controlled by Bi substitution and the susceptibility by the Ga and/or Al content. All these substitutions in $(Y, RE, Bi)_3(Fe, Ga, Al)_5O_{12}$ garnets, where RE stands for one or more rare earth ions, are adjusted by the flux melt composition and undercooling. Troublesome effects of morphological growth instabilities have been shown in another paper to be due to high lattice mismatch and can be avoided by the use of appropriate GGCMZ substrates [5, 19]. Structural and magnetic characterisation of $(Y, Bi)_3(Fe, Ga, Al)_5O_{12}$ layers have been carried out. Some examples illustrate the suitability of such layers for applications as magneto-optic sensors.

2 LPE of Bi, Ga, Al substituted yttrium iron garnets

It is well known that LPE takes place near the thermodynamic equilibrium. Therefore it is possible to deposit high quality films with excellently smooth surfaces. This is especially true in the case of garnet LPE [6], where GGG based substrates are dipped horizontally under axial rotation into a flux melt,

---

[*] Corresponding author: e-mail: pg@innovent-jena.de

Table 1 Mass ratios and liquidus temperatures of four typical solutions used for the deposition of BiYIG.

No.	mass ratio					liquidus temperature T_L
	PbO/B_2O_3	Y_2O_3/Fe_2O_3	Fe_2O_3/Bi_2O_3	Fe_2O_3/Ga_2O_3	Fe_2O_3/Al_2O_3	
BGY a	35.3	0.125	0.118	6.815	–	904 °C
BGY b	20.4	0.062	0.122	7.492	–	791 °C
BGY c	20.4	0.085	0.121	16.829	23.422	845 °C
BGY da	18.4	0.052	0.122	7.493	–	751 °C

which can be isothermally undercooled by more than 100 K. Under such conditions even the growth kinetics of garnets were studied quantitatively [7].

For the deposition of $(Y_{3-x}Bi_x)(Fe_{5-y-z}Ga_yAl_z)O_{12}$ LPE layers we used a lead oxide-boron oxide mixture with iron oxide excess as solvent in platinum crucibles under ambient atmosphere. The mass ratios and liquidus temperatures T_L of four different flux melts are given in Table 1 and the corresponding growth rates as a function of the growth temperature are presented in Fig. 1. The T_L values in Table 1 have been determined by extrapolation to zero growth rate.

In agreement with Klages and Tolksdorf [8] we found that the growth can be controlled essentially by PbO, B_2O_3, and Y_2O_3. An increasing B_2O_3 and a decreasing Y_2O_3 content results in a lower liquidus temperature T_L. Besides, the negative slope of growth rate against growth temperature decreases with decreasing Y_2O_3 content.

Figure 2 reveals that the Bi content reaches values of up to 1.8 f.u., when appropriate Ca, Mg, Zr substituted GGG substrates are used [5, 19]. Nevertheless such films are of high quality. Figure 3 illustrates the incorporation of diamagnetic Ga^{3+} (and Al^{3+}) ions as a function of the growth temperature. The increase of the Bi content and the decrease of the Ga content in dependence on undercooling is qualitatively consistent with the expectations for ions with distribution coefficients less than 1 in the case of Bi^{3+} ions and larger than 1 in the case of Ga^{3+} and/or Al^{3+} [7].

Figure 4 presents preliminary measurements of the saturation magnetisation M_s as a function of the Ga and Al content of flux melt a, c, and da in Tab. 1. Magnetic compensation appears near 1.25 f.u. as expected [9], although the negative slope for layers of flux melt c is surprisingly large.

For pure YIG the magnetisation is located nearly in the film plane because of the large demagnetising field. At an increased Ga and/or Al content of roughly 1 f.u., however, a transition from in plane to perpendicular magnetisation appears. Fig. 5a shows the magneto-optic hysteresis and fig. 5b the corresponding magnetic domain structure of a $(Y_{3-x}Bi_x)(Fe_{5-y}Ga_y)O_{12}$ layer with perpendicular magnetisation. The magnetic domain structure in fig. 5b remembers to the so-called *labyrinth structure* of magnetic bubble films [10].

The magneto-optic sensitivity of the layer of Figs. 5a and b is 0.12 deg/mT with a saturation field of 8.5 mT.

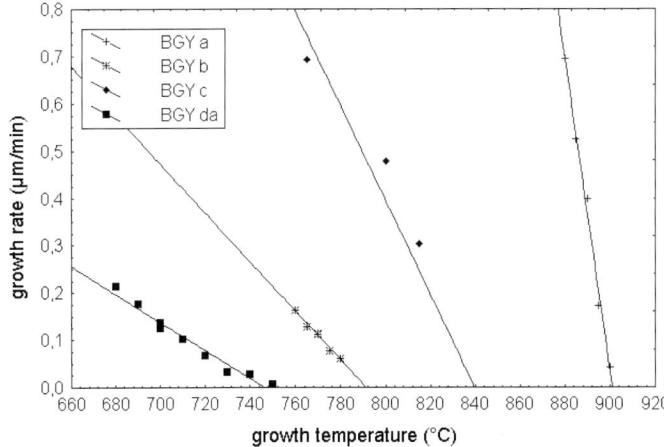

Fig. 1 Growth rate of garnet layers in dependence on growth temperature for the flux melt compositions in Table 1.

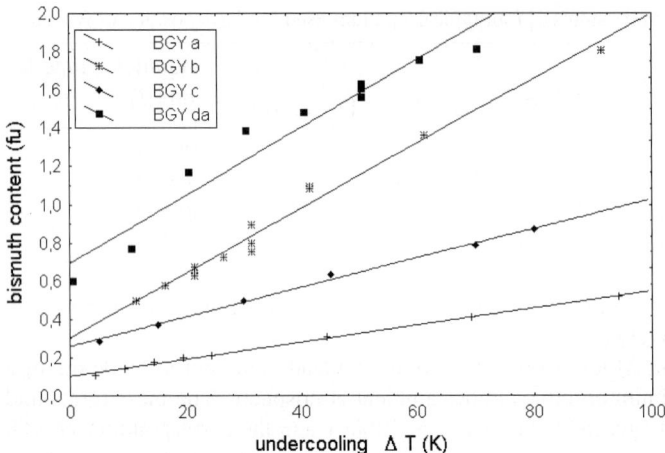

Fig. 2 Bi content in $(Y_{3-x}Bi_x)(Fe_{5-y-z}Ga_yAl_z)O_{12}$ layers in dependence on the undercooling for the flux melt compositions in Table 1.

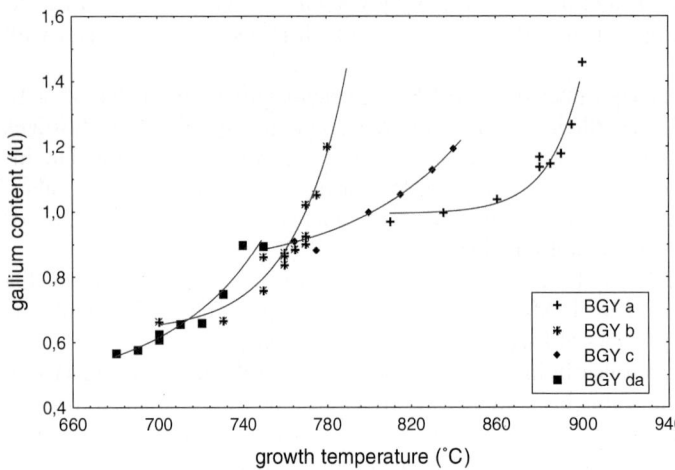

Fig. 3 Ga content in $(Y_{3-x}Bi_x)(Fe_{5-y-z}Ga_yAl_z)O_{12}$ layers as a function of the growth temperature for the flux melt compositions in Table 1.

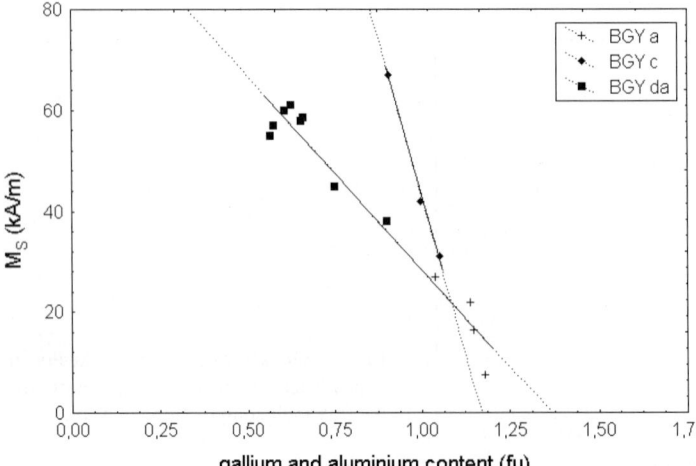

Fig. 4 Saturation magnetisation M_s as a function of the Ga and Al content of flux melt a, c, da (c.f. Table 1).

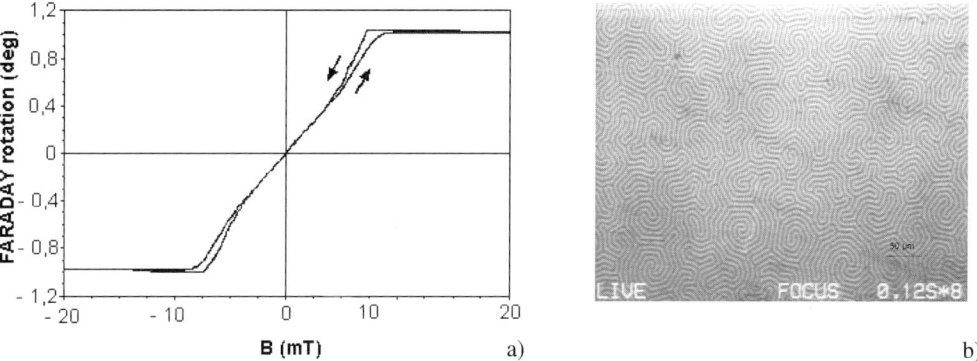

Fig. 5 a) Magneto-optical hysteresis of a $(Y_{3-x}Bi_x)(Fe_{5-y}Ga_y)O_{12}$ layer ($x = 0.6$ and $y = 1$) with perpendicular anisotropy measured at a wavelength $\lambda = 633$ nm; film thickness: 1.7 μm. b) Remanent magnetic domain structure of the layer of Fig. 5a.

3 Examples of applications

Already in the early nineties with flux grown garnet single crystals a field sensitivity of 100 pT/(Hz)$^{1/2}$ at 500 Hz was achieved which may be enhanced up to 1 pT/(Hz)$^{1/2}$ by using ferrite field concentrators [11].

Principle components of a magneto-optical field sensing device are an intense source of plane polarised light, the magneto-optical garnet film being placed in the magnetic field to be measured or imaged and a detection unit for probing the polarisation state after Faraday rotation. The commonly applied transmitting mode of operation of a Faraday rotator may be converted into a reflection mode by providing the garnet film with a reflection layer (e.g. aluminium). In this case the sensor is placed in front of the surface of a magnetic material or device to image its stray field distribution. The light beam crosses twice the garnet layer doubling the rotation angle in this way. The high quality of LPE grown garnet films allows observing the stray field with high spatial resolution using an optical microscope with large magnification.

Well known examples of field imaging are visualisations of the written information on several types of magnetic recording media from macroscopic patterns of credit cards down to the μm-range of e.g. hard discs or video tapes [12]. Potential areas of applications may be found in various fields of materials and device testing, for instance in the magnetic device industry for development and control of defined

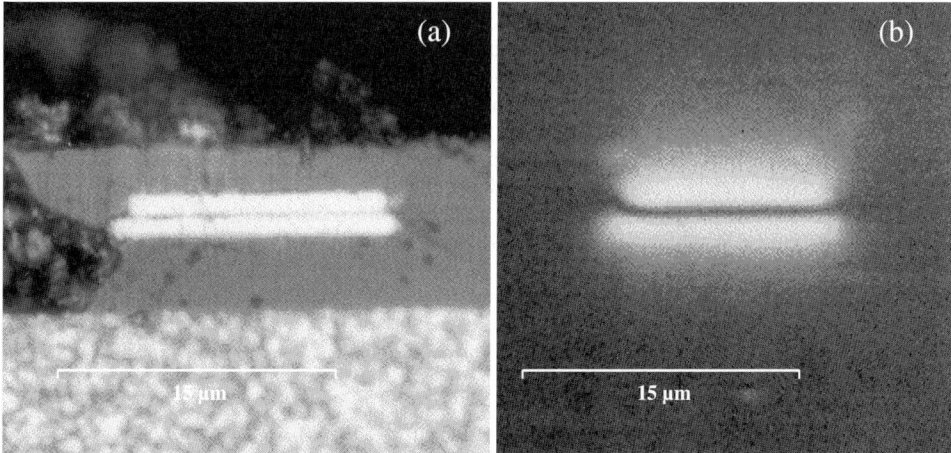

Fig. 6 (online colour at: www.interscience.wiley.com) Imaging of magnetic poles of a write head with typical magnetic field values in reflected light (a) and stray field image using a magneto-optical garnet film sensor (b).

stray field configurations (e.g. complicate multipole fields with narrow tolerances), in particular in the field of magnetic microcircuits. An important example is the detection of perturbations of magnetic scales, e.g. incremental sensors [5]. The high spatial resolution of magneto-optical imaging by means of garnet LPE films being limited only by the optical resolution of the microscope allows the imaging of the magnetic field of write heads with very small pole gap width (e.g. for hard discs) an example of which is shown in Fig. 6. There, an image of the head poles in reflected light (Fig. 6a) is compared with the stray field distribution made visible by means of a garnet sensor layer (Fig. 6b).

Other examples of application are the detection of precipitations or inclusions of non-magnetic phases in magnetic materials near to the surface of devices as shown by Taubert et al. [13] for barium ferrite glass ceramics. The same authors have demonstrated the validity of the method for investigation of weld joints. The quantity and distribution of otherwise difficult detectable magnetic δ-ferrite in austenitic ferritic steels or chrome-nickel steels may be determined even as fine precipitations [14].

Another important field of application is the imaging of current distributions [12], in particular due to field penetration in hard superconductors of second kind (see e.g. the review given by Jooss et al. [15]). An important advantage of the magneto-optical current detection e.g. in power electronics is the very rapid response time.

The high sensitivity of magneto-optical detection for small changes of magnetic fields is utilised in several areas of non-destructive evaluation (NDE) [16]. In the so-called flux leakage method, which is working without rf-excitation, stray fields arising from defects, e.g. cracks, in otherwise unperturbed magnetic material are detected magneto-optical [17]. While the flux leakage method is restricted to magnetic materials a more versatile method being applicable for all materials with metallic conductivity is the excitation of eddy currents in the specimen under investigation and the detection of the magnetic stray field by means of a magneto-optical imaging sensor instead of the commonly used inductive sensor coil [18].

Acknowledgements We like to express thanks to our colleagues M. Röder and Dr. M. Frigge from INNOVENT, and G. Bruchlos, Ch. Schmidt, U. Scheuermann, and C. Ulbrich from IPHT for measurements on the garnet layers, as well as to W. Tolksdorf for valuable discussions and hints.

References

[1] M. V. Chetkin, I. G. Morozova, and G. K. Tyutneva, Sov. Phys. Sol. State **9**, 2852 (1968).
[2] C. F. Buhrer, J. Appl. Phys. **40**, 4500 (1969).
[3] E. V. Berdennikova, R. V. Pisarev, and R. A. Petrov, Bull. Acad. Sci. USSR, Phys. Ser. **35**, 1081 (1971).
[4] J. M. Robertson, S. Wittekoek, T. J. A. Popma, and P. F. Bongers, Appl. Phys. **2**, 219 (1973).
[5] T. Aichele, A. Lorenz, R. Hergt, and P. Görnert, Cryst. Res. Technol. **38**, 575 (2003).
[6] P. Görnert, S. Bornmann, F. Voigt, and M. Wendt, phys. stat. sol. (a) **41**, 505 (1977).
[7] P. Görnert and F. Voigt, High temperature solution growth of garnets: Theoretical models and experimental results, in: Current Topics in Material Science, vol. 11, edited by E. Kaldis (Elsevier, 1984), p. 1.
[8] C.-P. Klages and W. Tolksdorf, J. Cryst. Growth **64**, 275 (1983).
[9] P. Görnert and C. G. d'Ambly, phys. stat. sol. (a) **29**, 95 (1975).
[10] A. H. Bobeck and E. Della Torre, Magnetic bubbles, North-Holland Publ. Comp., Amsterdam (1975).
[11] M. N. Deeter, G. W. Day, T. J. Beahn, and M. Manheimer, Electron. Lett. **29**, 993 (1993).
[12] W. Andrä, K. H. Geier, R. Hergt, and J. Taubert, Materialprüfung **36**, 7 (1994).
[13] J. Taubert, R. Hergt, R. Müller, C. Ulbrich, W. Schüppel, H. G. Schmidt, and P. Görnert, J. Magn. Magn. Mater. **168**, 187 (1997).
[14] J. Taubert, R. Hergt, G. Horn, and H. Heinemann, Schweißen und Schneiden (Welding and Cutting) E62 **49**, 233 (1997).
[15] Ch. Jooss, J. Albrecht, H. Kuhn, H. Kronmüller, and S. Leonhardt, Rep. Progr. Phys. **65**, 651 (2002).
[16] G. L. Fitzpatrick, D. K. Thome, R. L. Skaugset, E. Y. C. Shih, and W. C. L. Shih, in: Rev. Progr. Quantitative Nondestructive Evaluation (Plenum Press, New York, 1993).
[17] M. Shamonin, M. Klank, O. Hagedorn, and H. Dötsch, Appl. Opt. **40**, 3182 (2001).
[18] U. Radtke, R. Zielke, H.-G. Rademacher, H.-A. Crostak, and R. Hergt, Opt. Lasers Eng. **36**, 251 (2001).
[19] P. Hansen, C.-P. Klages, J. Schuldt, and K. Witter, Phys. Rev. B **31**, 5858 (1985).
[20] S. Kahl, I. Khartsev, A. M. Grishin, K. Kawano, K. Kong, R. A. Chakalov, and J. S. Abell, J. Appl. Phys. **91**, 9556, (2002).

Oxygen and fluorine doping in Sr_2MnGaO_5 brownmillerite

E. V. Antipov[*,1]**, A. M. Abakumov**[1,2]**, A. M. Alekseeva**[1]**, M. G. Rozova**[1]**, J. Hadermann**[2]**, O. I. Lebedev**[2]**, and G. Van Tendeloo**[2]

[1] Department of Chemistry, Moscow State University, Moscow 119992, Russia
[2] EMAT, University of Antwerp (RUCA), Groenenborgerlaan 171, 2020 Antwerp, Belgium

Received 1 September 2003, accepted 19 March 2004
Published online 26 April 2004

PACS 61.10.Nz, 61.14.Lj, 61.50.Nw, 61.66.Fn

Systematic study on crystal chemistry of $Sr_2MnGaO_{5+\delta}$ layered perovskites upon variation of anion stoichiometry and the Mn oxidation state (V_{Mn}) was performed. Starting from fully oxygenated $Sr_2MnGaO_{5.5}$ compound, the samples with $+3 \leq V_{Mn} \leq +4$ were prepared either by a reduction of oxygen content at controlled partial oxygen pressure or by a partial replacement of oxygen by fluorine. Varying δ is accompanied by structural transformations from *Imma* ($0.03 \leq \delta \leq 0.13$, $a \approx c \approx \sqrt{2}a_p$, $b \approx 4a_p$) to *Bmmm* ($0.41 \leq \delta \leq 0.46$, $a \approx c \approx \sqrt{2}a_p$, $b \approx 2a_p$) and to tetragonal *P4/mmm* ($\delta = 0.505$, $a \approx a_p$, $c \approx 2a_p$). The tetragonal $Sr_2MnGaO_{5-x}F_{1+x}$ oxyfluorides ($a \approx a_p$, $c \approx 2a_p$) were prepared by treatment of $Sr_2MnGaO_{5.5}$ with XeF_2. In the $Sr_2MnGaO_{4.78}F_{1.22}$ structure the MnO_6 octahedra are characterized by two short apical Mn–O distances and four long equatorial ones. This is interpreted as an "apically compressed" Jahn–Teller distortion, in contrast to the "apically elongated" one in $Sr_2MnGaO_{5+\delta}$.

© 2004 WILEY-VCH Verlag GmbH & Co. KGaA, Weinheim

1 Introduction

The brownmillerites $A_2MnGaO_{5+\delta}$ (A = Ca, Sr) [1–6] are an attractive structural matrix for new CMR materials. Sr_2MnGaO_5 can be oxidized at elevated oxygen pressure [2] up to $Sr_2MnGaO_{5.47}$ composition transforming from the orthorhombic brownmillerite structure with $a \approx c \approx a_p\sqrt{2}$, $b \approx 4a_p$ into a tetragonal perovskite with $a \approx a_p$, $c \approx 2a_p$. The structure of the oxidized compound is compressed along the direction of Mn–O apical bonds due to decreasing Jahn–Teller deformation of the MnO_6 octahedra. An appearance of extra oxygen atoms in the ($GaO_{1+\delta}$) layers alters the magnetic structure from antiferromagnetic (AFM) G-type with Mn magnetic moments aligned normal to the (MnO_2) planes ($\delta \approx 0$) to the AFM C-type where Mn magnetic moments are ordered antiferromagnetically within (MnO_2) planes but the planes are ferromagnetically coupled ($\delta \approx 0.5$) [7]. Thus the amount of anions in the Ga-layers are important for the magnetic interactions in this material. An alternative way to vary the oxidation state of the Mn cations is an insertion of fluorine atoms into the Ga-layers. Different amounts of oxygen and fluorine are required to create the same Mn oxidation state that can alter the structure and properties of the oxyfluorides in comparison with the oxygen-doped compounds. In this contribution we describe the preparation and structures of the $Sr_2MnGaO_{5+\delta}$ compounds with different oxygen content and the $Sr_2MnGaO_{5-x}F_{1+x}$ oxyfluorides.

---

[*] Corresponding author: e-mail: antipov@icr.chem.msu.ru, Phone: +007-095-9393375, Fax: +007-095-9394788

2 Experimental

The initial $Sr_2MnGaO_{5+\delta}$ oxide with $\delta \approx 0$ was prepared by a solid state reaction as described in [2]. Then the sample was treated in flowing oxygen at 415 °C to reach $\delta \approx 0.5$. The butch was separated into several samples for further reduction in a closed circle filled with Ar/O$_2$ mixture at fixed partial oxygen pressure. The partial oxygen pressure inside the circle was controlled by a cell with a unipolar oxygen conductivity. After annealing the sample was quenched. The compounds with $\delta \approx 0, 0.5$ were also treated with XeF$_2$. The operations with XeF$_2$ were carried out in a glove box filled with dried N$_2$. 0.4 g of $Sr_2MnGaO_{5+\delta}$ was mixed with XeF$_2$, ground, placed in a Ni crucible and sealed into a N$_2$-filled copper tube. The samples were annealed at 500–600 °C for 10–20 h and then furnace cooled to room temperature. The Mn oxidation state (V_{Mn}) was determined by an iodometric titration. Powder XRD study was performed with a focusing Guinier-camera FR-552 and a STADI-P diffractometer (CuK$_{\alpha1}$-radiation). The WINCSD program package and RIETAN-2000 program were used for the structure determination from X-ray powder data. The electron diffraction (ED) study was performed using JEOL 400EX and Philips CM20 transmission electron microscopes.

3 Results

The data on some representative oxygenated and fluorinated samples are given in Table 1. A difference between cell dimensions and V_{Mn} of the as-prepared samples **I** and **II** illustrates the influence of subtle changes in synthesis procedure on the final oxygen content. Both samples were prepared using the same

Table 1 Synthesis conditions, lattice parameters, cell volume, the degree of orthorhombic distortion $(c - a)/(c + a)$ and formal Mn oxidation state for the $Sr_2MnGaO_{5+\delta}$ (##1–6) and $Sr_2MnGaO_{5-x}F_{1+x}$ (##7–9) samples. Two times and four times increased unit cell volume are given in square brackets.

#	treatment conditions, P(O$_2$) in atm.	space group	cell parameters, Å	V, Å^3	$(c-a)/(c+a)$	δ, x	V_{Mn}
1	as-prepared **I**	Imma	a = 5.4023(5) b = 16.130(2) c = 5.5645(7)	484.9	0.0148	−0.03	+2.94
2	as prepared **II**	Imma	a = 5.3894(4) b = 16.193(1) c = 5.5348(4)	483.0	0.0133	0.025	+3.05
3	415 °C, lg(P(O$_2$)) = −5.19	Imma	a = 5.3700(5) b = 16.240(2) c = 5.4976(4)	479.5	0.0117	0.13	+3.26
4	415 °C, lg(P(O$_2$)) = −4.04	Bmmm	a = 5.3628(3) b = 8.0193(8) c = 5.4066(3)	232.5 [465.0]	0.0041	0.41	+3.82
5	415 °C, lg(P(O$_2$)) = −1.12	Bmmm	c = 5.3666(6) a = 7.959(2) b = 5.3861(5)	230.1 [460.1]	0.0018	0.46	+3.92
6	415 °C, lg(P(O$_2$)) = 0	P4/mmm	a = 3.7995(1) c = 7.9209(3)	114.3 [457.4]	0	0.505	+4.01
7	1:1 XeF$_2$, 500 °C, 10 h	P4/mmm	a = 3.8557(3) c = 7.7843(8)	115.7 [462.9]	0	1.22	+3.78
8	1:1 XeF$_2$, 600 °C, 20 h	P4/mmm	a = 3.865(1) c = 7.779(3)	116.2 [464.8]	0	1.46	+3.54
9	1:3 XeF$_2$, 600 °C, 20 h	P4/mmm	a = 3.8794(3) c = 7.7328(6)	116.4 [465.5]	0	1.61	+3.39

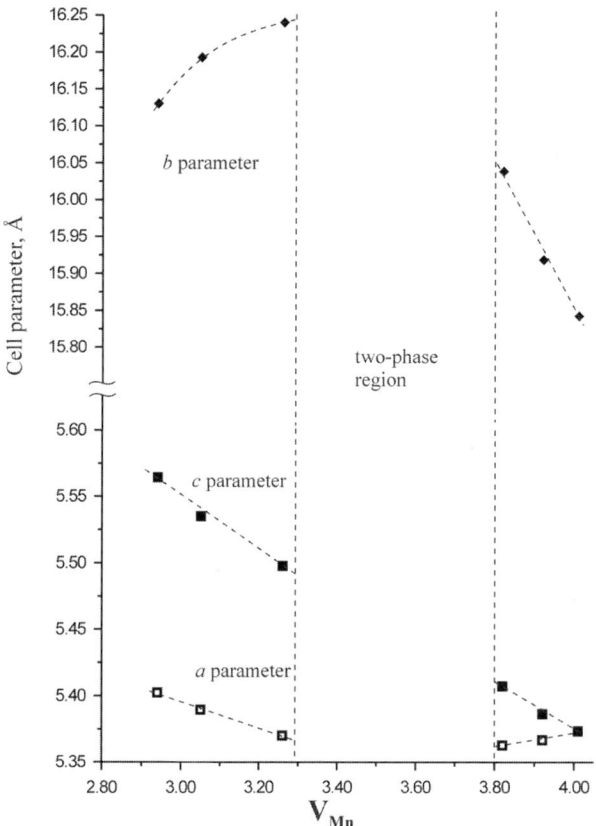

Fig. 1 Dependencies of cell parameters of $Sr_2MnGaO_{5+\delta}$ from δ. The cell parameters for all compounds are given in orthorhombic *Imma* setting.

synthesis procedure except for the ratio between the sample weight and the sealed silica tube volume. In the case of sample **II** this ratio was increased in 5 times in comparison with sample **I** which caused smaller oxygen loss and the formation of the more oxidized sample. The single phase oxygenated samples can be separated among two regions embracing oxidation states of Mn of $+2.94 \leq V_{Mn} \leq +3.26$ (a low-oxidized region) and $+3.82 \leq V_{Mn} \leq +4.01$ (a high-oxidized region). The samples with $+3.26 < V_{Mn} < +3.82$ consist of a mixture of two phases, which, according to their lattice parameters, belong to the low-oxidized and high-oxidized regions which may indicate either a presence of a two-phase region at $+3.26 < V_{Mn} < +3.82$ or non-equilibrium state of the samples. The XRD powder patterns of the samples from the low-oxidized region (samples ##1–3) were indexed in an orthorhombic brownmillerite-type unit cell ($a \approx c \approx a_p\sqrt{2}, b \approx 4a_p$, S.G *Imma*). The XRD patterns of the samples #4 and #5 from a high-oxidized region were indexed in a B-centered orthorhombic unit cell with $a \approx c \approx a_p\sqrt{2}, b \approx 2a_p$. The XRD pattern of the sample #6 can be indexed with a tetragonal primitive unit cell ($a \approx a_p, c \approx 2a_p$) For the oxygenated samples cell volume and degree of orthorhombic distortion decrease concomitantly with increasing oxygen content (Fig. 1). The behavior of the cell parameters *vs* oxygen content is not so straightforward. The *c* parameter decreases almost linearly with increasing oxygen content. In the low-oxidized region the *a* parameter decreases whilst the *b* parameter increases; this behavior changes to the opposite in the high-oxidized region (Fig. 1). It was found that the treatment of the reduced ($\delta \approx 0$) Sr_2MnGaO_5 phase with XeF_2 does not allow the preparation of single phase samples with a wide range of fluorine contents. Hence, the interaction of the oxidized $Sr_2MnGaO_{5+\delta}$ ($\delta \approx 0.5$) sample with XeF_2 was investigated. The XRD patterns of the fluorinated samples were indexed using a tetragonally distorted perovskite unit cell ($a \approx a_p, c \approx 2a_p$). Badly crystallised SrF_2 impurity was found in the samples in an amount growing with increasing amount of XeF_2 and with the temperature of fluorination. An increase of the fluorination temperature and a molar $XeF_2/Sr_2MnGaO_{5+\delta}$ ratio cause an increase of the *a* lattice parameter and a decrease of the *c* lattice parameter of the tetragonal phase. These changes occur simultaneously with a decrease of V_{Mn}. We should note that iodometric titration provides an estimate of V_{Mn} for the fluorinated phases in samples #8 and #9 because of the presence of impurity phases.

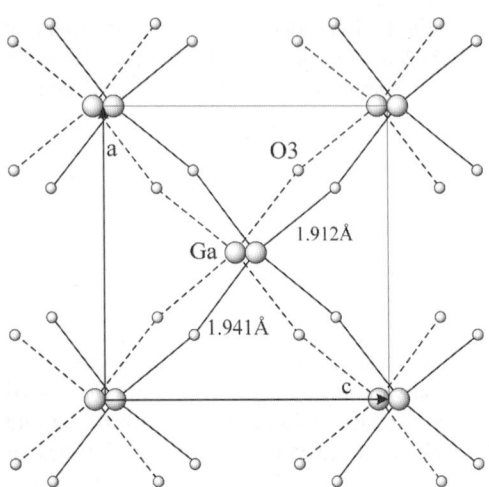

Fig. 2 Disorder in equatorial coordination environment of Ga atoms in $Sr_2MnGaO_{5.46}$.

3.1 Crystal structure of $Sr_2MnGaO_{5.46}$

The initial model for the refinement of the $Sr_2MnGaO_{5.46}$ crystal structure was calculated from the structure of tetragonal perovskite by transforming the atomic coordinates into an orthorhombic unit cell with $a \approx c \approx a_p\sqrt{2}$, $b \approx 2a_p$ and $Bmmm$ space symmetry. The refinement with Ga in (1/2, 0, 1/2) positions and O3 in (1/4, 0, 1/4) results in abnormally high atomic displacement parameters (ADP) for these atoms. The $Bm2m$, $Bmm2$ and $B222$ models were also tested, but no signigicant improvement was achieved. For subsequent refinement the Ga position was splitted into 8-fold (1/2, y, z) with occupancy factor $g = 0.25$ and the O3 position – into (x, 0, y) with $g = 0.365$ that corresponds to $\delta = 0.46$ found by iodometric titration. This led to reasonable values of isotropic ADPs for Ga and O3 atoms and to a decrease of the reliability factor from $R_I = 0.033$ to $R_I = 0.026$. The $Sr_2MnGaO_{5.46}$ structure is characterized by strong disorder in the $(GaO_{1+\delta})$ layers since the significant amount of extra oxygen in these layers changes the coordination environment of the Ga atoms (Fig. 2). Formally Ga atoms are surrounded by an octahedron built of O3 and O2 atoms. However, due the displacement of Ga and O3 atoms from their special positions, there are two equatorial squares of O3 atoms around Ga, each with the probability of 1/2. One equatorial square of oxygen atoms is displaced with respect to the other along the c axis, as it is shown in Fig. 2. Ga atoms are not situated exactly within the equatorial plane since they are statistically displaced towards one of the apical oxygen atoms O2. Thus five short (1.809–1.941 Å) and one long (2.206 Å) Ga–O separations were found. If one takes into account that the probability to find the O3 atom in each equatorial vertex of one of equatorial square is close to 3/4, the average coordination number of Ga should be close to 5, comprising three O3 atoms at the distances of 1.912–1.941 Å, one O2 atom at 1.809 Å and one O2 atom at 2.206 Å. These five oxygen atoms form distorted trigonal bipyramid around the Ga cations. The BVS calculation for this coordination environment gave +2.88 which is close to the formal charge of the Ga cation. The CN = 5 is not unusual for Ga^{3+} cations and the structures are known where Ga is situated in a trigonal bipyramid.

3.2 The crystal structure of $Sr_2MnGaO_{4.78}F_{1.22}$

The structure investigation was performed for the most pure sample #7, whose XRD pattern shows only traces of SrF_2. The reciprocal lattice of this compound was studied by electron diffraction. All reflections on the observed reciprocal lattice sections can be indexed on a tetragonal lattice with cell parameters as determined from XRD data. Since no extinctions were observed on the XRD and ED patterns, the $P4/mmm$ space group was chosen for the Rietveld refinement. The atomic coordinates were transformed from a perovskite structure and full occupancy was assigned to all anion positions giving the

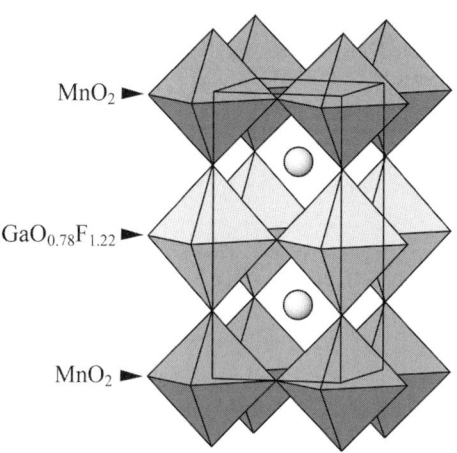

Fig. 3 Crystal structure of $Sr_2MnGaO_{4.78}F_{1.22}$. MnO_6 octahedra are darker shaded.

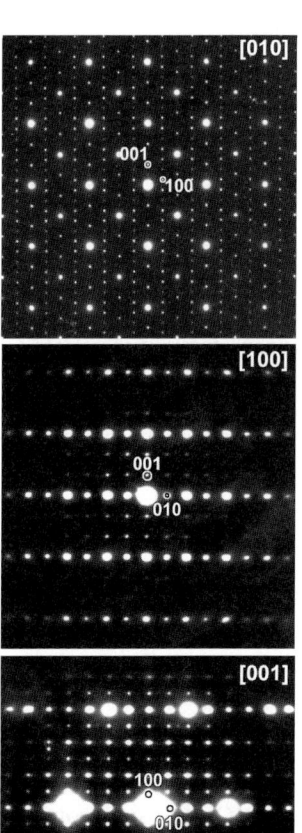

$Sr_2MnGa(O,F)_6$ composition. SrF_2 was introduced as a second phase into the refinement. The occupancy factors of the O(1), O(2) and O(3) positions were refined, which gives $gO(1) = 1.07(1)$, $gO(2) = 0.99(1)$, $gO(3) = 0.993(7)$ and shows that all anion positions are fully occupied. The reliability factors $R_I = 0.012$, $R_P = 0.019$ indicate a good agreement between experimental and calculated profiles. The main difference between the oxygen-doped $Sr_2MnGaO_{5.46}$ phase and the fluorinated $Sr_2MnGaO_{4.78}F_{1.22}$ phase is the full occupation of the anion positions in the latter structure. The Mn formal valence $V_{Mn} = +3.78$ known from iodometric titration allows to propose the $Sr_2MnGaO_{4.78}F_{1.22}$ formula for the fluorinated phase in the sample #7. In the $Sr_2MnGaO_{4.78}F_{1.22}$ structure the Mn and Ga atoms are situated in slightly distorted octahedra (Fig. 3). The bond valence sum calculations and evaluations of lattice energies support the location of fluorine in the (1/2, 0, 1/2) position, i.e. in the equatorial environment of the Ga atoms. The octahedral environment around the Mn atoms can be described as "apically compressed" with two short apical Mn–O distances of 1.876(8) Å and four long equatorial ones of 1.9278(1) Å. The octahedron around the Ga atoms is characterized by two apical Ga–O distances of 2.016(8) Å, which are longer than four equatorial Ga–(O, F) ones. The average interatomic distances are $\langle d(Mn-O)\rangle = 1.911$ Å and $\langle d(Ga-(O, F))\rangle = 1.957$ Å.

3.3 Electron diffraction study of anion ordering in $Sr_2MnGaO_{5.41}$

The reciprocal lattice for $Sr_2MnGaO_{5.41}$ was reconstructed using a large number of ED patterns (Fig. 4). All bright reflections on these patterns belong to the *Bmmm* sublattice, weaker reflections correspond to a superstructure indexed in monoclinic unit cell with parameters $a = 10.813$ Å, $b = 8.019$ Å, $c = 8.486$ Å, $\beta = 108.57°$, S.G. $P2/m$. Ordering of oxygen atoms and vacancies can be assumed as possible reason for

Fig. 4 Electron diffraction patterns of the $Sr_2MnGaO_{5.41}$ sample.

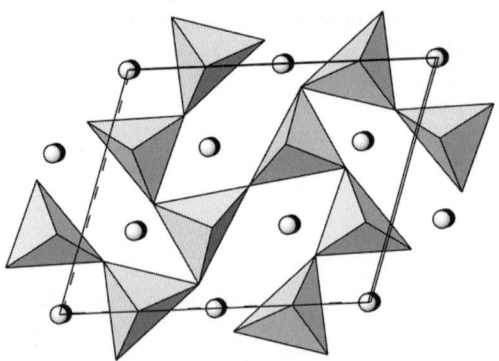

Fig. 5 Model of $Sr_2MnGaO_{5.41}$ structure: the Ga layer. Ga atoms are situated in trigonal bipyramids, Sr cations are shown as spheres.

the monoclinic superstructure. Since the average coordination number of Ga atoms in $Sr_2MnGaO_{5.41}$ is close to 5, we can propose an arrangement of GaO_5 pyramids for the $(GaO_{1+\delta})$ layer which is compatible with the monoclinic supercell and the $P2/m$ space group. However, tetragonal pyramid is not a typical coordination environment for the Ga^{3+} cation. In seldom cases of CN = 5 the Ga^{3+} cations are located in trigonal bipyramids rather than at tetragonal pyramids. Appropriate displacements of O and Ga atoms can be introduced to transform the coordination environment of Ga into trigonal bipyramids (Fig. 5). This model corresponds to the $Sr_2MnGaO_{5.5}$ composition, but the experimentally determined oxygen content implies a smaller oxygen amount, that leads to a partial occupation of at least one anion position in the Ga layer by oxygen and vacancies, so that part of the Ga atoms have a tetrahedral coordination.

4 Conclusions

An increase of oxygen content in the $Sr_2GaMnO_{5+\delta}$ brownmillerite occurs as an oxidative process resulting in several trends which can be summarised as follows:

– decrease of the average size of Mn cations on going from a formal oxidation state +3 to +4. The ionic radii of Mn^{3+} and Mn^{4+} in octahedral environment are equal to 0.65 Å and 0.54 Å, respectively;
– suppression of Jahn–Teller distortion of MnO_6 octahedra due to a transition from $t_{2g}^3 e_g^1$ to $t_{2g}^3 e_g^0$ electronic configuration;
– decrease of tilting distortion of the (MnO_2) layers;
– increase of the average Ga–O distance due to appearance of extra oxygen atoms in the $(GaO_{1+\delta})$ layers.

In contrast to the oxygen insertion, at our particular experimental conditions the fluorination is a reductive reaction of partial replacement of oxygen by fluorine. The overall scheme of the fluorination process can be expressed by the equation:

$$Sr_2MnGaO_{5.5} + (1 + x)/2 \; XeF_2 \rightarrow Sr_2MnGaO_{5-x}F_{1+x} + (0.5 + x) \; [O] + (1 + x)/2 \; Xe$$

Oxygen atoms released in this reaction are absorbed at the inner walls of the copper tube. Increasing the temperature of the fluorination and the amount of XeF_2 increases the degree of the anion replacement resulting in a gradual decrease of the formal Mn valence in the samples ##7 – 9.

It is interesting to compare the behavior of oxygen-doped $Sr_2MnGaO_{5+\delta}$ and fluorine doped $Sr_2MnGaO_{5-x}F_{1+x}$ compounds upon variation of the Mn oxidation state. In both cases the equatorial Mn–O distances in the MnO_6 octahedra increases simultaneously with decreasing V_{Mn} (decreasing δ or increasing F content): from 1.9001 Å for V_{Mn} = +4 to 1.9222 Å (V_{Mn} = +3.26, δ = 0.13) and 1.9411 Å (V_{Mn} = +3.39, x_F = 1.61, an estimation). These changes follow an increase of the Mn cation size upon reduction. However, the apical Mn–O distance varies differently in the $Sr_2MnGaO_{5-x}F_{1+x}$ solid solutions in comparison to that in the $Sr_2MnGaO_{5+\delta}$ ones. Elongation of the Mn–O_{ap} distance occurs together with a decrease of δ for the $Sr_2MnGaO_{5+\delta}$ compounds due to an increase of the degree of Jahn–Teller distortion: 1.956(3) Å (δ = 0.5), 1.983(6) Å (δ = 0.46), 2.225(8) Å (δ = 0.13), 2.372(5) Å (δ = 0.025) and 2.411(4) Å (δ = –0.03). For the $Sr_2MnGaO_{5-x}F_{1+x}$ solid solutions the c lattice parameter decreases together with decreasing V_{Mn} which may indicate a progressive apical compression of the MnO$_6$ octahedra. Indeed, d(Mn–O$_{ap}$) = 1.876(8) Å in the Sr$_2$MnGaO$_{4.78}$F$_{1.22}$ is clearly shorter than d(Mn–O$_{ap}$) = 1.956(3) Å in Sr$_2$MnGaO$_{5.5}$ [7]. At the same time, the apical Ga–O distance stays the same for these structures within the range of standard deviations: 2.024(3) Å for Sr$_2$MnGaO$_{5.5}$ [7] and 2.016(8) Å for Sr$_2$MnGaO$_{4.78}$F$_{1.22}$. It confirms that the compression of the unit cell of the Sr$_2$MnGaO$_{5-x}$F$_{1+x}$ solid solutions along the c axis occurs due to a shortening of the Mn–O$_{ap}$ bonds. Thus we can propose that the reduction of V_{Mn} by fluorination reverses the type of Jahn–Teller distortion of the MnO$_6$ octahedra from "apically elongated" to "apically compessed". We assume that the reversed Jahn–Teller effect in the fluorinated samples arises from an interplay between the decrease of the free energy from a stabilization due to a Jahn–Teller distortion and a simultaneous variation of the electrostatic lattice energy due to changes in the bond distances. In the oxygen doped Sr$_2$MnGaO$_{5+\delta}$ compounds the increase of the apical elongation of the MnO$_6$ octahedra is to some degree compensated by the decrease of the Ga–O$_{ap}$ distance due to a decrease of the coordination of Ga down to 4. For the Sr$_2$MnGaO$_{5-x}$F$_{1+x}$ solid solutions the Ga–O$_{ap}$ distance does not alter since no changes in the coordination number occur and the average ionic radius of the apical anions does not change. Thus, if the reduction of the Mn cations will be accompanied by an elongation of the Mn–O$_{ap}$ distances, it will result in abnormally long Sr–O separations and in a decrease of the lattice energy. From the crystal field theory, for a high-spin d^4 electronic configuration "apically elongated" and "apically compressed" octahedra are energetically equivalent, but the latter in the Sr$_2$MnGaO$_{5-x}$F$_{1+x}$ structure provides an additional stabilization due to the gain in the electrostatic energy.

Acknowledgements The American Chemical Society Petroleum Research Fund is acknowledged for partial support of this research (project 38459-AC5). A.M.A. is grateful to the INTAS for the Fellowship grant for Young Scientists YSF 2002-48. E.V.A. is grateful to Russian Science Support Foundation for the financial support.

References

[1] A. M. Abakumov, M. G. Rozova, B. Ph. Pavlyuk, M. V. Lobanov, E. V. Antipov, O. I. Lebedev, G. Van Tendeloo, D. V. Sheptyakov, A. M. Balagurov, and F. Bouree, J. Solid State Chem. **158**, 100 (2001).
[2] A. M. Abakumov, M. G. Rozova, B. Ph. Pavlyuk, M. V. Lobanov, E. V. Antipov, O. I. Lebedev, G. Van Tendeloo, O. L. Ignatchik, E. A. Ovtchenkov, Yu. A. Koksharov, and A. N. Vasil'ev, J. Solid State Chem. **160**, 353 (2001).
[3] A. J. Wright, H. M. Palmer, P. A. Anderson, and C. Greaves, J. Mater. Chem. **11**, 1324 (2001).
[4] A. J. Wright, H. M. Palmer, P. A. Anderson, and C. Greaves, J. Mater. Chem. **12**, 978 (2002).
[5] A. M. Abakumov, A. M. Alekseeva, M. G. Rozova, E. V. Antipov, O. I. Lebedev, and G. Van Tendeloo, J. Solid State Chem. **174**, 319 (2003).
[6] P. D. Battle, A. M. Bell, S. J. Blundell, A. I. Coldea, D. J. Gallon, F. L. Pratt, M. J. Rosseinsky, and C. A. Steer, J. Solid State Chem. **167**, 188 (2002).
[7] V. Yu. Pomjakushin, A. M. Balagurov, T. V. Elzhov, D. V. Sheptyakov, P. Fisher, D. I. Khomskii, V. Yu. Yushankhai, A. M. Abakumov, M. G. Rozova, E. V. Antipov, M. V. Lobanov, and S. J. L. Billinge, Phys. Rev. B **66**, 184412 (2002).

Effect of Fe doping on the structural and magneto transport properties in $Pr_{0.67}Sr_{0.33}MnO_3$ perovskite manganese

W. Boujelben[*,1], **M. Ellouze**[1], **A. Cheikh-Rouhou**[1], **R. Madar**[2], and **H. Fuess**[3]

[1] Laboratoire de Physique des Matériaux, Faculté des Sciences de Sfax, B. P. 802, 3018, Sfax Tunisia
[2] Laboratoire de Matériaux et du Génie Physique (UMR 5628 CNRS) ENSPG, B.P. 46, 38402 Saint Martin d'Hères Cedex, France
[3] Materials Science, University of technology, Petersenstrasse 23, 64287 Darmstadt, Germany

Received 1 September 2003, revised 10 March 2004, accepted 15 March 2004
Published online 11 May 2004

PACS 61.10.Nz, 75.30.Cr, 75.30.Kz, 75.47.Gk, 75.50.Pp, 75.60.Ej

In this work, we studied the crystallographic, magnetic and transport properties of $Pr_{0.67}Sr_{0.33}Mn_{1-x}Fe_xO_3$ polycrystalline compounds ($x = 0.01, 0.02, 0.04, 0.06, 0.08, 0.1, 0.2$ and 1). X-ray diffraction investigations show that our synthesized samples crystallize in the orthorhombic perovskite structure with Pnma space group. Magnetic measurements versus temperature for $Pr_{0.67}Sr_{0.33}Mn_{1-x}Fe_xO_3$ samples under magnetic applied field of 500 Oe show a paramagnetic (PM) to ferromagnetic (FM) transition with decreasing temperature for $x = 0.01-0.1$. The Curie temperature T_C decreases with increasing x, it is found to be 280 K for $x = 0.01$ and 136 K for $x = 0.1$. $Pr_{0.67}Sr_{0.33}Mn_{1-x}Fe_xO_3$ samples ($x = 0.01, 0.06, 0.08$ and 0.1) exhibit a semiconducting-metallic transition with decreasing temperature in the vicinity of T_C. Magneto-transport studies show a sharp drop in the resistivity as a function of the magnetic applied field.

© 2004 WILEY-VCH Verlag GmbH & Co. KGaA, Weinheim

1 Introduction

During the past few years, extensive studies have been performed on the perovskite manganese oxides $Ln_{1-x}A_xMnO_3$ (Ln is a rare-earth element and A is a divalent alkaline earth or a monovalent alkali metal) in order to understand the physical properties of such compounds [1]. Different substitutions in these systems lead to different crystal structures, spin states and transport properties. Electrical transport and ferromagnetism in these systems were traditionally interpreted within the frame work of double exchange (DE) interaction [2] which considers the exchange of electrons between neighboring Mn^{3+} and Mn^{4+} sites with strong on-site Hund's coupling. However Millis et al. [3] have showed that DE alone cannot explain the behaviors observed in these systems and suggested that other effects play a crucial role, such as charge ordering, average A-site cationic radius $\langle r_A \rangle$ [4], A-site cationic size mismatch [5], oxygen deficiency [6] and polaron effect due to the strong electron–phonon arising from Jahn–Teller distortion [3]. These systems have been given much attention because they could possibly be used for sensor applications [7].

Previous studies on $Pr_{1-x}Sr_xMnO_3$ showed that the Curie temperature T_C is optimized for Mn^{4+} content of about 33% 8]. Most of the studies on the Fe-doped substituted-lanthanum and neodymium manganites have been focused on $La_{1-x}Sr_xMnO_3$, $La_{1-x}Ca_xMnO_3$ and $Nd_{1-x}Sr_xMnO_3$ [9–12]. We have recently synthesized Fe-doped substituted-praseodymium manganites compounds of compositions $Pr_{0.67}Sr_{0.33}Mn_{1-x}Fe_xO_3$ (with $x = 0.01-0.2$ and 1).

To understand the physical properties of this system, a systematic study of the magnetic and magneto-transport properties of these bulk perovskite compounds have been performed and reported in this paper.

---

[*] Corresponding author: e-mail: Wahiba.boujelben@fss.rnu.tn, Phone: +00 216 74276400, Fax: +00 216 74274437

2 Experimental

The polycrystalline samples $Pr_{0.67}Sr_{0.33}Mn_{1-x}Fe_xO_3$ ($x = 0.01–0.2$ and 1) were prepared using the solid-state reaction by mixing Pr_6O_{11}, Mn_2O_3, $SrCO_3$ and Fe_2O_3 up to 99.9% purity in the desired proportions. The precursors were thoroughly mixed in an agate mortar, then fired in air at about 1000 °C for three days with an intermediate regrinding. After pulverizing again, the powder was pressed into pellets (of about 1mm thickness) and sintered at 1400 °C in air for two days.

Phase purity, homogeneity, and cell dimensions were determined by X-ray diffraction at room temperature (diffractometer using CuK_α radiation). Unit cell dimensions were obtained by least-squares calculations.

Magnetization measurements versus temperature were recorded by a vibrating sample magnetometer in the temperature range 10–350 K under a magnetic applied field of 500 Oe.

Resistivity measurements as a function of temperature and magnetic applied field were carried out on dense ceramic pellets using the standard four-probe technique.

3 Results and discussion

Figure 1 shows the X-ray diffraction patterns for $Pr_{0.67}Sr_{0.33}Mn_{1-x}Fe_xO_3$ compounds with $x = 0.02$, 0.06, 0.1 and 0.2 at room temperature. The patterns indicate that all the samples are single-phase without any detectable impurity or secondary phase. All the X-ray diffraction reflection lines were successfully indexed according to an orthorhombic perovskite structure with Pnma space group. No apparent structural changes by Fe doping can be identified. We list in Table 1 the unit cell volume and the lattice parameters for all our synthesized samples.

According to Ahn et al. [12], iron enters into the samples as Fe^{3+} and takes place of the Mn^{3+} cations. As Mn^{3+} and Fe^{3+} have almost the same ionic radius of 0.645 Å [13], a small change in the lattice parameters is observed by Fe doping. On the other hand, other works on $La_{0.7}Pb_{0.3}Mn_{1-x}Fe_xO_3$ [14], $Nd_{0.67}Sr_{0.33}Mn_{1-x}Fe_xMnO_3$ [15] and $La_{0.7}Sr_{0.3}Mn_{1-x}Fe_xO_3$ [10] samples show that the distortion of the structure is nearly independent of the amount of the Fe doping.

The $Pr_{0.67}Sr_{0.33}MnO_3$ sample (33% Mn^{4+}) is ferromagnetic below $T_C \sim 287$ K [16]. In order to study the effect of the Fe doping upon the magnetic properties, we have performed magnetic measurements versus temperature under a magnetic applied field of 500 Oe. We plot in Fig. 2 the magnetization evolution versus temperature for $Pr_{0.67}Sr_{0.33}Mn_{1-x}Fe_xO_3$ samples ($x = 0.01–0.2$). Our measurements show that polycrystalline samples with $x = 0.01–0.1$ exhibit a paramagnetic to ferromagnetic transition with decreasing temperature. A very small ferromagnetic component is observed for $x = 0$.

The Fe doping induces a weakness of the ferromagnetism at low temperature in our samples. This weakness is more important above 10% of Fe and indicates an antiferromagnetic Fe–O–Mn coupling. This result argues that the replacement of Mn by Fe destroys some ratios of Mn^{3+}–O^{2-}–Mn^{4+} bonds and the interactions between Fe–O–Mn bonds might be antiferromagnetic superexchange, supporting the previous authors results that most Fe ions are ferromagnetically coupled with the Mn ions in the La and Nd-based systems [10, 15]. In the high Fe-doping levels the strong competition between the Mn–O–Mn

Table 1 Crystallographic data of $Pr_{0.67}Sr_{0.33}Mn_{1-x}Fe_xO_3$ samples.

x	a (Å)	b (Å)	c (Å)	V (Å^3)
0.01	5.451(1)	7.703(3)	5.479(3)	230.09
0.02	5.451(7)	7.704(0)	5.479(3)	230.13
0.04	5.452(5)	7.706(0)	5.479(5)	230.23
0.06	5.453(2)	7.707(9)	5.480(0)	230.34
0.08	5.454(0)	7.708(7)	5.480(8)	230.43
0.10	5.454(7)	7.709(5)	5.480(9)	230.49
0.20	5.459(1)	7.717(0)	5.483(3)	231.00

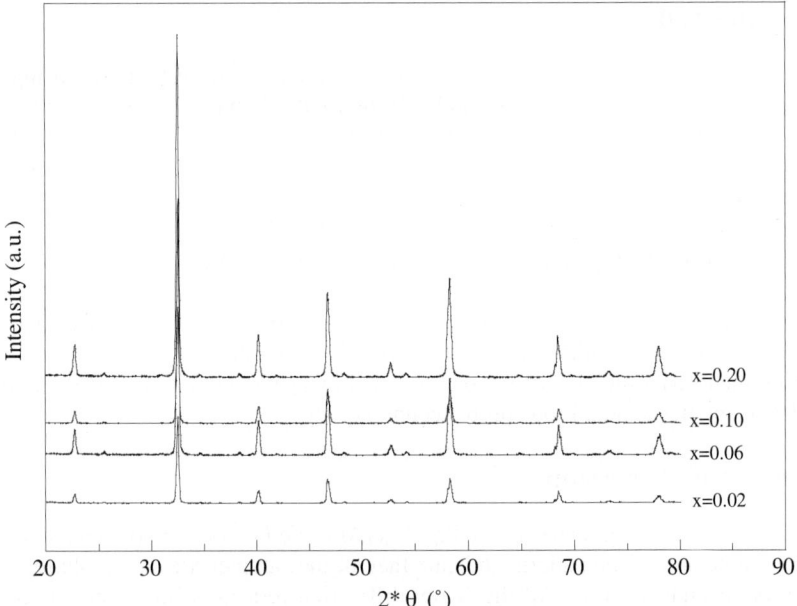

Fig. 1 X-ray diffraction patterns as a function of Fe doping for $Pr_{0.67}Sr_{0.33}Mn_{1-x}Fe_xO_3$ samples.

double exchange and Mn–O–Fe superexchange interactions might also result in a weakness of the ferromagnetism.

Another interesting magnetic characteristic is the rapid dawnshifting of the Curie temperature, T_C, by the Fe doping, as observed in Fig. 3. These values of T_C were determined as the temperature at which the

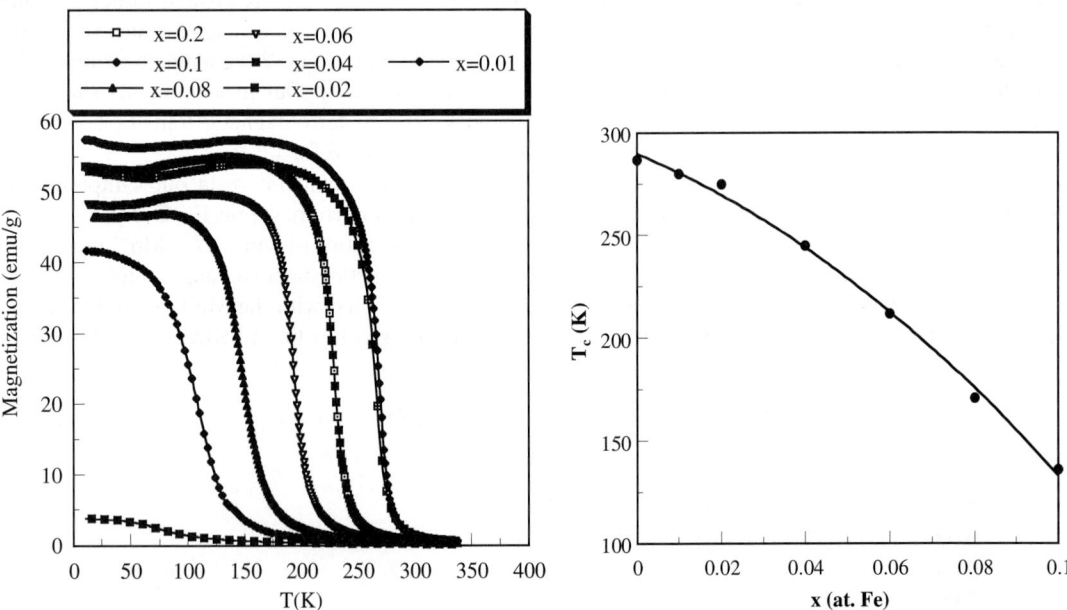

Fig. 2 Magnetization evolution as a function of temperature for $Pr_{0.67}Sr_{0.33}Mn_{1-x}Fe_xO_3$ samples under a magnetic applied field of 500 Oe.

Fig. 3 Curie temperature evolution as a function of x for $Pr_{0.67}Sr_{0.33}Mn_{1-x}Fe_xO_3$ samples.

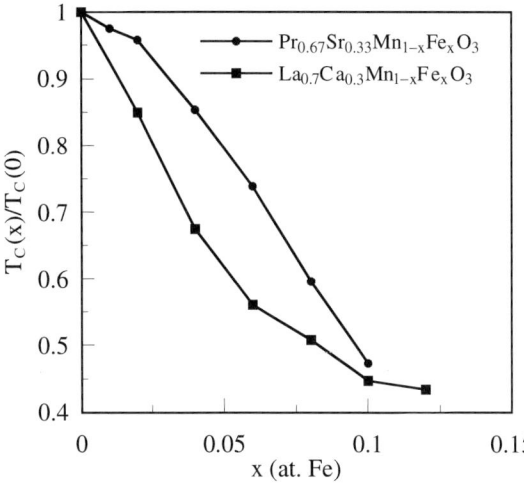
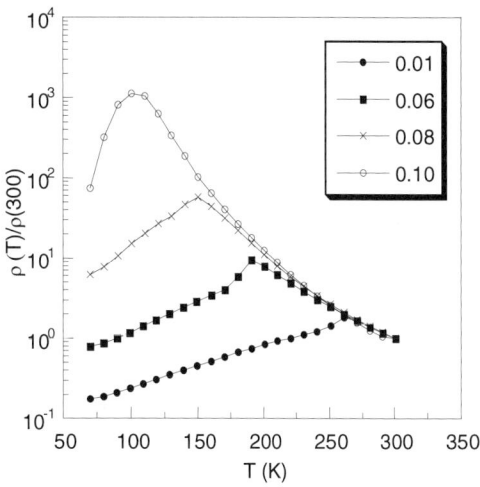

Fig. 4 $T_C(x)/T_C(0)$ evolution versus Fe doping in both $Pr_{0.67}Sr_{0.33}Mn_{1-x}Fe_xO_3$ and $La_{0.7}Ca_{0.3}Mn_{1-x}Fe_xO_3$ series.

Fig. 5 Temperature dependence of the resistivity at 0 T of $Pr_{0.67}Sr_{0.33}Mn_{1-x}Fe_xO_3$ ($x = 0.01, 0.06, 0.08$ and 0.1) samples.

$\frac{\partial M}{\partial T}(T)$ curves show a minimum. The feromagnetic transition temperature T_C decreases with increasing x. It is found to decrease from 280 K for $x = 0.01$ to 136 K for $x = 0.1$. Fe doping suppresses the double exchange ferromagnetic interactions between Mn^{3+} and Mn^{4+} by depopulating the available hopping e_g electrons of Mn^{3+} according to Ahn et al. [12].

Rao et al. [16] reported the Fe doping effect on the physical properties of $La_{0.7}Ca_{0.3}Mn_{1-x}Fe_xO_3$. A phenomenological cluster model has been proposed to elucidate the reduction of the Curie temperature T_C due to the Fe doping. According to this model, the relation between the reduction of T_C and x due to the Fe doping in our samples is given by $\frac{T_C(x)}{T_C(0)} = (1 - 1.49x)(1 - 6x)$. The initial reduction rate of T_C is thus $d\left(\frac{T_C(x)}{T_C(0)}\right)/dx\Big|_{x \to 0} = -7.32$, whereas the experimental value $\left(\frac{T_C(0.02)}{T_C(0)} - 1\right)/0.02 = -2$. The

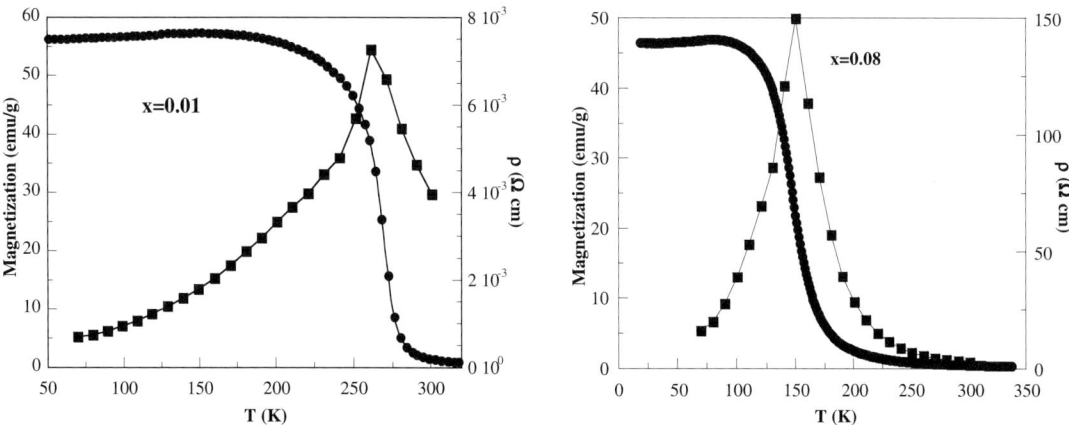

Fig. 6 Correlation between the temperature dependence of resistivity and magnetization for $Pr_{0.67}Sr_{0.33}Mn_{1-x}Fe_xO_3$ ($x = 0.01$ and 0.08).

© 2004 WILEY-VCH Verlag GmbH & Co. KGaA, Weinheim

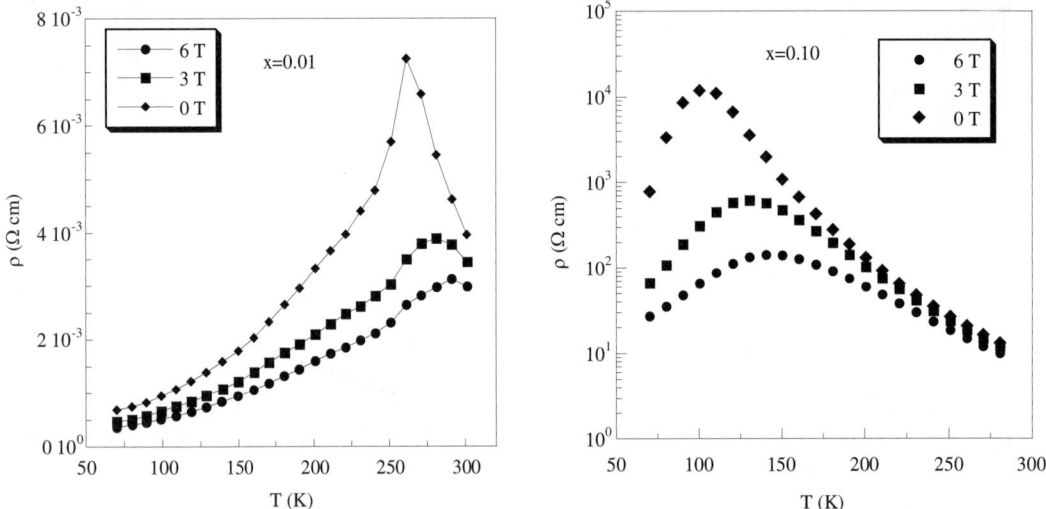

Fig. 7 Resistivity evolution versus temperature at several magnetic applied field of 0, 3 and 6 T for $Pr_{0.67}Sr_{0.33}Mn_{1-x}Fe_xO_3$ ($x = 0.01$ and 0.1).

difference between the experimental and the calculated values may be due to the praseodymium moment which can be polarized by the Mn moments. We plot in Fig. 4 the $T_C(x)/T_C(0)$ evolution versus Fe doping in both $Pr_{0.67}Sr_{0.33}Mn_{1-x}Fe_xO_3$ and $La_{0.7}Ca_{0.3}Mn_{1-x}Fe_xO_3$ [17] series.

Figure 5 presents the resistivity evolution, ρ, as a function of temperature for $Pr_{0.67}Sr_{0.33}Mn_{1-x}Fe_xO_3$ ($x = 0.01, 0.06, 0.08$ and 0.1) samples in the rare earth magnetic field. All samples exhibit a semiconducting–metallic transition with decreasing temperature (T_ρ corresponds to the maximum value of the resistivity). T_ρ decreases with increasing Fe content while the resistivity value at $T = T_\rho$ increases when increasing Fe doping. The evolution of the electrical transition temperature T_ρ is similar to that obtained for magnetic transition temperature T_C. The semiconducting–metallic transition occurs at the same temperature as the ferromagnetic-paramagnetic one, which indicates a strong correlation between magnetic and electrical properties.

We report in Fig. 6 the correlation between resistivity and magnetization evolutions versus temperature for samples corresponding to $x = 0.01$ and 0.08. The curves show a steep decrease of resistivity with the onset of the ferromagnetic magnetization.

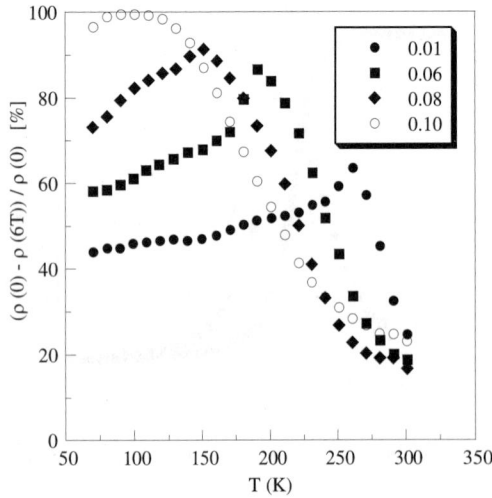

Fig. 8 Magnetoresistance evolution as a function of temperature for $Pr_{0.67}Sr_{0.33}Mn_{1-x}Fe_xO_3$ ($x = 0.01, 0.06, 0.08$ and 0.1) samples under 6 T.

In order to study the magneto-transport properties, we have carried out resistivity measurements versus temperature at several magnetic applied field up to 6 T. In Fig. 7, we plott the resistivity evolution versus temperature at 0, 3 and 6 T for $x = 0.01$ and 0.1. The effect of the magnetic field is to lower the resistivity values, meanwhile the resistivity peak becomes broader and shifts to higher temperatures.

The coexistence of ferromagnetism and metallic conductivity at low temperatures leads to the existence of magnetoresistance effect MR. Defining MR at a given temperature as $\mathrm{MR} = \dfrac{\Delta \rho}{\rho} = \dfrac{\rho(0) \, \rho(H)}{\rho(0)}$

where $\rho(H)$ and $\rho(0)$ are the resistivities at a magnetic field H and at zero field respectively. Figure 8 shows the temperature dependence of the MR for $Pr_{0.67}Sr_{0.33}Mn_{1-x}Fe_xO_3$ ($x = 0.01$, 0.06, 0.08 and 0.1) samples at 6 T.

For $x = 0.01$, a maximum of MR of about 64% was recorded at $T \approx 261$ K. With increasing Fe content, the maximum of MR increases up to 87%, 92% and 100% for $x = 0.06$, 0.08 and 0.1, respectively. However the corresponding T_ρ was significantly lowered.

4 Conclusion

We have studied the Fe doping effects on the structural, magnetic and electrical properties in $Pr_{0.67}Sr_{0.33}Mn_{1-x}Fe_xO_3$ powder samples. X-ray diffraction study show that Fe doping has a small effect on the lattice constant, however it leads to a decrease of T_C, a weakness of the ferromagnetism at low temperatures and an increase of the resistivity. The magnetotransport measurements show a sharp drop in the resistivity as a function of the magnetic applied field. The maximum of the magnetoresistance increases with increasing Fe content.

Acknowledgements This work has been supported by the Tunisian Secretary of State for Scientific Research and Technology.

References

[1] C. N. R. Rao and A. K. Raychaudhuri, Pekeris, Colossal magnetoresistance, charge ordering and other Novel properties of manganites and related materials (World Scientific, Singapore, 1998).
[2] C. Zener, Phys. Rev. **82**, 403 (1951).
[3] A. J. Millis, P. B. little Wood, and B. I. Shraiman, Phys. Rev. Lett. **74**, 5144 (1995).
[4] F. Damay, C. Martin, A. Martin, and B. Raveau, J. Appl. Phys. **81**, 1372 (1997).
[5] L. M. Rodriguez-Martinez and J. P. Atfield, Phys. Rev. B **54**, 15622 (1996).
[6] I. O. Troyanchuk, S. V. Trukhanov, H. Szymezak, and K. Baerner, J. Phys.: Condens. Matter **12**, L155 (2000).
[7] S. Jin. M. McCormack, T. H. Tiefel, and R. Ramesh, J. Appl. Phys. **76**, 6929 (1994).
[8] W. Boujelben, A. Cheikh-Rouhou, M. Ellouze, and J. C. Joubert, phys. stat. sol. (a) **177**, 503 (2000).
[9] S. K. Hasanaim, M. Madeem, W. H. Shah, M. J. Akhtar, and M. M. Hasan, J. Phys.: Conds. Matter **12**, 9007 (2000).
[10] J. H. Zhang, X. J. Fan, C. S. Xiong, and X.-G. Li, Solid State Commun. **115**, 531 (2000).
[11] Y. L. Chany, Q. Huang, and C. K. Ony, J. Appl. Phys. **91**, 789 (2002).
[12] H. Ahn, X. W. Wu, K. Liu, and C. L. Chien, J. Appl. Phys. **81**, 5505 (1999).
[13] J. M. D. Coey and M. Viret, Adv. Phys. **48**, 167 (1999).
[14] J. Gutienez, A. Pena, J. M. Barandiaran, J. L. Pizarro, T. Hemandez, L. Lezama, M. Inausti, and T. Roja, Phys. Rev. B **61**, 9028 (2000).
[15] Y. L. Chang, Q. Huang, and C. K. Ong, Phys. Appl. Phys. **91**, 789 (2002).
[16] W. Boujelben, A. Cheikh-Rouhou, M. Ellouze, and J. C. Joubert, Phase Transit. **71**, 127 (2000).
[17] G. H. Rao, J. R. Sun, A. Kattwinkel, L. Haupt, K. Bärner, E. Schmitt, and E. Gmelin, Physica B **269**, 379 (1999).

Deficiency effects on the physical properties of the lacunar $La_{0.5}Ca_{0.5-x}MnO_3$ manganese oxides

I. Walha[1], W. Boujelben[1], M. Koubaa[1,2], A. Cheikh-Rouhou[*,1], and A. M. Haghiri-Gosnet[2]

[1] Laboratoire de Physique des Matériaux, Faculté des Sciences de Sfax, B.P. 802, 3018 Sfax, Tunisia
[2] Institut d'Electronique Fondamentale, IEF/UMR 8622, Université Paris Sud, Bâtiment 220, 91405 Orsay Cedex, France

Received 26 January 2004, revised 21 March 2004, accepted 23 March 2004
Published online 11 May 2004

PACS 75.47.Gk, 75.60.Ej

We investigate the effects of calcium deficient on the structural and magnetotransport properties of non-stoichiometric $La_{0.5}Ca_{0.5-x}\square_x MnO_3$ manganese oxides. Powder samples have been elaborated using the solid-state reaction technique. Our synthesized samples crystallize in the orthorhombic perovskite structure with Pnma space group. Electrical measurements versus temperature in magnetic applied field up to 8 T on $La_{0.5}Ca_{0.5-x}\square_x MnO_3$ compounds have been performed. Resistivity measurements of the stoichiometric sample $La_{0.5}Ca_{0.5}MnO_3$ show a semiconducting behaviour in the whole temperature range of 20–300 K. An increase of the resistivity at very low temperature can be attributed to the charge ordering (CO) effect. Calcium defect leads to an important decrease of the resistivity at low temperature and consequently a destruction of the CO effect observed at very low temperature in the parent compound. Electrical measurements show that 5% of calcium deficiency induces a semiconducting–metallic transition with decreasing temperature. The electrical transition temperature increases with increasing deficient content. 5% of calcium deficient induces the same effect produced by an applied magnetic field of 8 T on the parent compound.

© 2004 WILEY-VCH Verlag GmbH & Co. KGaA, Weinheim

1 Introduction

The perovskite-type manganites of the general formula $Ln_{1-x}A_xMnO_3$ where Ln is a trivalent rare-earth and A is a divalent alkaline-earth or a monovalent alkali metal have been extensively investigated since the discovery of the colossal magnetoresistance effects in such materials [1–6]. In these materials, the magnetotransport could be explained on the basis of double exchange (DE) mechanism [7–9] between Mn^{3+}–Mn^{4+} pairs with conductivity occurring by electron hopping between manganese ions along Mn–O–Mn bonds. However, the double exchange cannot alone explain the rich variety of phenomena found in these compounds. The system $La_{1-x}Ca_xMnO_3$ has a rich phase diagram [10], the half-doped manganite, with $x = 0.5$, is charge-spin-orbital ordered [11, 12]. The effect of divalent alkaline-earth element substitution in the stoichiometric perovskite manganites $La_{1-x}Ca_xMnO_3$ have been extensively studied, however, only few studies have been carried out on the deficiency effects in lacunar systems [13, 14].

In order to study the vacancy effects in $La_{0.5}Ca_{0.5}MnO_3$, we investigate the structural and magnetotransport properties in the $La_{0.5}Ca_{0.5-x}\square_x MnO_3$ lacunar samples with $0 \leq x \leq 0.25$.

2 Experimental

Powder samples of $La_{0.5}Ca_{0.5-x}\square_x MnO_3$ ($0 \leq x \leq 0.25$) were prepared using the solid state reaction by mixing La_2O_3, MnO_2 and $CaCO_3$ up to 99.9% purity in the desired proportion according to the following reaction:

$$0.5 La_2O_3 + 2(0.5 - x) CaCO_3 + 2 MnO_2 \longrightarrow 2 La_{0.5}Ca_{0.5-x}\square_x MnO_3 + \delta CO_2.$$

---

[*] Corresponding author: e-mail: abdel.Cheikhrouhou@fss.rnu.tn

Table 1 Chemical analysis results for lacunar $La_{0.5}Ca_{0.5-x}\square_xMnO_3$ samples.

x	% Mn^{4+} theoretical	% Mn^{4+} experimental $La_{0.5}Ca_{0.5-x}\square_xMnO_3$	relative error (%)
0.05	60	59.75	0.5
0.10	70	68.86	1.6
0.15	80	77.92	2.6
0.20	90	87.63	2.6
0.25	100	96.73	3.2

The starting materials were thoroughly mixed in an agate mortar and then heated in air at 1000 °C for 60 hours. A systematically annealing at high temperature is necessary to ensure a complete reaction. In fact the powders are pressed into pellets (of about 1 mm thickness) and sintered at 1400 °C in air for 60 hours with intermediate regrinding and repelling. Finally, the pellets were rapidly quenched to room temperature in air.

Phase purity, homogeneity and cell dimensions were determined by X-ray diffraction at room temperature. Unit cell dimensions were obtained by least-squares calculations.

Resistivity measurements as a function of temperature and magnetic applied field were carried out on dense ceramic pellets by the standard four-probe technique.

3 Results and discussion A vacancy in the A site implies a partial conversion of Mn^{3+} to Mn^{4+} leading to an increase in the Mn^{4+} content above 50%. According to the general formula, the Mn tetravalent and Mn trivalent contents are $(0.5 + 2x)$ and $(0.5 - 2x)$ in the lacunar perovskite oxides $La_{0.5}Ca_{0.5-x}\square_xMnO_3$, respectively. The calcium vacancy leads to a change in the average ionic radius $\langle r_A \rangle$ of the A site. For electrostatic considerations, a vacancy must have an average radius $\langle r_V \rangle \neq 0$.

The Mn^{3+} and Mn^{4+} contents have been checked by chemical analysis. In Table 1 we list the chemical analysis results. The experimental results agree with the theoretical data.

X-ray diffraction patterns at room temperature of $La_{0.5}Ca_{0.5-x}\square_xMnO_3$ show that all our samples are of single phase (Fig. 1). Our synthesized samples crystallize in the orthorhombic perovskite structure with Pnma space group. Calcium deficiencies do not modify the $La_{0.5}Ca_{0.5}MnO_3$ structure. In Table 2, we list the crystallographic data for $La_{0.5}Ca_{0.5-x}\square_xMnO_3$ samples.

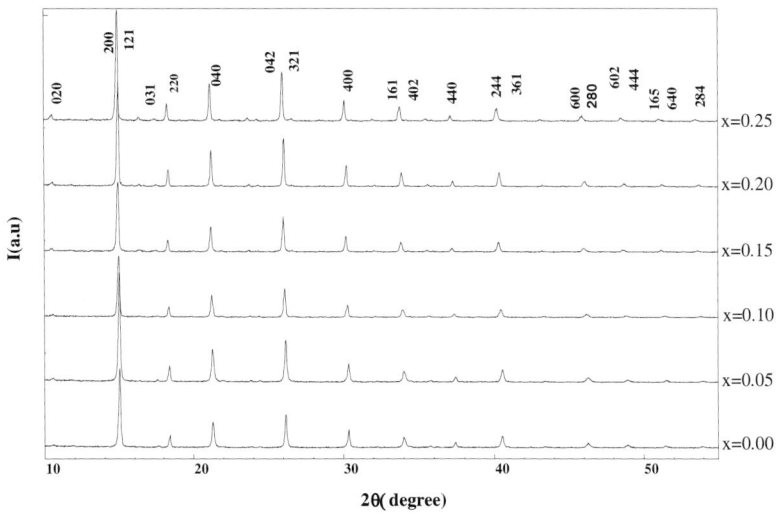

Fig. 1 X-ray powder diffraction patterns of $La_{0.5}Ca_{0.5-x}\square_xMnO_3$ samples.

Table 2 Crystallographic data of lacunar $La_{0.5}Ca_{0.5-x}\square_xMnO_3$ samples.

x	a (Å)	b (Å)	c (Å)	V (Å³)
0.00	5.429(1)	7.693(4)	5.421(8)	226.45
0.05	5.430(2)	7.696(1)	5.422(1)	226.62
0.10	5.438(9)	7.707(5)	5.426(5)	227.48
0.15	5.450(0)	7.718(8)	5.434(3)	228.60
0.20	5.459(8)	7.739(7)	5.449(3)	230.27
0.25	5.470(6)	7.752(4)	5.456(9)	231.43

With increasing calcium deficiency content, the cell parameters a, b, c and the unit cell volume (V) increase. Such a result has been also observed in a previous work on $Pr_{0.5}Sr_{0.5-x}\square_xMnO_3$ samples [15], with increasing strontium defect amount. These samples crystallize in the orthorhombic structure with Imma space group. The increase of Mn^{4+} content due to the increase of the calcium-defect amount cannot explain the increase of the unit cell volume, and consequently other parameters as the evolution of the average ionic radius $\langle r_A \rangle$ of the A site induced by the lacuna may explain such a behaviour.

In order to study the magnetotransport properties of our synthesized samples, we have performed resistivity measurement versus temperature at different applied magnetic fields up to 8 T.

In Fig. 2 we have plotted resistivity measurements of the stoichiometric sample $La_{0.5}Ca_{0.5}MnO_3$ at different applied magnetic fields up to 8 T.

Using the sign of the temperature coefficient of resistivity $\left(\frac{d\rho}{dT}\right)$ as a criterion $\left(\frac{d\rho}{dT} < 0 \text{ for a semiconductor-like system and } \frac{d\rho}{dT} > 0 \text{ for a metallic-like system}\right)$, we found that the parent compound at zero field shows a semiconducting behaviour in the whole temperature range of 20–300 K with a change in the slope at about 100 K. This behaviour can be explained by the charge ordering (CO) effect.

Applied magnetic field leads to an important decrease of the resistivity values at low temperature and consequently a destruction of the CO effect observed at low temperature. The applied magnetic field induces also a semiconducting–metallic transition when the temperature decreases. With increasing applied magnetic field the electrical transition temperature T_ρ increases and the resistivity peak becomes broader.

In order to study the effect of calcium deficiency on the electrical properties of $La_{0.5}Ca_{0.5}MnO_3$, we have performed electrical measurements at zero applied magnetic field of $La_{0.5}Ca_{0.5-x}\square_xMnO_3$ lacunar samples ($x = 0.05$ and 0.1) (Fig. 3).

Calcium deficiency leads to an important decrease of the resistivity at low temperatures and induces a semiconducting–metallic transition when temperature decreases and consequently a destruction of the CO effect observed at very low temperature in the parent compound.

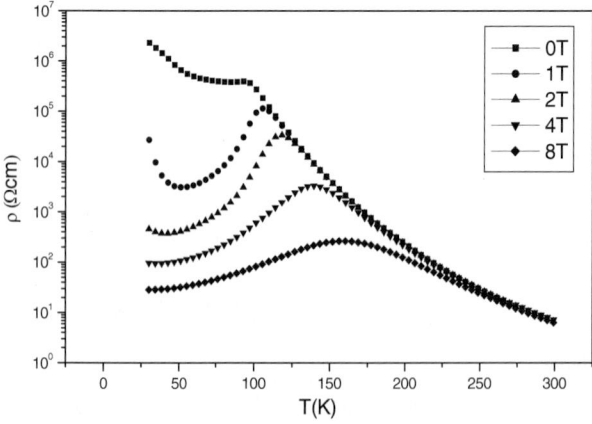

Fig. 2 Resistivity evolution as a function of temperature at different applied magnetic fields for the stoichiometric sample $La_{0.5}Ca_{0.5}MnO_3$.

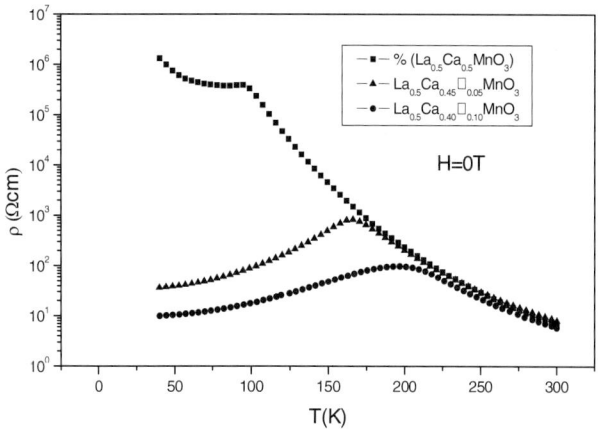

Fig. 3 Resistivity evolution as a function of temperature for $La_{0.5}Ca_{0.5-x}\square_x MnO_3$ samples ($0 \leq x \leq 0.1$) at $H = 0$ T.

The electrical transition temperature increases with increasing calcium deficiency content. It is important to notify that 5% of calcium deficient induces the same effect produced by an applied magnetic field of 8 T in the parent compound.

Figure 4 shows resistivity measurements versus temperature at applied magnetic fields of 0 and 8 T.

The effect of the magnetic field of 8 T is to lower the resistivity values; the resistivity peak becomes broader and shifts to higher temperatures.

Defining the MR at a given temperature as $\mathrm{MR} = \dfrac{\Delta \rho}{\rho} = \dfrac{\rho(0) - \rho(H)}{\rho(0)}$, where $\rho(0)$ and $\rho(H)$ are the resistivity values at zero and at applied magnetic field H, respectively, Fig. 5 shows the magnetoresistance evolution versus temperature at several applied magnetic fields for the sample $La_{0.5}Ca_{0.5}MnO_3$.

The MR remains constant at about 100% at low temperatures and than decreases with increasing temperature. The MR decrease depends strongly on the applied magnetic field. This decrease occurs rapidly for low field values and becomes more slowly with increasing magnetic field.

Figure 6 shows the magnetoresistance evolution versus temperature at an applied magnetic field of 8 T for lacunar samples $La_{0.5}Ca_{0.5-x}\square_x MnO_3$ ($x = 0.05$ and 0.1).

Contrary to the parent compound, the MR for calcium-deficient samples exhibits a maximum with decreasing temperature. The temperature corresponding to the maximum of the MR increases while the maximum of the MR decreases when the calcium deficiency content increases.

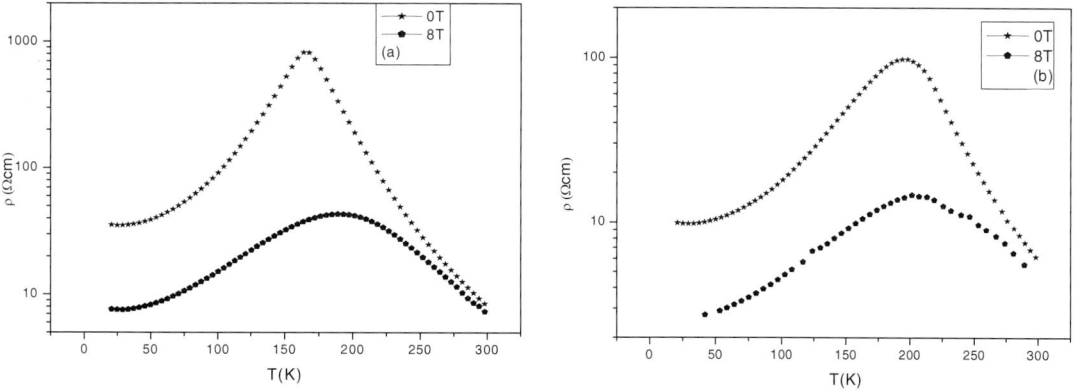

Fig. 4 Resistivity evolution as a function of temperature at $H = 0$ T and 8 T for both lacunar samples $La_{0.5}Ca_{0.45}\square_{0.05}MnO_3$ (a) and $La_{0.5}Ca_{0.4}\square_{0.1}MnO_3$ (b).

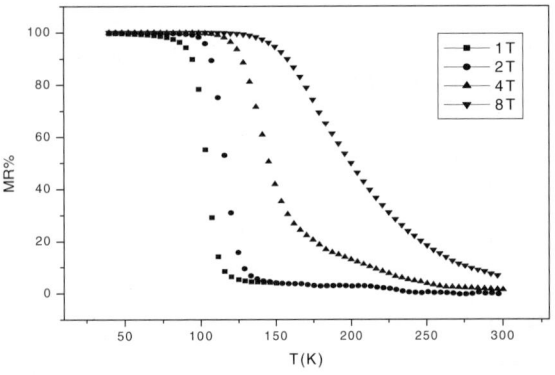
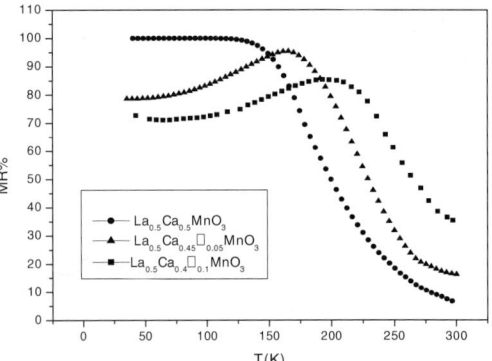

Fig. 5 Magnetoresistance evolution as a function of temperature for the sample $La_{0.5}Ca_{0.5}MnO_3$ at several magnetic applied fields.

Fig. 6 Temperature dependence of the magnetoresistance at $H = 8$ T for lacunar $La_{0.5}Ca_{0.5-x}\square_x MnO_3$ samples ($0 \leq x \leq 0.1$).

At 300 K, calcium deficient leads to an increase of the MR value. The MR is found to be 35% for $x = 0.1$, 20% for $x = 0.05$ and about 10% for $x = 0$.

A previous study on lacunar $Pr_{0.5}Sr_{0.5-x}\square_x MnO_3$ samples [16] shows that the effect of strontium vacancy on the electrical properties is in agreement with that observed in our lacunar $La_{0.5}Ca_{0.5-x}\square_x MnO_3$ samples.

4 Conclusion In this work we have investigated the calcium deficiency effect on the structural and magnetotransport properties of $La_{0.5}Ca_{0.5-x}\square_x MnO_3$ samples. Our studies show that the structural and electrical properties depend strongly on the vacancy content. Synthesized samples crystallize in the orthorhombic structure with Pnma space group. Calcium-deficient samples exhibit a semiconducting–metallic transition when the temperature decreases. The magneto-transport investigations of our samples show that 5% of calcium deficient induces the same effect produced by an applied magnetic field of 8 T on the parent compound $La_{0.5}Ca_{0.5}MnO_3$.

Acknowledgements This work has been supported by the Tunisian Secretary of State for Scientific Research and Technology.

References

[1] R. D. Sanchez, J. Rivas, C. V. Vazquez, A. L. Quintela, M. T. Causa, M. Tovar, and S. Oseroff, Appl. Phys. Lett. **68**, 134 (1996).
[2] H. L. Ju and H. Sohn, J. Magn. Magn. Mater. **167**, 200 (1997).
[3] F. Damay, C. Martin, M. Hervieu, A. Maignan, B. Raveau, G. André, and F. Boureé, J. Magn. Magn. Mater. **184**, 71 (1998).
[4] R. von Helmot, J. Weeker, B. Holzepfel, L. Schultz, and K. Samwer, Phys. Rev. Lett. **71**, 2331 (1993).
[5] B. Raveau, A. Maignan, and V. Caignaert, J. Solid State Chem. **117**, 424 (1995).
[6] W. Zhang, I. W. Boyd, N. S. Cohen, Q. T. Quentin, and A. Pankhaurst, Appl. Surf. Sci. **109**, 350 (1997).
[7] C. Zener, Phys. Rev. **81**, 440 (1951).
[8] P. W. Anderson and H. Hasegawa, Phys. Rev. **100**, 675 (1955).
[9] P. G. De Gennes, Phys. Rev. **118**, 141 (1960).
[10] P. E. Schiffer, A. P. Ramirez, W. Bao, and S. W. Cheong, Phys. Rev. Lett. **75**, 3336 (1995).
[11] E. O. Wollan and W. C. Koeler, Phys. Rev. **100**, 5455 (1955).
[12] B. Raveau, A. Maignan, and V. Caignaert, J. Solid State Chem. **130**, 162 (1997).
[13] W. Boujelben, A. Cheikh-Rouhou, and J. C. Joubert, J. Solid State Chem. **156**, 68 (2001).
[14] L. Laroussi, J. C. Joubert, E. Dhahri, J. Pierre, and A. Cheikh-Rouhou, Phase Transit. **70**, 29 (1999).
[15] S. Chaffai, W. Boujelben, M. Ellouze, A. Cheikh-Rouhou, and J. C. Joubert, Physica B **321**, 74 (2002).

^{55}Mn NMR study in magnetically ordered state of perovskite manganites

K. Shimizu[*,1], **W. Boujelben**[2], and **A. Cheikh-Rouhou**[2]

[1] Faculty of Education, Toyama University, 3190 Gofuku, Toyama 930-8555, Japan
[2] Faculté des Sciences de Sfax, B.P. 802, 3018 Sfax, Tunisia

Received 1 September 2003, accepted 31 December 2003
Published online 20 April 2004

PACS 71.30.+h, 75.47.Gk, 76.60.Lz

We present NMR results of $Pr_{1-x}Sr_xMnO_3$ and their lacunar samples, and also present results of the single crystalline layered manganites $La_{1.2}Sr_{1.8}Mn_2O_7$ and $La_{1.2}Sr_{1.6}Ca_{0.2}Mn_2O_7$. In $Pr_{1-x}Sr_xMnO_3$, a single resonance line is observed for $0.27 \leq x \leq 0.4$, which indicates the sample to be in a metallic state. In the Pr and Sr deficient samples, the NMR line arising from Mn^{2+} is observed. Introducing deficiency to Pr or Sr sites, a charge disproportionation of the type $2Mn^{3+} \rightarrow Mn^{2+} + Mn^{4+}$ probably occurs. For $La_{1.2}Sr_{1.8}Mn_2O_7$ and $La_{1.2}Sr_{1.6}Ca_{0.2}Mn_2O_7$ compounds, the NMR spectra under an external field of 1.5 T are broad and spread with several distinct lines in the frequency range 310–420 MHz. This is different from the results of $La_{1-x}Sr_xMnO_3$ and $Pr_{1-x}Sr_xMnO_3$ with a metallic state, where a single line has been observed. The distinct NMR lines are ascribed to Mn^{4+}, Mn^{3+} and metallic phase, taking account of the results for $La_{1-x}Sr_xMnO_3$ and $Pr_{1-x}Sr_xMnO_3$. In addition to the NMR line from the metallic phase, the observation of Mn^{3+} and Mn^{4+} lines suggests a phase separation at low temperature in $La_{1.2}Sr_{1.8}Mn_2O_7$ and $La_{1.2}Sr_{1.6}Ca_{0.2}Mn_2O_7$.

© 2004 WILEY-VCH Verlag GmbH & Co. KGaA, Weinheim

1 Introduction

Perovskite manganites with the general formula $R_{n(1-x)}A_{1+nx}Mn_nO_{3n+1}$ (R = rare-earth, A = alkaline-earth) have attracted considerable attention due to the discovery of colossal magnetoresistance (CMR) effects. In addition to their important technological applications, the perovskite manganites are of large basic physical interest. The complex interaction between charge, spin, and lattice results in competing ground states with distinct electronic and magnetic properties [1]. Many studies have been done on the magnetic and magneto-transport properties of the perovskite compounds. However, only a few studies have been done on the effects of site deficiency on the magnetic properties in the perovskite manganites, using the NMR technique [2, 3].

When electron holes are introduced into the parent compounds such as $LaMnO_3$ and $PrMnO_3$ by doping with divalent ions, e.g., Ca^{2+}, Sr^{2+} or Ba^{2+}, some of the Mn^{3+} ions are converted to Mn^{4+} depending on the doping level. In the mixed state of Mn^{3+} and Mn^{4+}, the double exchange interaction involves ferromagnetic coupling between Mn^{3+} ($t_{2g}^3 e_g^1$, $S = 2$) and Mn^{4+} (t_{2g}^3, $S = 3/2$) ions with transfer of electrons from Mn^{3+} to Mn^{4+} through the O p-orbital. The e_g electrons are itinerant, and Mn^{3+} is strongly affected by Jahn–Teller distortion. On the other hand, the t_{2g} electrons are well localized, and Mn^{4+} is hardly affected by Jahn–Teller distortion.

In the magnetically ordered state of the manganites, unlike the rare-earth ions for which the orbital momentum is not quenched, the dominant contribution to the hyperfine field at the Mn nuclei is the core polarization. The nuclear spin, I, of the Mn ion is coupled to the electron spin, S, by the hyperfine interaction IAS, where A is the hyperfine coupling tensor. Then, the nuclei experience a hyperfine field $H_n = 2\pi A\langle S \rangle/\gamma h$, where $\langle S \rangle$ is the average electronic spin of the Mn ion, γ is the nuclear gyromagnetic

---

[*] Corresponding author: e-mail: shimizu@edu.toyama-u.ac.jp

factor. The resonance frequency, ν, is given by $\nu = (\gamma/2\pi)H_n$. Thus, the spectra originating from Mn^{3+} and Mn^{4+} states are clearly separated experimentally in the insulating phase. In the metallic phase, the rate of electron hopping from a Mn^{3+} ion to an adjacent Mn^{4+} ion is much higher than the Larmor precession frequency of the Mn nuclear spin. Then, the nuclear spins experience a motionally averaged hyperfine field corresponding to an averaged Mn^{3+} and Mn^{4+} state. In perovskite manganites, the resonance frequency of the line at low temperature that originates from Mn^{4+} has been observed to be ~330 MHz, and that from Mn^{3+} falls into the range 350–435 MHz [4, 5]. Signals arising from the metallic phase are located at intermediate frequencies between those of Mn^{3+} and Mn^{4+}.

2 Experimental

$Pr_{1-x}Sr_xMnO_3$ and their lacunar samples were prepared using the solid state reaction method. Starting materials were intimately mixed in an agate mortar and then heated in air at 1000 °C for several days. The powders were sintered at 1400 °C in air with intermediate regrinding and repelleting. Finally, the pellets were rapidly quenched to room temperature. Phase purity and homogeneity were determined by powder X-ray diffraction at room temperature. The preparation of lacunar samples is almost the same as that of the stoichiometric ones. The characteristics have been reported, in detail, elsewhere [2, 3, 6, 7].

The single crystals of $La_{1.2}Sr_{1.8}Mn_2O_7$ and $La_{1.2}Sr_{1.6}Ca_{0.2}Mn_2O_7$ were grown using an image furnace by the floating zone method. The cell parameters decrease from $a = 3.875$ Å and $c = 20.122$ Å to $a = 3.871$ Å and $c = 20.076$ Å, upon increasing the Ca content from 0 to 0.2 [8]. For the compounds, the ab-plane is an easy plane and the c-axis is a hard axis. Their chemical, crystallographic and microstructural characterizations were reported, in detail, elsewhere [8, 9]. Magnetization measurements on $La_{1.2}Sr_{1.8}Mn_2O_7$ revealed a saturation magnetic moment $3.55 \pm 0.06 \mu_B$/Mn at low temperature, which is close to the full moment $3.6\mu_B$ for nominal 0.4 hole doping [8].

^{55}Mn spin-echo NMR measurements have been carried out using the frequency swept spin-echo apparatus. NMR spectra were obtained by measuring the integrated spin-echo intensity versus frequency. The typical spin-echo sequence of ~0.2 – τ – ~0.2 μs with $\tau = 2$ μs was used for zero-external magnetic field measurements. The shape of the spectrum taken with τ up to 6 μs remained almost unchanged. For measurements in external magnetic fields up to 1.5 T, the typical spin-echo sequence is ~1– τ – ~2 μs with $\tau = 4$ μs. The pulse widths were adjusted so as to obtain the maximum signal intensity at a frequency in a given external magnetic field. The pulse sequence was fixed while measuring. All the spectra are shown after the usual correction to be of frequency-square type for the dependence of the signal intensity on the frequency.

3 Results and discussion

3.1 Stoichiometric $Pr_{1-x}Sr_xMnO_3$

The zero-field ^{55}Mn NMR spectra of stoichiometric $Pr_{1-x}Sr_xMnO_3$ samples ($0.2 \leq x \leq 0.4$) taken at 4.2 K are shown in Fig. 1, together with that of $La_{0.75}Sr_{0.25}MnO_3$, for comparison. A single broad reso-

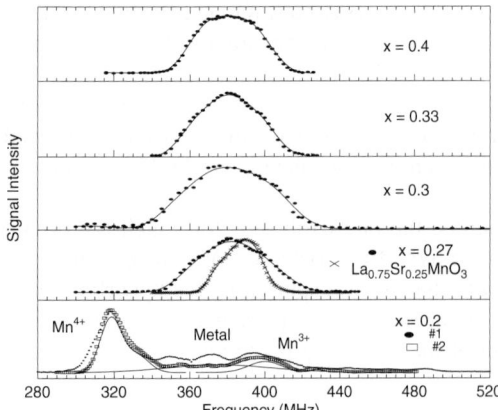

Fig. 1 Zero-field ^{55}Mn NMR spectra in $Pr_{1-x}Sr_xMnO_3$ at 4.2 K, together with $La_{0.75}Sr_{0.25}MnO_3$. Spectra of $Pr_{0.8}Sr_{0.2}MnO_3$ are shown for samples, #1 and #2. The solid lines for #2 denote the NMR lines from Mn^{4+}, Mn^{3+}, and metallic phase.

nance line has been observed around 380 MHz for $0.27 \leq x \leq 0.4$, which indicates these samples to be metallic. The results are consistent with the phase diagram in which the compound is of ferromagnetic metallic (FMM) state in the range $\sim 0.25 < x < \sim 0.5$ [11]. For $x = 0.2$, the results of two samples #1 and #2 are shown. Both spectra, which spread widely in the frequency range from ~ 300 to ~ 480 MHz, consist of a sharp line at 319 MHz and a broad line distributed in the range 330–480 MHz. This aspect is different from those of the samples for $0.27 \leq x \leq 0.4$. Taking account of the results of $La_{1-x}Sr_xMnO_3$, it is reasonable that the spectra consist of the lines from Mn^{4+}, the metallic phase, and Mn^{3+}. Those lines are shown by solid lines for sample #2 in the figure. The NMR spectra of $Pr_{0.8}Sr_{0.2}MnO_3$ indicate that a FMM phase is embedded in the ferromagnetic insulating (FMI) phase. The composition of $x = 0.2$ is close to the transition boundary from insulator to metal. The coexistence of FMM and FMI phases is probably due to the critical boundary between insulating and metallic phases.

For $x = 0.5$, the sample quenched rapidly in water (W) is ferromagnetic, whereas the sample cooled in air (A) is antiferromagnetic at 4.2 K [7]. ^{55}Mn NMR measurements have already been made on $Pr_{0.5}Sr_{0.5}MnO_3$ at 1.3 K [12] and $Pr_{0.51}Sr_{0.49}MnO_3$ at temperatures between 63 and 230 K [13]. Our results are different from them, which might be due to some difference in the experimental temperature and the composition. The NMR spectra of ferromagnetic and antiferromagnetic $Pr_{0.5}Sr_{0.5}MnO_3$ are shown in Fig. 2. Observed spectra are broad, and have a peak at 370 and 374 MHz, for samples W and A, respectively, which indicates that the compounds have almost a metallic state, although both spectra have small humps in the low frequency side. Concerning the temperature dependence of the resistivity, ρ, the sample A shows a semiconducting-like behavior, i.e., $d\rho/dT < 0$ below 160 K, where this sample is antiferromagnet. However, the resistivity ρ yet remains low, e.g., $\rho \sim 6 \times 10^{-2}$ Ω cm at 50 K. On the other hand, the sample W shows a metallic behavior, i.e., $d\rho/dT > 0$ in the temperature range between ~ 180 and ~ 80 K, and a semiconducting behavior below ~ 80 K. However, the resistivity ρ is still low, $\sim 2.3 \times 10^{-2}$ Ω cm at 4.2 K, as well as the sample A. The present NMR results suggest the antiferromagnetic sample A to have an antiferromagnetic metallic (AFMM) state. This metallic conduction has been proposed for $(La_{1-z}Nd_z)_{1-x}Sr_xMnO_3$ in the A-type antiferromagnetic state [14]. The weak signals around 325 MHz are due to Mn^{4+} in insulating phase. Thus, the NMR spectra of our both A and W samples indicate that a small amount of insulating phase coexists at 4.2 K in the metallic phase.

In Fig. 3, the resonance frequencies of $Pr_{1-x}Sr_xMnO_3$ are plotted against the hole concentration x, together with those of FMM $La_{1-x}Sr_xMnO_3$ [15]. In the case of $La_{1-x}Sr_xMnO_3$, the resonance frequency

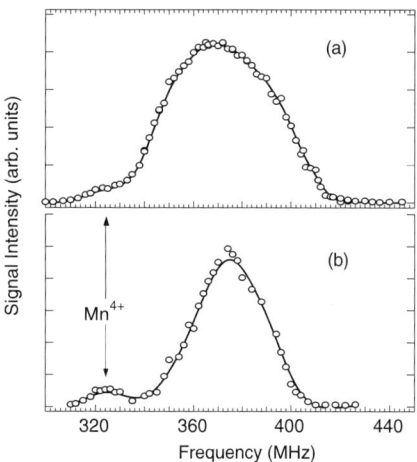

Fig. 2 ^{55}Mn NMR spectra, taken at 4.2 K, for (a) ferromagnetic (W) and (b) antiferromagnetic (A) $Pr_{0.5}Sr_{0.5}MnO_3$ [10].

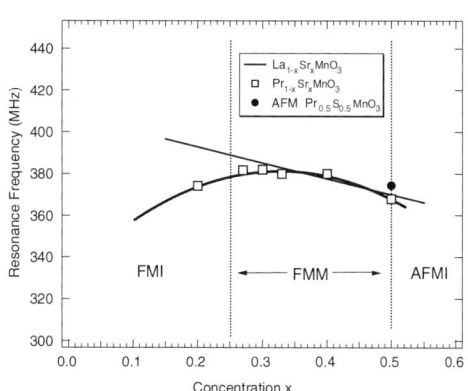

Fig. 3 Resonance frequency of $Pr_{1-x}Sr_xMnO_3$ plotted against x. The solid straight line indicates the results of $La_{1-x}Sr_xMnO_3$ [15]. The line for $Pr_{1-x}Sr_xMnO_3$ is only for eye guide. FMI, FMM, and AFMI denote ferromagnetic insulating, ferromagnetic metallic, and antiferromagnetic insulating states [11].

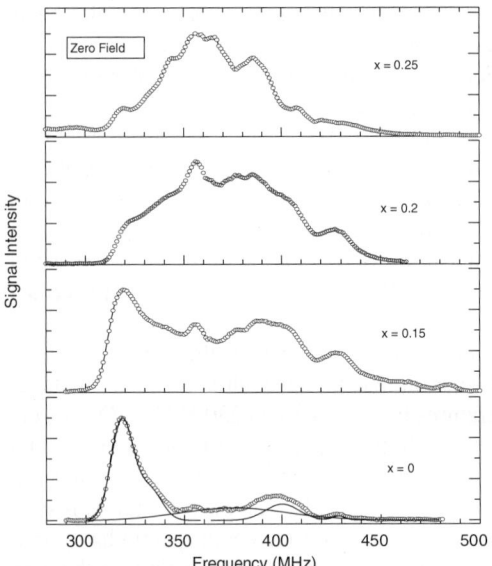
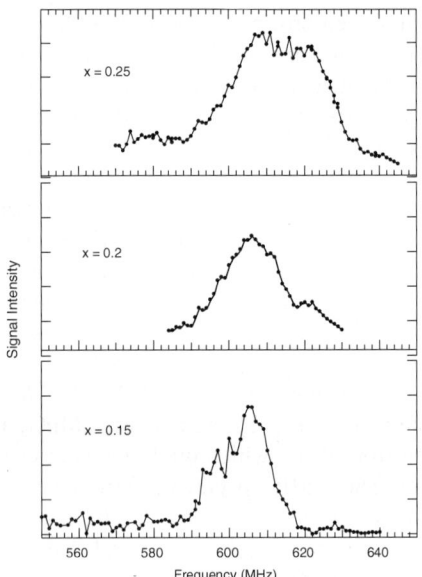

Fig. 4 Zero-field ^{55}Mn NMR spectra taken at 4.2 K for Pr deficient $Pr_{0.8-x}\square_x Sr_{0.2}MnO_3$.

Fig. 5 Zero-field NMR spectra of Mn^{2+} in $Pr_{0.8-x}\square_x Sr_{0.2}MnO_3$ ($x = 0.15, 0.2, 0.25$), taken at 4.2 K.

changes linearly against the hole concentration. The resonance frequency of $Pr_{1-x}Sr_xMnO_3$ deviates from that of $La_{1-x}Sr_xMnO_3$ with increasing Pr concentration. For comparison, the NMR spectrum of $La_{0.75}Sr_{0.25}MnO_3$ is shown in Fig. 1. The line shape for $La_{0.75}Sr_{0.25}MnO_3$ is narrow, and the full width at half height maximum of the line is about 1/2, compared with those of FMM Pr compounds. This may be related to higher hopping rate than those of FMM Pr compounds. The resonance line of $La_{0.75}Sr_{0.25}MnO_3$ appears clearly on the higher frequency side than that of the Pr compound. The Pr magnetic moment has been observed to be $0.273\mu_B$ in $Pr_{0.7}Sr_{0.3}MnO_3$ [6]. A coupling between Pr and Mn spins would cause a change in the hyperfine field at Mn site.

3.2 Lacunar $Pr_{0.8-x}\square_x Sr_{0.2}MnO_3$

Magnetic properties of lacunar compounds $Pr_{0.7}Sr_{0.2}\square_{0.1}MnO_3$ and $Pr_{0.7}Sr_{0.1}\square_{0.2}MnO_3$ have already been reported elsewhere [2]. In the present Pr deficient compounds $Pr_{0.8-x}\square_x Sr_{0.2}MnO_3$, a transition between semiconductor and metal occurs in the temperature range 180–190 K under zero external field, for $x = 0.2$ and 0.25. For $x = 0.15$, a semiconducting–metallic transition at about 130 K occurs. Then, the resistivity, ρ, undergoes an up-turn around 40 K with decreasing temperature, and exhibits semiconducting behavior at low temperature.

^{55}Mn NMR results in $Pr_{0.8-x}\square_x Sr_{0.2}MnO_3$ ($x = 0, 0.15, 0.2, 0.25$), which were taken at 4.2 K under zero-external field, are shown in Fig. 4. These complex spectra are due to superposition of domain wall signals on domain signals. In the samples with $x = 0$ and 0.15, a large resonance line is clearly observed around 320 MHz, which indicates these samples to have an insulating phase. The intensity of the line around 320 MHz decreases with increasing x. This result indicates that the metallic state region increases with increasing deficiency.

In the Pr deficient samples, furthermore, an NMR line at about 610 MHz that is attributed to Mn^{2+} is observed. These spectra are shown in Fig. 5. The Mn^{2+} line has also been observed in other lacunar compounds: both Sr and Pr deficient samples of $Pr_{0.7}Sr_{0.3-x}\square_x MnO_3$ ($x = 0.1, 0.2$) [2] and $Pr_{0.6}\square_{0.1}Sr_{0.3}MnO_3$ [3]. In $Pr_{0.8-x}\square_x Sr_{0.2}MnO_3$ ($x = 0.15, 0.2, 0.25$), the integral intensity of the Mn^{2+} line increases with increasing x, which indicates that number of Mn^{2+} ion increases, depending on the Pr deficiency. For the stoichiometric compound $Pr_{0.8}Sr_{0.2}MnO_3$, an Mn^{2+} signal was not observed under the same experi-

© 2004 WILEY-VCH Verlag GmbH & Co. KGaA, Weinheim

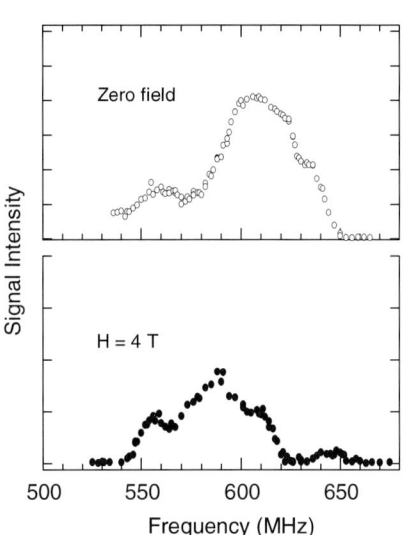
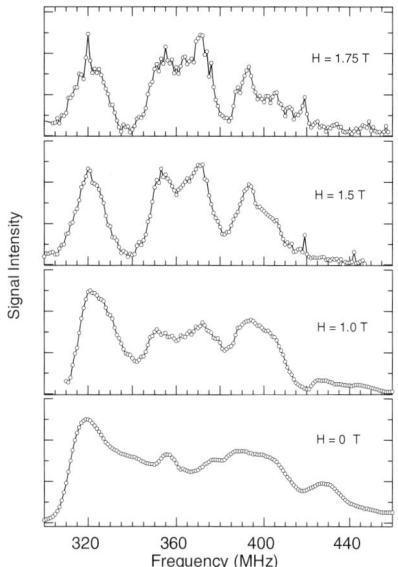

Fig. 6 Mn^{2+} spectra under zero-field and a field of 4 T for $Pr_{0.7}Sr_{0.2}\square_{0.1}MnO_3$ at 4.2 K.

Fig. 7 External magnetic field dependence of the NMR spectrum, taken at 4.2 K, for $Pr_{0.65}\square_{0.15}Sr_{0.2}MnO_3$.

mental condition. Introducing the deficiency to Pr or Sr sites, a charge disproportionation of the type $2Mn^{3+} \rightarrow Mn^{2+} + Mn^{4+}$ probably occurs [16].

The NMR line of Mn^{2+} in $Pr_{0.7}Sr_{0.2}\square_{0.1}MnO_3$ under an external magnetic field of 4T is shown in Fig. 6. The resonance frequency of the Mn^{2+} line shifts toward the lower frequency side under the magnetic field of 4 T, in comparison with the zero-field line. Since the direction of the hyperfine field at the Mn site due to the core polarization is antiparallel to that of the magnetic moment, this field dependence of the NMR line implies the magnetic moment of Mn^{2+} to lie parallel to the net magnetization. The NMR line of Mn^{2+} has also been observed in $Pr_{0.9}Ca_{0.1}MnO_3$. In this case, the external magnetic field dependence on the resonance frequency has revealed that the Mn^{2+} magnetic moment couples antiparallel to the magnetization [17].

As shown in Fig. 4, the zero-field spectrum of $Pr_{0.65}\square_{0.15}Sr_{0.2}MnO_3$ spreads widely in the frequency range 300–480 MHz. When applying an external magnetic field, the signal intensity decreases rapidly with increasing magnetic field, which indicates reduction of the domain wall enhancement. The NMR spectrum of $Pr_{0.65}\square_{0.15}Sr_{0.2}MnO_3$ was measured in external magnetic fields up to 1.75 T. All the spectra in fields exhibit distinct lines as shown in Fig. 7: a narrow line around 320 MHz, a line with two peaks in the range 340–380 MHz, and an inhomogeneous broad line in the range 380–430 MHz. The line around 320 MHz is ascribed to Mn^{4+}. Signals spreading in the range 380–430 MHz are ascribed to Mn^{3+}. Signals in the range 340–380 MHz are ascribed to a metallic phase. This assignment is consistent with the results of stoichiometric $Pr_{1-x}Sr_xMnO_3$. The appearance of Mn^{4+} and Mn^{3+} is characteristic of an insulating phase. Thus, in the sample with $x = 0.15$, metallic and semiconducting phases coexist. This is consistent with a semiconducting-like behavior observed in the resistivity measurements on the sample at low temperature [18].

3.3 Double-layered compounds $La_{1.2}Sr_{1.8}Mn_2O_7$ and $La_{1.2}Sr_{1.6}Ca_{0.2}Mn_2O_7$

The NMR spectra for $La_{1.2}Sr_{1.8-x}Ca_xMn_2O_7$ ($x = 0, 0.2$), taken under an external field of 1.5 T in order to wipe out the domain wall signal, are shown in Figs. 8 and 9. All the spectra are broad with several distinct lines. The spectrum shape taken under zero-field depends strongly on the rf radiation field for exciting and refocusing the spin-echo signal [19]. The zero-field spectrum taken at high rf power spreads widely in the frequency range 310–480 MHz, although the sample is in metallic state. This feature is in contrast with that of $La_{1-x}Sr_xMnO_3$ and $Pr_{1-x}Sr_xMnO_3$ with metallic state.

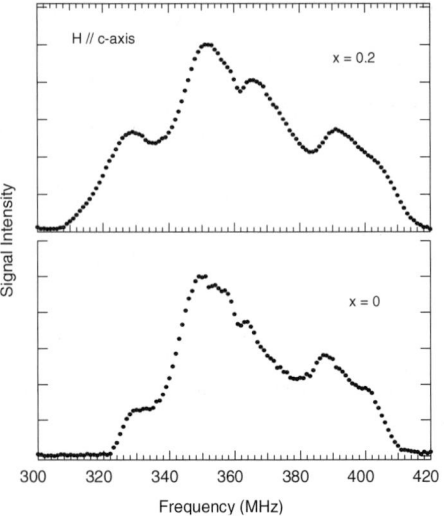
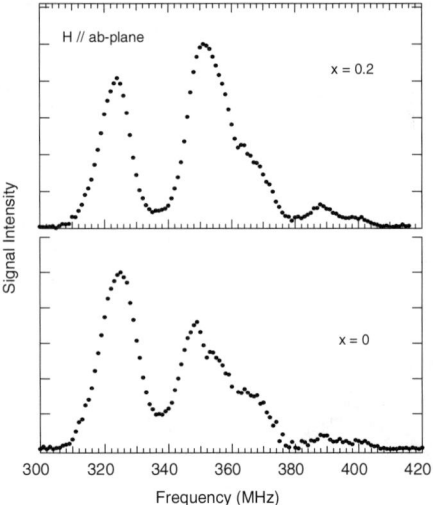

Fig. 8 NMR spectra of $La_{1.2}Sr_{1.8-x}Ca_xMn_2O_7$ for $x = 0$ (bottom), and 0.2 (top) at 4.2 K under a field of 1.5 T applied along the c-axis.

Fig. 9 NMR spectra of $La_{1.2}Sr_{1.8-x}Ca_xMn_2O_7$ for $x = 0$ (bottom), and 0.2 (top) at 4.2 K under a field of 1.5 T applied in the ab-plane.

When applying the field in the ab-plane, the line observed around 320–330 MHz, which has a full width of about 12 MHz at half height maximum, is assigned to Mn^{4+}. The line around 350 MHz is ascribed to Mn in the metallic phase. The aspect of the NMR spectrum of $La_{1.2}Sr_{1.7}Ca_{0.1}Mn_2O_7$ is almost the same as that of $La_{1.2}Sr_{1.6}Ca_{0.2}Mn_2O_7$ [20]. The NMR line shape of Mn^{3+} is anisotropic and inhomogenous due to the Jahn–Teller distortion, and is often practically smeared out [21, 22], while that of Mn^{4+} is symmetric and sharp in comparison with that of Mn^{3+}.

The effect of the direction of the external magnetic field, on the NMR line shape, is quite different for Mn^{3+} and Mn^{4+}. In the case of the field direction parallel to the ab-plane, the intensity of the Mn^{3+} line is weaker than that of Mn^{4+}, while the Mn^{3+} line is stronger than that of Mn^{4+} when the field is applied along the c-axis. We do not have, at present, an explanation for the dependence of the signal intensity on the field direction. For the comparison, the NMR spectrum of a polycrystalline $La_{1.2}Sr_{1.8}Mn_2O_7$, taken at 4.2 K, under the field of 1.5 T, is shown in Fig. 10. The anisotropy on the magnetic field direction is reduced, compared with that of the single crystal. Both Mn^{4+} and Mn^{3+} lines are observed clearly in the polycrystalline sample.

The NMR spectrum shape of the Ca-doped compound is similar to that of $La_{1.2}Sr_{1.8}Mn_2O_7$, but the NMR signal intensity of the former compounds is stronger than that of $La_{1.2}Sr_{1.8}Mn_2O_7$. As seen in Fig. 8,

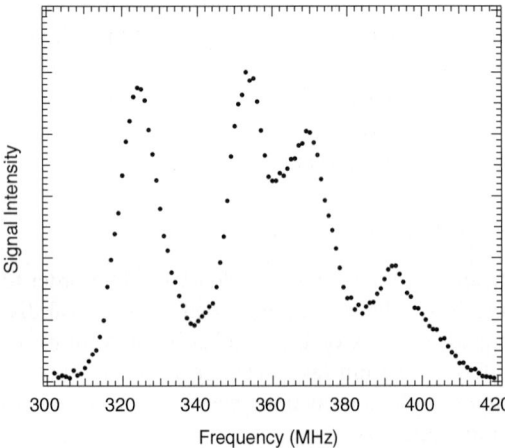

Fig. 10 NMR spectrum of polycrystalline $La_{1.2}Sr_{1.8}Mn_2O_7$ at 4.2 K under a field of 1.5 T.

the signal intensity of Mn^{4+} increases with increasing Ca content. The dependence of the signal intensity on the Ca content may correspond to the fact that the resistivity shifts towards higher values upon increasing the Ca content [23].

In the $La_{1.2}Sr_{1.8-x}Ca_xMn_2O_7$ ($x = 0$, 0.2), the Mn^{3+} and Mn^{4+} lines which are characteristic of an insulator are observed in addition to the line arising from the metallic phase, although this compound is of the FMM state at low temperature. This gives evidence for a phase separation in the compound [24].

Acknowledgements We are grateful to Dr. J.-P. Renard for continuous support and encouragement, and Dr. M. Velázquez and Prof. Revcolevschi for providing us the high quality samples of double-layered manganites.

References

[1] For example, C. N. R. Rao, and B. Raveau, Colossal Magnetoresistance, Charge Ordering and Related Properties of Manganese Oxides (World Scientific, Singapore, 1998).
[2] D. Abou-Ras, W. Boujelben, A. Cheikh-Rouhou, J. Pierre, J.-P. Renard, L. Reversat, and K. Shimizu, J. Magn. Magn. Mater. **233**, 147 (2001).
[3] W. Boujelben, A. Cheikh-Rouhou, J. Pierre, D. Abou-Ras, J. P. Renard, and K. Shimizu, Physica B **321**, 68 (2002).
[4] G. Matsumoto, J. Phys. Soc. Jpn. **29**, 615 (1970).
[5] A. Anane, C. Dupas, K. Le Dang, J. P. Renard, P. Veillet, A. M. de Leon Guevara, F. Millot, L. Pinsard, and A. Revcolevschi, J. Phys.: Condens. Matter **7**, 7015 (1995).
[6] W. Boujelben, M. Ellouze, A. Cheikh-Rouhou, J. Pierre, Q. Cai, W. B. Yelon, K. Shimizu, and C. Dubourdieu, phys. stat. sol. (a) **191**, 243 (2002).
[7] W. Boujelben, A. Cheikh-Rouhou, J. Pierre, and J. C. Joubert, J. Alloys Compd. **314**, 15 (2001).
[8] M. Velázquez, A. Revcolevschi, J. P. Renard, and C. Dupas, Eur. Phys. J. B **23**, 307 (2001).
[9] M. Velázquez, C. Haut, B. Hennion, and A. Revcolevschi, J. Cryst. Growth **220**, 480 (2000).
[10] K. Shimizu, W. Boujelben, A. Cheikh-Rouhou, J. Pierre, and J. C. Joubert, to be published in J. Magn. Magn. Mater.
[11] C. Martin, A. Maignan, M. Hervieu, and B. Raveau, Phys. Rev. B **60**, 12191 (1999).
[12] G. Allodi, R. de Renzi, M. Solzi, K. Kamenev, G. Balakrishnan, and M. W. Pieper, Phys. Rev. B **61**, 5924 (2000).
[13] M. M. Savosta, V. A. Borodin, M. Maryško, Z. Jirák, J. Hejtmánek, and P. Novák, Phys. Rev. B **65**, 224418 (2002).
[14] T. Akimoto, Y. Maruyama, Y. Moritomo, A. Nakamura, K. Hirota, K. Ohoyama, and M. Ohashi, Phys. Rev. B **57**, R5594 (1998).
[15] J.-P. Renard and A. Anane, Mater. Sci. Eng. B **63**, 22 (1999).
[16] M. F. Hundley and J. J. Neumeier, Phys. Rev. B **55**, 11511 (1997).
[17] G. J. Tomka, P. C. Riedi, Cz. Kapusta, G. Balakrishnan, D. McK. Paul, M. R. Lees, and J. Barratt, J. Appl. Phys. **83**, 7151 (1998).
[18] A. Cheikh-Rouhou, M. Koubaa, W. Boujelben, W. Cheikh-Rouhou, A. M. Haghiri-Gosnet, and J. P. Renard, to be published in J. Magn. Magn. Mater.
[19] K. Shimizu, M. Velázquez, J.-P. Renard, and A. Revcolevschi, J. Phys. Soc. Jpn. **72**, 793 (2003).
[20] K. Shimizu, M. Velázquez, J.-P. Renard, and A. Revcolevschi, to be published in J. Magn. Magn. Mater.
[21] Cz. Kapusta, P. C. Riedi, M. Sikora, and M. R. Ibarra, Phys. Rev. Lett. **84**, 4216 (2000).
[22] G. Papavassiliou, M. Fardis, M. Belesi, T. G. Maris, G. Kallias, M. Pissas, D. Niarchos, C. Dimitropoulos, and J. Dolinsek, Phys. Rev. Lett. **84**, 761 (2000).
[23] M. Velázquez, J. M. Bassat, J. P. Renard, C. Dupas, and A. Revcolevschi, J. Phys.: Condens. Matter **14**, 6667 (2002).
[24] E. Dagotto, T. Hotta, and A. Moreo, Phys. Rep. **344**, 1 (2001).

Neutron diffraction studies of magnetic and superconducting compounds

W. B. Yelon[*,1,2], **Q. Cai**[3], **W. J. James**[1,2], **H. U. Anderson**[1,4], **J. B. Yang**[1,5], **X. D. Zhou**[1,4], and **H. A. Blackstead**[6]

[1] Materials Research Center, University of Missouri, Rolla MO 65409, USA
[2] Chemistry Department, University of Missouri, Rolla MO 65409, USA
[3] Physics Department, University of Missouri, Columbia MO 65211, USA
[4] Department of Ceramic Engineering, University of Missouri, Rolla MO 65409, USA
[5] Department of Physics, University of Missouri, Rolla MO 65409, USA
[6] Physics Department, University of Notre Dame, Notre Dame, IN 46556, USA

Received 1 September 2003, accepted 24 February 2004
Published online 26 April 2004

PACS 61.12.Ld, 74.25 Nf, 74.70.Pq, 75.25+z, 75.47.Gk

Neutron diffraction is an extremely valuable tool for the investigation of magnetic and superconducting materials, because of its ability to directly observe periodic magnetic structures, determine magnetic moment directions and magnitudes, to observe light elements that are otherwise difficult to locate from X-ray diffraction due to the strong scattering of heavy elements, or to distinguish nearby elements in the periodic chart. This talk will focus on recent studies of superconducting and magnetic oxides that may provide insight into the interaction of magnetism and superconductivity, and into important changes in other transport properties (colossal magneto-resistance, ionic conductivity, etc.). These materials appear to show promise for a wide range of applications, and the neutron studies may not only help to understand their properties, but may also provide direction for synthesis of compounds that may overcome the limitations of those already discovered.

© 2004 WILEY-VCH Verlag GmbH & Co. KGaA, Weinheim

1 Introduction

Neutron scattering has proven to be an extremely valuable technique in the study of condensed matter. Neutron diffraction (ND), like its counterparts, X-ray and electron diffraction is a structural probe, but the special characteristics of the neutron permit the determination of structural features not readily accessible to the other probes. All atoms are normally well located in neutron diffraction studies, and vacancies can be seen, even for light atoms in systems containing heavy atoms. The absorption cross section for neutrons for all but a few elements (Sm, Gd, Eu, Cd, B) is very small to modest, allowing the easy development of special environments, such as furnaces, cryostats and pressure cells for neutron studies [1].

Unfortunately, compared to synchrotron X-ray and even modern laboratory sources, neutron sources are weak, producing peak fluxes between 10^{14} and 10^{15} n/cm^{2}s. To compensate, long counting times or large samples are usually required. Improvements in instrumentation, such as the use of multi-detectors or position sensitive detectors (PSD) can alleviate the problem and at the University of Missouri Research Reactor (MURR) we have employed a combination of focusing and PSD that allows high resolution data to be collected on 1 gm specimens in an hour or two (or more depending on the complexity of the structure) [2]. This leads, in turn, to the ability to study many samples within a chemical series or to follow the evolution of a material as a function of some parameter such as tem-

---

[*] Corresponding author: e-mail: yelonw@umr.edu, Phone: +10 573 341 4324, Fax: +10 573 341 2071

perature and gaseous environment. In this paper we report recent results on oxygen vacancies in simple ferrites and results on ruthenocuprate superconductors.

2 $La_{1-x}Sr_xFeO_{3-\delta}$

Recent years have seen an explosion in research into oxide materials, especially cuprates related to the high temperature superconductors YBCO and $La_{2-x}Sr_xCuO_{4-\delta}$, the compounds first discovered in this class of materials. In addition, the discovery of the colossal magnetoresistance (CMR) effect in manganites, has stimulated a broader investigation into the relationship between the magnetic interactions in these compounds and their transport properties [3–5].

High Tc superconductivity and CMR are "low temperature phenomena", observed only at temperatures low enough that the magnetic interactions overcome the thermally induced disorder. At high temperature other transport properties may be of considerable importance, and have received less study, especially from a fundamental perspective. In solid-state electrodes of similar materials, changes in chemistry and oxygen concentration can lead to remarkable changes in electrical conductivity. Optimizing their electrical properties at the desired operating temperature of a fuel cell may be an important step toward the realization of a hydrogen-based energy economy [6–8]. Virtually all of the perovskites exhibit these changes at some temperature (providing they do not first decompose) and manganites have been investigated for this application. Their optimum operating temperatures, however, appear to be too high for the present generation of materials. We have found in ferrites that there are important interrelationships between their electrical properties, structure and magnetic properties, which are helpful in understanding the ensemble of properties and in adjusting compositions in ways that bring us closer to practical realization of the cathode application. For simplicity we refer to the system $La_{1-x}Sr_xFeO_{3-\delta}$ as LSF and specify the La/Sr ratio with the nomenclature L6SF as representing 60% La, balance Sr. This ratio is fixed with temperature, but the oxygen concentration must be determined (and reported) at every point.

We have constructed a furnace that allows neutron diffraction measurements as a function of temperature and gaseous environment, but initially we investigated samples that had been heated in air or other environments and rapidly quenched to room temperature, thereby (hopefully) preserving the high temperature stoichiometry. Figure 1 shows the neutron diffraction diagrams of two of these samples (all

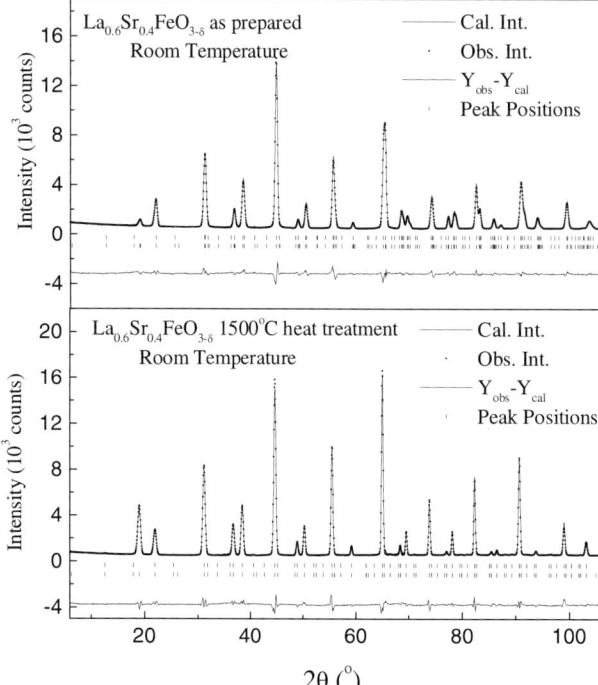

Fig. 1 Neutron diffraction diagrams for $La_{0.6}Sr_{0.4}FeO_{3-\delta}$ at room temperature for a) the (nearly stoichiometric) as prepared sample and b) An oxygen deficient sample prepared by heating to 1500 °C and quenching to room temperature.

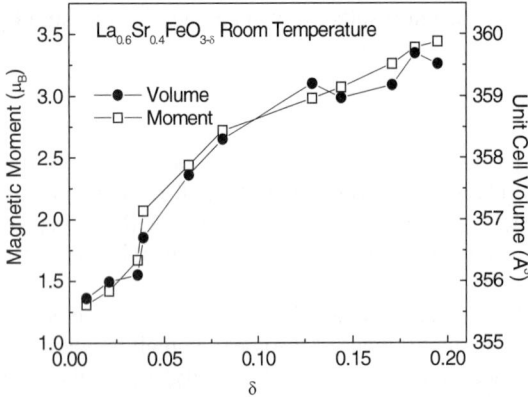
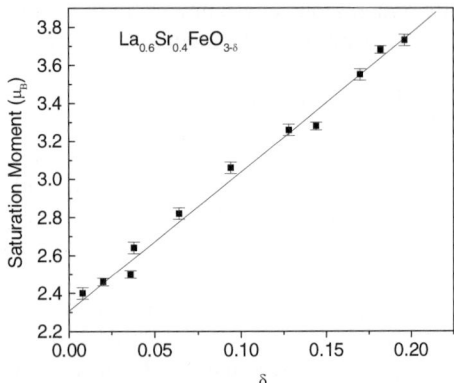

Fig. 2 Magnetic moments and unit cell volumes as a function of refined oxygen vacancy concentration, δ, for $La_{0.6}Sr_{0.4}FeO_{3-\delta}$ at room temperature

Fig. 3 Saturation moment per Fe atom for $La_{0.6}Sr_{0.4}FeO_{3-\delta}$ as a function of vacancy concentration, δ, from Rietveld refinement of data at low temperature.

derived from the same synthesis, guaranteeing constant La/Sr ratio). Samples heated only to low temperature show a weak first peak and large rhombohedral distortion, while those treated above about 1100 °C show a strong first peak and nearly cubic structure. The first Bragg peak is purely magnetic and the change in its intensity at room temperature reflects both the increase in Curie temperature with increasing oxygen vacancy concentration and the change in average valence of the Fe atoms. The neutron refinements also yield those concentrations and the unit cell volumes, which are shown in Fig. 2 as a function of refined vacancy concentration. It is clear that either the moment or the volume may be used to reliably determine the vacancy content. Either one may, in fact, be more reliable than the direct determination. The result for the 1500 °C sample shows a sample reduced to nearly pure Fe^{3+}, while the average valence of the Fe for other treatment temperatures can be readily determined (+3.4 for low temperature treatment) [9].

Neutron measurement at low temperature provides further insight into the magnetic properties of the samples. It is found (Fig. 3) that the low temperature moment increases linearly with vacancy concentration to about 3.8 μ_B for pure Fe^{3+}, the same value observed for $LaFeO_3$. Indeed, the linear behavior is consistent with a model in which the sample exists in a mixed 3+/–4 state, with a nearly zero moment on the Fe^{4+} ions. In the absence of other information this would have been the best assumption, but we have also collected Mössbauer effect data on these samples (Fig. 4). While the data for the samples treated at high temperature are consistent with a pure 3+ state, those treated at low temperature show high field sextets with about 80% of the total area and a low field sextet with the remaining 20%. This can be understood if the Fe^{4+} ions disproportionate into Fe^{3+} and Fe^{5+} states, enhancing the area of the 3+ sextets and reducing the area of the low field sextet. In support of this model is the observation that the isomer shift for the low field sextet is consistent with a 5+ charge state. The observation of more than one sextet at high field is probably a reflection of the variable environments of the Fe^{3+} atoms, with differing numbers of Fe^{5+} neighbors, while the sharper low field sextet suggests that Fe^{5+} ions have only Fe^{3+} neighbors, as is found in the charge ordered samples, we and others have investigated [10, 11].

The composition $La_{1/3}Sr_{2/3}FeO_3$ is expected to have an average Fe valence of 3.667. Battle et al. have found that this sample undergoes magnetic ordering around 200 K, showing a unit cell tripled along the threefold axis. This is explained as a 3+, 3+, 5+ stacking order, which is reflected in their and our neutron data (Fig. 5) which show new low angle reflections arising from the tripled cell. Battle's refinement yields a moment on the 3+ sites above 3 μ_B and one on the 5+ site around 2.8 μ_B, while our refinement shows considerably lower values. We believe this discrepancy arises from disorder; both samples are fully disproportionated but ours appears to have about 25% of the ions switched from their ideal positions, while Battle's shows only a small disorder (based on an ideal 3.8 μ_B moment for Fe^{3+}).

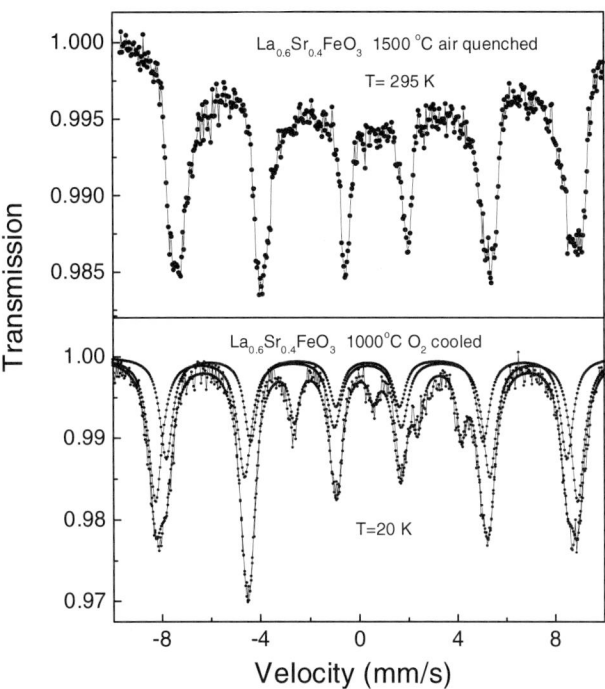

Fig. 4 Mössbauer data for a) oxygen deficient $La_{0.6}Sr_{0.4}FeO_{3-\delta}$ at room temperature, showing a single sextet, b) nearly stoichiometric $La_{0.6}Sr_{0.4}FeO_{3-\delta}$ at low temperature, showing high field sextets associated with Fe^{3+} and a low field sextet associated with Fe^{5+}.

We have recently begun to carry out measurements above room temperature. A sample of the L6SF quenched from 1500 °C was heated without air flow. Its volume increased linearly with temperature up to the highest temperature (Fig. 6), indicating that no oxygen was absorbed, and on returning to room temperature its volume and magnetic moment were virtually unchanged. When flowing air was introduced into the furnace, the sample volume began to decrease around 350 °C (consistent with TGA results) indicating oxygen uptake, while at 800 °C its volume was increasing markedly indicating oxygen loss. A stoichiometric sample heated without air flow showed a nearly linear volume increase up to about 600 °C above which the volume increase was more rapid, indicating oxygen loss. By using the volume difference between the extrapolated volume vs. temperature data for the stoichiometric and reduced samples, the oxygen content of these samples could be easily assessed.

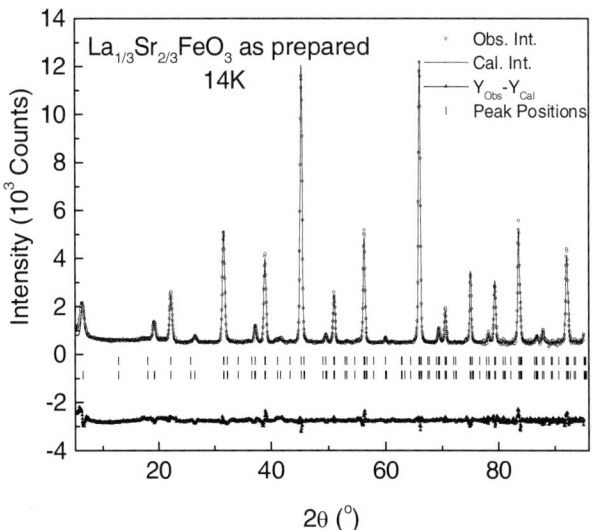

Fig. 5 Low temperature neutron diffraction data for $La_{1/3}Sr_{2/3}FeO_3$. The first Bragg peak, at about 6°, shows the tripling of the c-axis due to charge (and spin) ordering.

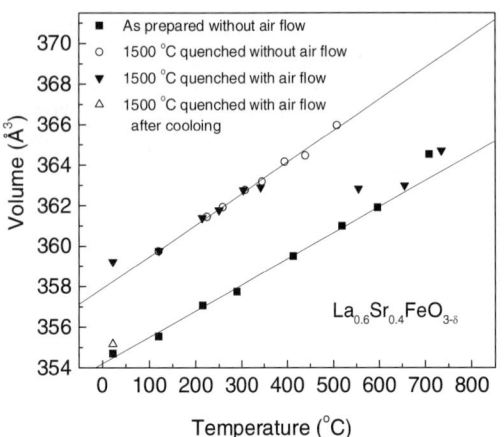

Fig. 6 (online colour at: www.interscience.wiley.com) Unit cell volumes for oxygen deficient and stoichiometric samples of $La_{0.6}Sr_{0.4}FeO_{3-\delta}$ as a function of temperature, showing the uptake or loss of oxygen through the volume change.

These results show that in situ ND can be an important tool in determining the oxygen vacancy kinetics in this important class of materials, while the combination of ND with other methods can provide unique insight into the charge states and magnetic properties.

3 Superconductivity and magnetism in rutheno-cuprates

It is widely believed that magnetism and superconductivity are incompatible, since they require different symmetries for the wavefunctions of the same electrons. The inclusion of paramagnetic ions in a superconductor is frequently deleterious, due to Abrikosov–Gorkov pair breaking effects. However, the A–G pair breaking due a paramagnetic ion can be largely neutralized, if there are crystal field effects of sufficient size. Crystal symmetry plays a determining role, since for example, cubic systems generally have small crystal fields. Ferromagnetism, especially of the itinerant type, is viewed as especially incompatible with superconductivity, since the same electrons would be involved. Although magnetism is observed on the rare earth sublattice of the YBCO type material (e.g. with Gd), the localized magnetic moments on those sites are thought to not interact with the superconducting electrons from other regions of the lattice. Nevertheless, NMR measurements demonstrate Korringa relaxation, showing that the rare earth site is relaxed by carrier scattering [12, 13]. We believe that the real anomaly of HTSC cuprate materials is the superconductivity of Gd123, since the Gd^{3+} ion is an S-state ion and exhibits essentially no crystal field effects. The two-layer O_6 materials are especially useful as test environments, since nearly all $LnA_2Ru_{1-x}Cu_xO_6$ (Ln = Lanthanide, A = Sr, Ba) homologues exist. Particularly, $GdBa_2Ru_{1-x}Cu_xO_6$ *does not* exhibit superconductivity, while $HoSr_2Ru_{1-x}Cu_xO_6$ reportedly does (Ho is a non-S-state ion). Thus, the issues of coexisting superconductivity and magnetism become especially acute in considering the superconducting rutheno-cuprates such as $GdSr_2RuCu_2O_8$ (referred to here as O_8), $Eu_{1.4}Ce_{0.6}Sr_2Cu_2O_{10}$ (referred to as O_{10}) and especially $YSr_2Ru_{1-x}Cu_xO_6$ (referred to as O_6) which show evidence for strong magnetic coupling, sometimes well above 100 K and yet show superconductivity onset around 40 K, (or higher when Sr is replaced by Ba) without significant change in the magnetic order [14]. In order to reconcile these facts, it is necessary to obtain detailed information about the magnetic ordering in these compounds. While this is typically thought of as an ideal powder ND problem, it is found that ND, alone, does not provide enough information to provide an unambiguous description of the type and direction of magnetic ordering. It is especially useful to combine these results with magnetic resonance (and other data) to obtain a comprehensive picture.

Neutron data have been collected by Lynn et al. on the Gd–O_8 compound, using a separated Gd isotope with relatively low neutron absorption [15]. They find that the lattice orders magnetically at 136 K, with the unit cell doubling in all directions, i.e. antiferromagnetic order along both the c-axis and in the basal plane. Only two magnetic reflections are observed, the (1/2 1/2 1/2) and the (1/2 1/2 3/2) peaks, and examination of the data in the Lynn paper suggests a ratio of about 4/1 for these two

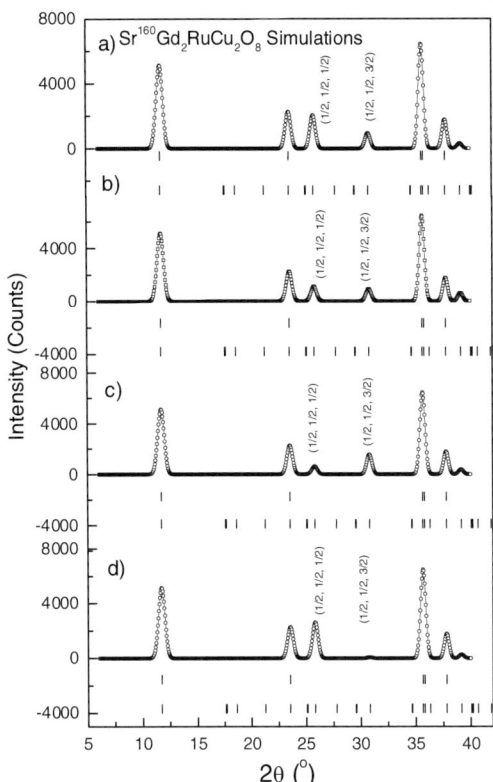

Fig. 7 Simulations of the low angle region of the Gd–O$_8$ compound for different magnetic models. (a) Ru order only with moments along the c-axis. (b) Ru order only with moments in the basal plane. (c) Type I (see text) Cu order only with moments along the c-axis. (d) Type II Cu order only with moments along the c-axis. Type I and type II Cu order models with basal moments are not shown since they are very similar to the c-axis results.

(although Lynn reports a lower ratio: 2.6/1). Since the ordering temperature is lower than that of SrRuO$_3$, which is a structural element of the O$_8$ cell, Lynn assumes that the magnetic ordering is associated with the Ru atoms only, even though SrRuO$_3$ is ferromagnetic not antiferromagnetic. His best fit with that model requires the Ru moments to lie along the c-axis and gives a ratio of the two magnetic reflections of about 2.2/1, in reasonable agreement with Lynn's estimate, but large compared to our estimate from his data. NQR data, however, are unambiguous in placing the Ru moments in the basal plane [16, 17]. Figure 7 shows a number of simulations we have carried out for the ND pattern for this system with different spin configurations. No attempt has been made to adjust the ratio of magnetic intensity to the nuclear; this can be done simply by scaling the moment. Models with Ru moments only give ratios (for the first two magnetic reflections) of 2.2/1 and 1/1 for the axial and basal moments respectively. Neither model can reconcile the NQR and ND observations. Since there are 4 Cu–O layers in the unit cell, two different stacking orders can be employed if Cu moments are allowed: + − − + (type I) and + + − − (type II) respectively, where the sign indicates the direction of the Cu moment on the sequential planes, relative to the first plane, regardless of moment direction. For Cu order only, type I ordering gives a large (1/2 1/2 3/2), intensity (regardless of moment direction), while type II ordering results in near zero (1/2 1/2 3/2) intensity. Thus the ND and NQR results can be reconciled by assuming that both the Cu–O and Ru–O layers order antiferromagnetically with moments lying in the basal plane direction. Data from Takagiwa et al. on the Y–O$_8$ compound shows no (1/2 1/2 3/2) peak in the ND spectrum despite excellent statistics [18], suggesting an even larger Cu contribution to the magnetic structure than in the Gd analog.

Neutron diffraction on the Eu–O$_{10}$ compound does not show any superlattice reflections at 15 K [19] even though the susceptibility data show magnetic ordering above 100 K [20, 21]. Thus, while the magnetic structure of these compounds has not been determined, ferromagnetic ordering cannot be ruled out.

This problem becomes even more perplexing in the O$_6$ compounds. In this case the Cu is partially substituted for Ru and no distinct Cu–O plane exists. Despite this, superconductivity is observed with Cu concentrations as low as 1% and resistivity transitions are seen at 5% Cu. In addition, magnetic order

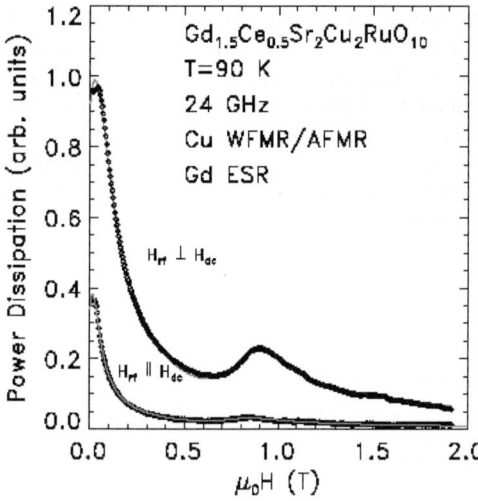 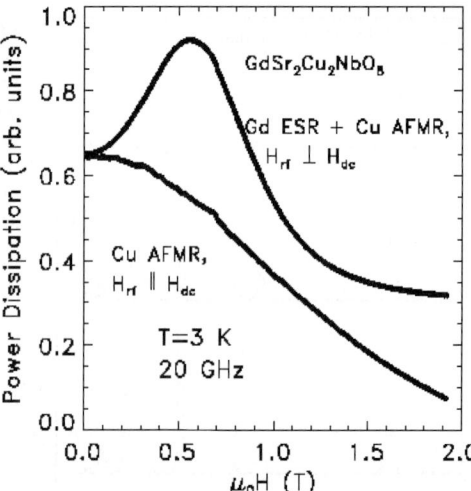

Fig. 8 (online colour at: www. interscience.wiley. com) Low field magnetic resonance (ESR-AFMR) for the Gd/Ce–O_{10} compound, showing two low field resonances, one related to the Gd ESR and the other believed to be due to ordered Cu.

Fig. 9 Low field magnetic resonance (ESR-AFMR) for $GdSr_2Cu_2NbO_8$ also showing two low field resonances. The absence of Ru in this compound confirms the identification of the zero-field feature as arising from Cu.

exists well above the onset temperature for superconductivity and is not quenched at Tc. Two neutron studies have been carried out on the Sr–O_6 compounds, and they agree that the magnetic ordering consists of ferromagnetic sheets, antiferromagnetically coupled [22, 23]. We believe that the neutron data show both Ru and Cu order, while Parkinson et al argue that the data can be understood with Ru ordering only. However, they find that the moment increases as the Cu content increases, a result that is hard to understand with non-magnetic Cu. Regardless, since the Cu and Ru are in the same layer and the Cu is dilute, it is difficult to imagine superconducting clusters of Cu–O_6 isolated from the magnetic Ru regions of sufficient size and concentration to support a resistivity transition.

Magnetic resonance (ESR, AFMR) data provide an important window on this problem. Measurements at the University of Notre Dame have shown two resonance modes at low fields in all of the superconducting rutheno-cuprates thus far studied, including the Gd and Eu–O_{10}s, the Gd and Er–O_8s, (Fig. 8) and the O_6, as well as in non-superconducting $GdBa_2$–O_8 and a non-superconducting O_8 in which the Ru is replaced by Nb (Fig. 9). In the superconducting Gd–O_8 compound the resonance is observed well above the magnetic ordering temperature of the Ru. While some authors have tried to attribute these features to Ru resonance, no similar features (or resonances of any kind not due to rare-earth) are seen in ruthenates without Cu including $SrRuO_3$ (which orders magnetically at 165 K), RuO_2, $Sr_3Ru_2O_7$, Ba_3RuNiO_9 and $GdBa_2RuO_6$. These non-resonant materials contain tetravalent Ru, in the first three, and pentavalent Ru, in the last two. It is difficult to avoid the conclusion that these resonance features are related to magnetically ordered Cu. This is not far fetched when it is recalled that oxygen deficient YBCO shows Cu ordering above 200 K. The only reasonable escape from the coexistence problem, then, is to assume that the superconducting electrons are hosted in the Sr–O layer and not the Cu–O_2 (O_8, O_{10}) or YRu/Cu–O_4 (O_6) layers.

4 Conclusions

ND has been an important tool in reaching a comprehensive picture for the systems described. The in-situ measurements of ferrites at high temperature should help to adjust compositions to produce optimum electrical properties for fuel cell application. The results on the rutheno-cuprates shed substantial doubt on the conventional model for cuprate plane superconductivity. Nevertheless, these data would be incomplete and largely incomprehensible without their connection to other microscopic and macroscopic probes used to investigate the same systems. The ability, at MURR, to collect high quality ND data in short times and as a function of important parameters has been essential to the success of this work.

Acknowledgement This work is partially supported by the U.S. Department of Energy under the contract UAF99-0038.

References

[1] H. Rauch and W. Waschkowski, Neutron scattering lengths in: Neutron Data Booklet, A.-J. Dianoux and G. Lander Eds. Inst., Laue-Langevin, Grenobole France 2002.
[2] W. B. Yelon, R. Berliner, and M. Popovici, Physics B **241**(3), 237 (1997).
[3] Y. Tokura, Fundamental features of colossal magnetoresistive manganese oxides, in: Y. Tokura (Editors), Contribution to Colossal Magnetoresistance Oxides, Monographs in Condensed Matter Science, (Gordon & Breach, London, 1999).
[4] A. J. Millis, Nature **392**, 147 (1998).
[5] E. Dagotto, T. Hotta, and A. Moreo, Phys. Reports **334**, 1 (2001).
[6] J. M. Ralph, J. T. Vaughey, and M. Krumpelt, in: Solid Oxide Fuel Cells VII, Proceedings of the seventh international symposium, Evaluation of potential cathode materials for SOFC operation between 500–800C, Edited by H. Yokokawa and S. C. Singhal, The Electrochemical Society, Inc. Pennington, New Jersey USA. p. 466–475.
[7] C. C. Chen, M. M. Nasrallah, and H. U. Anderson, J. Electrochemical Soc. **142**, 491–496 (1995).
[8] J. B. Goodenough, in Progress in Solid State Chemistry, Edited by H. Reiss (Pergamon, London, 1971) Vol. **5**, p. 145.
[9] J. B. Yang, W. B. Yelon, W. J. James, Z. Chu, M. Kornecki, Y. X. Xie, X. D. Zhou, H. U. Anderson, A. G. Joshi, and S. K. Malik, Phys. Rev. B **66**, 184415 (2002).
[10] P. D. Battle, T. C. Gibb, and P. Lightfoot, J. Solid State Chem. **84**, 271 (1990).
[11] J. B. Yang, X. D. Zhou, Z. Chu, W. M. Hikal, Q. Cai, J. C. Ho, D. C. Kundaliya. W. B. Yelon, W. J. James, H. U. Anderson, H. H. Hamdeh, and S. K. Malik, J. Phys. Condens. Matter **15**, 5093 (2003).
[12] M. Horvatic, T. Auler, C. Berthier, P. Butaud, W. G. Clark, J. A. Gillet, P. Segransen, and J. Y. Henry, Phys. Rev. B **47**, 3461 (1993).
[13] H. Alloul, T. Ohno, and P. Mendels, Phys. Rev. Lett. **63**, 1700 (1989).
[14] M. K. Wu, D. Y. Chen, F. Z. Chien, S. R. Sheen, D. C. Ling, C. Y. Tai, G. Y. Tseng, D. H. Chen, and F. C. Zhang, Zeitschrift für Physik B **102**, 37 (1997).
[15] J. W. Lynn, B. Keimer, C. Ulrich, C. Bernard, and J. L. Tallon, Phys. Rev. B **61**, R14964 (2000).
[16] Y. Tokunaga, H. Kotegawa, K. Ishida, Y. Kitaoka, H. Takagiwa, and J. Akimitsu, Phys. Rev. Lett. **86**, 5767 (2001).
[17] J. H. Han and J. Budnick, to be published.
[18] H. Takagiwa, J. Akimitsu, H. Kawano-Furukawa, and Y. Hoshizawa, J. Phys. Soc. Japan, **70**, 333 (2001).
[19] H. A. Blackstead, John D. Dow, D. R. Harshman, I. Felner, W. B. Yelon, W. J. Kossler, A. J. Greer, C. E. Stronach, E. Koster, and B. Hitti, to be published.
[20] L. Bauernfeind, W. Widder, and H. F. Braun, Physica C **254**, 151 (1995).
[21] I. Felner, U. Asaf, Y. Levi, and O. Millo, Phys. Rev. B **55**, R3374 (1997).
[22] H. A. Blackstead, J. D. Dow, D. R. Harshman, W. B. Yelon, M. X. Chen, M. K. Wu, D. Y. Chen, F. Z. Chein, and D. B. Pulling, Phys. Rev. B **63**, 214412 (2001).
[23] N. G. Parkinson, P. D. Hatton, J. A. K. Howard, C. Ritter, F. Z. Chien, and M. K. Wu, J. Mater. Chem. **13**, 1468 (2003).

YBa$_2$Cu$_3$O$_7$/La$_{2/3}$Ca$_{1/3}$MnO$_3$ superlattices showing simultaneously ferromagnetic and superconducting order

H.-U. Habermeier* and G. Cristiani

Max-Planck-Institut für Festkörperforschung, Heisenbergstr. 1, 70569 Stuttgart, Germany

Received 1 September 2003, accepted 3 March 2004
Published online 11 May 2004

PACS 74.25.Ha, 74.78.Fk, 75.70.Cn

All oxide thin film heterostructures and superlattices composed of materials of different functionalities such as ferromagnetic [FM], superconducting [SC], piezoelectric or ferroelectric open a new wide field of device oriented fundamental research by exploring the interaction of different long range ordered ground states. To demonstrate these principles we have prepared YBa$_2$Cu$_3$O$_7$ based superlattices [SL's] with La$_{2/3}$Ca$_{1/3}$MnO$_3$ as ferromagnetic [FM] part by pulsed laser deposition. The films are characterized with respect to their structual, magnetic and transport properties. Whereas simple heterostructures [single layer La$_{2/3}$Ca$_{1/3}$MnO$_3$ and single layer YBa$_2$Cu$_3$O$_7$ 50 nm thickness each] reproduce the intrinsic properties of the constituent material rather well [Curie temperature 250 K superconducting transition at $T = 75$ K] there are some novel effects emerging due to the coupling between the layers observed in the superlattices. The experimental findings are discussed within the frame of a model based on FM interlayer coupling and superconducting proximity effect.

© 2004 WILEY-VCH Verlag GmbH & Co. KGaA, Weinheim

1 Introduction

In a pioneering paper Ginzburg [1] addressed the question of coexistence of two different long-range ordering principles such as superconductivity and ferromagnetism. He concluded that these two antagonistic phenomena cannot coexist in a homogeneous system because superconductivity requires the formation of pairs of electrons with antiparallel spins whereas ferromagnetic order forces the spins to be aligned parallel. This situation can be completely different in inhomogeneous systems where the local environment can cause parallel spin alignment and superconductivity in different regions of the unit cell. This is realized either in natural or artificial layered structures such as superlattices and heterostructures. Examples for layered structures are the FM superconductors ErRh$_4$B$_4$ and HoMo$_6$S$_8$ [2, 3] with a Curie temperature, T_{Curie}, smaller than the transition temperature to superconductivity, T_C. The competing ordering mechanisms lead to superconductivity first and upon further cooling FM order is energetically more favorable and superconductivity is destroyed. Recently, the family of RuSr$_2$RECu$_2$O$_8$ and RuSr$_2$RE$_{2-y}$Ce$_y$Cu$_2$O$_{10}$ [4, 5] with RE = Gd, Eu has been discovered, and $T_{Curie} > T_C$ has been postulated, representing superconducting [SC] ferromagnets with simultaneous occurrence of ferromagnetism and superconductivity below T_C.

In the past, superlattices [SL's] based on metal superconductors and ferromagnets have been fabricated and studied, both experimentally and theoretically [6–9]. All these experiments have been carried out in systems which have in common that the superconductors are s-wave and their coherence length is always much larger compared to the FM metal layer thickness. They showed a non-monotonic – in some cases even an oscillatory – dependence of T_C with increasing FM layer thickness. This behavior has been controversially discussed and some authors claim an interplay of two competing mechanisms such as the

---

* Corresponding author: e-mail: huh@fkf.mpg.de

formation of a magnetically "dead layer" at the interface and its subsequent magnetization in the exchange field to account for the oscillation in T_C. In the case of SL's composed of cuprates and manganites this situation is changed fundamentally from the materials as well as physics point of view. According to TEM results [10, 11] intermixing is confined to roughly one unit cell at the interface thus not affecting the electronic system to the extend as in metallic SL's. Furthermore, the cuprate superconductors have a d-wave symmetry of the order parameter and an antiferromagnetic ground state. The coherence length is highly anisotropic with ~0.1 nm along the c-direction and ~1–2 nm along the a, b plane. The FM part has a nearly 100% spin polarization, ferromagnetism is caused by Zener double exchange rather than by an itinerant band mechanism. Due to comparable deposition conditions for the cuprates and the manganites high quality SL's can be fabricated by the conventional techniques such as sputtering and pulsed laser deposition. So far, several papers appeared dealing with all-oxide SC/FM superlattices [12–19] demonstrating the coexistence of both ordering phenomena, superconductivity and ferromagnetism and a suppression of T_{Curie} and T_C as well. In these papers especially the T_C depression has been studied and a characteristic length scale for the depression has been claimed much larger than predicted by the existing theories of the FM/SC proximity effect. On the other hand – as in the metallic systems – little attention was given to the changes of T_{Curie} in the SL's. In this paper the experimental results of a systematic study of the change of ordering temperature in FM/SC superlattices is presented, and discussed within the frame of a novel long-range proximity effect.

2 Experimental details and results

Using SrTiO$_3$ single crystal substrates, superlattices consisting of m unit cells of YBa$_2$Cu$_3$O$_{7-x}$ [YBCO] and n unit cells of La$_{2/3}$Ca$_{1/3}$MnO$_3$ [LCMO] repeated N times – $(m, n)_N$ SL's – and thus of different modulation lengths, Λ, have been deposited at 730 °C at an oxygen pressure of 5×10^{-3} Pa by pulsed laser deposition technique [20, 21]. To ensure complete oxygenation they were in-situ annealed in 1 atm oxygen for 1 hour at 530 °C. The deposition system (high vacuum chamber in conjunction with a KrF Excimer laser) is equipped with a FIR pyrometer temperature to control of the radiatively heated substrate surface and film temperature. A computer controlled target exchange system accommodating up to 5 different targets facilitates the desired SL formation. Thickness control of the individual layers is done by pulse counting after some calibration runs to ensure the stability of the ratio film thickness per pulse. X-ray diffraction analysis confirmed the phase purity of the c-axis oriented films. The X-ray diffraction shows diffraction peaks with a rather large FWHM of 2° – 4°. Superlattice peaks as expected due to the additional scattering planes in a superlattice are barely to be identified. A determination of the SL peak positions according to the standard formula $\lambda = \lambda/2(\sin \Theta_n - \sin \Theta_{n+1})^{-1}$ shows that for the first and second order peaks for our SL with the smallest modulation length $\lambda = 13$ nm the peak positions to be expected are buried in the main peak [14]. Cross-sectional high resolution TEM analysis reveals the SL formation and shows atomically flat interfaces. The epitaxial relation is clearly "cube-on-cube" [10]. The magnetic moments of the films are determined using standard SQUID magnetometry, transport properties are measured by standard 4-point probe techniques using evaporated gold contacts and a probing current of 0.1 mA. Figure 1 represents the $R(T)$ and $M(T)$ plots of a [YBCO 40 nm/LCMO 20 nm]$_5$ SL showing $T_{Curie} = 150$ K and $T_C = 70$ K, respectively. The normal state properties are expected to be a superposition of the linear dependence $R \sim T$ of the YBCO and the features of the metal-insulator transition of the LCMO. The metal-insulator transition appears as a change in the slope of the $R(T)$ around T_{Curie}. The main features in Fig. 1 are the reduction of the ordering temperatures to FM and SC, respectively, and a substantially higher resistivity as compared to the corresponding single layers. Optical measurements using spectroscopic ellipsometry as a bulk probe confirm this and rule out additional scattering at internal surfaces as the reason for the enhancement of resistivity [19]. For a systematic study of these effects we prepared [t_{YBCO}/t_{LCMO}]$_N$ – type SL's of different compositions, keeping either t_{YBCO} constant and change t_{LCMO} or vice versa. In Tabelle 1 the relevant data are given. The main results are summarized as follows: (i) Coexistence of ferromagnetism and superconductivity, (ii) composition dependent reduction of T_C and T_{Curie}, and (iii) reduction of conductivity.

© 2004 WILEY-VCH Verlag GmbH & Co. KGaA, Weinheim

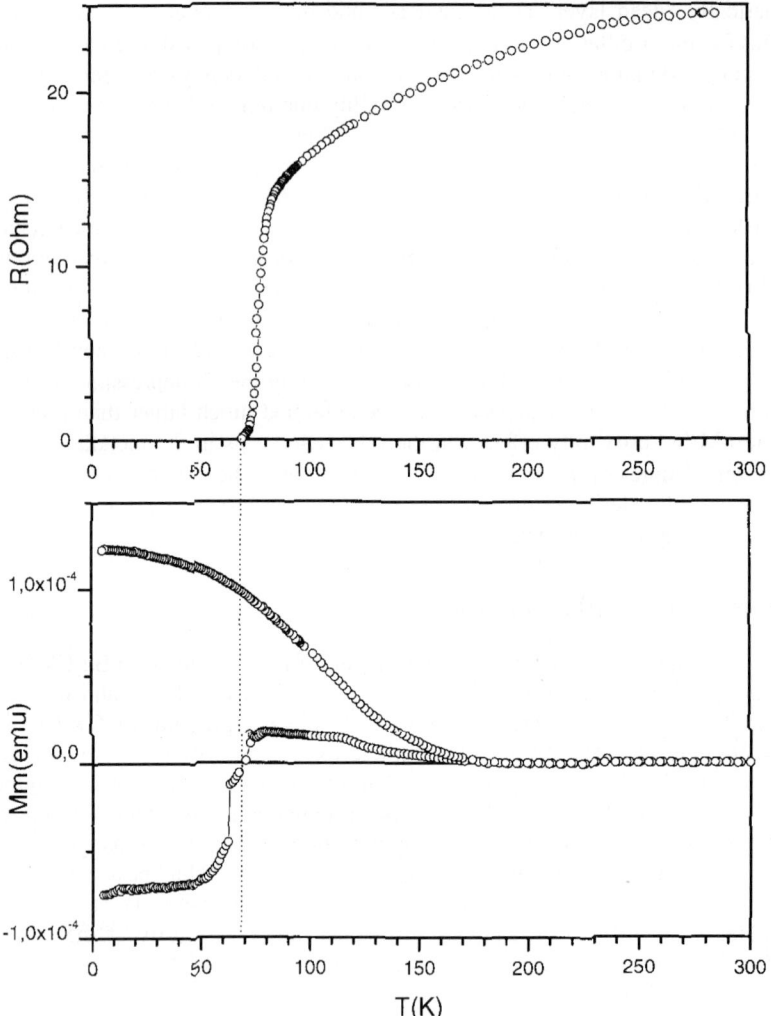

Fig. 1 Resistance (upper panel) and magnetic moment (lower panel for zero field cooled and field cooled measurement) of a [40 nm YBCO/20 nm LCMO]$_5$ superlattice as a function of temperature.

Table 1 Composition and transition temperature of various [YBCO/LCMO]$_5$ superlattices.

t_{YBCO} [nm]	t_{LCMO} [nm]	T_C [K]	T_{Curie} [K]
20	20	50	153
20	40	20	170
20	60	7	190
20	80	0	215
15	5	72	115
15	20	56	160
15	30	43	200
15	60	0	240

© 2004 WILEY-VCH Verlag GmbH & Co. KGaA, Weinheim

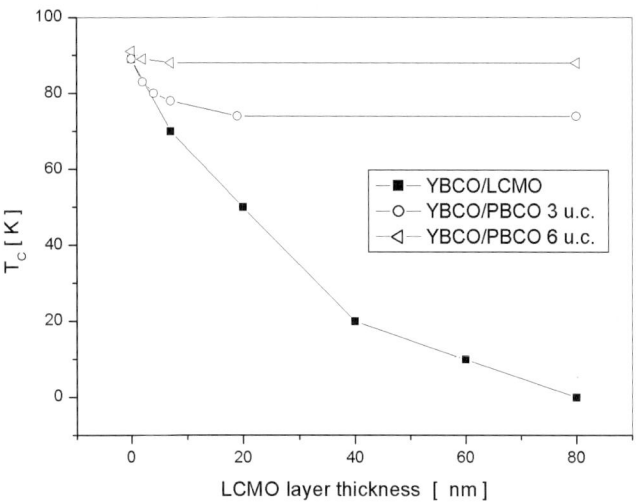

Fig. 2 Comparison of T_C vs. thickness of non-superconducting layers in YBCO/LCMO and YBCO/PrBCO superlattices (explanations in the text).

3 Discussion

The reduction of T_C in cuprate based superlattices with an insulating or normal spacer is not a new discovery; it has been investigated extensively in the past decade and a wealth of experimental data has been revealed and summarized in several review papers [22–24]. Amongst the mechanisms accounted for the T_C reduction material related extrinsic effects have been discussed such as interfacial lattice strains, incomplete oxygenation, disorder, grain boundaries and weak links due to the interface of the SC and non-SC layers [23]. Systematically ruling out these effects to be dominant in high quality SL's, intrinsic effects are treated as possible origin for the T_C reduction. Suggested explanations like interlayer coupling, Kosterlitz–Thouless transition, long-range Josephson coupling, and charge transfer mechanisms show that no conclusive quantitative interpretation of the T_C reduction in SL's exists. A quantitative agreement of the experimental data with theoretical ones has been achieved by Jansen and Block [25] for YBCP/PrBCO SL's on the basis of a microscopic approach based in indirect-exchange Cooper pairing between quasiparticles and oxygen anions. Qualitatively, all these arguments can be used in the case of the FM/SC SL's, quantitatively the T_C suppression is much more pronounced as in the case of e.g. YBCO/PrBCO. In Fig. 2 the experimental data for YBCO/LSMO SL's with $t_{YBCO} = 20$ nm and increasing LCMO thickness are represented and compared with the data reported by Lowndes [26] for the YBCO/PrBCO system for 3 unit cell and 6 unit cell thick YBCO layers in the SL. The data points for 80 nm spacer thickness are extrapolated from the plot given in Ref. [26]; this extrapolation seems to be justified according to the theory of Jansen and Block and the experimental results of our group and others. It is obvious that the reduction of T_C is much stronger compared to the YBCO/PrBCO case and additional interaction mechanisms must be effective. As already discussed in [19] magnetic correlations can play an important role in determining the temperature of the phase transitions, possibly due to a novel magnetic proximity effect where the charge carriers are coupled to different and competing kinds of magnetic correlations. The physical origin of the reduction of T_{Curie} is much less clear. Extrinsic effects such as incomplete oxygenation and epitaxial strain can be ruled out according to the arguments given in [10]. Massive charge transfer of holes into the LCMO layer can qualitatively reduce T_{Curie} according to the temperature/doping phase diagram of the La–Ca–Mn–O system. For bulk material – as well as relaxed thin films – at 33% Ca doping T_{Curie} is 260 K and an increase of the Mn^{4+} concentration – equivalent to increased doping – leads to a minimum T_{Curie} of 195 K at 50% doping. The Curie temperatures measured in some of the SL's [c.f. Tabelle 1] are much below this value consequently the argument of hole transfer to the LCMO can not quantitatively explain the reduction of T_{Curie}. Additionally, such a massive charge transfer would drive the YBCO into the non-superconducting insulating state in contrast to the experimental observations. In the literature there are several papers dealing with magnetic coupling in ferromagnetic/normal metal multilayers. The prevailing experimental evidence indicates that the exchange coupling with metal spacers is short range (1–5 nm) and a thickness dependent crossover from FM to AFM coupling occurs. These arguments can not simply be transferred to oxide SL's. The short-range spin diffusion length of several nm in metallic FM systems will confine the interaction effect due to the neighboring metal layer to a region of less than 3 nm close to the interfaces. In the case of the oxide SL's the interaction length must be apparently long range (10–30 nm). The close vicinity of T_{Curie} with the temperature for the spin gap opening in the YBCO normal state suggests the interrelation of the two temperatures. A reduced polarizability of the charge carrier spins below the spingap opening temperature can be the reason for this effect. A certain analogy to the model of Sa de Melo [27] can be seen which predicts a modification of the density of states and a weakening of the coupling via the appearance of a superconducting gap. Probably the spingap plays that role in the oxide SL's. A suggestive argument to support this view would be a simple relation between the pseudogap and the T_{Curie} reduction. Determining the pseudogap temperature for the SL's investigated from the generic phase diagram T vs. doping and plotting these data vs. the measured reduction of T_{Curie} a simple linear relation is found.

In summary, high quality cuprate/manganite superlattices have been prepared which show simultaneously the occurrence of superconductivity and ferromagnetism. The temperatures for the phase transitions can be varied systematically by tailoring the SL composition. The physical origin of the T_C reduction is seen in an interplay of charge transfer between adjacent layers and charge localization due to magnetic correlation giving rise to a long range magnetic proximity effect. For the reduction of T_{Curie} the role of the pseudogap temperature should be considered.

References

[1] V. L. Ginzburg, Zh. Eksp. Teor. Fiz. **32**, 202 (1956).
[2] M. B. Maple and Ø. Fischer, Superconductivity in Ternary Compounds II Topics Current Phys. Vol. 34, (Springer, Berlin, 1982).
[3] M. B. Maple, Physica C **341–348**, 47 (2000).
[4] L. Bauernfeind, W. Widder, and H.-F. Braun, Physica C **254**, 151 (1995).
[5] I. Felner, Physica C **341–348**, 25 (2000).
[6] P. Fulde and R. Ferrell, Phys. Rev. A **135**, 550 (1964).
[7] P. Korevaar, Y. Suzuki, R. Coehoorn, and J. Aarts, Phys. Rev. B **49**, 441 (1994).
[8] Th. Mühge et al., Phys. Rev. B **65**, 8945 (1997).
[9] A. S. Sidorenko et al., Ann. Phys. (Leipzig) **12**, 37 (2003).
[10] H.-U. Habermeier et al., Physica C **364–365**, 298 (2001).
[11] H.-U. Habermeier and G. Cristiani, IEEE Trans. Appl. Supercond. to appear July 2003.
[12] G. Jakob, V. Moschalkov, and Y. Bruynseraede, Appl. Phys. Lett. **66**, 2564 (1995).
[13] P. Przyslupski et al., IEEE Trans. Appl. Supercond. **7**, 2192 (1997).
[14] H.-U. Habermeier and G. Cristiani, Proceedings of SPIE **4811**, 111 (2002).
[15] Z. Sefrioui et al., Appl. Phys. Lett. **81**, 4568 (2002).
[16] H.-U. Habermeier and G. Cristiani, J. Supercond. **15**, 425 (2002).
[17] Z. Sefrioui et al., Phys. Rev. B **67**, 214511 (2003).
[18] P. Przyslupski et al. Physica C **387**, 40 (2003).
[19] T. Holden et al., accepted Phys. Rev. B (2003) – cond. Mat./0303284.
[20] D. Dijkamp and T. Venkatesan, Appl. Phys. Lett. **51**, 619 (1987).
[21] H.-U. Habermeier, Eur. J. Solid State Inorg. Chem. **28**, 201 (1991).
[22] J.-M. Triscone and Ø. Fischer, Rep. Prog. Phys. **60**, 1673 (1997).
[23] I. Bozovic and J. Eckstein, in Physical Properties of High Temperature Superconductors V. D. M. Ginsberg ed. (World Scientific, Singapore 1996) 99–207.
[24] H.-U. Habermeier, in: Crystal Growth in thin Solid Films: control of Epitaxy M. Guilloux-Viry and A. Perrin, ed. (Research Signpost Trivandrum, India, 2003) 207–244.
[25] L. Jansen and R. Block, Physica C **252**, 278 (1998).
[26] D. H. Lowndes, D. P. Norton, and J. D. Budai, Phys. Rev. Lett. **65**, 1160 (1990).
[27] C. A. R. Sa de Melo, Phys. Rev. Lett. **79**, 1933 (1997).

INTERNATIONAL CONFERENCE ON DEFECTS IN INSULATING MATERIALS
JULY 11 - 16, 2004

ICDIM-2004

Baltic Sea

Organizers:
University of Latvia TartuUniversity
Institute of Solid   Institute of Physics
State Physics
Latvia   Estonia
Chairmen
Prof. I. Tale   Prof. A. Lushchik

Riga Gulf

Important dates
2-nd announcement — 30 September, 2003
Abstracts — 1 March, 2004
Notice of acceptance — 30 April, 2004
Registration at regular fee — 15 May, 2004
Submission of Proceedings — 12 July, 2004

Sigulda

RIGA

Address: Institute of Solid State Physics
University of Latvia
8 Kengaraga Str., Riga LV-1063, Latvia
Website: http://www.fpd.lu.lv/
Conferences/icdim2004.htm
e-mail: icdim04@cfi.lu.lv

The proceedings of this conference will be published in physica status solidi (c).

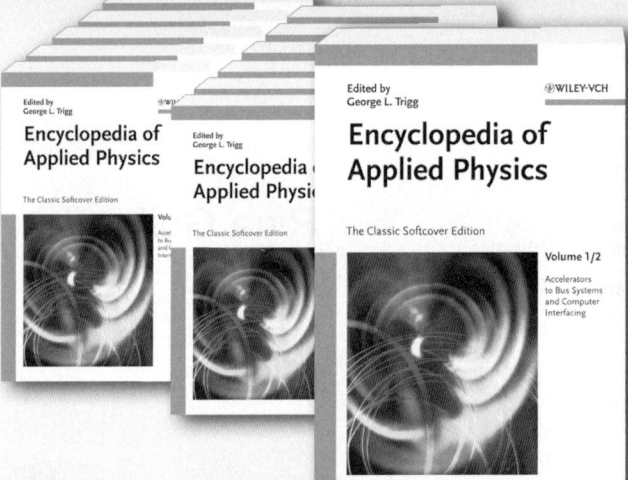

physica pss status solidi b

www.pss-b.com

basic research

The following pages have been reprinted from
phys. stat. sol. (b) **241**, No. 6, 1167–1250 (2004) as part of the
Proceedings of the 3rd International Conference on
Magnetic and Superconducting Materials (MSM '03)
Monastir, Tunisia, 1–4 September 2003.

Quantum dynamics of atomic magnets

B. Barbara*,1, R. Giraud1, and A. M. Tkachuk2

1 Laboratoire de Magnétisme Louis Néel, CNRS, BP166, 38042 Grenoble Cedex-09, France
2 All-Russia Scientific Center "S. I. Vavilov State Optical Institute", 199034 St. Petersburg, Russia

Received 31 December 2003, accepted 12 March 2004
Published online 14 April 2004

PACS 75.45.+j, 71.70.–d, 75.50.Lk

After the first evidence for tunneling of the total angular momentum of Ho^{3+} in $YLiF_4$ (insulating), where electronic and nuclear moments tunnel simultaneously (two-body entanglement), we show that pairs of Ho^{3+} ions also tunnel simultaneously (four-body entanglement). Such co-tunneling transitions, mediated by weak dipolar interactions, are discussed in terms of a four-spin representation including the rare-earth nuclear spin. This study of the entanglement of pairs of distant Ho^{3+} ions constitutes a first step to elucidate many-body quantum-spin relaxation. Finally, implications on the feasibility of two-spin qubits are discussed.

© 2004 WILEY-VCH Verlag GmbH & Co. KGaA, Weinheim

1 Introduction

Since the discovery of quantum tunneling of the magnetization in single molecule magnets (SMMs), followed by the first observation of Berry phases [1–3], the effects of environmental degrees of freedom on the tunneling mechanism of weakly-interacting mesoscopic spins have been investigated within the physical picture of a central spin immersed in a spin-bath generated by the other spins [4–6]. Most important features of the staircase hysteresis loop and the power-law relaxation were qualitatively understood within this single-spin picture [4, 7]. The spin-bath being generated by environmental spins coupled to the central spin by dipolar (and eventually exchange) interactions, one may expect to observe, not only single-spin tunnel transitions (shifted by the dipolar bias), but also multi-spin tunnel transitions. Such transitions have been recently observed. They involve pairs of Ho^{3+} ions diluted in a matrix of $YLiF_4$. This effect corresponds to simultaneous tunneling transition of pairs of Ho^{3+} ions (co-tunneling). It has been detected for the first time in [6], in the temperature range of 100 mK. In this paper we show that this phenomenon of co-tunneling, still present at a relatively high temperature (1.75 K), can be studied in detail by simple ac-susceptibility experiments [7]. We performed such experiments on a 0.1 at.% Ho-doped $LiYF_4$ single-crystal at frequencies up to 1 kHz. The results are analyzed using single-ion and two-ion models, including crystal-field, hyperfine and dipolar interactions. They all lead to the first description of co-tunneling in a magnetic system.

2 Single-ion and two-ion tunneling in ac-susceptibility

The measurements were performed with a Quantum Design SQUID magnetometer using a 4 Oe amplitude excitation field at $T \sim 1.75$ K and quasi-static fields applied along the easy c-axis. Figure 1 shows the complex susceptibility of a small 0.1% at. Ho-doped $LiYF_4$ single-crystal, measured at relatively low frequency (163 Hz). Numerous peaks (dips) in the in-phase (out of phase) response as a function of H are observed. They are typical of faster relaxation times, as this is shown in the scheme of Fig. 2. The largest

---

* Corresponding author: e-mail: barbara@grenoble.cnrs.fr, Phone: 33 (0)4 76 77 11 92, Fax: 33 (0)4 76 77 11 92

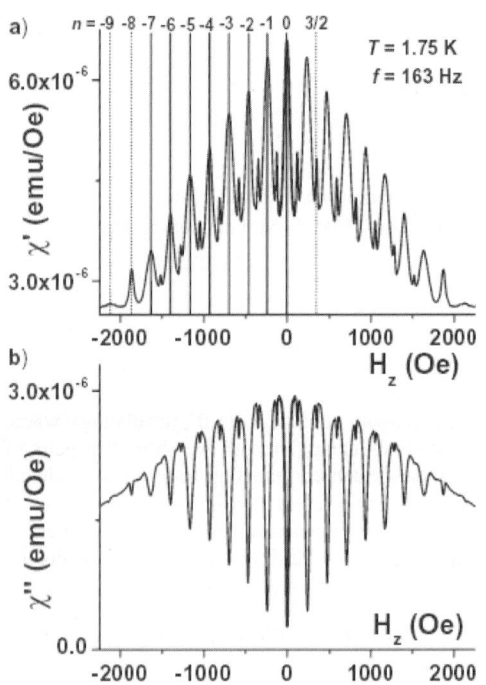

Fig. 1 Ac-susceptibility measured at $T = 1.75$ K with a frequency of 163 Hz. The non-monotonous ac-response, as a function of H gives evidence for a faster relaxation at well-defined values $H_n = n \times 230$ Oe. Large and broad peaks or dips correspond to single-ion resonances, while small peaks or dips correspond to two-ion tunneling. The solid lines show the positions of electro-nuclear level crossings calculated within the single-ion representation Fig. 3a, whereas the dotted lines are associated with the level crossings of the two-ion representation (Fig. 3b).

ones are clear signatures of single Ho^{3+} ion tunneling, in which the total angular momentum of the rare-earth and the nuclear spin tunnel at the same time [6]. This is seen in Fig. 3a, where the fields at which they occur coincide precisely with the crossing-fields of the electro-nuclear Zeeman diagram calculated within the single-ion picture. The analytical expression of these crossing, $H_n = n \times A_J / 2g_J\mu_B = n \times 230$ Oe, can easily be calculated from the level scheme. Much smaller and narrower peaks (dips) are also found between the large ones. They correspond to half-integer n values, with $-13 \leq 2n \leq 13$, as well as for $n = |8|$ and $n = |9|$. They do not correspond to the level crossings of the single-ion scheme of

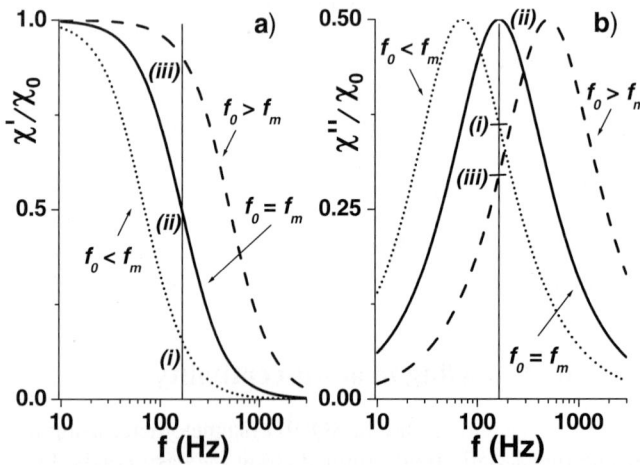

Fig. 2 Variation of the in-phase (a) and dissipative (b) responses in the ac-susceptibility for three different relaxationrates f_0, described within the Debye model assuming a single(dominant) relaxation time. The constant measurement frequency f_m is suggested by the vertical solid line, and χ_0 is the static susceptibility. Depending on the value of f_m, the dissipative reponse can be either increased (case (i)), or decreased (case (iii)) if the dynamics gets faster. In any case, the in-phase response follows the variations of f_0.

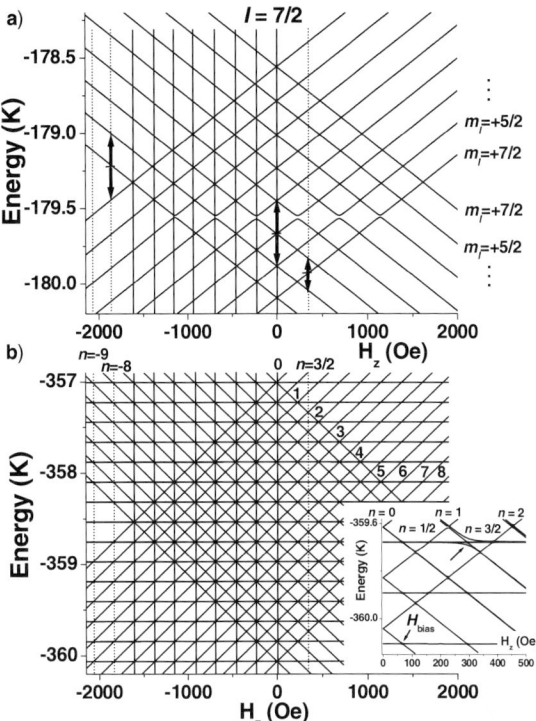

Fig. 3 (a) Electro-nuclear energy levels calculated within the single-ion representation. Equally-spaced levels in the Zeeman diagram allow for non-dissipative processes in the tunneling regime. As suggested by arrows, three various kinds of resonances may occur at well-defined values $H_n = n \times 230$ Oe. (b) Within the two-ion Zeeman diagram, each resonance involving a pair of magnetic moments now corresponds to a level crossing. Note that the finite slopes are twice larger than in (a). Their crossing describes a co-tunneling process (half-integer n, with $-14 \leq 2n \leq 14$). Other crossings only involve the flipping of a single magnetic moment (integer n, with $-14 \leq n \leq 14$). The inset shows how an isotropic interaction, with a given coupling strength, removes part of the degeneracies for integer n values, leading to biased tunneling.

Fig. 3a, but to the level crossings calculated within the two-ion representation (dotted lines in Fig. 3b) [7]. The single-ion level-scheme of Fig. 3a shows that these additional peaks/dips, situated just in between the single-ion peaks (dips), can be interpreted as the result of simultaneous transitions of two Ho^{3+} ions, with conservation of the total energy. This must corresponds to the co-tunneling of the electro-nuclear moments of these two ions (four-body entanglement).

In Fig. 4 we present some details of Fig. 1 obtained at the same temperature and higher frequency (801 Hz) [7]. Here too the smallest peaks corresponding to the co-tunneling of Ho^{3+} ion pairs. The better resolution of this experiment allows to determine their width, which is of the order of dipolar interactions (a few mT). This indicates an inhomogeneous line broadening of dipolar origin. The single-ion resonances ($n = 0$ and $n = 1$, in Fig. 4) have a Lorentzian-like shape, with a broad basis and a narrow half-height width.

The broad basis seems to be temperature-dependent, while the half-height width is of the order of dipolar interactions, as for co-tunneling. Note that, when approaching the single-ion resonance ($H \to H_n$), the tunneling rate increases, leading to an increase of the real susceptibility at low and high frequency (Fig. 1 and Fig. 4). Regarding the imaginary susceptibility, it also increases at low frequency (Fig. 1), but at high frequency it turns over a fast decreasing regime, producing a narrow dip (Fig. 4). This inversion corresponds to a different ac regime, in which the measurement frequency is (i) larger than the tunneling rate, when H is not too close to H_n, and (ii) smaller than the tunneling rate, when H is close to H_n (see Fig. 2). In all cases the tunneling rate, given by $\Gamma_t = \omega \chi'/\chi''$, shows a Lorentzian-like peak at single-ion and two-ion resonances, with half-height width of the order of dipolar interactions. The question of the exact shape of the inhomogeneous dipolar-field distribution is not really solved, due to the difficulty in distinguishing between different possible distributions. A Lorentzian distribution, already mentioned in [1] (for molecules), suggests that multi-spin correlations built by successions of tunneling transitions [4], are important. In the next section we will show the existence of three different types of multi-spin tunneling transitions, theoretically. One of them, the "dipolar-bias tunneling", takes place in the proximity of single-ion resonances and consequently should contribute to their lineshape.

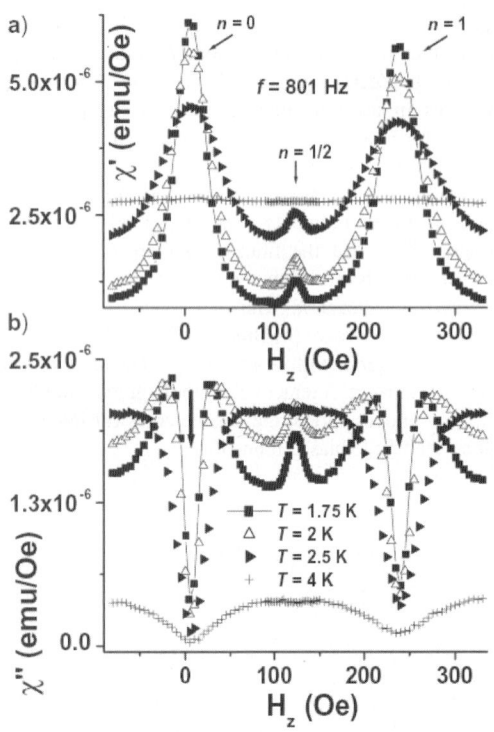

Fig. 4 Ac-susceptibility measured at a larger frequency (801 Hz). The decrease in the dissipative response for integer n (single-ion tunneling) is due to a change in the ac regime. In all cases the line-shape is of Lorentzian type with a half-height width of the order of dipolar interactions.

A follow up of this study might allow a better understanding of the general nature of Lorentzian-like lineshapes in this Ho^{3+} model system, with special emphasis on the range of multi-spin correlations built by multi-spin tunneling.

3 Two-body and four-body representations of tunneling

To underline the role of spin-spin interactions in $LiYF_4:Ho^{3+}$ we go beyond two-body/single-ion representation and introduce a four-body/two-ion one (two electronic angular moments and two nuclear spins). The former single-ion energy spectrum with equally spaced levels (Fig. 3a) now corresponds to a two-ion spectrum with pair-state energy levels (Fig. 3b). It was calculated on the basis of single-ion electro-nuclear states. The new features of this four-body representation, in the absence of coupling between Ho^{3+} ions, are (i) additional crossing fields with half-integer n, intermediate between single-spin resonances (integer n) and (ii) additional levels at constant energies (singlets).

These additional crossings calculated within the two-ion picture are avoided level crossing (gaps are not always discernible) associated with the *co-tunneling* of two parallel electro-nuclear moments from up to down or inversely, i.e. $[1/2, m_I; 1/2, m_{I'}] \leftrightarrow [-1/2, m_I; -1/2, m_{I'}]$. We recall here that such co-tunneling events can also be interpreted by simultaneous transitions of two ions between equally separated energy-levels of the single-ion picture (see dotted lines in Fig. 3a).

The singlets correspond to field-independent flips of two opposite electro-nuclear moments: $[1/2, m_I; -1/2, m_{I'}] \leftrightarrow [-1/2, m_I; 1/2, m_{I'}]$. In an antiferromagnetic medium these transitions would lead to the quantum reversal of "lines of spins" in which the sequence $+-+-+-\ldots$ becomes $-+-+-+\ldots$ (*spin-diffusion*). The introduction of an isotropic coupling removes partially the degeneracy at integer and half-integer n by shifting the singlet states by a field H_{bias} proportional to the coupling energy (inset of Fig. 3b; for the sake of clarity we have taken a coupling much larger than dipolar fields). The consequence is a shift of the single-ion resonance, in which a single Ho^{3+} electro-nuclear moment reverses, in the interaction field of other Ho^{3+} moments (the states of these moments should change, but very slightly). We call this *dipolar-bias tunneling*. In the two-ion level scheme it is related to the crossing of the singlet $[1/2, m_I; -1/2, m_{I'}]$ and doublet [±1/2, m_I; ±1/2, m_I] states. The distribution of dipolar interactions with zero mean-value and rms-deviation ~H_{dip}, leads to a distribution of dipolar-bias tunneling and thus to a resonance broadeing ~$g\mu_B H_{dip}$.

It is important to mention that the positions of avoided-level crossings associated with co-tunneling are not affected by interactions; they are related to the hyperfine or anisotropy constant only [8]. On the contrary the co-tunneling gap may depend strongly on interactions, but only if they are anisotropic. This is the case of non-secular dipole–dipole interactions, which must be an important source of multi-tunneling effects when multi-phonons transitions are not relevant. Finally, the same ideas can be applied to magnetic molecules such as Mn_{12}-ac and Fe_8. Recently Gaudin et al. [9] observed a weak Schottky-like specific heat anomaly, situated just in between two single-molecule resonances, i.e., according to the results of this paper, exactly where co-tunneling resonances of pairs of Fe_8 molecules are expected. This is the only case where a sign of co-tunnneling is given in a molecular system. The reason is probably the strong (anisotropic) dipolar interactions of this molecular system. Moreover, exchange-biased tunneling (similar to dipolar-biais tunneling, but with a shift due to isotropic exchange interactions) was recently observed in the molecular dimer Mn_4 molecules [10]. Exchange interactions being distributed about a non-zero value, exchange-bias resonances are shifted with respect of zero-field resonances by a field given by the strenght of the coupling, as for the metamagnetic transition of antiferromagnetic systems. This is quite different from dipolar-biais tunneling, where the dipolar distribution has zero mean-value and finite rms-deviation. The study of exchange bias tunneling may allow to get more insights in the understanding of antiferromagnetic moleculear or ionic clusters where similar transitions are observed [11, 12].

4 Role of anisotropic interactions and spin

In this first attempt to elucidate the origin of co-tunneling transitions, we will temporarily neglect the effects of phonons. In the absence of phonons, the Hamiltonian of two spins coupled by anisotropic interactions $H_{exc} = J_z S_1^z S_2^z + J_x S_1^x S_2^x + J_y S_1^y S_2^y$ can be written:

$$H_{exc} = J S_1^z S_2^z + \alpha J (S_1^+ S_2^- + S_1^- S_2^+) + \beta J(S_1^+ S_2^+ + S_1^- S_2^-) , \quad (1)$$

where $\alpha J = (J_x + J_y)/4$ and $\beta J = (J_x - J_y)/4$. The way the S^+ and S^- operators are tied in (1) immediately tells that spin-diffusions will require a breaking of the z-xy symmetry (↑↓ ⇒ ↓↑) and co-tunneling of the x-y symmetry (↑↑ ⇒ ↓↓). Our observation of co-tunneling with Ho^{3+} in $YLiF_4$ strongly suggests that both symmetries are broken in this system. This is not surprising because couplings arise from dipolar interactions only, and we know that these interactions are highly anisotropic. As an example the transformation of the term $(\boldsymbol{\mu}_i \boldsymbol{r}_{ij})(\boldsymbol{\mu}_j \boldsymbol{r}_{ij})/r_{ij}^5$ of the dipolar coupling gives Eq. (1) plus a few other terms mixing the xy and z components. Let us recall that the first part of (1) with $0 < \alpha < 1$ and $\beta = 0$, summed up to all spin pairs, corresponds to the first RVB Hamiltonian [13] allowing to create long living spin-spin excitations (spin-diffusions). Our studies differ from these examples of cross-spin relaxations by the fact that spin-diffusions ($\alpha \neq 0$) and co-tunneling ($\beta \neq 0$) occur by tunnelling "under a barrier".

We now illustrate in more details the role of anisotropic interactions taking Hamiltonian (1) in the simplest case of two effective spins 1/2. To be complete we add the Zeeman Hamiltonian:

$$H_Z = g_z \mu_B H_z (S_1^z + S_2^z) + g_x \mu_B H_x (S_1^+ + S_2^+ + S_1^- + S_2^-)/2. \quad (2)$$

The associated matrix (below) shows interesting features. In zero field, the states $|+ -\rangle$ and $|- +\rangle$ are mixed by the anisotropic term αJ (spin-diffusions), while the states $|+ +\rangle$ and $|- -\rangle$ are mixed by the anisotropic term βJ. The corresponding splittings will be calculated by diagonalisation of the matrix.

	$|+ +\rangle$	$|+ -\rangle$	$|- +\rangle$	$|- -\rangle$
$|+ +\rangle$	$g_z\mu_B H_z + J/4$	$-g_x\mu_B H_x$	$-g_x\mu_B H_x$	$\beta J/2$
$|+ -\rangle$	$-g_x\mu_B H_x$	$-J/4$	$J/2$	$-g_x\mu_B H_x$
$|- +\rangle$	$-g_x\mu_B H_x$	$J/2$	$-J/4$	$-g_x m_B H_x$
$|- -\rangle$	$J/2$	$-g_x\mu_B H_x$	$-g_x\mu_B H_x$	$-g_z\mu_B H_z + J/4$

In order to keep this discussion simple, we will assume in a first stage $H_x = 0$. The splitting linked with the zero-field crossing of the states $|+ +\rangle$ and $|- -\rangle$, is $\Delta = 2\{(g_z\mu_B H_z)^2 + (\beta J/2)^2\}^{1/2}$, while the split states are $\cos \varphi |+ +\rangle + \sin \varphi |- -\rangle$ and $\sin \varphi |+ +\rangle - \cos \varphi |- -\rangle$, with $\tan 2\varphi = \beta J/(4g_z\mu_B H_z)$ [14]. In this example the co-tunneling rate $\sim (\beta J)^2/(H_{dip})^2$ can be rather large. This is particularly true with dipolar interactions where $\beta \sim 1$. The maximum mixing, of course reached for $H_z = 0$ when $\varphi = \pi/2$ and $\Delta = \beta J$, decreases when H_z moves out of resonance. For large H_z the mixing goes to zero. In between, slow variation of H_z might be used to manipulate the quantum states of the two qubits formed by the pair of spins, while fast variations of H_z would allow to shift non-adiabatically from one state to another. The comparizon of the matrix (1) to different types of CNOT matrixes shows several interesting possibilities provided by anisotropic interactions and applied fields with different orientations θ ($\tan\theta = H_x/H_z$) [14].

The toy model given above, gives some qualitative results on the important role of anisotropic interactions in multi-spin tunneling and on their potentialities for quantum computing. In particular, it would possible to use pairs of rare-earth ions weakly coupled by dipolar interactions, for quantum calculation. An ensemble of two-ion qubits would require a sample made of chains of rare-earth ions (e.g. Ho^{3+} ions) with the possibility of adressing selectively given pairs of ions. This is one of our future challenge. Note, however that the toy model given above is too simple to be applied to a Ho^{3+} pair in $YLiF_4$, which involves both electronic angular momentum and nuclear spin. Furthermore, in the model, the states 1/2 and $-1/2$, can be connected in a single operation by the operator S^+ or S^-. This is not the case for Ho^{3+} ions where the two ground-states cannot mix directly, but through the excited singlet state 10 K above (top of the barrier) [6]. A more specific interpretation of the tunneling probabilty of a pair of Ho^{3+} ions, will be given elsewhere. Two different mechanisms will be discussed. The first one is based on magnetic interactions only, as above. The second one will involves co-tunneling transitions mediated by phonons. This is quite a realistic possibility in this system, where equidistant energy levels favor multi-spin and multi-phonon transitions. Furthermore, co-tunneling events were observed at relatively high temperature.

5 Conclusion

In conclusion, Ho^{3+} ions diluted in $LiYF_4$ constitute a model system to evidence and describe in detail the various cross-spin relaxations in the tunneling regime of an ensemble of weakly coupled mesoscopic spins. In the single-ion representation, intra-atomic Ho^{3+} hyperfine coupling gives equally separated energy levels allowing multi-spin transitions with conservation of the total energy (Fig. 3a). These transitions, co-tunneling and biased tunneling could be evidenced using simple ac-susceptibility experiments. In the case of co-tunneling, the angular moments of two Ho^{3+} ions tunnel coherently from up to down or down to up, with their nuclear spins, while with biased tunneling the angular momentum and the nuclear spin of only one Ho^{3+} ion tunnels, while the change of electro-nuclear state of the other Ho^{3+} ions remains small and does not involve tunneling. As shown in Fig. 3, co-tunneling processes constitute an intrinsic property of the mesoscopic many-body problem involving both resonant multi-phonons and multi-spins effects. In this last case, anisotropic interactions and in particular dipolar interactions play a crucial role. We believe that they are a key ingredient for the understanding of quantum relaxation as well as microscopic aspects of quantum phase transitions in complex systems such as heavy fermions or quantum spin-glasses.

Finally, our study shows that highly diluted rare-earth single-crystals have a great potential for making spin-qubits for quantum computation involving both electronic and nuclear spins.

References

[1] L. Thomas, F. Lionti, R. Ballou, D. Gatteschi, R. Sessoli, and B. Barbara, Nature (London) **383**, 145–147 (1996).
W. Wernsdorfer and R. Sessoli, Science **284**, 133 (1999).
[2] B. Barbara et al., J. Magn. Magn. Mater. **140–144**, 1825 (1995).
L. Thomas and B. Barbara, J. Low Temp. Phys. **113**, 1055 (1998).

L. Thomas, A. Caneschi, and B. Barbara, Phys. Rev. Lett. **83**, 2398 (1999).
T. Ohm, C. Sangregorio, and C. Paulsen, J. Low Temp. Phys. **113**, 1141 (1998).
[3] J. R. Friedman, M. P. Sarachik, J. Tejada, and R. Ziolo, Phys. Rev. Lett. **76**, 3830 (1996).
[4] N. V. Prokof'ev and P. C. E. Stamp, Rep. Prog. Phys. **63**, 669 (2000).
[5] For a review, see I. Tupitsyn and B. Barbara, in: Magnetoscience – From Molecules to Materials, edited by Miller and Drillon (Wiley-VCH Verlag, Weinheim, 2000).
[6] R. Giraud, W. Wernsdorfer, A. M. Tkachuk, D. Mailly, and B. Barbara, Phys. Rev. Lett. **87**, 057203 (2001).
[7] R. Giraud, A. M. Tkachuk, and B. Barbara, Phys. Rev. Lett. **91**, 25 (2003).
[8] B. Barbara, R. Giraud, W. Wernsdorfer, A. Tkachuk, P. Lejay, and H. Suzuki, to appear, Proc. ICM'03, Rome.
[9] G. Gaudin et al., J. Magn. Magn. Mater. **242–245**, 915 (2002).
[10] W. Wernsdorfer et al., Nature (London) **416**, 409 (2002).
W. Wernsdorfer, S. Bhaduri, R. Tiron, D. N. Hendrickson, and G. Christou, Phys. Rev. Lett. **89**, 197201 (2002).
[11] F. Varret, Y. Allain, and A. Miedan-Gros, Solid State Commun. **14**, 17–20 (1974).
F. Varret, J. Phys. Chem. Solids **37**, 257–263 (1976).
F. Varret and A. Ducouret-Cérèze, ICM88 Proceedings, J. Phys. Coll. **12** (C8), 847–848 (1988).
[12] Y. Shapira and V. Bindilatti, J. Appl. Phys. **92**(8), 4155 (2002).
[13] P. Fazekas and P. W. Anderson, Philos. Mag. **30**, 423 (1974).
[14] A. Abragam and B. Bleaney, Electron paramagnetic resonance of transition ions (Clarendon Press, Oxford, 1970).

Persistent current and Hall effect due to spin chirality

G. Tatara*

Graduate School of Science, Osaka University, Toyonaka, Osaka 560-0043, Japan

Received 1 September 2003, accepted 17 March 2004
Published online 23 April 2004

PACS 03.65.Vf, 03.67.–a, 73.23.Ra, 73.43.Cd, 73.50.Jt, 74.50.+r

We show that a spontaneous electric current is induced in a nano-scale conducting ring just by putting three ferromagnets (spin Josephson effect). The current is a direct consequence of the non-commutativity of the spin algebra, and is proportional to the non-coplanarity (chirality) of the magnetization vectors. This persistent current is shown to results in anomalus Hall effect in frustrated magnets. The application to a novel quantum logic gate is discussed.

© 2004 WILEY-VCH Verlag GmbH & Co. KGaA, Weinheim

1 Introduction

Persistent current in metallic rings is an equilibrium current which can be induced when the time-reversal symmetry is broken. Such a current appears in the presence of a magnetic flux through a normal ring. The effect is due to a U(1) phase factor attached by the flux to the electron wave function. Here we show theoretically that a permanent current is induced in a conducting normal ring just by attaching three ferromagnets, without magnetic flux through the ring [1]. This surprising effect can be seen in nano-scales at low temperatures. The key here is the non-commutativity of the SU(2) spin algebra, which breaks the time-reversal symmetry, and leads, in the presence of electron coherence, to a permanent electron current (spin Josephson effect).

We also show that this persistent current gives an intuitive explanation of the anomalous Hall effect due to spin Berry phase in frustrated magnets [2, 3]. It is shown that the system works as a quantum XOR gate, on which unitary transformation can be carried out; thus it works as a logic gate for quantum computing [4]. The realization of the superposition state of flux Qbit is realized simply by creating a superposition state of spin Qbits.

2 Persistent current

2.1 Spin chirality and time reversal symmetry breaking

The electron has spin 1/2 (i.e., has two components), and the spin obeys SU(2) algebra. The algebra is represented by three 2×2 Pauli matrices σ_i ($i = x, y, z$) satisfying the commutation relation

$$[\sigma_i, \sigma_j] = 2i\epsilon_{ijk}\sigma_k , \tag{1}$$

where ϵ_{ijk} is the totally antisymmetric tensor with $\epsilon_{xyz} = 1$. When a conduction electron in a conductor is scattered by some magnetic object, the electron wave function is multiplied by an amplitude $A(\mathbf{n}) = \alpha e^{i\beta \mathbf{n} \cdot \boldsymbol{\sigma}}$, which is generally spin-dependent and is represented by a 2×2 matrix in spin space.

---

* Corresponding author: e-mail: tatara@ess.sci.osaka-u.ac.jp, Phone: +81 66850 5544, Fax: +81 66850 5494

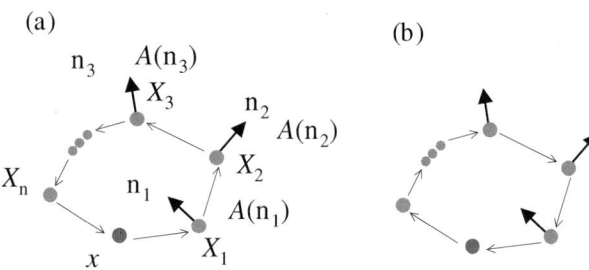

Fig. 1 A closed path contributing to the amplitude of the electron propagation from x to x. At X_i, the electron experiences a scattering represented by an SU(2) amplitude, $A(n_i)$. The contribution from one path (a) and the reversed one (b) are different in general due to the non-commutativity of $A(n_i)$'s.

Here α and β are complex numbers and $\boldsymbol{n}$ is a three-component unit vector characterizing the scattering object (such as the magnetization direction). We consider in this paper only classical, static scattering objects, and assume $\boldsymbol{n}$'s are constant vectors.

Let us consider two successive scattering events represented by $A(\boldsymbol{n}_1)$ and $A(\boldsymbol{n}_2)$ (Fig. 1). Due to the non-commutativity of σ_i, the amplitude depends on the order of the scattering event; $A(\boldsymbol{n}_1) A(\boldsymbol{n}_2) \neq A(\boldsymbol{n}_2) A(\boldsymbol{n}_1)$ in general. Various features in spin transport, which is under intensive pursuit recently, arise from this non-commutativity. It, however, does not affect the charge transport, since the charge is given as a sum of the two spin components (denoted by tr), and $\text{tr}\,[A(\boldsymbol{n}_1) A(\boldsymbol{n}_2)] - \text{tr}\,[A(\boldsymbol{n}_2) A(\boldsymbol{n}_1)] = 0$. Anomaly in the charge transport arises at the third order. We have, by virtue of Eq. (1) and the relation $\text{tr}\,[\sigma_i \sigma_j] = 2\delta_{ij}$,

$$\text{tr}\,[A(\boldsymbol{n}_1) A(\boldsymbol{n}_2) A(\boldsymbol{n}_3)] - \text{tr}\,[A(\boldsymbol{n}_3) A(\boldsymbol{n}_2) A(\boldsymbol{n}_1)] = 4\alpha^3 \sin^3 \beta \boldsymbol{n}_1 \cdot (\boldsymbol{n}_2 \times \boldsymbol{n}_3) \equiv iC_{123}, \quad (2)$$

where the cross denotes the vector product, i.e., $\boldsymbol{n}_1 \cdot (\boldsymbol{n}_2 \times \boldsymbol{n}_3) = \sum_{ijk} \epsilon_{ijk} n_1^i n_2^j n_3^k$. This relation indicates that in the presence of fixed $\boldsymbol{n}_i$'s with $\boldsymbol{n}_1 \cdot (\boldsymbol{n}_2 \times \boldsymbol{n}_3) \neq 0$, the symmetry under time-reversal (more appropriately, reversal of motion) is generally broken in the charge transport. In fact, relation (2) indicates that the contribution from one path, $x \to X_1 \to X_2 \to X_3 \to x$ (Fig. 1a), and its (time-)reversed one, $x \to X_3 \to X_2 \to X_1 \to x$ (Fig. 1b), are not equal, and this difference results in a spontaneous electron motion in a direction specified by the sign of C_{123}, namely, a permanent current. What is essential here is the non-commutativity of the SU(2) algebra. In fact, C_{123} vanishes if all $\boldsymbol{n}_i$'s lie in a plane, in which case the algebra is reduced to a commutative U(1) algebra. The degree of the symmetry breaking, $\boldsymbol{n}_1 \cdot (\boldsymbol{n}_2 \times \boldsymbol{n}_3)$, is given by the non-coplanarity, often called spin chirality.

2.2 Current

The spontaneous current above would be realized on a small conducting ring with three ferromagnets or magnetic dots with different magnetization direction, S_1, S_2 and S_3 (Fig. 4). The electron in a ring feels an effective spin polarization when it goes through the region (F_i) affected by the ferromagnets, and the effect will be modeled by the exchange (spin-dependent) potential, $V(x) = -\Delta \boldsymbol{n}_i \cdot \boldsymbol{\sigma}$ for $x \in F_i$. Here Δ represents the effective exchange field. The equilibrium charge current in the ring is calculated from

$$j(x) = \frac{\hbar e}{2m} \text{Im}\,(\nabla_x - \nabla_{x'})\,\text{tr}\,G(x, x', \tau = 0-)|_{x'=x}, \quad (3)$$

where $G(x, x', \tau) \equiv -\langle Tc(x, \tau) c^\dagger(x', 0)\rangle$ is the thermal Green function, e, m, c being the charge, the mass, the annihilation operator of electrons, respectively. $G(x, x', \tau)$ is calculated perturbatively from the Dyson equation, $G = g + gVG$, where g represents free Green function. As is seen from eq. (2), possible finite current arises at the third order in V. By summing the contribution of the two paths, $x \to X_1 \to X_2 \to X_3 \to x$ and the reversed one, we have $j(x) = -\frac{\hbar e}{m} B(x) \,\text{Re}\, C_{123}$. Here C_{123} is defined by Eq. (2) with $\alpha = i\Delta$, $\beta = \pi/2$, and

$$B(x) = \prod_{i=1}^{3} \int_{X_i \in F_i} dX_i \int \frac{d\omega}{2\pi} f(\omega) \nabla_{X_0} \text{Im}\,[g_{01} g_{12} g_{23} g_{34}]|_{X_4 = X_0 = x} \quad (4)$$

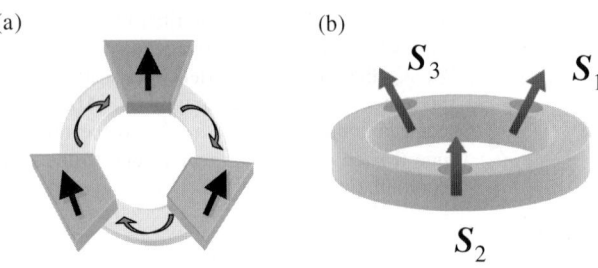

Fig. 2 Setup for the chirality-driven permanent current by use of (a): three ferromagnetic contacts and (b): three ferromagnetic dots, on a normal conducting ring.

describes the electron propagation through the ring, which is common to both paths. In Eq. (4), $f(\omega)$ is the Fermi distribution function and $g_{ij} = g^r(X_i - X_j, \omega)$ is the retarded Green's function of free electrons. Approximating the transport along the ring as one-dimensional and neglecting multiple circulation, we have $g^r(x, \omega) \simeq -i\pi (D/L) e^{ik_F|x|}$, where k_F is the Fermi wavenumber, D the density of states ($\sim 1/\epsilon_F$; $\epsilon_F = \hbar^2 k_F^2/2m$ being Fermi energy), and L the length of the ring perimeter. The final result is given by

$$j = -2e \frac{v_F}{L} \cos(k_F L) \left(\frac{J}{\epsilon_F}\right)^3 S_1 \cdot (S_2 \times S_3), \tag{5}$$

at zero temperature. Here $J \equiv \pi W \Delta/L$ with W being the width of the ferromagnets, and $v_F = \hbar k_F/m$ is the Fermi velocity.

The current is thus induced by the spin chirality $S_1 \cdot (S_2 \times S_3)$ of the ferromagnets. This quantity reduces to the Pontryagin index (density) for the case of smoothly varying field $S(x)$, which is also interpreted as Berry phase of the spin. The effect of spin Berry phase on the electron transport has so far been investigated in the limit of strong coupling to $S(x)$ where the electron spin adiabatically follows $S(x)$ [5]. In contrast, the present result (5) is obtained in the opposite limit; we have treated the coupling to S perturbatively (weak coupling) and made no assumption of smoothness on $S(x)$.

The appearance of the current is due to the symmetry breaking of the charge (U(1)) sector, as in the case of the current in Josephson junction. But note that here the U(1) symmetry breaking was due to the non-commutativity of spin (SU(2)) sector ("spin Josephson effect").

3 Anomalous Hall effect

The phenomenon predicted here is not restricted to artificial nano-structures, but will be present rather generally in metallic frustrated spin systems such as pyrochlore ferromagnets and spin glasses, where finite spin chirality is often realized [6]. The spin chirality was recently pointed out [7], in the adiabatic limit, to be the origin of the peculiar anomalous Hall effect observed in experiments [6]. The present chirality-driven persistent current affords an intuitive interpretation to it. The circulating current starts to drift when the electric field is applied, in the direction perpendicular to the electric field (Fig. 3), just as in the normal Hall effect. With the frequency of the circulating motion, read from Eq. (5) as

$$\Omega \simeq \frac{2\pi v_F}{L} \left(\frac{J}{\epsilon_F}\right)^3 S_1 \cdot (S_2 \times S_3), \tag{6}$$

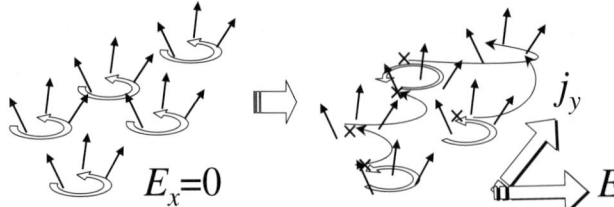

Fig. 3 Schematic picture of the chirality-induced Hall effect. Circulating permanent currents drift under applied electric field. The crosses denote impurity scattering.

we may estimate the Hall conductivity by $\sigma_{xy} = \sigma_0 \Omega \tau$. Here σ_0 is the classical (Boltzmann) conductivity, τ is the elastic lifetime, and the dirty case $\Omega \tau \ll 1$ is assumed. If the spin chirality is located uniformly on every triangle of size of inter-atomic distance (i.e., $S_1 \cdot (S_2 \times S_3) = \chi_0$ and $L \sim 1/k_F$), we have $\sigma_{xy}/\sigma_0 \simeq \chi_0 J^3 \tau / \epsilon_F^2$. This result agrees with the result based on the linear response theory [2] as we now show.

We start from the Hamiltonian with exchange interaction (J);

$$H = \sum_{k\sigma} \epsilon_{k\sigma} c_{k\sigma}^\dagger c_{k\sigma} + \frac{J}{N} \sum_{kk'} S_{k'-k} (c_{k'}^\dagger \sigma c_k) + H_{\text{imp}}, \tag{7}$$

where H_{imp} denotes the impurity scattering. Treating J perturbatively, Kubo formula calculation results in at the third order in J the Hall conductivity of [2]

$$\sigma_{xy} = \frac{N}{\pi V} \left(\frac{e}{m}\right)^2 (2\pi \nu J)^3 \tau^2 \chi_0 = (4\pi)^2 \sigma_0 J^3 \nu^2 \tau \chi_0. \tag{8}$$

The uniform chirality χ_0 is given by

$$\chi_0 \equiv \frac{1}{N} \sum_{X_i} S_{X_1} \cdot (S_{X_2} \times S_{X_3})$$

$$\times \left[\frac{(a \times b)_z}{ab} I'(a) I'(b) I(c) + \frac{(b \times c)_z}{bc} I(a) I'(b) I'(c) + \frac{(c \times a)_z}{ca} I'(a) I(b) I'(c) \right], \tag{9}$$

where X_i runs over all the positions of local spins, while $a \equiv X_1 - X_2$, $b \equiv X_2 - X_3$ and $c \equiv X_3 - X_1$ are the vectors representing sides of the triangle ($a \equiv |a|$ etc.). $I(r) \equiv \frac{\sin k_F r}{k_F r} e^{-r/2\ell}$ is an RKKY type interaction, and $I'(r) = \frac{dI(r)}{dr}$. It is seen that Hall current is driven by three spins which form a finite solid angle in spin space (i.e., finite local chirality $\chi_{123} \equiv S_{X_1} \cdot (S_{X_2} \times S_{X_3})$) spanning a finite area in coordinate space (as seen from $(a \times b)_z$ etc.). Contribution from largely separated three spins with the scale of r decays rapidly as $\sim e^{-3r/2\ell}/(k_F r)^3$, and the Hall effect is dominantly driven by chiralities of spins on small triangles.

3.1 Topological invariant and quantization

The above expression of the Hall conductivity looks similar to a topological invariant but not exactly so. Here we demonstrate that the Hall conductivity is indeed written in terms of a topological number of the localized spin configuration, if the spin texture is smoothly varying compared with the elastic mean free path, ℓ [3]. We expand $S_{X_1} = S_{X_3} - c \cdot \nabla S_{X_3} + o(\ell/\lambda)$, λ being the length scale where the spin texture varies, and neglect higher order term in ℓ/λ. By use of partial derivative, we obtain

$$\sigma_{xy} = \frac{e^2}{h} \alpha \int dX \frac{1}{4\pi} S \cdot (\partial_x S \times \partial_y S) + O(\ell/\lambda)^2. \tag{10}$$

Here the coefficient α represents the amount of non-adiabaticity, which is obtained as

$$\alpha = \frac{4\pi}{3Vm^2\tau} \left(\frac{J}{V}\right)^3 \int db \int dc\, I(|b-c|) I(b) I(c). \tag{11}$$

The integration over b and c is carried out, and we obtain $\alpha = 8\pi (J\tau)^3 \frac{\ell}{k_F L^2}$, where $L^2 = V$. The Hall conductivity is thus proportional to the density of vortices. The first term in Eq. (10) is a topological charge of a mapping $S^2 \to S^2$, i.e., an integer (n), if a boundary condition $S(\infty) = \text{const}$ is imposed; hence

$$\sigma_{xy} = \frac{e^2}{h} n\alpha + O(\ell/\lambda)^2. \tag{12}$$

The Hall conductivity is thus quantized in two-dimensions, and is proportional to the number of vortices in the sample. This topological invariant is therefore an non-adiabatic analog of the spin Berry phase.

4 Application to a quantum logic gate

The system of persistenct current by spin chirality works as a novel quantum operation gate, where spin Qbits and flux (current) Qbit are combined. For this, we consider the case where the magnetization is quantum spin of $S = 1/2$, which is carried by ferromagnetic dots on the ring. The spins in the dots can then be regarded as qubits. The current in one-dimensional ring is $\hat{J} = J_0 \hat{C}_3$, where $\hat{C}_3 \equiv (\hat{S}_1 \times \hat{S}_2) \cdot \hat{S}_3$ and $J_0 = -2e \frac{v_F}{L} \cos(k_F L) \left(\frac{\Delta}{\epsilon_F}\right)^3$. The state of the system is thus specified by a combination of states of the spin-qubits $\hat{S}_i$ and a current-qubit $\hat{J}$. The current takes a value according to the "volume" of the three spins, $(\hat{S}_1 \times \hat{S}_2) \cdot \hat{S}_3$.

Let us see how the quantum operation works. To remove an irrelevant degeneracy due to rotational symmetry, we fix S_3 in z-direction. Then the quantum operator $\hat{C}_3$ reduces to $\hat{C}_2 \equiv \frac{1}{2}(\hat{S}_1 \times \hat{S}_2)_z = \frac{i}{4}(\hat{S}_1^+ \hat{S}_2^- - \hat{S}_1^- \hat{S}_2^+)$. The eigenvalues λ and eigenstates (represented by $|S_1^z S_2^z\rangle$) of $\hat{C}_2$ are obtained as $\lambda = 0$ for $|++\rangle \equiv |0_+\rangle$ and $|--\rangle \equiv |0_-\rangle$, $\lambda = \frac{1}{4}$ for $\frac{1}{\sqrt{2}}(|+-\rangle + e^{-\frac{\pi}{2}i}|-+\rangle) \equiv |R\rangle$, and $\lambda = -\frac{1}{4}$ for $\frac{1}{\sqrt{2}}(|+-\rangle + e^{\frac{\pi}{2}i}|-+\rangle) \equiv |L\rangle$. Note that the current states $|R\rangle$ and $|L\rangle$ correspond to the entangled states as a result of "square-root swap" operation. We choose the axis of S_1 as in x-direction, and that of S_2 in y-direction. For instance, $|0\rangle = |x\rangle$ and $|1\rangle = |-x\rangle$ for S_1 is written as $|\pm x\rangle = \frac{1}{\sqrt{2}}(|+\rangle \pm |-\rangle)$. Quantum operations are implemented by use of projection into $S_z = 0$ subspace, which we write as P_0. (Note that $|R\rangle$ and $|L\rangle$ are eigenstates of $S_z = 0$.) After the projection, the mapping of spin states $|S_1, S_2\rangle$ to $P_0|S_1, S_2\rangle \equiv |C_2\rangle$ reads as a modified XOR operation

$$
\begin{array}{ll}
|S_1, S_2\rangle & C_2 \\
|00\rangle & \to |R\rangle \\
|01\rangle & \to e^{i\pi}|L\rangle \\
|10\rangle & \to |L\rangle \\
|11\rangle & \to e^{i\pi}|R\rangle
\end{array}
\tag{13}
$$

The extra factor of $e^{i\pi}$ can be removed by a single spin operation if one wants.

As is seen from the above consideration, our systems can be used as a preparation tool of an entangled state of two or more spins. For instance, in the case of three spins S_i ($i = 0, 1, 2$), with $S_0 \parallel z$, we can create an entangled state of $|S_1 S_2\rangle = \frac{1}{\sqrt{2}}|+-\rangle \mp i|-+\rangle$ by projecting the current state into $|R\rangle$ or $|\rangle$, respectively.

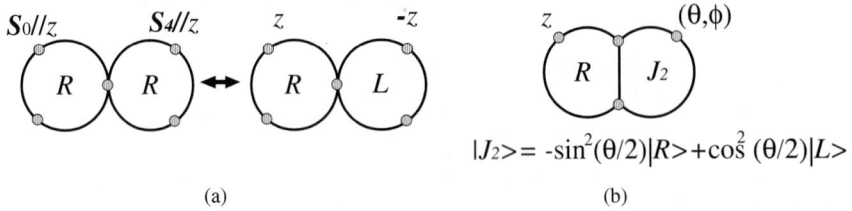

Fig. 4 Unitary operations by use of two rings coupled (a) with one spin in common and (b) with two spins in common. The current state J_2 in the second ring is a result of a unitary transformation of J_1 specified by (θ, ϕ). In (b), superposition state of current in the second ring is created from the R state in the first ring.

Unitary transformation can be carried out by couling rings (Fig. 4). For instance, coupling two rings to share two spins (Fig. 4b), In this case, the current J_1 and J_2 are both determined by the two spins shared by the two rings, but the state is controllable by (θ, ϕ). In fact, the current states of the first ring is translated into the current states of the second ring as (after projection P_0)

$$|R\rangle_1 = \frac{e^{i\pi/4}}{\sqrt{2}} \left[-\sin^2 \frac{\theta}{2} |R\rangle_2 + \cos^2 \frac{\theta}{2} |L\rangle_2 \right]. \tag{14}$$

Thus one can create from a current in ring 1 any superposition of $|R\rangle$ and $|L\rangle$ on the second ring.

4.1 Detection of current

The readout of the target bit is carried out by measuring the flux arising from the persistent current. Such measurement on a single ring has been successfully carried out in the case of conventional persistent current in a ring of gold [8] and GaAs–AlGaAs [9]. The flux due to this current is not large but may be detected with present lock-in technique.

We propose here an alternative detection of the flux state by use of a Hall like effect in the four terminal setup. In the presence of flux (or persistent current), the four-terminal conductance through a ring is expected to be asymmetric with respect to the flux, and a finite difference of the conductance arises when the voltage and current leads are reversed. The difference (which may be regarded as a "Hall conductance", G_H) is expected in semiconducting system [9] to be

$$G_H \simeq \frac{e^2}{h} (\Delta/\epsilon_F)^3 C_3 \ (\sim e^2/h \times O(10^{-2})) \quad \text{for} \quad C_3 \sim O(1).$$

This is of order of typical atomic size contacts of semiconductors, and would be measurable. The electric measurement, being very sensitive, detection of very small spin chirality C_3 would be possible, as well as the system with smaller coupling Δ. Much larger current would be obtained if we use a superconducting ring of p-wave order parameter (Sr_2RuO_4) or ferromagnetic semiconductors (Ga, Mn)As.

5 Summary

We demonstrated that a spontaneous charge current appears from the spin chirality when three ferromagnets are attached on a coherent conductive ring. This effect results in a Hall effect in frustrated magnets and also works as a quantum operation gate. The physics behind all these phenomena is "spin Josephson effect", which is a SU(2) analog of Josephson effect in superconductors. The spin chirality is equivalent to the quantum mechanical Berry phase carried by the spin. This Berry phase is a "fictitious magnetic flux", which does not affect the phenomena in the macroscopic world. In nanoscales, in contrast, it can be used to operate logic gates just in the same way as "real" magnetic flux can. This is a novel architecture of quantum logic gates.

Acknowledgements G. T. thanks the Mitsubishi foundation and Saneyoshi scholarship foundation for financial support.

References

[1] G. Tatara and H. Kohno, Phys. Rev. B **67**, 113316 (2003).
[2] G. Tatara and H. Kawamura, J. Phys. Soc. Jpn. **71**, 2613 (2002).
[3] G. Tatara, M. Yamanaka, and M. Onoda, J. Magn. Magn. Mater., to appear (2004).
[4] G. Tatara and N. Garcia, Phys. Rev. Lett. **91**, 076806 (2003).
[5] D. Loss, P. Goldbart, and A. V. Balatsky, Phys. Rev. Lett. **65**, 1655 (1990).
[6] Y. Taguchi, Y. Oohara, H. Yoshizawa, N. Nagaosa, and Y. Tokura, Science **291**, 2573 (2001).
[7] J. Ye, Y. B. Kim, A. J. Millis, B. I. Shraiman, P. Majumdar, and Z. Tesanovic, Phys. Rev. Lett. **83**, 3737 (1999).
[8] V. Chandrasekhar, R. A. Webb, M. J. Brady, M. B. Ketchen, W. J. Gallagher, and A. Kleinsasser, Phys. Rev. Lett. **67**, 3578 (1991).
[9] D. Mailly, C. Chapelier, and A. Benoit, Phys. Rev. Lett. **70**, 2020 (1993).

Energy-level diagrams of high-spin and low-spin molecules

H. De Raedt[*,1], **S. Miyashita**[2], and **K. Michielsen**[1,2]

[1] Department of Applied Physics-Computational Physics, Materials Science Centre,
University of Groningen, Nijenborgh 4, NL-9747 AG Groningen, The Netherlands
[2] Department of Applied Physics, Graduate School of Science, University of Tokyo, Bunkyo-ku,
Tokyo 113-8656, Japan

Received 1 September 2003, accepted 18 March 2004
Published online 14 April 2004

PACS 75.10.Jm, 75.45.+j, 75.50.Ee, 75.50.Xx

The magnetic energy-level diagrams for models of the Mn_{12} and V_{15} molecule are calculated using the Lanczos method with full orthogonalization and a Chebyshev-polynomial-based projector method. The effects of the Dzyaloshinskii-Moriya interaction on the appearance of energy-level repulsions and its relevance to the observation of steps in the time-dependent magnetization data are studied.

1 Introduction

Magnetic molecules such as Mn_{12} or V_{15} have attracted a lot of interest recently because these nanomagnets can be used to study e.g. quantum (de)coherence, relaxation and tunneling of the magnetization on a nanoscale [1–22]. As a result of the very weak intramolecular interactions between these molecules, experiments directly probe the magnetization dynamics of the individual molecules. In particular the adiabatic change of the magnetization at low-temperature is governed by the discrete energy-level structure [23–26]. As the magnetization dynamics of these molecules is determined by the (tiny) level repulsions, a detailed knowledge of the low-lying energy level scheme is necessary.

Magnetic anisotropy, a result of the geometrical arrangement of the magnetic ions within a molecule of low symmetry, mixes states of different total spin and enforces a treatment of the full Hilbert space of the system. Disregarding the single-ion anisotropy, the dominant contribution to the magnetic anisotropy due to spin-orbit interact ions is given by the Dzyaloshinskii-Moriya interaction (DMI) [27–33]. In principle this interaction can change energy-level crossings into energy-level repulsions. The presence of the latter is essential to explain the adiabatic changes of the magnetization at the resonant fields in terms of the Landau-Zener-Stückelberg (LZS) transition [23–26]. Thus a minimal magnetic model Hamiltonian should contain (strong) Heisenberg interactions, the single-ion anisotropy, the DMI and a coupling to the applied magnetic field [10, 34–42]. As the DMI mixes states with different magnetization, it is not possible to use the magnetization as a vehicle to block-diagonalize the Hamiltonian and effectively reduce the size of the matrices that have to be diagonalized. Therefore it is of interest to explore alternative routes to direct but accurate diagonalization of the full model Hamiltonian.

We have tested different standard algorithms to compute the low-lying eigenvalues of large matrices. The standard Lanczos method (including its conjugate gradient version) as well as the power method [47, 48] either converge too slowly, lack the accuracy to resolve the (nearly-)degenerate eigen-

---

[*] Corresponding author: e-mail: deraedt@phys.rug.nl, Phone: +31 50 363 4950, Fax: +31 50 363 4947

values, and sometimes even completely fail to correctly reproduce the low-lying part of the spectrum. It seems that model Hamiltonians for nanoscale magnets provide a class of Hermitian eigenvalue problems that are hard to solve. Extensive tests lead us to the conclusion that only the Lanczos method with full orthogonalization (LFO) [47, 48] and a Chebyshev-polynomial-based projector method (CP) [49] can solve these rather large and difficult eigenvalue problems with sufficient accuracy [49].

2 Manganese complex: Mn_{12}

In the Mn_{12}-acetate molecule the four inner Mn^{+4} ions have spin $S = 3/2$, the other eight Mn^{+3} ions have spin $S = 2$. The number of different spin states of this system is $4^4 \times 5^8 = 10^8$. If the total magnetization is a conserved quantity, it can be used to block-diagonalize the Hamiltonian, allowing the study of models of this size [39, 43]. However, to study the adiabatic change of magnetization, we have to treat all the states, and the dimension of the matrix becomes prohibitively large. A drastic reduction of the dimension of the matrix can be achieved by approximating the magnetic moment of an inner ion by an effective $S = 1/2$ moment. The schematic diagram of this simplified (but still complicated) model is shown in Fig. 1. The number of different spin states of this model is $2^4 \times 5^4 = 10^4$. The Hamiltonian for the magnetic interactions of the simplified Mn_{12} model can be written as [34]

$$\mathcal{H} = -J \left(\sum_{i=1}^{4} \boldsymbol{S}_{2i-1} \right)^2 - J' \sum_{\langle i,j \rangle} \boldsymbol{S}_{2i-1} \cdot \boldsymbol{S}_{2j} - K_z \sum_{i=1}^{4} (S_{2i}^z)^2 + \sum_{\langle i,j \rangle} \boldsymbol{D}^{i,j} \cdot [\boldsymbol{S}_{2i-1} \times \boldsymbol{S}_{2j}] - \sum_{i=1}^{8} \boldsymbol{h} \cdot \boldsymbol{S}_i, \quad (1)$$

where even (odd) numbered $\boldsymbol{S}_i$ are the spin operators for the outer (inner) $S = 2$ ($S = 1/2$) spins. The first two terms describe the isotropic Heisenberg exchange between the spins. The third term describes the single-ion easy-axis anisotropy of $S = 2$ spins. The fourth term represents the antisymmetric DMI in Mn_{12}. The vector $\boldsymbol{D}^{i,j}$ determines the DMI between the i-th $S = 1/2$ spin and the j-th $S = 2$ spin. We do not consider higher-order correction terms that restore the SU(2) symmetry [29–31, 44]. The last term describes the interaction of the spins with the external field $\boldsymbol{h}$. Note that the factor $g\mu_B$ is absorbed in our definition of $\boldsymbol{h}$. The first three terms in Hamiltonian (1) conserve the z-component of the total spin $M^z = \sum_{i=1}^{8} S_i^z$. The DMI on the other hand mixes states with different total spin and also states with the same total spin. Hence, the DMI can change level crossings into level repulsions and may explain the experimentally observed adiabatic changes of the magnetization.

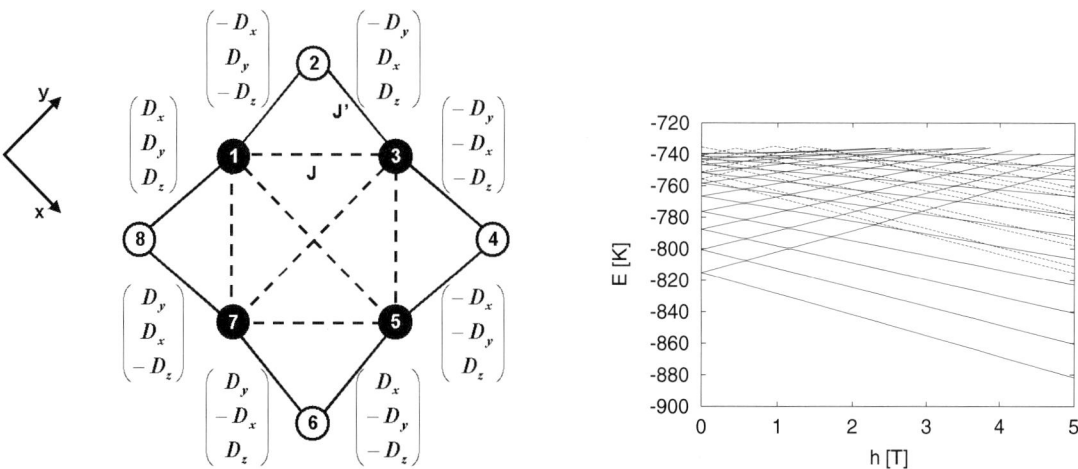

Fig. 1 Left: Schematic diagram of the magnetic interactions of the simplified model (1) of the Mn_{12} molecule. Right: The lowest 21 energy levels of the Mn_{12} model (1) as a function of the applied magnetic field $\boldsymbol{h}$. Solid lines: eigenstates with $|M^z| \approx 10$; dashed lines: eigenstates with $|M^z| \approx 9$.

The four-fold rotational-reflection symmetry (S_4) of the Mn_{12} molecule imposes some relations between the DM-vectors. It follows that there are only three independent DM-parameters: $D_x \equiv D_x^{1,8}$, $D_y \equiv D_y^{1,8}$, and $D_z \equiv D_z^{1,8}$, as indicated in Fig. 1. The above model satisfactorily describes a rather wide range of experimental data, such as the splitting of the neutron scattering peaks, results of EPR measurements and the temperature dependence of magnetic susceptibility [34]. The parameters of this model have been estimated by comparing experimental and theoretical data. In this paper we will use the parameter set B from Ref. [34, 40]: $J = 23.8$ K, $J' = 79.2$ K, $K_z = 5.72$ K, $D_x = 22$ K, $D_y = 0$, and $D_z = 10$ K. Although the amount of available data is not sufficient to fix all these parameters accurately, we expect that the general trends in the energy-level diagram will not change drastically if these parameters change relatively little.

Model (1) provides a good test case for diagonalization methods because it is small enough to be treated by full exact diagonalization but has all features of the larger problem. We find that the results obtained by full exact diagonalization, LFO and CP are the same to working precision (about 13 digits). For one set of model parameters, full exact diagonalization (using standard LAPACK algorithms) of the 10000×10000 matrix representing model (1) takes about 2 hours of CPU time on an Athlon 1.8 GHz/1.5Gb system. In Fig. 1 we show the results, obtained by LFO, for the lowest 21 energy levels of the Mn_{12} model as a function of the applied magnetic field. For each value of the h-field, the LFO calculation takes about 20 minutes and uses much less memory than the full diagonalization method.

Although the total magnetization is not a good quantum number, we can label the various eigenstates by their (calculated) magnetization. For large fields and/or energies, eigenstates with total spin 8, 9 and 10 appear. In Fig. 1 eigenstates with $|M^z| \approx 10(9)$ (within an error of about 10%) are represented by solid (dashed) lines (eigenstates with $|M^z| \approx 8$ appear for $h > 4$ but have been omitted for clarity). The standard $S = 10$ single-spin model for Mn_{12}, $\mathcal{H} = -D(S^z)^2 - hS^z$, is often used as a starting point to interpret experimental results [6, 7, 11–13, 37]. The energy levels of this model exhibit crossings at the resonant fields $h = \pm Dn$ for $n = -10, \ldots, 10$, in agreement with our numerical results for the more microscopic model (1). For the parameter set B, we find that $D \approx 0.55K$, in good agreement with experiments [6, 7]. The single-spin model commutes with the magnetization S^z and therefore it only displays level crossings, no level repulsions. Adding an anisotropy term of the form $S_+^4 + S_-^4$ only leads to level repulsions when the magnetization changes by 4, which does not agree with the observation of adiabatic changes of the magnetization for all $h = nD$ [6, 7, 11, 12]. In contrast, for the DMI the Hamiltonian has nonzero matrix elements for the pairs of states $|S, S_z\rangle$ and $|S \pm 1, S_z \pm 1\rangle$, but zero matrix elements for levels with the same value of the total spin.

In Fig. 1, for some values of h, level repulsions appear to be present. However, these are due to the fitting procedure used to plot the data and the number of h-values used (100) and disappear by using a higher resolution in h-fields (results not shown). Thus these splittings have no physical meaning. For the Mn_{12} system, the energy splittings at low field are extremely small. Their calculation requires extended-precision (128-bit) arithmetic [40]. Adding an extra transverse field by tilting the h-field by 5 degrees does not change this conclusion. Thus, it is clear that within the (very high) resolution in the h-field and the 13-digit precision of the calculation, there is no compelling evidence that the DMI gives rise to a level repulsion, at least not for the choice of model parameters (set B, see above) considered here. The algorithms developed for the work presented in this paper can be used for 33-digit calculations without modification and we leave the calculation of the splittings for future work.

3 Vanadium complex: V_{15}

In Fig. 2 we show the schematic diagram of the dominant magnetic (Heisenberg) interactions of the V_{15} molecule. The magnetic structure consists of two hexagons with six $S = 1/2$ spins each, enclosing a triangle with three $S = 1/2$ spins. All dominant Heisenberg interactions are antiferromagnetic. The number of different spin states of this model is $2^{15} = 32768$. The minimal Hamiltonian for the

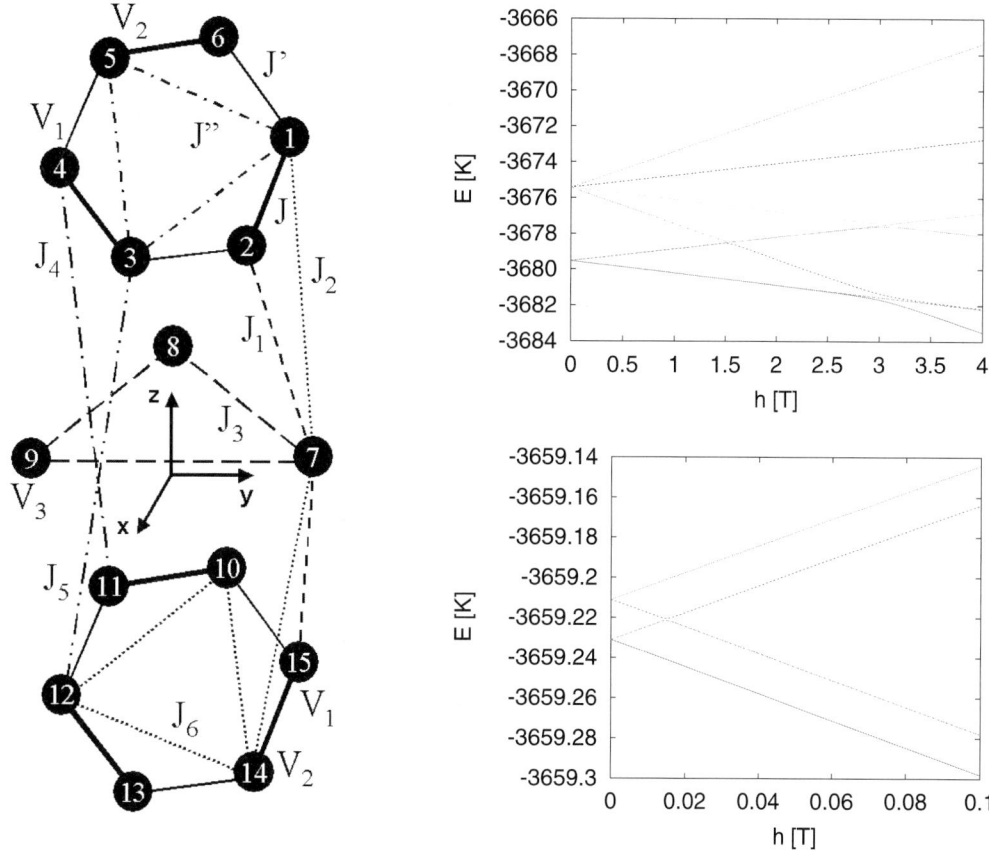

Fig. 2 (online colour at: www.interscience.wiley.com) Left: Schematic diagram of the magnetic interactions in model (2) of the V_{15} molecule. Right top: The lowest 8 energy levels of V_{15} model (2) with model parameters taken from Ref. [42] as a function of the applied magnetic field $\boldsymbol{h}$ parallel to the z-axis. Right bottom: Detailed view of the four lowest energy levels at $h \approx 0$.

magnetic interactions that incorporates the effects on magnetic anisotropy can be written as [22, 37, 38, 41]

$$\mathcal{H} = -\sum_{\langle i,j \rangle} J_{i,j} \boldsymbol{S}_k \cdot \boldsymbol{S}_l + \sum_{\langle i,j \rangle} \boldsymbol{D}^{i,j} \cdot [\boldsymbol{S}_i \times \boldsymbol{S}_j] - \sum_i \boldsymbol{h} \cdot \boldsymbol{S}_i \,. \qquad (2)$$

The various Heisenberg interactions $J_{i,j}$ are shown in Fig. 2. For simplicity, we assume that $\boldsymbol{D}^{i,j} = 0$ for sites i and j except for bonds for which the Heisenberg exchange constant is J (see Fig. 2) [37, 41]. Rotations about $2\pi/3$ and $4\pi/3$ around the axis perpendicular to and passing through the center of the hexagons leave the V_{15} complex invariant. This enforces constraints on the values of $\boldsymbol{D}^{i,j}$ [41, 42]. We have calculated the energy level diagram for the sets of model parameters given in Refs. [18], [37], and [41]. The level diagrams for these three choices are qualitatively similar [49]. Therefore we only present results for one set of model parameters.

Following Ref. [37] we take $J = -800$, $J_1 = J' = -54.4$ K, and $J_2 = J'' = -160$ K, $J_3 = J_4 = J_5 = J_6 = 0$. Then, in the absence of the DMI, we find for the energy gap between the ground state and the first excited state at $h = 0$ a value of 4.12478 K, in perfect agreement with Ref. [37]. As in Ref. [42] we take for the DMI parameters $D_x^{1,2} = D_y^{1,2} = D_z^{1,2} = 40$ K in the present lattice (Fig. 2). The amplitude is approximately 5% of the largest Heisenberg coupling. Using the rotational symmetry of the hexagon we have $D_x^{3,4} = 14.641$ K, $D_y^{3,4} = -54.641$ K, $D_z^{3,4} = 40$ K and $D_x^{5,6} = -54.641$ K, $D_y^{5,6} = 14.641$ K, $D_z^{5,6} = 40$ K. As the two hexagons are not equivalent we cannot use symmetry to reduce the number of free parameters. For simplicity, we assume that the (x, y) positions of the spins on the lower hexagons differ from those on the upper hexagon by a rotation about $\pi/3$. This yields for the remaining model parameters $D_x^{10,11} = -14.641$ K, $D_y^{10,11} = 54.641$ K, $D_z^{10,11} = 40$ K, $D_x^{12,13} = -40$ K, $D_y^{12,13} = -40$ K, $D_z^{12,13} = 40$ K, and $D_x^{14,15} = 54.641$ K, $D_y^{14,15} = -14.641$ K, $D_z^{14,15} = 40$ K.

In Fig. 2 we show the results for the eight lowest energy levels of V_{15} model (2) as a function of the applied magnetic field along the z-axis, using the parameters of Ref. [42]. At zero field, the DMI splits the doubly-degenerate doublet of $S = 1/2$ states into two doublets of $S = 1/2$ states. The difference in energy of the splitting is ≈ 0.0085 K, much smaller than the experimental estimate ≈ 0.05 K [22], but of the same order of magnitude as the values cited in Ref. [41]. The next four higher levels are $S = 3/2$ states. The energy-level splitting between the $S = 3/2$ and $S = 1/2$ states is ≈ 4.1 K, in reasonable agreement with the experimental value ≈ 3.7 K [45]. The transition between the states $|1/2, 1/2\rangle$ and $|3/2, 3/2\rangle$ takes place at $h \approx 2.8$ T in very good agreement with the experimental value 2.8 T. It should be noted that an energy gap does not necessarily implies an energy-level repulsion, as Fig. 2 demonstrates for the case when the magnetic field is applied in the z direction. Here we find that the levels simply cross at a finite value of the field, and the system does not allow for an adiabatic change of the magnetization between the states $|1/2, -1/2\rangle$ and $|1/2, 1/2\rangle$. If we apply the field in the x or y direction, the energy-level diagram exhibits degenerate repulsions as shown in Ref. [22]. If we apply in an intermediate angle, the energy structure changes smoothly from that for z direction to that for x (or y) direction. Therefore, although the DMI causes the avoided level crossing structure, it is anisotropic with respect to the direction of the field. This structure (by a factor of two at least) should lead to observable changes in the hysteresis loops but has not been seen in experiment [45]. Our numerical data for the model parameters given in Refs. [18], [37], and [41] suggest that the three-spin model reproduces the main features of the full V_{15} model. Within the three spin model we have studied the effects of higher-order correction terms that restore the SU(2) symmetry [29–31, 44]. While they cause the four $S = 3/2$ levels to be degenerate at $h = 0$, the low energy degenerate doublets do not change in an essential manner. In experiments only weak directional dependence was found. Thus, it seems that another type of mechanism for opening gaps is at work and, as we have shown elsewhere, hyperfine interactions seems to be a good candidate [50].

Acknowledgements We thank I. Chiorescu, and V. Dobrovitski for illuminating discussions. Support from the 'Nederlandse Stichting voor Nationale Computer Faciliteiten (NCF)' is gratefully acknowledged.

References

[1] L. Gunther and B. Barbara (Eds.), Quantum Tunneling of Magnetization, NATO ASI Ser. E, Vol. 301 (Kluwer, Dordrecht, 1995).
[2] A. Caneschi, D. Gatteschi, R. Sessoli, A. Barra, L. C. Brunel, and M. Guillot, J. Am. Chem. Soc. **113**, 5873 (1991).
[3] R. Sessoli, H.-L. Tsai, A. R. Shake, S. Wang, J. B. Vincent, K. Folting, D. Gatteschi, G. Christou, and D. N. Hendrickson, J. Am. Chem. Soc. **115**, 1804 (1993).
[4] D. Gatteschi, L. Pardi, A. L. Barra, and A. Müller, Mol. Eng. **3**, 157 (1991).
[5] G. Levine and J. Howard, Phys. Rev. Lett. **75**, 4142 (1995).
[6] J. R. Friedman, M. P. Sarachik, J. Tejada, and R. Ziolo, Phys. Rev. Lett. **76**, 3830 (1996).
[7] L. Thomas, F. Lionti, R. Ballou, D. Gattesehi, R. Sessoli, and B. Barbara, Nature **383**, 145 (1996).
[8] C. Sangregorio, T. Ohm, C. Paulsen, R. Sessoli, and D. Gatteschi, Phys. Rev. Lett. **78**, 4645 (1997).
[9] W. Wernsdorfer and R. Sessoli, Science **284**, 133 (1999).
W. Wernsdorfer, T. Ohm, C. Sangregorio, R. Sessoli, D. Mailly, and C. Paulsen, Phys. Rev. Lett. **82**, 3903 (1999).
[10] B. Barbara, L. Thomas, F. Lionti, A. Sulpice, and A. Caneschi, J. Magn. Magn. Mater. **177**, 1324 (1998).
[11] J. A. A. J. Perenboom, J. S. Brooks, S. Hill, T. Hathaway, and N. S. Dalal, Phys. Rev. B **58**, 330 (1998).
[12] I. Chiorescu, W. Wernsdorfer, A. Müller, H. Bögge, and B. Barbara, Phys. Rev. Lett. **84**, 3454 (2000).
[13] T. Pohjola and H. Schoeller, Phys. Rev. B **62**, 15026 (2000).

[14] Y. Zhong, M. P. Sarachik, J. Yoo, and D. N. Hendrickson, Phys. Rev. B **62**, R9256 (2000).
[15] I. Chiorescu, W. Wernsdorfer, A. Müller, H. Bögge, and B. Barbara, J. Magn. Magn. Mater. **221**, 103 (2000).
[16] I. Chiorescu, W. Wernsdorfer, A. Müller, H. Bögge, and B. Barbara, Phys. Rev. Lett. **84**, 3454 (2000).
[17] I. Chiorescu, R. Giraud, A. G. M. Jansen, A. Caneschi, and B. Barbara, Phys. Rev. Lett. **85**, 4807 (2000).
[18] D. W. Boukhvalov, V. V. Dobrovitski, M. I. Katsnelson, A. I. Lichtenstein, B. N. Harmon, and P. Kögerler, J. Appl. Phys. **93**, 7082 (2003).
[19] D. W. Boukhvalov, A. I. Lichtenstein, V. V. Dobrovitski, M. I. Katsnelson, B. N. Harmon, V. V. Mazurenko, and V. I. Anisimov, arXiv:cond-mat/0110488.
[20] W. Wernsdorfer, N. Allaga-Alcalde, D. N. Hendrickson, and G. Christou, Nature **416**, 407 (2002).
[21] A. Honecker, F. Meier, D. Loss, and B. Normand, Eur. Phys. J. B **27**, 487 (2002).
[22] I. Chiorescu, W. Wernsdorfer, A. Müller, S. Miyashita, and B. Barbara, arXiv: cond-mat/0212181.
[23] S. Miyashita, J. Phys. Soc. Jpn. **64**, 3207 (1997); ibid. **65**, 2734 (1996).
[24] V. V. Dobrovitskii and A. K. Zvezdin, Europhys. Lett. **38**, 377 (1997).
[25] L. Gunther, Europhys. Lett. **39**, 1 (1997).
[26] H. De Raedt, S. Miyashita, K. Saitoh, D. García-Pablos, and N. García, Phys. Rev. B **56**, 11761 (1997).
[27] I. E. Dzyaloshinskii, Zh. Eksp. Teor. Fiz. **32**, 1547 (1957) [Sov. Phys. – JETP **5**, 1259 (1957)].
[28] T. Moriya, Phys. Rev. **120**, 91 (1960).
[29] T. A. Kaplan, Z. Phys. B **49**, 313 (1983).
[30] L. Shekhtman, O. Entin-Wohlman, and A. Aharony, Phys. Rev. Lett. **69**, 836 (1992).
[31] L. Shekhtman, A. Aharony, and O. Entin-Wohlman, Phys. Rev. B **47**, 174 (1993).
[32] K. Yosida, Theory of Magnetism (Springer-Verlag, Berlin, New York, 1996).
[33] A. Crépieux and C. Lacroix, J. Magn. Magn. Mater. **182**, 341 (1998).
[34] M. I. Katsnelson, V. V. Dobrovitski, and B. N. Harmon, Phys. Rev. B **59**, 6919 (1999).
[35] M. Al-Saqer, V. V. Dobrovitski, B. N. Harmon, and M. I. Katsnelson, J. Appl. Phys. **87**, 6268 (2000).
[36] I. Rudra, S. Ramasesha, and D. Sen, Phys. Rev. B **64**, 014408 (2001).
[37] I. Rudra, S. Ramasesha, and D. Sen, J. Phys.: Condens. Matter **13**, 11717 (2001).
[38] S. Miyashita and N. Nagaosa, Prog. Theor. Phys. **106**, 533 (2001).
[39] C. Raghu, I. Rudra, D. Sen, and S. Ramasesha, Phys. Rev. B **64**, 064419 (2001).
[40] H. De Raedt, A. H. Hams, V. V. Dobrovitsky, M. Al-Saqer, M. I. Katsnelson, and B. N. Harmon, J. Magn. Magn. Mater. **246**, 392 (2002).
[41] N. P. Konstantinidis and D. Coffey, Phys. Rev. B **66**, 174426 (2002).
[42] I. Rudra, K. Saito, S. Ramasesha, and S. Miyashita, unpublished.
[43] N. Regnault, Th. Jolicoeur, R. Sessoli, D. Gatteschi, and M. Verdaguer, Phys. Rev. B **66**, 054409 (2002).
[44] A. Zheludev, S. Maslov, I. Tsukada, I. Zaliznyak, L. P. Regnault, T. Masuda, K. Uchinokura, R. Erwin, and G. Shirane, Phys. Rev. Lett. **81**, 5410 (1998).
[45] I. Chiorescu, private communication
[46] V. V. Kostyuchenko and A. K. Zvezdin, Phys. Solid State **45**, 903 (2003).
[47] J. H. Wilkinson, The Algebraic Eigenvalue Problem (Clarendon Press, Oxford, 1965).
[48] G. H. Golub and C. F. Van Loan, Matrix Computations (John Hopkins University Press, Baltimore, 1996).
[49] H. De Raedt, S. Miyashita, and K. Michielsen, submitted to Phys. Rev. B, http://arXiv.org/abs/cond-mat/0306275.
[50] S. Miyashita, H. De Raedt, and K. Michielsen, Prog. Theor. Phys. **110**, 889 (2003).

On the hyperfine and magnetic properties of the alloy system $Fe_{3-x}Mn_xSi$

S. H. Mahmood[*,1], **A.-F. Lehlooh**[1], **A. S. Saleh**[1], and **F. E. Wagner**[2]

[1] Physics Department, Yarmouk University, Irbid, Jordan
[2] Technical University of Munich, Garching, Germany

Received 1 September 2003, accepted 4 March 2004
Published online 23 April 2004

PACS 75.30.Kz, 75.50.Bb, 76.80.+y

The results of a comprehensive Mössbauer study of the alloy system $Fe_{3-x}Mn_xSi$ are reported. The study covers a wide range of temperatures (4.2 K to 300 K), and a wide range of concentrations between $x = 0$ and $x = 2.5$. Alloys with $x < 0.75$ show ferromagnetic behavior up to RT, and the spectra were fitted with several sextets indicating several configurations around the Fe atoms at B and A, C sites. Alloys with $x > 1.5$ show antiferromagnetic ordering, and those with $x > 2.0$ develop new non-cubic phases. In the concentration region $1.0 < x < 1.5$ the alloys exhibit a complex magnetic behavior. Thus, the alloys were studied over a wide temperature range where transition from ferromagnetic ordering to paramagnetic phase is observed. The Curie Temperature (T_c) for the alloys in this concentration range is found to drop down sharply as x increases. The subspectrum characteristic of the B site disappears at $x \sim 1.3$, indicating that the B site is completely filled with Mn at this concentration.

© 2004 WILEY-VCH Verlag GmbH & Co. KGaA, Weinheim

1 Introduction

Both alloys Fe_3Si and Mn_3Si are known to have a $D0_3$ crystallographic structure, in which the Si atoms selectively occupy site D [1]. Fe_3Si orders ferromagnetically below 850 K, where as Mn_3Si orders antiferromagnetically below 22 K [2]. The two alloys combine to form the solid solution $Fe_{3-x}Mn_xSi$ with Mn atoms preferentially substituting Fe atoms in the B site of the $D0_3$ structure for x values up to 0.75, and then partially occupying A, C sites for higher x values.

For $x < 0.75$, the alloys exhibit ferromagnetic (FM) ordering, and onset of antiferromagnetic (AFM) ordering is observed for $x > 1.8$ [1, 3]. The alloys in the concentration region $0.75 < x < 1.75$ exhibit complex magnetic properties, showing a paramagnetic to ferromagnetic transition at T_c, followed by a re-entrant transition at a lower temperature [2]. This re-entrant spin-glass behavior is a result of the competition between the ferromagnetic ordering and the antiferromagnetic ordering. Due to the fact that the atomic form factors of Fe and Mn are similar, it is difficult to study atomic order using X-Ray Diffraction (XRD). Thus, neutron diffraction (ND) is used to investigate atomic order, and the magnetic structure in these alloys. Also, Mössbauer spectroscopy (MS) was extensively used to study these alloys [4–7], due to the fact that it is based on a local effect, which is capable of detecting minute changes in the chemical environment of the Fe atoms in the various sites. Investigation of a single crystal of Fe_3Si using a polarized neutron beam has shown that the moment of the Fe atoms on site B is 2.3, and that of atoms on A and C is ~1.2 μ_B [8]. Also, MS studies [1, 4, 5] have shown that the hyperfine field at B site is about 31.5 T, and at A and C sites is 20.0 T.

The structural study of the alloy system $Fe_{x-3}Mn_xSi$ conducted by Yoon and Booth [1] indicated the presence of a single cubic phase of the $D0_3$ type (for $x = 0$ and $L2_1A$ type (for $x > 0$) with lattice parame-

---

[*] Corresponding author: e-mail: science@yu.edu.jo

ter varying from 5.65 to 5.68 Å for x values up to 1.6. The results also show that Mn atoms occupy the B site initially, and then start occupying the A and C sites before completely filling the B site. Alloys with $x > 1.8$ were found to contain a second phase that was identified as a tetragonal phase consistent with the structure of the $(FeMn)_5Si_2$ alloys, or a hexagonal phase consistent with the Mn_5Si_3 structure. Alloys with x values in between have shown either one or two phases depending on the heat treatment.

The magnetic studies indicate that the alloys with $x < 0.75$ are ferromagnetic with a magnetization decreasing monotonically as Mn concentration increases. The ND data indicate that the moment on the B site is almost constant for $x < 0.75$, implying that the Mn atoms carry a magnetic moment approximately equal to that of Fe atoms in this site (~2.3 μ_B). The slight decrease, observed in the moment on the B site, is associated with the NNN interaction. The Fe atoms in A and C sites show a moment which decreases from 1.2 μ_B for $x = 0$ to ~ 0.4 μ_B for $x = 0.75$ due to the predominant NN interaction [1].

However, for $x > 0.75$, the observed decrease in the magnetic moment on the B site, and the decrease in T_c are attributed to NN interactions of atoms in site B with Mn atoms occupying A and C sites. Also, in this concentration regime the next nearest neighbors NNN or higher interactions become comparable to the FM interactions. Some of these interactions are antiferromagnetic, leading to a canted moment. At still higher Mn concentrations the FM interaction is completely obscured, leading to an AFM structure [1].

This work involves a comprehensive study of the alloy system $Fe_{3-x}Mn_xSi$ for x values between 0 and 2.5 using Mössbauer spectroscopy at temperatures ranging from 4.2 K to 300 K. A special focus is made for the alloys with x between 1.0 and 1.3 due to the complex magnetic behavior of the system in this concentration region. The study is intended to provide more information on the hyperfine interactions, magnetic phases and transition temperatures of this system.

2 Experimental

The samples were prepared by arc melting the proper amounts of spec-pure materials in an argon atmosphere. The buttons were flipped and re-melted several times to insure homogeneity. Mass losses in the samples were <1%. The buttons were crushed, and the powders were annealed in vacuum at 800 °C for two weeks.

Mössbauer spectroscopy on the powders was performed using a standard constant acceleration spectrometer. For more details the reader is referred to [4, 5].

3 Results and discussion

Low temperature (4.2 K) Mössbauer spectra for the alloy system $Fe_{3-x}Mn_xSi$ for x varying from 0 up to 2.5 are shown in Fig. 1. The spectra for alloys with $x < 0.5$ are not very much different from those at RT [4, 5]. However the high field component resulting from Fe at the B-site shows some structure and was fitted with three sextets with hyperfine fields $B_{hf} = 33$, 31 and 27 T. This structure could be associated with next nearest neighbors NNN effects or with Si anti-structure atoms at A, C sites. The other components corresponding to Fe at A, C sites are similar to those at RT but occur at higher hyperfine fields (the main peak occurs now at 21.6 T, when it was at 20 Tesla or less at RT), and are associated with the different configurations as discussed earlier [4]. For the alloy with $x = 0.5$ a new peak appears at 6.0 T which may be associated with a configuration in which Fe atoms at A, C sites are surrounded by 4 Mn atoms and 4 Si atoms. As x increases this peak grows in intensity at the expense of other peaks, indicating a higher probability of occurrence for this configuration.

The spectrum for the sample with $x = 0.75$ exhibits ferromagnetic order with a high field component at $B_{hf} \sim 29.0$ T, and five other components at 21.6, 18, 14.3, 10.5 and 6.2 T. The intensity of the high field component drops down to 15% from its original value of 30% at $x = 0$. As x increases further the high field component continues to drop in intensity until it disappears at around $x = 1.3$. The other five components change in such a way that those at the low field end of the spectrum grow in intensity at the expense of those at the high field end. Two components only are observed at $x = 1.3$, and only one component (at $B_{hf} \sim 5.8$ T) remains at $x = 1.5$. These developments in the spectrum, as described, are consistent with the increase in the concentration of Mn. The high field component decreases gradually in inten-

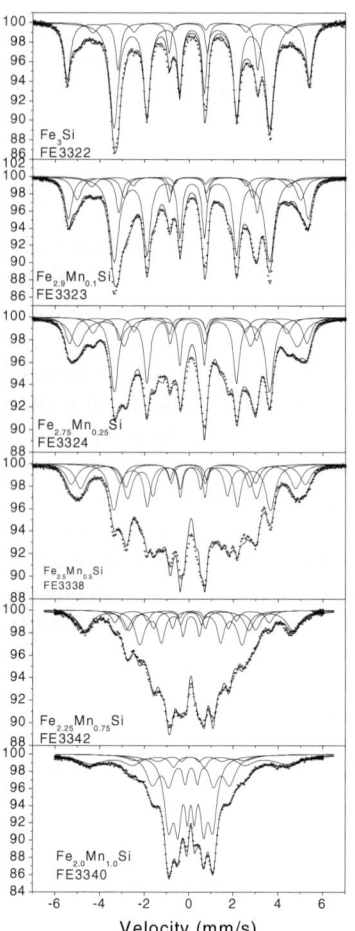
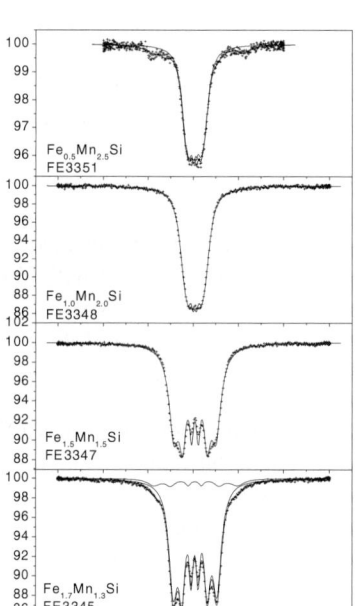

Fig. 1 Mössbauer spectra for the alloy system $Fe_{3-x}Mn_xSi$ at 4.2 K.

sity and shifts to lower fields as x increases until the B site is completely occupied with Mn at about $x \sim 1.3$ when this component disappears.

The other components corresponding to Fe atoms in the A, C sites follow a pattern in which configurations with more Mn atoms surrounding the A, C sites become more probable than those with smaller number of Mn atoms. This trend continues until the B site is completely occupied with Mn atoms, when only one configuration is possible, that is (4 Mn + 4 Si) surrounding the A, C sites, giving rise to the component with $B_{hf} \sim 6.0$ T.

The spectra for alloys with $x > 1.5$ are different from the above spectra for $x < 1.5$. For the sample with $x = 2$, only one component is observed with $B_{hf} \sim 3$ T which could be attributed to an antiferromagnetic phase (with low magnetic moment) with Fe at the A, C sites. For the sample with $x = 2.5$, a new magnetic component with low intensity emerges at $B_{hf} \sim 12.7$ T. This component is associated with a second non-cubic phase since it has a relatively large quadrupole shift (0.33 mm/s). This result is in agreement with the work of Yoon and Booth [1], which reports that alloys with $x > 1.8$ contain a second non-cubic phase.

4 Temperature effect

The alloys with $x < 0.75$ are known to exhibit ferromagnetic behavior up to RT, and those with $x > 1.5$ develop other magnetic and structural phases. In the concentration range $1.0 < x < 1.5$ the temperature-dependence of the spectra shows interesting features and a critical transition temperature is clearly ob-

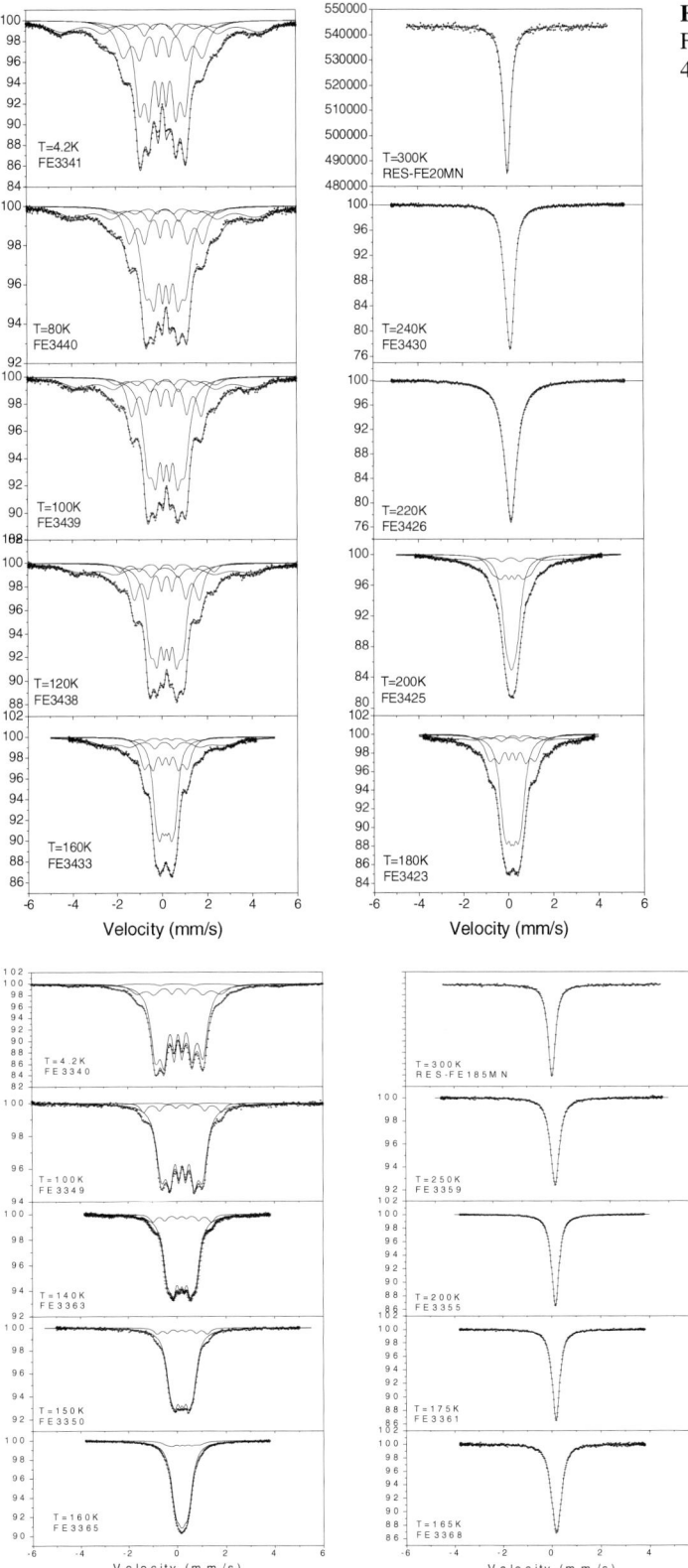

Fig. 2 Mössbauer spectra for the alloy Fe$_2$MnSi at different temperatures between 4.2 K and 300 K.

Fig. 3 Mössbauer spectra for the alloy with $x = 1.15$ at different temperatures.

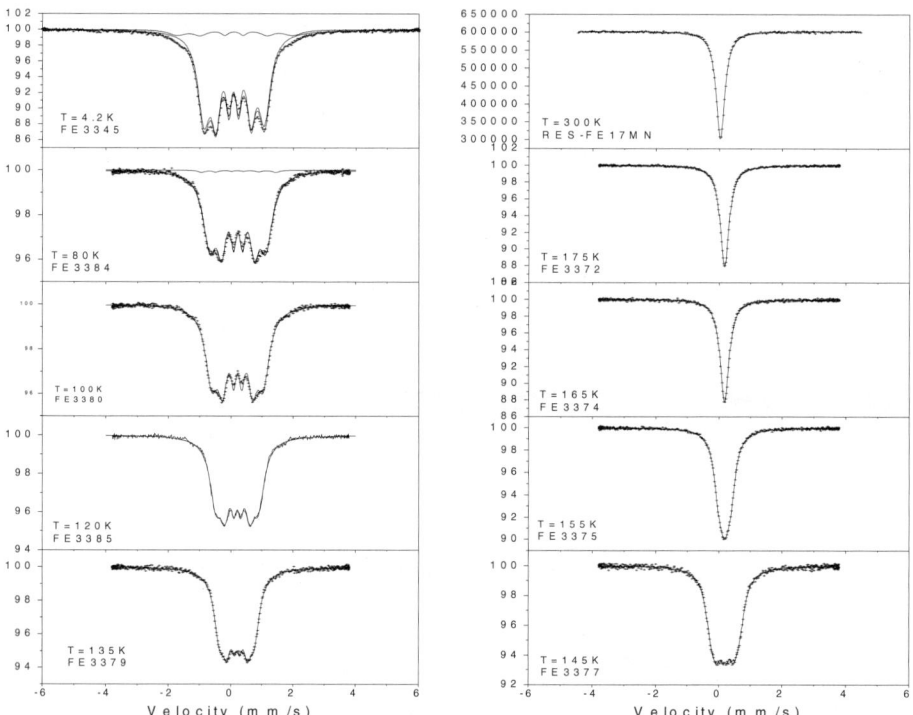

Fig. 4 Mössbauer spectra for the alloy with $x = 1.3$ at different temperatures.

servable. Thus, Mössbauer spectra for the samples with $x = 1.0$, 1.15, 1.3 were collected over the range of temperatures from 4.2 K to 300 K. The spectra for $x = 1.0$ (Fig. 2) were best fitted with four magnetic components. The value of B_{hf} for each component decreases monotonically, with increasing temperature. At temperatures ($T > 200$ K) the components coalesce into a broad singlet which is fitted with a single component with a small B_{hf} (~1–2 T). The presence of a weak high-field component is a clear indication that the B site is still partially occupied by Fe atoms at this concentration. The gradual decrease in B_{hf} for each component and the broadening of the absorption lines as T increases show that the magnetic interaction weakens as the temperature gets closer to Curie temperature T_c. A plot of B_{hf} versus T was extrapolated, yield a $T_c \sim 240$ K for this alloy. This value is consistent with that of 219 K obtained for this alloy using magnetization measurements [9].

A similar analysis was done on the spectra for the sample with $x = 1.15$ (Fig. 3). The high field component becomes weaker, and all three components shift to lower fields as the temperature increases, and finally they coalesce into a single component with $B_{hf} \sim 1$ T. The extrapolation of the B_{hf} versus T curve yields $T_c \sim 175$ K. This result agrees with that of 183 K reported on the alloy with $x = 1.14$ [10].

The low temperature spectra for the sample $x = 1.3$ (Fig. 4) consist almost of one single component with $B_{hf} \sim 6.0$ T, a result which indicates that the B site is completely filled with Mn, and the alloy has a single magnetic phase with Fe atoms at A, C sites having identical environments. The hyperfine field for this component decreases from 6.0 T at 4.2 K to less than 1.0 T as the temperature increases. Analysis of the temperature dependence of the spectra implies that $T_c \sim 165$ K. This value is close to that of 157 K reported for the alloy with $x = 1.5$.

5 Conclusions

The results of the present study indicate that:
1. The alloy system $Fe_{3-x}Mn_xSi$ orders ferromagnetically for x values up to 1.5, and antiferromagnetically for higher Mn concentrations.

2. As x increases, Mn atoms replace Fe atoms in B sites for low Mn concentrations, and then start ocuupying A, C sites as well as x approaches 1.0, until the B site is fully occupied by Mn at $x \sim 1.3$.
3. The alloy system is single phase for x below 2.0, and develop a non-cubic phase for $x = 2.5$.
4. Curie temperature for the alloy system drops down sharply from about 240 K at $x = 1.0$ to about 165 K at $x = 1.3$.

References

[1] S. Yoon and J. G. Booth, J. Phys. F: Met. Phys. **7**, 1079 (1977).
[2] A. Chakravarti, R. Ranganathan, and S. Chatterjee, J. Magn. Magn. Mater. **138**, 329 (1994).
[3] J. G. Booth, J. E. Clark, J. D. Ellis, P. J. Webster, and S. Yoon, Proc. Int. Conf. 73 on Magnetism, Moscow, Vol. 4, 577 (1974).
[4] S. H. Mahmood, A.-F. D. Lehlooh, A. S. Saleh, and N. Darlington, Proceedings of the Fifth International Symposium on Advanced Materials, A. Q. Khan Research Labs, Kahuta, Pakistan, 321 (1997).
[5] G. A. Al-Nawashi, S. H. Mahmood, A.-F. D. Lehlooh, and A. S. Saleh, Physica B **321**, 167 (2002).
[6] J. T. T. Kumaran and C. Bansal, Solid State Commun. **69**, 779 (1989).
[7] J. T. T. Kumaran and C. Bansal, Solid State Commun. **74**, 1125 (1990).
[8] A. Paoletti and L. Passari, Nuovo Cimento **32**, 1449 (1964).
[9] T. Nagano, S. Uwanuyu, and M. Kawakami, J. Magn. Magn. Mater. **140**, 123 (1995).
[10] T. Ersez, G. T. Etheridge, and T. J. Hicks, J. Magn. Magn. Mater. **148**, L381 (1995).

Effects of d-wave symmetry in high-T_C grain boundary Josephson junctions

Antonio Barone[*,1], **Filomena Lombardi**[2], **Antonia Monaco**[3], **Ettore Sarnelli**[3], **Francesco Tafuri**[4], and **Gianluca Testa**[3]

[1] Coherentia-INFM, Dip. Scienze Fisiche, Università di Napoli Federico II, Napoli, Italy
[2] Department of Microelectronics and Nanoscience, MINA, Chalmers University of Technology and Goteborg University, 41296 Goteborg, Sweden
[3] Istituto di Cibernetica "E. Caianiello" del CNR, Pozzuoli (Napoli), Italy
[4] Coherentia – INFM, Dip. Ingegneria dell' Informazione, Seconda Università di Napoli, Aversa (CE), Italy

Received 1 September 2003, accepted 15 March 2004
Published online 23 April 2004

PACS 73.40.Gk, 73.40.Rw, 74.50.+r, 74.72.Bk

Recent experimental results on the physics of high temperature superconductivity (HTS) related to coherent phenomena and unconventional pairing symmetry are discussed. Some consequences of the d-wave order parameter symmetry on the phenomenology of HTS junctions of different type are investigated. The attention is focused on recent experiments based on biepitaxial $YBa_2Cu_3O_{7-\delta}$ Josephson structures where junctions with a π shift in the phase have been reproducibly obtained. Anisotropy measurements confirm the relevance of effects of d-wave order parameter symmetry on the properties of the junctions, and prove that intrinsic d-wave effects can be predominant over extrinsic effects. The possibility to obtain some kind of tunnel-like barriers and to tune d-wave effects through a suitable geometry has been demonstrated opening some perspectives towards novel device concepts within the framework of π circuitry and "quiet qubit".

1 Introduction

The fundamental role of the Josephson effect as a very powerful probe of the underlying physics of unconventional pairing symmetry is well known since many years [1]. Evidence of d-wave symmetry [2] for high temperature superconductors was well established through extensive experimental investigations. YBCO based junction structure configurations were mainly considered in such a context although various other cuprates superconductors have been employed in a variety of experiments. In section 2 grain boundary (GB) Josephson structures are discussed in the most interesting configurations which provide a proper control of different interface orientations. Thus the fundamental issue of the angular dependence of the Josephson critical current is extensively discussed in connection with the recent experimental observation [3] of the actual signature of the d-wave symmetry of the order parameter inferred from structures consisting of single junctions not inserted in any loop (see below). Finally aspects of the role of midgap states on the zero-voltage current are summarized. In particular the observation of a nanometric dependence of the Josephson current is discussed together with the striking experimental evidence of a 0–π junction transition.

---

[*] Corresponding author: e-mail: barone@na.infn.it

2 Unambigous evidence of angular dependence of the critical current in single HTS Josephson junctions

This section is devoted to describe a very exciting aspect of an unconventional order parameter (OP) symmetry [2], and in particular the angular dependence of the critical current [3]. This effect can be directly correlated to the OP symmetry as shown below, and is of great importance since it is the first direct consequence of d-wave OP measured on single all high-T_C Josephson junctions (not inserted in any loop). The two main ingredients required to observe such a peculiar feature, are the availability of a flexible technique able to provide different angle orientations on the same chip and some kind of directionality of the tunneling or transport mechanism across the barrier. Grain boundary Josephson junctions [4], which still represent the optimal compromise situation for HTS Josephson junctions to investigate fundamental physical aspects and for simple basic electronic applications, offer both the required features. The biepitaxial technique is the answer to the first requirement. This is based on the epitaxial growth of YBCO on the various materials [5] and the barrier-interface orientation can be simply modified through an opportune photolithographic processes and mask design [6]. Unfortunately the original biepitaxial technique provided junctions with poor barrier properties (in the sense specified below) and the angular information on the OP was lost in the transport processes. The "modified biepitaxial" junctions, developed at the University of Napoli and described below [3, 6, 7], finally gave the missing link because of their characteristic low barrier transparency which favours directional transport. Section 2 includes three subsections. The first is devoted to a brief description of the junctions; the second to the transport properties and the third to anisotropy transport issues respectively.

2.1 Junctions description

The GB Josephson junctions are obtained at the interface between a (103) YBCO film grown on a (110) SrTiO$_3$ substrate and a c-axis film deposited on a (110) CeO$_2$ seed layer. The presence of the CeO$_2$ produces an additional 45° in-plane rotation of the YBCO axes with respect to the in-plane directions of the substrate [3, 6]. As a consequence, the GBs are the product of two 45° rotations, a first one around the c-axis, and a second one around the b-axis (Fig. 1). This configuration produces the desired 45° misorientation between the two electrodes to enhance d-wave OP effects.

Furthermore, the degrees of freedom of the fabrication process allow to select any possible in plane orientation for the GB interface. Details about the fabrication process can be found elsewhere [6]. We will define the GB interface angle θ with respect to [001] in-plane SrTiO$_3$ direction. In the two limiting configurations, $\theta = 0°$, $\theta = 90°$, the GBs are characterized by a (100) 45° tilt or twist respectively of the c-axis with respect to the interface, plus a 45° tilt around the c-axis. In this paper they will be referred to as tilt-tilt and twist-tilt GBs and are indicated in Fig. 1 along with an intermediate situation. We also recall that if we rather use an MgO seed layer, no in-plane rotation occurs. MgO-based

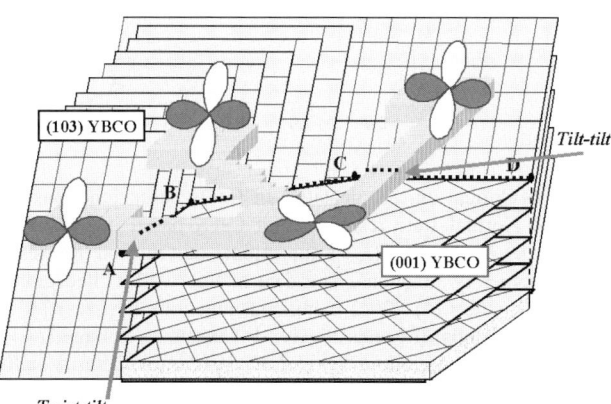

Fig. 1 (online colour at: www.interscience. wiley.com) Sketch of the grain boundary structure with different interface orientations.

junctions have been proved to be of high quality and their properties to be very weakly influenced by OP symmetry [6]. It is reasonable to assume that these MgO-based and CeO_2-based junctions can be considered complementary from the OP symmetry point of view in a circuit design perspective [6, 8, 9].

2.2 General transport properties

In our GBs, the crucial feature of relatively lower barrier transmission seems to be associated with a c-axis tilt. This is somehow the conclusion of different studies realized on our MgO-based and CeO_2-based junctions. Such a property is of particular interest when combined with the 45° in-plane rotation of one of the electrodes, which may produce π-junction behavior. Specific resistivities $R_N A$ (where R_N is the normal state resistance and A is the junction area respectively) in our biepitaxial grain boundary junctions typically range from 10^{-7} to 10^{-5} (Ω cm^2) at $T = 4.2$ K, on average at least one order of magnitude higher than the values extracted from measurements on other (in particular bicrystal) types of GB junctions [4]. The critical current density J_C typically ranges from 10^2 to 10^3 A/cm^2. These values are compared with data in literature taken from [4] in Figs. 2a and b. The data available in

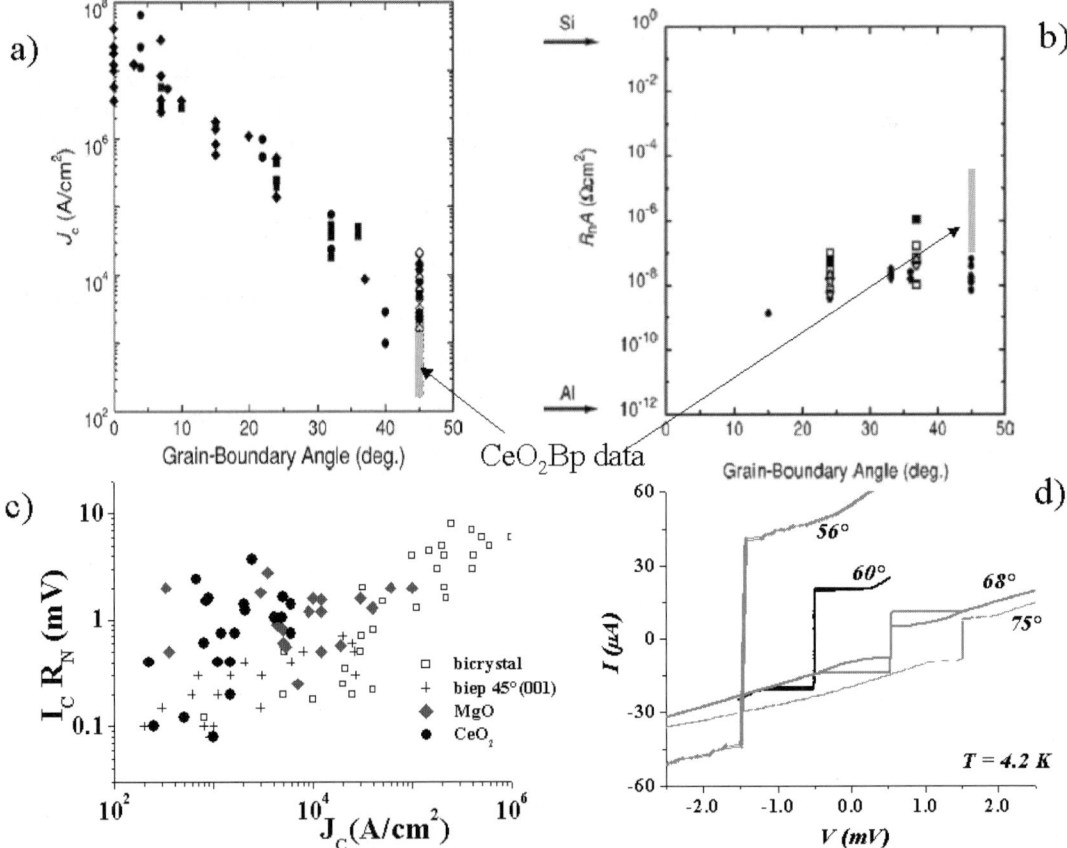

Fig. 2 (online colour at: www.interscience.wiley.com) a) Critical current densities of CeO_2 based junctions are compared with data in literature taken from Ref. [4]; b) Normal state resistances of CeO_2 based junctions are compared with data in literature taken from Ref. [4]; c) $I_C R_n$ product (I_C is the critical current and R_n is the normal state resistance) as a function of J_C for different types of junctions. The data of this work (filled circle) are compared with values from bicrystal junctions (open squares) [4], (001) biepitaxial junctions (crosses) [4, 5] and our junctions employing MgO as a seed layer (filled diamonds) [6] respectively; d) Current vs voltage (I–V) characteristics of biepitaxial junctions for various interface orientations ($T = 4.2$ K). Curves are horizontally shifted for clarity.

literature mostly concern to GB junctions where the angle refers to an in-plane tilt. Our grain boundary is more complicated as discussed in the previous subsection. Nevertheless for sake of simplicity we have assumed 45° as angle representative of our structures for this particular aspect. As a matter of fact, our J_C and R_NA are characterized by the lowest and the highest values respectively, indicating a trend toward the tunnel-like behavior. In Fig. 2c we report the $I_C R_N$ product (I_C is the critical current) as a function of J_C for different types of junctions. The data relative to CeO_2-based junctions (filled circles) are compared with values from bicrystal junctions (open squares) [4], (001) biepitaxial junctions (crosses) [4, 5] and our junctions employing MgO-based (filled diamonds) [6] respectively. The comparison confirms the relatively lower barrier transparency of our CeO_2-based biepitaxial junctions, and seems to suggest that the off-plane rotation (characteristic of both of the MgO-based and CeO_2-based junctions) is probably an important feature to increase the $I_C R_N$ product.

In Fig. 2d typical $I-V$ curves for different angles orientations are shown. We can clearly observe a tuning of $I-V$ curves and of the junctions parameters within the values given above that we will discuss in in some detail in the next subsection.

2.3 Anisotropy measurements: evidence of d-wave induced effects

In Fig. 3 we report the dependence of I_C on the angle (θ) for 4 μm wide junctions. A clear oscillatory dependence of the critical current I_C on θ was observed [3] as expected in structures dominated by d-wave induced effects.

Minima in the critical current are observed for $\theta = 0°$, 34°, and 90° respectively. These values correspond to configurations in which the tunneling direction (the normal to the barrier) points towards a node of the OP on either of the two sides. The minima at $\theta = 0°$ and 90° arise from the position of the nodes in the c-axis oriented side of the junction. The minimum at $\theta = 34°$ occurs when the projection in the $a'-b'$ planes of the normal to the barrier points towards a node of the OP on the (103) side. In the Sigrist–Rice (S–R) phenomenological approach, the Josephson current density of an all d-wave junction is given by [10]:

$$J_c = J_o (n_x^2 - n_y^2)_L (n_x^2 - n_y^2)_R \times \sin(\phi) . \tag{1}$$

In this expression J_o is the maximum Josephson current density, ϕ is the difference between the phase of the OP in the two electrodes, and n_x, n_y are the projections of the unit vector $\boldsymbol{n}$ onto the crystallographic axes $\boldsymbol{x}$ and $\boldsymbol{y}$ in the left (L) and right (R) electrode respectively. The S–R formula represents a significant reference to test the main physics expressed by the $J_C(\theta)$ measurements, especially in our non coplanar configuration where contributions due to Andreev bound states are quite hard to evaluate exactly [11–15]. Specifying the above expression to our non coplanar configuration, assuming a more efficient tunneling in the lobes directions [3], we obtain the curve plotted as open circles in Fig. 3.

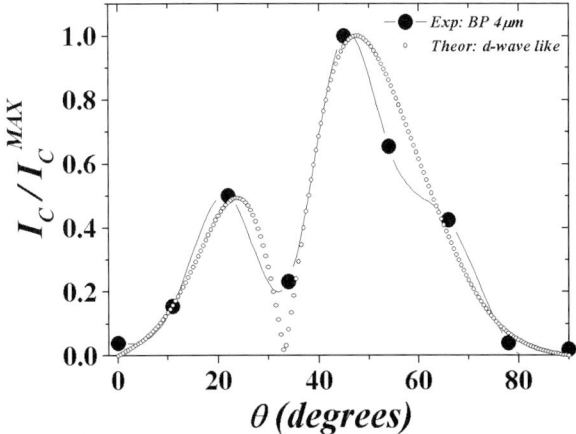

Fig. 3 I_C dependence on the angle is reported; experimental data (filled circles) are compared with theoretical predictions based on $d_{x^2-y^2}$-wave pairing symmetry (open circles) [3].

The coplanar configuration is clearly responsible for the position of the minimum at around 35° rather than 45° which is the expected value for the (001) tilt traditional biepitaxial junctions. The experimental behavior is well reproduced by the S–R-like theoretical prediction. The $I_C(\theta)$ dependence in agreement with the S–R formula apparently seems to be mostly determined by the OP symmetry.

In the past previous attempts to observe d-wave induced behavior in all-HTS single junctions have been unsuccessful for various reasons [2]. Common to most attempts was the high barrier transparency. Our junction configuration, instead, with low barrier transmission probabilities preserves the directionality of the Cooper pairs, whether the transport be by tunneling or some other mechanism, and it is in principle more sensitive to the angular dependence of the order parameter symmetry. The anisotropy measurements have demonstrated for the first time that "intrinsic" d-wave effects are dominant in the phenomenology of the Josephson junctions themselves (not inserted in any loop) independently of the interface details [3].

We finally mention that the magnetic properties of CeO_2-based junctions have been investigated in some detail thorough Scanning SQUID Microscopy. Results are presented elsewhere [7]. We remark that interesting behaviors have been observed consistently with d-wave induced effects. In particular it has been shown the possibility to control the presence of spontaneous currents along faceted GBs and the presence of the half flux quantum effect in suitably designed corners. The phenomenology of the observed effects can be reasonably explained on the basis of d-wave induced effects on junctions characterized by low barrier transparencies [7].

In conclusion the tunnel-like behavior and the low barrier transparencies are crucial features to isolate intrinsic d-wave effects such as the half flux quantum effect and the oscillatory dependence of the critical current on the interface orientation.

The availability of such tunnel-like junctions is a crucial pre-requisite for quantum circuitry and qubit proposals [8]. As a matter of fact they can combine the peculiar d-wave feature required for the "quiet" qubits and the junction quality in priciple adequate to have access to a quantum regime.

3   0 and π junctions. Evidence of midgap states

The unconventional order parameter symmetry leads to additional peculiar behaviours in the Josephson current of high critical temperature grain boundaries. In particular, the sign change of the order parameter for rotation of 90° can add an extra phase of π between the two superconducting electrodes. By suitably designing the GB misorientation angle, it is then possible to fabricate both 0 and π junctions, characterized by a minimum of the free energy at $\varphi = 0$ and π, respectively. In Ref. [2], the S–R formula has been used to calculate the angles where 0 and π junctions are theoretically expected. Results have been experimentally verified in both tricrystal experiments [2] and dc SQUIDs on tetracrystal substrates [16] by inserting the junctions in superconducting loops.

The S–R formula is however only an approximated equation. It doesn't take into account the presence of midgap state (MGS) at the interface of the two superconducting electrodes and describes the Josephson current fairly enough only for quasiparticle trajectories perpendicular to the barrier. In order to include both the intrinsic phase of the pair potential and the formation of localised states at the interface, a tunneling cone has to be considered. A more complete approach, describing the effect of the d-wave symmetry of the order parameter on grain boundary Josephson junctions, has been extensively reported [13, 14, 17].

Experimentally, the effect of midgap states on the junction properties has been widely reported on quasiparticle currents as a zero bias conductance peak in different superconducting /insulator/ normal metal (or superconducting) structures [18, 19]. More difficult is the observation of their influence on the zero voltage state current, which requires very smooth interfaces with small meandering.

Recent experiments, performed by using submicron [001] 45° bicrystal grain boundary junctions, have demonstrated that the Josephson current shows a non-monotonic temperature dependence, in agreement with theoretical predictions [20]. This behaviour is due to a competition between the dc

Josephson current I_{CONT}, characterized by a minimum of the free energy at the phase $\varphi = 0$, and the opposite sign midgap state mediated current I_{MGS}, which has an energy minimum at $\varphi = \pi$ [13]. At low temperatures, in the case of low-transparency barriers, the midgap state current, proportional to the square root of the barrier transparency $D^{1/2}$ can become larger than I_{CONT}, proportional to D, leading to a negative zero voltage state current. By measuring $I-V$ characteristics, a non-monotonic temperature dependence should then be observed [13, 14, 17].

The effect has been first observed in Ref. [20], where, however, the free energy showed a transition from a 0 state to a double degenerate ground state, similar to that expected in 45° asymmetric junctions. The double degenerate state, not expected in theoretical predictions, was mainly due to an anomalously large second harmonic component of the Josephson current and explained by the authors as the result of phase fluctuations.

Very recently, the first direct evidence of a 0–π junction transition has been achieved by using two phase sensitive tests [21]. First, a non-monotonic temperature dependence of the critical current has been observed in a dc SQUID made by two sub-micron 45° [001] symmetric bicrystal grain boundary junctions. Moreover, the SQUID itself showed a transition from 0 to π (second phase sensitive test), as indicated by the half flux quantum shift in the magnetic field dependence of its critical current. Both results give evidence of a 0–π transition in only one of the two junctions. Indeed, if both junctions showed such transition, the loop would not becomes frustrated and no half flux quantum would then be spontaneously generated.

It is worth noting that, in order to have a better control of such transition and to exploit it for applications, it is necessary to change suitably the transparency of the barrier and to optimize the junction interface, making the faceting negligible.

4 Conclusions

Experiments concerning the d-wave order parameter in High T_C superconductors have been discussed. The "intrinsic" character of recent experimental data stemming for the angular dependence of the Josephson supercurrent, confirms unambiguously the d-wave nature of the OP. The availability of such tunnel-like junctions is a crucial pre-requisite for quantum circuitry and qubit proposals. They can combine the peculiar d-wave feature required for the "quiet" qubits and the junction quality in priciple adequate to have access to a quantum regime. Moreover, results of a 0–π transition observed in SQUID configuration structure are briefly analized.

Acknowledgements This work has been partially supported by the ESF Network "Pi-shift", by MIUR under the project DG236RIC "NDA" and by the TRN "DeQUACS". The authors would like to thank M. Blamire, E. Ilichev, D.J. Kang, J. Kirtley, F. Miletto Granozio and C. Tsuei for useful discussions.

References

[1] V. B. Geshkenbein, A. I. Larkin, and A. Barone, Phys. Rev. B **36**, 235 (1987); Josephson Effect (J. Wiley, New York, 1982).
[2] C. C. Tsuei and J. R. Kirtley, Rev. Mod. Phys. **72**, 969 (2000).
 D. J. Van Harlingen, Rev. Mod. Phys. **67**, 515 (1995).
[3] F. Lombardi, F. Tafuri, F. Ricci, F. Miletto Granozio, A. Barone, G. Testa, E. Sarnelli, J. R. Kirtley, and C. C. Tsuei, Phys. Rev. Lett. **89**, 207001 (2002).
[4] H. Hilgenkamp and J. Mannhart, Rev. Mod. Phys. **74**, 485 (2002) and references therein.
[5] K. Char, M. S. Colclough, S. M. Garrison, N. Newman, and G. Zaharchuk, Appl. Phys. Lett. **59**, 773 (1991).
[6] F. Tafuri, F. Carillo, F. Lombardi, F. Miletto Granozio, F. Ricci, U. Scotti di Uccio, A. Barone, G. Testa, E. Sarnelli, and J. R. Kirtley, Phys. Rev. B **62**, 14431 (2000).
[7] F. Tafuri, J. R. Kirtley, F. Lombardi, and F. Miletto Granozio, Phys. Rev. B **67** 174516 (2003).
[8] L. B. Ioffe, V. B. Geshkenbein, M. V. Feigel'man, A. L. Fauchere, and G. Blatter, Nature **398**, 679 (1999).
 G. Blatter, V. B. Geshkenbein, and L. B. Ioffe, Phys. Rev. B **63**, 174511 (2001).

[9] A. Blais and A. M. Zagoskin, Phys. Rev. A **61**, 42308 (2000).
A. M. Zagoskin, cond-mat/9903170 (1999).
[10] M. Sigrist and T. M. Rice, J. Phys. Soc. Jpn. **61**, 4283 (1992); Rev. Mod. Phys. **67**, 503 (1995).
[11] A. F. Andreev, Zh. Eksp. Teor. Fiz. **48**, 1823 (1964); Sov. Phys. JETP 19, 1228 (1964).
[12] C.-R. Hu, Phys. Rev. Lett. **72** 1526 (1994).
[13] T. Lofwander, V. Shumeiko, and G. Wendin, Supercond. Sci. Tech. **14**, R53 (2001) and references therein.
[14] Y. Tanaka and S. Kashiwaya, Phys. Rev. B **56**, 892 (1997).
[15] A. Golubov and F. Tafuri, Phys. Rev. B **62**, 15200 (2000).
[16] R. R. Schulz, B. Chesca, B. Goetz, C. W. Schneider, A. Schmehl, H. Bielefeldt, H. Hilgenkamp, and J. Mannhart, Appl. Phys. Lett. **76**, 912 (2000).
[17] Yu. S. Barash, Phys. Rev. B **61**, 678 (2000).
[18] M. Covington et al., Phys. Rev. Lett. **79**, 277 (1997).
[19] L. Alff et al., Phys. Rev. B **58**, 11197 (1998).
[20] E. Ilichev, V. Zakosarenko, R. P. J. IJsselsteijn, H. E. Hoenig, V. Schultze, H. G. Meyer, M. Grajcar, and R. Hlubina, Phys. Rev. B **60**, 3096 (1999).
E. Ilichev, M. Grajcar, R. Hlubina, R. P. J. IJsselsteijn, H. E. Hoenig, H. G. Meyer, A. Golubov, M. H. S. Amin, A. M. Zagoskin, A. N. Omelyanchuk, and M. Yu. Kupriyanov, Phys. Rev. Lett. **86**, 5369 (2001).
[21] G. Testa, A. Monaco, E. Esposito, E. Sarnelli, D.-J. Kang, E. J. Tarte, S. H. Menneima, and M. G. Blamire, cond-mat/0310727 (2003).

First-principles study on the creation of holes in high T_c cuprates

C. Ambrosch-Draxl*,1, E. Ya. Sherman1, H. Auer1, and T. Thonhauser2

1 Institut für Theoretische Physik, University Graz Universitätsplatz 5, A-8010 Graz, Austria
2 Department of Physics, Pennsylvania State University, University Park, Pennsylvania 16802, USA

Received 1 September 2003, accepted 24 February 2004
Published online 14 April 2004

PACS 71.15.Mb, 74.25.Jb, 74.72.Jt, 74.81.–g

We investigate the charge redistribution in high T_c cuprates as a function of pressure, composition, and doping. To this extent we have performed first-principles calculations based on density functional theory for several representatives of the Hg based cuprates. In particular, we focus on the creation of holes in the copper-oxygen planes. Conclusions are drawn about the similarities and differences between the three parameters influencing the superconducting transition temperature.

© 2004 WILEY-VCH Verlag GmbH & Co. KGaA, Weinheim

1 Introduction

In the high temperature superconductors, in particular the high T_c cuprates, the Cooper pairs are formed by holes in the CuO_2 planes. The number of holes depends on the particular compound, but also on other parameters like the doping concentration or pressure. Despite fifteen years of intensive research on the mechanism of superconductivity in these materials not much is known about the details, how the holes within the copper-oxygen planes are created. Whereas for the experimental side the doping concentration or the pressure value is the key quantity, most of the models for superconductivity and related properties rely on the number of holes as an input parameter. In order to fully understand superconductivity in these materials, however, a profound understanding of how the charge is redistributed as a function of doping or pressure is inevitable. Without this knowledge also no improvement or tailoring of the material is possible. For these reasons we have performed first-principles calculations based on density functional theory (DFT) for several representatives of the cuprate families, where we concentrate on the mercury based compounds. Within the single-layered materials, $HgBa_2CuO_4$ exhibits the highest superconducting transition temperature upon doping. Moreover, the three-layer compound of this family reaches the highest T_c under pressure. We investigate how the charges are distributed when one or two CuO_2 planes are present in the unit cell. Moreover, we study the effect of pressure on the charge redistribution in the compounds. For the single-layer material, we compare the effect of pressure with the influence of doping, which has been studied recently over the full doping range up to the overdoped case [1, 2]. Possible consequences on the superconducting transition temperature will be discussed.

---

* Corresponding author: e-mail: claudia.ambrosch@uni-graz.at, Phone: +43 316 380 5235, Fax: +43 316 380 9820

2 Method

All calculations have been carried out within the full-potential linearized augmented plane-wave (LAPW) method as implemented in the WIEN2k code [3]. The computational details for the particular compounds are given elsewhere [1]. Here we focus on the aspects which are needed to analyse the charges with respect to their atomic and orbital character.

In augmented planewave methods [4] the unit cell is partitioned into two different regions, which are the atomic spheres, centered at the nuclear positions, and the interstitial region in between these spheres. Since in the interstitial the wavefunctions and hence the charge density are much less oscillating than close to the nuclei, planewaves are a good choice for a basis set. Every planewave is then augmented to an atomic-like basis function in each of the atomic spheres α:

$$\phi_{k+G}(S_\alpha + r) = \sum_{lm} [A^\alpha_{lm}(k+G) u^\alpha_l(r, E_l) + B^\alpha_{lm}(k+G) \dot{u}^\alpha_l(r, E_l)] Y_{lm}(\hat{r}). \quad (1)$$

Here, G denotes the reciprocal lattice vector characterizing the planewave, and k is a point in the first Brillouin zone (BZ). S_α denotes the position vector of the atomic nucleus α, and the radius of the corresponding muffin-tin sphere is R_α. $Y_{lm}(\hat{r})$ are the spherical harmonics. The radial function $u^\alpha_l(r, E_l)$ is the regular solution of the equation

$$\left\{-\frac{d^2}{dr^2} + \frac{l(l+1)}{r^2} + V^\alpha(r) - E_l\right\} r u^\alpha_l(r, E_l) = 0, \quad (2)$$

and $\dot{u}^\alpha_l(r, E_l)$ is its energy derivative. The matching coefficients $A^\alpha_{lm}(k+G)$ and $B^\alpha_{lm}(k+G)$ are determined by the requirement that the basis functions and their first spatial derivatives have to be continous at the sphere boundaries.

This type of basis functions for solving the Kohn-Sham equations of DFT allows to analyze the charge densities inside the spheres not only according to their atomic origin, but also with respect to their l and m character. E.g. inside the copper spheres the d charge can be split up into $d_{x^2-y^2}$, d_{z^2}, d_{xy}, d_{xz}, and d_{yz} contributions, wheras the oxygen p charge can be further analyzed with respect to its p_x, p_y, and p_z character. However, there is a drawback in the charge being analyzed within the sphere, i.e., these charges are not unique quantities but depend on the chosen sphere sizes. Nevertheless, the partial charges can be used to study trends, e.g. as a function of doping, when the sphere sizes are kept the same for all calculations.

For the investigation of pressure, there is no standard way of chosing the atomic radii when the unit cell volume is decreased. The most widely used procedure is to keep the radii fixed, although the amount of charge (volume) lying within the spheres with respect to the total charge (volume) becomes larger, and therefore complicate the interpretation of the partial charges. Furthermore, the spheres have usually to be chosen rather small in order to avoid overlapping spheres due to short bond lengths. In this case a considerable amount of charge can leak out of the spheres.

3 Crystal structure optimization

In order to calculate properties from first-principles, including lattice relaxation effects upon applying pressure, we first have optimized the crystal structures with respect to all degrees of freedom, i.e., the lattice parameteres a and c, and the atomic positions. a and c have been varied in a range which allowed to study pressure up to 15 GPa. Thereby the experimental crystalline data have been taken as a starting point. For each set of lattice parameters the internal coordinates have been relaxed. The *optimized crystal structure* corresponds to the minimum of the energy function, obtained from a fit of all total-energy values. From this analytic expression for the energy also the pressure can be calculated. For pressure values of 0, 5, 10, and 15 GPa new self-consistent calculations have been performed. This procedure has been applied to the Hg based systems with one and two CuO_2 layers, furtheron denoted as Hg1201 and Hg1212, respectively. The corresponding unit cells are depicted in Fig. 1. When going from Hg1201 to Hg1212, one additional CuO_2 plane is added as well as one Ca

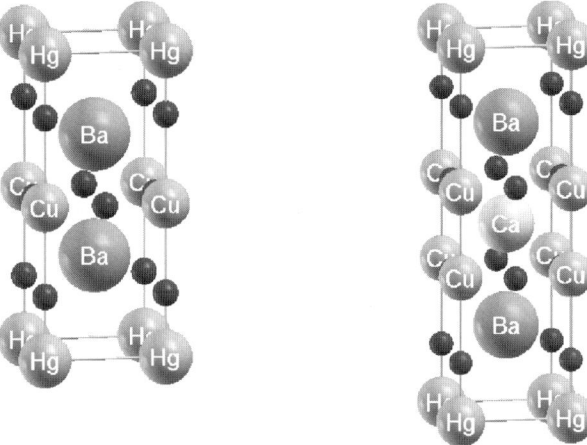

Fig. 1 Crystal structures of Hg1201 (left) and Hg1212 (right). The black spheres mark the oxygen atoms.

atom separating the two planes. Since the latter does not contribute to the states near the Fermi level, but serves to stabilize the crystal structure, only the additional charges stemming from copper (Cu) and plane oxygen (simply denoted as O) are important for our studies, which will be discussed in the next section.

4 Results

Table 1 shows the Cu $d_{x^2-y^2}$ and O p_x charges in Hg1201 and Hg1212 as a function of pressure. The copper charges exhibit a significant decrease, which is linear and more pronounced in the lower pressure range. This decrease appears despite the fact that due to the constant atomic sphere radii a bigger fraction of the volume is inside the spheres. Thus we can even expect the effect to be a bit underestimated. While the same trend, i.e. decreasing charges with increasing pressure, is observed for both compounds, the opposite behavior is found when increasing the number of Cu–O layers instead of applying pressure: When going from Hg1201 to Hg1212 the copper charges increase by nearly 0.008 e for ambient pressure, where this difference becomes bigger for higher pressures. But although the number of pressure-induced holes per copper atoms is smaller in the two-layer compound, the effect is still larger in Hg1212 since there are two layers in the unit cell. This finding is concomitant with the higher T_c. The oxygen charges are not much affected by pressure. Due to the very small sphere sizes which had to be chosen for treating the high pressure values, they even slightly decrease. When performing calculations with bigger oxygen spheres, which is possible for somewhat smaller pressure

Table 1 Cu $d_{x^2-y^2}$ and O p_x charges in Hg1201 and Hg1212 as a function of pressure. The sphere radii of 1.9 (1.2) for Cu (O) have been kept constant for all pressure values.

	Hg1201		Hg1201	
pressure [GPa]	Cu $d_{x^2-y^2}$	O p_x	Cu $d_{x^2-y^2}$	O p_x
0	1.4336	0.7330	1.4419	0.7330
5	1.4141	0.7375	1.4283	0.7362
10	1.3945	0.7412	1.4178	0.7400
15	1.3866	0.7466	1.4137	0.7443

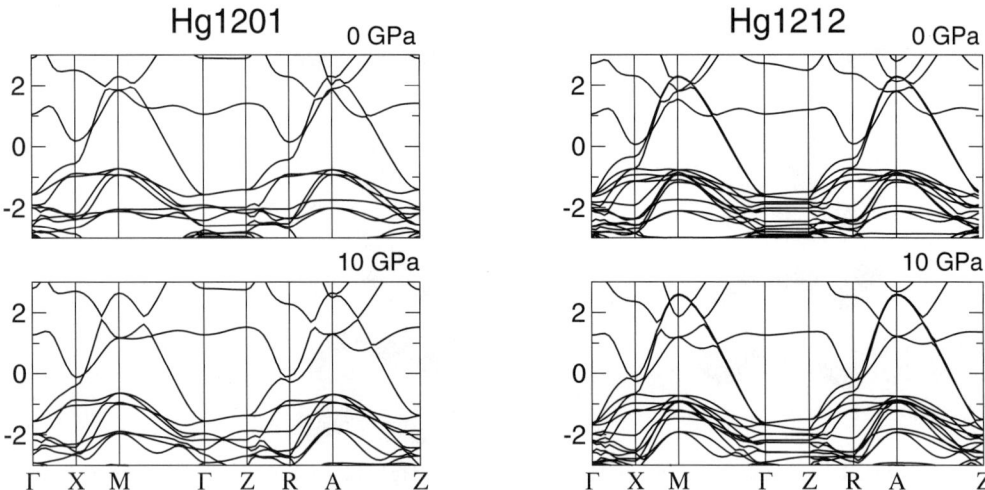

Fig. 2 Band structures for Hg1201 and Hg1212 for ambient pressure and 10 GPa, respectively. The energies are given in eV, the Fermi level has been set to zero.

values, they nearly stay the same or exhibit a marginal decrease. One way to crosscheck this finding is to look at the corresponding bands. Indeed, in the regions of the Brillouin zone, where the CuO_2 band is partially depopulated when applying pressure, it is dominated by the copper character, which explains the different behavior of Cu and O partial charges.

The band structures for both compounds are depicted for ambient pressure and 10 GPa in Fig. 2. When adding an additional CuO_2 plane the number of the corresponding bands are doubled. Focusing on the dispersive bands crossing the Fermi level between Γ and M, they are nearly degenerate in a big fraction of the BZ, e.g. along $\Gamma-M$ and $A-Z$, where in particular no visible splitting can be observed along k_z. Only between Γ and X and the corresponding line $Z-R$ there is significant splitting of a few tenth of an eV. The total band width is the same for both compounds at the same pressure, but increases with pressure by approximately 0.2 eV. The splitting of the highest valence band discussed above becomes more pronounced between Γ and M compared to the line $Z-R$ where it is hardly affected by pressure. This shoulder, however, moves up and nearly reaches the Fermi level close to R in Hg1201 and to X in Hg1212.

5 Discussion

Since the number of CuO_2 layers, pressure, and doping strongly influence the superconducting transition temperature, comparison should be made of how these three *control parameters* influence the number of charge carriers in the CuO_2 planes. Let us focus first on comparing pressure and doping. This can be done for the single layer compound Hg1201, where doping has been studied recently [1, 2]. In both cases, the number of holes in the CuO_2 planes is increased, where in case of pressure the copper charges are affected, while with doping also the oxygen p_x charges decrease significantly. As stated above, the effect of pressure is similar for the one and the two-layer material, where in the latter the amount of pressure-induced holes is smaller in one layer, but larger in total due to the second CuO_2 plane. The latter effect is also important comparing compounds with different composition: Going from Hg1201 to Hg1212, the amount of holes at the copper sites is larger in Hg1201, but due to twice the amount of Cu atoms, the total number of holes is bigger in Hg1212. Summarizing these results, we find that the number of holes induced by any of the three *control parameters* exhibits the same behavior as the superconducting transition temperature [5–7] and thus indicates the direct correlation between the hole content and T_c.

Acknowledgements The work was carried out within projects P13430 and P14004 of the Austrian Science Fund (FWF). TT acknowledges support by NSF Grant No. DMR–02–05125. Part of the calculations were performed at the Materials Simulation Center, a Penn-State MRSEC and MRI facility. We also appreciate a financial contribution from the Heinrich-Jörg-Stiftung of the University of Graz.

References

[1] C. Ambrosch-Draxl, P. Süle, H. Auer, and E. Ya. Sherman, Phys. Rev. B **67**, 100505(R) (2003).
[2] C. Ambrosch-Draxl, E. Ya. Sherman, H. Auer, and T. Thonhauser, J. Supercond. (in print).
[3] P. Blaha, K. Schwarz, G. K. H. Madsen, D. Kvasnicka, and J. Luitz, 2001, WIEN2k, Vienna University of Technology, An Augmented Plane Wave + Local Orbitals Program for Calculating Crystal Properties.
[4] C. Ambrosch-Draxl, Physica Scripta T (in print).
[5] S. N. Putilin, E. V. Antipov, O. Chmaissem, and M. Marezio, Nature **362**, 226 (1993).
[6] L. Gao, Y. Y. Xue, F. Chen, Q. Xiong, R. L. Meng, D. Ramirez, J. H. Eggert, and H. K. Mao, Phys. Rev. B **50**, 4260 (1994).
[7] Q. Xiong, Y. Y. Xue, Y. Cao, F. Chen, Y. Y. Sun, J. Gibson, and C. W. Chu, Phys. Rev. B **50**, 10346 (1994).

Magnetic resonant excitations in High-T_c superconductors

Y. Sidis[1], S. Pailhès[1], B. Keimer[2], P. Bourges[*,1], C. Ulrich[2], and L. P. Regnault[3]

[1] Laboratoire Léon Brillouin, CEA-CNRS, CE-Saclay, 91191 Gif sur Yvette, France
[2] Max-Planck-Institut für Fertkörperforschung, 70569 Stuttgart, Germany
[3] CEA Grenoble, DRFMC, 38054 Grenoble cedex 9, France

Received 1 September 2003, revised 26 February 2004, accepted 27 February 2004
Published online 14 April 2004

PACS 74.72.–h, 78.70.Nx, 75.40.Gb

The observation of an unusual spin resonant excitation in the superconducting state of various High-T_c copper oxides by inelastic neutron scattering measurements is reviewed. This magnetic mode is discussed in light of a few theoretical models and likely corresponds to a spin-1 collective mode.

© 2004 WILEY-VCH Verlag GmbH & Co. KGaA, Weinheim

More than fifteen years after the high temperature superconductivity discovery, antiferromagnetic (AF) fluctuations pairing mechanism [1] is still highly controversial. However, inelastic neutron scattering (INS) measurements have successfully brought to light the existence of unusual AF excitations that develop below T_c and could be the hallmark of an unexpected spin-1 collective mode, tightly bound to the superconducting (SC) state. Whatever the role of that mode for superconductivity, it has to be derived from the same microscopic model used to discuss superconductivity. We here review its characteristic features in a few cuprates and discuss its possible origin in light of different theoretical models.

In optimally doped $YBa_2Cu_3O_{6+x}$ (YBCO) (T_c = 93 K) where it has been discovered [2] (Fig. 1a), the spin excitation spectrum is dominated in the SC state by a sharp magnetic excitation at an energy of ~ 40 meV and at the planar antiferromagnetic wave vector $q_{AF} = (\pi/a, \pi/a)$, the so-called magnetic resonance peak [2–5]. Its intensity decreases with increasing temperature and vanishes steeply at T_c, without any significant shift of its characteristic energy E_r. In the underdoped regime, E_r monotonically decreases with decreasing hole concentration [6, 7] so that $E_r \simeq 5\ k_B T_c$ (Fig. 2). Besides, it is possible to vary T_c without changing the carrier concentration through impurity substitutions of Cu in the CuO_2 planes. This is the case in $YBa_2(Cu_{1-y}Ni_y)_3O_7$ ($y = 1\%$, $T_c = 80$ K), where the magnetic resonance peak shifts to lower energy with a preserved $E_r/k_B T_c$ ratio (Fig. 1b) [8].

In optimally doped $Bi_2Sr_2CaCu_2O_{8+\delta}$ (BSCO) (T_c = 91 K), a similar magnetic resonance peak has been observed at 43 meV (Fig. 1d) [9]. Furthermore, E_r shifts down to 38 meV in the overdoped regime (T_c = 80 K) [10], preserving a constant ratio with T_c: $E_r \simeq 5.4\ k_B T_c$ (Fig. 2). Thus, whatever the hole doping, the energy position of the magnetic resonance peak always scales with T_c. In contrast to YBCO, where the resonance peak is resolution limited in energy, the resonance peak in BSCO exhibits an energy width of ~ 13 meV. In addition, the momentum width of the excitation is twice broader. A similar energy and momentum broadening has been also reported in $YBa_2(Cu_{1-y}Ni_y)_3O_7$ [8] (Fig. 1b) and can therefore be ascribed to disorder, such as impurities or inhomogeneities. Furthermore, the observation of a spatial distribution of the SC coherence peaks in $Bi_2Sr_2CaCu_2O_{8+\delta}$ by Scanning Tunneling Microscopy measurements [11] provides evidence in favor of an intrinsic disorder in this system.

---

[*] Corresponding author: e-mail: bourges@llb.saclay.cea.fr

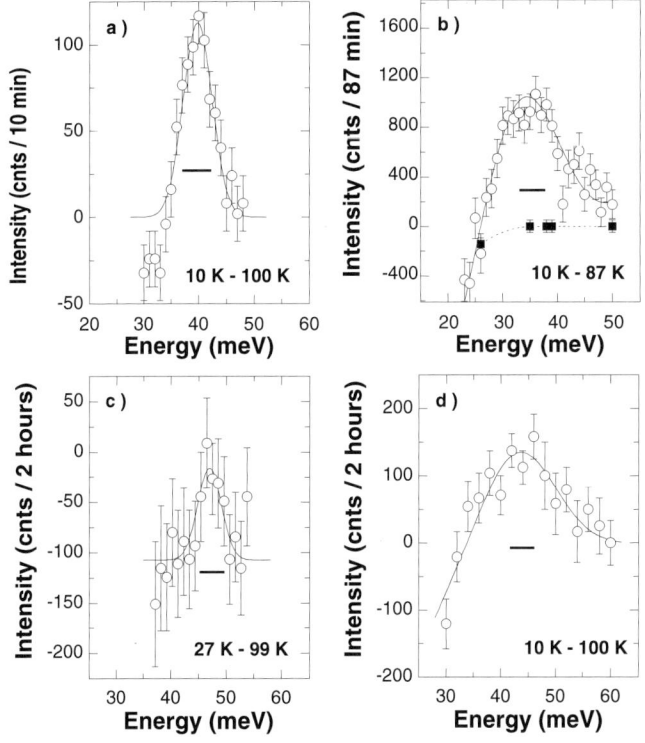

Fig. 1 Difference spectrum of the neutron intensities at low temperature, measured at the wave vector $(\pi/a, \pi/a)$ and $T \geq T_c$: a) YBa$_2$Cu$_3$O$_{6.95}$: $T_c = 93$ K, $V = 10$ cm^3 [4], b) YBa$_2$(Cu$_{1-y}$Ni$_y$)$_3$O$_7$: $T_c = 80$ K, $V \sim 2$ cm^3 [8], c) Tl$_2$Ba$_2$CuO$_{6+\delta}$: $T_c \sim 90$ K, $V = 0.11$ cm^3 [12], d) Bi$_2$Sr$_2$CaCu$_2$O$_{8+\delta}$: $T_c = 91$ K, $V = 0.06$ cm^3 [9]. Data are fitted to a Gaussian profile. The solid bar indicates the energy resolution.

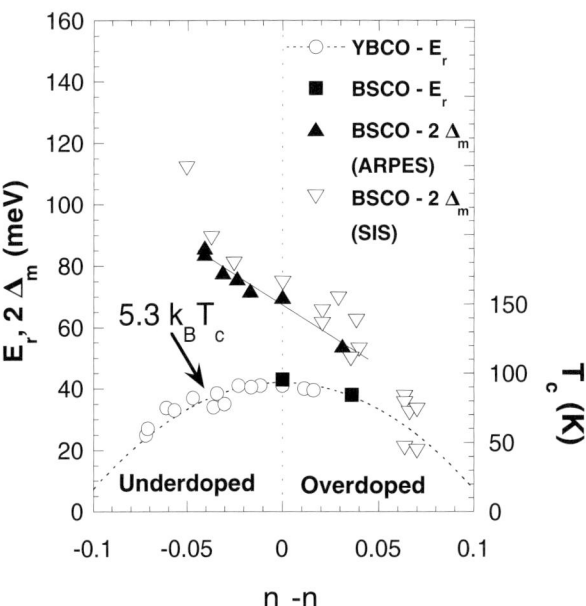

Fig. 2 Doping dependence of the energy of the magnetic resonance peak at $(\pi/a, \pi/a)$, E_r, and of twice the maximum of the single particle gap measured at low temperature, $2\Delta_m$ (usually assigned to twice the superconducting gap). E_r has been measured by INS in YBCO [2–7], and BSCO [9, 10]. $2\Delta_m$ is determined from angle resolved photo-emission spectroscopy (ARPES) [13] or superconducting-insulator-superconductor (SIS) tunneling [14] data performed in BSCO. The doping level is estimated through the empirical relation: $T_c/T_c^{max} = 1 - 82.6(n_h - n_{opt})^2$ [15].

Further, the magnetic resonance peak has been observed in optimally doped $Tl_2Ba_2CuO_{6+\delta}$ (T_c = 90 K) at E_r = 47 meV (Fig. 1c) [12]. This yields a ratio $E_r/k_BT_c \simeq 6$ slightly larger than $YBa_2Cu_3O_{6+x}$. Nevertheless, as in $YBa_2Cu_3O_7$, the excitation is limited by the resolution in energy and displays a momentum width of 0.25 Å^{-1} (half width at half maximum). Meanwhile, the energy integrated intensity of the magnetic resonance peak is almost the same in both systems: 0.7–0.8 μ_B^2 eV^{-1}/CuO$_2$ plane. Thus, the magnetic resonance peak appears as a common excitation to the SC state of all High-T_c superconductors, investigated so far by INS measurements, whose maximum T_c can be as high as ~ 90 K. Furthermore, the existence of this excitation does not depend on the number of CuO$_2$ planes per unit cell: one for $Tl_2Ba_2CuO_{6+\delta}$ and two for $YBa_2Cu_3O_{6+x}$ and $Bi_2Sr_2CaCu_2O_{8+\delta}$.

While the magnetic resonance peak exists in cuprates whose maximum T_c is about 90 K, it has not yet been reported in the mono-layer system $La_{2-x}Sr_xCuO_4$ with a maximum of T_c of ~ 40 K. Furthermore, the magnetic excitations in that compound are rather strong even in the normal state and located at incommensurate planar wave vector $Q_{mag} = (\pi/a(1 \pm \delta_{inc}), \pi/a)$ and $(\pi/a, \pi/a(1 \pm \delta_{inc}))$ [16]. This is in a marked contrast with the systems mentioned above, for which the normal state magnetic fluctuations (if observable) remain centered around $(\pi/a, \pi/a)$. However, passing through T_c, the incommensurate spin fluctuations of $La_{2-x}Sr_xCuO_4$ are enhanced and become narrower in momentum space, in an energy range which is about $5k_BT_c$ [17]. This phenomenon, usually referred to as a "coherence effect" and the resonance peak could eventually share a common origin.

In underdoped $YBa_2Cu_3O_{6+x}$ (x = 0.6, T_c = 63 K, E_r = 34 meV), INS measurements provide evidence for incommensurate-like spin fluctuations at 24 meV and low temperature (seemingly similar to those observed $La_{2-x}Sr_xCuO_4$) [18]. These incommensurate-like spin fluctuations are also observed at higher oxygen concentrations: x = 0.7 [19], x = 0.85 [20]. As a function of temperature and energy [20], the incommensurability (δ_{inc}) increases below T_c with decreasing temperature and decreases upon approaching E_r in the SC state (Fig. 3a). The simultaneous disappearance of δ_{inc} at E_r and T_c indicates that the resonance peak and the incommensurate-like spin fluctuations are intrinsic features of the SC state and that they can be viewed as continuously connected (Fig. 3a). In other words, these results lead to an unified description of both the incommensurate spin excitations and the magnetic resonance peak in terms of a unique collective spin excitation mode with a downward dispersion [20]. Recently, the actual symmetry of this dispersion for an optimally doped YBCO sample has been looked at carefully [21] and was found basically circular within the 2D copper-oxygen plane. In addition, a second magnetic mode with much weaker intensity is reported dispersing upward above the $(\pi/a, \pi/a)$ peak [21]. The deep underdoped state YBCO$_{6.5}$ has been also recently re-investigated in partly detwinned sample with an ortho-II structure [22]. In contrast to the dispersive mode picture, it is claimed [22] that the low energy magnetic excitations are essentially one-dimensional as expected for hydrodynamic stripes. Therefore, the detailed doping dependence of the spin fluctuations needs to be clarified to reconcile these conclusions by studying fully detwinned samples. Indeed, one needs to determine the specific role of the Cu–O chains in YBCO for the magnetic anisotropy.

The main difference between mono-layer and bilayer systems shows up in the momentum dependence of the magnetic resonance peak along the (0 0 1) direction. In $Tl_2Ba_2CuO_{6+\delta}$, the excitation remains purely bidimensional. In contrast in bilayer systems, the two CuO$_2$ planes correlate antiferromagnetically within the bilayer. That interlayer AF coupling is responsible in insulating parent compounds for both acoustic and optic magnons, whose counterpart in the metallic state are the odd (o) and even (e) excitations. The neutron scattering cross section then reads [6]:

$$\frac{d^2\sigma(Q,\omega)}{d\Omega\, d\omega} \propto \sin^2(Q_z d/2)\, \text{Im}\,[\chi_o(Q,\omega)] + \cos^2(Q_z d/2)\, \text{Im}\,\chi_e(Q,\omega)], \tag{1}$$

where Im$[\chi_{o,e}(Q,\omega)]$ corresponds to the imaginary part of the dynamical magnetic susceptibility in each channel and d (= 3.3 Å) stands for the distance between CuO$_2$ planes within the bilayer. Intuitively, one could expect a splitting of the magnetic resonance peak under the interlayer AF coupling, leading to a magnetic resonance in each channel. For a long time, the magnetic resonance peak was observed only in the odd channel. However, we could recently observe a resonant mode in each channel in slightly overdoped YBCO through 10% substitution of Y by Ca [23]. They occur at two different energies E_r^o = 36 meV and E_r^e = 43 meV and the even mode exhibits an intensity one third times less than the odd one. The question why the even mode is now sizeable in this overdoped regime and not in previous studies remains open as it could be simply due to improvement of neutron instruments. However, it might also be related to the electronic transport between closely spaced CuO_2 layers which becomes coherent in the overdoped regime, as demonstrated by recent experiments showing well-defined bonding and antibonding bands.

Considering the different models for the magnetic resonance peak, we focus on models where electron-electron interactions play the central role, despite the still possible existence of electron-phonon couplings in cuprates. Essentially, there is no indication of an effect of the lattice on the magnetic resonance peak. Secondly, we consider here models where the downward dispersion of the resonant mode would naturally emerge. For instance, approaches [24, 25] which associate the resonance peak to a pre-existing soft mode reminiscent of nearby (commensurate or incommensurate) AF phase would yield a collective mode dispersing predominantly upward.

The existence of a spin 1-collective in d-wave superconductors such as high-T_c cuprates, is derived from an itinerant description of the magnetic properties of the system and of strong correlation effects ([26–31] and references therein). This leads to a particle-hole (p–h) bound state, usually referred as a spin exciton. In these strong coupling models (see also Refs. [32, 33] in the framework of the $t-J$ model), the generalized spin susceptibility $\chi(q,\omega)$ is expressed as a function of the non interaction spin susceptibility $\chi_0(q,\omega)$ and the magnetic interaction. $\chi(q,\omega)$ has an RPA-like form:

$$\chi(q,\omega) = \frac{\chi_0(q,\omega)}{1 + J(q)\chi_0(q,\omega)}, \qquad (2)$$

where $J(q) = 2J(\cos(q_x) + \cos(q_y))$ is the intra-plane AF super-exchange coupling and $\chi_0(q,\omega)$ describes in an itinerant system the continuum of spin flip particle–hole (p–h) excitations, given by the Lindhard function in the normal state or the BCS function in the SC state. In the SC state, due to the opening of the SC gap, the continuum becomes gapped (Fig. 3b) below a threshold energy at $\omega_c = 2\Delta_{k_s}$, where Δ_k is the momentum dependent superconducting d-wave energy gap and k_s is the so-called hot-spot wave vector defined as both k_s and $k_s + Q_{AF}$ are lying on the Fermi surface. In addition to the p–h excitations within the continuum, a spin triplet p–h bound state can form below the continuum due to the AF interaction. In Eq. (2), the dynamical Stoner criterion, i.e. $1 + J(q)\,\mathrm{Re}\,[\chi_0(q,\omega)] = 0$, is then fulfilled for an energy smaller that the threshold of the continuum at wave vector q. This spin 1-collective mode is characterized by a downward dispersion controlled by the momentum dependences of the continuum threshold and the magnetic interaction (Fig. 3b). The mode vanishes when approaching the continuum by changing wave vector from $(\pi/a, \pi/a)$. The mod-

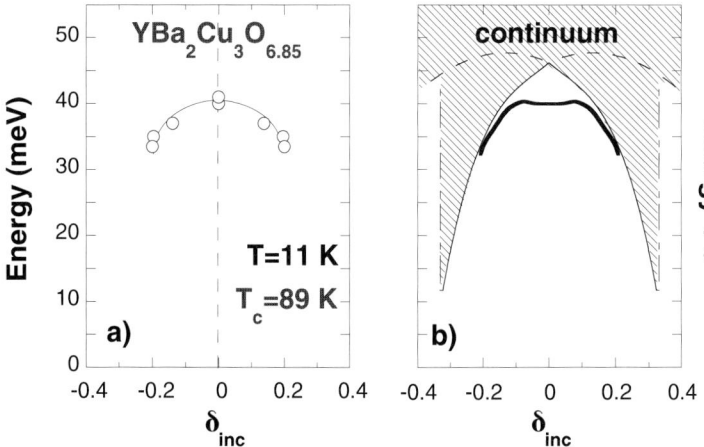

Fig. 3 a) Dispersion of the magnetic resonance peak as a function $q = (\pi/a(1 \pm \delta_{\mathrm{inc}}), \pi/a)$, measured in $YBa_2Cu_3O_{6.85}$ [20], b) dispersion of the spin exciton [31] (thick line) below the continuum which is represented by the dashed area. The maximum of the dashed lines correspond to $2\Delta_m$, and their crossing defines $\omega_c = 2\Delta_{k_s}$.

el also predicted an excitation dispersing upward within the continuum [29, 31] which might correspond to the recently observed high energy mode [21].

From the experimental point of view, a major challenge for future INS experiments would be to observe the magnetic continuum. So far, one can deduce the continuum threshold, ω_c, from (i) the measurement of the maximum of the SC gap, Δ_m (determined by angle resolved photo-emission spectroscopy [13], by the measurement of the B_{1g} mode in Raman scattering [34] or by tunneling data [14], see Fig. 2) and ii) from the Fermi surface topology [35]. At optimal doping in BSCO, one can show that the magnetic resonance peak lies well below the continuum: $\Delta_m \simeq 35$ meV, $\omega_c \simeq 1.8\Delta_m$ and $E_r \simeq 1.2\Delta_m$. From optimal doping to the overdoped regime, one may expect ω_c to reach the limit $\sim 2\Delta_m$. Simultaneously, the ratio $2\Delta_m/k_B T_c$ evolves from 7–8 to 5–6, while according to INS data, the ratio $E_r/k_B T_c$ seems to be preserved (Fig. 2). These evolutions suggest that in the overdoped regime the binding energy of the spin exciton, $\omega_c - E_r$, weakens, leading to the possible disappearance of the spin 1-collective mode. Further INS experiments, in the deeply overdoped regime are required to test such a possibility.

For a bilayer system [26, 27, 36], the interlayer AF coupling, $J_\perp$, is treated in perturbation, such that the odd (o) and even (e) spin susceptibilities are given by:

$$\chi_o(q,\omega) = \frac{\chi(q,\omega)}{1 - J_\perp \chi(q,\omega)} \quad \text{and} \quad \chi_e(q,\omega) = \frac{\chi(q,\omega)}{1 + J_\perp \chi(q,\omega)}. \tag{3}$$

In Fig. 4, we calculated these susceptibilities with a realistic value of $J_\perp$ ($\sim 0.1J$) [37] and with other parameters identical to those of Ref. [31]. Because of $J_\perp$, the odd spin exciton is likely pushed to lower energy, while the even one merges into the continuum. Actually, this explains why the odd channel mode is mainly observed as well as why the resonant mode appears to shift to lower energy in a bilayer system in contrast with a mono-layer system. Going further, their respective spectral weights $W_{o,e}$ are predicted to be approximately proportional to their binding energies [27] as $W_{o,e} \sim \omega_c - E_r^{o,e}$. This is found in the calculated susceptibilities in Fig. 4 and sketched in the figure inset. Using this property in overdoped YBCO-Ca [23], one could directly estimate of continuum threshold at $\omega_c(Q_{AF}) \simeq 49$ meV.

If the spin exciton scenario provides a plausible explanation for the resonance peak around the optimal doping, more theoretical work is needed to fully account for its evolution as a function of hole doping. In particular, there is no explicit relationship between the energy of the collective mode

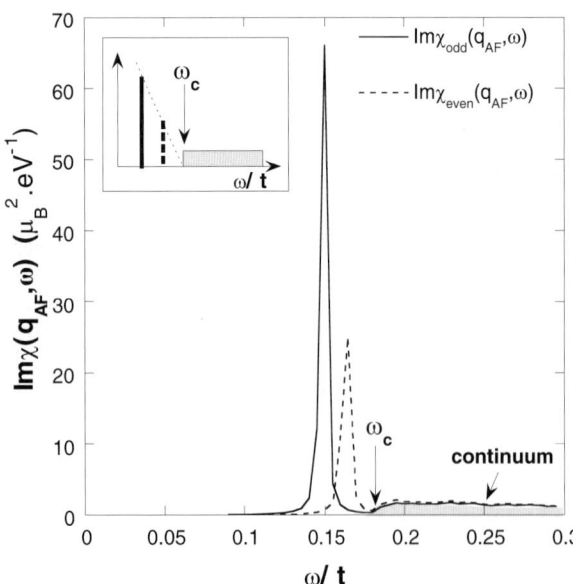

Fig. 4 Calculated spin susceptibility for both odd and even symmetry according to Eq. (3) with an interlayer coupling of $J_\perp \simeq 0.1\ J$ and with the band structure parameters and $t/J = 2$ of Ref. [31] and $t = 250$ meV. The dotted areas show the weak electron-hole continuum intensity.

at $(\pi/a, \pi/a)$ and the value of T_c. The phenomenological relationship $E_r/k_B T_c \simeq 5-6$ remains to be explained. Furthermore, the same model must simultaneously describe the unusual features observed below T_c as well as the spin excitation spectrum of the normal state. In the underdoped regime, the magnitude and the energy dependence of spin fluctuations observed by INS are still difficult to reproduce quantitatively. Thus, most likely, the model needs to go beyond a purely itinerant picture in order to capture the deep underdoped state.

However the spin-exciton interpretation for the magnetic neutron resonance and its downward dispersion is not unique. Indeed, besides such an approach the spin 1-collective mode corresponds to a p–h bound state, it has also been proposed that the magnetic resonance peak could be a p–p pair, the π-excitation, in SO(5) model (spin itinerant picture) or a magnon-like excitation (spin localized picture) of a disordered stripe phase.

The SO(5) model [38] considers as a starting point that, in the AF insulating state of cuprates, there exists a super-symmetry that allows the system to switch from the AF state to the d-wave SC state. This symmetry involves the existence of a Goldstone mode, the π-excitation [39]. This excitation can be depicted as an excitation that creates a p–p pair carrying a charge $2e$, a spin $S = 1$ and with total momentum $(\pi/a, \pi/a)$. Upon doping the symmetry is broken, the π-excitation survives, but becomes massive. While it exists already in the normal state, it can only be observed by INS in the SC state due to the p–h mixing. Its characteristic energy is roughly linear with hole doping and its intensity in the SC state scales with $|\Delta_m|^2$. Moreover, a downward dispersion is obtained due to by the phase slips of the SC order parameter induced by the propagation of the π-excitation [40]. Such an approach accounts for several features observed by INS, especially in the underdoped regime, where T_c and E_r increase with hole doping. The π-excitation should decrease in magnitude, but remains almost at the same energy, as observed experimentally. However this scenario cannot explain the decrease of E_r in the overdoped regime. Furthermore, recent calculations have shown that, if the π-excitation existed, it should be observed at high energy, above $2\Delta_m$ [41]: this casts some doubt about the interpretation of the resonance as a π-triplet excitation.

Alternatively, the stripe model considers that in a $S = 1/2$ AF Heisenberg system, doped holes segregate to form lines of charges, separating AF domains in anti-phase. The metallic state is viewed as a disordered stripe phase, where charged lines can fluctuate. While there is not a general interpretation of the magnetic resonance peak in stripe models, it has been for instance proposed that the resonance peak and the incommensurate spin fluctuations observed by INS in the SC state could be viewed as magnon-like excitations reminiscent of the ordered stripe phase [42, 43] or could correspond to the eigen magnetic modes of the liquid stripe phase [44]. Magnons, developing symmetrically around the magnetic incommensurate wave-vector, Q_{mag}, of the stripe ordered phase and merging at $(\pi/a, \pi/a)$, actually describe correctly the spin dynamics observed in stripe-ordered nickelates [45] as predicted in the spin-only model [42, 43]. In cuprates, the lack of symmetric peaks around the incommensurate wave-vector Q_{mag} (see Fig. 3) does not seem to validate these approaches. In addition, this model, if interesting, fails to explain why the resonance peak and the incommensurate spin fluctuations exist basically only in the SC state. Independently, it could be also interesting to understand how, in bilayer compounds, the adjacent CuO_2 planes succeed in accommodating the Coulomb repulsion between charged lines and the AF interlayer coupling. This is a central issue to account for as the magnetic resonance peak exists mostly in the odd channel.

Finally, INS experiments have shown the existence of an unusual enhancement of spin fluctuations in the SC state around the vector $(\pi/a, \pi/a)$ and at an energy E_r which is found experimentally to scale with T_c. Combined with the observation at lower energy of incommensurate spin fluctuations, that develop also below T_c, INS data point toward the existence of a dispersive spin 1-collective mode deep inside the SC state. The observation of that mode, first discovered in $YBa_2Cu_3O_7$, has been then extended to other systems with one or two CuO_2 planes per unit cell, such as $Bi_2Sr_2CaCu_2O_{8+\delta}$ and $Tl_2Ba_2CuO_{6+\delta}$. This establishes the magnetic resonance peak as a generic excitation of the SC state of cuprates whose maximum T_c can be as high as 90 K. In the strongly under- and overdoped regimes ($T_c \leq 50$ K) or in other cuprate families with lower maximum T_c (such as $La_{2-x}Sr_xCuO_4$), the observation of the spin 1-collective mode (if any) still requires more experimental work. In any case, the observation of such an excitation, thanks to inelastic neutron scattering, is one of the most persuasive experimental indications of the crucial role of magnetic interactions in the physics of high-T_c copper oxides.

References

[1] See e.g. D. Scalapino, Phys. Rep. **250**, 329 (1995).
[2] J. Rossat-Mignod, L. P. Regnault, C. Vettier, P. Bourges, P. Burlet, J. Bossy, et al., Physica C **185–189**, 86 (1991).
[3] H. A. Mook, M. Yethiraj, G. Aeppli, T. E. Mason, and T. Armstrong, Phys. Rev. Lett. **70**, 3490 (1993).
[4] H. F. Fong, B. Keimer, P. W. Anderson, et al., Phys. Rev. Lett. **75**, 316 (1995); Phys. Rev. B **54**, 6708 (1996).
[5] P. Bourges, L. P. Regnault, Y. Sidis, and C. Vettier, Phys. Rev. B **53**, 876 (1996).
[6] H. F. Fong, P. Bourges, Y. Sidis, L. P. Regnault, J. Bossy, A. S. Ivanov, et al., Phys. Rev. B **61**, 14774 (2000).
[7] P. Dai, H. A. Mook, R. D. Hunt, F. Doğan, Phys. Rev. B **63**, 054525 (2001).
[8] Y. Sidis, P. Bourges, H. F. Fong, B. Keimer, L. P. Regnault, J. Bossy, et al., Phys. Rev. Lett. **86**, 4100 (2001).
[9] H. F. Fong, P. Bourges, Y. Sidis, L. P. Regnault, A. S. Ivanov, G. D. Gu, et al., Nature **398**, 588 (1999).
[10] H. He, Y. Sidis, Ph. Bourges, G. D. Gu, A. Ivanov, N. Koshizuka, B. Liang, C. T. Lin, L. P. Regnault, E. Schoenherr, and B. Keimer, Phys. Rev. Lett. **86**, 1610 (2001).
[11] K. M. Lang, V. Madhavan, J. E. Hoffman, E. W. Hudson, H. Eisaki, S. Uchida, and J. C. Davis, Nature **415**, 412 (2002).
[12] H. He, P. Bourges, Y. Sidis, C. Ulrich, L. P. Regnault, S. Pailhès, et al., Science **295**, 1045 (2002).
[13] J. Mesot, M. R. Norman, H. Ding, M. Randeria, J. C. Campuzano, et al., Phys. Rev. Lett. **83**, 840 (1999).
[14] J. F. Zasadzinski, L. Ozyuzer, N. Miyakawa, et al., Phys. Rev. Lett. **87**, 067005 (2001) (cond-mat/0102475).
[15] J. L. Tallon, C. Bernhard, H. Shaked, R. L. Hitterman, and J. D. Jorgensen, Phys. Rev. B **51**, 12911 (1995).
[16] G. Aeppli, T. E. Mason, S. M. Hayden, H. A. Mook, and J. Kulda, Science **278**, 1432 (1997).
[17] T. E. Mason, A. Schröder, G. Aeppli, H. A. Mook, and S. M. Hayden, Phys. Rev. Lett **77**, 1604 (1996).
[18] H. A. Mook, P. Dai, S. M. Hayden, G. Aeppli, T. G. Perring, and F. Doğan, Nature **395**, 580 (1998).
[19] M. Arai, T. Nishijima, Y. Endoh, T. Egami, S. Tajima, K. Tomimoto, et al., Phys. Rev. Lett. **83**, 608 (1999).
[20] P. Bourges, Y. Sidis, H. F. Fong, L. P. Regnault, J. Bossy, A. Ivanov, and B. Keimer, Science **288**, 1234 (2000).
[21] D. Reznik, P. Bourges, L. Pintschovius, Y. Endoh, et al., submitted to Phys. Rev. Lett. (cond-mat/0307591).
[22] C. Stock, W. J. L. Buyers, R. Liang, D. Peets, Z. Tun, D. Bonn, W. N. Hardy, and R. J. Birgeneau, Phys. Rev. B **69**, 014502 (2004) (cond-mat/0308168).
[23] S. Pailhès, Y. Sidis, P. Bourges, C. Ulrich, V. Hinkov, L. P. Regnault, A. Ivanov, B. Liang, C. T. Lin, C. Bernhard, and B. Keimer, Phys. Rev. Lett. **91**, 237002 (2003) (cond-mat/0308394).
[24] D. K. Morr and D. Pines, Phys. Rev. Lett. **81**, 1086 (1998).
[25] S. Sachdev, C. Buragohain, and M. Vojta, Science **286**, 2479 (1999).
[26] D. Z. Liu, Y. Zha, and K. Levin, Phys. Rev. Lett. **75**, 4130 (1995).
[27] A. J. Millis and H. Monien, Phys. Rev. B **54**, 16172 (1996).
[28] A. Abanov and A. V. Chubukov, Phys. Rev. Lett. **83**, 1652 (1999).
[29] M. N. Norman, Phys. Rev. B **61**, 14751 (2000).
[30] D. Manske, I. Eremin, and K. H. Bennemann, Phys. Rev. B **63**, 054517 (2001).
[31] F. Onufrieva, and P. Pfeuty, Phys. Rev. B **65**, 014502 (2002) (cond-mat/9903097).
[32] J. Brinckmann and P. A. Lee, Phys. Rev. B **65**, 014502 (2002).
[33] I. Sega, P. Prelovšek, and J. Bonča, Phys. Rev. B **68**, 054524 (2003).
[34] For a recent review, see e.g. M. Cardona, Physica C **318**, 30 (1999).
[35] H. Ding, M. R. Norman, T. Yokoya, T. Takeuchi, M. Randeria, et al., Phys. Rev. Lett. **78**, 2628 (1997).
[36] T. Li and Z. Gan, Phys. Rev. B **60**, 3092 (1999).
T. Li, Phys. Rev. B **64**, 012503 (2001).
[37] D. Reznik, P. Bourges, H. F. Fong, L. P. Regnault, J. Bossy C. Vettier, D. L. Milius, I. A. Aksay, and B. Keimer, Phys. Rev. B **53** R14741 (1996).
[38] S. C. Zhang, Science **275**, 1089 (1997).
[39] E. Demler, H. Kohno, and S. C. Zhang, Phys. Rev. B **58**, 5719 (1998).
[40] J. P. Hu and S. C. Zhang, Phys. Rev. B **64**, 100502 (2001).
[41] O. Tchernychyov, M. R. Norman, and A. V. Chubukov, Phys. Rev. B **63**, 144507 (2001).
[42] C. D. Bastista, G. Ortiz, and A. V. Balatsky, Phys. Rev. B **65**, 180402 (2002).
[43] F. Krüger and S. Scheidl, Phys. Rev. B **67**, 134512 (2003).
[44] N. Hasselmann, A. H. Castro Neto, C. Morais Smith, and Y. Dimashko, Phys. Rev. Lett. **82**, 2135 (1999).
[45] P. Bourges, Y. Sidis, M. Braden, K. Nakajima, and J. M. Tranquada, Phys. Rev. Lett. **90**, 147202 (2003).

Coexistence of superconductivity and magnetism in low dimensional conductors

A. BenAli[1], S. Charfi-Kaddour[1], C. Pasquier[2], R. Bennaceur[1], and M. Héritier*,[2]

[1] Laboratoire de Physique de la Matière Condensée, Faculté des Sciences de Tunis, Tunisia
[2] Laboratoire de Physique des Solides, UMR CNRS-Paris XI, Bat. 510, 91405 Orsay, France

Received 1 September 2003, accepted 17 March 2004
Published online 23 April 2004

PACS 74.25.Dw, 75.30.Fv, 75.50.Ee

Coexistence of superconductivity and antiferromagnetism in low dimensional conductors can be explained by a phase segregation scenario: (i) In quasi 1D or 2D conductors in which the (T, pressure) phase diagram exhibits a border line between superconducting and spin density wave (SDW) phases, it is favourable to create a phase segregation in the direct space by forming alternatively superconducting and magnetic slabs (ii) a phase segregation may also occur in reciprocal space. A $d_{x^2-y^2}$-wave superconducting order is established, with lines of zero gap on the Fermi surface, as in cuprates. In the node region, superconductivity is weak enough to leave room for a confined SDW gap. These models can explain consistently various recent experimental data observed in low dimensional conductors.

© 2004 WILEY-VCH Verlag GmbH & Co. KGaA, Weinheim

1 Introduction

Competition between a magnetic state and a superconducting state is one of the key features in the domain of strongly correlated electrons. This point is extensively studied in layered superconductors, such as the high-T_c copper oxide superconductors and heavy fermion compounds. Such a competition is also common in organic materials [1]. In all these systems, coexistences of antiferromagnetic phases and superconducting phases have been observed. These phenomena are unexpected in conventional superconductors, since magnetic orders are usually known to break the Cooper pairs. However, an easy way to understand these phenomena is to assume a phase segregation between the antiferromagnetic phase and the superconducting phase, thus preserving each type of ordering in its own part of the phase space. Such a segregation can occur either in the direct space or in the reciprocal phase. We believe that these two types of phase coexistence or phase segregation have been recently observed in low dimensional conductors and can explain consistently various recent experimental data in cuprate high T_c superconductors, as well as in organic superconductors.

2 Segregation in the direct space

Recently, the phase diagram of the Bechgaard salts has been revisited, with careful determination of the applied pressure [2]. The pressure dependence of the resistivity and of the critical current in $(TMTSF)_2PF_6$ revealed the simultaneous formation of two different phases: a Spin Density Wave phase (SDW) coexisting with a metal phase and, at lower temperature, a SDW phase coexisting with a superconducting (SC) one. Such a coexistence corresponds, in fact, to a segregation in the direct space. It is,

---

* Corresponding author: e-mail: heritier@lps.u-psud.fr

indeed, quite plausible that such a segregation is produced on a macroscopic scale λ, much larger than the ordering correlation length ξ, which is much more favorable to the carrier localisation energy, but also to the interface energy between regions of different orders. We have to study the relative stabilities of different phases: the metal, the spin density wave (SDW), the superconducting phase (SC), but also possible phases in which coexist either a SDW phase and a metallic phase, or, at lower temperature, a SDW phase and a SC one. To discuss the relative stabilities of these phases, we describe the quasi-one dimensional electron gas of the organic conductor by an open Fermi surface with strong nesting properties. The non-interacting electron dispersion relation $\varepsilon(k)$ includes a perfect nesting term $\varepsilon_0(k)$, with a transverse tunneling integral t_b much lower than the longitudinal one t_a, and an anti-nesting term $-2\,t'_b \cos 2\,k_y b$, where b is the lattice parameter in the transverse direction.

Simple and general arguments prove that, near enough to the critical line for the formation of an homogenous SDW phase, a spatially heterogenous phase has a lower free energy than the homogenous SDW phase, especially when the Fermi surface exhibits strong nesting properties. The origin of such a phenomenon is due to the following features: (i) the relevant quantity for the SDW order stability is the departure from perfect nesting energy parameter t'_b. Applying a pressure increases t'_b and, therefore, the SDW free energy, up to a critical value $t'_b{}^*$ at which the homogenous SDW phase disappears; (ii) the SDW stability decreases very strongly near $t'_b{}^*$; (iii) the relevant quantity to stabilize the SDW phase near $t'_b{}^*$ is b, the unit cell parameter along the transverse direction. Increasing b strongly lowers the magnetic free energy; (iv) it is always favourable to create a heterogenous phase: one part has a cell parameter $b + \delta b_1$ and is magnetic ; the other part is metallic and has a cell parameter $b - \delta b_2$. The energy cost for such a deformation is proportional to $(\delta b)^2$, while the deformation allows a first order gain of magnetic free energy. In a Fermi liquid approach for an almost perfectly nested Fermi surface, we obtain the phase diagramm schematized in Fig. 1. The shaded areas correspond to inhomogenous phase in which coexist a SDW and a metal, and, at lower temperature a SDW and a superconducting phase. The pressure range of coexistence estimated from this model, using the usual values of the parameters is about 1 kbar, which is in very good agreement with the experimental data of Vuletic et al.

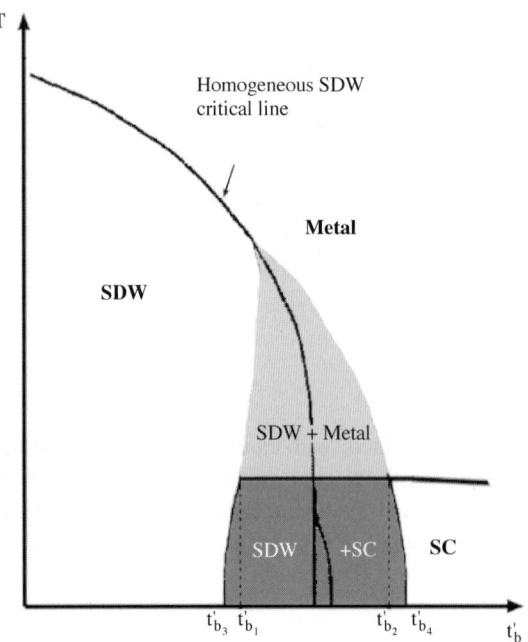

Fig. 1 Phase diagram of a quasi-one-dimensional conductor, such as (TMTSF)$_2$PF$_6$ in the (Temperature, t'_b) plane, where t'_b is the amplitude of the antinesting term (see the text). The t'_b axis can be considered as a pressure axis. The shaded areas are regions of coexistence of SDW and metal (lighter area) or of SDW and SC (darker area).

3 Segregation in the reciprocal space

3.1 Introduction

During the last few years, several authors have given experimental evidences for a d-wave symmetry of the order parameter in various high T_c superconductors, in particular in Bi2212. The presence of nodes in the superconducting gap allows the existence of quasiparticles, even at low temperature, offering the opportunity of testing the quasiparticle properties. Experimental results have brought a large number of new insights concerning quasiparticle behaviour, in particular in a magnetic field. When a magnetic field is applied perpendicular to the CuO_2 plane, the thermal conductivity shows a sharp first order transition to a field independent regime indicating the elimination of the nodes [3]. Thermal transport measurements from other groups have confirmed these findings, although the phenomenon seems sample dependent. Indications of a field induced transition are also present in other experiments, such as superfluid density measurements [4]. Moreover, neutron scattering experiments reported the generation of AF moments in the SC phase by a magnetic field perpendicular to the planes [5].

3.2 Theoretical model

A two-dimensional system of quasiparticles exhibits original properties when a magnetic field is applied perpendicularly to its plane. The orbital effect of the field makes the quasiparticle motion strictly one-dimensional. The result is a SDW instability induced by the field and the opening of a gap at the Fermi level [6]. The intensity of the relevant magnetic field is determined by the Fermi surface geometry and such a field can be available in the laboratory only when this geometry exhibits good nesting. This is the case for various high T_c superconductors, in particular Bi2212. We interpret the data as manifestations of a field induced phase transition from a $d_{x^2-y^2}$ SC phase to a phase in which $d_{x^2-y^2}$ SC coexists with a confined density wave which develops in the gap node of the Fermi surface [7], as schematized in Fig. 2.

Our HTCS system is built of two subsystems: subsystem I is the Fermi surface region covered by the superconducting gap and subsystem II is a virtual normal quasiparticle region created by the magnetic field. The confined field-induced SDW (FISDW) will develop because of the orbital effect of the field in this region. In region II, we necessarily have open Fermi surface sheets, since, indeed the Fermi surface, in this part of the k-space, is strictly similar to that of a quasi-one-dimensional conductor. We can calculate the spin susceptibility of subsystem II in the presence of a magnetic field assuming that a Fermi liquid approach is valid. This is certainly justified at low temperature in a region of the phase diagram where the quasiparticles are well defined. Details of calculation will be given elsewhere. A first order transition to a state in which a $d_{x^2-y^2}$ SC coexists with a confined FISDW is expected. Such a theoretical description seems to explain various experimental data, for which, as far as we know, no clear theoretical interpretation has been given so far.

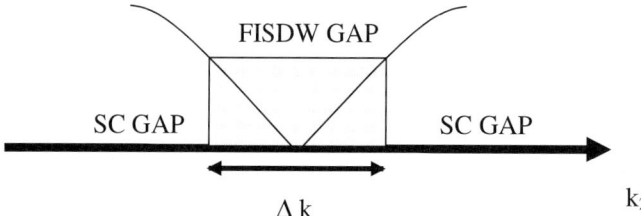

Fig. 2 In a quasi-two-dimensional d-wave superconductor, a phase segregation can be formed in the reciprocal space: the superconducting order is developed in the k-space region where the order parameter is large, but it leaves room for a SDW order parameter in the region of the superconducting nodes.

3.3 Experimental evidences

Such a theoretical model, indeed, predicts various original phenomena, which have been actually observed in quasi-two-dimensional d-wave superconductors, such as the cuprates.

3.3.1 Superfluid density

Since the superconducting order is destroyed in subsystem II when a magnetic field is applied perpendicularly to the conducting planes, we expect a field induced decrease of the superfluid density, as observed by the penetration length measurements of Sonier et al. [4], which clearly indicates a decrease of the superfluid density above a critical field of a few Teslas.

3.3.2 Field induced magnetic moments

The present theory also predicts the formation of a spin density wave outside of the vortex cores. The magnetic field penetrates in the sample by forming vortices, up to a distance from the vortex core given by the penetration length λ, which is much larger than the superconducting coherence length ξ in these extreme type II superconductors. Therefore, on this length λ, the formation of oscillating magnetic moments, with a wave vector corresponding to the Fermi surface nesting vector, is expected. Such a prediction seems consistent with the recent observation of field induced antiferromagnetism in neutron experiments by Lake et al. [5].

3.3.3 Disappearance of low temperature quasiparticles

Since, according to this model, a SDW gap should be formed in the region of the k-space where the superconducting order parameter exhibits lines of zero, we should expect that the field induced magnetic order indeed opens a gap on the whole Fermi surface. The superconducting gap nodes should disappear, inducing a disappearance of the low temperature quasiparticle density. This might explain the phenomenon observed by Krishana et al. [3] in the thermal conductivity of Bi2212. According to these authors, when a magnetic field, perpendicular to the conducting planes and larger then a threshold field, is applied, a plateau of thermal conductivity, independent of the applied field, is observed. This has been interpreted as due to the disappearance of low temperature quasiparticles. The opening of the SDW gap in the region of the superconducting nodes might explain this phenomenon. However, one should be cautious in this interpretation of the data. It is still controversial, since the effect observed by Krishana et al. [3] seems definitely sample dependent. This sample dependence might reflect the fact that a very high sample quality is required to observe this quantum orbital effect. When the field-induced SDW involves the entire Brillouin zone, the condition for the existence of well defined quantized orbits is $\omega_c \tau \gg 1$, where ω_c is the cyclotron frequency and τ is the electron relaxation time. Such a condition is already difficult to fulfill in poor quality samples. In our case, where the field-induced SDW is confined in a part of the Brillouin zone, the condition for the observation of the quantum effect should be even more difficult to fulfill, particularly in cuprate samples. This might explain the difficulty of observing reproducible quantum effects in the thermal conductivity of cuprates. One should also insist on the fact that these various experimental data have not been observed in a single sample, but rather on various high Tc superconducting cuprates. However, it seems clear that all these phenomena are generic of this class of quasi-two-dimensional d-wave superconductors.

4 Conclusion

We have proposed a model explaining coexistence phenomena of antiferromagnetism and superconductivity in different systems of strongly correlated fermions by two kinds of phase segregation. In the first one, it is proposed that separate domains in real space are formed, in which an antiferromagnetic phase and a superconducting phase order separately in its own domain. Such a segregation model relies on very general and simple arguments and can certainly be extended to other coexistence phenomena. In the second one, the two different orderings avoid themselves by a segregation in the reciprocal space. There is a partition of the k-space in which, around the superconducting nodes, a SDW order is established, while the rest of the Brillouin zone, a superconducting order persists. Such a description seems consistent with various experimental observations, which, as far as we know are still not well understood.

Acknowledgement We acknowledge the financial support from CMCU to project 01/F1303.

References

[1] Hito et al., J. Phys. Soc. Jpn. **65**, 2987 (1996).
    S. Lefebvre et al., Phys. Rev. Lett. **85**, 5420 (2000).
[2] T. Vuletic et al., Eur. Phys. J. B **25**, 319 (2002).
[3] K. Krishana et al., Science **277**, 83 (1997).
[4] J. E. Sonier et al., Phys. Rev. Lett. **83**, 4156 (1999).
[5] B. Lake et al., Science **291**, 175 (2001).
[6] L. P. Gork'ov and A. G. Lebed, J. Phys. Lett. (France) **45**, L-433 (1984).
    M. Héritier et al., J. Phys. Lett. (France) **45**, L-943 (1984).
[7] A. BenAli et al., Europhys. Phys. J. B **14**, 53 (2000).

Quasi-one dimensional organic conductors: interplay between a magnetic field and the dimensionality

S. Haddad[*,1], **S. Charfi-Kaddour**[1], **C. Nickel**[2], **M. Héritier**[2], and **R. Bennaceur**[1]

[1] Faculté des Sciences de Tunis, LPMC, Campus Universitaire, 1060 Tunis, Tunisie
[2] Laboratoire de Physique des Solides, Bât. 510, Université Paris-Sud, 91405 Orsay, France

Received 1 September 2003, accepted 15 March 2004
Published online 14 April 2004

PACS 64.60.Ak, 71.10.Pm, 72.80.Le

We briefly review some key properties of the quasi-one dimensional organic conductors. We focus on some puzzling issues of the normal state in particular the problem of dimensional crossover. We show that several features of the normal state are consistent with the Luttinger liquid description. We stress on the competition between the magnetic field and the dimensionality.

© 2004 WILEY-VCH Verlag GmbH & Co. KGaA, Weinheim

1 Introduction

On the routes towards high temperature superconductors with always higher transition temperatures, organic superconductivity has been discovered for the first time in 1979 in the quasi-one dimensional organic conductor (TMTSF$_2$PF$_6$). From then on, a large number of theoretical and experimental works have been devoted to study the properties of quasi-one dimensional organic compounds. A wealth of new concepts in physics have then been obtained such as the dimensionality crossover and the one dimensionalization effect under a magnetic field.

In spite of this extensive study, these compounds are still the subject of open questions [1, 2]. In this paper we review the crossover from the high temperature one dimensional (1D) regime to the low temperature bidimensional (2D) regime. We also consider the deconfinement-confinement transition under a magnetic field. The paper is organized as follows: In Section 2, we present some physical properties of the TMTSF compounds and their sulfur analog TMTTF. We will emphasize on the points related to the dimensional crossover. In Section 3, we consider the renormalization group method as a theoretical approach dealing with the above issues. Concluding remarks are summarized in Section 4.

2 The (TM)$_2$X family: some physical properties

2.1 Crystal structure

The organic Bechgaard salts (TMTSF)$_2$X and the Fabre salts (TMTTF)$_2$X are typical examples of quasi 1D systems. The crystal structure of these compounds is built up from the nearly planar organic molecules: the selenium based tetramethylselenofulvalene (TMTSF) or its sulfur analog tetramethyltetrathiafulvalene (TMTTF) (Fig. 1).

The organic molecules TM form zig-zag stacks along the crystallographic a-axis. The chains of molecules are separated by the inorganic ions X (X = PF$_6$, ClO$_4$, Br etc.) which maintain the charge neutrality of the salts. The nature of the TM molecules and the X anions is crucial for the properties of the (TM)$_2$X salts.

---

[*] Corresponding author: e-mail: shaddad@lagrange.physique.usherb.ca

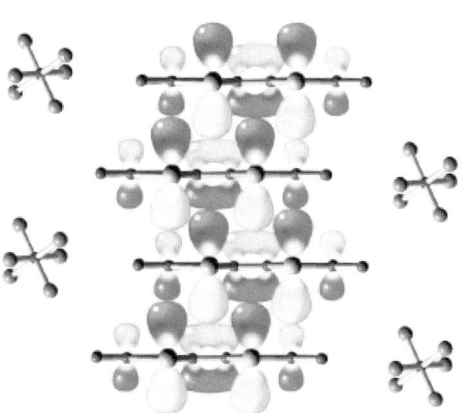

Fig. 1 Structure of the quasi-1D organic salts $(TM)_2X$. The stacks of the organic molecules TM (TMTSF or TMTTF) are arranged along the vertical axis (a-axis) corresponding to the most conducting axis. The horizontal axis, along which the stacks of TM molecules are separated by the X ions, is the least conducting axis (c-axis), after [1].

The transverse hopping integrals in the directions perpendicular to the molecule chains axis (a-axis) are smaller by at least one order of magnitude than the longitudinal hopping integral t_a along the chains. Indeed, the values of these integrals are 2000 K : 200 K : 10 K for respectively the integrals along the a-axis, b-axis and c-axis. This strong anisotropy of the hopping integrals is at the origin of the quasi-1D character of the $(TM)_2X$ salts. One should then expect that the hopping processes are substantially important for the nature of the electronic states of these compounds.

2.2 $(TM)_2X$ compounds: from 1D to 2D

At high temperature, the magnitude of the thermal fluctuations bypass the interchain hopping integrals. Therefore, the transverse particle motion is erased in this temperature regime and the particle are restricted to move along the chains. The compound behaves then as isolated 1D chains and the Fermi surface may be regarded as that of a 1D electron gas with a linear dispersion relation:

$$\varepsilon(\boldsymbol{k}) = v_F \left(|k| - k_F\right),$$

which defines the spectrum of the Luttinger model. The nature of the high temperature 1D regime is the subject of a controversial debate. However, there is a general consensus that it cannot be viewed as a Fermi liquid.

Due to the 1D character of the $(TM)_2X$ family, the electron–electron interactions are relevant. The magnitude of such interactions may justify the use of either a Luttinger liquid or a Fermi liquid.

Optical conductivity [3] and photoemission [4] data can be well explained by the theoretical predictions for the Luttinger liquid indicating the presence of very strong electron interactions in these compounds [5]. The behavior of the latter is correctly described by a Luttinger liquid down to 200 K. The deviations from a Fermi liquid description is also confirmed by transport measurements [6] showing striking different temperature dependencies for the in and out plane resistances for $T > 80$ K in $(TMTSF)_2PF_6$. This could not be explained within a Fermi liquid approach [1]. Moser et al. [6] argued that the high temperature phase ($T > 130$ K) is compatible with a Luttinger liquid. Recently, Mihaly et al. [7] have carried out Hall effect and transport measurements on the normal phase of $(TMTSF)_2PF_6$.

They suggested that their results may be interpreted within a Fermi liquid description and proposed that the 1D Luttinger liquid regime should be pushed above the room temperature. However, their Hall effect data could be explained within a weakly coupled one-dimensional Luttinger chains [8]. Furthermore, Moser et al. [9] have reported Hall effect measurements performed on $(TMTSF)_2PF_6$. The results are suggestive of a Luttinger liquid regime above 130 K. On the other hand, the transport measurements of Mihaly et al. show, as found by Moser et al. [6], opposite temperature dependence of the in plane and out plane resistances, which indicates a non-Fermi liquid behavior.

One may then conclude, as it has been argued in references [1, 2], that in the high temperature regime the $(TM)_2X$ compounds cannot be viewed as Fermi liquid conductors. The features of the normal phase in this temperature range may be described within the Luttinger liquid.

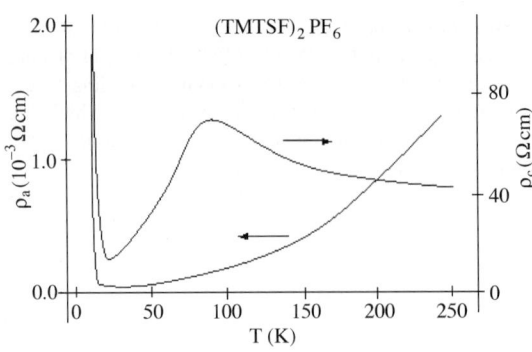

Fig. 2 Longitudinal and transverse resistivities as a function of temperature in $(TMTSF)_2PF_6$, after [6].

By decreasing the temperature, the thermal fluctuations are reduced. The transverse hopping along the b-axis becomes, then, coherent and the system undergoes a crossover from the 1D Luttinger liquid to a 2D regime. The signature of this dimensional crossover has been observed in the transport along the c-axis, which shows the same temperature dependence as the longitudinal transport below a crossover temperature (Fig. 2, where the crossover temperature is about 100 K). The latter, which depends strongly on pressure, sets the crossover between two different regimes: a high temperature Luttinger liquid regime and an incipient 2D Fermi liquid. The 3D Fermi liquid is fully restored below 10 K since the transports along the three axes show the same behavior with temperature.

In the low temperature regime, the $(TM)_2X$ compounds undertake a phase transition to a long range ordered state. A variety of ground states may be obtained including charge and spin density waves, Mott insulator and superconductivity. By varying the anisotropy of these compounds, the relative stability of the different ordered phases may be substantially changed. This may be achieved by applying a hydrostatic pressure which furthers the transverse hopping processes.

The generic pressure-temperature phase diagram of the $(TM)_2X$ family is depicted in Fig. 3. The crossover temperature is denoted by T_1^*.

It should be stressed that the dimensional crossover does not occur in the close vicinity of T_1^*, but it is spread over a broad domain of temperature including T_1^* which is so-called *transient regime* [10]. The presence of such regime for the crossover has been confirmed both experimentally [6] and theoretically [2, 10].

Although there is a general consensus on the non-Fermi liquid nature of the high temperature 1D phase, the nature of the low temperature phase (10 K $< T <$ 100 K) is still a puzzling question. This phase shows a non-Fermi liquid behavior suggesting the presence of 1D effects even below the crossover.

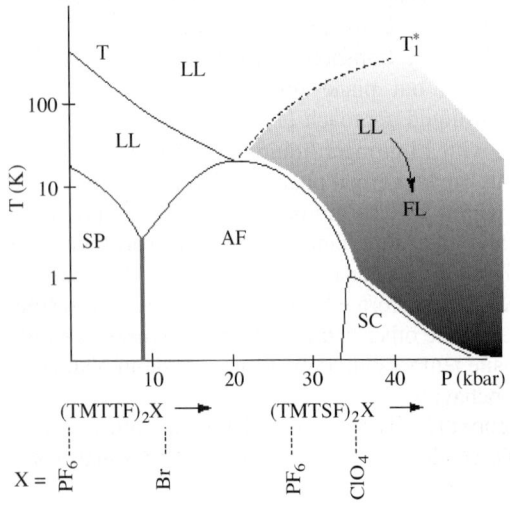

Fig. 3 Generic phase diagram of the $(TM)_2X$ as a function of pressure. LL (FL) denotes Luttinger (Fermi) liquid. The ordered states SP, AF and SC correspond respectively to the Spin-Peierls, the Antiferromagnetic and the superconducting states, after [1].

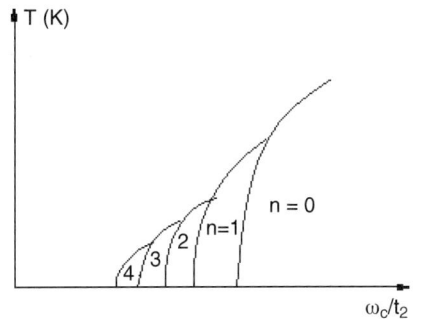

Fig. 4 Generic temperature-magnetic field phase diagram. The horizontal axis gives the magnitude of the field compared to the imperfect nesting parameter t_2.

2.3 Interplay between dimensionality and the magnetic field

It has been found that, applying a magnetic field along the least conducting axis (c-axis) has dramatic effects on the phase diagram of the Bechgaard salts. A magnetic field of the order of 10 T induces an anomalous behavior of the longitudinal transport in $(TMTSF)_2X$ family [11–13]. This unusual behavior has been interpreted as a result of the changes in the effective dimensionality of the system caused by the applied magnetic field H [14]. The system is expected to be effectively 1D at $t_c < T < \omega_c$, here t_c is the hopping integral along the c-axis and ω_c is the magnetic energy given by $\omega_c = ev_F Hb$, where e is the electron charge, v_F is the Fermi velocity and b is the distance between the chains in the b direction.

An experimental evidence of this *magnetic field induced one-dimensionalization effect* has been given in Ref. [15] from magnetoresistance measurements in $(TMTSF)_2PF_6$.

Besides the decrease of the effective dimensionality of the system towards the 1D state, the transverse magnetic field induces, above the critical pressure at which the superconducting state vanishes, a cascade of phase transition between field induced spin density wave (FISDW) states [16]. This phenomenon is exclusive to the Bechgaard salts $(TMTSF)_2X$. Each FISDW phase is characterized by the quantized Hall resistance $\rho_{xy} = h/2Ne^2$ in the sequence $N = \ldots 5, 4, 3, 2, 1, 0$ as the field increases (Fig. 4).

The FISDW cascade was explained within the Quantized Nesting Model (QNM) [17, 18] based on a 2D Fermi liquid approach where the nesting properties of the open Fermi surface are a fundamental issue.

A system of infinite number of chains weakly coupled by interchain one-particle hopping processes t_1 and t_2 respectively to the first and to the second-nearest neighbors along the b direction can be described by the following dispersion relation:

$$\varepsilon(\boldsymbol{k}) = v_F(|k| - k_F) - 2t_1 \cos k_\perp b - 2t_2 \cos 2k_\perp b, \qquad (1)$$

k is the longitudinal momentum while $k_\perp$ is the transverse momentum along the b direction.

At zero magnetic field and for $t_2 = 0$, the electron–hole symmetry is satisfied at the nesting vector $\boldsymbol{Q}_0 = (2k_F, \pi/b)$:

$$\varepsilon(\boldsymbol{k}) = -\varepsilon(\boldsymbol{k} + \boldsymbol{Q}).$$

However, the presence of small deviations in $\varepsilon(\boldsymbol{k})$, parameterised by the t_2 term, breaks down this symmetry. As shown in QNM, an applied magnetic field along the c direction may overcome the effect of t_2 process and restore the perfect nesting. This is possible thanks to a variation of the nesting vector $\boldsymbol{Q}$ which is characterized by a quantized longitudinal component $Q_x = 2k_F + NG$, where $G = \omega_c/v_F$ is the magnetic wave vector. The competition between the magnetic field and the t_2 term is substantially important for the formation of the FISDW phases. In the next section we will focus on this issue.

3 (TM)$_2$X family: a renormalization group approach

3.1 Why renormalization group method

As discussed in the previous section, a wealth of features suggest that the high temperature 1D phase has a non-Fermi liquid nature. It turns out that correlations are crucial in this phase which runs out any mean field treatment.

In the 1D regime, the response functions in either the Cooper (electron–electron) channel or the Peierls (electron–hole) channel give a logarithmic divergence as $\ln E_0/T$, where E_0 is a bandwidth cutoff, which is of the order of the Fermi energy.

The divergence is not only present in the lowest order but is also found in the higher orders of the perturbative expansion of the vertex functions. To handle these divergences, the renormalization group (RG) method is an appropriate approach within which it is possible to sum up divergences of the perturbative series.

The RG method proposed by Bourbonnais and Caron [20] provides not only a coherent description of the 1D features of the (TM)$_2$X salts but it also accounts for the low temperature 2D behavior. In particular, one may derive within this method the long-range ordered states. It is worth to note that in the RG method, the transition temperatures from the metallic regime to these ordered states have the same expression as those obtained by mean field theories but with renormalized parameters.

The latter take into account the *history* of the system in the 1D regime, which is neglected in the mean field theories. In the RG method, the transverse hopping parameter t_1 is treated perturbatively. However, it can no more be considered as a perturbation below the 1D–2D crossover temperature.

To get a detailed analysis of the nature of the low temperature regime and on the scale of the dimensional crossover, a non-perturbative method is necessary. The chain-dynamical mean field theory is a very promising approach where the 1D–2D crossover has been studied in a non-perturbative manner [2, 21].

In the following, we study the effect of the magnetic field on the imperfect nesting parameter t_2 based on the renormalization group (RG) method.

3.2 Renormalization formulation

Within the RG method the scaling of the physical parameters is governed by a differential equation. The renormalization group equation of t_2 is generally written as [20]

$$\frac{d\ln \tilde{t}_2}{dl} = 1 - f(l), \qquad (2)$$

$\tilde{t}_2 = t_2/E_0$ and $f(l)$ is a function depending on the scattering strengths. The scaling parameter l is related to the temperature as $l = \ln E_0/T$. The scaling process is carried out until $\tilde{t}_2$ reaches unity at a scaling parameter l_2^* [22]. At the temperature $T_2^* = E_0 \exp(-l_2^*)$, the t_2 hopping becomes coherent.

The energy scale of t_2 is of the order of 20 K, it belongs then to the 2D Fermi liquid regime. One should, therefore, derive the scaling equation of t_2 in the 2D regime and in the presence of the magnetic field.

The problem is characterized by three cutoff parameters: the bandwidth cutoff E_0, the 1D–2D crossover temperature T_1^* and the magnetic energy ω_c. So, a three step renormalization procedure is required.

In the course of the first step, which corresponds to the high temperature 1D regime, the scaling parameter ω_c/E_0 scales from 1 to T_1^*. In this step, the scaling of t_2 is given by Eq. (2).

The second step is carried out on the 2D regime ($T < T_1^*$). The scaling equation of t_2 reduces, at the one loop level to [20]

$$\frac{d\ln \tilde{t}_2}{dl} = 1.$$

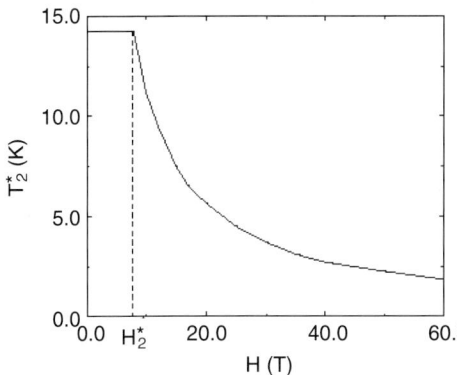

Fig. 5 Field renormalization of T_2^*. H_2^* is the critical field under which the hopping process t_2 is unaffected by the magnetic field.

The scaling energy is in this case $E(l) = T_1^* \exp(-l)$. This scaling step is stopped at $l_c = \ln T_1^*/\omega_c$ and a third step should be started where the scaling parameter is varied from 1 to T/ω_c. The flow equation of t_2 is also given by [23]

$$\frac{d \ln \tilde{t}_2}{dl} = 1.$$

This last step of the renormalization procedure is stopped at l_2^* defined as $\tilde{t}_2(l_2^*) = 1$, which corresponds to $T_2^* = \omega_c \exp(l_2^*)$. The present three step renormalization procedure has been discussed in details by the present authors in Ref. [24].

3.3 Results

We have solved the scaling equations of t_2 and the coupling constants for a bare value $t_2(0) = 25$ K. Carrying out the different steps of the renormalization procedure, we have obtained the dependence of the T_2^* temperature on the magnetic field as depicted in Fig. 5.

The latter shows that, below a critical field H_2^*, T_2^* is field independent. For $H > H_2^*$, T_2^* decreases as a power law with the increasing field. The effective parameter t_2^* is, then, *reduced* as the field increases, which *improve the nesting properties*. The field renormalization of t_2 should be taken into account when deriving, within the QNM, the transition temperatures of the FISDW phases. This may improve the quantitative agreement between theory and experiments.

4 Conclusion and outlook

In these notes, we have reviewed a few of the key features of quasi-1D organic conductors in particular those dealing with the normal states. Great interest is now focused on these states where the Fermi liquid description may break down. Clear deviations from the Fermi liquid behavior have been observed and several properties suggest a Luttinger liquid description of the high temperature 1D phase. Even below the 1D–2D crossover temperature, the incipient 2D Fermi liquid cannot be viewed as a canonical Fermi liquid.

Several aspects of the dimensional crossover remain puzzling: at which energy scale does it take place? How the system evolves from a 1D Luttinger liquid to a 2D or 3D Fermi liquid? Why the signature of the crossover is absent in some probes?

Some properties of the quasi-1D organic compounds may be explained within the perturbative renormalization group approach. Recent non-perturbative theoretical method, as the chain dynamical mean field theory, have been proposed to handle the above issues. However a complete theory of the dimensional crossover in quasi-1D systems is still lacking.

References

[1] C. Bourbonnais and D. Jérome, in: Advances in Synthetic Metals, Twenty Years of Progress in Sciences and Technology, edited by P. Bernier, S. Lefrant, and G. Bidan (Elsevier, New York, 1999), p. 206.
D. Jérome, in: Organic Conductors: fundamentals and applications, edited by J.-P. Farges (Dekker, New York, 1999), p. 405.

[2] S. Biermann et al., cond-mat/0201542, Proceedings of the NATO ASI Field Theory of Strongly Correlated Fermionsand Bosons in Low-Dimensional Disordered Systems, Windsor, August 2001, and references therein.

[3] Schwartz et al., Phys. Rev. B **58**, 1261 (1998).
V. Vescoli et al., Science **281**, 1191 (1998).

[4] V. Vescoli et al., Eur. Phys. J. B **13**, 503 (2000).

[5] T. Giamarchi, Phys. Rev. B **44**, 2905 (1991).
T. Giamarchi, Physica B **230–232**, 975 (1997).

[6] J. Moser et al., Eur. Phys. J. B **1**, 39 (1998).

[7] G. Mihaly et al., Phys. Rev. Lett. **84**, 2670 (2000).

[8] V. Lopatin, Phys. Rev. B **57**, 6342 (1998).
A. Lopatin, A. Georges, and T. Giamarchi, cond-mat/000806.

[9] J. Moser et al., Phys. Rev. Lett. **84**, 2674 (2000).

[10] C. Bourbonnais, Proceedings of the Int. summer school on High Magnetic Fields: Application in Condensed Matter Physics and Spectroscopy, Cargese, May 2001, edited by C. Berthier, L. P. Levy, and G. Martinez, to be published by Springer, 2002, cond-mat/0204345.

[11] K. Behnia et al., Phys. Rev. Lett. **74**, 5272 (1995).

[12] D. Jérome, in: Correlated Fermions and Transport in Mesoscopic Systems, edited by T. Martin, G. Montambaux, and J. Trâm Thanh Vân (Editions Frontières, Gif-sur-Yvette, 1996), p. 95.

[13] E. I. Chashechkina and P. M. Chaikin, Phys. Rev. Lett. **80**, 2181 (1998).

[14] T. Zheleznyak and V. M. Yakovenko, Eur. Phys. J. B **11**, 385 (1999).

[15] L. Balicas, Phys. Rev. B **59**, 12830 (1999), and references therein.

[16] P. M. Chaikin, J. Phys. I **6**, 1875 (1996).

[17] L. P. Gor'kov and A. G. Lebed, J. Phys. (Paris) Lett. **45**, L433 (1984).

[18] M. Héritier, G. Montambaux, and P. Lederer, J. Phys. Lett. **45**, L943 (1984).
M. Héritier, G. Montambaux, and P. Lederer, J. Phys. C **19**, L293 (1986).

[19] Virosztek, L. Chen, and K. Maki, Phys. Rev. B **34**, 3371 (1986).

[20] C. Bourbonnais and L. G. Caron, Int. J. Mod. Phys. B **5**, 1033 (1991).
C. Bourbonnais, Strongly Interacting Fermions and High T_c Superconductivity, in: Les Houches, Session LVI, 1991, edited by B. Douçot and J. Zinn-Justin (Elsevier Science, Amsterdam, 1995), p. 307.

[21] S. Biermann et al., Phys. Rev. Lett. **87**, 276405 (2001).
A. Georges et al. Phys. Rev. B **61**, 16393 (2000).

[22] J. Kishine and K. Yonemitsu, J. Phys. Soc. Jpn. **67**, 5301 (1998).

[23] S. Haddad, M. Héritier, and R. Bennaceur, Eur. Phys. J. B **11**, 429 (1999).

[24] S. Haddad et al., Eur. Phys. J. B **34**, 33 (2003).

Effects of magnetic field on the cuprate high-T_c superconductor $La_{2-x}Sr_xCuO_4$

B. Lake[*,1,2], **G. Aeppli**[3,4], **N. B. Christensen**[5], **K. Lefmann**[5], **D. F. McMorrow**[3,5], **K. N. Clausen**[5], **H. M. Rønnow**[4,6], **P. Vordewisch**[7], **P. Smeibidl**[7], **M. Mankorntong**[8], **T. Sasagawa**[8], **M. Nohara**[8], **H. Takagi**[8], and **N. E. Hussey**[8,9,10]

[1] University of Oxford, Clarendon Laboratory, Parks Road, Oxford OX1 3PU
[2] Oak Ridge National Laboratory, P.O. Box 2008 MS 6430, Oak Ridge, Tennessee 37831-6430, USA
[3] University College London, Department of Physics and Astronomy, London WC1E 6BT, UK
[4] N.E.C. Research Institute, 4 Independence Way, Princeton, New Jersey 08540, USA
[5] Materials Research Dept., Bldg 108, Risoe National Lab., Frederiksborgvej 399, 4000 Roskilde, Denmark
[6] CEA Grenoble (MDN/SPSMS/DRFMC), 17 Ave. des Martyrs, 38054 Grenoble cedex 9, France
[7] BENSC, Hahn-Meitner-Institut, Glienicker Strasse 100, 14109 Berlin, Germany
[8] Department of Advanced Materials Science, Graduate School of Frontier, Sciences, University of Tokyo, Hongo 7-3-1, Bunkyo-ku, Tokyo 113-8656, Japan
[9] Department of Physics, University of Loughborough, Loughborough, LE11 3TU, UK
[10] H. H. Wills Physics Laboratory, University of Bristol, Bristol BS8 1TL, UK

Received 1 September 2003, accepted 5 March 2004
Published online 14 April 2004

PACS 74.25.Qt, 74.72.Dn, 75.25.+z

This article discusses neutron scattering measurements on the cuprate, high transition temperature superconductor $La_{2-x}Sr_xCuO_4$ (LSCO) in an applied magnetic field. LSCO is a type-II superconductor and magnetic flux can penetrate the material via the formation of vorticies. Phase coherent superconductivity characterized by zero resistance is suppressed to the lower field-dependent irreversibility temperature ($T_{irr}(H)$) and occurs when the vortices freeze into a lattice. Because superconductivity is destroyed within the vortex cores, an investigation of the vortex state provides information about the ground state that would have appeared had superconductivity not intervened. Our measurements reveal that both optimally doped LSCO ($x = 0.16$, $T_c = 38.5$ K) and underdoped LSCO ($x = 0.10$, $T_c = 29$ K) have an enhanced antiferromagnetic response in a field. Measurements of the optimally doped system for $H = 7.5$ T show that inelastic sub-gap spin fluctuations first disappear with the loss of finite resistivity at T_{irr}, but then reappear at a lower temperature with increased lifetime and correlation length compared to the normal state. In the underdoped system elastic antiferromagnetism develops below T_c in zero field, and is significantly enhanced by application of a magnetic field; phase coherent superconductivity is then established within the antiferromagnetic phase at T_{irr}.

© 2004 WILEY-VCH Verlag GmbH & Co. KGaA, Weinheim

$La_{2-x}Sr_xCuO_4$ (LSCO) is derived from the parent compound La_2CuO_4 which is an insulating antiferromagnet consisting of alternating CuO_2 and LaO layers. The Cu^{2+} ions possess spin-1/2 and give rise to long-range magnetic order below a Néel temperature of $T_N = 325$ K [1]. The magnetism is essentially two-dimensional with strong antiferromagnetic exchange interactions between nearest neighbours within the CuO_2 planes and weak interactions between these planes. Introduction of Sr doping, reduces the size of the magnetic signal and for $x > 0.02$ the long-range commensurate order gives way to incommensurate magnetism visible as four peaks surrounding the original Bragg peak position. The material becomes

---
[*] Corresponding author: e-mail: bella.lake@physics.ox.ac.uk, Phone: +44 (0)1865 282226, Fax: +44 (0)1865 272400

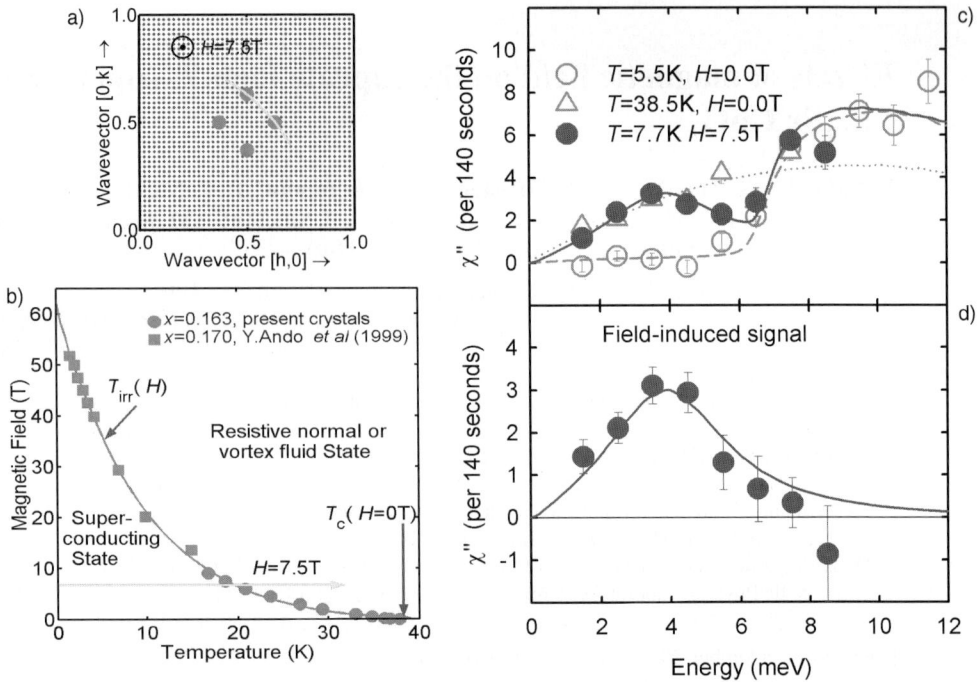

Fig. 1 (online colour at: www.interscience.wiley.com) (a) shows the reciprocal space for the CuO_2 planes, the large dots give the magnetic peak positions for LSCO $x = 0.16$ and the small dots give the reciprocal vortex lattice for a magnetic field of 7.5 T applied perpendicular to the planes. (b) shows the irreversibility line for LSCO, $x = 0.16$, as a function of temperature and field; (c) gives the magnetic susceptibility as a function of energy at the magnetic peak position for different fields and temperatures; (d) shows the field-induced signal – low temperature signal measured for $H = 7.5$ T minus zero-field signal.

superconducting for Sr dopings, $0.06 < x < 0.25$, [2, 3] and the incommensurate magnetism persists in this phase. There are in fact two regimes within the superconducting doping range. The underdoped regime, $0.06 < x < 0.125$, where there is long-range magnetic order, and the optimally doped and overdoped regime, $0.125 < x < 0.25$, where the long-range order is lost but magnetic fluctuations remain at the peak positions. Figure 1a shows the two-dimensional reciprocal lattice of the CuO_2 planes, the large filled circles represent the peak positions for 16% Sr doped LSCO.

When a magnetic field is applied to LSCO it behaves like a type-II superconductor and magnetic flux is able to penetrate the material due to the formation of vorticies. At high temperatures the vorticies are mobile while at low temperatures they form a lattice. The vortex freezing temperature is also the temperature below which phase coherent superconductivity characterised by zero resistance occurs for a given applied field, which in this article is called the irreversibility temperature, $T_{irr}(H)$, (Fig. 1b). The reciprocal vortex lattice for a field of $H = 7.5$ T applied perpendicular to the superconducting CuO_2 planes is represented by the small dots in Fig. 1a; for this field the separation between the vorticies is $a_v(7.5\ T) = 166$ Å. The size of the vortex cores is typically given by the superconducting pair coherence length ξ which for LSCO is $\xi \sim 20$ Å.

This article discusses the effects of an applied magnetic field on two LSCO samples, first an optimally doped sample with $x = 0.16$ and a superconducting transition temperature of $T_c = 38.5$ K and second an underdoped crystal with $x = 0.10$ and $T_c = 29$ K. The experimental technique in both cases is neutron scattering which is able to track changes in the magnetism of these materials with applied field and temperature. The $x = 0.10$ sample has long-range magnetic order and elastic neutron scattering is used to probe it; the $x = 0.16$ sample has no long-range order, however it does have magnetic fluctuations which show up in inelastic neutron scattering.

© 2004 WILEY-VCH Verlag GmbH & Co. KGaA, Weinheim

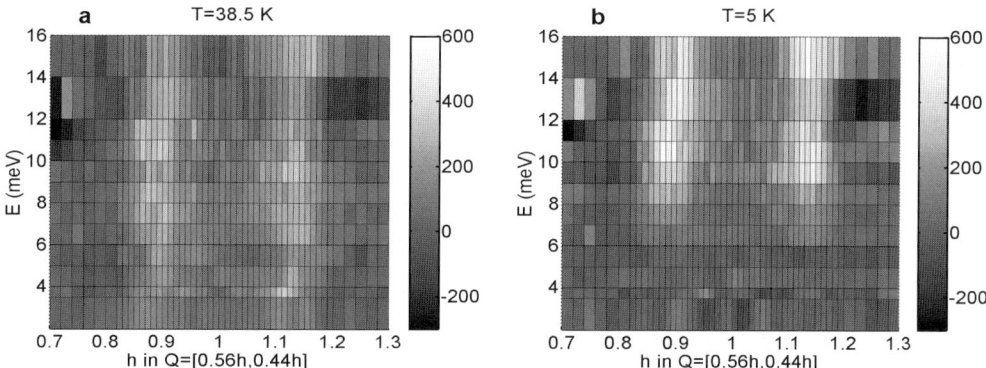

Fig. 2 (online colour at: www.interscience.wiley.com) The magnetic signal of LSCO $x = 0.16$ measured in zero field in the normal and superconducting states. The data is displayed as a function of energy transfer and wavevector for the trajectory shown by the curved line in figure 1a that passes through two of the magnetic peaks. (a) shows the data collected in the normal state at $T_c = 38.5$ K the white streaks indicate the magnetic peak positions. (b) shows the data collected in the superconducting state at $T = 5$ K, where the spin-gap is revealed by the absence of scattering below 6.7 meV.

The experiments on optimally doped LSCO, $x = 0.16$, took place on the RITA-1 triple-axis spectrometer at Risø National Laboratory, Denmark, just before the closure of the reactor in 2000. In zero field the main feature is the spin gap that appears in the superconducting state [5–7]. Figure 2 compares the magnetic response of the material as a function of energy in the normal state at $T_c = 38.5$ K with the superconducting state at 5 K. In the normal state magnetic signal is observed at the incommensurate peak positions down to the lowest energies. In contrast at 5 K a gap appears in the magnetic excitation spectrum at $\Delta = 6.7$ meV below which the magnetic signal drops to zero. The spin-gap has similar origins to the superconducting energy gap observed in the quasi-particle excitation spectrum measured by for example angle resolved photoemission [8]. The latter is the energy required to break up the charge pairing that occurs in the superconducting state while the spin gap refers to the pairing energy of the spins of these quasi-particles.

Figure 1c shows the magnetic susceptibility at the incommensurate peak position as a function of energy for various fields and temperatures [9] where the applied fields were supplied by an 8 T vertical field magnet manufactured by Oxford Instruments. The open circles give the magnetic signal at 5 K in zero field and show that there is complete loss of signal below the spin gap energy in the superconducting state, compared to the normal state at these energies (open triangles). If the measurement is now repeated at $T = 5$ K in an applied field of 7.5 T, where the field direction is perpendicular to the copper oxide planes, (filled circles) the data reveal magnetic signal induced below the spin-gap energy. This signal appears as a peak with a maximum at an energy of ~4.3 meV (Fig. 1d) and when fitted to a damped harmonic oscillator it has an inverse lifetime of $\Gamma = 4.3$ meV which is much less than the normal state value of 9 meV suggesting slower magnetic fluctuations. Further measurements of the wavevector-dependence of the field-induced signal reveal a correlation length at low temperatures of 75 Å that is considerably greater than the normal state value of 24 Å. These two results – an enhanced lifetime and an enhanced correlation length – suggest a greater tendency towards long-range magnetic order for LSCO in an applied field compared to its normal state.

As a first guess one might assume that the field-induced magnetism originates from the vortex cores, however the correlation length which gives the size of the magnetic regions in the material is considerably larger than the diameter of the vorticies ($\xi \sim 20$ Å). Furthermore the size of the ordered spin moment obtained by a phonon normalization is 0.22 μ_B/Cu^{2+} in absolute units [9]; this value is averaged over the Cu^{2+} ions throughout the material and is clearly too large to come from the vortex cores alone which make up less than 5% of the total sample for a field of 7.5 T. A potential scenario is for vortices to nucleate magnetism but where these magnetic regions extend beyond the vortex cores into the surrounding superconducting regions. Evidence for this comes from high-field nuclear magnetic resonance experi-

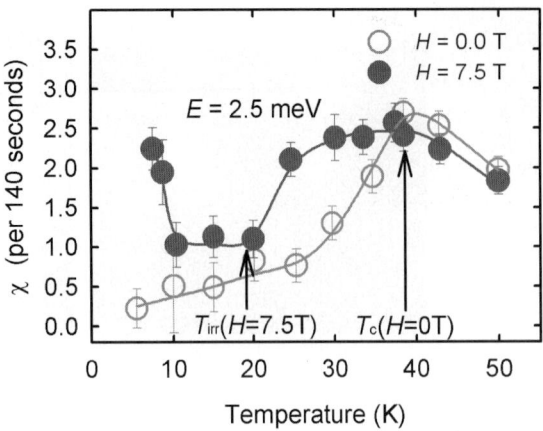

Fig. 3 (online colour at: www.interscience.wiley.com) Temperature dependence of the magnetic signal in LSCO, $x = 0.16$ measured below the spin-gap at 2.5 meV, the data was collected at the magnetic peak position. The open circles shows the data collected in zero field where the signal drops rapidly below the $T_c(H = 0\,\text{T})$ due to the opening up of the spin-gap. The blue circles shows the same measurement in an applied field of $H = 7.5\,\text{T}$, magnetic signal lingers below $T_c(H = 0\,\text{T})$ and does not drop away until $T_{irr}(H = 7.5\,\text{T}) = 19\,\text{K}$ is reached. As the temperature is cooled further the magnetic susceptibility rises again for $T < 10\,\text{K}$.

ments which have detected an enhanced resonance rate both within and surrounding the vortices in optimally doped $YBa_2Cu_3O_{7-\delta}$; which has been attributed to field-induced magnetic excitations [10, 11]. There are also interesting results from scanning tunnelling microscopy (STM) measurements of optimally doped $Bi_2Sr_2CaCu_2O_{8+\delta}$ in an applied magnetic field. These measurements reveal a field-induced checkerboard pattern around the vorticies, and while this modulation is clearly associated with the formation of vortices it extends beyond the cores to cover a region of ~100 Å [12]. The signal measured by STM is electronic and the relationship between this and the field-induced magnetic signal measured by neutrons has yet to be established. Nevertheless it seem likely that these two phenomena are linked in which case the field-induced magnetism would also originate from a region around the vortex cores.

Other interesting aspects of the field-induced signal can be deduced from the temperature dependence of the subgap magnetic signal which is shown in Fig. 3. In the absence of an applied field the signal drops away rapidly as the sample is cooled below the zero-field superconducting transition temperature $T_c(H = 0\,\text{T})$ due to the opening up of the spin gap in the superconducting state. In an applied field however the signal lingers below $T_c(H = 0\,\text{T})$ and does not drop away until $T < 19\,\text{K}$, a temperature that coincides with the irreversibility temperature for the applied field. At lower temperatures, $T < 10\,\text{K}$, the magnetic signal starts to increase rapidly suggesting an onset to long-range order. Unfortunately the available temperatures did not go low enough to fully explore this region, however it is clear that rising magnetic susceptibility occurs within the state of phase coherent superconductivity possibly via a mechanism where magnetism nucleated by the vorticies starts to extend outwards beyond the cores as temperature is lowered and different magnetic region begin to couple to each other through the intervening superconducting medium.

Since an applied magnetic field has the effect of inducing magnetism in optimally doped LSCO in the form of magnetic fluctuations without long-range order, this raises the question of what the effect of field would be in an underdoped sample which has long-range magnetic order in zero field. In the next set of experiments, elastic neutron scattering measurements were performed on underdoped LSCO $x = 0.10$. These measurements took place on the V2/FLEX triple-axis spectrometer at the Hahn-Meitner Institute in Berlin and the magnetic field was provided by the VM1 15 T vertical field magnet. Figure 4a and 4b show a scan through one of the incommensurate peaks in zero field and for $H = 14.5\,\text{T}$ [13]. In the normal state at $T_c = 29\,\text{K}$ (filled circles) no signal is observed, but on cooling down below T_c in zero field (open circles), signal appears in the superconducting state – a phenomenon that has been well documented in a number of La_2CuO_4 based superconductors [3]. If a magnetic field is now applied perpendicular to the CuO_2 planes at low temperatures, a strong enhancement of the elastic magnetic signal is observed (a factor of three larger than the zero field signal for an applied field of $H = 14.5\,\text{T}$). The magnetic peaks are resolution limited in both zero and non-zero field putting a lower limit of 400 Å on the magnetic correlation length. The large correlation length is significant because it suggests that the magnetic regions are not only greater than the vortex cores ($\xi \sim 20$ Å) but are also, unlike the case of the optimally doped sample, greater than the separation of the vorticies ($a_v(5\,\text{T}) = 200$ Å for $H = 5\,\text{T}$). The

© 2004 WILEY-VCH Verlag GmbH & Co. KGaA, Weinheim

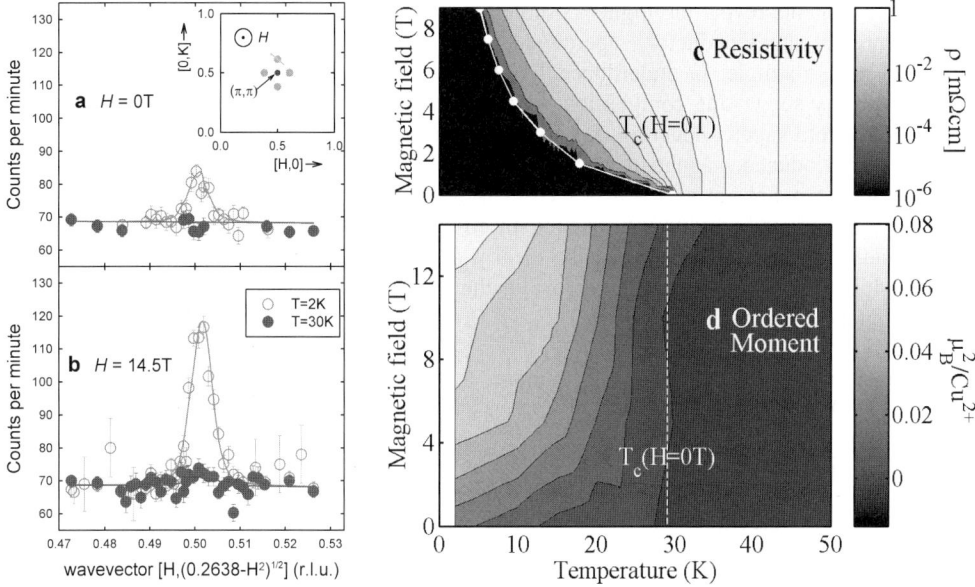

Fig. 4 (online colour at: www.interscience.wiley.com) (a) and (b) show the elastic magnetic scattering in underdoped $La_{2-x}Sr_xCuO_4$, $x = 0.10$, measured in the normal and superconducting states. The data is plotted as a function of wavevector through one of the magnetic peaks (see inset diagram). (a) shows the data collected in zero field. (b) shows the data collected in an applied field of 14.5 T. (c) and (d) compare the resistivity and magnetic signal as a function of temperature and magnetic field. (c) shows the transport data. The white dots mark the irreversibility temperature T_{irr}. (b) shows the ordered moment squared per Cu^{2+} ions obtained from a phonon normalisation.

large correlation length coupled with the large value of the average ordered spin moment per Cu^{2+} site of 0.24 μ_B/Cu^{2+} at $H = 5$ T, imply that as for the optimally doped sample many more sites are involved in the magnetic ordering than the 3% of Cu^{2+} ions that form the vortex cores for this field. Other experiments have subsequently taken place that confirm this picture of field-induced long-range magnetic order in oxygen-doped La_2CuO_{4+y} [14, 15].

Figure 4c and 4d compare the resistivity and the ordered spin moment as a function of temperature and magnetic field. The data show that irrespective of the size of the field that is applied the Néel temperature for the magnetic ordering lies close to the zero-field superconducting transition temperature $T_c(H = 0\,\text{T}) = 29$ K (first contour in Fig. 4d), and not the irreversibility temperature at which phase coherent superconductivity occurs. It is also clear that the state of phase coherent superconductivity characterised by zero resistance that occurs below T_{irr} always forms within the magnetically ordered phase – i.e. as the sample is cooled first it becomes magnetically ordered and than at a lower temperature it becomes superconducting. In many ways then the results for the underdoped sample are opposite to those for the optimally doped sample where phase coherent superconductivity is achieved at a higher temperature than the field-induced magnetism.

This duality between underdoped and optimally doped LSCO is extremely interesting and suggests a gradual evolution of the magnetic ordering temperature as a function of doping resulting in a crossover from above to below the irreversibility temperature at some doping intermediate between $x = 0.10$ and $x = 0.16$ [16]. Doping dependent changes also occur in the correlation length and energy of the field-induced magnetism. In the optimally doped sample, the field-induced signal lies at a non-zero energy transfer and has a finite correlation length whereas the field-induced signal in the underdoped sample is elastic and has a large resolution limited correlation length. This implies a phase diagram where the energy of the subgap field-induced signal decreases progressively as the doping decreases while at the same time the correlation length and hence the size of the magnetic regions increases.

These experiments reveal the highly complex interaction between superconductivity and magnetism. In a conventional superconductor, superconductivity and magnetism are incompatible but in LSCO the situation is not so simple. In underdoped LSCO although a magnetic field suppresses phase coherent superconductivity suggesting competition between these phases, field-induced magnetic order is turned on at the zero-field superconducting transition temperature suggesting co-operation. In optimally doped LSCO the field-induced magnetism occurs within the state of phase coherent superconductivity again suggesting that they are not necessarily incompatible. A number of questions remain to be answered in particular do superconductivity and magnetism co-exist or phase separate, and to what extent do they compete and/or co-operate? There is also clearly a very strong doping dependence which need to be explored which could help reveal the nature of the quasi-particle pairing and mechanism for superconductivity in the cuprate high-T_c's.

Acknowledgements We thank P. Dai, P. Gammel, S. E. Hayden, P. Hedegard, S. Kivelson, M. Marchevsky, H. Mook, N. P. Ong, C. Renner, S. Sachdev, J. Zaanen, and S.-C. Zhang for useful discussions.

References

[1] D. Vaknin et al., Phys. Rev. Lett. **58**, 2802 (1987).
[2] Y. S. Lee et al., Phys. Rev. B **60**, 3643 (1999).
[3] H. Kimura et al., Phys. Rev. B **59**, 6517 (1999).
[4] R. Gilardi et al., Phys. Rev. Lett. **88**, 217003 (2002).
[5] B. Lake et al., Nature **400**, 43 (1999).
[6] K. Yamada et al., Phys. Rev. Lett. **75**, 1626 (1995).
[7] S. Petit, A. H. Moudden, B. Hennion, A. Vietkin, and A. Revcolevschi, Physica B **234–236**, 800 (1997).
[8] B. Lake et al., Science **291**, 1759 (2001).
[9] T. Yoshida et al., LANL Preprint server cond-mat/0206469 (2003).
[10] V. F. Mitrovic et al., Nature **413**, 501 (2001).
[11] V. F. Mitrovic et al., Phys. Rev. B **67**, 220503 (2003).
[12] J. E. Hoffman et al., Science **295**, 466 (2002).
[13] B. Lake et al., Nature **415**, 299 (2002).
[14] B. Khaykovich et al., Phys. Rev. B **66**, 014528 (2002).
[15] B. Khaykovich et al., Phys. Rev. B **67**, 054501 (2003).
[16] E. Demler, S. Sachdev, and Y. Zhang, Phys. Rev. Lett. **87**, 067202 (2001).

Normal state understanding within a Fermi liquid approach

I. Sfar[1,2], **S. Charfi-Kaddour**[*,1], **M. Héritier**[2], and **R. Bennaceur**[1]

[1] LPMC, Département de Physique, Faculté des Sciences de Tunis, Campus Universitaire 1060 Tunis, Tunisia
[2] Laboratoire de Physique des Solides, UMR CNRS-Paris XI, Bat. 510, 91405 Orsay, France

Received 1 September 2003, revised 7 March 2004, accepted 8 March 2004
Published online 23 April 2004

PACS 74.20.Mn, 74.72.–h

Using the Hubbard model in the weak coupling limit within a Fermi liquid approach, we have studied spin fluctuation effects in quasi-2D superconductors such as high critical temperature superconductors. We show that, due to nesting properties of the Fermi surface, the magnetic and the transport properties are different from the behaviours we should observe in usual metal, because of strong magnetic fluctuation effect. We distinguish different regions in the normal state with different dependences as a function of temperature for several properties, such as the resistivity, NMR relaxation rate and spin susceptibility. We show that the pseudo-gap region is strongly related to the existence of a pseudo-gap in the density of states at the Fermi level.

1 Introduction

The high value of T_c found in the cuprate oxide superconductors, as well as many quite singular properties of their normal state phases have led many theorists to propose various original models [1–3], which have been strongly debated for the past fifteen years. Some important fundamental questions are addressed to the physicist trying to understand the nature of High Critical Temperature Superconductivity:

(i) Is this phenomenon simply resulting from a "conventional" mechanism, taking advantage of exceptionally favourable conditions (such as a singular density of state, a quasi-two-dimensional structure, a quite strong electron-phonon coupling etc…) or are we confronted to an essentially novel phenomenon, in which, in particular, magnetic interactions or magnetic fluctuations might play a crucial and quite original role ?

(ii) A subsequent question immediately arises: what is the relevant theoretical description of the "normal state" in these superconductors, stable above the critical temperature? The standard basis of the quantum theory of solids, used in particular in the BCS theory of conventional superconductivity, is the Landau theory of normal Fermi liquids, which describes the correlated electrons, in the limit of low energy and low temperature excitations, as a gas of weakly coupled well defined and long lived "quasi-particles", near enough the Fermi level.

Experimental data impose severe constraints on the various possible responses to this question. In particular, the generic phase diagram of the High Critical Temperature Superconductors (HCTS) exhibits a quite unusual behaviour above the critical temperature. Three different regimes separated by crossover lines are observed. With increasing the hole content, first a "pseudo-gap" regime, then a "strange metal" behaviour and finally a usual Fermi liquid are found in strong contrast with the usual behaviour of a normal Fermi liquid phase. These observed data in (HCTS) have cast some doubt about the validity of the Landau theory of Fermi liquid and lead many theorists to discard it and to propose the electron corre-

---

[*] Corresponding author: e-mail: samia.kaddour@fst.rnu.tn, Phone: +00 216 98 925 884, Fax: +00 216 71 885 073

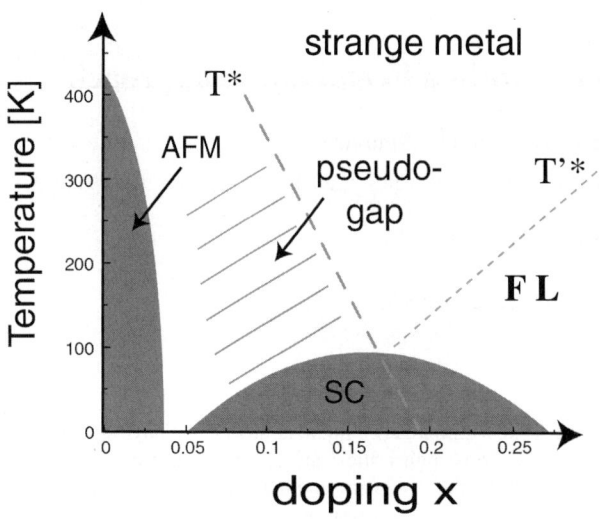

Fig. 1 (online colour at: www.interscience.wiley.com) Generic phase diagram of high critical temperature superconductors.

lations are too strong to allow the existence of well defined quasi-particles: Coulomb correlations are believed to be so strong that the so-called "normal" phase in no longer Fermi liquid [1]. In this non-Fermi liquid picture, the under-doped materials, such as La_2CuO_4 or $YBa_2Cu_3O_{6.5}$ is a Mott-Hubbard or a Charge Transfer insulator, where the electrons are localised by Coulomb correlations.

Some authors [2], indeed, have proposed a phenomenological model describing the electron system as a Marginal Fermi liquid. Although these phenomenological models cannot provide a microscopic description of the phenomenon, such marginal Fermi liquid theories seem to give a satisfactory account of many abnormal properties. Undoubtedly, it would be quite interesting if we were able to find a microscopic theoretical description able to account for such a marginal Fermi liquid behaviour. It is important, also, to emphasise an essential property: the normal state phase clearly exhibits a crossover behaviours, with characteristic energy scales which depend on the doping for all the HCTS.

In this work, we would like to discuss a weak correlation limit, which, surprisingly, seems to account for most of the normal phase anomalies. While the experimental data are definitely incompatible with the usual properties of standard Fermi liquids, it can be shown that this is no longer the case if we consider, instead of a standard Fermi liquid, a model of two-dimensional Fermi liquid in close vicinity of an antiferromgnetic instability. It is well known, from many experiments, that HCTS exhibits nesting properties of the Fermi surface [3–7], as well as Van Hove singularity [8] in the neighbourhood of the Fermi level. Such anomalous features strongly modify the usual standard Fermi liquid behaviour and can reproduce a marginal Fermi liquid behaviour. Therefore, such a simple theoretical model might provide a possible interpretation of a lot of experimental data such as magnetic susceptibility, neutron scattering, resistivity...

2 Normal state properties

Because of the observed anomalies and their implication on the validity of the Fermi liquid picture, a great interest has been devoted to the discussion of the normal state properties. In this section, we remind briefly the most important experimental data reported on the HCTS properties. There are three distinct regions where the normal state properties are very different. The first region is the pseudo-gap regime which corresponds to the under-doped region, below a cross-over temperature defined as T^*. The second region roughly corresponds, near T_c, to optimal doping (i.e. where T_c is maximum), this region becomes broader as T increases. Finally, the third region corresponds to the strongly over-doped region below a second cross-over temperature, defined as T'^*.

In the pseudo-gap region of the phase diagram, a large variety of experiments definitely show evidence of a strong suppression of spectral weight in the low energy part of the excitation spectrum. This phenomenon has been named in the literature as the pseudo-gap [9]. The behaviours, as a function of temperature, of the magnetic susceptibility, of the nuclear relaxation rate and of the transport measurements clearly show deviations from the usual properties of conventional Fermi liquids.

In a conventional metal, the Pauli susceptibility is, in principle, temperature independent. Such a behaviour has been observed, indeed, in $YBa_2Cu_3O_7$. However, for lower oxygen content, a different behaviour has been found [10]. A general feature appears, which is the first observed signature of the pseudo-gap. The magnetic susceptibility displays a reduction at low temperature. The closer the antiferromagnetic phase, the more pronounced the susceptibility reduction. This pseudo-gap effect in the magnetic susceptibility is also observed is other cuprates [11]. It is generally believed that this effect is restricted to the under-doped region.

Usually, in a metallic state, the NMR spin-lattice relaxation rate divided by the temperature $1/(T_1T)$ is temperature independent, as predicted by the Korringa law. In the cuprates, the NMR data for the copper nucleus show an important temperature dependence of $1/(T_1T)$ with a maximum value at a temperature close to the pseudo-gap temperature T^*. This enhancement of $1/(T_1T)$ depends on the hole doping. This effect could reflect the existence of the antiferromagnetic (AF) fluctuations. Moreover, the pseudo-gap effect observed by NMR experiments in the under-doped samples disappeared completely in the over-doped samples [11].

A considerable amount of inelastic neutron scattering (INS) experiments have been performed [12]. The experimental data provide a clear evidence of the persistence of antiferromagnetic fluctuations. Magnetic fluctuations are observed, depending on the compounds, to be either commensurate peaked at AF wave vector $Q_{AF} = (\pi, \pi)$ or incommensurate away from Q_{AF}.

One major feature of the non Fermi liquid behaviour of HCTS is the electrical resistivity temperature dependence. Many experiments have shown a linear T dependence of the resistivity in the (a, b) plane, obeyed over a quite large temperature range, with a very small $T = 0$ intercept. As well as the linearity, one should note that the resistivity is high: these are poor metals. A linear $\rho(T)$ is of course expected from electron-phonon interaction in a normal metal at temperature larger than the characteristic Debye temperature θ_D. However, one knows, from studies of low T_c samples that the range of validity of the linear law extends at much lower temperatures. Careful analysis of the resistivity have been made in the whole phase diagram. In single and double layered $Bi_2Sr_2Ca_{n-1}Cu_nO_y$, in the under-doped region, Konstantinovic et al. [13] observe a downward deviation of the resistivity from the high temperature law: $\rho(T) = \rho_0 + \alpha T$ starting at a characteristic temperature T^*, which corresponds fairly well to the opening of the pseudo-gap. This temperature T^* shifts systematically to higher temperatures as doping decreases. This phenomenon is also observed in $La_{2-x}Sr_xCuO_4$ and $YBa_2Cu_3O_y$ compounds.

3 Theoretical approach: magnetic fluctuation effects on the normal state

The electronic systems under discussion are described by a simple two-dimensional one-band Hubbard model, where U is the intra-atomic Coulomb interaction.

$$H = \sum_{k,s} \varepsilon_k c^+_{k,s} c_{k,s} + \frac{U}{N} \sum_{k,p,q,s} c^+_{k+q,s} c_{k,s} c^+_{p-q,-s} c_{p,-s}$$

where $c_{k,s}$ indicates electron annihilation operator corresponding to a Bloch state with wave vector k and a spin s, ε_k is the energy band dispersion law and N is the number of sites. In the Hubbard approximation, U is usually the "bare" local on-site Coulomb integral. However, the sensible value of U to be put in the Hubbard hamiltonian treated in the RPA is obviously not the "bare" one, but an effective one which is reduced by screening or other many body effects [14].

A careful study has been done to determine the shape of the Fermi surface. It seems clear that in all the cuprates, the Fermi surface exhibits a nesting property which is more or less perfect, depending on the compound. The best nesting is probably observed in the Bi-2212 compound [15].

Let us consider a metal with a nearly perfect nesting with a Q wave vector, i.e. in which the following relationship is almost exactly verified:

$$\varepsilon_{k+Q} = -\varepsilon_k - \omega_0.$$

Here, ω_0 is considered as the deviation from perfect nesting. The spin susceptibility for non interacting electrons is given by the following relation:

$$\chi^0(q,\omega) = \sum_k \frac{f(\varepsilon_{k+q}) - f(\varepsilon_k)}{\omega + \varepsilon_k - \varepsilon_{k+q}},$$

where f is the Fermi–Dirac distribution. The imaginary part of this susceptibility at the Q wave vector is then

$$\mathrm{Im}\,\chi^0(Q,\omega) = \frac{1}{2}N^0\left(-\frac{\omega}{2} - \frac{\omega_0}{2}\right)\frac{\sinh\dfrac{\omega}{2kT}}{\cosh\dfrac{\omega}{2kT} + \cosh\dfrac{\omega_0}{2kT}},$$

where N^0 is the density of states of the non interacting electrons.

3.1 Over-doped region

We have considered the RPA approximation to calculate the magnetic and transport properties in this region where

$$\chi^{RPA} = \frac{\chi^0}{1 - U\chi^0}.$$

Actually, the NMR relaxation rate is given by the relation

$$\frac{1}{T_1 T} = \lim_{\omega \to 0}\sum_q \frac{\mathrm{Im}\,\chi^{RPA}(q,\omega)}{\omega} \cong \lim_{\omega \to 0}\frac{\mathrm{Im}\,\chi^{RPA}(Q,\omega)}{\omega},$$

where Q is the best nesting vector corresponding the less imperfect nesting.

We will try first to understand the T'^* line by taking into account the effect of moderate spin fluctuations on the normal state properties. Our calculations of the spin relaxation rate reveal that $1/(T_1 T)$ have a peak as a function of the temperature. The temperature corresponding to this maximum is in fact very sensitive to nesting which is related to the strength of the spin fluctuations. Indeed, this peak vanishes rapidly when ω_0 increases. It is commonly known that, by moving from an antiferromagnetically correlated system to a more usual metallic system, we gradually recover the Korringa law. As a consequence, we should have a maximum of $1/(T_1 T)$ shifted to higher temperatures and the temperature T'^* is then an increasing function of hole doping.

Since the spin fluctuations play an essential role in the normal state, these fluctuations should also contribute significantly to the resistivity. A simple way to determine this contribution is to calculate the following resistivity due to AF fluctuations using the approach considered for nearly magnetic systems [16]:

$$\rho(T) = \frac{1}{T}\int_0^{2k_F} q^3 dq \int_0^\infty \mathrm{Im}\,\chi^{RPA}(q,\omega)\omega e^{\omega/T}\left(e^{\omega/T} - 1\right)^{-2} d\omega.$$

Starting from optimal doping, as the doping increases, the antiferromagnetic fluctuations effects are progressively reduced. These fluctuations less and less affect the quasi-particles and the usual T^2 electron-electron Fermi liquid damping will be progressively dominating. This is in agreement with the observed gradual crossover of $\rho(T)$ at low temperatures from $\rho(T) \sim T$ near optimal hole concentration towards $\rho(T) \sim T^2$ in the over-doped region.

© 2004 WILEY-VCH Verlag GmbH & Co. KGaA, Weinheim

3.2 Under-doped region

The nesting property induces the existence of strong antiferromagnetic fluctuations, which are not strong enough to induce a SDW state, but are able to yield important consequences on many physical properties by inducing a pseudo-gap in the density of states. This pseudo-gap, obtained by different authors [3, 5, 6], is the key point to elucidate the unusual properties in the under-doped region and the T^* line. This pseudo-gap have quite important consequences in the under-doped region for the following reason. When the hole doping concentration is small, the spin fluctuations are rather strong because of the good Fermi surface nesting. In that case, the pseudo-gap of the density of states is deep. In this situation, the Fermi level sits inside the pseudo-gap and the density of states at the Fermi energy, $N(\varepsilon_F)$, is very low. Moreover, it is well known that the properties of a metal are directly related to the density of states at the Fermi energy. Therefore, all these electronic properties will be considerably affected because of the strong reduction of $N(\varepsilon_F)$, compared to a normal metal value.

In order to take into account the electron interactions in the calculation of the spin susceptibility and this important modification of the density of states, we consider N^{corr}, the density of states of the antiferromagnetically correlated system, instead of N^0 which are defined by

$$N^0(\omega) = -\sum_k \frac{1}{\pi} \mathrm{Im}\, G^0(k, \omega) \quad \text{and} \quad N^{corr}(\omega) = \sum_k \frac{1}{\pi} \mathrm{Im}\, G(k, \omega)$$

where G^0 is the non interacting Green function and N^{corr} is the density of states calculated by the Green function G of the correlated system taking into account self-energy corrections. Then the imaginary part of the susceptibility $\mathrm{Im}\,\chi(Q,\omega)$ becomes:

$$\mathrm{Im}\,\chi(Q,\omega) = \frac{1}{2} N^{corr}\left(-\frac{\omega}{2} - \frac{\omega_0}{2}\right) \frac{\sinh\frac{\omega}{2kT}}{\cosh\frac{\omega}{2kT} + \cosh\frac{\omega_0}{2kT}}.$$

Therefore, we should expect an important effect on the Pauli susceptibility $\chi(Q = 0)$ which depends directly on the density of states at the Fermi level. The energy scale of the observed pseudo-gap in the Pauli suceptibility is then the width of the pseudo-gap. Since the spin fluctuations are reduced as the hole doping grows, the pseudo-gap width of the density of states is reduced and consequently T^* decreases as a function of doping. In the other hand, the nuclear relaxation rate and the electrical resistivity which are related to $\mathrm{Im}\,\chi$ will be also reduced at low temperature ($T < T^*$) due to the pseudo-gap.

3.3 Strange metal region

In the strange metal region, in the under-doped region, if we consider the temperature at which the pseudo-gap is going to be filled, the density of states will become almost constant for higher temperatures and the pseudo-gap effect will vanish. In that case, we should have an almost constant magnetic susceptibility. The imaginary part of the susceptibility corresponding to the best nesting will behave roughly as follows:

$$\mathrm{Im}\,\chi(Q,\omega) \approx \frac{\omega}{T} \quad \text{for} \quad \omega < T.$$

In the overdoped region, we show that the resisitivity is linear for temperatures higher than a certain temperature depending on ω_0. Indeed, in this region, the thermal excitations are enough to allow the spin fluctutations, those corresponding to imperfect nesting with the Q wave vector. Hence, the $q = Q$ scattering is so large in these conditions that other fourier components may be neglected, which put an additional constraint in the final scattered state and lead to a marginal Fermi liquid behaviour at high temperature.

Our microscopic approach leads to theoretical predictions which verify the basic hypotheses of the phenomenological marginal Fermi liquid. They are, indeed, in good agreement with the experimental observations. We can conclude that the T^* observed in the under-doped region is a characteristic energy related to the width of the density of state pseudo-gap and that the T'^* line is related to energy corresponding to the deviation from perfect nesting.

We should note that important studies have been carried in this field to explain the pseudo-gap effect and photoemission experiments [17].

4 Conclusion

We have proposed microscopic interpretations of the normal state properties on the basis of weak electron coupling theories. We have tried to give a coherent interpretation of the normal state experimental observations. We have shown that the pseudo-gap regime is governed by the pseudo-gap formed in the single particle density of states by the magnetic short range order. This pseudo-gap in the density of states gets deeper and wider as the antiferromagnetic phase is approached, i.e. at low doping. In this regime, the width of the density of states pseudo-gap is the pertinent energy scale which is direcly connected to T^* and the normal state properties are directly affected by this pseudogap. In the strange metal region, due to thermal activation, the normal state behaves now like a marginal Fermi liquid where the spin fluctuations are responsible for the temperature dependence of the normal state properties. For higher doping, the usual Fermi liquid behaviour is gradually recovered due to a weakening of the spin fluctuations.

The weak coupling limit used in this simple theoretical calculation is certainly too schematic. It is known that the electron correlations are probably not very weak. In a more realistic description, one should certainly improve this point, for example by using a Dynamical Mean Field Theory to take into account more precisely the correlation effects. However, we believe that the properties of a Fermi liquid in close vicinity of a magnetic instability and properly modified by the existence of a two-dimensional Van Hove singularity provide a qualitatively correct description of the normal state.

Acknowledgement We acknowledge the financial support from CMCU to project 01/F1303.

References

[1] P. W. Anderson, Science **235**, 1196 (1987).
P. W. Anderson et al., Phys. Rev. Lett. **58**, 2790 (1987).
[2] C. M. Varma et al., Phys. Rev. Lett. **63**, 1996 (1989); Phys. Rev. Lett. **64**, 497 (1990).
[3] J. Kampf and J. R. Schrieffer, Phys. Rev. B **41**, 6399 (1990); Phys. Rev. B **42** 7967 (1990).
[4] A. Viroszteck and J. Ruvalds, Phys. Rev. B **42**, 4064 (1990).
J. Ruvalds et al., Phys. Rev. B **51**, 3797 (1995).
[5] S. Charfi-Kaddour et al., J. Phys. I (France) **2**, 1853 (1992).
[6] S. Wembter and L. Tewordt, Phys. Rev. B **48**, 10514 (1993).
[7] N. Bulut, D. J. Scalapino, and S. R. White, Phys. Rev. B **47**, 2742 (1993).
[8] J. Friedel, J. Phys. (France) **48**, 1787 (1987); **49**, 1435 (1988).
J. Labbé and J. Bok, Europhys. Lett. **3**, 1225 (1987).
[9] B. Battlog, in: Physics of high temperature superconductors, edited by S. Maekawa and M. Sato (Springer-Verlag, Berlin, Heidelberg, New York, 1992), p. 219.
Z. X. Shen and J. R. Shrieffer, Phys. Rev. Lett. **78**, 1771 (1997).
[10] H. Alloul et al., Phys. Rev. Lett. **63**, 1700 (1989).
[11] H. Yasuoka et al., in: 'Strong Correlation and Superconductivity', edited by Fukuyama, S. Maekawa and A. P. Malozemoff (Springer, New York, 1989).
Ishida et al., Phys. Rev. B **58**, R5960 (1998).
[12] J. Rossat-Mignod et al., Physica B **169**, 58 (1991).
H. A Mook et al., Phys. Rev. Lett. **70**, 3490 (1992).
J. M. Tranquada, P. M Gehring, G. Shirane, S. Shamoto, and M. Sato, Phys. Rev. B **46**, 5561 (1992).
[13] Z. Konstantinovic et al., Physica B **259–261** (1999) 567; Physica C **341–348**, 859 (2000).

[14] J. Kanamori, Prog. Theor. Phys. **30**, 275 (1963).
     N. Bulut et al., Phys. Rev. B **47**, 2742 (1993).
     N. F. Berk and J. R. Schrieffer, Phys. Rev. Lett. **17**, 433 (1966).
[15] D. S. Dessau et al., Phys. Rev. Lett. **71**, 2781 (1993).
[16] R. Jullien et al., Phys. Rev. B **9**, 1441 (1974).
[17] A. V. Chubukov and J. Schmalian, Phys. Rev. B **57**, R 11085 (1998).

Normal state properties of BEDT compounds

N. Joo[1,2], **D. Meddeb**[1], **S. Charfi-Kaddour**[*,1], **R. Bennaceur**[1], and **M. Héritier**[2]

[1] LPMC, Département de Physique, Faculté des Sciences de Tunis, Campus Universitaire 1060 Tunis, Tunisia
[2] Laboratoire de Physique des Solides, UMR CNRS-Paris XI, Bat. 510, 91405 Orsay, France

Received 1 September 2003, revised 4 March 2004, accepted 4 March 2004
Published online 14 April 2004

PACS 74.20.Mn, 74.25.Fy, 74.70.Kn

The κ-(BEDT-TTF)$_2$X organic superconductors are described by a two parameter 2D surface model, in which bandwidth and departure from perfect nesting can be varied. We have studied the spin fluctuation effect on the normal state properties in a Fermi liquid approach using the RPA approximation. The calculated NMR relaxation rate exhibits a peak in $1/(T_1T)$, which strongly decreases when the departure from the perfect nesting of the Fermi surface and the bandwidth decrease. Our calculations show also that the spin fluctuations induce an unusual resistivity temperature dependence. These results are in a good agreement with experimental data done in κ-(BEDT-TTF)$_2$X, at least qualitatively.

© 2004 WILEY-VCH Verlag GmbH & Co. KGaA, Weinheim

1 Introduction

The synthetic molecular coumpounds based on the BEDT-TTF (bisethylendithio-tetrathiafullvalene) molecule have attracted great interest in the field of condensed matter physics because a variety of novel electronic phases are included in these systems [1]. They have layered structure of alternating sheets of conducting BEDT-TTF radical cations and insulating anions X which the valency is -1. The κ-type molecular arrangement of (BEDT-TTF)$_2$X [X = Cu[N(CN)$_2$]Cl, Cu[N(CN)$_2$]Br and Cu(NCS)$_2$] gives rise to various kinds of electronic ground states such as insulating, metallic and superconducting phases.

At ambiant pressure, the κ-(BEDT-TTF)$_2$Cu[N(CN)$_2$]Br [2] and κ-(BEDT-TTF)$_2$Cu(NCS)$_2$ [3] are superconducting with transition temperatures of 11.5 K and 9.5 K, respectively, while κ-(BEDT-TTF)$_2$Cu[N(CN)$_2$]Cl has an antiferromagnetic ordering [4–6]. This latter compound converts into a 13 K superconductor under a pressure of about 300 bar.

In addition to the fairly high critical temperature of superconductivity occuring in these organic compounds, original properties of the normal state, exhibiting deviations from the conventional Landau Fermi liquid behaviour, have been observed. Moreover, important experimental data, inconsistent with the BSC theory seem to indicate a non conventional superconductivity. Some similarities of the phase diagram and some unusual properties in the normal state imply analogies to the high-T_c cuprates with carrier doping playing the role of pressure in organics [4, 5], the AF spin-fluctuations was expected to be the origin of the superconductivity [7–11].

The κ-(BEDT-TTF)$_2$X salt family can be described by a unified pressure-temperature (P, T) diagram (Fig. 1), where the ground state at low temperature and at low pressure is antiferromagnetic and superconducting at higher pressure. A metal insulator transition occurs at $P = 200$ bar [12]. Above the critical temperature, the normal phase is not a conventional metal. Changing the anion can be considered as equivalent to varying the pressure. The lower pressure, i.e. the stronger relative correlations, corresponds

---

[*] Corresponding author: e-mail: samia.kaddour@fst.rnu.tn, Phone: +00 216 98 925 884, Fax: +00 216 71 885 073

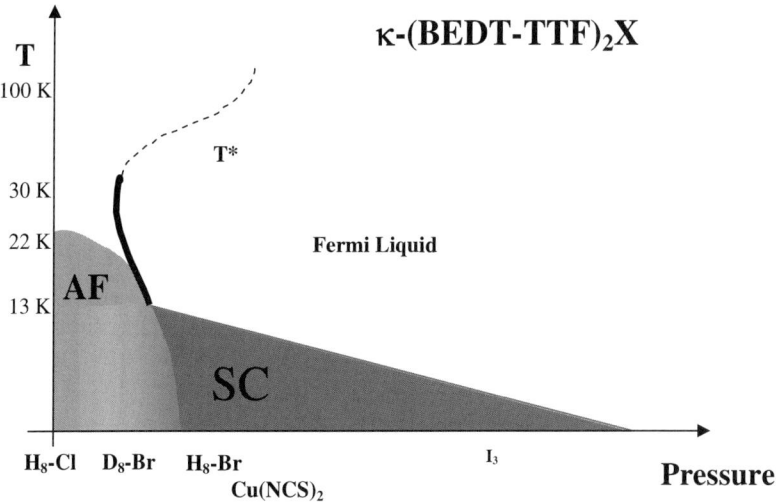

Fig. 1 (online colour at: www.interscience.wiley.com) Unified phase diagram of κ-(BEDT-TTF)$_2$X salt family.

to the anion X = Cu[N(CN)$_2$]Cl, which is usually considered as an antiferromagnetic Mott insulator at ambient pressure. This case requires more sophisticated treatment of the electron correlations and is out of the scope of this paper. Changing the anion to X = Cu[N(CN)$_2$]Br and X = Cu(NCS)$_2$ corresponds to lower the electron-electron correlations (or higher pressure).

In the normal state of the κ-(BEDT-TTF)$_2$X salts, the spin-lattice relaxation rate observed by ^{13}C NMR experiments shows an enhancement below 150 K and takes a cusp around T^* [13–15]. This temperature T^* occurs at 50 K for X = Cu[N(CN)$_2$]Br and at 55 K for X = Cu(NCS)$_2$ with a lower amplitude for the second compound. Moreover, the NMR experiments under pressure done by Mayaffre et al. [14] have shown a vanishing of the $1/T_1T$ enhancement at 4 kbar for X = Cu[N(CN)$_2$]Br. The enhancement and the anomaly at T^* have been interpreted as AF spin fluctuations and a pseudogap formation.

A pseudogap effect was also observed in the magnetic susceptibility below 100 K by different authors [14–17]. It was observed that the susceptibility for X = Cu[N(CN)$_2$]Br drops more rapidly at low temperature than the one for the X = Cu(NCS)$_2$.

Concerning the transport properties, above the critical temperature, a broad resistivity hump is observed around T = 80 K for the bromine salt and T = 110 K for the Cu(NCS)$_2$ salt. This hump disappears rapidly as a moderate pressure is applied [18, 19].

Ultrasonic velocity measurements in the κ-(BEDT-TTF)$_2$X in their normal state have revealed compressibility anomalies where an important softening is identified around 40–50 K [20]. In order to characterise the origin of this anomaly, the authors have studied its behaviour under the application of hydrostatic pressure. The observed behaviour is found to mimic those of the transport and magnetic properties of these materials which have been attributed to the magnetic fluctuations.

In our previous work [11], we have suggested that spin fluctuations, which are present in the system due to an imperfect nesting property of the Fermi surface, can induce anomalies of the magnetic susceptibility and explain the behaviour of the $1/T_1T$. We have found the pressure induces changes in the Fermi surface topology and nesting become worse. Consequently, the spin fluctuation are reduced and the usual behaviour is recovered. Our point of view is supported by experimental observations of the Fermi surface topology made by magneto-transport measurements [21, 22]. In this paper, we expose the effect of these spin fluctuations on the transport properties to try to make a link with the magnetic properties.

Some experiments and theories support another point of view in which the Mott-Insulator transition plays an important part in the origin of the pseudogap observed in the normal state [23, 24]. Moreover, it was also suggested that a second order transition takes place at T^* [17].

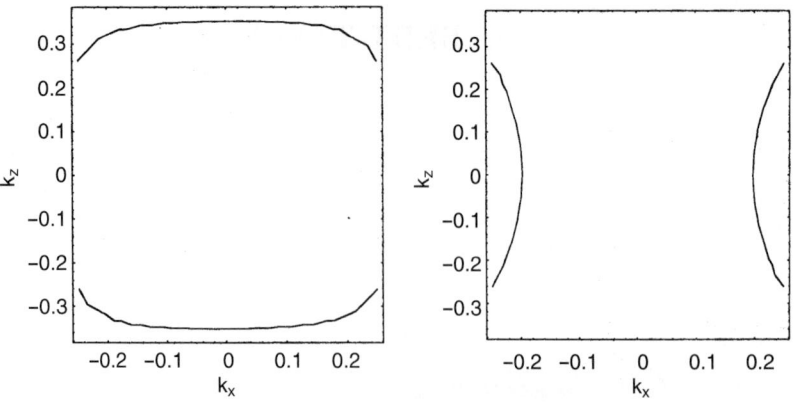

Fig. 2 Fermi surface corresponding to the band (1) and (2) for $t_1/t_2 = 0.3$.

2 The model

Band structure calculations and several theoretical studies [25, 26] have been carried out for the κ-(BEDT-TTF)$_2$X family and lead to almost the same shape of the Fermi surface which is in a reasonable qualitative agreement with magneto-transport data [21]. The κ-(BEDT-TTF)$_2$X organic compounds have large and structurally complex unit cell. We have considered a simplified dispersion relation, that reproduces band structure calculations and fits well magneto-transport data, which is described by a two parameter (t_1 and t_2) 2D surface model as follows:

$$\varepsilon_k^{(1)} = 2t_1\left(\cos k_z - \cos 0.7\pi\right) + 2t_2\sqrt{(1+\cos k_x)(1+\cos k_z)},$$
$$\varepsilon_k^{(2)} = 2t_1\left(\cos k_z - \cos 0.7\pi\right) - 2t_2\sqrt{(1+\cos k_x)(1+\cos k_z)}.$$

(1)

For both bands, the shape of the Fermi surface displays a nesting property which is very sensitive to the ratio t_1/t_2. This property is optimal around $t_1/t_2 = 0.3$ (see Fig. 2). According to the data obtained by Caulfield et al. [21] and references therein, the family under study corresponds to values of t_1/t_2 higher than the 0.3. The nesting property of the Fermi surface is more imperfect when the value of t_1/t_2 is higher.

The electronic systems under discussion are described by a simple two-dimensional Hubbard model. Since there are two conduction bands, we consider intraband intra-atomic Coulomb interaction U, which couples electrons occupying the same band, supposed to be identical for both bands, and repulsive on-site inter-band Coulomb repulsion U' coupling electrons occupying different conduction bands.

$$H = \sum_{k,s,i} \varepsilon_k^{(i)} c_{i,k,s}^+ c_{i,k,s} + \frac{U}{N} \sum_{k,p,q,s,i} c_{i,k+q,s}^+ c_{i,k,s} c_{i,p-q,-s}^+ c_{i,p,-s}$$
$$+ \frac{U'}{N} \sum_{k,p,q,s,i,j\neq i} c_{i,k+q,s}^+ c_{i,k,s} c_{j,p-q,-s}^+ c_{j,p,-s} + \frac{U'}{N} \sum_{k,p,q,s,i,j\neq i} c_{i,k+q,s}^+ c_{j,k,s} c_{i,p-q,-s}^+ c_{j,p,-s}$$

(2)

where $c_{i,k,s}$ indicates electron annihilation operator corresponding to the i band with a wave vector k and a spin s, N is the number of sites. In the Hubbard approximation, U is usually the "bare" local on-site Coulomb integral. The sensible value of U to be put in the Hubbard Hamiltonian treated in the RPA is obviously not the bare one, but an effective one which is reduced due to the screening. It is well known that many body effects such as those described by Kanamory–Bruchnev theory screen the Coulomb repulsion in metals and lead to a reduced value U [27].

Besides screening, low dimensional fluctuations may affect the RPA parameters and possibly invalidate such a mean field treatment, particularly in the quasi-1D band. However, the energy bands of the BEDT-TTF salts are not strictly one-dimensional and transverse coupling between neighbouring chains exists with a typical energy scale $t_\perp$ even in the quasi-1D band. It is proved that $t_\perp$ destabilizes the Luttinger liquid fixed point [28]. In the weak correlation regime, for a given $t_\perp$ and a temperature much lower than T_F ($T_F = E_F/k_B$ where E_F is the Fermi energy and k_B the Boltzman constant), as the temperature decreases, transverse one-particle coherent motion starts to develop. A crossover from Luttinger liquid to Fermi liquid takes place at a deconfinement temperature T_x. Since we are studying the metallic region of the BEDT-TTF salts, they are not in the strong interaction limit, but rather weak or intermediate limit, this cross-over temperature is certainly larger than the temperature discussed in this paper, which justifies a Landau-Fermi liquid approach with renormalized parameters [11 and references therein].

Therefore, U and U' used in our RPA calculation are definitely not the "bare" values, which would be given by quantum chemistry but effective reduced values. In the spirit of Landau theory of normal Fermi liquids, we consider U and U' as phenomenological quantities. We determine their values as the best choice to account for the properties of the normal state. In our study, we have considered $U = U'$.

3 Normal state properties

To take into account the spin fluctuation effects, we have calculated the magnetic susceptibility, using the RPA approximation where

$$\chi^{RPA} = \frac{\chi_0}{1 - U\chi_0} \qquad (3)$$

and

$$\chi^0 = \frac{1}{2\pi} \int d^2k \, \frac{f(\varepsilon_{k+q^i}) - f(\varepsilon_{k^i})}{(\omega + \varepsilon_{k^i} - \varepsilon_{k+q^i})}.$$

The real part of magnetic susceptibility for both bands exhibits a series of maximum values, as it shown in Fig. 3. The corresponding values are associated to good nesting vectors of the Fermi surface and will contribute to an enhancement of the susceptibility in the metallic phase due to the Stoner factor. Therefore the Korringa law, well established for usual metal in which $1/T_1T$ is temperature independent, is not valid in the present case.

Indeed, we have calculated the spin-lattice relaxation rate using the following expression:

$$\frac{1}{T_1T} \propto \lim_{\omega \to 0} \sum_q \frac{\text{Im}\,\chi^{RPA}(q,\omega)}{\omega}. \qquad (4)$$

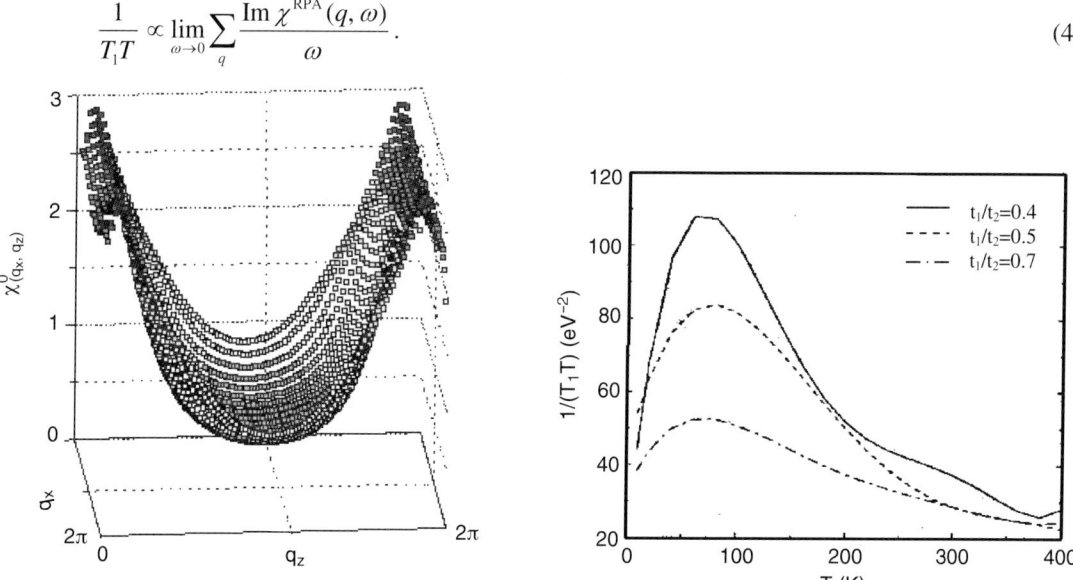

Fig. 3 Real part of the spin susceptibility $\chi^0(\omega = 0, q_x, q_z)$ for $t_1/t_2 = 0.3$ and $T = 100$ K for the band (1).

Fig. 4 Temperature dependence of $1/(T_1T)$ for different values of t_1/t_2 for $U = 0.3$ eV and $t_1 = 0.1$ eV.

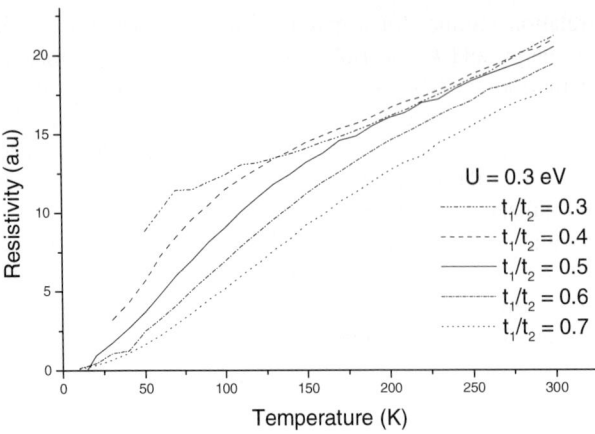

Fig. 5 Resistivity versus temperature for different values of t_1/t_2.

Our calculations of the relaxation rate [11], displayed in Fig. 4, reveal that $1/T_1T$ have a peak as a function of temperature which is very sensitive to the nesting property related to the value of t_1/t_2 and the intensity of the Coulomb interaction related to U/W (W is the bandwidth). The temperature T^*, corresponding to the maximum of $1/T_1T$, is around 70–80 K in our calculation, in a rather good agreement with NMR experiments leading to T^* around 50–60 K [13–15]. The peak is suppressed when t_1/t_2 increases. This means that the existence of the peak is related to well defined spin fluctuations. Such a condition is guaranteed by a rather good nesting property of the Fermi surface. Indeed, by increasing t_1/t_2, we gradually destroy the quality of the nesting property and we move from an antiferromagnetically correlated system to a more usual metallic system where Korringa law is recovered.

We have performed a full calculation of the density of states taking into account the spin fluctuations and we show that it appears a pseudogap near the Fermi level particularly of the second band. This pseudogap depends strongly on temperature since the spin fluctuations involved are strongly reduced for temperatures higher than 150 K. The existence of this pseudogap induces a reduction of the density of states near the Fermi level and consequently a reduction of the magnetic susceptibility. We have obtained particularly that the static susceptibility $\chi(Q = 0)$ is reduced at low temperature:

$$\chi(T = 50 \text{ K}) = 0.66 \quad \chi(T = 300 \text{ K})$$

for $t_1/t_2 = 0.4$ and $U = 0.3$ eV [29], which corresponds to value of $U/W = 0.53$.

Transport measurements performed on BEDT TTF salts in the normal state do not exhibit a T^2 behaviour but show a peak at 50–100 K which disappears under pressure. To understand the transport properties, we have calculated the resistivity due to an exchange of antiferromagnetic fluctuations using the formulation given by R. Jullien et al. [30]. We have considered the reduced resistivity ρ/ρ_0 in the conducting plane as follows:

$$\frac{\rho}{\rho_0} = \frac{1}{T} \int_0^{2\pi}\int_0^{2\pi} \sqrt{(q_x^2 + q_z^2)} \left(\int_0^{2\infty} 2 \operatorname{Im}\chi^{RPA}(q,\omega,T)\,\omega n(\omega)[1+n(\omega)]\,d\omega \right) dq_x\, dq_z$$

where $\operatorname{Im}\chi^{RPA}$ is the imaginary part of the RPA susceptibility defined in Eq. (3), $n(\omega)$ is the Bose–Einstein distribution function. We have obtained a temperature dependence with an unusual behaviour which is more pronounced when the nesting is better, i. e. when the spin fluctuations are stronger. The temperature dependence starts to deviate from a quadratic behavior at around 50 K and the enhancement of the resisitivity due to spin fluctuation occurs particularly between 50 and 200 K.

In the same range of temperatures, between 50 and 200 K, the behaviour of the $1/T_1T$ is different from what we should observe in a ususal metal (Korringa law). We can conclude that abnormal magnetic and transport properties of the normal state of these compounds come both from the spin fluctuations present in the system.

5 Conclusion

According to our results, we can confirm that antiferromagnetic fluctuations are enhanced by nesting as it can be seen from the shape of the Fermi surface. It turns out that when we apply a pressure in these compounds, we modify not only the ration U/W but also the ratio t_1/t_2 and consequently the nesting of the Fermi surface. Indeed, comparing the effects of pressure in experiments and the effect of increasing t_1/t_2 according to our calculations, we find that they are comparable if the pressure acts in the sense of increasing t_1/t_2. Our results are in agreement with experiments [21, 22]. We have explained how in a Fermi liquid approach, we can obtain a strange metallic region due to antiferromagnetic fluctuations. We have also demonstrated that this unusual behaviour in the transport and magnetic properties disappears gradually when the nesting property is destroyed progressively.

Acknowledgement We acknowledge the french-tunisian cooperation CMCU (project 01/F1303).

References

[1] T. Ishiguro, K. Yamaji, and G. S Saito, Organic superconductors (Springer, Berlin, 1998).
   J. M. Williams et al. Organic superconductors (Prentice Hall, New Jersey, 1992); Phys. Rev. **126**, 1470 (1962).
[2] A. M. Kini et al., Inorg. Chem. **29**, 2555 (1990).
[3] H. Urayama et al., Chem. Lett. **1988**, 55 (1988).
[4] J. M. Williams, A. M. Kini, H. H. Wang, K. D. Carlson, U. Geiser, L. K. Montgomery, L. K. Pyrka, D. M. Watkins, J. M. Kommers, S. J. Boryschuk, A. V. Striedy Crouch, W. K. Kwok, J. E. Schirber D. L. Overmyer, D. Jung, and M.-H. Whangbo, Inorg. Chem. **29**, 3262 (1990).
[5] Sushko et Andres. Phys. Rev. B **47**, 330 (1993).
[6] R. H. McKenzie, Comments Condens. Matter Phys. **18**, 309 (1998).
[7] H. Kino and H. Kontani, J. Phys. Soc. Jpn. **67**, 3691 (1998).
[8] H. Kondo et al., J. Phys. Soc. Jpn. **67**, 3695 (1998); J. Phys.: Condens. Matter **11**, L363 (1999).
[9] K. Kuroki and H. Aoki, Phys. Rev. B **60**, 3060 (1999).
[10] J. Schmalian, Phys. Rev. Lett. **81**, 4232 (1998).
[11] R. Louati, S. Charfi-Kaddour, A. Ben Ali, R. Bennaceur, and M. Héritier, Phys. Rev. B **62**, 5957 (2000).
[12] D. Fournier et al., Phys. Rev. Lett. **90,** 127002 (2003).
[13] S. M. de Soto et al., Phys. Rev. B **52**, 10364 (1995).
[14] H. Mayaffre et al., Europhys. Lett. **28**, 205 (1994).
[15] A. Kawamoto et al., Phys. Rev. Lett. **74**, 3455 (1995).
   K. Kanoda, Hyperfine Interact. **104**, 235 (1997).
[16] V. Kataev, Solid State Commun. **83**, 435 (1992).
[17] T. Sazaki et al., Phys. Rev. B **65**, 06505 (2002).
[18] Y. V. Sushko et al., J. Phys. I (France) **1**, 1375 (1991).
[19] I. D. Parker et al., J. Phys.: Condens. Matter **1**, 4479 (1989).
[20] K. Frikach et al., Phys. Rev. B **61**, R6491 (2000).
[21] J. Caulfield et al., J. Phys.: Condens. Matter **6**, 2911 (1994).
[22] J. Biggs et al., J. Phys.: Condens. Matter **14**, L495 (2002).
[23] D. Fournier et al., Phys. Rev. Lett. **90**, 127002 (2003).
[24] P. Limelette et al., Phys. Rev. Lett. **91**, 16401 (2003).
[25] E. Demirap and W. A. Goddard, Phys. Rev. B **56**, 11907 (1997).
[26] Y. N. Xu et al., Phys. Rev. B **52**, 12946 (1995).
[27] J. Kanamori, Prog. Theor. Phys. **30**, 275 (1963).
   N. Bulut et al., Phys. Rev. B **47**, 2742 (1993).
   N. F. Berk and J. R. Schrieffer, Phys. Rev. Lett. **17**, 433 (1966).
[28] V. N. Prigodin and Y. A. Firsov, Sov. Phys. JETP **49**, 369 (1979).
[29] D. Meddeb et al., in preparation.
[30] R. Jullien, M. Béal-Monod, and B. Coqblin, Phys. Rev. Lett. **30**, 1057 (1973).

Ba@Pr or Pr@Ba in R123 HTSC – To be or not to be SC

M. Akhavan[*]

Magnet Research Laboratory (MRL), Department of Physics, Sharif University of Technology,
P.O. Box 11365-9161, Tehran, Iran, and The Abdus Salam International Centre for Theoretical Physics,
Strada Costiera, 11, 34014 Trieste, Italy

Received 1 September 2003, accepted 6 March 2004
Published online 14 April 2004

PACS 74.25.Fy, 74.25.Ha, 74.72.Bk

In spite of the vast amount of experimental and theoretical knowledge accumulated in HTSC, the nature of the interaction driving charge carriers to form Cooper pairs below T_c is still unknown. To elucidate the controversial insulating/SC Pr123, the substitution of Pr at Ba site (Pr@Ba) with Ba at Pr site (Ba@Pr) in R123 is compared and reviewed. This might shed some light on the microscopic origin of HTSC.

© 2004 WILEY-VCH Verlag GmbH & Co. KGaA, Weinheim

1 Introduction

The substitution effect of Ba@Pr, Pr@Ba, and Pr@R in $RBa_2Cu_3O_{7-\delta}$ (R123) ($R = Y$ or rare earth elements, except Ce, Tb, Pm) have been the subject of many investigations. Suppression of superconductivity by the substitution of Pr@R has attracted great interest since the discovery of the (R123) series with $T_c \approx 90$ K. For a recent review on the role of Pr in HTSC see Ref. [1]. The insulating behavior of Pr123 compound in the orthorhombic phase and isostucture with other superconducting (SC) R123 is one of the most studied subjects in recent years [2]. Pr@Ba has been presented the reason for both the insulating Pr123 [3], and the SC Pr123 [4]. There have been many controversies as the reason for the anomalous electronic and magnetic behaviors [5]. Does Pr have a unique chemical or electronic characteristic among the light rare earth elements, which causes such a strange behavior, and/or how the electronic structure of Pr123 compounds changes with Pr doping? The ultimate question is under what condition, if ever, the Pr123 compound forms a SC state. [Hereafter, x denotes the doping concentration of Pr@Ba or Ba@Pr, and δ donates the oxygen concentration].

The purpose of this review is to highlight some of the similarities and differences when R123 HTSC systems accept Pr/Ba mis-substitution (i.e. Ba@Pr or Pr@Ba). This we hope would elucidate the highly controversial Pr123 issue. Due to the vast amount of work on Pr123, and the page limitation for this paper, it has become impossible to mention all the works on the subject. Therefore, the author apologizes in advance for omission from an incomplete reference list. A comprehensive review is due elsewhere.

2 Sample preparation

The main difficulties in the experimental verification of the models describing the Pr123 arise from the complex doping mechanism in R123 series, especially with Pr/Ba variant compounds, and from the strong influence of crystal preparation on the material properties. Therefore, it is necessary to have detail knowledge of the sample preparation procedures and their limitations/drawbacks.

The samples prepared under solid-state reaction procedures at 930 °C, close to the melting point of the compound, are found to be inhomogeneous, but single phase with marginal $BaCuO_2$ and $PrBaO_3$ impu-

---

[*] e-mail: akhavan@sharif.edu, Phone: +98 21 6164510, Fax: +98 21 6012983

rity phases. To prevent the formation of magnetic impurity phase PrBaO$_3$, the initial sintering is carried out in the presence of Ar gas [6].

The epitaxial thin film samples are usually grown by pulsed-laser ablation of the pressed and fired ceramic targets (prepared by the bulk preparation procedures described above) on heated SrTiO$_3$ (100), LaAlO$_3$ (100), or MgO (100) substrates at 650 ± 100 °C. The film growth is usually carried out in a low oxygen pressure, and after the deposition, the films are slowly cooled to room temperature in a high oxygen pressure. Under such a low substrate temperature, just due to the presence of seed crystals, nucleation of the 123 phase over other competing phases such as BaCuO$_2$, PrBaO$_3$, Pr$_2$BaCuO$_5$ is formed, but since the atoms arrange themselves after arriving at the substrate, and the diffusion length is small comparing to the cell parameters, considerable Pr/Ba site-switching is possible. Therefore, the presence of comparable Ba/Pr inhomogeneity in thin films with that of solid-state procedure is expected.

There have been reports on oxygen annealed traveling solvent flouting zone (TSFZ) Pr123 single crystal grown in a deoxidized atmosphere [7]. One major and unavoidable problem with the TSFZ method is the fluctuation in local composition on a microscopic scale, inevitably present in a multi-component liquid like 123, leading to the nonuniform distribution of the Pr/Ba substitution ratio over the crystal. This, in turn, results in crystals having larger c-axis parameter (11.735 Å), very large change of T_c under applied pressure ($T_c > 100$ K under pressure of 8 GPa with still potential to increase), insulating behavior in the normal state, and large difference between the T_c of different samples grown by the same method [8]. As a result of the drawbacks in the TSFZ technique, large scattering in both structural and SC properties are exhibited even within a single sample.

In addition, in general, preparing high quality Pr123 crystals of a size (≥ 10 mg) suitable for most precise characterization experiments (e.g., NMR) requires extensive work. For one reason, Al from Al$_2$O$_3$ crucible causes contamination of the Cu (1) sublattice as well as on the magnetic structure of the Cu(2) sublattice at low temperature. Using MgO crucible improves the situation, and BaZrO$_3$ is ideal, but crucible of large enough sizes is not available. Another reason is the existence of a finite solid solution range for the R/Ba sublattices for the light rare earths like Pr. So far, many unsuccessful attempts have been made to prepare SC Pr123. Of course, the local Pr/Ba ratio of different samples can be influenced by an additional heat treatment, but again as in other methods, the composition of the resulting crystal is inhomogeneous.

To conclude, in light of the extent of the controllable parameters for the polycrystalline samples prepared by the solid-state reaction technique over the limitations in the thin film production such as substrate temperature and diffusion length, and especially, the non-uniformity of the composition in the TSFZ single crystal or crystal pulling technique, it is justifiable that the homogeneity of the polycrystalline Pr123 samples and their Pr/Ba mis-substitution can be comparable with that in the thin films and not any more than in the TSFZ single crystals. Therefore, for the present purpose of studying Pr@Ba and Ba@Pr in R123 HTSCs, the data extracted from the polycrystalline samples are very reliable.

3 Structural properties

To clarify the role of Pr played in R123, and in particular, to differentiate the effect of Pr/Ba mis-substitution on the R or Ba site, it is necessary to count not only how much Pr is in the compounds, but also the site it occupies. The knowledge of the crystal parameters as well as different bond distances and the valences of different substituting atoms in the compound elucidate the mechanism involved.

Many experiments have been carried out for Pr@Ba on samples of Pr123 [6, 9–12], Y123 [13, 14], Sm123 [12], Eu123 [15, 16], Gd123 [17], and other rare earth elements R@Ba such as La and Eu@Ba in Pr123 [18], La@Ba in Y123 [14], or in Pr123 [4], Nd@Ba in Nd123 [13, 19], Ho,Y and Ca@Pr in Pr123 [20], Ba@Pr in Pr123 [21, 22]. Through these experiments different structural and valence modifications related to Pr are reported.

A striking feature arising out of Pr@Ba substitution is the appearance of an orthorhombic-tetragonal transition (OT), or in some cases, orthorhombic-tetragonal-orthorhombic transition (OTO). For the case of Pr@Ba [11] in Pr123, the lattice parameters $a < b \approx 3.9$ Å $\approx c/3$ for $0 \leq x < 0.4$, orthorhombic O(I) symmetry; $a = b \approx 3.9$ Å $> c/3$ for $0.4 \leq x \leq 0.65$ tetragonal T symmetry; and then to a different type of orthorhombic O(II) symmetry for $0.65 < x < 1$ with $a < b \approx 5.5$Å ($\approx \sqrt{2} \times 3.9$ Å).

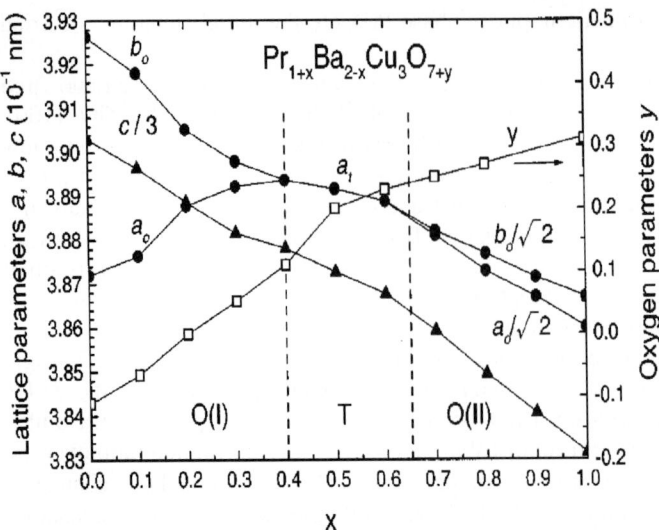

Fig. 1 Variation of orthorhombicity/tetragonal lattice parameters a, b, c, oxygen parameter $\delta(y)$ and O(I)–T–O(II) phase boundaries for the oxygenated $Pr_{1+x}Ba_{2-x}Cu_3O_{7+\delta}$ system.

Figure 1 shows the variation of O–T–O lattice parameters a, b, c, δ, and O(I)–T–O(II) phase boundaries of the oxygenated $Pr(Ba_{2-x}Pr_x)Cu_3O_{7+\delta}$ system. The correlation between crystal symmetry O(I)–T–O(II) and oxygen distribution in the $CuO_{1+\delta}$ plane in fact results in the increase of Pr–O bond length d_{Pr-O} with Pr@Ba due to the accompanying increasing δ in the $CuO_{1+\delta}$ plane.

Ha [14] has performed a similar comparative study of Nd, La and Pr@Ba in Y123 system with similar findings. The XRD pattern indicates the OT transition identified by merging the double peaks around $2\theta = 47°$. The x domain can be divided into two parts at around $x = 0.2$ at which the ratio a/b starts to change rapidly. As x increases, oxygen content slightly increases for $x \leq 0.2$, but it does rapidly increase for high x content. The rapid otorhombic distortion of the a/b is accompanied by the increase of oxygen content above seven. Meanwhile, the average Cu valence state shows a rather steep reduction for $x \leq 0.2$ followed by a slow decrease, and reaching ~2.28 for La@Ba and ~2.20 for Pr@Ba with $x = 0.6$. For Pr@Ba, the valence state of copper reduces more rapidly and the oxygen content is smaller than that for the La@Ba substitution. Similar results have been obtained for Sm123 and Y123 [12]. Here, the oxygen content increases from 6.94 ($x = 0$) to 7.10 ($x = 0.35$) and OT transition occurs at $x = 0.3$–0.35. When Sm is changed to Nd or Eu the same oxygen and OT transition trend prevails.

Another comparative study is followed further by Xu et al. [18] on La, Eu, and Pr@Ba in Eu123 compound. Again, the occupancy of O(5) site increases as more Ba^{2+} is replaced by R^{3+}. Once the OT transition occurs, the O(1) and the O(5) sites are equivalent. Figure 2 shows the lattice parameters, cell volume, and orthorhombicity of the system. It shows that a-parameter stays almost the same; b-parameter for La and Pr are consistently greater than Eu; c-parameter and cell volume decreases with x. The OT transition occurs for the three elements between $x = 0.10$ and $x = 0.15$. Occurrence of this transition is due to disorder of the oxygen at the chain and antichain sites. However, the larger R^{3+}(La) allows for higher anti-chain site occupancy before the transition, most likely due to less strain on the bonds. This is in contrast to the observation in fully oxygenated R123, where otorhombicity decreases as the ionic radius increases. It is also shown that R^{3+}@Ba has stronger effect on the bond length between Cu(1) and O(1) than the same atoms on R site. Thus, this further evidences that the R went to the Ba site. Even, in the case of Pr@Ba in Gd123, Pr exclusively occupies Ba [23]. Of course, the neutron data supports the idea that the ionic size is dominant factor in site preference of the R@Ba [18].

Figure 3 shows that as x increases, the buckling angle of CuO_2 plane decreases, resulting in the flattening of the CuO_2 plane, and the distance between the CuO_2 planes decreases. The trend of variation follows the ionic size differences between the substituting La^{3+}, Pr^{3+}, Eu^{3+} ions and Ba^{2+}. The striking feature is that from the point of view of structural distortion arising from these substitutions the Pr follows the ion size relation, between La and Eu, exhibited in Fig. 3, but from the electronic point of view, Pr

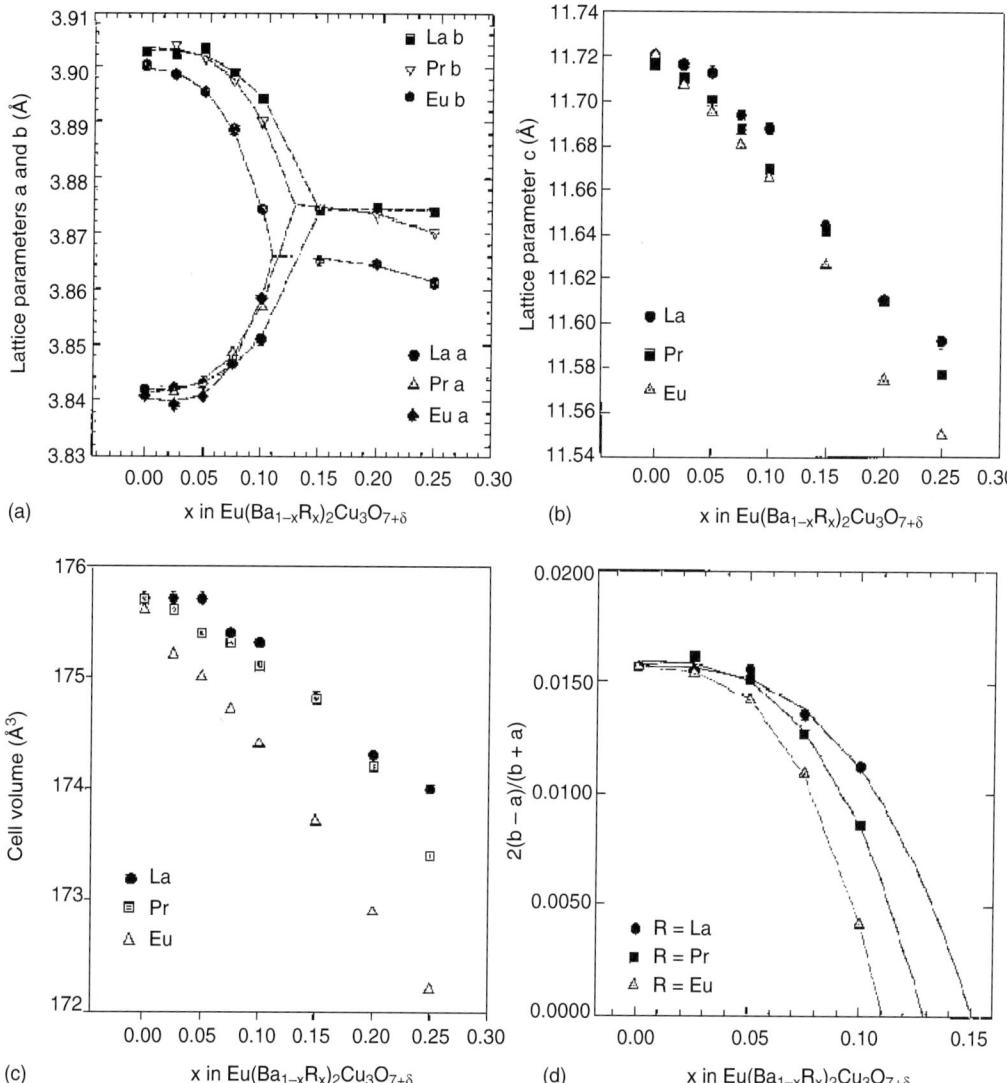

Fig. 2 Lattice parameters of $Eu(Ba_{1-x}R_x)_2Cu_3O_{7+\delta}$ for R = La, Pr, and Eu. (a) Lattice parameters a and b. (b) Lattice parameter c. (c) Cell volume. (d) Orthorhombicity (after Xu et al. [18]).

has a stronger effect than both La and Eu. This is exhibited from the values of critical doping concentration for suppression of SC $x_c(Pr) = 0.35 < x_c(Eu) = 0.55 < x_c(La) = 0.65$.

The valency of different atoms for the case of Pr@Ba or Ba@Pr is important, as it would allow one to distinguish between different models. It should be noted that the standard X-ray and neutron diffraction techniques cannot fully resolve this issue due to the very similar X-ray scattering factor and neutron-scattering length of Pr and Ba. All reported valences based on these techniques are rough estimates and suggestive, or indirect values by comparative results. Obviously, there is debate over the valence of Pr at different sites in Pr123.

Supports for models based on tetravalent Pr@Ba, is provided by recent high-energy X-ray Compton scattering experiments [24]. However, such a scenario contradicts the interpretation of transport measurement of Nd123 and Pr123 with Pr@Ba substitution [25]. Similar reduction in T_c for similar amounts of Nd and Pr@Ba indicates that both ions are incorporated in the trivalent oxidation state. Moreover,

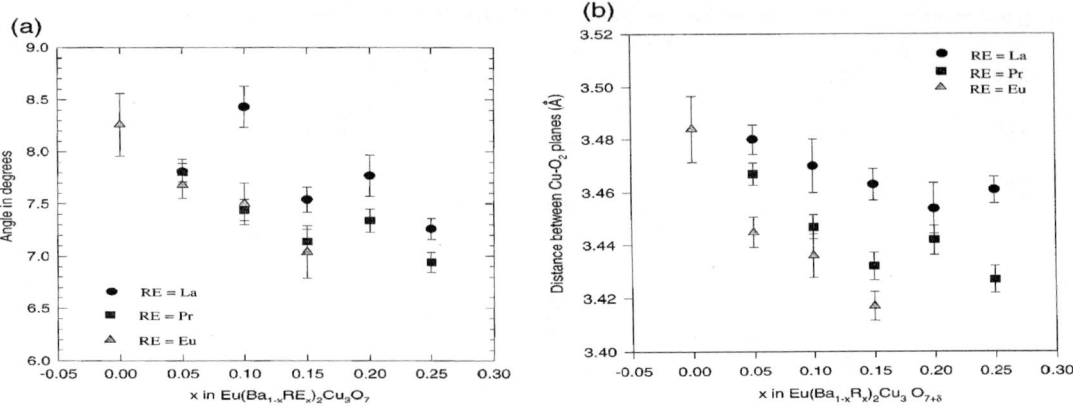

Fig. 3 (a) The buckling angle of CuO_2 plane and (b) the distance between two adjacent CuO_2 plane layers of $Eu(Ba_{1-x}R_x)_2Cu_3O_{7+\delta}$ as a function of substitution level x (after Xu et al. [18]).

recent X-ray-absorption results for Pr@Ba in Pr123 show that the Pr valence is intermediate [26]. The valency of Pr is nominally 3+, but with strong hybridization of Pr_{4f}–O_{2p} with neighboring oxygen results in a characteristically mixed or fluctuating valence.

Different valences exhibited by Pr, is a result of its different bondings at different sites. For Pr@Ba, by increasing the amount of Pr, more oxygen is brought into the structure in order to maintain charge neutrality of the system. The increase in the oxidation state of Cu(1) increases the bonding length of Cu(1)–O(4). Conversely, the ionic radius $r_{Pr3+} \ll r_{Ba2+}$ resulting in an overall decrease in the c-axis lattice parameters as x increases. These competing effects cancel each other resulting in a very small change in Cu(1)–O(4) bond lengths. Furthermore, this valence fluctuation effect induces metal (SC)–insulator transition.

In addition to Pr doping at R or Ba sites, it might be possible to induce transition between the SC (metallic) and insulating states by applying external stresses. A metal (SC)–insulator transition might occur in Nd123 under uniaxial pressure along the c-axis, which would shift all structural, ionic, and energy parameters to the directions favorable for the Nd^{IV} state. Similarly, an insulator–metal (SC) transition might be induced in Pr123 by applying biaxial pressure in the ab-plane or negative chemical pressure along the c-axis by Ba@Pr. Because of the coupling between the electronic state and the crystal structure the transition would be 1st order, accompanied by discontinuous changes in structural parameter such as c-axis, d_{R-O}, and buckling angle. This could well be considered as the reason for SC exhibition reported in some Ba rich Pr123 samples with some Ba@Pr.

For Ba@Pr [27] (or similarly, Ca@Y [28]), we expand the surrounding O_8 cage, resulting in an increase in d_{R-O}. The expansion of the O_8 cage will not be isotropic because of the Cu–O bonds, having a significant covalent character, are stiffer and more reluctant to expand than the other bonds. As a consequence, the expansion along the c-axis dominates that in the ab-plane, resulting in a decrease in buckling angle. The larger c-axis parameter has indeed been observed in the SC Pr123 samples [8]. The above effects, which are purely geometrical in origin, are accentuated by the lower valence of Ba(Ca) as compared with Pr(Y).

4 Physical properties

4.1 Normal state transport

For the case of Ba@Pr in Pr123, the resistivity versus doping concentration x for the temperatures range of 70 to 300 K is shown in Fig. 4. All the samples have insulating behavior. By increasing x, resistivity increases for low doping values ($x < 0.08$) and then, decreases sharply with the increase of x up to $x = 0.3$ [21]. This is caused by two mechanisms: (1) carrier scattering from lattice disorders and (2) increase of

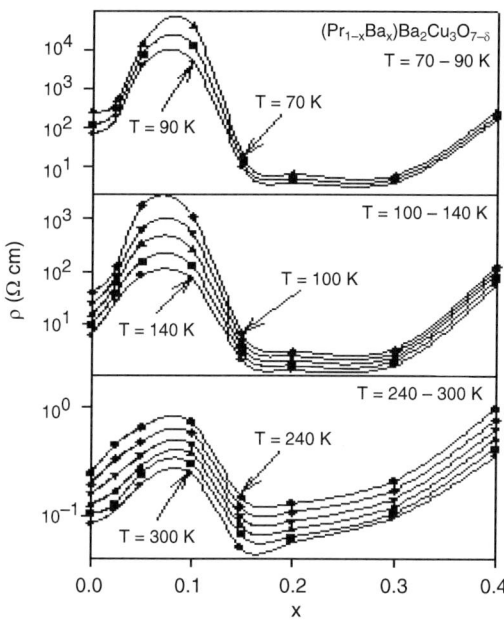

Fig. 4 Resistivity versus Ba@Pr doping concentration x for temperatures of 70 to 300 K for $(Pr_{1-x}Ba_x)Ba_2Cu_3O_{7-\delta}$ (after Khosroabadi et al. [21]).

conductivity with the increase of hole carrier content due to Ba@Pr. For lower x values the first factor is dominant, while for larger x values the second factor dominates. Similar results have been reported by Yang et al. [22] and Khosroabadi et al. [27].

In contrast to above, the normal state resistivity of Y123 for Pr@Ba shows metallic behavior for low x values. The metal–insulator transition (MIT) and superconductor-insulator transition (SIT) are at $x = 0.7$ [14]. In addition, Fig. 5 shows the plot of resistivity for Pr@Ba in Gd123. It shows relatively sharp superconducting transition width with no indication of any secondary phase like two-step transitions suggesting single phase Gd123 compound are obtained, consistent with the XRD results. The MIT occurs at 0.2 corresponding to OT transition and SIT occurs at $x = 0.35$.

When La, Pr, or Eu are substituted for Ba in Eu123 system [18], the suppression of T_c in all substitutions have a similar trend. For low $x \leq 0.025$, T_c is not affected. As x increases, the decrease of T_c is monotonic. The suppression for Pr@Ba is fastest, while for La@Ba it is slowest with x.

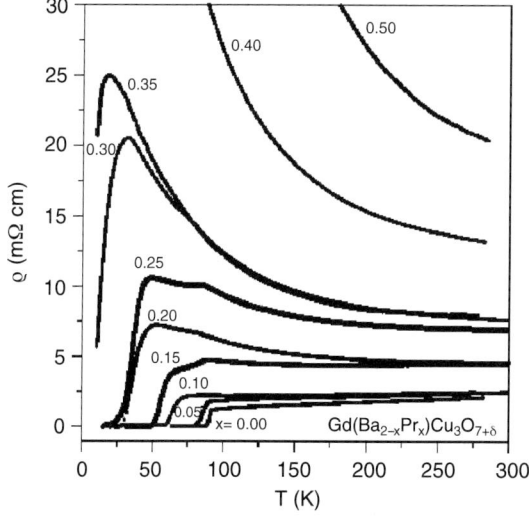

Fig. 5 Resistivity versus temperature of $Gd(Ba_{2-x}Pr_x)Cu_3O_{7+\delta}$ samples for Pr@Ba doping concentration $0.00 \leq x \leq 0.50$ (after Mohammadizadeh et al. [23]).

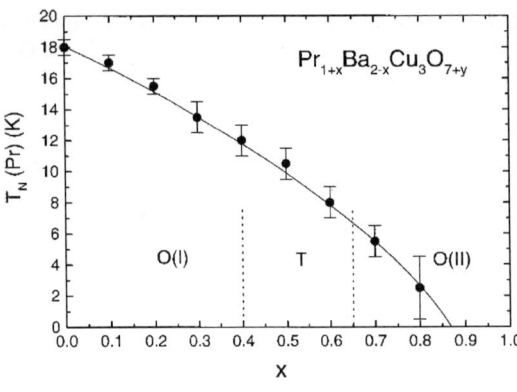
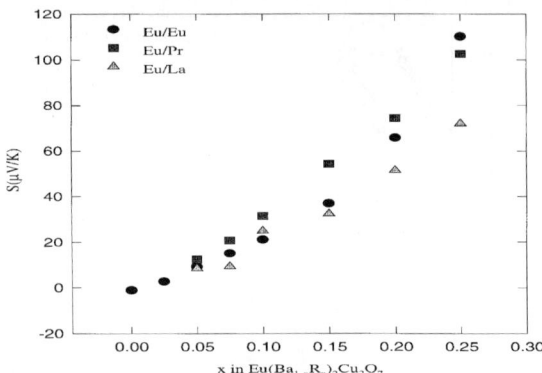

Fig. 6 Variation of $T_N(Pr)$ for the oxygenated $Pr_{1+x}Ba_{2-x}Cu_3O_{7+\delta}$ system (after Luo et al. [11]).

Fig. 7 Seebeck coefficient of $Eu(Ba_{1-x}R_x)_2Cu_3O_{7+\delta}$ for R = Eu, Pr, and La (after Xu et al. [18]).

Another comparative study is reported in Ref. [12]. The Pr@Ba in Sm123 affects the SC properties much stronger and, as a result, the Sm123 system becomes non-SC at $x = 0.3$. It can be seen that the observed drop of T_c for Pr@Ba is even faster than it should be in the case of Pr@Sm in Sm123. This observation is verified for Pr@Ba and Pr@Gd in Gd123 system, where T_c suppression occurs at $x = 0.45$ for Pr@Gd [29] and $x = 0.35$ for Pr@Ba [30]. This fact clearly indicates that for Pr@Ba, Pr doping affects the stronger SC suppression by two mechanisms: (1) the hybridization of Pr_{4f} states with O_{2p}, (2) nonisovalent impurity. This conclusion agrees with the observation in Nd123, where substitution of Ca^{2+} for Pr failed to restore SC in the insulating 0.7Pr@Nd system, whereas the same substitution restored SC in the again insulating 0.7Pr@Ba system to $T_c = 31$ K [31].

4.2 Magnetic properties

For the case of Pr@Ba in Pr123, the temperature dependence of molar magnetic suceptibility $\chi_m(T)$ and $\chi_m^{-1}(T)$ for insulating phase 0.2Pr@Ba can be well fitted by a Curie-Weiss law $\chi_m(T) = \chi_0 + C/(T-\theta_p)$ with negative paramagnetic intercept $\theta_p = -11.7$ K and an effective magnetic moment $\mu_{eff} = 2.86\ \mu_B$ for Pr ion, if the small saturation Cu moment of ~0.3–0.5 μ_B can be neglected [11]. μ_{eff} variation in the range of 2.86 to 3.18 μ_B/Pr indicates Pr^{3+} nature. Moreover, μ_{eff}/Pr-ion does not substantially differ for oxygenated and deoxygenated Pr123 ($\approx 2.85\ \mu_B$), while for the case of 0.3Ca@Pr in Pr123 it increases to 3.01 μ_B [20]. To get a clearer picture, the Cu AF should also be considered thoroughly. There are deviations from the Curie-Weiss law in the high temperature, which indicates that Cu moments are ordered around $T_N(Cu) = 320$–330 K. The low temperature deviation indicates that Pr^{3+} moments in 0.2 Pr@Ba are ordered AF around $T_N(Pr) \sim 15.5$ K. The variation of $T_N(Pr)$ for the oxygenated $Pr(Ba_{2-x}Pr_x)Cu_3O_7$ system is shown in Fig. 6. $T_N(Pr)$ decreases monotonically from 18 K for $x = 0$ to zero for $x = 0.85$. Similar behavior was observed for the isostructural Pr (BaLa) with non-magnetic La@Ba [32].

It follows again that when 20% more Pr^{3+} ions are incorporated for Ba in Pr123 compound, these extra moments make no contribution for $T_N(Pr)$ ordering due to the random and dilute distribution of Pr in Ba matrix with longer (Pr@Ba)–O(1)/O(2)/O(3) bond lengths of 2.76–2.98 Å compared to shorter Pr–O(1)/O(2) bond lengths of 2.44–2.45 Å between Pr and CuO_2 bilayers. There is a correlation between $T_N(Pr)$ and d_{Pr-O} in the adjacent CuO_2 layer Pr(BaPr): $d_{Pr-O} \approx 2.44$ Å for $x = 0$ to $d_{Pr-O} \approx 2.47$ Å for $x = 0.7$. The monotonic increase of d_{Pr-O} with decreasing T_N indicates that Pr AF order is closely correlated with Pr_{4f}–$O_{2p\pi}$ orbital hybridization that is essential in the superexchange interaction mechanism. This superexchange interaction is highly anisotropic and the long-range Pr ordering is thus very much quasi-two-dimensional. It is interesting to note that similar variation of $T_N(Pr)$ with d_{Pr-O} exist for the tetragonal 1212 compounds $MA_2PrCu_2O_{7+\delta}$ (M = Hg, Tl, Pb/Cu, Cu; A = Sr, Ba, Ba/Pr) [11].

In contrast to above, for the case of Ca@Pr (similar to Ba@Pr), the AC and DC magnetization measurements have resulted in no sign of SC even though weaker Pr_{4f}–O_{2p} hybridization and extra holes doped by Ca (Ba) are expected [20]. It appears that hole-localization mechanism is active here.

4.3 Thermal properties

In the case of Pr@Ba in Sm123 for low x, the temperature dependence of the thermopower S is almost temperature independent at $T > 150$–200 K [12]. With increasing doping level x the $S(T)$ dependence at high temperature becomes stronger. Pr@Y or Pr@Ba affects the $S(T)$ to a different extent. With increasing x, $S(300)$ values rises extremely strongly for Pr@Ba. This result can be considered as the direct evidence for the hole filling/localization effect in Sm123 system realized due to a higher valence of Pr compared to that of Ba and the anomalous behavior of Pr.

Another report on the result of thermopower measurements is shown in Fig. 7 for Eu123 with Eu, Pr, and La@Ba at room temperature [18]. The values of $S(300)$ are almost equal for different R@Ba substitution. When Ba is replaced with R ion, the Seebeck coefficient increases with doping concentration x. The increase in S indicates the decrease of the concentration of the mobile-holes. It is noteworthy that data of thermopower measurements and neutron diffraction show that the suppression of SC by Pr@Ba is electronic, not magnetic in nature [33].

5 Concluding remarks

Blackstead et al. [3] present another argument suggesting that, preparing Pr123 by the conventional solid-state method, due to the tendency of Pr to sit on the Ba-site, where it acts as a pair-breaker, therefore destroying SC. Extensive XRD, neutron, and EXAFS [34] allow us to conclude that Pr does not substitute for Ba in Pr123 compound. Kramer et al. [35] report that Pr@Ba and Nd@Ba have similar influence on the material properties. Therefore, an independent mechanism not active in Nd123 must be present in Pr123 (i.e. Pr_{4f}–O_{2p} hybridization) to suppress the MIT and SC in the CuO_2 plane.

In contrast to Blackstead's view, Grevin et al. [4] has even reported small SC fractions in some Pr rich Pr123 samples in which the largest effect has been detected below 90 K for 0.3Pr@Ba. The SC effect has been identified by 0.2% diamagnetic observation, with no report of resistivity data on the samples. The whole set of results regarding Pr@Ba is interpreted in the framework of a phase separation mechanism in the CuO_2 planes induced by the Ba/Pr substitution. In this model, the pairs of Pr are formed around Cu(1) sites. This is claimed to locally keep the charge transfer active and break the Pr_{4f}–O_{2p} hybridization. The weak SC fraction observed find their origin in the existence of such clusters when their size reaches a threshold value corresponding to the existence of several elementary CuO_2 cells: defect-induced SC.

Of course, such a model that a small localized structural distortion on the Ba site promotes the hole localization at the R site is unrealistic, particularly because the 4f electrons form bands, which are not expected to be disturbed by a few percent of disorder. Besides, on the contrary, for the case of Pr@Ba in Gd123 compound [17], observation of an anomaly in resistivity data at about 80–90 K has been related to SC in a part of the samples containing Ba@Gd; for Pr@Ba, the insulating behavior is observed for all x.

In summary, many different Pr/Ba doping combinations have been implemented for explaining the anomalous insulating/SC Pr123 compound. Through these experiments, so far, no definite and convincing data have been presented for the SC Pr123, while many experimental data have been presented for the insulating Pr123. It is concluded that Pr@Ba affects all the properties of R123 much stronger compared to Pr@R. this manifests itself in an increasing oxygen content, OTO transitions, stronger rise of thermopower value, and faster drop of transition temperature. Also, no definite sign of SC is observed for Pr@Ba. In contrast to above, Ba@Pr affects the electronic properties of Pr123 in such a way that SC is possibly revived if the solubility limit could be enhanced by some sample preparation techniques.

Acknowledgments In preparing this manuscript, I am indebted to the works of my students: M. Kariminezhad, H. Khosroabadi, P. Maleki, M. Mirzadeh, M. Modarreszadeh, M. R. Mohammadizadeh, Z. Mokhtari, P. Taheri,

Z. Yamani in the Magnet Research Laboratory. This work was supported in part by the Office of Vice President for Research at Sharif University of Technology.

References

[1] M. Akhavan, Physica B **321**, 265 (2002), and references therein.
[2] Z. Yamani and M. Akhavan, Phys. Rev. B **56**, 7894 (1997).
[3] H. A. Blackstead, D. B. Chrisey, J. D. Dow, J. S. Horwitz, A. E. Klanzinger, and D. B. Pulling, Phys. Lett. A **207**, 109 (1995).
[4] B. Grevin, Y. Berthier, P. Mendels, and G. Collin, Phys. Rev. B **61**, 4334 (2000).
[5] Z. Yamani and M. Akhavan, phys. stat. sol. (a) **163**, 157 (1997).
[6] M. W. Pieper, F. Weikhorst, and T. Wolf, Phys. Rev. B **62**, 1392 (2000).
[7] K. Oka, Z. Zou, and J. Ye, Physica C **300**, 200 (1998).
[8] J. Ye, Z. Zou, A. Matsushita, K. Oka, Y. Nishihara, and T. Matoumoto, Phys. Rev. B **58**, R619 (1998).
[9] G. Cao, Y. Qian, Z. Chen, and Y. Zhang, J. Phys. Chem. Solids **56**, 981 (1995).
[10] M. Tagami and Y. Shiohara, J. Cryst. Growth **171**, 409 (1997).
[11] H. M. Luo, B. N. Lin, H. C. Chiang, Y. Y. Hsu, T. I. Hsu, T. J. Lee, H. C. Ku, C. H. Lin, H.-C.I. Kao, J. B. Shi, J. C. Ho, C. H. Chang, S. R. Hwang, and W.-H. Li, Phys. Rev. B **61**, 14825 (2000).
[12] V. E. Gasumyants, M. V. Elizarova, and R. Suryanarayanan, Phys. Rev. B **61**, 12404 (2002).
[13] Ch. Bertrand, Ph. Galez, R. E. Gladyshevskii, and J. L. Jorda, Physica C **321**, 151 (1999).
[14] D. H. Ha, Physica C **302**, 299 (1998).
[15] Z. Klencsar, E. Kuzmann, A. Vertes, P. C. M. Gubbens, and A. M. v. d. Kraan, Physica C **329**, 1 (2000).
[16] Z. Klencsar, E. Kuzmann, Z. Homonnay, A. Vertes, K. Vad, J. Bankuti, T. Racz, M. Bodog, and I. Kotsis, Physica C **304**, 124 (1998).
[17] M. R. Mohammadizadeh and M. Akhavan, Phys. Rev. B **68**, 104516 (2003).
[18] Y. Xu, M. J. Kramer, K. W. Dennis, H. Wu, A. O'Connor, R. W. McCallum, S. K. Malik, and W. B. Yelon, Physica C **333**, 195 (2000).
[19] N. Watanabe, K. Kuroda, K. Abe, N. Koshizuka, M. Tagami, and Y. Shiohara, Physica C **300**, 301(1998).
[20] Z. Tomkowicz, P. Lunkenheimer, G. Knebel, M. Balanda, A. W. Pacyna, and A. J. Zaleski, Physica C **331**, 45 (2000).
[21] H. Khosroabadi, P. Taheri, M. Modarreszadeh, and M. Akhavan, submitted to Physica C (2003).
[22] H. D. Yang, I. P. Hong, S. Chatterjee, P. Nachimuthu, J. M. Chen, and J.-Y. Lin, Physica C **341–348**, 411 (2000).
[23] M. R. Mohammadizadeh and M. Akhavan, Eur. Phys. J. B **33**, 381 (2003).
[24] A. Shukla, B. Barbiellini, A. Erb, A. Manuel, T. Buslaps, V. Honkmaki, and P. Surotti, Phys. Rev. B **59**, 12127 (1999).
[25] M. J. Kramer, K. W. Dennis, D. Falzgraf, and R. W. McCallum, Phys. Rev. B **56**, 5512 (1997).
[26] U. Staub, L. Soderholm, S. Skanthakumar, S. R. Wasserman, A. G. O. Conner, M. J. Kramer, B. D. Patterson, M. Shi, and M. Knapp, Phys. Rev. B **61**, 1548 (2000).
[27] H. Khosroabadi, M. Modarreszadeh, P. Taheri, and M. Akhavan, phys. stat. sol. (c) **1**, No. 7 (2004), this conference.
[28] G. Botter, I. Mangelschots, E. Kaldis, P. Fischer, Ch. Kruger, and F. Fauh, J. Phys.: Condens. Matter **8**, 8889 (1996).
[29] M. Akhavan, Physica C **250**, 25 (1995).
[30] M. R. Mohammadizadeh, H. Khosroabadi, and M. Akhavan, Physica B **321**, 301 (2002).
[31] L. Colonescu, J. Berthon, R. Suryanarayanan, and I. Zelenay, Physica C **291**, 85 (1997).
[32] R. K. Li, Z. Y. Chen, and Y. T. Qian, Physica C **172**, 335 (1990).
[33] R. Nagarajan, V. Pavate, and C. N. R. Rao, Solid State Commun. **84**, 183 (1992).
[34] V. G. Harris, D. J. Fatemi, V. M. Browning, M. S. Osofsky, and T. A. Vanderrah, J. Appl. Phys. **83**, 6783 (1998).
[35] M. J. Kramer, S. I. Yoo, R. W. McCallum, W. B. Yelon, H. Xie, and P. Allenspach, Physica C **219**, 145 (1994).

physica pss status solidi c

www.pss-c.com

conferences and critical reviews

The Fe$_8$ molecular magnet: A proving ground for the semiclassical theory of spin

Anupam Garg[*]

Department of Physics and Astronomy, Northwestern University, 2145 Sheridan Road, Evanston, Illinois 60208, USA

Received 31 August 2003, accepted 31 December 2003
Published online 23 April 2004

PACS 75.50.Xx, 75.10.Dg, 75.60.Jk

Tunnel splitting oscillations in the molecular magnet Fe$_8$ are best interpreted in terms of interfering spin paths. However, this picture is incomplete and one must also consider *discontinuous* paths to explain the observations. This is one example of how the Fe$_8$ system has motivated workers to better understand spin coherent state path integrals and the semiclassical limit for spin. As a second example, we discuss the recently derived Bohr-Sommerfeld rule for spin.

© 2004 WILEY-VCH Verlag GmbH & Co. KGaA, Weinheim

1 General remarks Among the various molecular magnets studied over the last decade, the Fe$_8$ system is remarkable because it provides dramatic and unambiguous evidence for quantum dynamics of the spin, specifically the tunneling of spin between *deep* levels, as shown by the oscillatory variation of the tunnel splitting with the applied magnetic field [1]. These oscillations are most visually understood in terms of interfering instantons, or semiclassical spin trajectories in the spin-coherent-state path integral [2, 3], and this explanation is widely known. What is not so widely known is that the experimental data contain equally unambiguous evidence for yet another aspect of spin instantons, namely, that these need not be continuous [4]. As a general point of formalism, this has been known since at least 1979, when it was shown by Klauder that such trajectories were necessary for understanding Larmor precession [5]. To our knowledge, however, the Fe$_8$ system is the first non trivial one where such trajectories arise, and it is the only one where both continuous and discontinuous instantons must be considered. More generally, the Fe$_8$ system has provided the impetus for revisiting the semiclassical limit for spin. This is a long-standing problem, and although one knows how to incorporate the first quantum corrections in specific cases (spin waves, e.g.), one is only now learning how to do this in general.

In this paper we will discuss two issues in spin semiclassics: the business of discontinuous instantons mentioned above, and the Bohr-Sommerfeld rule for spin.

2 Instantons in Fe$_8$ To a first approximation, the anisotropy Hamiltonian in Fe$_8$ is given by

$$\mathcal{H}_0 = k_1 J_z^2 + k_2 J_y^2 - g\mu_B J_z H , \qquad (1)$$

where $\mathbf{J} = (J_x, J_y, J_z)$ is a spin operator with magnitude $j = 10$, describing the spin of one molecule in the ground manifold, and $k_1 > k_2 > 0$. Thus the x, y, and z directions are easy, medium, and hard respectively. Measurements yield $g \simeq 2$, $k_1 \simeq 0.338$ K, and $k_2 \simeq 0.246$ K.

---
[*] e-mail: agarg@northwestern.edu, Phone: +1 847 491 3229, Fax: +1 847 491 9982

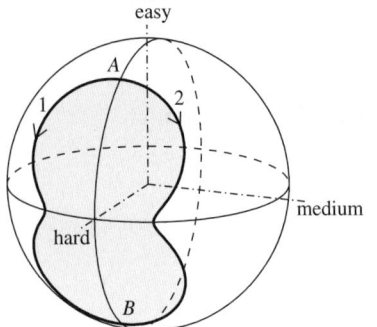

Fig. 1 Instantons for the Hamiltonian $\mathcal{H}_0$ projected onto the real unit sphere.

In Eq. (1), we have also included a term to describe a magnetic field along the hard axis. This field causes the classical minimum energy directions to cant away from $\pm\hat{x}$ toward the $\hat{z}$ axis. However, the two classical minima still have equal energy, so one may conceive of tunneling (more precisely resonance) between them. In the instanton picture, one thinks of the tunneling as taking place along the least action trajectory in imaginary time, τ. Along such a trajectory, the Euler-Lagrange equations are obeyed, and the energy is constant. It follows that a trajectory that runs between two minima of the energy can not lie on the real sphere, i.e., can not have real values of $\theta(\tau)$ and $\phi(\tau)$, where θ and ϕ are the usual spherical polar coordinates. However, solutions can be found if θ and ϕ are allowed to become complex, i.e., if the trajectory is allowed to lie on the complex unit sphere. Indeed, one may think of this trajectory as the tunnel that the system digs in order to go from one minimum to the other.

In the present problem, there are two instantons by symmetry, and in Fig. 1, we show their projections onto the real unit sphere. By general quantum mechanical principles, we must add the amplitudes for transition along each instanton in order to obtain the net tunneling amplitude, Δ. In other words,

$$\Delta = \sum_k D_k \, e^{-S_k^{cl}}. \tag{2}$$

Here S_k^{cl} is the action for the k-th instanton, and D_k is a prefactor arising from the integral over the Gaussian fluctuations around it. Now, $D_1 = D_2$ by symmetry once again, but the actions are not equal. Since the kinetic part of the action is given by

$$S_{kin} = iJ \int (1 - \cos\theta) \, \dot\phi \, d\tau, \tag{3}$$

it follows that the two instantons have a relative phase given by

$$\Lambda = J\mathcal{A}, \tag{4}$$

where $\mathcal{A}$ is the area of the loop formed by the two instantons. Hence, $\Delta \propto \cos(\Lambda/2)$. At $H = 0$, $\mathcal{A} = 2\pi$, since the instantons divide the sphere into two equal halves, and as H increases, $\mathcal{A}$ decreases smoothly toward 0. Whenever Λ passes through an odd multiple of π, Δ vanishes.

It is this interference phenomenon that is seen in the experiments (see Fig. 2). At first sight this is a great success for the theory, and it is rather pleasing that the phenomenon has such a direct geometrical explanation. There is, however, a fly in the ointment. Only four quenching points (field values where Δ vanishes) are seen in the experiment. The argument of the previous paragraph predicts ten quenches, since $\Lambda = 20\pi$ at $H = 0$, and so must pass through 10 odd multiples of π (19π, 17π, ..., π) with increasing H. This argument is a topological one, and so is, a priori, very general, and depends very little on the details of the Hamiltonian. Why doesn't it work?

The answer is that the Hamiltonian of Eq. (1) is not quite correct, and that a better fit to the data is obtained with

$$\mathcal{H}_1 = k_1 J_z^2 + k_2 J_y^2 - C[(J_z + iJ_y)^4 + \text{h.c.}] - g\mu_B J_z H. \tag{5}$$

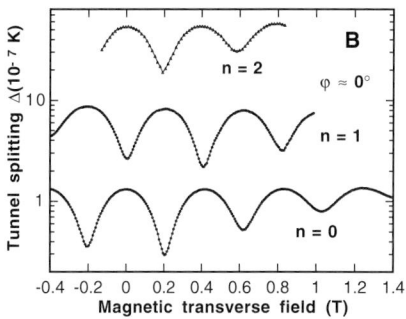

Fig. 2 Measured splittings for Fe_8 for $\mathbf{H} \parallel \hat{z}$ between the states $m = -10$ and $m = 10 - n$. Figure courtesy of Dr. Wolfgang Wernsdorfer.

This differs from $\mathcal{H}_0$ in the 4th order term with $C \simeq 29\,\mu\text{K}$. Since the dimensionless ratio $\lambda_2 = CJ^2/k_1$ is only 8.58×10^{-3}, it is surprising that this term can make such a huge qualitative difference. The reason that it does is that there are two new solutions to the Euler-Lagrange equations for any $C \neq 0$, which will in general be discontinuous at the end points. In Eq. (2), therefore, k runs from 1 to 4. We show the real part of the action for all four instantons in Fig. 3, and the time dependence for the one labelled $k = 3$ in Fig. 4. (Here, $h = H/H_c$, with $H_c = 2k_1 J/g\mu_B$.) As can be seen, this instanton has the least action for H just beyond the fourth quench point, so it dominates the sum in Eq. (2), and since it has no interfering partner, Δ ceases to oscillate with further increase in H.

As an aside we mention that the experimenters proposed [1] the Hamiltonian $\mathcal{H}_1$ because the interval between the quenches was found to be 50% larger than predicted [3] on the basis of Eq. (1). By trying various forms of fourth order anisotropy compatible with the symmetry of the molecule, and comparing their data with the results of exact numerical diagonalization, they arrived at Eq. (5). The coincindental fact that numerical diagonalization of $\mathcal{H}_1$ also yielded four quenches and not ten went unremarked. However, as we have shown, the *number* of quenches is much more significant than the *interval* between them.

Let us now briefly discuss why discontinuous instantons must exist. (The full argument is given in Refs. [4, 5].) The Euler-Lagrange equations are

$$iJ \sin\theta \, \frac{d\theta}{d\tau} = \frac{\partial H_{\text{sc}}}{\partial \phi}, \qquad iJ \sin\theta \, \frac{d\phi}{d\tau} = -\frac{\partial H_{\text{sc}}}{\partial \theta}, \qquad (6)$$

where $H_{\text{sc}}(\theta, \phi)$ is the expectation value of $\mathcal{H}$ in the spin coherent state $|\theta, \phi\rangle$, i.e., the state with maximal spin projection along the direction with coordinates (θ, ϕ). Suppose that the instanton runs from direction (θ_i, ϕ_i) to (θ_f, ϕ_f). If we impose these boundary conditions on the path $\theta(\tau)$, $\phi(\tau)$, there will in general be no solution to Eq. (6) since it forms a second order system, but we have four boundary conditions. In fact, it turns out that to have a properly formulated least-action principle, we must add to the action a term that depends on the explicit boundary values. (This term can actually be discovered by writing the quantum mechanical propagator as a path integral in the usual way. One

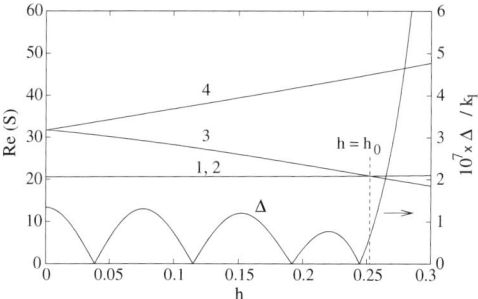

Fig. 3 Real parts of the action for the four Fe_8 instantons, and the tunnel splitting (right scale).

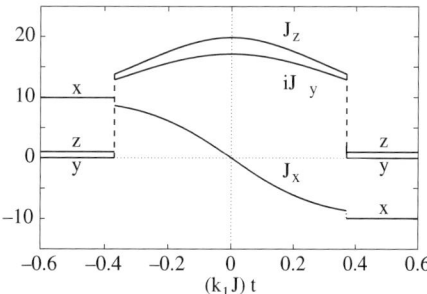

Fig. 4 Time dependence of the components of the spin vector $\mathbf{J}$ for instanton no. 3 for $h = 0.1$.

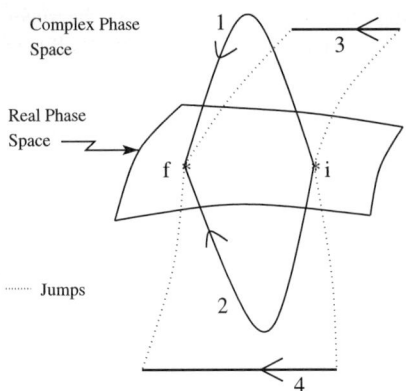

Fig. 5 Instanton paths for the Fe$_8$ Hamiltonian $\mathcal{H}_1$ on the real and complex sphere.

divides the total time interval T into a large number of infinitisimal time slices, and between each time slice, one inserts a resolution of unity in the form of an integral over the overcomplete set of spin coherent states.) Once this term is added, one discovers that δS, the first variation of the action, vanishes, if the combination $\bar{z} = \tan(\theta/2) e^{-i\phi}$ is varied at the initial point. Likewise, $\delta S = 0$ if we vary $z = \tan(\theta/2) e^{i\phi}$ at the final point. This means that the correct boundary conditions on the Euler-Lagrange equations are

$$\tan \tfrac{1}{2}\theta(0) e^{i\phi(0)} = \tan \tfrac{1}{2}\theta_i e^{i\phi_i}, \qquad \tan \tfrac{1}{2}\theta(T) e^{-i\phi(T)} = \tan \tfrac{1}{2}\theta_f e^{-i\phi_f}. \qquad (7)$$

A classical (i.e., least action) path now exists as the number of conditions now equals the order of the system of differential equations. However, note that $\theta(0)$ and $\phi(0)$ need not be real (and likewise for $\theta(T)$ and $\phi(T)$). Thus we see that the dynamics must be defined on the complexified unit sphere from the very start in order to have a sensible least action principle. Further, the starting and ending points of the least action path need not lie on the real sphere (a submanifold of the complex sphere), which is equivalent to saying that the path can be discontinuous at the ends. For an instanton problem, where the path runs between minima of the energy, a continuous path (or paths) can always be found by letting $T \to \infty$, so the natural inclination is to stop there. However, in general, discontinuous paths also exist, and the Fe$_8$ example shows that we must be alert to their presence. A schematic of the four instantons in this example is shown in Fig. 5. A portion of real phase space (the real unit sphere), a two-dimensional surface, is shown embedded in complex phase space. The latter is a four-dimensional manifold, but appears three-dimensional in this perspective drawing. The initial and final points i and f lie on the real sphere. Paths 1 and 2 (which interfere with each other) start and end at these points but lie on the complex sphere otherwise. Paths 3 and 4 (which do not interfere) have jumps at the end points and lie entirely in complex phase space.

At a deeper level, these discontinuous paths exist because spin coherent states are not orthogonal. Thus, in the limit in which the width of a time slice in the discrete path integral goes to zero, it is possible for the system to "evolve" from $|\theta, \phi\rangle$ to $|\theta', \phi'\rangle$ in zero time. This is not possible in the Feynman path integral, and so the class of permissible paths there must be continuous, though it need not be differentiable anywhere, giving rise to the so-called Brownian paths. In the spin case, by contrast, the paths need not even be continuous. What is surprising is that these discontinuities are relevant even at the level of classical or least-action paths.

The above arguments apply to *any* path integral based on coherent states, not just spin. In particular, they apply to a massive particle with a geometrical position coordinate, and here too discontinuous classical paths must be considered. Of course, in this case one can resort to the Feynman path integral, but there may be problems where the coherent state integral leads to a simpler analysis. For spin there is no choice but to use coherent state based integrals.

3 Bohr-Sommerfeld quantization In the previous section, we have skipped over the calculation of the prefactors D_k in Eq. (2). To do this we must know how to compute the prefactor (analogous to the so-called van Vleck determinant) in the general propagator. Although this was done in 1987, the result remained obscure till recently [6–8]. (The seminal paper by Solari [6] had a grand total of five citations when Ref. [7] was written!) The difficulty can once again be traced to the end points, which now leads to an anomaly in the fluctuation determinant. By carefully examining the discrete form, however, the anomaly can be resolved, and the final answer can be expressed in classical terms.

One of the uses to which one can put the semiclassical propagator is to derive the Bohr-Sommerfeld (BS) quantization formula. We first state this rule for massive particles in the form in which it naturally appears when we start with the coherent-state propagator:

$$\oint_{H_{\rm sc}=E} p\,dq + \frac{\hbar}{2}\int \nabla^2 H_{\rm sc}\,dt = (2n+1)\pi\hbar. \tag{8}$$

This differs from the textbook form in that it involves the semiclassical Hamiltonian $H_{\rm sc}$, and not the classical one, which we call H_c. We define

$$H_{\rm sc}(p,q) = \langle p,q|\mathcal{H}|p,q\rangle, \tag{9}$$

where $|p,q\rangle$ is the coherent state

$$|p,q\rangle = e^{-i(xP-pX)/\hbar}|0;\omega_r\rangle. \tag{10}$$

Here, P and X are quantum mechanical operators for momentum and position, and $|0;\omega_r\rangle$ is the ground state of a harmonic oscillator of mass m and reference frequency ω_r. The second term in Eq. (8) is the "Solari correction". It involves the Laplacian $\nabla^2 H_{\rm sc}$, defined as

$$\nabla^2 H_{\rm sc} = \frac{1}{2}\left(m\omega_r\frac{\partial^2}{\partial p^2} + \frac{1}{m\omega_r}\frac{\partial^2}{\partial q^2}\right)H_{\rm sc}(p,q). \tag{11}$$

Table 1 Positive part of energy spectrum of the LMG model from the BS rule (12) and numerical diagonalization for $j = 15$. Entries marked * are by extrapolation, and those marked ** are from mapping to x, p variables.

$r = 0.6$		$r = 1.0$		$r = 5.0$	
BS	exact	BS	exact	BS	exact
15.10*	15.09	15.33**	15.31	37.98	38.05
14.26	14.26	14.77**	14.80	37.98	38.05
13.38	13.38	14.12	14.09	31.37	31.44
12.46	12.46	13.28	13.27	31.37	31.44
11.52	11.51	12.38	12.37	25.35	25.42
10.54	10.54	11.41	11.41	25.35	25.42
9.55	9.54	10.40	10.39	20.02	20.14
8.53	8.53	9.34	9.33	20.02	20.05
7.50	7.50	8.24	8.24	15.66	16.13
6.45	6.45	7.12	7.11	15.01*	15.24
5.39	5.39	5.97	5.96	12.82	12.63
4.33	4.33	4.80	4.79	10.46	10.47
3.25	3.25	3.61	3.61	7.96	7.93
2.17	2.17	2.41	2.41	5.36	5.35
1.09	1.09	1.21	1.21	2.69	2.69
0.00	0.00	0.00	0.00	0.00	0.00

Using the standard Weyl prescription for relating $\mathcal{H}$ to H_c, one can show that Eq. (8) is equivalent to the textbook rule in which the action integral is taken over the orbit $H_c(p,q) = E$. The difference between this and the integral in Eq. (8) precisely cancels the Solari correction.

The rule for spin can be written in analogous form:

$$\left(j + \frac{1}{2}\right) \oint_{H_{sc}=E} (1 - \cos\theta)\, d\phi + \frac{1}{4j} \int \nabla_\Omega^2 H_{sc}\, dt = (2n+1)\pi. \tag{12}$$

Now, $H_{sc}(\theta, \phi) = \langle \theta, \phi | \mathcal{H} | \theta, \phi \rangle$, and ∇_Ω^2 is the angular part of the Laplacian. This rule can also be written in terms of a classical Hamiltonian which can be uniquely related to the quantum mechanical Hamiltonian $\mathcal{H}$ using the spherical harmonic tensor operators [9]. This time a non trivial correction involving the gradient of the Hamiltonian remains.

We test Eq. (12) on the Lipkin-Meshov-Glick (LMG) model of certain collective excitations in nuclei. The same model was used by Shankar [10] to test a previous BS rule. It has the Hamiltonian

$$\mathcal{H}_{LMG} = J_z^{op} + \frac{r}{2j}\left[(J_x^{op})^2 - (J_y^{op})^2\right], \tag{13}$$

with J_x^{op} being the x component of the spin operator, etc. The model has a symmetry which causes states to occur in pairs at energy E and $-E$, and for $r > 1$ it has two degenerate maxima divided by separatrices at energy $E = j$. In Table 1 we compare the answers given by the rule with numerical diagonalization. It is evident that except in the vicinity of the separatrix, where the energy levels are tunnel split by large amounts, the rule is extremely good. Energies near the separatrix will also be accurately obtained once tunneling is included via complex orbits.

This BS rule can be used to study single spin Hamiltonians in other contexts, spinning molecules and nuclie, e.g. One could also use it to find the low lying energies of Heisenberg-like models for the interacting atomic spins of molecules like Fe_8 and Mn_{12} themsleves.

Acknowledgements The research described here has been done, in part, in collaboration with Ersin Keçecioğlu and Michael Stone, and supported by various National Science Foundation grants, most recently, No. DMR-0202165.

References

[1] W. Wernsdorfer and R. Sessoli, Science **284**, 133 (1999).
[2] M. Wilkinson, Physica **21D**, 341 (1986).
  D. Loss, D. P. DiVincenzo, and G. Grinstein, Phys. Rev. Lett. **69**, 3232 (1992).
  J. von Delft and C. L. Henley, Phys. Rev. Lett. **69**, 3236 (1992).
[3] A. Garg, Europhys. Lett. **22**, 205 (1993).
[4] E. Keçecioğlu and A. Garg, Phys. Rev. Lett. **88**, 237205 (2002); Phys. Rev. B **67**, 054406 (2003).
[5] J. R. Klauder, Phys. Rev. D **19**, 2349 (1979), and in: Path Integrals, and Their Applications in Quantum, Statistical, and Solid State Physics, edited by G. J. Papadopoulos and J. T. Devreese (Plenum, New York, 1978).
[6] H. G. Solari, J. Math. Phys. **28**, 1097 (1987).
  E. A. Kochetov, J. Math. Phys. **36**, 4667 (1995).
[7] M. Stone, K.-S. Park, and A. Garg, J. Math. Phys. **41**, 8025 (2000).
[8] A. Garg, E. Kochetov, K.-S. Park, and M. Stone, J. Math. Phys. **44**, 48 (2003).
[9] A. Garg and M. Stone, Bohr-Sommerfeld Quantization of Spin Hamiltonians, cond-mat/0304125.
[10] R. Shankar, Phys. Rev. Lett. **45**, 1088 (1980).

Spin-waves in ferromagnetic double layers: effect of a lateral patterning

N. Sergeeva[1,2], **S.-M. Chérif**[*,1], **A. Stachkevitch**[1], **M. Kostylev**[2], and **Y. Roussigné**[1]

[1] Laboratoire P.M.T.M (CNRS-UPR 9001), Université Paris 13, 93430 Villetaneuse, France
[2] St. Petersburg Electrotechnical University, St. Petersburg 197376, Russia

Received 31 August 2003, accepted 31 December 2003
Published online 1 April 2004

PACS 75.30.Ds, 75.40.Gb, 75.50.Bb, 75.70.Ak, 75.75.+a

Brillouin light scattering (BLS) spectroscopy is used to investigate the thermally activated spin-waves in a Fe/Au/Fe sandwich where various arrays of micron-size stripes were elaborated. The experimental results of the unpatterned and patterned double layer are compared and discussed. The measured frequencies of the observed discrete modes related to the quantisation of the surface acoustic SA mode are compared to those obtained from analytical and numerical calculations of eigen modes in stripe-patterned double layers.

© 2004 WILEY-VCH Verlag GmbH & Co. KGaA, Weinheim

1 Introduction

Brillouin light scattering (BLS) spectroscopy is one of the most powerful methods to derive the parameters which monitor the physical properties of magnetic films containing magnetic layers. It relies upon the inelastic scattering of light by thermally excited spin-waves. Layered magnetic structures have been widely investigated both theoretically and experimentally [1]. During the last few years, the BLS technique has also proved to be powerful to investigate the influence of the finite lateral dimension of single layer micrometric magnetic elements [2–6]. The most striking observed effect is related to the quantisation of the surface mode of the magnetic excitations due to the lateral confinement. If small single layer elements have been investigated to a certain extent using BLS, this is not the case of small elements elaborated in double or multilayers. To our knowledge, such a study is still missing. This is the main goal of this paper.

The present study deals with the investigation of the dynamic magnetic properties of patterned iron–gold–iron layers. BLS spectroscopy was used to investigate the thermally activated spin-waves in a Fe/Au/Fe sandwich where various arrays of micron-size stripes were elaborated even if in this paper we only focus on results relative to the array of 1.5 μm wide stripes. The experimental results of the unpatterned and patterned double layer are compared and discussed. The measured frequencies of the observed discrete modes related to the quantisation of the so-called surface acoustic SA mode, are compared to those obtained from analytical and numerical calculations of eigen modes in stripe-patterned double layers.

2 Experiment

The patterned arrays were prepared from a Fe (20 nm)/Au (3 nm)/Fe (20 nm) sandwich deposited onto a chemically cleaned Si(100) substrate. The substrate was plated with a 25 nm Au buffer and the sample was top protected by a thin Au layer (3 nm). The fabrication of the stripe arrays was achieved using a

---

[*] Corresponding author: e-mail: cherif@lpmtm.univ-paris13.fr, Phone: 33/1 49 40 34 73, Fax: 33/1 49 40 39 38

technique combining e-beam lithography and ion beam etching [7]. BLS measurements (p-s scattering) were carried out in the usual geometrical set up where the transferred in-plane wave vector $\mathbf{q}_\parallel$ is perpendicular to the magnetic field **H** which is applied along the axis of the stripe (Fig. 1). Light of a single-mode Ar$^+$ laser using a power of 100 mW at wavelength λ of 5140 Å was focused onto the sample and the frequency spectrum of the backscattered light was analysed using a computer controlled Fabry–Perot interferometer. The value of $\mathbf{q}_\parallel$ was changed by varying the angle of light incidence θ measured against the surface normal of the sample: $\mathbf{q}_\parallel = (4\pi/\lambda) \sin\theta$.

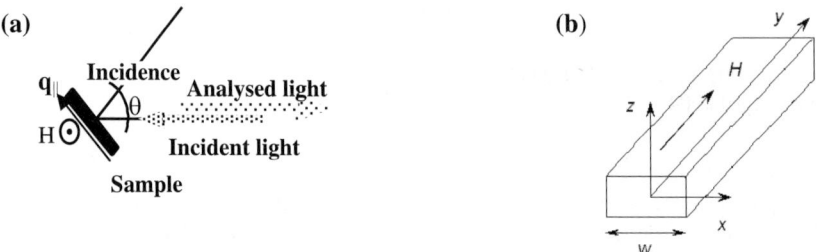

Fig. 1 The used back-scattering geometry (a) and the stripe axes (b).

3 Results and discussion

The BLS spectra were recorded with H = 1 kOe and using an angle of incidence θ varying between 3° and 55° (i.e. $\mathbf{q}_\parallel$ varies between 0.12 and 2×10^5 cm^{-1}). Let us first comment upon the results related to the unpatterned area. Figure 2 shows an experimental Brillouin spectrum (a) and a calculated one (b): they were obtained with an angle of incidence $\theta = 20°$, which leads to an in-plane wavevector $\mathbf{q}_\parallel = 0.835 \times 10^5$ cm^{-1}. The spectrum exhibits two lines. One is the so-called acoustic mode SA, which is independent on interlayer exchange coupling ; the other one, the so-called optic mode SO, is sensitive to interlayer exchange coupling. The numerical method of calculation of the BLS spectra, which is based on the evaluation of the appropriate spin dependent response functions (Green functions) has been reported elsewhere [8]. The agreement between the experimental spectrum (Fig. 2a) and the calculated one (Fig. 2b) is very satisfactory: it was obtained by fitting the magnetisation ($4\pi M = 17.0$ kOe); the interlayer exchange coupling A_{12} constant ($A_{12} = 0.1$ erg/cm^2) and using the published values of the gyromagnetic factor ($\gamma = 1.84 \times 10^7$ Hz/Oe, i.e. g = 2.1) and of the magnetic exchange (D = 2.15×10^{-9} Oe/cm^{-2}) in bulk iron.

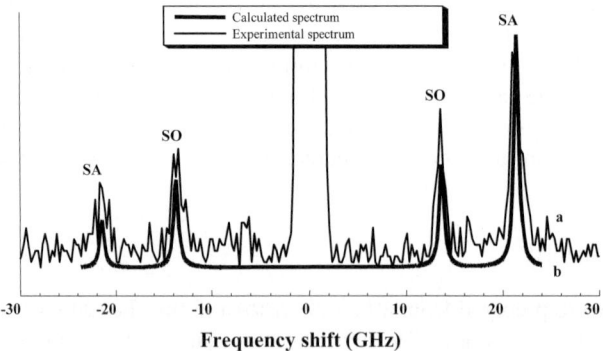

Fig. 2 Brillouin spectra of the unpatterned Fe/Au/Fe double layer: (a) experimental, (b) calculated.

© 2004 WILEY-VCH Verlag GmbH & Co. KGaA, Weinheim

Let us now investigate the patterned area. Figure 3 exhibits a typical BLS spectrum obtained from the 1.5 µm-width stripe array (upper panel) for θ = 15°, together with the one from the continuous unpatterned area (lower panel). One can observe that the BLS spectrum from the patterned area is drastically changed: several peaks are now observed. They are related to the discretisation of the magnetic excitations due to the lateral confinement within the stripe width.

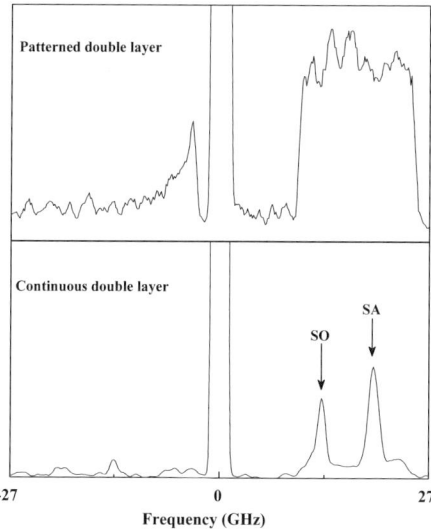

Fig. 3 BLS spectrum obtained in the array of stripes with width of 1.5 µm (upper panel) and for the continuous double layer (lower panel) with an angle of incidence θ = 15° and an external magnetic field H = 1 kOe applied along the long stripe axis (y axis). The scanning speed is reduced by a factor of 4 in the anti-Stokes side of the BLS spectrum from the array of stripes, increasing the sensitivity by the same factor

The dispersion of the spin-waves for both the array of stripes (closed symbols) and the continuous film (open symbols) is presented in Fig. 4 together with the calculated curves (solid and dashed horizontal lines). Since the discrete, dispersionless spin wave modes converge towards the dispersion of the SA mode of the continuous film, it is natural to assume that these modes result from a width dependent quantisation of the in-plane wave vector characterising the SA mode. This is not surprising since the SA mode is similar to the Damon–Eshbach DE mode of the combined system. Conversely the SO mode seems to be not significantly affected. The spin-waves frequencies were calculated by using the approach first proposed in Ref. [9]. Such an approach takes into account peculiarities of dynamic demagnetisation at the stripe lateral boundaries. It allows one finding effective dipolar-origin pinning conditions for the dynamic magnetisations at the boundaries. In our work the approach was extended to the case of a double-layered stripe and it was shown that the eigen-frequencies ω_n of standing spin waves having the in-plane wave vector along the stripe width as in [9] are described by the relation :

$$\left(\frac{\omega_n}{\omega_M}\right)^2 = \left(\frac{\omega_H}{\omega_M} + 1 + \frac{\lambda_n}{4\pi}\right)\left(\frac{\omega_H}{\omega_M} - \frac{\lambda_n}{4\pi}\right), \quad \text{with} \quad \omega_H = \gamma H \quad \text{and} \quad \omega_M = \gamma 4\pi M. \tag{1}$$

Here λ_n are the eigen-values of the integral operator of dynamic dipolar field of the structure. All the peculiarities of the spin-wave dispersion in the double-layered stripes are determined by the kernel of the integral operator. In the particular case of two identical magnetic stripe layers and of vanishing interlayer magnetic coupling it has the following form:

$$g^{(\pm)}(x,x') = \ln \frac{(x-x')^4}{[L^2 + (x-x')^2]^2} \mp \ln \frac{[(d+L)^2 + (x-x')^2]^2}{[L^2 + (x-x')^2][(d+2L)^2 + (x-x')^2]} \tag{2}$$

Here L is the magnetic-layer thickness, d is the spacer thickness, and x and x' are the coordinates along the stripe width. The upper (lower) sign corresponds to the resonances of the SA (SO) mode.

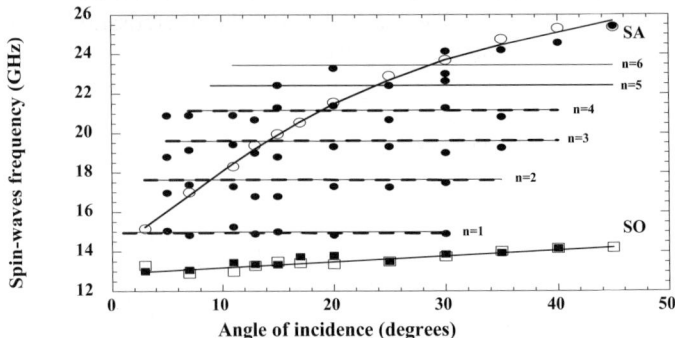

Fig. 4 Spin-wave dispersion for the Fe(20 nm)/Au(3 nm)/Fe(20 nm) double layer obtained with an external magnetic field H = 1 kOe applied along the stripe axis and the in-plane wave vector $\mathbf{q}_\parallel$ perpendicular to it: continuous double layer (open symbols), 1.5 μm-wide stripes (closed symbols). The calculated spin wave frequencies of the quantised modes are represented by the solid lines (obtained from Eq. (1) (see text)) and by the dashed lines (obtained by using the usual Damon–Eshbach dispersion relation for a continuous layer with an appropriate quantisation condition for the in-plane wave vector: $\mathbf{q}_{\parallel,n} = (n + \beta)(1 - \delta_{n,0})\pi/W$ (with $\beta = -1/3$).

The calculations of eigen-functions have shown that for the lower resonances, where the confinement effects *a-priori* should be maximum, the effective dipolar-origin pinning at the lateral boundaries is nearly total for the SA resonances and is much smaller for the SO ones. This leads to a noticeable shift of the frequency positions of the resonances relative to the spectrum of the unpatterned double layer, the shape of BLS-spectrum being modified correspondingly. The calculated frequencies are shown in Fig. 4 by horizontal solid lines. They are in a good agreement with the experimental data. They also well agree with the frequencies obtained using the simplified approach, as follows (see Fig. 4, dashed lines) : namely, the dependence of the frequencies upon the stripe width W is obtained by using the usual Damon–Eshbach dispersion relation for a continuous layer with an appropriate quantisation condition for the in-plane wave vector [1] :

$$\mathbf{q}_{\parallel,n} = (n + \beta)(1 - \delta_{n,0})\pi/W \qquad \text{(with } \beta = -1/3\text{) where n is an integer .} \qquad (3)$$

This demonstrates that for the total thickness of the sandwich (43 nm), which is much larger than that of the spacer, the structure behaves nearly like a homogeneous layer of double thickness of a single iron layer (20 nm).

References

[1] S. O. Demokritov and E. Tsymbal, J. Phys. C **6**, 7145 (1994).
[2] Y. Roussigné, S.-M. Chérif, C. Dugautier, and P. Moch, Phys. Rev. B **63**, 134429 (2001).
[3] J. Jorzick et al., J. Appl. Phys. **89**, 7091 (2001).
[4] S.-M. Chérif, Y. Roussigné, and P. Moch, IEEE. Trans. Magn. **38**, 2529 (2002).
[5] J. Jorzick, S. O. Demokritov, B. Hillebrands, M. Bailleul, C. Fermon, K. Guslienko, A. N. Slavin, D. Berkov, and N. L. Gorn, Phys. Rev. Lett. **88**, 47204 (2002).
[6] G. Gubbiotti, P. Candeloro, L. Businaro, E. Di Fabrizio, A. Gerardino, R. Zivieri, M. Conti, and G. Carlotti, J. Appl. Phys. **93**, 7595 (2003).
[7] S.-M. Chérif and J.-F. Hennequin, J. Magn. Magn. Mater. **165**, 504 (1997).
[8] Y. Roussigné, F. Ganot, C. Dugautier, P. Moch, and D. Renard, Phys. Rev. B **52**, 350 (1995).
[9] K. Yu. Guslienko, S. O. Demokritov, B. Hillebrands, and A. N. Slavin, Phys. Rev. B **66**, 132402 (2002).

Brillouin light scattering study of magnetic dots

S.-M. Chérif*, **S.-Y. Roussigné**, and **P. Moch**

Laboratoire P.M.T.M (CNRS/UPR 9001), Université Paris 13, 93430 Villetaneuse, France

Received 31 August 2003, accepted 31 December 2003
Published online 25 March 2004

PACS 75.30.Ds, 75.40.Gb, 75.50.Bb, 75.50.Ak, 75.70.+a

Brillouin light scattering investigation of the thermally activated spin-waves in micrometer-size permalloy square dots shows evidence of a frequency discretization of the surface spin-wave mode which is due to the lateral confinement of the magnetic excitations. The observed frequencies are compared to those derived from the analytical expression in a continuous layer assuming that the finite size of the dots is responsible of the quantization of the wave vectors and of the lowering of the internal magnetic field.

© 2004 WILEY-VCH Verlag GmbH & Co. KGaA, Weinheim

1 Introduction

Brillouin light scattering (BLS), which allows access to inelastic scattering of the electromagnetic radiation by thermal spin-waves, provides a powerful non-destructive tool in order to investigate the parameters monitoring the physical properties of patterned magnetic thin films. The steadily growing applications in magnetic recording media technology like storage devices and sensor technologies partly motivate the interest in the determination of those parameters. The magnetic properties of such artificial structures are very different from those of a two-dimensional continuous ferromagnetic layer, because of their reduced size, of their specific shape and of their periodic structure. In patterned films, when the size of the magnetic objects becomes comparable to the wavelength of the propagating spin-waves, a quantization of the magnetic excitations is observed [1–5]: it arises from the lateral confinement of the in-plane wave vector $q_\parallel$ of the surface mode (so-called DE mode) of the continuous film. In the case of saturated wires, we have published exact numerical calculations [1]: they show that the dependence of the frequencies upon the width W does not simply follow the naive model which uses the usual dispersion relation derived for a continuous layer [6]:

$$\nu_{DE} = (\gamma/2\pi)\,[H\,(H + 4\pi M) + (2\pi M)^2\,(1 - e^{-2q_\parallel t})]^{1/2} \tag{1}$$

and which states that:

$$q_{\parallel,n} = n\pi/W \quad \text{(where n is an integer)}. \tag{2}$$

In Eq. (1) γ is the gyromagnetic factor, H is the static magnetic field, M is the saturation magnetization and t the layer thickness. We found that the use of expression (1) leads to a significantly different quantization condition, namely:

$$q_{\parallel,n} = (n + \beta)\,(1 - \delta_{n,0})\,\pi/W \quad \text{(with } \beta = -1/3\text{)}. \tag{3}$$

Equation (3) reveals an effective pinning of the spin waves. Notice that the lowest mode (n = 0), which is easily obtained analytically in the above assumed dipolar approximation, behaves in a singular

---

* Corresponding author: e-mail: cherif@lpmtm.univ-paris13.fr, Phone: 33/1 49 40 34 73, Fax: 33/1 49 40 39 38

way. Our experimental results very well agree with the calculated ones using equation (3): this agreement includes the results concerning n = 0, which have given rise to alternative significantly different interpretations [2, 4, 5]. In addition, we have recently published calculations including the exchange terms, related to the frequencies and the magnetization profiles of the spin waves in magnetic stripes [7].

In this work, we report on a BLS study of the dynamic properties of spin-waves in periodic arrays of 1 and 2 μm-size square permalloy dots. Quantization effects are clearly evidenced. An attempt to interpret the results in terms of Eq. (1) is presented, i) noticing that the appropriate field H is the internal field, which, compared to the applied field, is significantly reduced by the demagnetizing field and, ii) using a quantization condition close to expression (3).

2 Experiment

The studied dots were elaborated starting from a continuous permalloy layer of 29 nm thickness, with the help of electron beam lithography and of ion beam sputtering. The required patterns were obtained by irradiation of a polymethylmethacrylate (PMMA) resist, previously deposited on the top of the permalloy film, in a scanning electron microscope (SEM). The irradiated resist was then dissolved in a soft chemical solution. The sample was further transferred in a ultra high vacuum (UHV) chamber where the whole area was ion beam sputtered. A detailed description of our combinated technique is presented elsewhere [8]. The (500 × 500 μm^2) patterned areas consisted of periodic arrays of 1 and of 2 μm-size square dots. At the end of the fabrication process, large parts of the continuous unpatterned permalloy film were still available, thus allowing for a direct comparison of the experimental results between the patterned and the unpatterned regions. As an example of the investigated structures, Fig. 1 (left) shows an atomic force microscopy (AFM) image of the 1 μm-size dots. Magnetic force microscopy (MFM) was additionally used and showed (right) that in the zero field state the dots mainly exhibit a flux closure configuration : domains are formed in order to minimize the external magnetic flux and, consequently, the demagnetizing field.

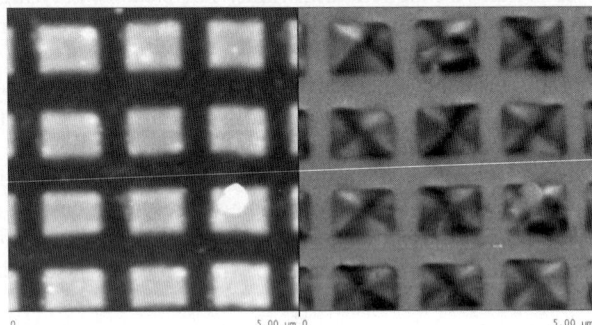

Fig. 1 AFM (left) and MFM (right) image of the 1 μm-size permalloy dots.

The dominant configuration seems to correspond to the so-called Landau state in which four domains are formed. A similar behaviour was also observed on the 2 μm-size dots. The BLS measurements were achieved using a (3+3)-pass tandem Fabry–Pérot interferometer.

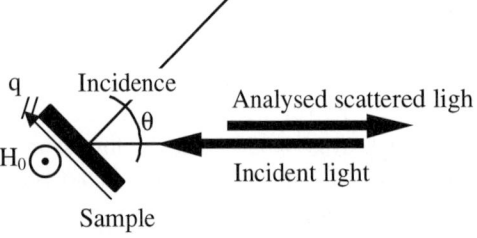

Fig. 2 The used back-scattering geometry.

The samples were illuminated by a single-mode Ar$^+$ ion laser, using a power of 50 mW at the wavelength $\lambda = 5145$ Å. The backscattering geometrical set up was chosen such as the in-plane transferred wave vector $\mathbf{q}_\parallel$ remained perpendicular to the external applied field $\mathbf{H}_0$ (Fig. 2). Its amplitude $\mathbf{q}_\parallel = 2k_i \sin\theta = (4\pi/\lambda)\sin\theta$ (where θ is the angle of incidence) could sweep the range [0.12–2.2 × 10^5 cm^{-1}] (by varying θ from 3° to 65°).

3 Results and discussion

The BLS spectra were recorded under an applied magnetic field H_0 of 600 Oe. This value is large enough to almost fully saturate the magnetization of the dots along $\mathbf{H}_0$, as shown by micromagnetic calculations using the OOMMF code [9]. Figure 3 shows the obtained BLS anti-Stokes spectra, with $\mathbf{H}_0$ applied along an edge, at different values of θ, i) for the 1 µm-size (left panel) and, ii) for the 2 µm-size dots (right panel).

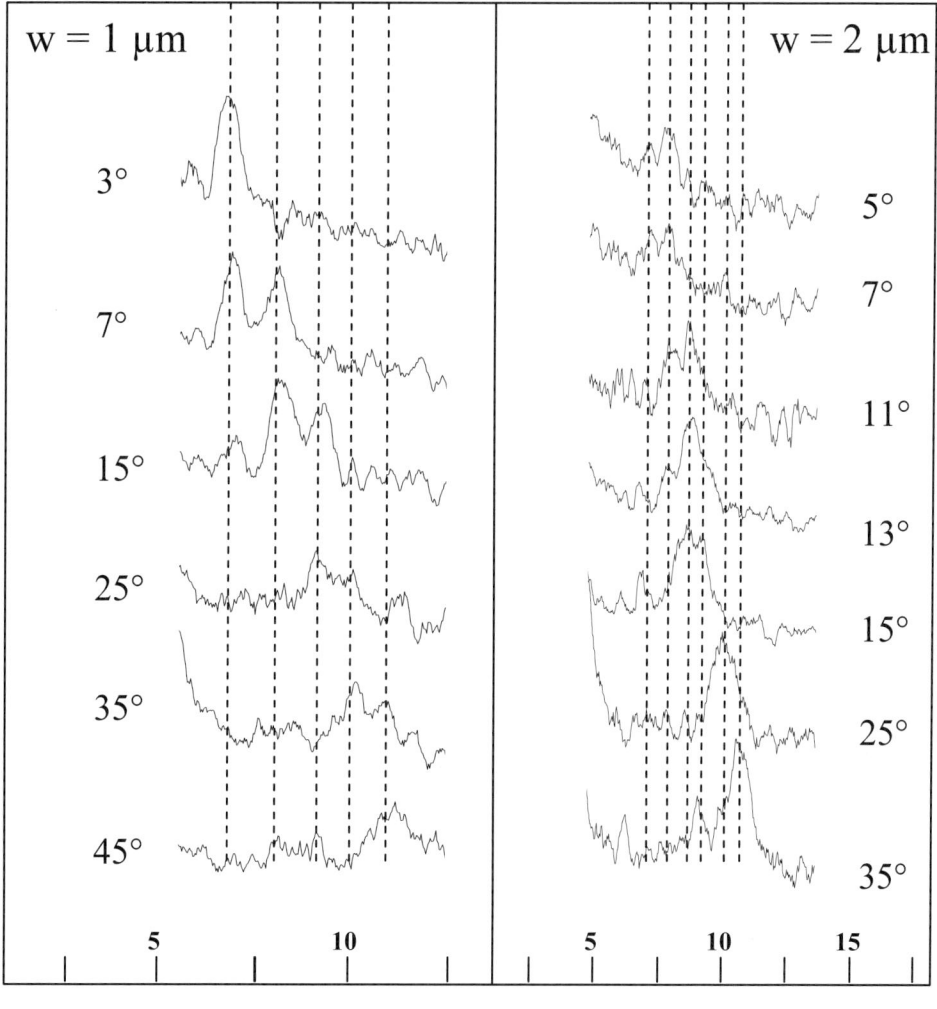

Fig. 3 Anti-Stokes parts of Brillouin spectra obtained from the 1 µm (left panel) and 2 µm (right panel) dots arrays at various angles of incidence with H = 600 Oe.

The lateral confinement within the edge of each dot leads to the expected sampling of the DE mode of the continuous layer into a series of discrete lines related to quantized values of the in-plane wave vector $q_{\parallel}$. Up to five discrete modes are observed in the investigated θ range at the measured frequencies: (6.5, 7.9, 9, 10, 10.7 GHz) and (6.6, 7, 7.7, 8.6, 9.1 GHz) for the 1 and 2 μm dots, respectively. The uncertainty on these determinations is estimated to ± 0.1 and ± 0.2 GHz, respectively: such rather large values follow from the the non negligible width of the observed lines and from the weakness of the scattered intensity. These Brillouin spectra exhibit the typical observed behaviour in the absence of dispersion: by changing the angle of incidence θ, the measured frequencies do not vary, but, indeed, their relative intensities are modified. On the other hand, the frequency splitting between neighboring discrete modes decreases with increasing W.

Considering the high aspect ratio W/t (35 and 70, respectively) of the 1 and 2 μm studied dots, it is reasonable to expect that the spin waves are still satisfactorily described by the quantization of the wave vector $q_{\parallel}$, which specifies the magnetic modes of the continuous layer and their dispersion through Eq. (1). Our experimental results agree with such an approach assuming :

$$q_{\parallel,n} = (n + \beta)(1-\delta_{n,0})\pi/W \quad \text{(with } \beta = -1/4) \tag{4}$$

which is very close to Eq. (3), and H = 570 Oe. The γ and M values ($\gamma = 1.87 \times 10^7$ Hz/Oe and $4\pi M = 7800$ Oe, respectively) are obtained from the analysis of the Brillouin spectra of the unpatterned area. The calculated frequencies of the first discrete modes are: (6.5, 7.71, 8.89, 9.76, 10.44 GHz) and (6.5, 7.15, 7.88, 8.49, 9.01 GHz) for the 1 and 2 μm dots respectively. The 570 Oe calculated value of the internal field is not completely satisfactorily: within the experimental uncertainty it provides a nearly vanishing demagnetizing field, in contrast to a direct evaluation which leads to a mean value of about 150 Oe, and, consequently, to an internal magnetic field H of 450 Oe. However, we have to point that our conclusions differ from a recently published study [5] which, using a different analysis concerning the lowest observed line, leads to a significantly larger reduction of the magnetic field.

To summarise, we observe the quantization of the spin wave modes in square permalloy dots but it does not seem to follow the behaviour derived from a simple extension of the model appropriate for the wires. In view of these difficulties we are presently performing analytical estimations including the exchange contributions, through the development of previously published models [10, 11]. On the other hand, numerical calculations using a finite element method are in progress in our laboratory [12].

References

[1] Y. Roussigné, S.-M. Chérif, C. Dugautier, and P. Moch, Phys. Rev. B **63**, 134429 (2001).
[2] J. Jorzick et al., J. Appl. Phys. **89**, 7091 (2001).
[3] S.-M. Chérif, Y. Roussigné, and P. Moch, IEEE. Trans. Magn. **38**, 2529 (2002).
[4] J. Jorzick, S. O. Demokritov, B. Hillebrands, M. Bailleul, C. Fermon, K. Guslienko, A. N. Slavin, D. Berkov, and N. L. Gorn, Phys. Rev. Lett. **88**, 47204 (2002).
[5] G. Gubbiotti, P. Candeloro, L. Businaro, E. Di Fabrizio, A. Gerardino, R. Zivieri, M. Conti, and G. Carlotti, J. Appl. Phys. **93**, 7595 (2003).
[6] R. W. Damon and J. R. Eshbach, J. Phys. Chem. Solids **19**, 308 (1961).
[7] Y. Roussigné, S.-M. Chérif, and P. Moch, J. Magn. Magn. Mater. **263**, 289 (2003).
[8] S. M. Chérif and J.-F. Hennequin, J. Magn. Magn. Mater. **165**, 504 (1997).
[9] http://math.nist.gov/oommf
[10] K. Yu. Guslienko and A. N. Slavin, Mater. Sci. Forum **373–376**, 217 (2001).
[11] B. A. Kalinikos and A. Slavin, J. Phys. C **19**, 7013 (1986).
[12] Y. Roussigné, S.-M. Chérif, and P. Moch, to be published.

Field-induced phase transitions (FIPT) in molecular magnets

B. Barbara[1], V. V. Kostyuchenko[2], A. S. Mischenko*,[3], and A. K. Zvezdin*,[4]

[1] Laboratoire Louis Néel, Grenoble, France
[2] Institute of Microelectronics and Informatics, RAS, Yaroslavl, Russia
[3] M.V. Lomonosov Moscow State University, Moscow, Russia
[4] General Physics Institute of RAS, Moscow, Russia

Received 31 August 2003, accepted 31 December 2003
Published online 18 March 2004

PACS 75.50.Gg, 75.50.Xx

Ultrahigh magnetic fields ($B = 100–1000$ T) are valuable means for many-sided research in solid state physics. A brief description of the explosion method and some results of its application are presented. The magnetic coil compensation method has been used to study the first order phase transitions in a broad range of fields and temperatures. The Faraday rotation method has been applied to measure the second order magnetic phase transitions. A number of magnetic materials were studied experimentally and theoretically: first order phase transitions in RCo_2 compounds, second order transitions in $KMnF_3$, MnF_2, etc. and steplike quantum magnetization process (transition from ferrimagnetic to ferromagnetic phase) in some mesoscopic magnets – $Mn_{12}Ac$, Mn_6Rad_6 and V_{15}. These magnetic measurements provide us with a direct and unique method of determination of the exchange interactions between magnetic ions in magnetic molecules. The knowledge of the magnetic interacion energies is important for molecular engineering in order to design new magnetic nanoscale materials with desireable properties. The intramolecule exchange integrals of the magnetic nanoclusters $Mn_{12}Ac$, Mn_6Rad_6 and V_{15} that have been determined by this method are presented.

© 2004 WILEY-VCH Verlag GmbH & Co. KGaA, Weinheim

1 Magnetic cumulation

The magnetic cumulation technique and magnetic explosion generators have been studied since about 1950 independently in Russia and in the USA. The Russian project was headed by A.D. Sakharov and the American – by F. Villing and E. Teller [1, 2]. The principle of magnetic cumulation is in the folowing. A hollow metallic cylinder is placed inside a magnet coil thus embraced by initial magnetic field flux. A chemical high explosive charge surrounds the cylinder all over its surface. When the charge is blown up it compresses the hollow cylinder and magnetic field flux at the same time. More dense magnetic lines correspond to higher intensity and energy of the magnetic field. From the physical point of view it can be described by the well known law of electromagnetic induction: the reduction of a magnetic contour's radius remains the magnetic flux almost constant by means of current and magnetic field increase. The average reproducible magnetic field value achieved in the first pool of experiments was about 5×10^6 Oe in a 10 mm cavity. Larger fields were inaccessible because of the instability of magnetic cumulation and limits on the initial magnetic field values. The solution to the Rayleigh – Taylor instability problem turned out to be rather simple: the compression should be made by several cylindrical coaxial liners. Thus every time the compression becomes unstable, a liner is changed with a new one and the stable compression continues. The new method has allowed to achieve magnetic fields up to 1.3×10^7 Oe in a 4 cm^3 cavity. The magnetic field sweep time is tens of microseconds.

---

* Corresponding authors: e-mails: zvezdin@fpl.gpi.ru, smischenko@yahoo.com, Phone: +7 (095) 939 38 83

The ultra high magnetic fields described in the previous paragraph provide us with highly valuable means for conducting of research in different fields of solid state- and nanophysics. This report is a brief overview of some ultra high field experiments and corresponding theoretical considerations carried out so far.

2   The first order FIPT in the RCo_2 itinerant metamagnets

Field induced phase transitions are valuable means for studying various aspects of solid state physics such as the exchange interactions in a broad range of bulk and mesoscopic materials [3].

A number of the RCo_2 compounds experience a first order phase transition from weak to strong magnetic state under the influence of internal or external magnetic field (itinerant metamagnetism). These metamagnetic transitions have been intensively studied both experimentally [4, 5] and theoretically [6]. The internal field can be changed indirectly via pressure and concentration of chemical components [4]. The intermetallic rare-earth compounds RCo_2 (cubic Lave phase $C15$) with heavy rare earths are ferrimagnetics. There are two magnetic subsystem in these compounds. One of the subsystems is formed by magnetic component of rare earth f-electrons, the other is formed by magnetic cobalt $3d$-electrons which are hybridized with $5d$-electrons of rare earth ($4d$-electrons in case of Y). The d-electron subsystem is magnetically unstable. The phase transition is accompanied by a jump of magnetisation and peak of the magnetic susceptibility at helium temperatures in ultrastrong magnetic field (from 60 to 380 Tesla depending on the compound). The peak has been measured at the MC-1 experimental setup by means of a magnetic coils compensation method [5]. The method is rather easy and is highly effective in study of the first order phase transitions accompanied by changes of magnetic structure as it can be applied to any samples regardless their mechanical, optical and thermal properties.

3   The second order FIPT and the Faraday rotation in the transparent antiferromagnets

However, the second order transitions are accompanied only by a kink on a M vs B dependence, where M is magnetization and B is magnetic field and a jump on a magnetic susceptibility curve at the corresponding B value. The coil compensation method is not accurate enough to locate the jumps of susceptibility in megagauss range of fields. An alternative approach is the well known Faraday effect – the rotation of plane of light polarization at propagation through magnetized media. A number of samples ($KMnF_3$, MnF_2) were measured with the Faraday technique and the kinks of magnetization are clearly seen [5, 7]. It is quite obvious that the method can be used with transparent materials only.

4   The full magnetization process in ferrimagnets. Difference between bulk and mesoscopic samples

Now let us describe how the ultra high magnetic field setup has been used to study some fundamental properties of molecular magnets (or mesoscopic molecules). Some of them are shown on slide 15. The antiferromagnetic indirect interaction between the magnetic ions embedded within the molecules results in ferrimagnetic ground state of the molecules. However, there is a significant difference between the magnetization process of bulk and mesoscopic ferrimagnetic materials that can be seen from the slide 18: the former has a monotonous increase of magnetization with the external magnetic field and a plateau on its magnetic susceptibility curve while the latter has specific quantum steps on magnetization and oscillations of susceptibility with the field [3].

5   The full magnetization process in the $Mn_{12}Ac$ nanocluster

The $Mn_{12}Ac$ nanocluster attracts much attention particularly because it is one of the first objects exhibiting the macroscopic quantum tunneling [8–10]. A detailed description of this molecular ferrimagnet can be found elsewhere [8, 11, 12]. The theoretical dependence of its magnetization, magnetic susceptibility and spin crossover mechanism are described in [13]. The curves have the characteristics of a ferrimagnetic mesoscopic behavior – quantum steps on the magnetization and oscillations on the susceptibility curves. Recent experiments in ultra high magnetic fields have proved the theoretical picture [5, 14]. Note that the highest peak on the usceptibility was found at about 500 Tesla. The experimental results has allowed us to reveal the values of the important parameters of exchange interactions within the molecule.

Table 1 The exchange parameters of $Mn_{12}Ac$.

Exchange integral	Literature, cm^{-1} [12]	MC-1 experiment, cm^{-1}
J_1	–150	–61
J_2	–60	–30
J_3	–60	5.6
J_4	$|J_4| < 30$	–15

6 Ferrimagnetic wheel Mn$_6$Rad$_6$

Another example of the mesoscopic family is a ring ferrimagnet Mn_6Rad_6 studied in [15]. The experimental magnetization vs magnetic field curve and a relevant Hamiltonian describing the molecule's properties best of all is presented in [16]. Our analysis has shown that a simple Heisenberg exchange mechanism is not enough and a special three-spin interaction must be added to fit the experimental data properly. Besides, the obtained data has allowed us to estimate the parameters in the Hamiltonian. The results of the MC-1 experiment is shown on fig.1. All essential physics appeared in the megagauss magnetic fields region [14].

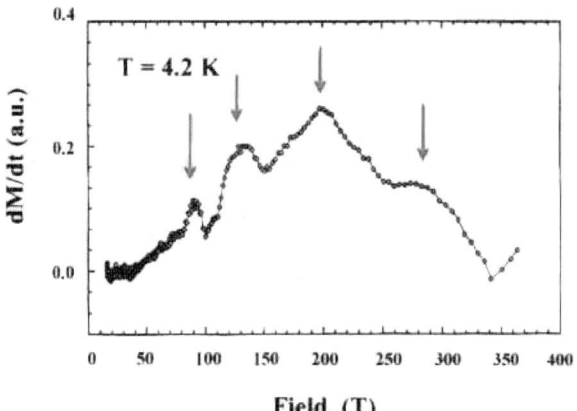

Fig. 1 The MC-1 measurement of Mn_6Rad_6. Theoretically obtained critical fields are shown by arrows.

Table 2 The exchange parameters of Mn_6Rad_6.

Exchange integral	Literature, cm^{-1}	MC-1 experiment, cm^{-1}
J_1	~10^2 [16]	27.0
J_3	– // –	1.4

7 Mesoscopic ferrimagnet V$_{15}$

The last application of the megagauss technique we will speak about is a study of a V_{15} molecular magnet. Its unique feature is that it has the Kramers doubly degenerate ground state and nevertheless some kind of the macroscopic quantum tunneling was observed on it [17]. The magnet consists of 15 vanadium $s = \frac{1}{2}$ ions surrounded by organic ligands. Its structure and scheme of the exchange interactions are described elsewhere among some other interesting features [17, 18]. The vanadium ions form a quasispherical three-layer structure. Indirect antiferromagnetic interaction results in a S = ½ net spin in the ferrimagnetic ground state. The ferrimangetic order changes to a ferromagnetic one after a number of quantum steps on the magnetization curve in megagauss magnetic fields (see Fig. 2) [19]. The experimental results allow us to estimate the values of the exchange integrals [14, 20, 21].

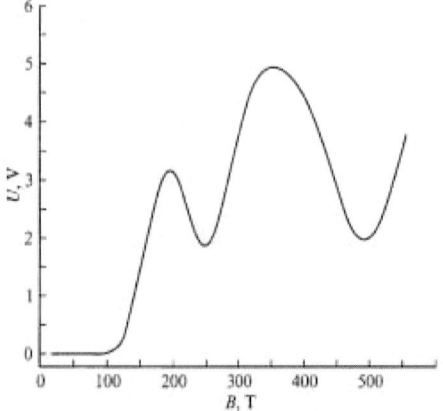

Fig. 2 Uncompensated signal from the measuring coils U (proportional to the variation of the sample's magnetization with time) vs. magnetic field B. The MC-1 measurement of Vn_{15}.

Table 3 The exchange parameters of V_{15}.

Exchange integral	Literature [22], K	MC-1 experiment, K
J	−800	−490
J'	−150	−80
J_1	−150	−80
J''	−300	−161
J_2	−300	−161
J_0	−2.5	– // –

8 Conclusion

As a conclusion, let's ask a rhetorical question – why is it of importance to study the full magnetization process of the magnetic nanoclusters? These magnetic measurements provide us with a direct and unique method of determination of the exchange interactions between magnetic ions within a nanocluster. On the other hand, the knowledge of the magnetic interacion energies is important for molecular engineering in order to design new magnetic nanoscale materials with desireable properties.

The magnetic cumulation method has proved to be a powerful tool for many-sided research in solid state physics. The magnetic coil compensation method has been used to study the first order phase transitions in a broad range of fields and temperatures. The Faraday method has been applied to measure the second order magnetic phase transitions. A number of magnetic materials were studied with the method: first order phase transitions in RCo_2 compounds, second order transitions in $KMnF_3$, MnF_2 and steplike magnetization process (transition from ferrimagnetic to ferromagnetic phase) in some molecular magnets – $Mn_{12}Ac$, Mn_6Rad_6 and V_{15}.

References

[1] A.I. Pavlovskii, Priroda **8**, 39 (1990).
[2] A.D. Sakharov, Uspekhi UFN **88**(4), 725–734 (1966).
[3] A.K. Zvezdin, Field-induced phase transitions in ferrimagnets, Handbook of Magnetic Materials, Vol. 9, ed. K.H.J. Buschow, 1995, pp. 405–543.
[4] T. Goto, K. Fukamichi, T. Sakakibara, and H. Komatsu, Solid State Commun. **72**(9), 945–947 (1989).

[5] I.S. Dubenko, A.K. Zvezdin, A.S. Lagutin, R.Z. Levitin, A.S. Marcosyan, V.V. Platonov, and O.M. Tatsenko. JETP Lett. **64**, 202–206 (1996).
A.K. Zvezdin, I.A. Lubashevskii, R.Z. Levitin, V.V. Platonov, and O.M. Tatsenko, Uspekhi UFN **168**(10), 1141–1146 (1998):
A.K. Zvezdin, I.A. Lubashevsky, R.Z. Levitin, G.M. Musaev, V.V. Platonov, and O.M. Tatsenko, Itinerant electron magnetism: Fluctuation Effects (Kluwer Acad. Publishers, 1998), pp. 285–302.
[6] E.P. Wohlfarth and P. Rhodes, Philos. Mag. **7**, 1817 (1962):
A.K. Zvezdin, JETP Lett. **58**(9), 719–725 (1993):
L.R. Evangelista and A.K. Zvezdin, JMMM **140–144**, 1569–1570 (1995)
S.N. Utochkin and A.K. Zvezdin, JMMM **140–44**, 787–788 (1995).
[7] A.A. Mukhin, V. Platonov, V.I. Popov, O.M. Tatsenko, and A.K. Zvezdin, Physica B **246/247**, 195–9 (1998).
[8] B. Barbara, L. Thomas, F. Lionti, I. Chiorescu, and A. Sulpice, JMMM **200**, 167–181 (1999).
[9] A. Caneschi, D. Gatteschi, C. Sangregorio, R. Sessoli, L. Sorace, A. Cornia, M.A. Novak, C. Paulsen, and W. Wernsdorfer, JMMM **200**, 182–201 (1999).
[10] Julio F. Fernandez, Fernando Luis, and Juan Bartolome, Phys. Rev. Lett. **80**(25), 5659–5662 (1998).
[11] D. Gatteschi, L. Pardi, A.L. Barra, A. Mueller, and J. Doering, Nature **354**, 465 (1991).
[12] R. Sessoli, Hin-Lien Tsai, A.R. Shake et al., J. Am. Chem. Soc. **115**, 1804 (1993).
[13] A.K. Zvezdin and A.I. Popov. JETP **82**(6), 1140–1144 (1996).
[14] A.K. Zvezdin, V.V. Kostyuchenko, V.V. Platonov, V.I. Plis, A.I. Popov, V.D. Selemir, and O.M. Tatsenko. Uspekhi UFN **172**(11), 1303–6 (2002).
[15] V.V. Kostyuchenko, I.M. Markevtsev, A.V. Filippov, V.V. Platonov, V.D. Selemir, O.M. Tatsenko, A.K. Zvezdin, and A. Caneschi, arXiv: cond-mat/0209670 v1 (30 Sept. 2002).
[16] A. Caneschi, D. Gatteschi, J. Laugier, P. Rey, R. Sessoli, and C. Zanchini, J. Am. Chem. Soc. **110**, 2795–2799 (1988).
[17] I. Chiorescu, W. Wernsdorfer, A. Mueller, H. Boegge, and B. Barbara, Phys. Rev. Lett. **84**, 3454 (2000).
[18] A.S. Mischenko, A.K. Zvezdin, and B. Barbara, Sov. Phys. – Solid State **45**(2), 292–297 (2003).
[19] V.M. Platonov, O.M. Tatsenko, V.I. Plis, A.I. Popov, A.K. Zvezdin, and B. Barbara, Sov. Phys. – Solid State **44**, 11, 2104–2106 (2002).
[20] A.K. Zvezdin, V.I. Plis, A.I. Popov, and B. Barbara, Sov. Phys. – Solid State **43**(1), 185–189 (2001).
[21] V.V. Kostyuchenko and A.K. Zvezdin, Sov. Phys. – Solid State **45**(5), 903–906 (2003).
[22] D. Gatteschi, L. Pardi, A.L. Barra, and A. Muller, Molecular Engineering **3**, 157 (1993).

Spin-accumulation contribution to the magnetic nanobridge magnetoresistance

A. K. Zvezdin[1,2], **K. A. Zvezdin**[1,2], **and D. Pullini**[*,2]

[1] General Physics Institute of RAS, Vavilova st. 38, 199991 Moscow, Russia
[2] Research Centre of FIAT, Strada Torino, 50, 10043 Orbassano (TO), Italy

Received 31 August 2003, accepted 31 December 2003
Published online 18 March 2004

PACS 75.47.–m, 75.75.+a

We present here the investigation of the spin-accumulation effect in magnetic nanocontacts [1] and nanobridges [3]. Spin-accumulation effect is employed to explain the huge values of magneto-resistance experimentally discovered in these structures. We present the solution of the 2D spin-diffusion problem in a magnetic nanobridge. Dependences of the magneto-resistance created by the domain wall and of the distribution of non-equilibrium spin density on the nanobridge geometry and the material parameters are presented.

© 2004 WILEY-VCH Verlag GmbH & Co. KGaA, Weinheim

Very recently it was discovered that domain walls in magnetic nanocontacts and nanowires create huge values of magnetoresistance, which can reach more than thousands percents. This is the reason why such structures are considered to be very promising for the upcoming spintronics devices. Up to now, all the experiments were carried out on the nanocontacts with uncontrolled geometry [1].

In [2] it was demonstrated that geometry of nanocontacts strongly affects the position of domain wall inside nanocontact and magnetization distribution around it, i.e. it affects the final structure magnetoresistivity. It follows that to carry out more meaningful experiments and to build up nanocontact based high performing nano-devices, it is necessary to fabricate nanocontacts of strictly defined geometry.

In the present article the investigation of spin-accumulation effect in magnetic planar nanobridge of strictly controlled geometry [3] (Fig.1) is reported.

An electric current flowing through the plane magnetic nanobridge cause non-equilibrium spin-concentration in the domain wall region and it leads to an additional resistance.

The equations which describe the spin density distribution and the electrical potential in ferromagnetic with electric current with density J take the form:

$$\mathbf{j}_\alpha = eD_\alpha \nabla n_\alpha - \sigma_\alpha \nabla U , \qquad (1)$$

$$\operatorname{div} \mathbf{j}_\alpha = \frac{en_\alpha}{\tau_S} , \qquad (2)$$

$$\sum_\alpha n_\alpha = n_+ + n_- = 0 . \qquad (3)$$

Here $\mathbf{j}_\alpha$ – current density of the current of the charges with spin index $\alpha \in \{+,-\}$, which corresponds to two spin polarisations; n_α, σ_α, D_α, μ_α – density, conductivity, electro-diffusion coefficient, and electrochemical potential correspondently, U – electrical potential, e – electron charge, τ_S – longitudinal time of the spin relaxation of the electrons.

---

[*] Corresponding author: e-mail: d.pullini@crf.it; daniele.pullini@crf.it

The n_α value represents the addition to the equilibrium density of the electrons with the spin state α. Electrochemical potentials μ_α describe the non-equilibrium destribution of the majority and minority electrons in a domain:

$$\mu_\alpha = \zeta_\alpha - eU , \qquad (4)$$

here ζ_α – chemical potential of the electron subsystem of spin state α.

The non-equilibrium density n_α and the chemical potential are to one another connected by the following relation

$$n_\alpha = g_\alpha \zeta_\alpha , \qquad (5)$$

where

$$g_\alpha = \sigma_\alpha / e^2 D_\alpha \qquad (6)$$

– Einstein relation. For metals g_α value represents the density of states (DOS) at Fermi state.

Further we use the following symmetrized values:

$$\mu_t = \mu_+ + \mu_-, \quad \mu_S = \mu_+ - \mu_-,$$
$$\xi_t = \xi_+ + \xi_- , \qquad (7)$$
$$j_t = j_+ + j_-, \quad j_S = j_+ - j_-$$

and

$$\sigma_\pm = \frac{\sigma}{2}(1 \pm \beta) = \frac{1}{2\rho}(1 \pm \beta), \quad g_\pm = \frac{g}{2}(1 \pm \delta) , \qquad (8)$$

where σ, ρ, g – conductivity, unit resistivity, and DOS of the nanowire material, β and δ – unit-less parameters of the asymmetry of above mentioned characteristics.

Following (1)–(6) the spin potential μ_S is equal to

$$\mu_S = n_+ (g_+^{-1} + g_-^{-1})$$

i.e. it is proportional to the non-equilibrium spin-concentration. The distribution of μ_S in the connecting bridge and shores can be found from the solution of the diffusion equation:

$$\Delta \mu_S = \frac{\mu_S}{L_S^2} . \qquad (9)$$

Here $L_S = (D_S \tau_S)^{1/2}$ – is the spin-diffusion length,

$$D_S = \frac{1}{e^2} \frac{g_+^{-1} + g_-^{-1}}{\sigma_+^{-1} + \sigma_-^{-1}} .$$

We can write the functional which corresponds to this equation:

$$2E(\mu_S) = \iint_{NB} \left(\left(\frac{\mu_S}{L_S}\right)^2 + \left(\frac{d\mu_S}{dx}\right)^2 + \left(\frac{d\mu_S}{dy}\right)^2 \right) dxdy , \qquad (10)$$

here integration is done over the all nanobridge (Fig. 1). For this functional the differential equation (9) is the equation of Euler-Lagrange.

According to the nanobridge geometry we split it into three regions –*I*, *B*, and *P* (Fig. 1) and look for the μ_S distribution in each of them.

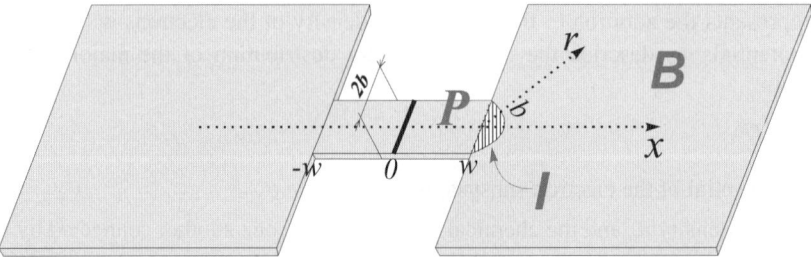

Fig. 1 Nanobridge geometry.

The connecting bridge I is supposed to be enough long and narrow, and the μ_S distribution in P area can be considered as one-dimensional and can be written as:

$$\mu_S(x) = \frac{\mu_1 - \mu_0 e^{-w/2}}{2\sinh(w/2)} e^x + \frac{-\mu_1 + \mu_0 e^{w/2}}{2\sinh(w/2)} e^{-x}, \quad 0 \leq x \leq w/2. \tag{11}$$

Here μ_0 and μ_1 are values of the spin-potential correspondently in the bridge centre and near the connection with the shores.

In the region B desired spin-potential distribution is supposed to be radial-symmetrical and can be written as:

$$\mu_S(r) = \frac{a}{\pi} K_0(r), \quad r \geq b/2. \tag{12}$$

It is obvious that $a = \pi\mu_1/K_0(b/2) + O(b^2)$. To express the unknown coefficients μ_0 and μ_1 through each other the boundary conditions are used. Finally the last independent coefficient is found by the energy minimum:

$$E(\mu_S) = E_P + E_I + E_B, \tag{13}$$

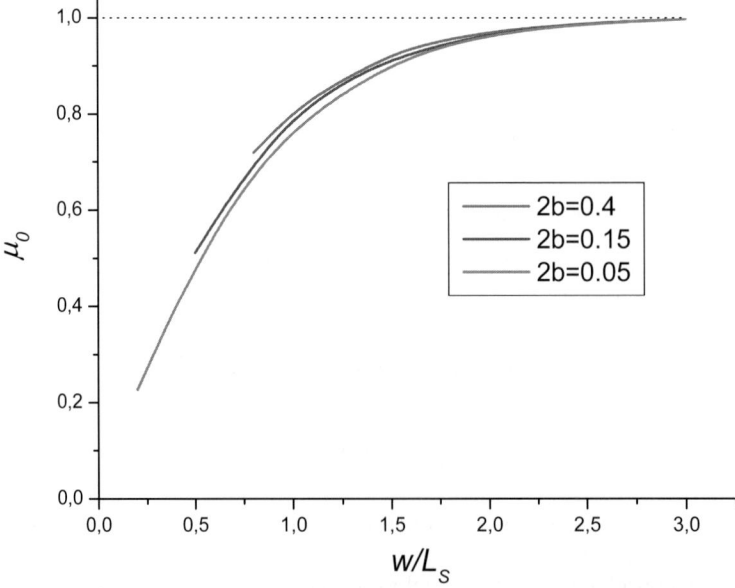

Fig. 2 Dependence of *MR* on the nanobridge length w for various nanobridge thickness.

$$\frac{dE(\mu_s(\mu_1))}{d\mu_1} = 0 \tag{14}$$

This approach leads to the expression for the total voltage created by the domain wall:

$$\Delta U_b = \frac{\beta}{e}\mu_0, \tag{15}$$

here β stands for scattering asymmetry.

Concluding, it can be stated that the 2D spin-accumulation model in nano-bridge is constructed. The distribution of the non-equilibrium spin concentration in the system is calculated. The voltage caused by the presence of the domain wall is calculated. The dependence of spin potential and magnetoresistivity on the geometry factors are obtained (Fig. 2). According to the calculation the MR ratio can reach up to 20 ÷ 30 percents.

References

[1] N. Garcia, M. Munoz, and Y.-W. Zhao, Phys. Rev. Lett. **82**, 2923 (1999).
[2] A.K. Zvezdin, A.F. Popkov, K.A. Zvezdin, and L.L. Savchenko, Phys. Met. Metallogr. **9**, 165 (2001).
[3] A.A. Zvezdin and K.A. Zvezdin, JETP Lett. **75**, No. 10, 613 (2002).
[4] A.K. Zvezdin and K.A. Zvezdin, Bull. Lebedev Physics Institute **8**, 3 (2002).
[5] M. Dzero, L.P. Gorkov, A.K. Zvezdin, and K.A. Zvezdin, Phys. Rev. B 0004XXR (2003).

Tunable refractive index of magnetic fluids and its applications

Chin-Yih Hong[*,1], **H. E. Horng**[2], and **S. Y. Yang**[2]

[1] Department of Mechanical and Automation Engineering, Da-Yeh University, Chang-Hwa 515, Taiwan
[2] Institute of Electro-optical Science and Technology, National Taiwan Normal University, Taipei 116, Taiwan

Received 31 August 2003, accepted 31 December 2003
Published online 18 March 2004

PACS 75.50.Mm, 78.20.Ci, 78.20.Ls

Magnetic fluid is a type of colloid consisting of magnetic nano-particles dispersed in a liquid carrier with the aid of surfactants and Brownian motion. Thus, the optical properties of magnetic fluids can be manipulated through careful selection of magnetic particles and carriers. In this work, we give an example by designing a refractive index around 1.465 at 1.557 µm wavelength with magnetic fluids composed of various carriers and particles, that could play an important role in optical fiber communication. In addition, we also show how to achieve a desired flexibility in the tunable refractive index with externally varying fields by adopting suitable magnetic fluid films. These results reveal the feasibility of developing index-match or index-tunable devices using magnetic fluids.

© 2004 WILEY-VCH Verlag GmbH & Co. KGaA, Weinheim

1 Introduction With the vast advancement in opto-electronics and communications, research on novel devices, whose properties can be manipulated by varying the refractive index, has become important because of the tunability and flexibility of these devices. For example, long-period fiber gratings, transmissive devices that couple the co-propagation core and cladding modes of the fibers, are very useful for loss filtering and are usually used to flatten the spectra of erbium-doped-fiber amplifiers in optical fiber communication [1]. The core mode transmission losses of the long-period fiber gratings are expected to be modulated by varying the surrounding refractive index. Stegell et al. used several types of liquid media for obtaining various surrounding refractive indices and indicated that the coupling wavelength between the core and the cladding modes shifted to shorter wavelengths as the refractive index of the liquid surrounding the long-period fiber grating increased [2]. Once the refractive index of the liquid medium exceeds that of the cladding, the coupling wavelength increases for higher refractive indices of liquids, and a high coupling efficiency is also achievable [3]. These results imply that the desired transmission characterization of a long-period fiber grating can be obtained by embedding the fiber grating in a suitable liquid medium.

In addition to long-period fiber gratings, the photonic crystals, in which the dielectric function is spatially periodic thus causing their optical properties to be dominated by strong diffraction effects, exhibit a strong dependence of the photonic band gap on the index contrast between the micro-arrayed columns and the interstitial region [4–6]. For instance, by reducing the index contrast between the semiconductor columns and the liquid media in the interstitial regions, the forbidden band can be shifted to a lower-frequency region. This means that the forbidden band of the photonic crystal can be tunable by using liquid media of various refractive indices for the interstitial region.

At present, the liquid media used for these devices include several kinds of liquids such as organic solutions, polymers or liquid crystals. However, due to the limitations of liquid specimens, only a few values, depending on the liquids used, are available for the refractive index. Thus, the optical properties

---

[*] Corresponding author: e-mail: cyhong@mail.dyu.edu.tw, Phone: +886-4-8511213, Fax: +886-4-8511213

of the devices are restricted. In this work, we reveal that continuous values of the refractive index can be obtained by using magnetic fluids of various concentrations. Furthermore, for a given concentration, the tunability of the refractive index of a magnetic fluid is also verified. The possibility of designing the required tunable refractive index with magnetic fluids needs more investigation.

2 Experimental details The magnetic fluid was injected into a glass cell to form a magnetic fluid film, which was then covered with a triangular prism made of silicon or other material with a high known refractive index. Thus, there was an interface, PM, between the prism and the magnetic fluid. Then, the critical angle of total reflection at interface PM was precisely measured to determine the refractive index of the magnetic fluid films [7]. The wavelength of the light emitted from the laser diode was 1.557 µm. A solenoid was used to provide an external magnetic field with a resulting deviation of the applied magnetic field within the sample region of around 0.5%. The angle, θ_H, between the applied field and the magnetic fluid film was varied by adjusting the solenoid. The temperature was maintained at 24.3 °C ± 0.1 °C with a circulating water system. The uncertainty of the temperature was 0.1 °C. An optical microscope and a CCD camera were adapted to record the images of the structural patterns in the magnetic fluid films under external fields.

3 Results and discussion Since magnetic fluids consist of magnetic particles dispersed in liquid carriers with the aid of surfactants and Brownian motion [8], the refractive indices of magnetic fluids can be manipulated through the careful selection of magnetic particles and carriers. Before dispersing the magnetic particles into the carrier of the magnetic fluid, the fluid exhibits the intrinsic refractive index of the carrier. With the dispersion of the magnetic particles, the refractive index of the magnetic fluid increases linearly and continuously with the particle concentration. Figure 1 shows examples of the concentration-dependent refractive indices of the magnetic fluids under zero field. The carriers here are kerosene, heptane and water, and the magnetic particle is Fe_3O_4 with an average diameter of around 10 nm [9]. The surfactant used for both kerosene and heptane is oleic acid, whereas lauric acid is used for water. These three curves exhibit a similar behavior except the existence of a shift, depending on the carrier of the magnetic fluid.

According to the results in Fig. 1, the magnetic fluid with a required refractive index, say 1.465 which is close to that of optical fibers, is available by setting the concentration, ϕ, at 0.71%, or at 0.75% and 1.45% in terms of volume fraction for the kerosene, heptane- and water-based Fe_3O_4 magnetic fluids, respectively. Other desired values of the refractive indices for various applications may also possibly be obtained by using magnetic fluids with suitable concentrations. It is noted that the highest value of around 1.6 for the refractive index of the magnetic fluid in Fig. 1 is due to the limitations of the experiment. Higher refractive indices of the concentrated magnetic fluids can be probed by using a high-index prism in the experimental setup [7].

When an external magnetic field is applied to a magnetic fluid film at a fixed temperature, say 8.0 °C, the refractive index of the water-based Fe_3O_4 magnetic fluid film does not change until the applied magnetic field H exceeds a critical field strength $H_{c,n}$ (= 22 Oe for 8.0 °C), as shown by the data points in Fig. 2. Over $H_{c,n}$, the refractive index n_{MF} of the magnetic fluid film increases under a higher field strength and then almost becomes saturated. As the temperature was raised, a similar behavior was obtained for the experimental n_{MF}–H curve, except that a higher $H_{c,n}$ resulted at a higher temperature. Through careful inspection, the critical $H_{c,n}$ was found to increase from 22 to 50 Oe with the increasing temperature from 8.0 to 60.0 °C. The temperature dependence of the critical field $H_{c,n}$ is plotted in Fig. 3. In addition, the experimental n_{MF}–H curve at a higher temperature shifted toward the region with lower values for n_{MF}'s. This means that the n_{MF} under a certain field higher than $H_{c,n}$ reduces when the temperature goes up. A negative thermo-optical coefficient dn_{MF}/dT with an order of magnitude less than 10^{-4} °C^{-1} was found for the magnetic fluid film. The temperature shows a compensation effect on the variation in the refractive index of the magnetic fluid with respect to the magnetic field.

Since the variation in the refractive index of the magnetic fluid film with the field strength is suggested to be due to the structural formation [7], we then investigated the structures in the magnetic fluid film under external magnetic field at various temperatures to clarify the phenomena observed in Figs. 2 and 3.

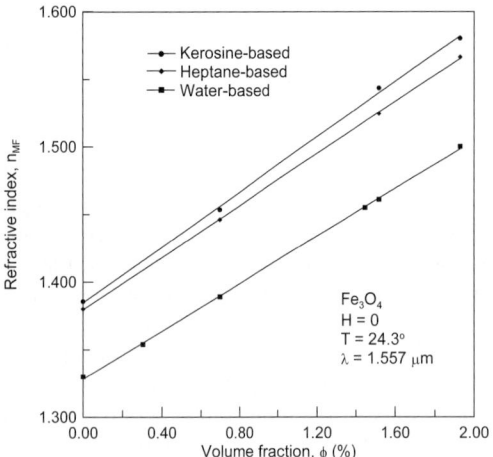
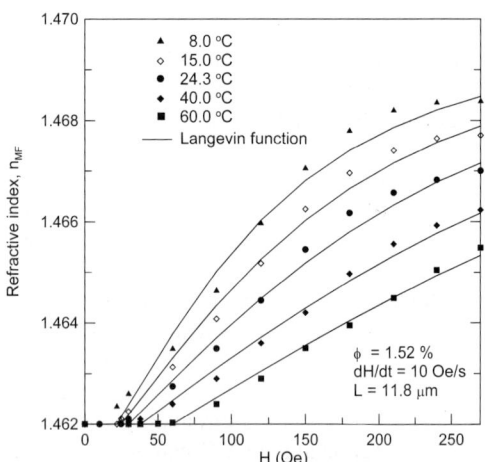

Fig. 1 Refractive index versus concentration of the magnetic fluids under zero field for the carriers of kerosine, heptane and water using Fe_3O_4 as magnetic particles. The wavelength of the light used is 1.557 μm, and the working temperature is 24.3 °C.

Fig. 2 Magnetic field dependent refractive index of the magnetic fluid film under various temperatures from 8.0 to 60.0 °C. The solid lines are obtained from Eq. (1) for the corresponding $n_{MF}(H,T)$ data.

Under zero field strength at a given temperature, the magnetic fluid film is at the monodispersion state, at which the magnetic particles Fe_3O_4 are dispersed homogeneously in the fluid. A typical image of the monodispersion state of the magnetic fluid film is shown as the bottom-right inset in Fig. 3. It is noted for this image that because the diameter of the magnetic particles is around 9 nm, nothing can be observed via using an ordinary optical microscope. When the field strength is increased from zero, the film is still at the monodispersion state, while the magnetic particles start to agglomerate to form magnetic columns as the field strength exceeds a critical field (denoted by H_o). These particle columns are distributed in the film randomly, and hence the state is assigned as the disordered column state. The image of the disordered columns in the magnetic fluid film under 120 Oe at 8.0 °C is shown as the upper-left inset in Fig. 3. With the investigation on the structural evolution from the monodispersion to the disordered column state in the magnetic fluid film at various temperatures, the critical field H_o as a function of temperature can be determined experimentally. The results are shown in Fig. 3. A larger H_o is required for a higher temperature. This phenomenon was also found by Rosensweig et al. [10] and Horng et al. [11]. The result is attributed to the enhancement in the kinetic energy of magnetic particles at higher temperatures. Thus, higher field strength is needed to compensate for the kinetic energy, making the particles to be able to agglomerate to form columns.

Another remarkable result obtained when we compare the H_o–T curve with the $H_{c,n}$–T curve is that these two curves overlap with each other, as shown in Fig. 3. Since the down-right side of the H_o–T curve in Fig. 3 corresponds to the monodispersion state and the up-left part denotes the disordered columns, the overlap of these two curves gives strong evidence to that the variation in the refractive index of the magnetic fluid film under various fields is due to the formation of columns.

It is well known that two major physical energies are important for the formation of particle columns in the magnetic fluid films under external magnetic fields [10] one is the thermal energy of particles, the other is the magnetic energy of particles under external magnetic fields. The compensation effect between the thermal energy and the magnetic energy can be described with a Langevin function [12]. With the fact that the variation in n_{MF} with the field strength H and the temperature T is attributed to the column formation, the trend of $n_{MF}(H,T)$ should be similar to a Langevin function. Hence, the experimental data shown in Fig. 2 are compared with Langevin-function-like $n_{MF}(H,T)$ expressed as

$$n_{MF}(H,T) = [n_s - n_o]\left[\coth\left(\alpha\frac{H - H_{c,n}}{T}\right) - \frac{T}{\alpha((H - H_{c,n})}\right] + n_o, \quad \text{for} \quad H > H_{c,n}, \tag{1}$$

© 2004 WILEY-VCH Verlag GmbH & Co. KGaA, Weinheim

where n_o (= 1.4620 here) is the refractive index of the magnetic fluid film under fields lower than $H_{c,n}$ that depends on the type of carrier liquid and the concentration of magnetic fluid, n_s denotes the saturated value of the refractive index of the magnetic fluid and is obtained as 1.4704 by fitting the experimental n_{MF}–H data at 8 °C to Eq. (1) in which the n_{MF}–H curve becomes saturated at H > 200 Oe, H is the field strength in Oe, T represents the temperature in Kelvin, and α is the fitting parameters. The n_{MF}–H curves calculated from Eq. (1) at various temperatures are plotted with the solid lines in Fig. 2. It was found that the data points in Fig. 2 almost follow the Langevin-function type of n_{MF}(H,T) curves.

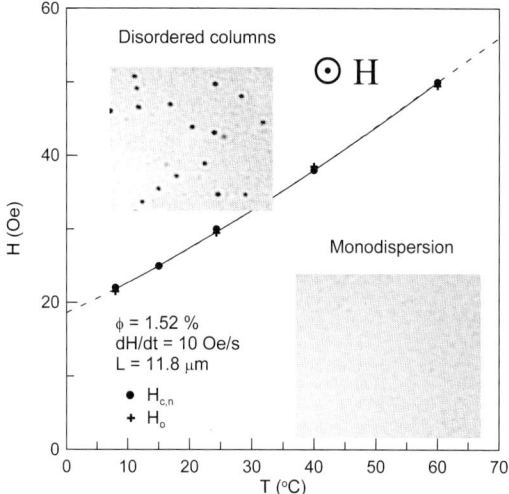

Fig. 3 Critical fields $H_{c,n}$ and H_o as functions of temperature for the magnetic fluid film. $H_{c,n}$ denotes the critical field under which the refractive index of the magnetic fluid film starts to increase with the raising field strength, and H_o represents the field strength required for the formation of the magnetic columns in the film. The down-right inset shows the structure of the magnetic fluid film at the monodispersion state, whereas the up-left one exhibits the cross-section image of the structural pattern for the disordered columns (shown as black dots) in the magnetic fluid film.

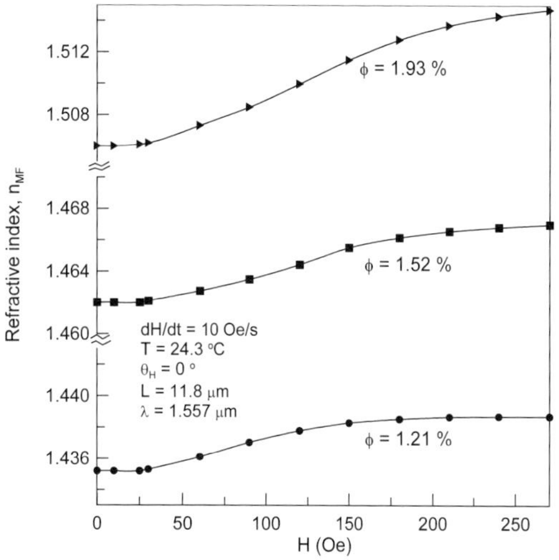

Fig. 4 Refractive index versus magnetic field strength for various concentrations of magnetic particles in the magnetic fluid films.

In addition to the temperature dependent n_{MF}–H curve, the magnetic-field dependent refractive index of magnetic fluid films with various concentrations, ϕ, were also measured. The results are shown in Fig. 4. A similar trend was obtained for each n_{MF}–H curve in Fig. 4: the n_{MF} does not change until the field strength exceeds a critical value (~ 30 Oe), and then increases with rising field strength, reaching a saturated value under higher strengths. But, the difference between saturated value and zero-field n_{MF} becomes larger as ϕ increases from 1.21% to 1.93% of volume fraction. This is due to enhanced of phase separation in a magnetic fluid film with a higher concentration when an external field is applied [13]. Furthermore, as the concentration increases, the n_{MF}–H curve moves to a region with higher refractive indices. This shift of the n_{MF}–H curve is attributed to the variation in the volume fraction of the magnetic particles Fe_3O_4 in the magnetic fluid. It is noteworthy that the n_{MF} under zero field increased from 1.4352 to 1.5062 as ϕ rose from 1.21% to 1.93%, whereas a variation in n_{MF} of magnitude of 10^{-3} resulted from rising field strength from zero to 200 Oe. This indicates that the shift in the n_{MF}–H curve with varying concentration is more sensitive than the change in field strength. Hence, one can adjust the concentration to obtain a refractive index roughly around the desired value and then fine-tune the refractive index by applying a magnetic field.

According to our previous study [14], film thickness also plays a role in the magnetically modulated structures of magnetic fluid films. Hence, the influence of the film thickness on the refractive index of a magnetic fluid film under external fields was also examined. Figure 5 plots the n_{MF}–H curves for the water-based Fe_3O_4 magnetic fluid films with various values of thickness ranging from 11.8 to 200 μm. When a magnetic field applied along a fixed direction was increased, a larger variation in the n_{MF} was observed for the thicker films when H increased from zero up to 270 Oe. On the other hand, the n_{MF} under a given field strength for a certain film thickness became larger as the magnetic field was applied parallel to the film surface ($\theta_H = 90°$) with respect to that under the perpendicular field ($\theta_H = 0°$). Figure 6 shows clearly the shift of the n_{MF}–H curve toward the region with higher n_{MF}'s when the direction of the externally magnetic field is rotated from perpendicular to parallel to the plane of the film.

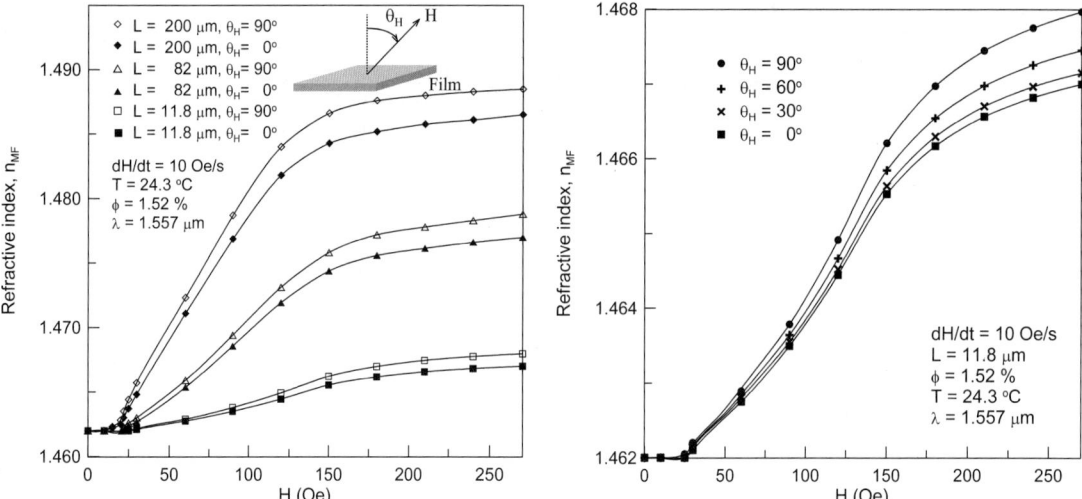

Fig. 5 Magnetic-field dependent refractive index of the magnetic fluid films with various thicknesses. The solid and the hollow symbols are for perpendicular ($\theta_H = 0°$) and parallel ($\theta_H = 90°$) fields, respectively. The angle θ_H is illustrated in the inset.

Fig. 6 Refractive index of the magnetic fluid film versus the external magnetic field strength. The direction of the applied magnetic field is changed from perpendicular to the film ($\theta_H = 0°$) aligned with the film surface ($\theta_H = 90°$).

4 Conclusion The refractive index of a magnetic fluid film increases as an external magnetic field is applied. This magnetically induced variation in the refractive index can be controlled by the temperature, concentration of magnetic material in the fluid, film thickness or the direction of the applied field. Thus, we have not only clarified a control mechanism for the refractive index of the magnetic fluid film, but also laid the groundwork for further applications based on the tunable refractive index of magnetic fluid films.

Acknowledgements This work is supported by National Science Council of ROC under Grant Nos. NSC90-2212-E-212-031 and NSC90-2112-M-003-020.

References

[1] A. M. Vengasarkar, P. J. Lemaire, J. B. Judkins, V. Bhatia, T. Erdogan, and J. E. Sipe, J. Lightwave Technol. **14**, 58 (1996).
[2] D. B. Stegall and T. Erdogan, IEEE Photo. Technol. Lett. **11**, 343 (1999).

[3] Olivier Duhem, Jean-Francois Henninot, Marc Warenghem, and Marc Douay, Appl. Opt. **37**, 7223 (1998).
[4] H.-B. Lin, R.J. Tonucci, and A.J. Campillo, Appl. Phys. Lett. **68**, 2927 (1996).
[5] Jane F. Bertone. Peng Jiang, Kevin S. Hwang, Daniel M. Mittleman, and Vicki L. Colvin, Phys. Rev. Lett. **83**, 300 (1999).
[6] Katsumi Yoshino, Shigenori Satoh, Yuki Shimoda, and Yoshiaki Kawagishi, Jpn. J. Appl. Phys. **38**, L961 (1999).
[7] S.Y. Yang, Y.F. Chen, H.E. Horng, Chin-Yih Hong, W.S. Tse, and H.C. Yang, Appl. Phys. Lett. **81**, 4931 (2002).
[8] R.E. Rosensweig, Ferrohydrodynamics (Cambridge U.P., New York, 1985).
[9] H.E. Horng, I.J. Jang, K.L. Kung, Y.D. Yao, H.C. Yang, and Chin-Yih Hong, Czech. J. Phys. **46**, Suppl. S4, 2023 (1996).
[10] R.E. Rosensweig and J. Popplewell, Int. J. Appl. Electromagn. Mater. **2**, Suppl., 83 (1992).
[11] S.Y. Yang, Y.H. Ke, W.S. Tse, H.E. Horng, Chin-Yih Hong, and H.C. Yang, J. Magn. Magn. Mater. **252**, 290 (2002).
[12] R.E. Rosensweigs, "Ferrohydrodynamics" (Dover Publication Inc., New York, 1998).
[13] S.Y. Yang, H.E. Horng, Chin-Yih Hong, H.C. Yang, M.C. Chou, C.T. Pan, and Y.H. Chao, J. Appl. Phys. **93**, 3457 (2003).
[14] C.-Y. Hong, C.H. Ho, H.E. Horng, C.-H. Chen, S.Y. Yang, Y.P. Chiu, and H.C. Yang, Magnetohydrodynamics **35**, 297 (1999).

Relationship between the giant magnetoimpedance effect and the relaxation of magnetic permeability

J. Íñiguez[*,1], **V. Raposo**[1], **D. García**[1], **O. Montero**[1], **P. Hernández-Gómez**[2], and **C. de Francisco**[2]

[1] Dpto. Física Aplicada, Universidad de Salamanca, Plaza de la Merced s/n, 37071 Salamanca, Spain
[2] Dpto. Electricidad y Electrónica, Universidad de Valladolid, Prado de la Magdalena s/n, 47071 Valladolid, Spain

Received 31 August 2003, accepted 31 December 2003
Published online 18 March 2004

PACS 75.40.Gb, 75.50.Kj, 75.60.–d

Temperature dependence of the magnetoimpedance and its relaxation has been measured in amorphous ferromagnetic wires. The presence of several peaks in the frequency dependence of the magneoimpedance is ascribed to the magnetoelastic resonance of standing waves in the circular direction. A core-shell model that includes a term accounting for the magnetic looses interprets the analysis of the phase angle of the complex impedance. It is also observed a time evolution of the impedance that reveals the existence of magnetic after effect with a maximum at temperatures close to 350 K.

1 Introduction

As a result of the skin effect at high enough frequencies, the current distribution in amorphous ferromagnetic samples is not uniform across the transversal area of the conductor. Although the electrical conductivity is considerably smaller than the conventional conductive materials, its magnetic permeability is several orders of magnitude higher. The great influence of the external magnetic field on the permeability values suggests remarkable expectations for its application as magnetic sensors. With compositions of minimum magnetostriction, the GMI reaches values up to 800% with fields of a few nT for frequencies of several MHz [1].

In a lot of experimental papers we can find, a circular magnetic permeability independent of the work frequency is usually assumed. On the other hand, in the radio-frequencies range, usual for the GMI experiments, we consider necessary to take into account the dispersion of the magnetic permeability, according to the amorphous manufacturing procedure.

In this work we present a theoretical study of this analysis as well as a simulation and a set of measurements that show us the convenience of considering these effects in different GMI experiences.
For amorphous magnetic ribbons and wires the study is focused on the analysis of the relationship between the circular magnetic permeability and the domain walls dynamics. According to the bibliography [2], typical values of parameters involved in such a dynamics goes to relaxation/resonance frequencies corresponding to a few MHz, reason because of this analysis can not be omitted in GMI experiences. In fact, there is in the bibliography a bit of information concerning to magnetic relaxation in amorphous samples, but the results are not conveniently related to GMI relaxation [3–5].

Our experimental set-up measures the GMI ratio and the magnetic relaxation of the sample. According to our results it could be plausible to connect both experiences in terms of a conventional magnetic after effect theory.

---

[*] Corresponding author: e-mail: nacho@usal.es, Phone: +34 923 294 400 - 1301, Fax: +34 923 294 584

2 Experimental

Two different amorphous wires were studied in our experimental set-up. $Co_{68.15}Fe_{4.35}Si_{12.5}B_{15}$ amorphous wires were prepared at the laboratory by in-rotating-water quenching-technique. The diameter of the wires was 125 µm and pieces about 10 cm in length were taken for the experiments. $Fe_{77.5}Si_{7.5}B_{15}$ amorphous wires with 125 µm in diameter were provided by Goodfellow.

Hysteresis loops were measured by the conventional fluxmetric method. For GMI measurements, both real and imaginary parts of the sample impedance were obtained using the four-point technique by means of an LCR bride model HP 4285A from Agilent Technologies. The frequency dependence was studied in the range from 75 kHz to 30 MHz, while the AC current was maintained constant to 1 mA. The magnetic field was generated by a Helmholtz coils pair connected to a programmable bipolar power supply from Kepco which produces a maximum magnetic field strength of 15 kA/m. The system was placed inside a magnetic screening to avoid the influence of any external field.

In magnetic after effect (MAE) measurements, the impedance of the sample was measured by an LCR bridge or a lock-in amplifier, depending on the frequency range measured. The sample was introduced in a cryostat to measure the temperature dependence of the GMI. In order to analyze the MAE, the sample was submitted to the appropriate magnetic field during 30 seconds and then suddenly removed by disconnecting the power supply by a relay. MAE when removing magnetic fields from 80 to 10000 A/m was studied by recording impedance data after the switching off. (see Fig. 1).

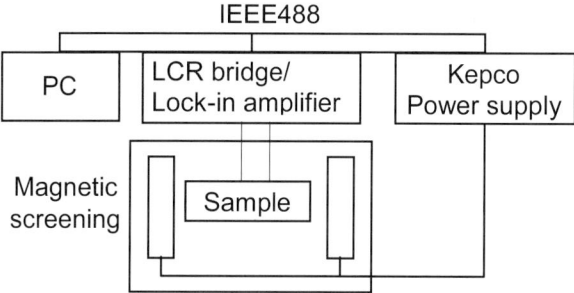

Fig. 1 Schematic block diagram of the measuring set-up.

3 Results and discussion

As expected, $Co_{68.15}Fe_{4.35}Si_{12.5}B_{15}$ wire presents a very high GMI ratio, close to 800% corresponding to the high circular permeability of the sample. This percent is reached at frequencies close to 7 MHz and it is maintained up to 30 MHz. On the other hand $Fe_{77.5}Si_{7.5}B_{15}$ sample has a lower GMI ratio due to the different domain structure with radial domains in the outer shell and thus smaller circular permeability [6, 7]. In this case the presence of magnetic resonance in the sample is clear, and it is observed by several drops of the sample resistance as a function of frequency and the dispersive behavior of the phase angle of the complex impedance (Fig. 2).

The presence of resonances can be justified according to the dynamics of Bloch walls and/or by the existence of magnetoelastic resonances in the sample. As the resonances are only found in samples with high magnetostriction constant, the most probable cause of it will be the second one. In that case the frequency of resonance is given by:

$$f = \frac{n}{2\pi r}\sqrt{\frac{E}{\rho}} \qquad (1)$$

where r is the radius of the sample, E the Young's module ρ the density an n an integer. Using bibliography values for E we obtain a resonance frequency of 10.7 MHz, quite close to the experimental one (13 MHz) confirming the assumption that the magnetostriction is the major cause of the existing resonances.

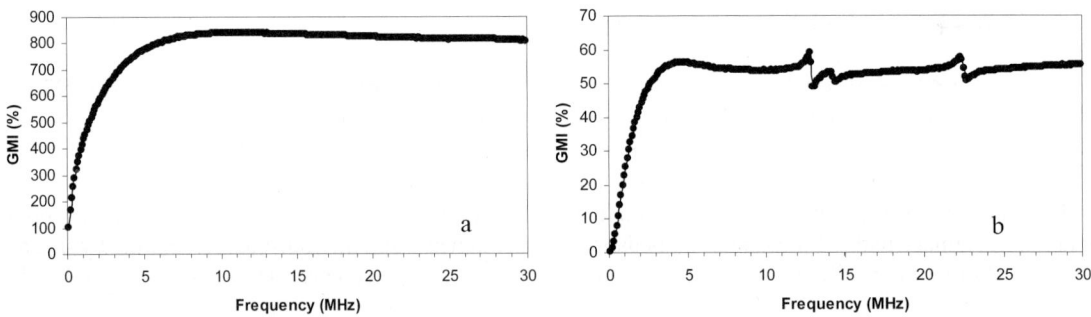

Fig. 2 Frequency dependence of the GMI ratio for the $Co_{68.15}Fe_{4.35}Si_{12.5}B_{15}$ (a) and $Fe_{77.5}Si_{7.5}B_{15}$ (b) samples.

Another interesting feature is the phase angle of the complex impedance. The phase angle contains information of both real and imaginary components of the impedance and reveals to be a very critical parameter to evaluate the validity of permeability models in amorphous wires. The conventional skin effect theory with merely a constant value of permeability is not enough to explain the frequency dependence of the impedance in amorphous wires. For such systems we must consider at least two different values of the permeability, one for the core and another for the outer shell [3]. This model can explain with a quite good agreement the behavior of the phase angle in some wires, but in those presenting high values of circular permeability, magnetic looses must be also considered. Figure 3 shows the frequency dependence of the phase angle of the complex impedance of FeSiB and CoFeSiB for zero external field. FeSiB sample presents a uniform increase of the phase angle, corresponding to the classical skin effect theory, while the Co-based sample has a more complex behavior. For this wire it is necessary to consider two regions with different magnetic permeabilities and magnetic looses. If we take into account only two regions without looses, we can reproduce the presence of a peak in the phase angle at low frequencies, but it will asymptotically go to 45 degrees at high frequencies. To reproduce the experimental behavior it is necessary to consider also the magnetic looses of the circular hysteresis loop, which reduces the phase angle at high frequencies as shown in Fig. 3. In effect, if we introduce a complex magnetic permeability ($\mu = \mu' + j\mu''$) in the equations proposed by [3] we will find another contribution for the resistance proportional to $\omega\mu''$ responsible of the increase of the resistance at high frequencies. In the FeSiB sample it is also necessary to consider the dispersive behavior of the permeability to account the resonances and the corresponding drops of the permeability.

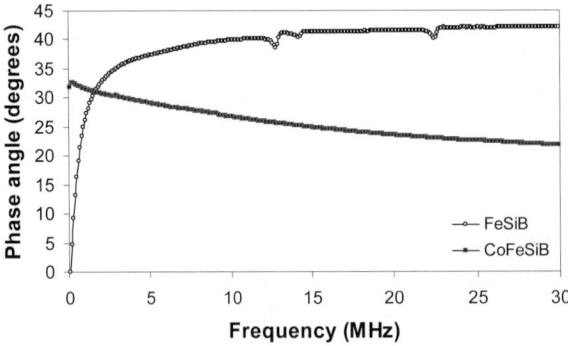

Fig. 3 Experimental frequency dependence of the phase angle of the impedance.

Complementary to the study of the frequency dependence is the time dependence of the GMI. If we submit the sample to a step change in the applied magnetic field the time evolution, usually called magnetic after-effect reveals the dynamics of the process. Figure 4(a) shows the temperature dependence of the isochronal spectrum for the resistance defined as:

$$MAE(\%) = \frac{R(t_2) - R(t_1)}{R(t_1)} \times 100 \qquad (2)$$

with $t_1 = 1$ s and $t_2 = 2, 4, 8,..., 128$ s. The corresponding isothermal curves are presented in Fig. 4(b). The presence of magnetic after effect in amorphous wires has been previously reported in ribbons [4, 5]. The Presence of MAE is usually ascribed to the rearrangement of the magnetic moments in the magnetically coupled grains [8]. MAE was only found in the CoFeSiB sample, where the GMI percent was higher, and no significant changes were found for the FeSiB wire. The presence of 3% MAE near room temperature has to be considered in order to apply GMI for sensor applications working in usual conditions.

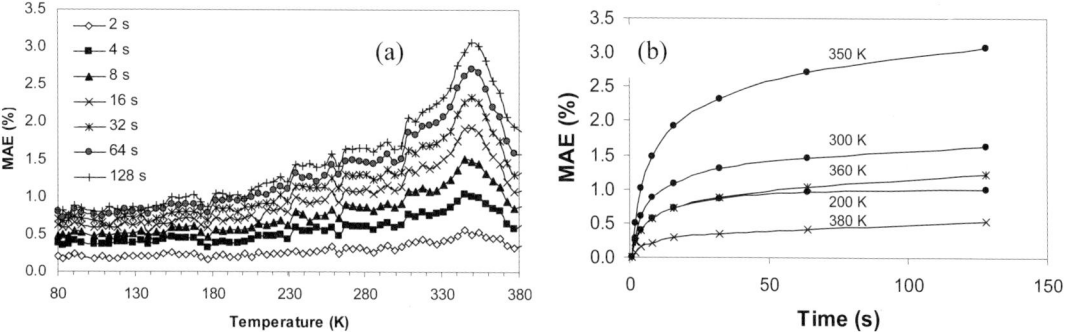

Fig. 4 Isochronal spectrum of the magnetic after effect for the $Co_{68.15}Fe_{4.35}Si_{12.5}B_{15}$ wire at 37.5 kHz (a). Isothermal curves (b).

4 Conclusion

The GMI in amorphous wires can be analyzed by a core-shell model that includes a term of magnetic looses. The dispersive behavior of the permeability and the presence of magnetoelastic resonances must be considered in magnetostrictive samples assuming several contributions to the permeability. The time dependence of the impedance shows an important amount of MAE in some samples with a maximum above room temperature.

Acknowledgements This work was supported in part by the MCyT of Spain under project MAT2001-0082-C04-03 and by the Junta de Castilla y León under project SA010/03.

References

[1] V. Raposo, A. G. Flores, M. Zazo, and J. Íñiguez, J. Magn. Magn. Mater. **254–255**, 204 (2003).
[2] A. H. Morrish, "The Physical Principles of Magnetism" (John Wiley and Sons., Inc., New York, 1965).
[3] D.-X. Chen, L. Pascual, E. Fraga, M. Vázquez, and A. Hernando, J. Magn. Magn. Mater. **202**, 385 (1999).
[4] M. Knobel, M. L. Sartorelli, and J. P. Sinnecker, Phys. Rev. B **57**, 3362 (1997).
[5] M. L. Sartorelli, M. Knobel, J. Schoenmaker, J. Gutierrez, and J. M. Barandiarán Appl. Phys. Lett. **71**, 2208 (1997).
[6] H. Chiriac, C.S. Marinescu, and T.A. Ovári, J. Magn. Magn. Mater. **196–197**, 162 (1999).
[7] J.P. Sinnecker, M. Vázquez, A. Garcia-Arribas, and M. Knobel, IEEE Trans. Magn. **33**(5), 3343 (1997).
[8] G. Buttino, A. Cecchetti, and M. Poppi, J. Magn. Magn. Mater. **241**, 183–198 (2002).

Substrate surface engineering for tailoring properties of functional ceramic thin films

H.-U. Habermeier[*]

Max-Planck-Institut für Festkörperforschung, Heisenbergstr. 1, 70569 Stuttgart, Germany

Received 31 August 2003, accepted 31 December 2003
Published online 5 April 2004

PACS 68.55.Jk, 74.76.Bz, 75.70.–i

Using oxide substrates for functional ceramic thin film deposition beyond their usual application as chemical inert, lattice-matched support for the films represents a novel concept in ceramic thin film research. The substrates are applied as a functional element in order to controllably modify the atom arrangement and the growth mode of ceramic prototype materials such as cuprate superconductors and colossal magnetoresistance manganites. One example is the use of epitaxial strain to adjust the relative positions of cations and anions in the film and thus modify their physical properties. The other makes use of vicinal cut $SrTiO_3$ which enables the fabrication of regular nanoscale step and terrace structures. In $YBa_2Cu_3O_{7-x}$ thin films grown on vicinal cut $SrTiO_3$ single crystals a regular array of antiphase boundaries is generated causing an anisotropic enhancement of flux-line pinning. In the case of La–Ca–Mn–O thin films grown on vicinal cut substrates it could be demonstrated that magnetic in-plane anisotropy is achieved.

© 2004 WILEY-VCH Verlag GmbH & Co. KGaA, Weinheim

1 Introduction

The physical properties of perovskite-type functional ceramic thin films are known to sensitively depend on details of deviations from their ideal composition and/or crystal structure. Therefore, substrate-induced lattice strain, substrate surface morphology and growth-induced defects are some examples for extrinsic effects playing an important role in determining thin-film properties and consequently the application potential of functional ceramics [1–5]. The sensitivity of the physical properties on the structure arises from the origin of the functionality at the level of sub-unit cells. In the case of the doped rare earth manganites, e.g. the bonding distance and bonding angle of the Mn–O–Mn building block determines the charge transfer of an electron from Mn to Mn via the oxygen, thus the bandwidth, metallicity, charge-ordering and the appearance of ferromagnetism [6–8]. In high temperature superconductor (HTS) cuprates such as $La_{1-x}Sr_xCuO_4$ the distance of the apex oxygen from the CuO_2 planes affects the Cu4s–O2p hybridization and thus doping and T_C [9, 10].

The multi-component chemical composition of the ceramic materials combined with their complex crystal structure represents a much higher degree of sophistication for a thin-film technology compared to that of metals and semiconductors. Consequently, to understand the film growth process and modify it intentionally in order to open a path for defect control, is a tremendous challenge. In contrast to metals and semiconductors where the deposition temperature for epitaxial growth from the vapor phase is around 20% (metals) to 40% (semiconductors) of the melting temperature, T_M, for ceramics such as the HTS cuprates much higher values of around (0.7–0.8) T_M are required. This reflects the chemical dissimilarity of the constituent cations and their quite different diffusion coefficient at the substrate surface at growth conditions. The general problem of the vapor deposition of oxides with complex chemical

---

[*] e-mail: huh@fkf.mpg.de

composition and large unit cells to fabricate single crystal type epitaxially grown thin films had not been addressed prior to 1986, the year of the discovery of the HTS cuprates. The role of the substrates in the efforts exploring the epitactic growth of ceramic thin films has been treated so far simply as that of a mechanical support combined with chemical stability and compatibility with the prerequisites given by the epitaxy relations. In order to pave a new way for all-oxide electronics and novel device concepts an advanced oxide epitaxy technology is required based on nanoscale substrate engineering as well as on atomic layer control of oxide films. Furthermore, the opportunities buried in tailoring the substrate surface morphology have to be explored in order to intentionally modify growth conditions and defect arrangements.

In this paper, a novel concept in ceramic thin-film research is introduced, the use of substrates as a functional element. The function of a substrate can originate either from well designed topological properties affecting the growth and defect structure of a ceramic thin film or from its electronic and/or magnetic properties. In a first step, we restrict ourselves to functions derived from the substrate surface and structural peculiarities of artificially tailored substrates. Two examples may serve for this case study. First, an intentional lattice mismatch is used to tailor the epitaxial strain of the film, second, vicinal cut surfaces with a nanoscale step-and-terrace structure are applied to generate macroscopically aligned antiphase boundaries

2 Some basic considerations for ceramic thin film growth

The key requirements of a mature technology for future device applications of functional ceramics are the ability to deposit compact homogeneous layers and multilayer structures with flat surfaces in the view of the short correlation lengths e.g. for superconducting or ferromagnetic thin films. The difficulties achieving structurally and morpholologically perfect ceramic thin films of a specific type are connected with the complex parameter space for deposition spanned by deposition rate, kinetic energy of the particles impinging the substrate surface, surface energies of substrate, film and interface, respectively, substrate flatness and termination. Additionally, the thermo-dynamic requirements for phase stability as given e.g. in the oxygen-pressure/ temperature phase diagram [11] for HTS thin film deposition has to be fulfilled. In general, the rate of growth and the morphology of a film depend on several factors. The most important one, governing the nucleation and growth mode, is the relative supersaturation, μ, determining the chemical potential, σ, as driving force for epitaxy. They are related by

$$\sigma = k_B T_s \ln\mu = k_B T_s \ln(\Phi \Delta T / R T_s^2) \tag{1}$$

where k_B is the Boltzmann constant, Φ denotes the molar heat of solution, ΔT the undercooling, R and T_s are the gas constant and the absolute temperature of the substrate. When the thermodynamic driving force for crystallization from the vapor phase is small, (small supersaturation) dislocation-controlled growth (spiral growth) is observed. The growth

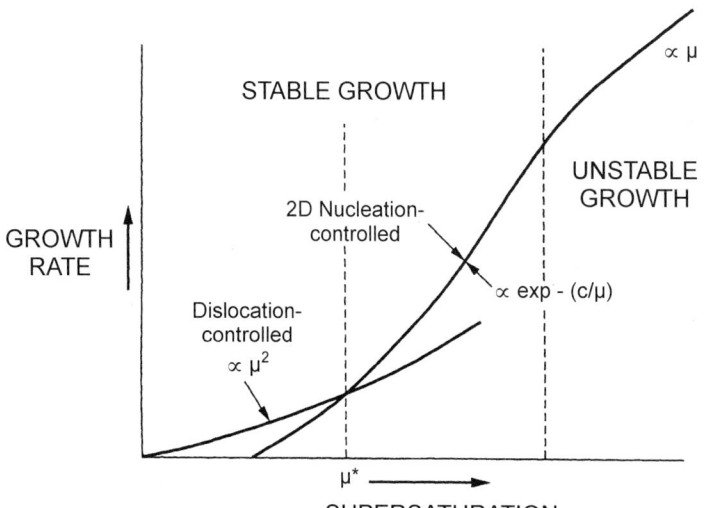

Fig. 1 Schematic dependence of the growth regimes on super-saturation (taken from Ref. [12] ex-planations are given in the text).

mechanism is spiral type in which nuclei are continuously added to the edge of a spirally expanding step. Surface steps are energetically favorable attachment sites. Growth of materials with highly anisotropic surface energies and thus growth velocities such as $YBa_2Cu_3O_{7-x}$ (growth velocity in a, b-direction much larger than in c-direction) has an extended region for spiral growth. Increase of the effective supersaturation favors two-dimensional nucleation (island formation) and a crossover from dislocation-dominated to nucleation-dominated growth is expected at a critical supersaturation μ^*. Further increase of μ causes a transition to unstable growth, uncontrolled nucleation of growth centers on top of each others lead to a dendritic growth type [12]. These regimes are schematically shown in Fig. 1. In addition to supersaturation the deposition temperature, T_s, plays an important role in determining the growth kinetics and the surface morphology. Defining a normalized bonding energy, E_b, for the substrate surface atoms

$$E_b = 4\Phi_{ss}/2k_BT_s \qquad (2)$$

(Φ_{ss} denotes the potential energy of a solid–solid nearest neighbor pairs of atoms in the substrate unit cell) it is obvious that increasing deposition temperature implies a smaller E_b and the surface becomes rougher. Thus a higher density of kink sites on the surface is offered for the oncoming vapor particles, thus leading to a more rapid growth. For the spontaneous nucleation of a unit cell a critical volume of the deposited material is required. Surface diffusion that supplies a nucleus with the necessary material, however, is quite different for the cationic constituents of the material. For $YBa_2Cu_3O_{7-x}$ e.g. the surface diffusion coefficients at deposition temperature, $T_s = 800$ °C, vary by 4 orders of magnitude from Y (10^{-13} m^2/s) to Cu (10^{-9} m^2/s), thus facilitating the formation of micro-precipitates and a phase separation if the stoichiometry is not perfect [13].

3 The "perfect" film on a "perfect" substrate

As briefly sketched in the previous section, supersaturation and substrate temperature determine whether the nucleation is dislocation controlled or island growth controlled. Substrates for the growth of ceramic thin films are more than just a mechanical support for the film. The ideal substrate has to fulfill the requirements of perfect lattice match to ensure epitaxy, match of the thermal expansion coefficients to avoid cracking, lack of structural phase transition between deposition temperature and operating temperature to prevent additional stress, chemical inertness with respect to the film forming species and, finally, lack of interdiffusion. The misfit between the substrate and the film at epitaxial growth temperature not only affects the selection of the epitaxy relations it also influences the surface nucleation and growth modes.

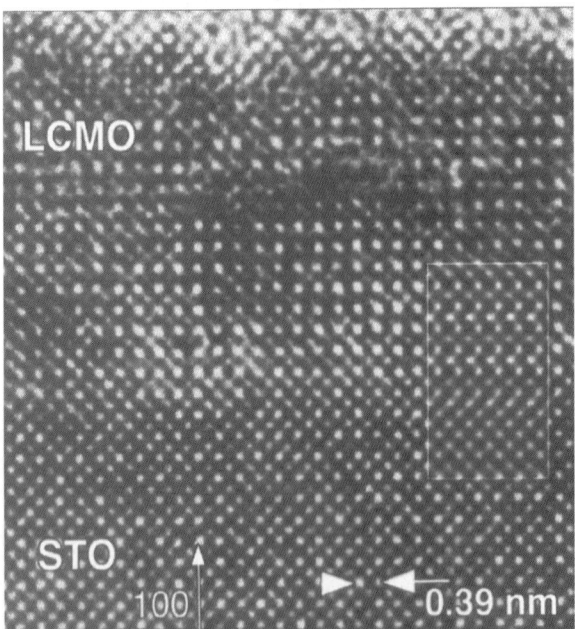

Fig. 2 High resolution cross-sectional TEM of an $SrTiO_3$/$La_{2/3}Ca_{1/3}MnO_3$ interface showing perfect epitaxy.

Misfit reduces the step-flow regime and enhances two-dimensional nucleation. Additionally, the stored elastic energy in the pseudomorphically grown layer adjacent to the substrate acts as a driving force for cluster generation. Misfit acts in the same direction as supersaturation. Strain caused by lattice mismatch and cooling due to thermal expansion coefficient differences can be accommodated by misfit dislocation formation, twinning, and redistribution of oxygen in the lattice. If these mechanisms cannot relieve the strain, cracking may occur. The difficulties in achieving the "perfect" ceramic thin film are connected with problems of the phase stability and oxygen supply superimposed to the peculiarities of the kinetically controlled thin film growth process and the substrate related effects. In Fig. 2 a TEM cross section micrograph of a $La_{2/3}Ca_{1/3}MnO_3$ thin film is given showing the perfect epitaxy on the $SrTiO_3$ substrate [14].

4 Tailoring epitaxial strain in HTS thin films

It is a well known phenomenon that a lattice mismatch between substrate and film will result in a pseudomorphically strained layer with subsequent stress relieve by different accommodation mechanisms such as generation of misfit dislocations, stacking faults or undulations of the lattice planes or a combination of these. As a rule of thumb, a lattice mismatch in the order of 1–2 % is accommodated by the generation of stressed pseudomorphically grown films up to a critical thickness, t_c, which decreases with increasing lattice mismatch. For YBCO deposited onto $LaGaO_3$ single crystal substrates, t_c has been determined to be around 50 nm [15]. A careful analysis of the pressure and strain dependence of T_c for different HTS materials by Locquet et al. [9] shows that in most cases the uniaxial dT_c/dp values have different signs and thus add up to an increase of T_c for compressive strain and a decrease of T_c for tensile strain. Indeed, Locquet et al. could demonstrate, that biaxial strain in $La_{1.9}Sr_{0.1}CuO_4$ (LSCO) thin films pushes T_c up from the bulk value of 25 K to 50 K.

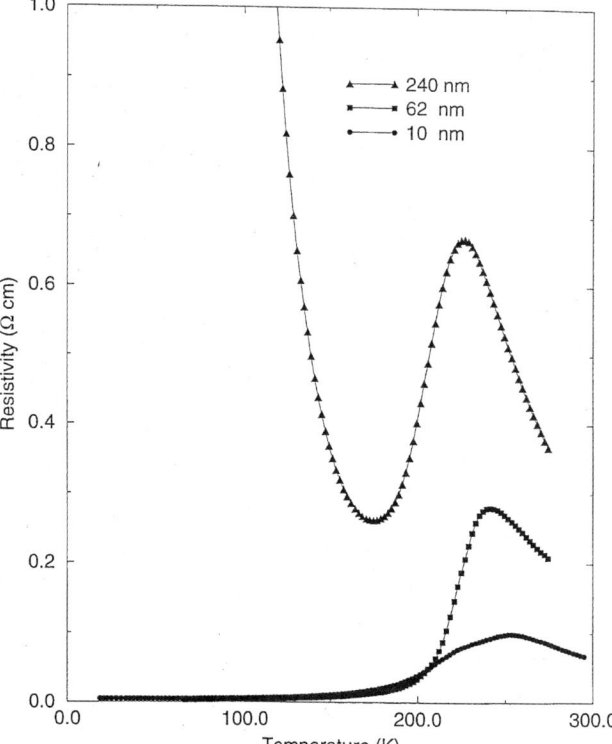

In the case of manganite thin films compressive strain can cause an even more dramatic effect on the temperature dependence of resistivity. In Fig. 3 the ρ (T) curves $La_{.88}Sr_{.1}MnO_3$ thin films of different thickness are represented. Whereas the thick film (240 nm) shows the bulk-like behavior with a metal–insulator transition, T_{MI}, at 220 K followed by a transition to a charged ordered insulator at 170 K, charge ordering is destroyed in the homogeneously strained thin films. Compressive epitaxial strain causes here the transition from a ferromagnetic insulator to a ferromagnetic metal [4].

Fig. 3 Temperature dependence of resistivity of $La_{.88}Sr_{.1}MnO_3$ thin films of different thickness.

Similarly, in thin films of the charge-ordered [CO] $Pr_{1/2}Ca_{1/2}MnO_3$ the robustness of the CO state depends strongly on the strains and thus on the film thickness as shown by Prellier et al. [8].

5 YBCO thin films deposited on vicinal cut $SrTiO_3$ single crystals

When (001)-oriented $SrTiO_3$ substrates with an intentional miscut (angle $\alpha < 15^0$) towards the [010] direction are annealed at 950 °C a regular step and terrace structure is generated with the step height of typically one unit cell of SrTiO$_3$ [a = 0.3905 nm] and a step width corresponding to w = a / tanα. In contrast to films grown onto closely lattice matched (001)-oriented perovskite-type oxide substrates the films deposited onto the vicinal cut substrates show a terrace-like surface morphology with steps along the [100] direction, indicating a change from the usual Stranski–Krastanov growth to a step-flow growth mode. The substrate-mediated modification of the growth mode YBa$_2$Cu$_3$O$_{7-x}$ influences the microstructure of the films and causes an artificially introduced anisotropy of their transport and pinning

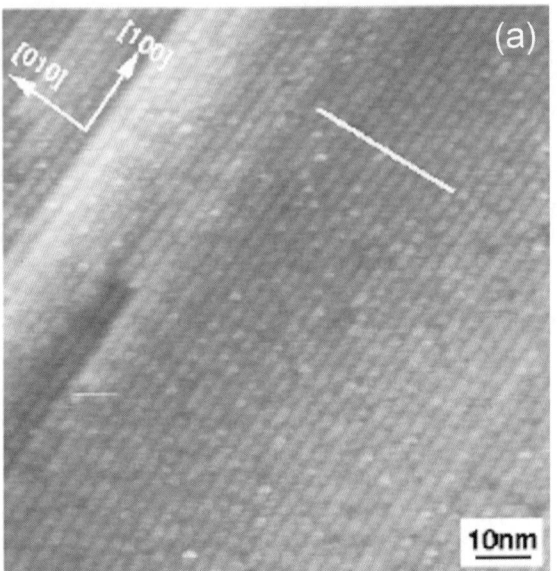

Fig. 4 STM image of a 10° vicinal cut SrTiO$_3$ single crystal substrate after UHV annealing.

properties. We find a close correlation between the film morphology and the film properties as revealed by transport measurements and Raman spectroscopy. In Fig. 4 the surface of a 10° miscut SrTiO$_3$ surface clearly demonstrates a remarkable nanoscale terrace structure as a result of the UHV annealing. Along the [010] direction regularly spaced terraces with a period of 2.3 nm and a step height of $\approx$ 0.39 nm corresponding to one unit cell of SrTiO$_3$ is found. The terrace-like structure is basically preserved after YBCO film deposition, indicating a change of the usual Stranski–Krastanov island growth a step-flow growth for YBCO. Compared to the substrate, the terraces of the film are wider by roughly a factor of three ($\approx$ 7 nm). The step heights, however, are multiples of 0.2 nm, deviating from the full integer of the unit cell height of 1.2 nm of YBCO. This implies that the unit cells grown on the upper and the lower part of a single unit cell step of the substrate can be shifted vertically forming an antiphase boundary (APB). The planar APBs are oriented perpendicular to the film plane forming a regular nanoscale array with an APB distance of around 7 nm. A detailed analysis of the defect structure due to the nanoscale surface step structure of the substrate and its implications to transport properties and flux pinning is given in by Haage et al. [16]. The predominant modifications of the properties of such films are summarized as follows [17]: (i) Strong anisotropy of the electrical dc resistivity; (ii) thickness dependent enhancement of the critical current; (iii) partial detwinning of the films with a better perfection of the CuO chains along the step edges; (iIV) anisotropy of the dimensionality of the fluctuation conductivity above T$_c$ with a large temperature range for the 3-dimensional fluctuations along the step edges and a small range for 3-dimensional fluctuations perpendicular to the step edges.

Similarly, manganite thin films grown on vicinal cut SrTiO$_3$ substrates show an in-plane magnetic anisotropy with the easy axes along the substrate steps. Over a large angular range the angular dependence of the magnetic switching field is found to obey the 1/cosθ law, indicating that the magnetic reversal is completed by a 180° domain nucleation and sweeping along the easy axis [18].

6 Conclusions and outlook

The experiments described above demonstrate the new possibilities in ceramic thin film research if the substrate is not only treated as a support material for the films but also additionally regarded as a functional integrated part of the system film/substrate. Making use of epitaxial strain opens the possibility to externally affect the atom arrangement and thus the properties of the films. Modifying the growth mode from Stranski–Krastanov to step flow in the case of the vicinal cut substrates is a new possibility for tailoring the defect structure and thus flux-line pinning sites in cuprates and in-plane magnetic anisotropy in manganites. Recently, a further technique for controlled substrate surface modification has been introduced using either ion implantation with a focused ion beam microscope [19] or laser surface treatment to regularly etch μm scale grooves or trenches into the substrate. The physical concept behind these experiments is the search of matching effects in HTS flux-line pinning, fabrication of flux guides and formation of regular arrays of manganite ferromagnetic quantum dots for in plane spin valve devices.

References

[1] Interfaces in High T_c Superconducting Systems, S. L. Shinde and A. Rudman (eds.) (Springer, Berlin, 1993).
[2] J. M. Philips, in: H. Weinstock and R. W. Ralston (eds.) (The New Superconducting Electronics, NATO ASI Series, Kluwer, Doordrecht/Boston/London, 1993), p. 59.
[3] T. H. Hylton and M. R. Beasley, Phys. Rev. B **41**, 11669 (1990).
[4] F. S. Razavi, M. Gross, H.-U. Habermeier, O. Lebedev, S. Amelinckx, G. V. van Tendeloo, and A. Vigliante, Appl. Phys. Lett. **76**, 155 (2000).
[5] H.-U. Habermeier, F. S. Razavi, O. Lebedev, G. M. Gross, R. Praus, and P. X. Zhang, phys. stat. sol. (b) **215**, 679 (1999).
[6] A. Ramirez, J. Phys.: Condens. Matter **9**, 8171 (1997).
[7] J. Coey, M. Viret, and S. Von Molnar, Adv. Phys. **48**, 167 (1999).
[8] W. Prellier, Ch. Simon, A. M. Haghiri-Gosnet, B. Mercey, and B. Raveau, Phys. Rev. B **62**, R16337 (2000).
[9] J.-P. Locquet, J. Perret, J. Fompeyrine, E. Machler, J. W. Seo, and G. V. van Tendeloo, Nature **394**, 453 (1998).
[10] E. Pavarini, I. Dasgupta, T. Sasha-Dasgupta, O. I. Jepsen, and O. K. Andersen, Phys. Rev. Lett. **87**, 47003 (2001).
[11] R. Feenstra, T. B. Lindemer, J. D. Budai, and M. D. Galloway, J. Appl. Phys. **69**, 6569 (1991).
[12] I. D. Raistrick, M. Hawley, in: Interfaces in High T_c, S. L. Shinde and A. Rudman (eds.) Superconducting Systems (Springer, Berlin, 1993), p. 28.
[13] R. E. Somekh, Z. H. Barber, and J. Evetts, in: Concise Encyclopedia of Magnetic and Superconducting Materials, J. Evetts (ed.) (Pergamon Press, Oxford, 1992), p. 431.
[14] O. I. Lebedev, G. van Tendeloo, S. Amelinckx, B. Leibold, and H.-U. Habermeier, Phys. Rev. B **58**, 8065 (1998).
[15] M. Ece, E. Garcia-Gonzalez, H.-U. Habermeier, and B. Oral, J. Appl. Phys. **77**, 1646 (1995).
[16] T. Haage, J. Zegenhagen, J. Q. Li, H.-U. Habermeier, M. Cardona, Ch. Jooss, R. Warthmann, A. Forkl, and H. Kronmüller, Phys. Rev. B **56**, 8404 (1997).
[17] H.-U. Habermeier, Proc. SPIE **3481**, 204 (1998).
[18] Z.-H. Wang, G. Cristiani, and H.-U. Habermeier, Appl. Phys. Lett. **82**, 3731 (2003).
[19] J. Albrecht, S. Leonhardt, R. Spolanek, U. Täffner, H.-U. Habermeier, and G. Schütz, Surf. Sci. **547**, L847 (2003).

Optical and magneto-optical study of the Au/Co/Au/Cu multilayer grown on vicinal Si (111) surfaces

W. Cheikh-Rouhou Koubaa[*,1], **B. Bartenlian**[1], **P. Beauvillain**[1], **A. Brun**[2], **P. Georges**[2], **T. Maroutian**[1], and **V. Mathet**[1]

[1] Institut d'Electronique Fondamentale, UMR CNRS 8622, Université Paris Sud, 91405 Orsay, France
[2] Laboratoire Charles Fabry UMR CNRS 8501, Université Paris Sud, 91405 Orsay, France

Received 31 August 2003, accepted 31 December 2003
Published online 5 April 2004

PACS 42.65.Ky, 68.37.Ef, 75.70.Cn, 78.20.Ls, 78.67.Pt

The appearance of well-defined steps on vicinal Si (111) surfaces has a very pronounced effect on the rotational anisotropy of the Second Harmonic Generation (SHG) from these surfaces. This effect is related to the lowering of the surface symmetry and an enhancement of the non-linear susceptibility at these steps. In this paper, we present a study of the growth parameters dependence on the optical and magneto-optical response of Au / Co / Au / Cu ultrathin films epitaxied on vicinal Si (111) surfaces, using Scanning Tunneling Microscopy (STM), linear and non-linear Kerr effect.

1 Introduction

The strong activity and success in magnetic thin films research is based on the current development in surface and material science, enabling the manipulation and control of thin films structure on the atomic level [1]. Some of the most exciting recent discoveries, like the giant magnetoresistance (GMR) and the oscillatory exchange coupling, are related to the properties of multilayers of alternating ferromagnetic and paramagnetic layers. As the interfaces between these layers appear to play an essential role for these phenomena [2] and consequently for the device properties based on them, a detailed study of the magnetic interface properties is needed. In the last few years, there has been a resurgence of interest in Second Harmonic Generation from crystals, since it is an exclusively surface and buried interfaces sensitive technique. The surface specific character of SHG arises from the selection rule which states that, in the electric dipole approximation, second-order non linear optical processes in the bulk of centrosymmetric media are forbidden [3, 4]. Only at an interface or a surface, where the inversion symmetry is broken, a non-linear polarization at the second harmonic frequency can be generated. In this study, single crystals miscut at small angles with respect to a low index orientation were used as substrates, in order to explore the influence of the growth parameters on the interface magnetic anisotropy and to illustrate the lowering of the surface symmetry induced by the presence of steps on the vicinal substrates.

2 Experimental procedures

The used substrates are Si wafers with their polished face miscut from the (111) orientation, the (111) planes being the cleavage planes of silicon. In fact, each Si–Si bond in the bulk is directed along one of the <111> axes, so that a (111) plane cuts as few bonds per area as possible in the crystal [5]. All substrates were taken from two wafers, cut 2° and 6°, respectively, off the [111] axis towards the high symmetry [−1−12] direction. This direction of misorientation allows a step mixture formation of mono and

---

[*] Corresponding author: e-mail: wissem.cheikh-rouhou@fss.rnu.tn

three atomic height [6, 7]; whereas the opposite direction, i.e. [11–2] gives rise to a step bunching phenomenon (Fig. 1) [8]. This difference between the two directions of misorientation is due to the number of dangling bonds in each configuration [9]. Before thermal annealing in a UHV chamber, the substrates were chemically treated in order to obtain a hydrogenated Si surface, which is especially suitable for molecular beam epitaxy. Upon annealing, the surface structure changes with the appearance of the famous 7×7 reconstruction, checked by Reflective High Energy Electron Diffraction (RHEED).

Fig. 1 (a) Misorientation towards [–1–12] direction: mono- and triple-atomic step formation, and (b) misorientation towards [11–2] direction: step bunching.

We have grown on the silicon substrates an ultrathin Cu layer with two different conditions: 2 monolayers (ML) at 100 °C and 4 ML at 200 °C. In order to study the influence of Au and Co on the magnetic properties, different thicknesses of Au buffer (3, 5, 10 and 15 ML) and of Co film (from 2.5 to 15 ML) were grown, on the same substrate, using a moveable shutter. Finally, 15 ML Au were added to protect the Co layer. The topographic study of the samples, at different stages of the growth, was carried out using a combined in situ Scanning Tunneling Microscope and Atomic Force Microscope (STM/AFM). In order to correlate the surface morphology with the magnetic properties of our ultrathin layers, we have investigated linear magneto-optical Kerr effect in both polar and longitudinal configurations, as well as second harmonic generation exploring all growth parameters (miscut angle, Cu growth conditions and Au and Co thicknesses). The SHG measurements were done using a femtosecond Ti: sapphire oscillator with a diameter of 30 μm and an average power of 20 mW. In order to do azimuthal measurements, a special sample holder was designed to perform the experiments exactly on the same point of the sample.

3 Results and discussion

Upon thermal annealing in UHV-snip-of the hydrogenated Si (111) surfaces, STM images (Fig. 2a) show a mixture of monoatomic and triple-height steps with a relative proportion of the latter increasing with the miscut angle of the surface. All terraces exhibit a 7×7 reconstruction as shown in Fig. 2b for a Si (111) surface with 2° misorientation towards the [–1–12] direction.

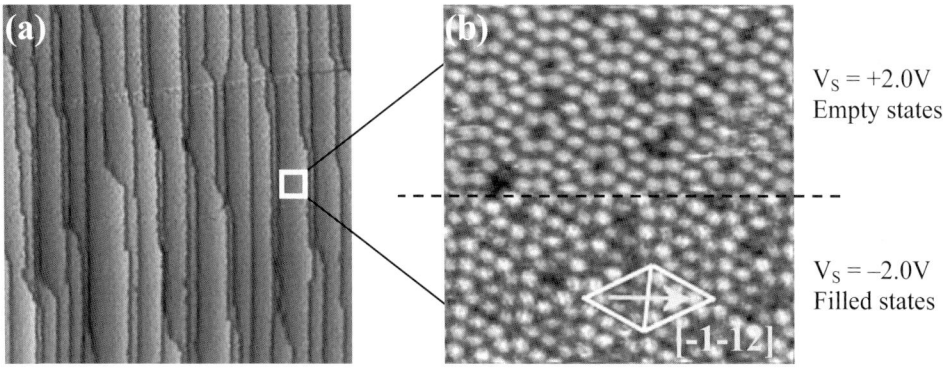

Fig. 2 STM images on the 2° off Si (111) surface.

Copper deposition on Si (111) substrates leads to three-dimensional growth, characterized by small islands formation, which present a quasi-hexagonal shape with typically 10 nm lateral dimension. These islands are locally arranged along crystallographic directions of the Si substrate (along the step edges and two equivalent directions <–110>). For Cu growth at 200 °C, the crystallinity is better pronounced and the surface is flatter [10]. This morphological difference between the two Cu buffer layers will surely affect the magnetic properties of the metallic multilayer.

Hysteresis loops were recorded using polar and longitudinal Magneto-Optical Kerr effect experiment (MOKE) exploring all growth parameters. The magnetic perpendicular Au/Co interface anisotropy is found to be strongly dependent on Cu growth conditions and only slightly dependent upon Au buffer thickness. The magnetization reversal from perpendicular to planar occurs at 7 ML of Co for the multilayer grown on 2 ML of Cu deposited at 100 °C, and it lies in the film plane already at 4 ML of Co for a Cu deposition at 200 °C. In this case, a uniaxial magnetic anisotropy is observed with an easy magnetization axis parallel to the monoatomic steps of the vicinal surface. As expected, this uniaxial in-plane magnetic anisotropy, linked to the symmetry breaking of atoms at the step edge of the vicinal surface, increases with the step density and thus with the miscut angle [11].

These magnetic properties are directly related to the surface and buried interfaces roughness, so SHG measurements, which have been noticed as a versatile and sensitive interface probe, can reveal the effect of the surface misorientation. The reflected SHG intensity was recorded, while the samples were rotated about their normal at a constant rate. At the surface, the inversion symmetry is broken and a second-order dipole response can exist. This contribution to SHG arises from a layer a few angstroms thick instead of the entire region within the escape depth of the second harmonic radiation, as in bulk-originating contributions.

Before investigating non-linear measurements on the metallic multilayer, we first study the influence of surface misorientation on the azimuthal reflectivity. Figure 3 displays the SH anisotropy of oxidized Si (111) wafers (2° and 6° off), observed by rotating the sample while detecting the reflected S-polarized SHG signal under P-polarized excitation. The results show a very strong effect of the misorientation angle on the observed anisotropy: for the vicinal surfaces the threefold symmetry is clearly broken. A more detailed analysis of the experimental results shows the interference of a step-induced C_{1v} symmetry with the underlying C_{3v} symmetry of the Si (111) [10].

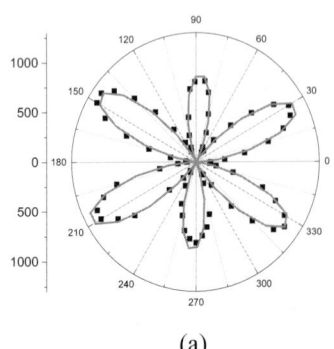 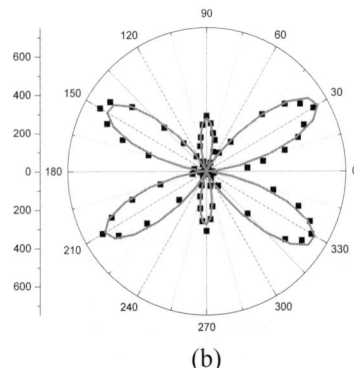

(a)   (b)

Fig. 3 Azimuthal SHG reflectivity on Si (111) surface for (a) 2° off and (b) 6° off.

In order to check the propagation of the vicinal character of the Si substrate through the metallic multilayer, we have studied the evolution of azimulhal reflectivity as a function of Au thickness in Au/Cu/Si (111) 2° off. This experiment shows that the vicinal character of the Si substrate is macroscopically duplicated even for thick Au layer [12].

After the study of the influence of Au thickness on the SH experiment, we have investigated azimuthal reflectivity on the whole metallic multilayer as a function of growth conditions for all polarization configurations. For all these measurements, we have applied a sufficient in-plane magnetic field in order to saturate the magnetization in transverse configuration. In this configuration, we observe a magnetic signal only for P-polarized SHG light. In this case, the azimuthal measurements show only three main lobes and not six as observed in Fig. 3, due to the presence of an isotropic term in the expression of the intensity [10]. In Fig. 4 are represented the SH reflectivity in S_{in}–P_{out} configuration on Au/15 ML Co/15 ML Au/Cu/Si (111) 6° off for 2 ML Cu grown at 100 °C and 4 ML at 200 °C. It shows that the uniaxial character is stronger in multilayer deposited on 4 ML Cu (200 °C), as well as in-plane magnetic anisotropy, which favors easy magnetization along the step direction. This result is in good agreement with longitudinal MOKE experiments.

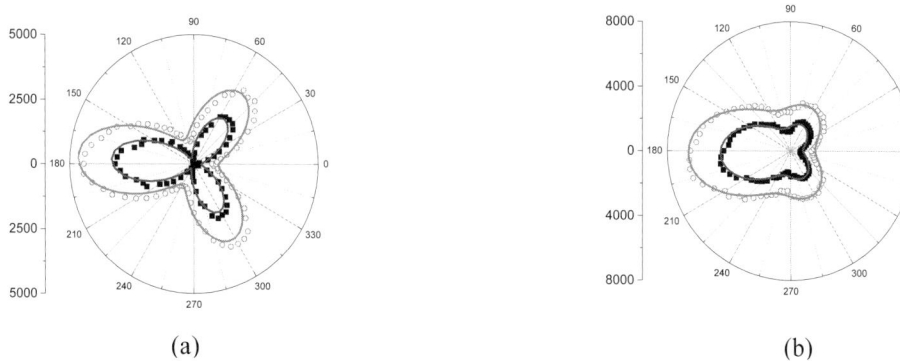

Fig. 4 Azimuthal SHG reflectivity on Au/15 ML Co/15 ML Au/Cu/Si(111) 6° for (a) 2ML of Cu at 100 °C and (b) 4ML of Cu at 200 °C.

4 Conclusion

We investigated linear and non-linear magneto-optical properties of SiO_2/Si interface as well as Au/Co/Au/Cu multilayer deposited on vicinal Si (111) substrate, with changing film thickness, growth conditions and misorientation angle of the vicinal surface. We have observed by the azimuthal dependence of second harmonic experiments that the miscut angle strongly affects the anisotropic reflectivity. Furthermore, this vicinal character is macroscopically duplicated in the Au/Co/Au magnetic multilayer grown on a Cu buffer. The uniaxial symmetry induces an in-plane magnetic anisotropy with easy axis parallel to the monoatomic steps of the silicon substrate. This anisotropy is strongly dependent on Cu growth conditions. This work clearly shows the powerful character of SHG technique to reveal the uniaxial character of the metallic multilayer.

References

[1] L. Falivor et. al., J. Mater. Res. **5**, 1299 (1990).
[2] W. P. Meiklejohn and C. P. Bean, Phys. Rev. **102**, 1413 (1956).
[3] Y. R. Shen, Chemistry at Surfaces and Interfaces, Eds. R.B. Hall and A.B. Ellis (VCH. Deerfield Beach, FL, 1986).
[4] G. L. Richmont, J. M. Robinson, and V. L. Shannon, Prog. Surf. Sci. **28**, 1 (1988).
[5] J. Dabrowski and H-J. Müssig, Silicon Surfaces and Formation of Interfaces (World Scientific, Singapore/New Jersey/London/Hong Kong, 2000).
[6] J. Wei, X.-S. Wang, J. L. Goldberg, N. C. Bartelt, and E. D. Williams, Phys. Rev. Lett. **68**, 3885 (1992).
[7] H. Hibino and T. Ogino, Phys. Rev. Lett. **72**, 657 (1994).
[8] J. Tershoff, Y. H. Phang, Z. Zhang, and M.G. Lagally, Phys. Rev. Lett. **75**, 2730 (1995).

[9] D. J. Chadi and J. R. Chelikowsky, Phys. Rev. B **24**, 4892 (1981).
[10] W. Cheikh-Rouhou, L. C. Sampaio, B. Bartenlian, P. Beauvillain, A. Brun, J. Ferré, P. Georges, J.-P. Jamet, V. Mathet, and A. Stupakewicz, Appl. Phys. B **74**, 665 (2002).
[11] R. K. Kawakami, E. Escorcia-Aparicio, and Z. Qiu, Phys. Rev. Lett. **77**, 2570 (1996).
[12] W. Cheikh-Rouhou, L. C. Sampaio, B. Bartenlian, P. Beauvillain, A. Brun, J. Ferré, P. Georges, P. Gogol, J.-P. Jamet, V. Mathet, and A. Stupakewicz, J. Magn. Magn. Mater. **240**, 532 (2002).

Crystal chemistry of non-perovskite manganese oxides – implications for magnetic properties

P. Strobel[*,1], **A. Ibarra-Palos**[1], **M. Pernet**[1], **S. Zouari**[2], **W. Cheikh-Rouhou**[2], and **A. Cheikh-Rouhou**[2]

[1] Laboratoire de Cristallographie, CNRS, BP 166, 38042 Grenoble Cedex 9, France
[2] Faculté des Sciences, Université de Sfax, Tunisie

Received 31 August 2003, accepted 31 December 2003
Published online 23 April 2004

PACS 61.50.Nw, 61.66.Fn, 75.47.Lx, 75.50.Ee

This paper reviews crystal structures of ternary manganese oxides and their implications for magnetic properties. We point out the critical role of the counter-cation (A) size in driving given stoichiometries to different structure types, in particular $AMnO_3$ compositions to ilmenite or perovskite structure, or A_2MnO_4 ones to spinel or K_2NiF_4 type. Mn^{2+} lies near the size borderline. It occupies the large site in $A_2Mn_2O_7$ pyrochlores, contrary to Mn^{4+}, and occupies separate crystallographic sites in known mixed-valent Mn^{2+}–Mn^{3+} or $^{4+}$ oxides. The bond distances, Mn–O–Mn bonding angles and Mn sublattice topology are given for the main A–Mn–O structures. Only the perovskite and K_2NiF_4 structures give Mn–O–Mn angles near 180°, while these are close to 90° in spinels, to 135° in pyrochlores and marokites. The magnetic properties of non-perovskite manganese oxides are briefly discussed in view of these structural features.

© 2004 WILEY-VCH Verlag GmbH & Co. KGaA, Weinheim

1 Introduction

Manganese with valence 2 to 4 enters a wide range of ternary oxides. Just as cuprates are intimately associated with high-T_c superconductivity, ternary manganese oxides have attracted considerable attention in the last decade for their giant magnetoresistance properties (GMR). Interestingly, this applies not only to the widely studied "manganites" (or rather manganates(III, IV), to come closer to chemical nomenclature rules), but also to pyrochlore-type $Tl_2Mn_2O_7$ with a quite different structure.

The "manganites" and other perovskite-related manganese oxides have given rise to a number of excellent reviews [1, 2]. The aim of this paper is to review the general trends in the stoichiometry and structure of the other manganese ternary oxides, and their relevance to magnetic properties. To this end, we will focus on Mn–O and Mn–O–Mn bonding geometries. The latter are seldom included in crystal structure reviews and need to be recalculated in order to discuss superexchange conditions in a given structure. The importance of the counter-cation size on the structures formed will also be highlighted. Given the format of this paper and the abundance of recent literature, perovskites, K_2NiF_4-type phases and their combinations (Ruddlesden–Popper phases) will be purposely given very limited coverage, to leave room for a description of structural families which have received less attention.

2 Manganese cation properties and crystal chemistry

Manganese is a very versatile element, which can have oxidation states ranging from +2 to +7. Fundamental properties of manganese ions with various valences v_{Mn} are summarized in Table 1. Several pa1rticularities should be noted:

---

[*] Corresponding author: e-mail: strobel@grenoble.cnrs.fr

1. Manganese ions with valence ≥ 5 form very short, highly covalent Mn–O bonds, resulting in stable, isolated tetrahedral MnO_4^{n-} species. Mn^{5+} and Mn^{6+} are known only in few compounds with very basic cations such as Ba or K. Together with diamagnetic Mn^{7+}, they will not be addressed further here.

2. Due to its d5 electron configuration, Mn^{2+} has no coordination preference; it can occupy almost indifferently crystallographic sites with coordination numbers 4, 6, or even 8, depending on other structural constraints. Mn^{3+} and Mn^{4+}, on the contrary, are always found in octahedral sites in complex oxides. In addition, the $t_{2g}^3 e_g^1$ configuration of Mn^{3+} makes it a typical Jahn-Teller cation. This results in a distortion of the MnO_6 coordination octahedron along four equatorial shorter bonds and two axial longer bonds. This effect is cooperative and clearly reflected in Mn^{3+}-O distances in most structures.

3. Within the Mn^{2+}–Mn^{4+} series, the ionic radius difference is much larger between Mn^{2+} and Mn^{3+} (0.185 Å) than between Mn^{3+} and Mn^{4+} (0.115 Å). This has important consequences on the crystal chemistry and physical properties of mixed-valence manganese oxides, as shown in Section 4 below).

Table 1 Properties of Mn cations as a function of valence v_{Mn}.

v_{Mn}	electron configuration	spin	preferred coordination	r_i(Å)[1]	superexchange interaction [2]
+2	$t_{2g}^3 e_g^2$	5/2	no preference	0.83	180°: AF strong 90°: AF moderate
+3	$t_{2g}^3 e_g^1$	2	octahedral (*Jahn-Teller*)	0.645	180°: AF weak
+4	$t_{2g}^3 e_g^0$	3/2	octahedral	0.53	180°: AF moderate 90°: ferromagn.
+5	t_{2g}^2	1	tetrahedral	0.33	
+6	t_{2g}^1	1/2	tetrahedral	0.255	
+7	(d^0)	0	tetrahedral	0.25	

[1] ionic radius based on $r_i(O^{2-}) = 1.40$ Å [3].
[2] according to Goodenough–Kanamori rules; AF = antiferromagnetic.

3 Crystal chemistry of ternary Mn oxides

The structures adopted by ternary oxides are determined primarily by (i) the *stoichiometry*, especially the cation/anion ratio ($R_{c/a}$), (ii) the *size* of the constituting cations. Mn^{n+} with n = 2 to 4 can be considered as "small" cations, with ionic radii comparable to those found in Mg^{2+} and the first-row transition metals. Cations with $r_i > \approx 0.90$ Å (typically Na, K, Ca, Sr, Ba, Y, the rare earths, Bi, Tl^{3+}) are considered as "large" counter-cations. Table 2 summarizes the structures encountered in manganese ternary oxides [4, 5] and shows important examples of the effect of A- cation size on structure, especially for $R_{c/a} = 0.667$ (2/3) and 0.75 (3/4). Note that Mn^{2+} ($r_i = 0.83$ Å) has a size close to the borderline, and actually occupies the *large* site in pyrochlores ($R_{c/a} = 0.57$).

There is one notable exception to the small/large A cation separation, namely the layered structures. The presence of an interlayer space provides more structural flexibility than in 3-D structures, as shown by several examples such as :

(i) the $Zn_2Mn_3O_8$ structure, based on a CdI_2-type framework, where Zn can be replaced by cations with sizes ranging from Co^{2+} ($r_i = 0.65$ Å) to Ca^{2+} ($r_i = 1.00$ Å), also including Mn^{2+};

(ii) the layered rocksalt-superstructures of $AMnO_2$ and A_2MnO_3 (A = alkali metal), where A can be lithium (small) as well as much larger potassium;

(iii) the phyllomanganates A_xMnO_2, where A can encompass numerous cations ranging in size from Mg^{2+} ($r_i = 0.72$ A) to K^+ and even Cs^+ ($r_i = 1.67$ Å).

The stability of the *spinel* and *perovskite* structures is such that they are found even for off-stoichiometry compositions, giving rise to cation- or anion-deficient compositions (see examples in Table 2).

Table 2 Main structural types in ternary Mn oxides as a function of stoichiometry (R = rare earth, cub = cubic, orh = orthorhombic, mon = monoclinic).

Rc/a	small counter-cation	large counter-cation
0.57 (4/7)	**pyrochlore** $(Mn^{2+})_2Sb_2O_7$ with Mn^{2+} in *large site* $Zn(Mn^{4+})_3O_{7-}$ layered (mon.)	– cub. $R_2(Mn^{4+})_2O_7$ (+ In, Tl, Bi)
0.625 (5/8)	– **layered** $(Mn^{2+})_2(Mo^{4+})_3O_8$ $Zn_2(Mn^{4+})_3O_8$ – $(Mn^{2+})_2Ti_3O_8$ – *defect spinel*	$Ca_2Mn_3O_8$
0.667 (2/3)	– **ilmenite**: $TiMn^{2+}O_3$ (+ Ge) $MgMn^{4+}O_3$ – $Li_2(Mn^{4+})_4O_9$ – *defect spinel*	– **perovskite** (distorted) $RMn^{3+}O_3$, $CaMn^{4+}O_3$ – $BaMn^{4+}O_3$ ("hexagonal perovskite")
0.7 (7/10)		$Ca_4(Mn^{4+})_3O_{10}$ – Ruddlesden–Popper
0.71 (5/7)		$Ca_3(Mn^{4+})_2O_7$ – Ruddlesden–Popper $Na_2(Mn^{4+})_3O_7$ – layered (triclinic)
0.75 (3/4)	**spinel**: $Mn^{2+}(M^{3+})_2O_4$, $(Mn^{2+})_2TiO_4$, $M^{2+}(Mn^{3+})_2O_4$	$Ca(Mn^{3+})_2O_4$ – marokite $Ca_2(Mn^{4+})O_4$ – **K_2NiF_4**
0.80 (4/5) 0.86 (6/7)		$Ca_2Mn_2O_5$) *defect perovskite* $Ca_4Mn_2O_7$)
0.87 (7/8)	$Mg_6(Mn^{4+})O_8$ - murdochite (cub., also Ni)	
1.0 (1/1)	$AMn^{3+}O_2$ [A = Li, Na, K]) $Li_2Mn^{4+}O_3$, $Na_2Mn^{4+}O_3$)	ordered NaCl superstructures | with Mn^{2+}: $BaMn_2O_3$ (orh), Ba_2MnO_3

Turning now to mixed-valent manganese compounds, a number of specific examples are given in Table 2b. In all cases involving Mn^{2+}, this cation occupies a different crystallographic site than $Mn^{3+/4+}$, whereas this is not the case for Mn^{3+}–Mn^{4+} phases. The comparison of AMn_7O_{12} phases with A = Ge (a $Mn^{2+/3+}$ phase) and Na (a $Mn^{3+/4+}$ one) clearly illustrates this point. This feature is a consequence of the gap in ionic size between Mn^{2+} and Mn^{3+} noted in Section 1.

Another distinction arises from the counter-cation size. Ilmenite-type oxides, for instance, could be written $M^{3+}Mn^{3+}O_3$ as well as $M^{2+}Mn^{4+}O_3$, in a similar way as the mixed $A^{3+}Mn^{3+}O_3$–$A^{2+}Mn^{4+}O_3$ perovskite systems. But because of the size similarity, known $M^{3+}Mn^{3+}O_3$ phases with small M have a disordered occupation of M and Mn, and solid solutions with their $M^{2+}Mn^{4+}O_3$ analogues would give rise to a mixing of not only Mn^{3+} and Mn^{4+}, but of M^{3+} as well. Thus the only structure with small counter-cation appropriate to yield mixed-valence Mn compounds on a specific site is the spinel, where the counter-cation (e.g. Li) is structurally constrained to a different, tetrahedral site.

Table 2b Ternary manganese oxides with mixed valence.

Rc/a	small counter-cation	large counter-cation
0.625 (5/8)	$Mn_5O_8 = (Mn^{2+})_2(Mn^{4+})_3O_8$ –> *separate* Mn^{2+}/Mn^{4+} sites	$BaMn_4O_8$ – hollandite (tunnel structure) –> unique $Mn^{3+/4+}$ site
0.667 (2/3)	$GeMn_7O_{12}$ – tetragonal –> *separate* Mn^{2+}/Mn^{3+} sites	$NaMn_7O_{12}$ – double perovskite (also Ca) $K_{\approx 1/3}MnO_2$ – layered (pseudo-hex.) –> unique $Mn^{3+/4+}$ site
0.75 (3/4)	$Li[Mn^{3.5+}]_2O_4$ – spinel –> unique $Mn^{3+/4+}$ site Mn_3O_4 – distorted spinel –> *separate* Mn^{2+}/Mn^{3+} sites	$NaMn_2O_4$ – layered (hex./mon) –> unique $Mn^{3+/4+}$ site

4 Bond lengths and bond angles in ternary Mn oxides

The bonding properties of a number of typical ternary oxides of manganese are given in Table 3. Mn–O bond lengths are clearly dominated by the Jahn-Teller effect in all Mn^{3+} compounds, with long/short distance ratios ranging from 1.11 in $LaMnO_3$ to 1.25 in marokite (the latter gives especially irregular Mn-O distances). $LiMn_2O_4$, with has an exact 1:1 mixture of Mn^{3+} and Mn^{4+}, possesses purely cubic Mn sites at room temperature, but undergoes a phase transition near 275 K to a complex superstructure with partial charge ordering [6]. The Mn^{4+} oxides have a much more regular Mn–O environment.

Regarding Mn–O–Mn angles and Mn–Mn connectivity, other differences show up. The most favourable cases for a 180° (or close to it) interaction are the structures in which the Mn–Mn neighbours follow perpendicular axes. This occurs in the *perovskite and K_2NiF_4 structures only*, hence the interest of these structures – and of their intermediaries, the Ruddlesden–Popper phases – for magnetic properties. As shown by the $LaMnO_3$ case, however, Mn–O–Mn angles can be severely bent from 180° due to the A–O and Mn–O bondlength mismatch, which results in Mn–O octahedra tilting ($GdFeO_3$-type orthorhombic structure instead of the ideal perovskite one). This tilting varies with the size of A cation, hence the importance of A-cation size and size variance on the magnetic properties in the $AMnO_3$ "manganites", which has been widely studied [7].

In the ilmenite structure, Mn and the counter-cation order in alternating layers along the trigonal axis. As a result, Mn layers are rather isolated from one another, and no 180° connectivity is to be expected. The same applies to purely layered compounds such as $CaMn_3O_8$ or phyllomanganates (birnessite).

$LiMnO_2$ is a particular case: because of the Jahn–Teller distortion, the thermodynamically stable phase does not adopt a $NaFeO_2$-type layered structure as its cobalt or nickel analogues, but a corrugated chain-like ordering of Li and Mn atoms with Mn-O-Mn angles close to 90 and 180°. Its magnetic susceptibiliy variation at low temperature is complex and reminiscent of 2-dimensional magnetic systems [8].

Finally, the important spinel and pyrochlore families possess a similar manganese sublattice (16d site in space group F-d3m). This sublattice is *tetrahedral*, and gives rise to strong magnetic frustration. The oxygen positions, however, are different, resulting in Mn–O–Mn angles close to 90 and 135° in the spinels and pyrochlore-type compounds, respectively.

Table 3 Structure and Mn–O bonding characteristics in the main ternary structures.

formula	v_{Mn}	structure space group	Mn–O bond lengths (Å)	Mn–O–Mn angles (°)	Mn sublattice	magnetic properties[1]
$LaMnO_3$	+3	perovskite (dist.) Pnma	2x1.92 + 2x1.97 + 2x2.165	155	cubic	GMR if mixed valence
$MnCoO_3$	+4	ilmenite R-3	3 x 1.88 + 3 x 2.18	84 + 92 + 124 + 131	layered triangular	
$LiMnO_2$	+3	Pmnm	4 x 1.89 + 2 x 2.29	92-95 + 171	≈tetrahedral	short-range
$CaMn_2O_4$	+3	marokite Pbcm	4x1.90-1.96 +2x 2.37-2.46	90-98 + 135	Mn chains	antiferro. T_N 220 K
$ZnMn_2O_4$	+3	dist. spinel $I4_1/amd$	4 x 1.93 + 2 x 2.21	95	tetrahedral	magnetic frustration
$LiMn_2O_4$	+3.5	spinel Fd3m	6 x 1.96	96	ibid.	ibid.
$Tl_2Mn_2O_7$	+4	pyrochlore Fd3m	6 x 1.90	134	= spinel	frustration + GMR
$Ca_2Mn_3O_8$	+4	layered C2/m	6 x (1.895–1.91)	98 + 100	≈tetrahedral	
Ca_2MnO_4	+4	K_2NiF_4 $I4_1/acd$	4 x 1.86 + 2 x 1.94	162	layered ≈ squares	

5 Magnetism in manganese compounds with intermediate angles

Pyrochlores – In addition to the difference in Mn–O–Mn angle, the pyrochlores exhibit another important difference with respect to to perovskite-type oxides, the absence of mixed valence, which was confirmed by spectroscopic studies [9]. Figure 1 shows the magnetoresistive effect in samples prepared by high pressure synthesis in our laboratory [10]. The occurrence of magnetoresistance in the absence of mixed valence is remarkable, as it seems to be unrelated to double exchange. In fact, the mechanism underlying magnetoresistance in this compound is obviously quite different from that in the mixed-valent $AMnO_3$ "manganites", and remains to be elucidated.

Spinels – The magnetic properties of $LiMn_2O_4$ are complex, with partial long-range antiferromagnetic order below ca. 60 K [11]. We studied several substitutions on the manganese site in order to relieve the frustration. In the $Li(Mn_{2-x}Ti_x)O_4$ system, for instance, T_N and θ_N were actually found to decrease with increasing dilution of Mn (Fig. 2). Even in the x = 1 case (50 % Mn dilution), the frustration remains, probably because the substitution is random. In $Li(Mn_{1.5}M_{0.5})O_4$ compositions with other magnetic as well as non-magnetic M substituents, the Weiss constants are less negative than in unsubstituted $LiMn_2O_4$, and cation ordering, which has been shown to occur for M = Mg, Ni or Cu [12], does not suppress the frustration either [13]. The M = Ni case shows particularly strong magnetic interactions starting at 120 K resembling the ferrimagnetic behaviour previously observed in $NiMn_2O_4$ [14].

Marokite – The structure of marokite, $CaMn_2O_4$, is much more distorted (see Mn–O distances in Table 3). The presence of the large counter-cation Ca leads to double chains of Mn–O octahedra sharing edges and vertices. This Mn-only compound possesses both 90 and 135° Mn–O–Mn angles. The Mn chains allow the occurrence of long-range antiferromagnetic ordering at 220 K. The magnetic structure, for which doubts between two possible structures persisted, was recently redetermined from new neutron diffraction data [15]. It requires a doubling of the unit cell along the *a* axis.

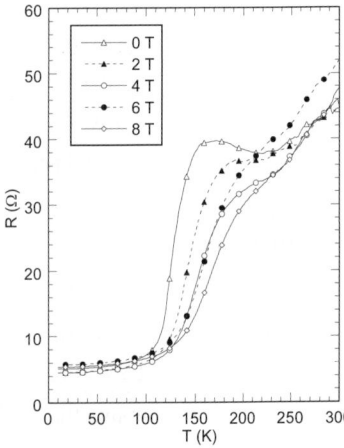
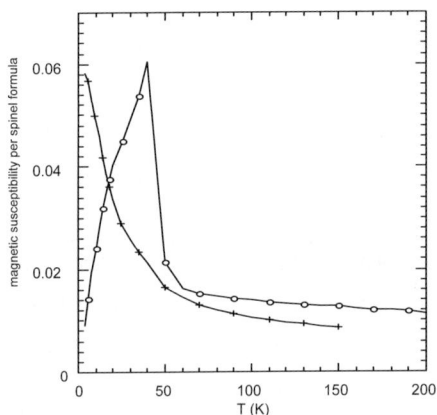

Fig. 1 Magnetoresistance effect in $Tl_2Mn_2O_7$.

Fig. 2 Magnetic susceptibilities of $LiMn_2O_4$ (o) and $LiMnTiO_4$ (+) [A. Ibarra, Thesis, Grenoble].

6 Conclusions

Manganese ternary oxides display a remarkable variety of structures, depending on stoichiometry and the size of the counter-cation. We first showed the peculiar role of Mn^{2+}, which does not occupy similar crystallographic sites as Mn^{3+} or Mn^{4+} due to its significantly larger size. Most of the structures formed do not give rise to 180° Mn–O–Mn bonds such as those existing in perovskite and K_2NiF_4-type structures. The magnetic properties of oxides with intermediate superexchange angles are complex, especially when the manganese lattice is frustrated. This situation arises in two important structural families, the spinels and the pyrochlores. As a result, mixed-valence in spinels, which is easy to achieve, does not give rise to physical properties comparable to those of the so-called "manganites".

References

[1] C.N.R. Rao and B. Raveau (eds.), Colossal Magnetoresistance (World Scientific, Singapore, 1998).
[2] J.M.D. Coey, M. Viret, and S. von Molnar, Adv. Phys. **48**, 167 (1999).
[3] R.D. Shannon, Acta Crystallogr. **32**, 751 (1976).
[4] O. Müller and R. Roy, The Major Ternary Structural Families (Springer, Berlin, 1974).
[5] ICSD Database, FIZ Karlsruhe (Germany), version 1.2.0 (2003).
[6] G. Rousse, C. Masquelier, J. Rodriguez, and M. Hervie, Electrochem. Solid State Lett. **2**, 6 (1999).
[7] G.H. Rao, K. Barner, and I.D. Brown, J. Phys.: Condens. Matter **10**, L757 (1998).
[8] J.E. Greedan, N.P. Raju, and I.J. Davidson, J. Solid State Chem. **128**, 209 (1997).
[9] H.D. Rosenfeld and M.A. Subramanian, J. Solid State Chem. **125**, 278 (1996).
[10] W. Cheikh-Rouhou, P. Strobel, C. Chaillout, S.M. Loureiro, R. Senis, B. Martinez, X. Obradors, and J. Pierre, J. Mater. Chem. **9**, 743 (1999).
[11] A.S. Wills, N.P. Raju, and J.E. Greedan, Chem. Mater. **11**, 1510 (1999).
[12] P. Strobel, A. Ibarra, M. Anne, C. Poinsignon, and A. Crisci, Solid State Sci. **5**, 1009 (2003).
[13] P. Strobel, A. Ibarra-Palos, M. Anne, and F. Le Cras, J. Mater. Chem. **10**, 429 (2000).
[14] S. Asbrink, A. Waskowska, M. Drozd, and E. Talik, J. Phys. Chem. Solids **58**, 725 (1997).
[15] S. Zouari, L. Ranno, A. Cheikh-Rouhou, O. Isnard, M. Pernet, P. Wolfers, and P. Strobel, J. Alloys Compd. **353**, 5 (2003).

Magnetic and electrical properties of the lacunar $La_{0.7}Ca_{0.3-x}\square_x MnO_3$ and $La_{0.7-x}\square_x Ca_{0.3}MnO_3$ oxides

I. Kamoun[1], **W. Boujelben**[1], **A. Cheikh-Rouhou**[*,1], **H. Roussel**[2], and **R. Madar**[2]

[1] Laboratoire de Physique des Matériaux, Faculté des Sciences de Sfax, B. P. 802, 3018 Sfax-Tunisie
[2] Laboratoire des Matériaux et de Génie Physique (UMR 5628 CNRS), ENSPG, B. P. 46, 38402 Saint Martin d'Hères, France

Received 31 August 2003, accepted 31 December 2003
Published online 23 April 2004

PACS 75.47.Gk, 75.60.Ej

Previous studies on lacunar $Ln_{0.7-x}A_{0.3}MnO_3$ and $Ln_{0.7}A_{0.3-x}MnO_3$ with Ln = Pr, Nd and A = Sr, Ba perovskite manganese oxides have shown that their magnetic and electrical properties depend strongly on the nature of the deficient-element. In order to confirm such behaviour, we investigate the effects of lanthanum and calcium deficient upon the physical properties of $La_{0.7}Ca_{0.3}MnO_3$ oxide. $La_{0.7}Ca_{0.3-x}\square_x MnO_3$ and $La_{0.7-x}\square_x Ca_{0.3}MnO_3$ powder samples have been elaborated by the conventional ceramic method at 1400 K. All our synthesized samples are single phase and exhibit a paramagnetic to ferromagnetic transition with decreasing temperature. With increasing vacancy content, the Curie temperature decreases in the calcium-deficient samples and increases in the lanthanum-deficient ones. With increasing x from 0 to 0.2, T_C is found to decrease from 218 to 180 K in the first series and to increase from 218 to 248 K in the second one. Resistivity measurements versus temperature showed that all lanthanum-deficient samples exhibit a semiconducting–metallic transition with decreasing temperature, however, such transition has been observed only for low x values (x ≤ 0.1) in the calcium-deficient samples.

© 2004 WILEY-VCH Verlag GmbH & Co. KGaA, Weinheim

1 Introduction

The manganites of general formula $Ln_{1-x}A_x MnO_3$ where Ln is a trivalent rare earth element (Ln = La, Pr, Nd) and A is a divalent alkaline-earth (Ca, Ba, Sr) or a monovalent alkali metal (Na, K) have been extensively studied during the last ten years in reason of the colossal magnetoresistance that they exhibit in the vicinity of the Curie temperature T_C [1–6]. Stoichiometric compounds have been widely studied, however only few investigations have been performed on lacunar materials [7–9]. Previous studies on deficient samples have shown that their physical properties depend strongly on the rare-earth or alkali-earth deficient amount. In $Pr_{0.7}Sr_{0.3}MnO_3$, strontium-defect leads to a decrease of the Curie temperature and destroys the metallic behaviour observed in the parent compound at low temperature, whereas the praseodymium-defect leads to an increase of the Curie temperature, moreover these lacunar samples are metallic at low temperature. In this paper, we investigate the effects of lanthanum and calcium defect upon the structural, magnetic and electrical properties of $La_{0.7}Ca_{0.3-x}\square_x MnO_3$ and $La_{0.7-x}\square_x Ca_{0.3}MnO_3$ lacunar samples (0 ≤ x ≤ 0.2).

2 Experimental

Polycrystalline $La_{0.7-x}\square_x Ca_{0.3}MnO_3$ and $La_{0.7}Ca_{0.3-x}\square_x MnO_3$ powder samples (0 ≤ x ≤ 0.2) were elaborated by the conventional solid–solid reaction. Starting from stoichiometric mixtures of La_2O_3, $CaCO_3$

---

[*] Corresponding author: e-mail: Abdel.Cheikhrouhou@fss.rnu.tn and MnO_2 of high-purity (99.9%) powders according to the following reactions:

$$(0.7-x)La_2O_3 + 0.6\ CaCO_3 + 2MnO_2 \longrightarrow 2La_{0.7-x}\square_xCa_{0.3}MnO_3 + \delta\ CO_2,$$

$$0.7La_2O_3 + 2(0.3-x)CaCO_3 + 2MnO_2 \longrightarrow 2La_{0.7}Ca_{0.3-x}\square_xMnO_3 + \delta\ CO_2.$$

The starting materials are intimately mixed in an agate mortar and then heated in air at 1000 °C for 48 hours. A systematically annealing at high temperature is necessary to achieve a complete reaction. The powders are pressed into pellets (of about 1 mm thickness) and sintered in air at 1400 °C for 48 hours with several periods of grinding and sintering. Finally, these pellets are rapidly quenched to room temperature in air. This step is done in order to retain the structure adopted at a given annealing temperature.

Phase purity, homogeneity, and cell dimensions are determined by powder X-ray diffraction at room temperature. The unit cell dimensions were obtained by least-squares calculations.

Magnetization measurements versus temperature were recorded by a vibrating sample magnetometer in the temperature range 10–350 K in an applied field of 500 Oe. Resistivity measurements versus temperature, in the earth magnetic field, were performed on dense ceramic pellets using the conventional four-probe method.

3 Results and discussions

A vacancy in the A site implies a partial conversion of Mn^{3+} to Mn^{4+} leading to an increase of the Mn^{4+} content above 30%. This vacancy leads also to a change in the average ionic radius $<r_A>$ of the A site. According to the general formula, the Mn tetravalent and Mn trivalent contents are respectively $(0.3 + 2x)$ and $(0.7-2x)$ in the calcium-deficient samples and $(0.3 + 3x)$ and $(0.7-3x)$ in the lanthanum-deficient ones. The Mn^{3+} and Mn^{4+} contents have been checked by chemical analysis. We list in Table 1 the chemical analysis results for $La_{0.7-x}\square_xCa_{0.3}MnO_3$ and $La_{0.7}Ca_{0.3-x}\square_xMnO_3$ compounds. As our samples have been elaborated in air, they must be stoichiometric in oxygen and consequently the increase of Mn^{4+} in our samples is due only to the deficiency in the A cation site. It is important to notify the agreement between the theoretical and experimental results. The relative error does not exceed 3%, which confirms really the existence of lacuna in our samples.

Table 1 Chemical analysis results for $La_{0.7-x}\square_xCa_{0.3}MnO_3$ and $La_{0.7}Ca_{0.3-x}\square_xMnO_3$ samples.

x	%Mn^{4+} theoretical	%Mn^{4+} experimental	Relative error (%)	x	%Mn^{4+} theoretical	%Mn^{4+} experimental	Relative error (%)
	$La_{0.7-x}\square_xCa_{0.3}MnO_3$				$La_{0.7}Ca_{0.3-x}\square_xMnO_3$		
0.00	30	29.75	0.8	0.00	30	29.75	0.8
0.05	45	44.25	1.6	0.05	40	39.26	1.8
0.10	60	58.65	2.2	0.10	50	48.95	2.1
0.15	75	73.05	2.6	0.15	60	58.65	2.2
0.20	90	87.15	3.1	0.20	70	68.15	2.6

X-ray diffraction patterns showed that all our synthesized samples are single phase and crystallize in the orthorhombic structure with Pmna space group. We plot in Fig. 1 the X-ray diffraction patterns of $La_{0.7}Ca_{0.3}MnO_3$, $La_{0.7}Ca_{0.1}\square_{0.2}MnO_3$ and $La_{0.5}\square_{0.2}Ca_{0.3}MnO_3$ compounds and we list in Table 2 the crystallographic data for all our synthesized samples.

The stoichiometric sample $La_{0.7}Ca_{0.3}MnO_3$ (30% Mn^{4+}) is ferromagnetic below $T_C = 218$ K. This value is smaller than $T_C = 263$ K given by Wang et al. [10]. The difference may be explained by the elaborating method. In order to study the deficiency effects in the A site on the magnetic properties, we performed magnetization measurements versus temperature in a magnetic applied field of 500 Oe. Figures 2 and 3 show the temperature dependence of the magnetization for $La_{0.7}Ca_{0.3-x}\square_xMnO_3$ and $La_{0.7-x}\square_xCa_{0.3}MnO_3$ samples respectively.

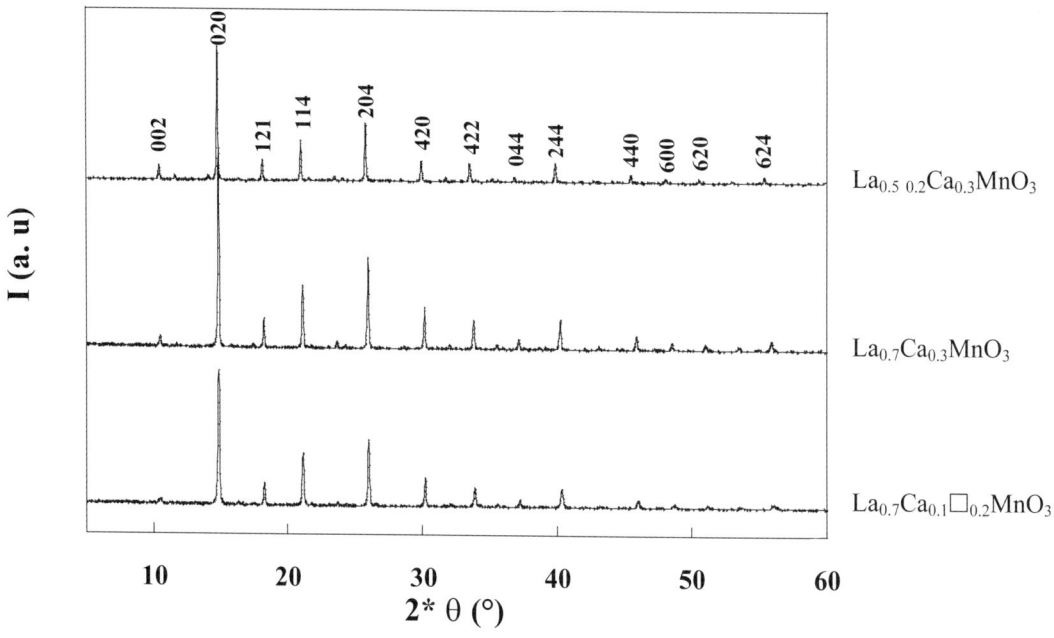

Fig. 1 X-ray diffraction patterns at room temperature for $La_{0.7}Ca_{0.3}MnO_3$, $La_{0.7}Ca_{0.1}\square_{0.2}MnO_3$ and $La_{0.5}\square_{0.2}Ca_{0.3}MnO_3$ samples.

Table 2 Crystallographic data for $La_{0.7-x}\square_xCa_{0.3}MnO_3$ and $La_{0.7}Ca_{0.3-x}\square_xMnO_3$ samples.

$La_{0.7-x}\square_xCa_{0.3}MnO_3$					$La_{0.7}Ca_{0.3-x}\square_xMnO_3$				
x	a(Å)	b(Å)	c(Å)	v(Å³)	x	a(Å)	b(Å)	c(Å)	v(Å³)
0.00	5.454(9)	5.467(2)	7.694(0)	229.46	0.00	5.454(9)	5.467(2)	7.694(0)	229.46
0.05	5.454(9)	5.466(9)	7.692(3)	229.40	0.05	5.466(3)	5.481(1)	7.725(8)	231.48
0.10	5.454(5)	5.466(4)	7.692(1)	229.35	0.10	5.466(5)	5.481(5)	7.726(0)	231.51
0.15	5.449(7)	5.460(2)	7.690(0)	228.83	0.15	5.484(5)	5.502(9)	7.754(8)	234.04
0.20	5.435(5)	5.446(7)	7.654(9)	226.63	0.20	5.502(8)	5.519(1)	7.782(1)	236.35

All our deficient samples exhibit a paramagnetic to ferromagnetic transition with decreasing temperature. The deficiency effects on the Curie temperature are very spectacular. In fact, with increasing x, T_C is observed to decrease in the calcium-deficient samples and to increase in the lanthanum-deficient ones. It is found to decrease from 218 K for x = 0 to 180 K for x = 0.2 in the first series and to increase from 218 K for x = 0 to 248 K for x=0.2 in the second one (Fig. 4). Such result has been observed by Boujelben et al. [11] in the strontium-deficient $Pr_{0.7}Sr_{0.3-x}\square_xMnO_3$ and in the praseodymium-deficient $Pr_{0.7-x}\square_xSr_{0.3}MnO_3$ samples. The T_C decrease in the calcium-deficient samples can be explained by the increase of the Mn^{4+} content, which leads to a weakness of the double exchange interactions [12], however, the T_C increase with the Mn^{4+} content above 30% in the lanthanum-deficient compounds cannot be explained by the same phenomena. There are surely other parameters which lead to the increase of T_C.

The average ionic radius $\langle r_A \rangle$ of the A site may explain such results. In fact, as a vacancy must have an average radius $\langle r_v \rangle \neq 0$, T_C decreases with increasing calcium-deficient in the lacunar $La_{0.7}Ca_{0.3-x}\square_x MnO_3$ samples and the T_C increase with increasing lanthanum-deficient in the lacunar $La_{0.7-x}\square_x Ca_{0.3} MnO_3$ samples can be explained according to Hwang et al. [13] with a vacancy radius $\langle r_v \rangle$ smaller than Ca^{2+} and larger than La^{3+}.

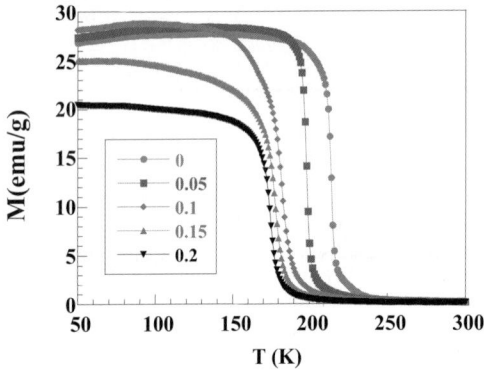

Fig. 2 Magnetization evolution as a function of temperature for lacunar $La_{0.7}Ca_{0.3-x}\square_x MnO_3$ samples at H = 500 Oe

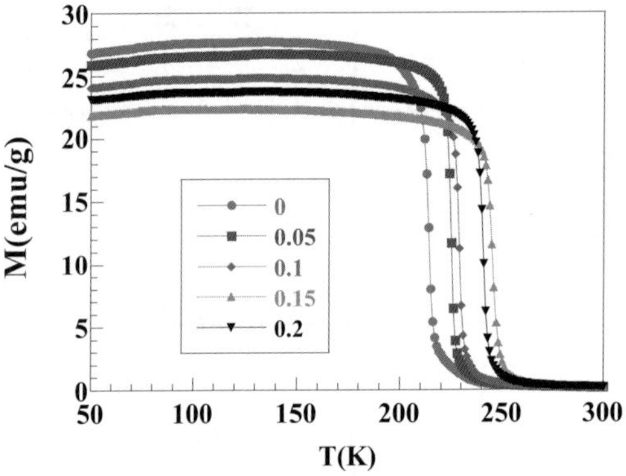

Fig. 3 Magnetization evolution as a function of temperature for lacunar $La_{0.7-x}\square_x Ca_{0.3} MnO_3$ samples at H = 500 Oe

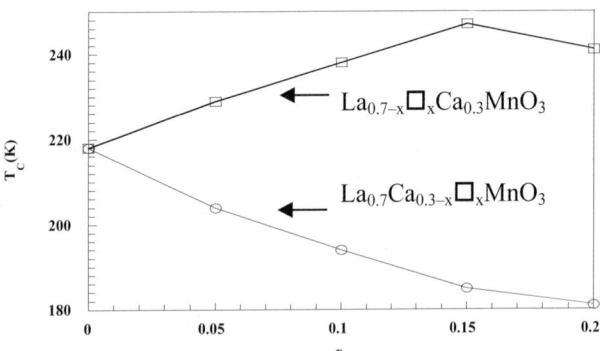

Fig. 4 Curie temperature T_C as a function of x for $La_{0.7}Ca_{0.3-x}\square_x MnO_3$ and $La_{0.7-x}\square_x Ca_{0.3} MnO_3$ samples.

We reproduce in Figs. 5 and 6 the temperature dependence of the resistivity $\rho(T)$ of our lacunar samples in the earth magnetic field.

Calcium and lanthanum deficiencies do not have the same effects on the electrical properties. In fact, all lanthanum-deficient samples $La_{0.7-x}\square_x Ca_{0.3} MnO_3$ exhibit a semiconducting-metallic transition with decreasing temperature. However, such transition has been observed only for low x values ($x \leq 0.1$) in the calcium-deficient $La_{0.7}Ca_{0.3-x}\square_x MnO_3$ compounds. Such results have been also obtained in previous work on lacunar $Pr_{0.7}Sr_{0.3-x}\square_x MnO_3$ and $Pr_{0.7-x}\square_x Sr_{0.3} MnO_3$ compounds [11].

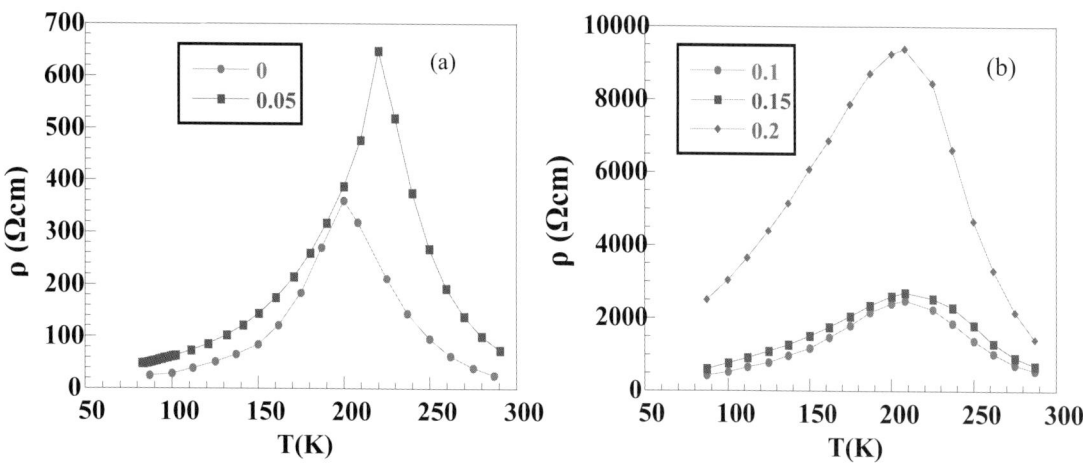

Fig. 5 Resistivity evolution as a function of temperature for lanthanum-deficient $La_{0.7-x}\square_x Ca_{0.3}MnO_3$ samples.

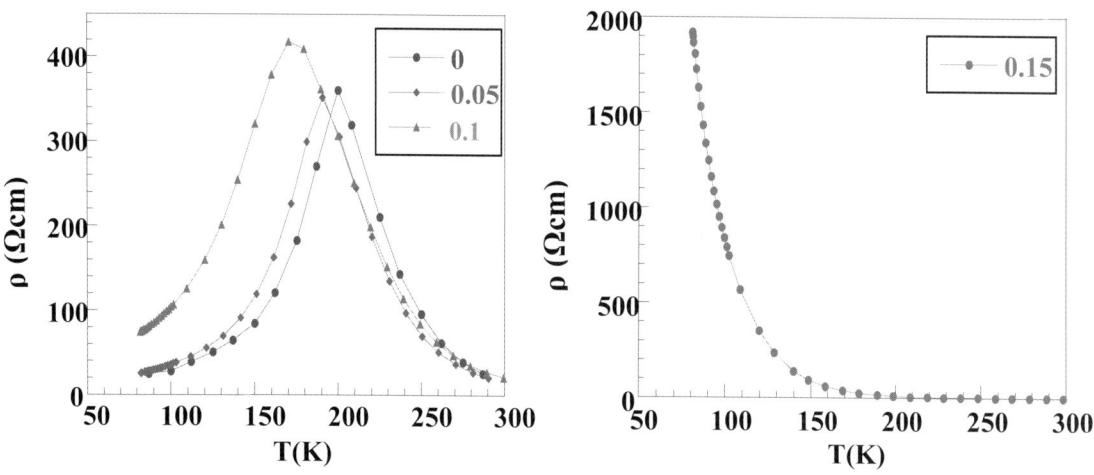

Fig. 6 Resistivity evolution as a function of temperature for calcium-deficient $La_{0.7}Ca_{0.3-x}\square_x MnO_3$ samples.

3 Conclusions

In this work we have studied the effect of lanthanum and calcium deficiencies on the physical properties of $La_{0.7}Ca_{0.3}MnO_3$ perovskite manganite oxide. Calcium and lanthanum-deficient have different effects on the magnetic and electrical properties. Calcium-deficient leads to a decrease of the Curie temperature with increasing calcium-defect. However, all lanthanum-deficient samples exhibit a ferromagnetic behavior at low temperatures, moreover T_C increases with increasing lanthanum vacancy content. The electrical measurements show that all lanthanum-deficient samples exhibit a semiconducting–metallic transition with decreasing temperature while such transition is observed only for low x values ($x \leq 0.1$) in the calcium-deficient ones.

Acknowledgement This work has been supported by the Tunisian Secretary of State for Scientific Research and Technology.

References

[1] R. D. Sanchez, J. Rivas, C. V. Vazquez, A. L. Quintela, M. T. Causa, M. Tovar, and S. Oseroff, Appl. Phys. Lett. **68**, 134 (1996).
[2] H. L Ju and H. Sohn, J. Magn. Magn. Mater. **167**, 200 (1997).
[3] F. Damay, C. Martin, M. Hervieu, A. Maignan, B. Raveau, G. André, and F. Boureé, J. Magn. Magn. Mater. **184**, 71 (1998).
[4] R. Von Helmot, J. Weeker, B. Holzepfel, L. Schultz, and K. Samwer, Phys. Rev. Lett. **71**, 2331 (1993).
[5] B. Raveau, A. Maignan, and V. Caignaert, J. Solid State Chem. **117**, 424 (1995).
[6] W. Zhang, I. W. Boyd, N. S. Cohen, Q. T. Quentin, and A. Pankhaurst, Appl. Surf. Sci. **109**, 350 (1997).
[7] W. Boujelben, A. Cheikh-Rouhou, and J. C. Joubert, J. Solid State Chem. **156**, 68 (2001).
[8] L. Laroussi, J. C. Joubert, E. Dhahri, J. Pierre, and A. Cheikh-Rouhou, Phase Transitions **70**, 29 (1999).
[9] S. Chaffai, W. Boujelben, M. Ellouze, A. Cheikh-Rouhou, and J. C. Joubert, Physica B **321**, 74 (2002).
[10] K. Y. Wang, W. H. Song, J. M. Dai, S. L. Ye, and S. G. Wang, J. Appl. Phys. **90**, 6263 (2001).
[11] W. Boujelben, A. Cheikh-Rouhou, J. Pierre, and J.-C Joubert, J. Alloys Compd. **68**, 315 (2001).
[12] C. Zener, Phys. Rev. **82**, 403 (1951).
[13] H. Y. Hwang, W. Cheong, P. G. Radaelli, M. Marezio, and B. Batlogg, Phys. Rev. Lett. **75**, 914 (1995).

Observations of magnetic domain structures and phase segregation in single-crystal $Nd_{1/2}Sr_{1/2}MnO_3$ using X-ray scattering

Mohammad E. Ghazi[*,1] **and P. D. Hatton**[2]

[1] Department of Physics, Shahrood University of Technology, Shahrood, P.O. BOX 36155-316, Iran
[2] Department of Physics, University of Durham, South Road, Durham DH1 3LE, UK

Received 31 August 2003, accepted 31 December 2003
Published online 18 March 2004

PACS 61.50.Ks, 61.72.Dd, 75.60.Ch

We report high-resolution synchrotron X-ray scattering measurements on the structure and phase transitions of a single-crystal of $Nd_{1/2}Sr_{1/2}MnO_3$. Detailed measurements of the peak intensity of the Bragg peaks as a function of temperature have allowed us to determine the onset temperatures and hysteresis associated with a series of structural and magnetic phase transitions. Measurement of the anisotropic peak profiles and widths allows a direct probe of the correlated volume of magnetic domains in each phase. Upon cooling we observe a dramatic decrease in the correlated volume, at the $T_c \approx 252$ K corresponding to the transition from a paramagnetic to a ferromagnetic state. Below approximately 200 K the formation of a mixture of antiferromagnetic and ferromagnetic phases is observed via a dramatic increase in the width of Bragg reflections. This width increase continues until the first-order structural phase transition at $T_{CO}=152$ K. This structural phase transition was observed to display a large hysteresis width of 10 K upon warming.

© 2004 WILEY-VCH Verlag GmbH & Co. KGaA, Weinheim

1 Introduction

Much effort has been devoted to the synthesis of manganites and the understanding of colossal magnetoresistance (CMR) [1]. Typical manganite compounds $R_{1-x}A_xMnO_3$, R = La, Pr, Bi, Nd; A = Sr, Ca have perovskite-type structure. Experiments have revealed a rich phase diagram with a variety of different structures and magnetic properties as a function of stoichiometry, temperature and applied magnetic field. In the case of $Nd_{1-x}Sr_xMnO_3$, for a hole concentration of $0.48 < x < 0.52$, first undergoes a transition to the ferromagnetic metallic state (FMM) at about 250 K, and then becomes an A-type antiferromagnetic (AFM) metal at about 200 K. Upon further cooling, it becomes a CE-type AFM at about 160 K and displays both charge and orbital ordering in this phase.

$Nd_{0.5}Sr_{0.5}MnO_3$ shows a phase transitions from a paramagnetic (PM) insulator to a FMM at Curie temperature, $T_C \approx 255$ K via the double-exchange (DE) mechanism. Then it undergoes a first-order transition to a CE-type AFM insulating state at $T_{CO} \approx 158$ K [2] which displays charge ordering (CO) below T_{CO} [3, 4]. Recently, using neutron powder diffraction on $Nd_{1/2}Sr_{1/2}MnO_3$, Ritter et al. [5] observed phase segregation at low temperature, contains two different crystallographic structures and three magnetic phases: orthorhombic *(Imma)* FM, orthorhombic *(Imma)* A-type AFM, and monoclinic *(P2₁/m)* CE-type AFM phases. For certain compounds of CMR materials, a spatially inhomogeneous distribution of domain structures has also been observed using electron microscopy [6, 7]. Here, we report the first experimental evidence using high-resolution X-ray scattering on a *single crystal* of $Nd_{1/2}Sr_{1/2}MnO_3$, rather than powder material, to support these features of the phase segregation in CMR materials.

---

[*] Corresponding author: e-mail: ebrahim_ghazi@yahoo.com, Phone: +00 98 273 32204-9, Fax: 0098 273 3336007

2 Experimental details

The single crystal of $Nd_{1/2}Sr_{1/2}MnO_3$ was grown by the standard floating zone. The lattice parameters a = 5.431 Å, b = 7.625 Å, and c = 5.477 Å with the space group of *Imma* [8, 9] were used for indexing the Bragg reflections. Preliminary measurements were carried out at the University of Durham using a four-circle triple-axis diffractometer on a rotating anode X-ray generator. The detailed studies utilising synchrotron radiation were performed at station 16.3 at Daresbury Laboratory. Si (1 1 1) single crystals were used as monochromator and analyser. In order to avoid probing the twin structure of the sample, a beam size of 0.2 x 0.1 mm^2 was used to search around the crystal until the Bragg reflections were found to be singlet. The mosaic width of the sample was determined to be ~ 0.05° (FWHM) as measured on the Bragg reflection (5, –2, 3) at room temperature. Such an experimental arrangement provides extremely high instrumental resolution and ensures that the widths of the Bragg reflections are entirely due to the size of coherent domains within the sample.

3 Results and discussion

The evolution of the integrated intensities of Bragg peak (5, –2, 3) as a function of temperature is displayed in Figure 1. Upon cooling, the intensity of the Bragg peak firstly drops at $T \approx 200$ K, and then a sharp and abrupt change takes place at $T_{CO} \approx 152$ K. Similar behaviour was also observed on warming run with the transition at $T_{CO} \approx 162$ K. The transition into the charge ordered phase with hysteresis width of ~10 K is in agreement with the resistivity measurements [10] and belongs to the class of the first-order phase transition as observed by the transport measurements [2, 3]. By carefully tracking the evolution of Bragg peaks around T_{CO}, we found that the transition boundary wherein the high temperature phase co-existed with the low temperature phase only within ~ 1.3 K [11]. A structural phase transition from a FM phase to an *A*-type AFM phase at $T \approx 200$ K has been reported by Kajimoto *et al* using neutron powder diffraction [3]. As X-rays are not sensitive to changes of magnetic phases but to the changes of domain sizes in a crystal, the decrease in the integrated intensities of Bragg peak suggests that the FM domains break into smaller domains at $T = 200$ K due to the formation of *A*-type AFM domains. The transition happens at $T \approx 200$ K, upon cooling and warming runs respectively, is a second-order transition, as the data shows no hysteresis.

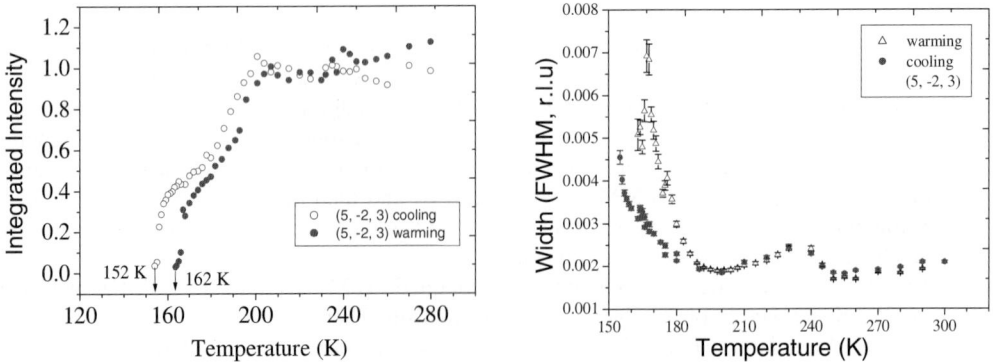

Fig. 1 The evolution of the integrated intensities of the Bragg reflection (5, –2, 3) as a function of temperature.

Fig. 2 Scans through the longitudinal direction of the Bragg peak (5, –2, 3) displaying changes of the peak width (FWHM) with temperature on cooling and warming.

It is also noteworthy that both curves in Fig. 1 measured in cooling and warming runs split at $T \approx 180$ K, which suggests the existence of a non-equilibrium state between domains in the temperature range 180 to 160 K due to the first order phase transition at T_{CO}. This suggestion of a change in the size of the domains is also supported by measurements of the Bragg peak width along the longitudinal direction as displayed in Fig. 2. It is clear that both curves, measured upon cooling and warming runs, split at $T \approx 180$ K and this splitting can be ascribed to the hysteresis behaviour due to the first-order phase transition at T_{CO}. Below 180 K the peak widths diverge as the temperature approaches T_{CO}, indicating a decrease of the length scale of the long-range ordered structure. The width of the Bragg reflection shows unusual peak broadening around the transitions from PM to FM and FM to AFM charge-ordered phases. Such a peak broadening of the Bragg peak has also been observed on $La_{1/2}Ca_{1/2}MnO_3$ [12] in the whole FM range in contrast to $Nd_{1/2}Sr_{1/2}MnO_3$. The peak width increases at $T \approx 252$ K and then decreases at $T \approx 234$ K, indicating a transition starting at 252 K, which is completed at 234 K. At 252 K, the PM domains start to break up into smaller domains due to the formation of the FM phase, as demonstrated by an abrupt change in both resistivity and magnetization [2]. This change is continuous, and it suggests the existence of a mixture of both domains, PM and FM domains, in the temperature range 252–234 K. Such a transition has been reported by both transport and neutron powder diffraction measurements [3–5].

In order to further study the evolution of the domain structures with temperature, we also performed scans along the 3 crystallographic axis, H, K, and L, individually. Figure 3 shows normalised integrated intensity of the (5, –2, 3) Bragg reflection in each crystallographic direction. The corresponding structural correlation lengths were obtained from widths measurement and by $\xi = 2\pi\rho$ where δ is the HWHM. The correlated domain volume was determined from the product of these 3 correlation lengths along the H, K, and L directions and is displayed in Fig. 4. We realise that this determined domain size may not be the actual size of the domains [13], but rather the size of long-range correlations within a crystallite. However, it will reflect the domain structure, size and dislocation density and the effect of magnetic and crystallographic transitions. For temperatures above 252 K, (the region marked by PM in Fig. 3), the PM domains were observed to grow as the temperature was lowered, indicating long-range order was forming. Upon cooling, we observed a drop of about 70 % in the volume of the correlated domains from 252 K (PM state) to 234 K (FM state). In the interval 234–200 K, the size of the domains increases with decreasing temperature and the integrated intensities increase as well.

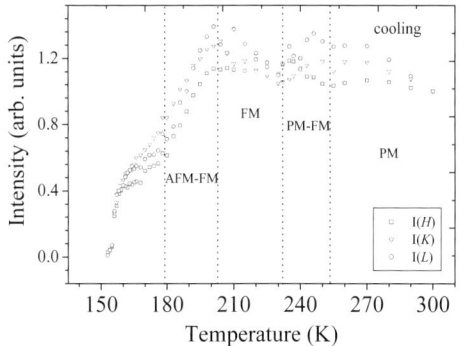 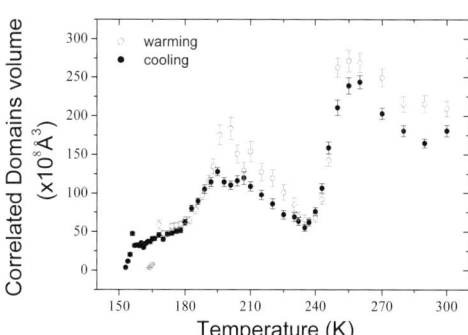

Fig. 3 The normalized integrated intensity of the Bragg peak (5, –2, 3) versus temperature. The data were taken by scanning through the H, K, and L directions. PM-FM and AFM-FM represent mixture of paramagnetic and ferromagnetic and mixture of antiferromagnetic and ferromagnetic, respectively.

Fig. 4 The temperature dependence of the domain size on warming and cooling runs.

The development of a long- range order of the FM state stops at about 200 K, where the integrated intensity starts to drop and the peaks profiles became increasingly broad upon cooling. An A-type AFM phase has been demonstrated to co-exist with a FM phase below 200 K on neutron powder diffraction patterns [3, 5]. Using transmission electron microscopy, Fukumoto et al. also observed the transformation of microstructures relating to the microscopic-scale electronic phase separation [6]. On further cooling, a turning point on the transition curves of Fig. 3 was observed at $T \approx 182$ K. Such a discontinuous change is more pronounced on the curves along the H and L directions than that along the K direction.

The origin of this is not exactly known yet, but it might be the precursor of orbital ordering. It has been demonstrated that orbital ordering results from a cooperative JT effect distorting the $M^{3+}O_6$ octahedra, associated with long Mn^{+3}–O bonds along the a and c axes, which could suppress the domains formed at high temperatures and change the distribution of domains. Kajimoto *et al.* proposed that orbital ordering could cause the observed phase segregation [3]. A two-component feature (PM and FM) has been observed at T ≈ 250 K in a single crystal of $Nd_{1/2}Sr_{1/2}MnO_3$ using inelastic neutron scattering [14, 15]. The coexistence of multiple phases ascribed to micro-domain structures in the temperature range of $T_N < T < T_C$ has also been observed in $Pr_{1/2}Ca_{1/2}MnO_3$, and $La_{1/2}Ca_{1/2}MnO_3$ using electron microscopy and neutron diffraction on powder samples [5, 6, 16].

In conclusion, we have reported the transformation of domain structures in *single crystal* $Nd_{1/2}Sr_{1/2}MnO_3$ using synchrotron X-ray scattering. The phase transitions observed in the temperature interval $T_N < T < T_C$ are unlikely to be caused by the structural phase transition, instead, it is evidence for phase segregation due to the formation of different magnetic domains. Such a mixture of magnetic phases has previously been observed in powdered material. Our study is the first to observe such effects in a bulk single crystal. As $Nd_{1/2}Sr_{1/2}MnO_3$ is a typical charge ordered CMR material, the observed phase segregation, especially in the temperature range of $T_{CO} < T < 200$ K, could therefore also be sensitive to the application of magnetic field [16, 17].

Acknowledgments We are grateful for the facilities and beamtime made available at the Daresbury Synchrotron Radiation Source and beamline teams S. P. Collins, and B. M. Murphy. The authors are thankful for the great help during the experiment from Dr. C. H. Du and Prof. S.-W. Cheong at Rutgers University and Bell Laboratories for providing crystal.

References

[1] Colossal magnetoresistive oxides, edited by Y. Tokura (Gordon and Breach Science Publishers, Vol. 2).
[2] H. Kuwahara, Y. Tomioka, A. Asamitsu, et al., Science **270**, 961 (1995).
[3] R. Kajimoto, H. Yoshizawa, H. Kawano, et al., Phys. Rev. B **60**, 9506 (1999).
[4] H. Kawano, R. Kajimoto, H. Yoshizawa, et al., Phys. Rev. Lett. **78**, 4253 (1997).
[5] C. Ritter, R. Mahendiran, M. R. Ibarra, et al., Phys. Rev. B **61**, R9229 (2000).
[6] S. Mori, C. H. Chen, and S. W. Cheong, Phys. Rev. Lett **81**, 3972 (1998).
[7] N. Fukumoto, S. Mori, N. Yamamoto, et al., Phys. Rev. B **60**, 12963 (1999).
[8] V. Caignaert, F. Millange, M. Hervieu, et al., Solid State Commun. **99**, 173 (1996).
[9] P. M. Woodward, D. E. Cox, T. Vogt, et al., Chem. of Mater. **11**, 3528 (1999).
[10] Y. Tokura, H. Kuwahara, Y. Moritomo, Y. Tomioka, and A. Asamitsu, Phys. Rev. Lett **76**, 3184 (1996)
[11] C. H. Du, Y. Su, M. E. Ghazi, et al., in Magnetic and Superconducting Materials (MSM-99), edited by M. Akhavan, J. Jensen, and K. Kitazawa (World Scientific Press, London, 2000, 1999), p. 936.
[12] P. G. Radaelli, D. E. Cox, M. Marezio, et al., Phys. Rev. B **55**, 3015 (1997).
[13] Strictly speaking, in order to obtain the actual correlated domain size the peak profiles should be deconvoluted from the resolution function. In this study, our resolution function was so very narrow compared to the measured Bragg widths that such a deconvolution is unnecessary.
[14] R. Kajimoto, H. Yoshizawa, H. Kawano, et al., J. Phys. Chem. Solids **60**, 1177 (1999).
[15] R. Kajimoto, T. Kakeshita, Y. Oohara, et al., Phys. Rev. B **58**, R11837 (1998).
[16] C. H. Chen and S. W. Cheong, Phys. Rev. Lett. **76**, 4042 (1996).
[17] S. Mori, T. Katsufuji, N. Yamamoto, et al., Phys. Rev. B **59**, 13573 (1999).

Jahn-Teller distortion ordering in single-crystal $Nd_{1/2}Sr_{1/2}MnO_3$

Mohammad E. Ghazi[*,1] and **P. D. Hatton**[2]

[1] Department of Physics, Shahrood University of Technology, Shahrood, P.O. Box 36155-316, Iran
[2] Department of Physics, University of Durham, Durham DH1 3LE, UK

Received 31 August 2003, accepted 31 December 2003
Published online 18 March 2004

PACS 60.10.–i, 64.60.–i, 71.30.+h

Charge, spin and orbital degrees of freedom play important roles in the electrical and magnetic properties of manganites. $Nd_{0.5}Sr_{0.5}MnO_3$ is a paramagnetic insulator with *Imma* orthorhombic symmetry at room temperature. Upon cooling it becomes a ferromagnetic metal below 240 K and at 160 K undergoes a first-order phase transition. Below this temperature it is a CE-type antiferromagnetic insulator. In this paper we report synchrotron X-ray scattering investigations of superlattice reflections observed in the low temperature phase of $Nd_{0.5}Sr_{0.5}MnO_3$. Upon cooling below 160 K we observed additional peaks occurring at non-integer positions in reciprocal space. These reflections had a wavevector (1/2, 0, 0) and are due to the structural modulation arising from the Jahn-Teller distortion associated with charge ordering of Mn^{3+} and Mn^{4+} ions and orbital ordering in the sample. The peaks observed below 160 K have an intensity of approximately 10^{-3} that of the Bragg reflections. The intensity profiles of these reflections show that the transition into the ordered phase is strongly first-order. The structural phase transition was observed to display a hysteresis width of 10 K. The measured widths of these peaks show that the charge modulation is correlated primarily in the *ac* plane rather than the *K*-direction (long-axis).

© 2004 WILEY-VCH Verlag GmbH & Co. KGaA, Weinheim

1 Introduction

Charge, orbital and spin ordering in direct space has recently attracted considerable attention due to their probable role in the observed colossal magnetoresistance (CMR) in manganites. The $Ln_{1-x}A_xMnO_3$ manganite compounds have perovskite structures and provide an ideal material to study the physics of strongly correlated electronic systems. Every manganese Mn ion in the CMR manganite is surrounded by six oxygen ions and forms an octahedron. The 3d state on the Mn-site placed in such an crystal field are subject to the partial lifting of the degeneracy into lower-lying t_{2g} states and higher lying e_g sates. The electronic configuration of Mn^{3+} (d^4) ions is (t^3_{2g}, e^1_g) and Mn^{4+} (d^3) ions is (t^3_{2g}) with the t_{2g} electrons localized at the Mn sites. The e_g electrons are hybridised with the oxygen 2p orbitals and participate in a Jahn-Teller (JT) distortion of the MnO_6 octahedra. In high temperature Mn^{4+} ions are distributed randomly and charge ordering (CO) is formed in low temperature with the periodic ordering of Mn^{3+} and Mn^{4+} ions. The fraction of Mn^{4+} ions can be controlled with the concentration of A ions.

$Nd_{0.5}Sr_{0.5}MnO_3$ is a paramagnetic (PM) insulator with *Imma* orthorhombic symmetry at room temperature [1–3]. Upon cooling it becomes a ferromagnetic metal (FMM) below $T_C \sim 250$ K via the double-exchange (DE) mechanism and at ~ 160 K undergoes a first-order phase transition to a CE-type antiferromagnetic (AFM) insulator state [4]. The measurements of the lattice parameters showed an abrupt change at this temperature (~ 160 K) [5]. Neutron and electron diffraction have demonstrated the presence of charge, spin and orbital ordering in this phase [6]. Upon cooling the crystal from the FM state, and before transition to this ordered state, the system undergoes to another transition, which is a mixed

---

[*] Corresponding author: e-mail: ebrahim_ghazi@yahoo.com, Phone: +00 98 273 32204-9, Fax: 0098 273 336007

domain state of *A*-type AFM and *CE*-type AFM [4]. Phase segregation due to the formation of different magnetic domains on a single crystal of $Nd_{0.5}Sr_{0.5}MnO_3$ has been observed by Ghazi et al. using synchrotron X-ray scattering [7], by Mori et al. [8] and Fukumoto et al. [9] using electron microscopy, and by Ritter et al. using neutron powder diffraction [10].

In addition to the DE interactions and JT distortions, the ordering of the two-fold e_g orbitals of Mn ions have an essential effect on physical properties in hole-doped manganites [4]. The charge, spin, and the $d_{3x^2-r^2/3y^2-r^2}$ type orbital ordering [6] satellites appear at the same temperature in $Nd_{1/2}Sr_{1/2}MnO_3$. In this half-doped system, the CO is believed to be the ordering of Mn^{3+} and Mn^{4+} ions in *CE*-type AFM phase. It is known the CO state can be melted and transferred to FM state by magnetic field [11, 12] and even by substitution of ^{16}O with ^{18}O [13]. In this paper the results obtained by a very high-resolution X-ray scattering study of JT distortion ordering in a single crystal of $Nd_{1/2}Sr_{1/2}Mn$ (associated with the charge and orbital ordering) will be described.

2 Experimental details

The single crystal of $Nd_{0.5}Sr_{0.5}MnO_3$ was grown using the floating zone method at Bell Laboratories and then polished to get a shiny and even surface with an area of ~ 2×1 mm^2 using 1 µm diamond paste. The crystal was indexed in the orthorhombic structure with $a_o \approx c_o \approx \sqrt{2}a_c$ and $b_o = 2a_c$, where subscripts c and o represent the cubic and orthorhombic perovskite structure respectively. The lattice parameters a = 5.431 Å, b = 7.625 Å, and c = 5.477 Å with the space group of *Imma* were used for indexing of the reflections. The crystal was first aligned and studied at the University of Durham using a four-circle triple-axis diffractometer on a rotating anode X-ray generator and a closed-cycle displex cryostat. It was determined that the longitudinal direction in the scattering plane is the $\langle 1\ 0\ 1\rangle$ direction. By cooling the sample, satellite ordering was observed at low temperatures.

Further experiments were conducted using synchrotron radiation at the *XMaS* diffractometer situated on BM28 at the European Synchrotron Research Facility in Grenoble (ESRF) and at beamline 16.3 at the SRS in Daresbury. At Daresbury, a double bounce Si (111) monochromator was used to select the incident X-ray beam wavelength of 1.000 Å, whilst at the ESRF a focusing mirror was used to focus the beam to a small spot. Single crystals were used as analysers to provide a good wavevector resolution and decrease the background at the SRS and ESRF. The mosaic width of the sample was determined to be ~ 0.05° (FWHM) as measured on the Bragg reflection (5, –2, 3) at room temperature. Measurements at the SRS were performed along the longitudinal and transverse directions with a Si (111) single crystal as an analyser, while at the ESRF they were performed along the *H*-, *K*-, and *L*-directions in reciprocal space by using a single crystal Cu (220) as an analyser.

3 Results and discussion

Upon cooling the crystal below 160 K additional peaks were observed at non-integer positions in reciprocal space. These peaks had a wavevector (1/2, 0, 0) and are due to the structural modulation arising from the JT distortion (due to the presence of the JT active ions Mn^{3+}) associated with charge ordering of Mn^{3+} and Mn^{4+} ions and orbital ordering in the sample. This causes the undistorted $Mn^{4+}O_6$ octahedra in the *ac* plane to displace in opposite directions with respect to the neighbouring (along the *a* direction) $Mn^{4+}O_6$ octahedra [14, 15] and a doubling of the unit cell along the *a*-axis. The peaks observed below 160 K have an intensity of approximately 10^{-3} that of the Bragg reflections. The search for superlattice reflections with wavevector (0, 2 K+1, 0), which directly probes the CO, was not successful probably due to the crystal orientation and the peak intensity. Figure 1 shows the integrated intensity (normalised) of the superlattice reflection (2.5, 0, 2) as a function of temperature in the *H*-, *K*-, and *L*-directions in reciprocal space. As the Figure indicates, the intensity of the peak increased as the temperature decreased from T_{CO} ~ 160 K and the variation of the intensity with temperature and observation of the hysteresis curve (Fig. 2) are in accord with a first-order phase transition. The observed peak could be indexed using the wavevectors (*h*/2, *k*, *l*) and (*h*, *k*, *l*/2). These wavevectors are associated with the structural modulation due to the charge ordering and the presence of two wavevectors 90° apart is due to twinning. The peak intensity is approximately constant over the whole temperature range, except near to the transition temperature, where the intensity suddenly vanishes.

A property of a first-order phase transition is the presence of a hysteresis loop upon reversing the temperature in the transition region. The normalized integrated intensity profile of the superlattice reflection (2.5, 0, 2) and Bragg reflection (5, –2, 3) on cooling and warming runs is displayed in Fig. 2. When the sample was warmed from low temperature, the superlattice satellite (2.5, 0, 2) disappeared at ~ 162 K, the same temperature as the high-temperature Bragg peak appeared, and on the cooling run this peak appeared at ~ 152 K, the temperature at which the Bragg peak disappeared. Figure 2 clearly shows that the transition has a marked hysteresis of ~ 10 K on cooling and warming runs in agreement with the resistivity measurements [9]. Between 200 K and 160 K no superlattice reflections were observed, indicating that charge ordering does not occur in the A-type AFM state.

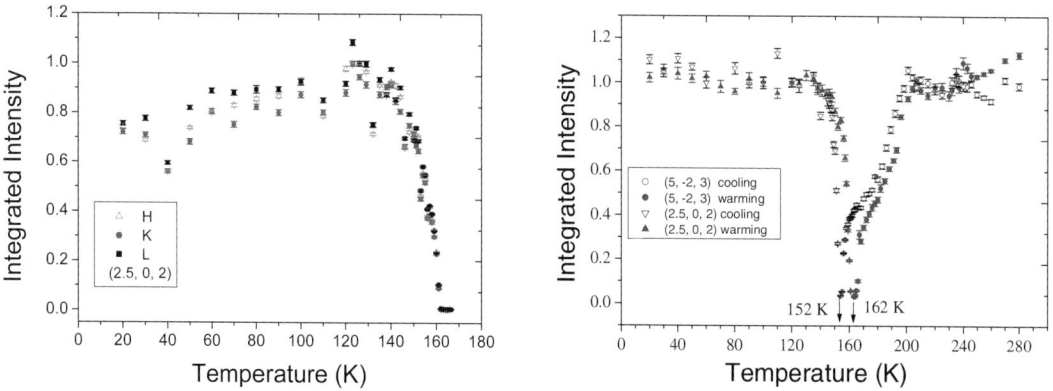

Fig. 1 The integrated intensity (normalized) of the superlattice reflection (2.5, 0, 2) as a function of temperature in the H-, K-, and L- directions in reciprocal space.

Fig. 2 Integrated intensities (normalised) of the superlattice reflection (2.5, 0, 2) and Bragg peak (5, –2, 3) as a function of temperature in cooling and warming runs along the longitudinal direction in reciprocal space.

Scans along the H-, K-, and L-directions in reciprocal space were performed for the JT distortion ordering peak (2.5, 0, 2) at every temperature and the results have been fitted with a Gaussian line shape. Figure 3 shows the width (FWHM) of this peak in reciprocal lattice units obtained from the fitting. The error bars resulted from this fitting are also included in the graph. The width in the K-direction is added to the Figure with a reduced scale to ease comparison with the other directions. The width along the K-direction, which is the long-axis direction in this setting, is much larger than the other directions, probably due to the structural anisotropy e.g. anisotropy in the Mn–O–Mn bond angles and Mn–O bond lengths in the ac plane and b-direction [1, 6]. As the correlation length is proportional to the inverse of the width, the ordered regions are correlated over much longer distances in the ac planes, compared to those in normal to these planes. The width is approximately constant and independent of temperature unlike the charge stripes in nickelate crystals [16], indicating the CO in this crystal is stable. The width abruptly increases at the charge ordering transition temperature.

In conclusion, below T_{CO} satellite reflection with wavevector (1/2, 0, 0) were observed due to the JT distortion of the MnO_6 octahedra. These satellites appear at the same temperature as the CO temperature. These results show the transition into the charge-ordered phases is strongly first-order as it was seen in the superlattice peak intensity profile and the presence of the hysteresis loop in cycling of the temperature. The measured widths of these peaks show that the charge modulation is correlated primarily in the ac plane rather than the K-direction (long-axis). Between 200 K and 160 K no superlattice reflections were observed, indicating that charge ordering does not occur in the A-type AFM state.

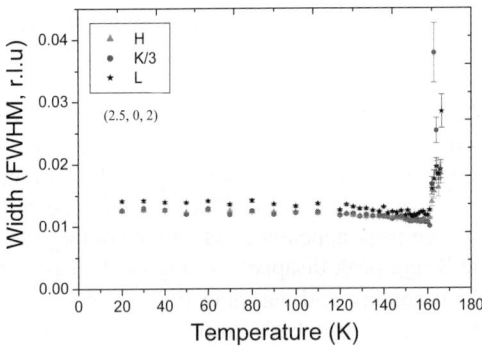

Fig. 3 The width (FWHM) of the superlattice reflection (2.5, 0, 2) as a function of temperature in the H-, K-, and L-directions in reciprocal space.

Acknowledgements We are pleased to acknowledge the facilities and beamlines teams at SRS and ESRF, Dr. S. Brown, Dr. D. Paul, Dr. S. P. Collins and B. M. Murphy. We also thank Dr. S. Wilkins, and Mr. P. Spencer for their help and assistance during the experiments and Prof. S-W. Cheong at Rutgers University and Bell Laboratories for providing crystal.

References

[1] V. Caignaert, F. Millange, M. Hervieu, et al., Solid State Commun. **99**, 173 (1996).
[2] P. M. Woodward, T. Vogt, D. E. Cox, et al., Chem. of Mater. **10**, 3652 (1998).
[3] P. M. Woodward, D. E. Cox, T. Vogt, et al., Chem. of Mater. **11**, 3528 (1999).
[4] H. Kawano, R. Kajimoto, H. Yoshizawa, et al., Phys. Rev. Lett. **78**, 4253 (1997).
[5] S. Shimomura, K. Tajima, N. Wakabayashi, et al., J. Phys. Soc. Jpn. **68**, 1943 (1999).
[6] R. Kajimoto, H. Yoshizawa, H. Kawano, et al., Phys. Rev. B **60**, 9506 (1999).
[7] M. E. Ghazi and P. D. Hatton, MSM03 (2003).
[8] S. Mori, C. H. Chen, and S. W. Cheong, Phys. Rev. Lett. **81**, 3972 (1998).
[9] N. Fukumoto, S. Mori, N. Yamamoto, et al, Phys. Rev. B **60**, 12963 (1999).
[10] C. Ritter, R. Mahendiran, M. R. Ibarra, et al., Phys. Rev. B **61**, R9229 (2000).
[11] A. Biswas, A. Arulraj, A. K. Raychaudhuri, et al., J. Phys.: Condens. Matter **12**, L101 (2000).
[12] H. Kuwahara, Y. Tomioka, A. Asamitsu, et al., Science **270**, 961 (1995).
[13] G. M. Zhao, K. Ghosh, and R. L. Greene, J. Phys.: Condens. Matter **10**, L737 (1998).
[14] S. Shimomura, N. Wakabayashi, H. Kuwahara, et al., Phys. Rev. Lett. **83**, 4389 (1999).
[15] P. G. Radaelli, D. E. Cox, M. Marezio, et al., Phys. Rev. B **55**, 3015 (1997).
[16] C. H. Du, M. E. Ghazi, Y. Su, et al., Phys. Rev. Lett. **84**, 3911 (2000).

Fe doping effects on the structural and magnetic properties in $Pr_{0.5}Sr_{0.5}Mn_{1-x}Fe_xO_3$ with $0 \leq x \leq 0.3$

A. Ammar[*], S. Zouari, and **A. Cheikh-Rouhou**

Laboratoire de Physique des Matériaux, Faculté des Sciences de Sfax, B. P. 802, 3018 Sfax, Tunisie

Received 31 August 2003, accepted 31 December 2003
Published online 25 March 2004

PACS 61.66.Dk, 61.72.Ww, 75.50.Bb, 75.60.Ej

Structural and magnetic properties of $Pr_{0.5}Sr_{0.5}Mn_{1-x}Fe_xO_3$ ($0.0 \leq x \leq 0.3$) powder samples have been investigated. Powder X-ray diffraction showed that all our synthesized samples are single phase and crystallize in the orthorhombic symmetry with Imma space group. Fe doping leads to a weakness of the ferromagnetic (FM) ordering at low temperature. With increasing Fe content, the Curie temperature T_c decreases.

© 2004 WILEY-VCH Verlag GmbH & Co. KGaA, Weinheim

1 Introduction

The doped manganese perovskites $Ln_{1-x}A_xMnO_3$ (where Ln is a rare earth ion and A is a divalent one) have stimulated great interest in reason of the very large negative magnetoresistance effects that they exhibit [1–4]. The substitution of the trivalent rare-earth element by a divalent one leads to a mixed valence Mn^{3+} and Mn^{4+} ions. The double exchange interaction between these ions [5] is responsible for the metallic and ferromagnetic properties. However, Millis et al. [6] proposed that Jahn–Teller distortions and associated strong electron–phonon coupling play a key role in the metal to insulator transition.

In order to reveal the mechanism of the magnetotransport, some authors [7–11] have investigated the Fe doping effects on the structural, magnetic and magneto-transport properties of $La_{1-x}Ca_xMn_{1-y}Fe_yO_3$ and found that the dopant Fe leads to a weakness of the ferromagnetism at low temperatures and modifies the magnetoresistance effects.

Earlier studies have shown that in this Fe doping range, a direct replacement of Mn^{3+} by Fe^{3+} with identical ion size occurs [12, 13]. Consequently, the otherwise strong lattice effects can be bypassed, and the effects due to changes in the electronic structure become accessible.

In this work, we investigated systematically the Fe doping effects on the structural and magnetic properties of $Pr_{0.5}Sr_{0.5}Mn_{1-x}Fe_xO_3$ samples ($0.0 \leq x \leq 0.3$).

2 Experimental procedure

Polycrystalline samples $Pr_{0.5}Sr_{0.5}Mn_{1-x}Fe_xO_3$ with $0 \leq x \leq 0.3$ were prepared using the standard solid-state reaction method by mixing Pr_6O_{11}, MnO_2, $SrCO_3$ and Fe_2O_3 with high purety up to 99.9% in the desired proportion according to the following reaction:

$$(1-x)\, MnO_2 + \frac{0.5}{6} Pr_6O_{11} + 0.5\, SrCO_3 + \frac{x}{2} Fe_2O_3 \longrightarrow Pr_{0.5}Sr_{0.5}Mn_{1-x}Fe_xO_3 + \delta\, CO_2$$

---

[*] Corresponding author: e-mail: abdallahammar@yahoo.fr, Phone: +00 216 98 632 927

The starting materials were intimately mixed in an agate mortar and then heated in air at 1000 °C for 48 h. A systematic annealing at high temperature is necessary to ensure a complete reaction. The powders are pressed into pellets (of about 1 mm thickness) and sintered at 1400 °C in air for 72 h with intermediate regrinding and repelling. Finally the pellets are rapidly quenched in air. This step was done in order to retain the structure adopted at a given annealing temperature.

Phase purity, homogeneity and cell dimensions are determined by powder X-ray diffraction at room temperature using a SIEMENS D500 Diffractometer ($CuK_{\alpha 1}$ radiation). Unit cell parameters are obtained by least squares calculations. The Mn^{4+} content was checked by redox titration using standard potassium permanganate and ferrous sulfate solution.

Magnetization measurements versus temperature were recorded by a vibrating sample magnetometer in the temperature range 10–300 K, in a magnetic applied field of 500 Oe.

3 Results and discussion

3.1 X-ray diffraction analysis

The X-ray powder diffraction analysis show that our synthesized samples $Pr_{0.5}Sr_{0.5}Mn_{1-x}Fe_xO_3$ ($0 \leq x \leq 0.3$) are single phase and crystallyze in the orthorhombic structure with Imma space group. However we observe for the high x values, the presence of diffraction peaks with very small intensities, witch are attributed to traces of Mn_3O_4. We report in Fig. 1 the X-ray diffraction patterns of our synthesized samples. The room temperature lattice constants are listed in Table 1. As the Fe^{3+} takes the place of Mn^{3+} with the same ionic radii [9, 14], the dopant Fe has a very small influence on the unit cell parameters.

Table 1 Crystallographic data of $Pr_{0.5}Sr_{0.5}Mn_{1-x}Fe_xO_3$ samples ($0.00 \leq x \leq 0.30$).

x	a (Å)	b (Å)	c (Å)	V (Å)	% Mn^{4+}
0.00	5.391(6)	5.434(0)	7.733(8)	226.58(5)	52.2
0.05	5.387(0)	5.434(9)	7.729(5)	226.30(3)	51.8
0.10	5.380(2)	5.435(9)	7.715(4)	225.64(6)	51.5
0.15	5.374(5)	5.437(3)	7.708(4)	225.26(1)	50.7
0.20	5.371(6)	5.438(7)	7.702(0)	225.01(0)	50.2
0.25	5.367(9)	5.441(3)	7.699(8)	224.89(8)	48.7
0.30	5.355(9)	5.444(6)	7.696(9)	224.44(7)	47.3

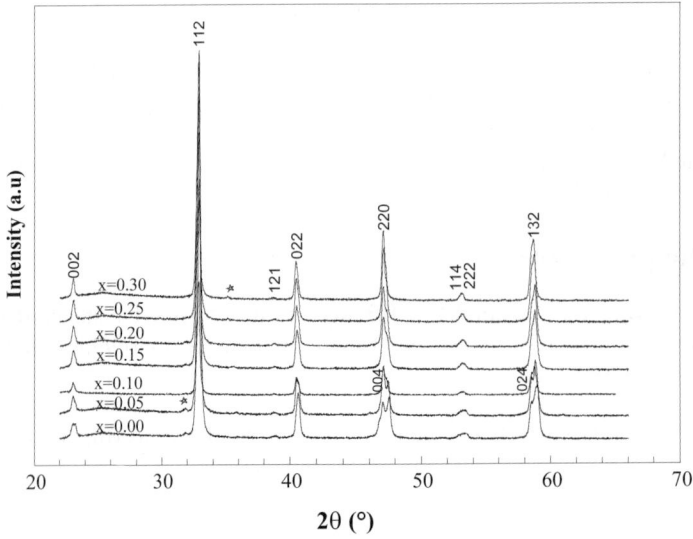

Fig. 1 Powder X-ray diffraction patterns for $Pr_{0.5}Sr_{0.5}Mn_{1-x}Fe_xO_3$ ($0.00 \leq x \leq 0.30$).

3.2 Magnetic properties

Figure 2 shows the temperature dependence of the magnetization for polycrystalline $Pr_{0.5}Sr_{0.5}Mn_{1-x}Fe_xO_3$ samples. Our samples exhibit a paramagnetic to an ordered state transition with decreasing temperature.

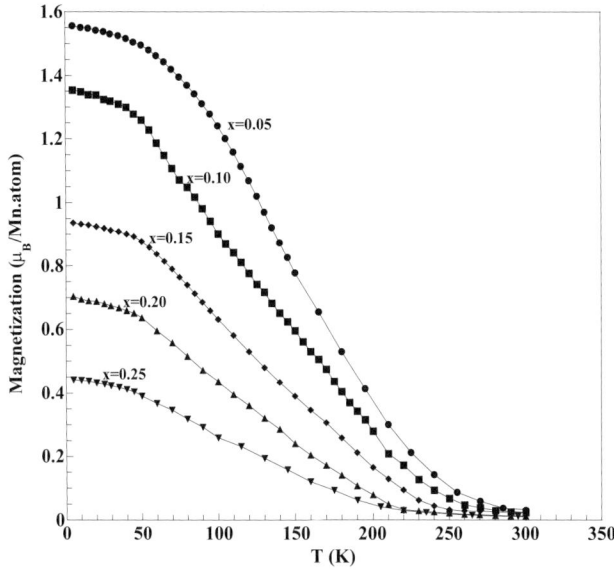

Fig. 2 Evolution of the magnetization as a function of temperature at H = 0.05 T for $Pr_{0.5}Sr_{0.5}Mn_{1-x}Fe_xO_3$ samples ($0.00 \leq x \leq 0.30$).

We observe a great weakness of the magnetization for $x \geq 0.05$. This behavior, explained by the interruption of the double exchange interaction between Mn^{3+} and Mn^{4+} ions, is responsible for the ferromagnetic behavior. With increasing Fe content, the Curie temperature T_C decreases from 256 K for x = 0 to 209 K for x = 0.25 (Fig. 3).

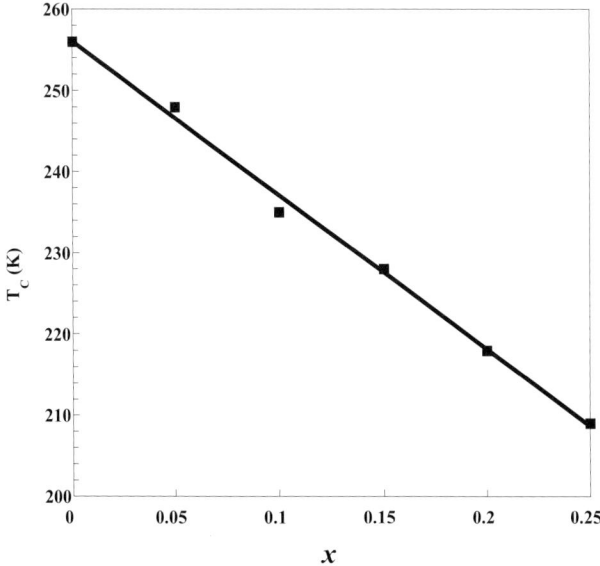

Fig. 3 Curie temperature dependence as a function of x for $Pr_{0.5}Sr_{0.5}Mn_{1-x}Fe_xO_3$ samples.

4 Conclusion

In summary, we have studied the Fe doping effect on the structural and magnetic properties in $Pr_{0.5}Sr_{0.5}Mn_{1-x}Fe_xO_3$ ($0.00 \leq x \leq 0.30$) compounds. All our synthesized samples crystallize in the orthorhombic symmetry with Imma space group. Fe doping leads to a weakness of the ferromagnetic (FM) ordering at low temperature and a decrease of the Curie temperature T_c with increasing Fe content.

Acknowledgements This work has been supported by the Ministry of High Education, Scientific Research and Technology. Magnetization measurements have been performed in Laboratoire de Magnétisme Louis Néel, Grenoble.

References

[1] G. H. Jonker and J. H. Van Santen, Physica **16**, 337 (1950).
[2] G. H. Jonker, Physica **22**, 707 (1956).
[3] P. V. Vanétha, R. S. Singh, S. Natarajan, and C. N. R. Rao, J. Solid. State Chem. **137**, 365 (1998).
[4] G. Ji, X. J. Fan, J. H. Zhang, C. S. Xiong, and X. G. Li, J. Phys. D: Appl. Phys. **31**, 3036 (1998).
[5] C. Zener, Phys. Rev. **82**, 403 (1951).
[6] A. J. Millis, P. B. Littlewood, and B. I. Shraiman, Phys. Rev. Lett. **74**, 5144 (1955).
[7] A. Tkachuk, K. Rogacki, D. E. Brown, B. Dabrowski, A. J. Fedro, C. W. Kimball, B. Pyles, X. Xiong, D. Rosenmann, and D. D. Dunlap, Phys. Rev. B **57**, 8509 (1998).
[8] S. B. Ogale, R. Shreekala, R. Bathe, S. K. Date, S. I. Palil, B. Hannoyer, F. Petit, and G. Marest, Phys. Rev. B **57**, 7841 (1998).
[9] K. H. Ahn, X. W. Wu, K. Liw, and C. L. Chien, Phys. Rev. B **54**, 15299 (1996).
[10] Z. W. Li, A. H. Morrish, X. Z. Zhou, and S. Dai, J. Appl. Phys. **83**, 7198 (1998).
[11] J. H. Zhang, X. J. Fan, C. S. Xiong, and X. G. Li, Sold. State Commun. **115**, 531 (2000).
[12] M. F. Hundley, M. Hawley, R. H. Heffner, Q. X. Jia, J. J. Neumeier, J. Tesmer, J. D. Thompsen, and X. D. Wu, Appl. Phys. Lett. **67**, 860 (1995).
[13] J. M. D. Coey, M. Viret, L. Ranno, and K. Ounadjela, Phys. Rev. Lett. **75**, 3910 (1995).
[14] R. D. Shannon, Acta Crystallogr. A **32**, 751 (1976).

Magnetism and giant magnetoresistance in $La_{0.7}Sr_{0.3}Mn_{1-x}M_xO_3$ (M = Cr, Ti) systems

N. Kallel[*,1,2], **K. Fröhlich**[3], **M. Oumezzine**[1], **M. Ghedira**[1], **H. Vincent**[2], and **S. Pignard**[2]

[1] Laboratoire de Physico-Chimie des Matériaux, Département de Physique,
Faculté des Sciences de Monastir, BP 22, 5019 Monastir, Tunisia
[2] Laboratoire des Matériaux et du Génie Physique, Ecole Nationale Supérieure de Physique de Grenoble, BP 46, 38 402 Saint-Martin d'Hères Cedex, France
[3] Institut of Electrical Engineering, Slovak Academy of Sciences, Dúbravská 9, 842 39 Bratislava, Slovak Republic

Received 31 August 2003, accepted 31 December 2003
Published online 25 March 2004

PACS 61.10.Nz, 61.72.Ww, 72.60.+g, 75.30.Kz, 75.47.De

Magnetic and transport measurements have been performed on $La_{0.7}Sr_{0.3}Mn_{1-x}M_xO_3$ (M = Cr, Ti) in order to study the influence of the substitution on the Mn-site on the magnetotransport properties in the $La_{0.7}Sr_{0.3}MnO_3$ perovskite-like compound. Polycrystalline samples were prepared by a conventional solid-state reaction method in air. The two series of samples show a decrease of the magnetic characteristics (magnetic transition temperature and saturation magnetization) when the level of chromium or titanium increases. The substitution of manganese by other metal (Cr, Ti) also induces strong changes in the transport properties: the metallicity of $La_{0.7}Sr_{0.3}MnO_3$ observed below the room temperature is destroyed and the resistivity increases many orders of magnitude when the ratio of doping metal increases. For low substituting ratios, an electrical transition is observed; the temperature corresponding to the semiconductor–metal transition shifts to a lower values with increasing x, similarly to the T_C behaviour. For substituting ratio higher than 10% of Ti and 20% of Cr, the electrical transition disappears and the samples exhibit a semiconductor behaviour in the whole range of temperature. The magnetoresistive (MR) properties have been investigated between 0 and 0.5 T for $La_{0.7}Sr_{0.3}MnO_3$, $La_{0.7}Sr_{0.3}Mn_{1-x}Cr_xO_3$ (x = 0.05, 0.1, and 0.2), and $La_{0.7}Sr_{0.3}Mn_{1-x}Ti_xO_3$ (x = 0.1, 0.2, and 0.3) samples.

© 2004 WILEY-VCH Verlag GmbH & Co. KGaA, Weinheim

1 Introduction

Since ten years a great deal of attention has been focused on the colossal magnetoresistance (CMR) properties of manganese perovskites [1–3]. In most cases the large resistance changes are achieved only in a strong field in the Tesla range, thus severely limiting their practical utility. Reducing the field scale and increasing the operating temperature has been the goal of a number of research groups. There has been some recent progress in reducing the field scale by exploiting the intrinsically high degree of spin polarization in these materials [4]. The insulator-to-metal transition, paramagnetic-to-ferromagnetic transition, and the CMR properties depend crucially on the doping level as well as on the nature of the doping element. It has been thought that the spin structure and electronic transport properties of $Ln_{1-x}A_xMnO_3$ are correlated via the double exchange (DE) mechanism, i.e. the hopping of e_g electrons between Mn^{3+} and Mn^{4+} ions mediated by oxygen anions. However, Millis et al. [5, 6] argued that the double exchange alone could not quantitatively account for some features of the CMR effect in $Ln_{1-x}A_xMnO_3$, such as the magnitude of the resistivity and the magnitude of CMR. A strong Jahn–Teller effect of Mn^{3+} ions should play an important role. The lattice distortion may not only influence the effective transfer integral but it may also result in a superexchange interaction between the manganese ions

---

[*] Corresponding author: e-mail: Nabil.Kallel@fsm.rnu.tn, Phone: +216 97 600 841, Fax: +216 73 500 278

and in the complicated magnetic structure of the compound. Additionally, the properties of the manganites might be influenced also by other factors such as the chemical and magnetic inhomogeneity, which are generally related to each other. This kind of in homogeneity is a competitive result between the ferromagnetic coupling due to double exchange interaction and the antiferromagnetic coupling caused by intrinsic orbital interaction. Because Mn acts at the heart of DE interaction, the influence of the substitution at Mn sites with other elements has spurred considerable interest in recent years [7]. In this article, we will discuss our investigation of the effect of chrome and titanium substitution on Mn sites of the prototype colossal magnetoresistance material $La_{0.7}Sr_{0.3}MnO_3$ by magnetization, electrical transport, and magnetoresistance measurements.

2 Experimental $La_{0.7}Sr_{0.3}Mn_{1-x}Cr_xO_3$ ($0 \leq x \leq 0.2$) and $La_{0.7}Sr_{0.3}Mn_{1-x}Ti_xO_3$ ($0 \leq x \leq 0.3$) polycrystalline compounds were prepared by a conventional solid-state reaction method in air. The starting reagents, La_2O_3, $SrCO_3$, Cr_2O_3, TiO_2 (dried before use at 373 K in order to remove any absorbed water) and MnO_2 were weighted in stoichiometric proportions. The mixed powders were first heated in air at 1173 K during 72 h to achieve decarbonation. After grinding, they were heated again at 1473 K for 24 h in air to ensure homogenization. Intermediate cooling and mechanical grinding steps were repeated in order to get an accurate homogenization and complete reaction. The powders were pressed into pellet forms under 4tons/cm^2 and sintered at 1673K for two days in air with several periods of grinding and repelleting. Finally, these pellets were quenched to room temperature. This step was carried out in order to conserve the structure at an annealed temperature. Structural characterizations have been reported elsewhere: all samples are single phased and can be indexed in a rhombohedral structure of $R\bar{3}c$ space group [8, 9]. The magnetic properties were explored with a Vibrating Sample Magnetometer, working at temperatures between 10 and 400 K and fields between 0 and 8 Tesla. The temperature dependence of resistivity ρ was measured by the conventional four-probe method in the temperature range 77–350 K with a typical sample size of 5 x 5 x 2 mm^3. The magnetoresistive properties have been investigated between 0 and 0.5 T for $La_{0.7}Sr_{0.3}MnO_3$, $La_{0.7}Sr_{0.3}Mn_{1-x}Cr_xO_3$ (x = 0.05, 0.1, and 0.2), and $La_{0.7}Sr_{0.3}Mn_{1-x}Ti_xO_3$ (x = 0.1, 0.2, and 0.3) compounds.

3 Results and discussion

3.1 Magnetic properties Magnetization (M) versus temperature (T) curves measured in a field of 500 Oe are shown in Fig. 1. All the samples present a single magnetic transition from a ferromagnetic to paramagnetic state as T increases. However, it is obvious that the magnetic transition temperature T_C and the saturation magnetization (M_s) decrease for both series as x increases. The T_C values of Cr-doped samples are higher than those of Ti-doped samples and the decrease with x is more sensitive to the Ti-substitution compared to the Cr-one: T_C decreases from 369 K to 266 K (resp. 327 K) when 10% of Mn is substituted by Ti (resp. Cr). This softer influence of Cr on the magnetic properties compared to Ti can be analysed by considering the electronic configurations of both ions. Ferromagnetism in undoped $La_{0.7}Sr_{0.3}MnO_3$ is linked to the ferromagnetic Mn^{3+}–O–Mn^{4+} superexchange interactions leading to a maximum magnetic transition temperature of 380 K [10] with the ratio $Mn^{4+}/(Mn^{4+}+Mn^{3+}) = 0.3$. With Cr substitution Cr^{3+}–O–Mn^{3+} and Cr^{3+}–O–Cr^{3+} interactions appear; as Cr^{3+} and Mn^{4+} exhibit the same t_{2g} electronic configurations Mn^{3+}–O–Cr^{3+} interaction is also ferromagnetic whereas Mn^{4+}–O–Cr^{3+} and Cr^{3+}–O–Cr^{3+} interactions are antiferromagnetic like Mn^{4+}–O–Mn^{4+} (two cations with empty e_g orbitals establish a weak antiferromagnetic superexchange interaction) [11, 12]. The slow decrease of T_C and M_s in Cr-doped compounds as x increases is therefore in good agreement with the gradual development of Cr^{3+}–O–Cr^{3+} and Mn^{4+}–O–Cr^{3+} antiferromagnetic interactions and the weakening of Mn^{3+}–O–Mn^{4+} and Mn^{3+}–O–Cr^{3+} ferromagnetic interactions (as Mn^{3+} is progressively replaced by Cr^{3+}). On the other hand, Ti^{4+} ion is not magnetic and does not possess any 3d electrons; the substitution of Mn by Ti causes a sudden break of ferromagnetic Mn^{3+}–O–Mn^{4+} interactions without any ferromagnetic compensation leading to a much stronger decrease of T_C and M_s than for the Cr substitution. A specific behaviour of the Ti-doped samples is the broadness of the magnetic transitions compared to the Cr-samples; this broadness increases with x leading to a non well-defined Curie temperature T_C. This particular behaviour can be linked to magnetic disorder caused by Ti substitution and chemical inhomogeneities at grain boundaries, as observed in other granular manganites [13, 14].

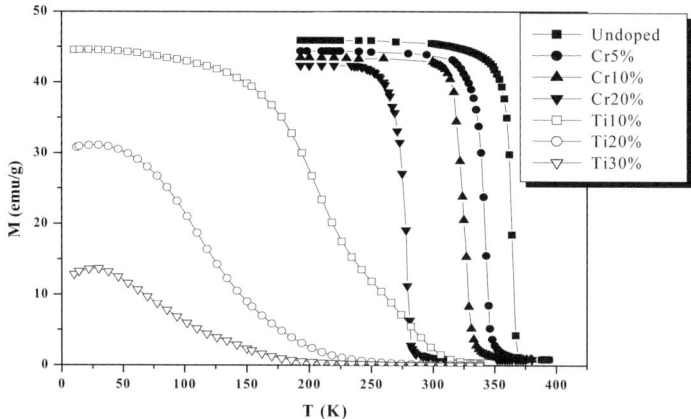

Fig. 1 Magnetization (M) vs. temperature (T) measured in a field of 500 Oe.

Figure 2 shows two micrographs achieved with backscattered electrons (BSE scanning mode) of $La_{0.7}Sr_{0.3}Mn_{0.9}Cr_{0.1}O_3$ and $La_{0.7}Sr_{0.3}Mn_{0.9}Ti_{0.1}O_3$.

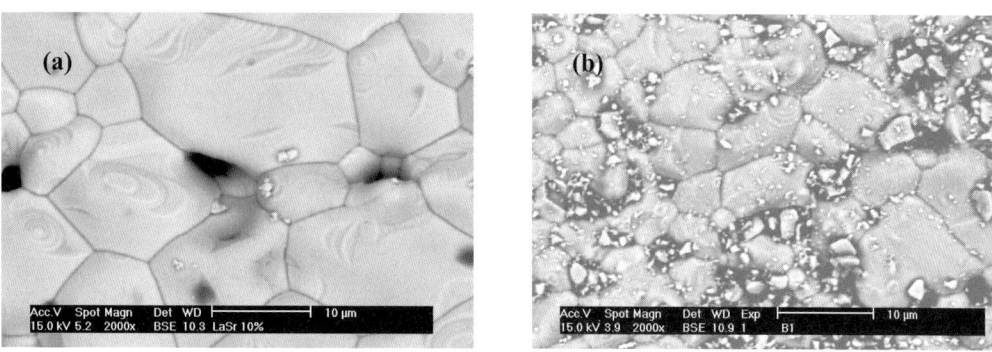

Fig. 2 Micrographs of $La_{0.7}Sr_{0.3}Mn_{0.9}Cr_{0.1}O_3$ (a) and $La_{0.7}Sr_{0.3}Mn_{0.9}Ti_{0.1}O_3$ (b).

It is clear that the grain size is bigger in Cr-doped sample which confirms that the role of grain boundaries is enhanced in the case of Ti-doped compounds leading to broader magnetic transitions.

3.2 Magnetotransport properties The magnetoresistance ratio (MR) and the resistivity curves versus temperature for $La_{0.7}Sr_{0.3}MnO_3$, $La_{0.7}Sr_{0.3}Mn_{1-x}Cr_xO_3$ (x = 0.05, 0.1, and 0.2), and $La_{0.7}Sr_{0.3}Mn_{1-x}Ti_xO_3$ (x = 0.1, 0.2, and 0.3) are shown in Fig. 3. The ratio of MR is calculated according: MR(%) = $\frac{\Delta\rho}{\rho(0)} = \frac{\rho(0T) - \rho(0.5T)}{\rho(0T)}$. As a general feature, values of resistivity increase with x whatever the substituting cation. The resistivity curves of undoped, Cr-doped (x = 0.05; 0.1) and Ti-doped (x = 0.1) samples exhibit a large peak of the metal/semiconductor transition; for the undoped sample we don't see the transition in the range of working temperature because the temperature transition is above 370 K. The temperature T_{M-SC}, corresponding to the metal/semiconductor transition, shifts to a lower values with increasing x similarly to the T_C. This electrical transition disappears for substituting ratio higher than 10% (Cr(x = 0.2) and Ti(x = 0.2, 0.3)).

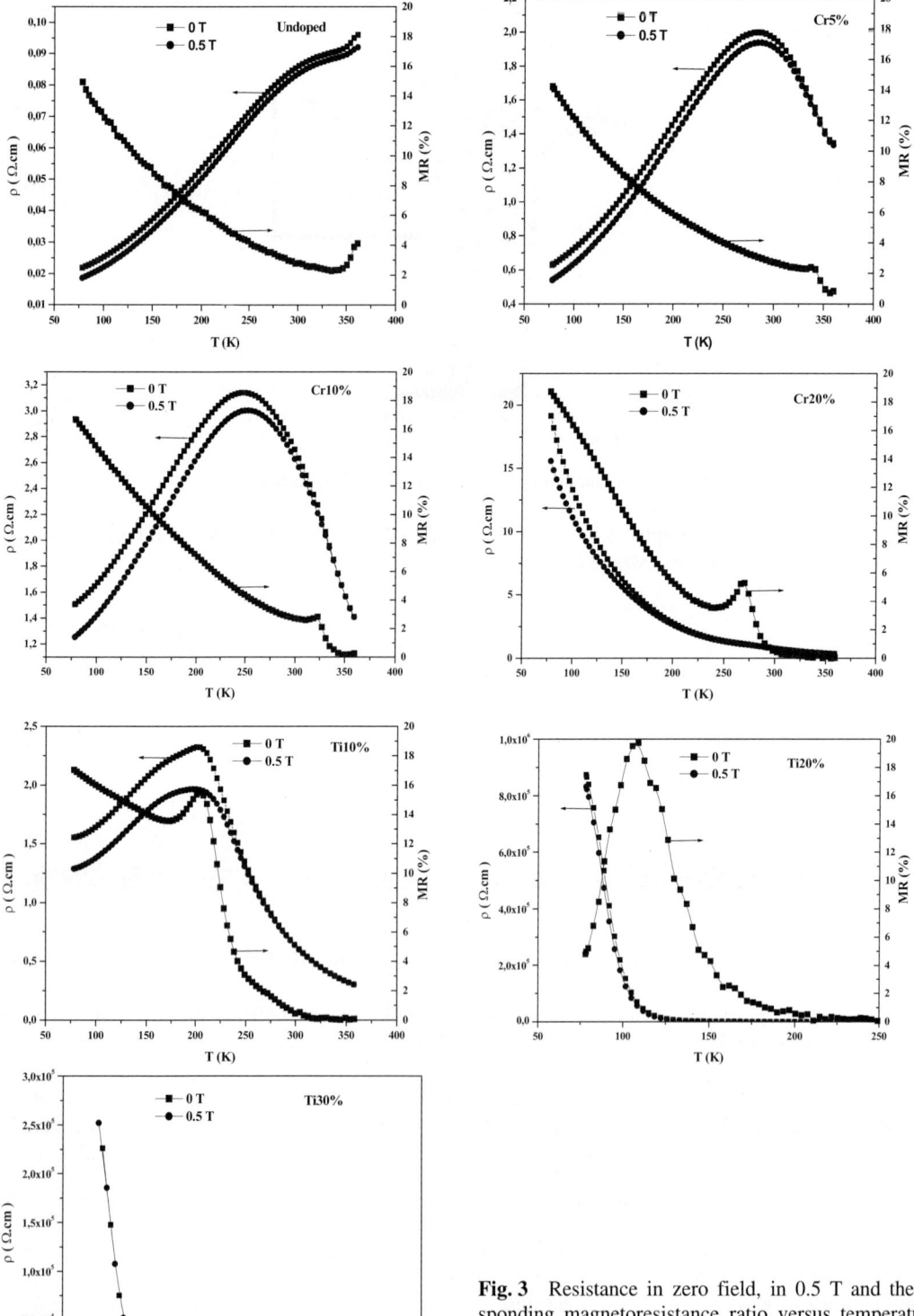

Fig. 3 Resistance in zero field, in 0.5 T and the corresponding magnetoresistance ratio versus temperature for $La_{0.7}Sr_{0.3}MnO_3$, $La_{0.7}Sr_{0.3}Mn_{1-x}Cr_xO_3$, and $La_{0.7}Sr_{0.3}Mn_{1-x}Ti_xO_3$.

This electrical transition disappears for substituting ratio higher than 10% (Cr(x = 0.2) and Ti(x = 0.2, 0.3)). In a magnetic field of 0.5 T the resistivity decreases in the whole range of temperature and the peaks of resistivity shift towards higher temperatures, resulting in a MR effect. Magnetic transition temperatures (T_C), metal/semiconductor (T_{M-SC}) transitions and MR obtained at 78 K under H = 0.5 T are resumed in Table 1.

Table 1 Temperatures of magnetic (T_C) and metal–semiconductor (T_{M-SC}) transitions and MR.

Sample	T_C (K)	T_{M-SC} (K) (H = 0 T)	MR (%) (at 78 K)
$La_{0.7}Sr_{0.3}MnO_3$	369	>360K	15
$La_{0.7}Sr_{0.3}Mn_{0.95}Cr_{0.05}O_3$	346	284	15
$La_{0.7}Sr_{0.3}Mn_{0.9}Cr_{0.1}O_3$	327	249	17
$La_{0.7}Sr_{0.3}Mn_{0.8}Cr_{0.2}O_3$	286	No transition	19
$La_{0.7}Sr_{0.3}Mn_{0.9}Ti_{0.1}O_3$	266	198	17
$La_{0.7}Sr_{0.3}Mn_{0.8}Ti_{0.2}O_3$	~170	No transition	5
$La_{0.7}Sr_{0.3}Mn_{0.7}Ti_{0.3}O_3$	~140	No transition	0

The magnetoresistance shows a local maximum for $T \approx T_C$ bound with the magnetic transition. At lower temperatures the MR effect increases in a monotonous way with decreasing temperature and reaches a value around 20% at low temperature for all samples (for Ti20% this monotonous increase of the MR is not visible because it should occur below 77 K). It is to be noticed that the sample. $La_{0.7}Sr_{0.3}Mn_{0.7}Ti_{0.3}O_3$ exhibits no MR effect. The transport properties in our doped samples are strongly dependent on the existence of DE interactions between Mn cations and Cr or Ti dopants and its coexistence with the Mn^{3+}–O–Mn^{4+} double exchange. As said before, Cr^{3+} exhibits the same electronic configuration as Mn^{4+} so that a double exchange between Cr and Mn is possible through the following dynamic charge transfer: $Mn^{3+}+Cr^{3+} \leftrightarrows Mn^{4+}+Cr^{2+}$ as evidenced in other Cr-doped compounds [15]. With increasing x, the Mn^{3+}–O–Mn^{4+} DE is progressively replaced by the Cr^{3+}–O–Mn^{3+} charge transfer mechanism; this latter is not equivalent to the former one because of the strong difference between ionic radii of Mn^{3+}/Mn^{4+} and Cr^{3+}/Cr^{2+}; this implies a strong deviation from 180° of the bond angles Mn^{3+}–O–Cr^{3+} and Mn^{4+}–O–Cr^{2+} compared to this of Mn^{3+}–O–Mn^{4+} and therefore a weakening of the charge transfer as x increases. Direct consequences are (i) the increase of the resistivity with increasing x for a given temperature and (ii) the disappearance of the electrical transition in the sample Cr(20%) because of the stabilization of the insulating phase in the ferromagnetic regime (low temperatures). When substituting Ti for Mn, the density of holes (Mn^{4+}) decreases so that parts of the Mn^{3+}–O–Mn^{4+} network are broken. As there are no e_g electrons hoping from Mn^{3+} to Ti^{4+}, the double exchange interaction is strongly weakened even for small doping values. This implies much higher resistivities than in Cr-doped samples for x ≥ 0.20 and the disappearance of MR in the Ti(30%) sample. The MR variation with the temperature is typical for polycrystalline samples presenting a local phase separation. When decreasing the temperature, MR appears with the magnetic ordering leading to a local maximum near T_C (in the case of well defined T_C). For temperatures far below T_C, the continuous increase of the MR as T decreases is probably linked to the coexistence of insulating and metallic regions as reported in other Mn-doped compounds [16]: the effect of the external applied field is very strong in the insulating regions which present a decrease of the resistivity leading to a strong MR (this can be compared to a local metallic/insulating transition caused by the external field). This effect is also present in the low-doped and even in the undoped compounds: in these cases, metallic regions dominates leading to a global metallic behaviours but insulating clusters have to be supposed to explain the monotonous increase of the MR at low temperatures.

4 Conclusion In summary, we have presented a systematic research on the magnetic, and transport properties of the perovskites $La_{0.7}Sr_{0.3}Mn_{1-x}M_xO_3$ (M = Cr, Ti). It has been shown that the substitution of the trivalent Cr ion for Mn^{3+} ion weakens the double exchange interactions. As a result T_C and T_{M-SC} decrease obviously and the resistivity increases with increasing Cr^{3+} content. Substitution by the non magnetic tetravalent Ti ion for the Mn^{4+} ion strongly reduces carrier mobility and makes magnetic transitions broader: direct consequences are the very strong increase of the resistivity and the disappearance of the MR for a substituting ratio of 30%.

References

[1] S. Jin, T. H. Tiefel, M. McCormack, R. A. Fastnacht, R. Ramesh, and L. H. Chen, Science **264**, 413 (1994).
[2] S. Jin, H. M. Bryan, T. H. Tiefel, M. McCormack, and W. W. Rhodes, Appl. Phys. Lett. **66**, 382 (1995).
[3] Y. Moritomo, A. Asamisu, H. Kuwahara, and Y. Tokura, Nature **380**, 141 (1996).
[4] W. E. Pickett and D. J. Singh, Phys. Rev. B **53**, 1146 (1996).
[5] A. J. Millis, P. B. Littlewood, and B. I. Shraiman, Phys. Rev. Lett. **74**, 5144 (1995).
[6] A. J. Millis, B. I. Shraiman, and R. Mueller, Phys. Rev. Lett. **77**, 175 (1996).
[7] K. Y. Wang, W. H. Song, B. J. Gao, J. M. Dai, S. L. Ye, S. G. Wang, J. Fang, J. L. Chen, B. J. Gao, J. J. Du, and Y. P. Sun, phys. stat. sol. (a) **188**, No. 3, 1121 (2001).
[8] N. Kallel, J. Dhahri, S. Zemni, M. Oumezzine, M. Ghedira, and H. Vincent, phys. stat. sol. (a) **184**, 319 (2001).
[9] N. Kallel, G. Dezanneau, , J. Dhahri, M. Oumezzine, H. Vincent, J. Mag. Mag. Mater. **261**/1-2, 56-65 (2003).
[10] A. Urushibara, Y. Moritomo, T. Arima, A. Asamitsu, G. Kido, and Y. Tokura, Phys. Rev. B **51**, 14103 (1995).
[11] J. B. Goodenough, A. Wold, R. J. Arnott, and N. Menyuk, Phys. Rev. **124**, 373 (1961).
[12] G. H. Jonker, Physica **22**, 707 (1956).
[13] C. Vazquez-Vazquez, M. C. Blanco, M. A. Lopez-Quintela, R. D. Sanchez, J. Rivas, and S. B. Oseroff, J. Mater. Chem. **8**, 991 (1998).
[14] N. Zhang, W. P. Ding, W. Zhong, D. Y. Xing, and Y. W. Du, Phys. Rev. B **56**, 8138 (1997).
[15] Y. Sun, W. Tong, X. Xu, and Y. Zhang, Phys. Rev. B **63**, 174438 (2001).
[16] J. Blasco, J. Garcia, J. M. de Teresa, M. R. Ibarra, J. Perez, P. A. Algarabel, and C. Marquina, Phys. Rev. B **55** (14), 8905 (1997).

Effect of Fe doping on the physical properties of LaKMn$_{1-x}$Fe$_x$MoO$_6$ double perovskite with $0 \leq x \leq 0.2$

S. Megdiche[*,1], **M. Ellouze**[1], **A. Cheikh-Rouhou**[1], and **R. Madar**[2]

[1] Laboratoire de Physique des Matériaux, Faculté des Sciences de Sfax, B. P. 802, 3018 Sfax, Tunisie
[2] Laboratoire des Matériaux et de Génie Physique, (UMR 5628 CNRS) ENSPG, B. P. 46, 38402 Saint Martin d'Hères, France

Received 31 August 2003, accepted 31 December 2003
Published online 25 March 2004

PACS 61.66.Dk, 61.72.Ww, 71.30.+h, 75.47.Pq, 75.60.Ej

Structural and magnetic properties of LaKMn$_{1-x}$Fe$_x$MoO$_6$ ($0 \leq x \leq 0.2$) double perovskite oxides have been investigated. Powder samples have been elaborated using the conventional solid-state reaction at 1100 °C. Structural studies at room temperature show that all our samples are single phase and crystallize in the orthorhombic structure with P222 space group. Magnetization measurements showed that our synthesized samples exhibit a paramagnetic to ferromagnetic transition with decreasing temperature. With increasing Fe content, the Curie temperature T_C decreases, it is found to decrease from 180 K for $x = 0$ to 110 K for $x = 0.2$. Fe doping induces also a decrease of the spontaneous magnetization M_{sp}.

© 2004 WILEY-VCH Verlag GmbH & Co. KGaA, Weinheim

1 Introduction

Transition-metal oxides with ordered double perovskite structure have begun to attract the interest of material scientists from the practical point of view as a magnetoresistance device. More than 300 compounds with double perovskite structure, A$_2$B'B"O$_6$ have been synthesized [1]. Among them A$_2$FeMoO$_6$ (A = Ca, Sr and Ba) are known to be ferrimagnetic with high critical temperature T_C [2]. Up today, many researchers [2–9] have reported the structural, magnetic and electrical properties of the double perovskite transition–metal oxides, especially for the Fe – based compounds. To our knowledge, no studies have been carried on A'A"B'B"O$_6$ where A' and A" are respectively trivalent rare earth and monovalent alkali elements. Previous studies showed that LaKMnMoO$_6$ exhibits a ferromagnetic behaviour at low temperature, the Curie temperature is found to be 180 K [10]. In this paper, we report the effect of Fe doping on the structural and magnetic properties of LaKMn$_{1-x}$Fe$_x$MoO$_6$ powder double perovskite compounds.

2 Experimental

Polycrystalline LaKMn$_{1-x}$Fe$_x$MoO$_6$ samples were prepared using the standard ceramic processing technique by mixing La$_2$O$_3$, K$_2$CO$_3$, Fe$_2$O$_3$, MnO$_2$ and MoO$_3$ up to 99.9% purity in the desired proportions according to the reaction

$$0.5\ La_2O_3 + 0.5\ K_2CO_3 + 0.5\ x\ Fe_2O_3 + (1-x)\ MnO_2 + MoO_3 \rightarrow LaKMn_{1-x}Fe_xMoO_6 + \delta\ CO_2.$$

The starting materials were intimately mixed in an agate mortar and heated in air at about 950 °C for 72 hours. A systematically annealing at high temperature is necessary to ensure a complete reaction. In fact the powder is pressed into pellets (of about 1mm thickness) and sintered at 1100 °C in air for 48

---

[*] Corresponding author: e-mail: smaiha@yahoo.fr, Phone: 00 216 74 274 923, Fax: 00 216 74 274 437

hours. Finally these pellets were rapidly quenched at room temperature in air. This step was done in order to retain the structure achieved at high temperature. In fact the magnetic properties depend strongly on the quenching method.

Phase purity, homogeneity and cell dimensions were determined by powder X – ray diffraction at room temperature using a SIEMENS diffractometer with Fe radiation ($\lambda = 1.936$ Å). The data were analyzed by CELEREF program, where the space group and the cell parameters were determined.

Magnetic measurements as a function of temperature at 500 Oe in the whole temperature range 50–300 K have been carried on using a Vibrating Sample Magnetometer (VSM). Magnetization measurements versus magnetic applied field at different temperatures have been performed using an extraction magnetometer.

3 Results and discussions

3.1 Crystallographic study

X-ray diffraction patterns at room temperature of $LaKMn_{1-x}Fe_xMoO_6$ ($0 \leq x \leq 0.2$) samples are shown in Fig. 1. Our samples are single phase and crystallize in the orthorhombic system with P222 space group. The lattice parameters a, b, c and the unit cell volume V of our samples are listed in Table 1. The unit cell volume remains constant; in fact, both Mn^{3+} and Fe^{3+} have approximately the same average ionic radius (r $[Mn^{3+}]$ = 0.580 Å and r $[Fe^{3+}]$ = 0.50 Å) [11]. This result may be also explained by the change in the oxidation states of manganese from Mn^{2+} to Mn^{3+} and molybdenum from Mo^{6+} to Mo^{5+}.

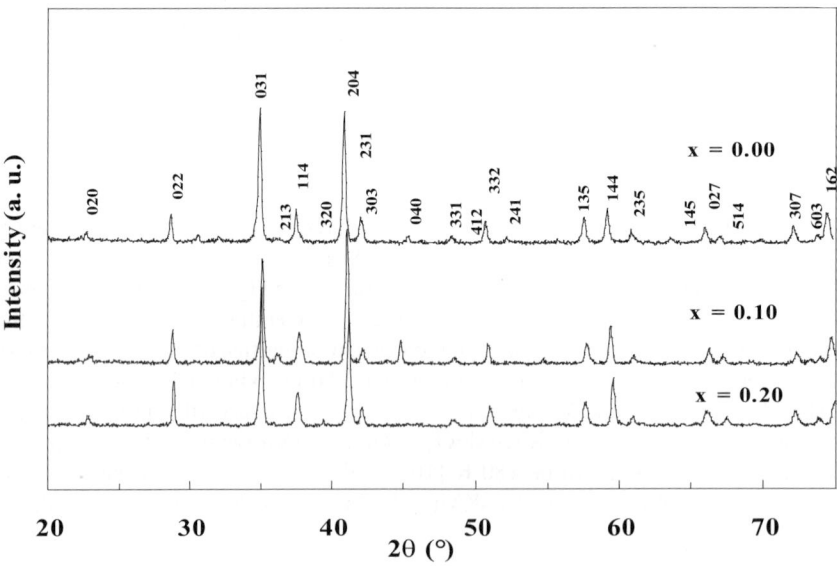

Fig. 1 The X-ray diffraction patterns at room temperature for $LaKMn_{1-x}Fe_xMoO_6$ ($0 \leq x \leq 0.2$) samples.

Table 1 Crystallographic data of $LaKMn_{1-x}Fe_xMoO_6$ ($0 \leq x \leq 0.2$) compounds.

x	Symmetry	a (Å)	b (Å)	c (Å)	V (Å³)
0.00	Orthorhombic	10.437(1)	9.972(1)	13.070(3)	1360.3
0.05	Orthorhombic	10.437(9)	9.970(1)	13.069(7)	1360.1
0.10	Orthorhombic	10.438(8)	9.965(4)	13.062(1)	1358.8
0.15	Orthorhombic	10.439(8)	9.962(4)	13.057(5)	1358.0
0.20	Orthorhombic	10.441(1)	9.959(1)	13.053(1)	1357.3

3.2 Magnetic properties

Itoh and al. [8] have reported that Sr_2MnMoO_6 show a paramagnetic behaviour. This magnetic state has been explained by the absence of the 4d electrons in the hexavalent Mo^{6+} ($4d^0$) ions [9]. Our parent compound $LaKMnMoO_6$ exhibits a ferromagnetic behaviour at low temperature [10]. This result may be probably explained by the existence of both La and K elements at the A sites. In fact, in both Sr_2MnMoO_6 and $LaKMnMoO_6$ the oxidation state of Mn and Mo are respectively the same. The Curie temperature T_C is found to be 180 K [10]. Magnetization measurements versus temperature for $LaKMn_{1-x}Fe_xMoO_6$ ($0 \leq x \leq 0.2$) samples are plotted in Fig. 2. All our synthesized samples exhibit a paramagnetic to ferromagnetic transition with decreasing temperature. The substitution of manganese by iron in small amount does not destroy the ferromagnetic behaviour observed in the parent compound at low temperature. For high iron content (x > 0.2), the substituted samples are not magnetic at low temperatures. The Fe substituted phases must change the electronic properties. Therefore change in Mn and Mo oxidation states is expected (Mn^{2+} oxidation to Mn^{3+} and Mo^{6+} reduction to Mo^{5+}).

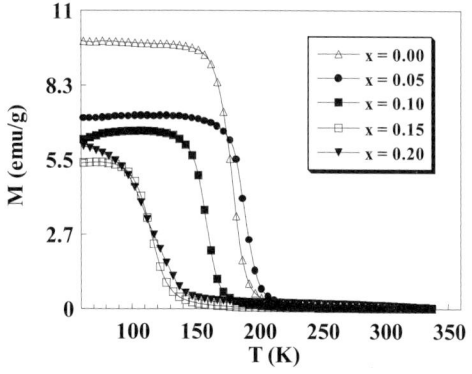 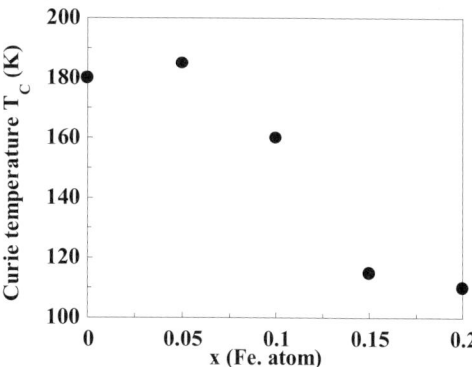

Fig. 2 Magnetization evolution versus temperatures for $LaKMn_{1-x}Fe_xMoO_6$ compounds ($0 \leq x \leq 0.2$).

Fig. 3 Curie temperature T_C evolution versus Fe content in $LaKMn_{1-x}Fe_xMoO_6$ compounds ($0 \leq x \leq 0.2$).

The Curie temperature T_C as a function of Fe content is plotted in Fig. 3. We use the maximum slope of the curve M (T) as a criterion to determine T_C. Fe doping leads to a decrease of the Curie temperature, T_C is found to decrease from 180K for x = 0 to 110 K for x = 0.2 with a small increase for x = 0.05 which can be probably explained by a disordered Mn(Fe)/Mo distribution.

To confirm the ferromagnetic behavior at low temperatures, we performed magnetization measurements versus magnetic applied field up to 7 T at several temperatures for $LaKMn_{1-x}Fe_xMoO_6$ with x = 0.05, 0.1, and 0.2 (Fig. 4-a, 4-b and 4-c, respectively). Magnetization at low temperatures (T < T_C) increases sharply with magnetic applied field for H<1T and seems to saturate for high magnetic field values. For x = 0.2, the magnetization versus magnetic applied field seems to indicate a spin canted state at low temperatures. The Fe doping leads to a weakness of the ferromagnetism at low temperatures.

We plot in Fig. 5 the spontaneous magnetization (M_{sp}) as a function of temperature for $LaKMn_{1-x}Fe_xMoO_6$ samples (x = 0.05, 0.1 and 0.2). With increasing Fe content, the spontaneous magnetization at 10 K decreases. It decreases from 3.25 μ_B/mole for x = 0 to 1.3 μ_B/mole for x = 0.2.

4 Conclusion

In this work we investigated the structural and magnetic properties of Fe doped $LaKMn_{1-x}Fe_xMoO_6$ ($0 \leq x \leq 0.2$) double perovskite powder samples. All our synthesized samples crystallize in the orthorhombic system with the P222 space group. With substitution of Mn by Fe, the unit cell volume remains constant. Our samples exhibit a paramagnetic to ferromagnetic transition with decreasing temperature. The Curie temperature T_C decreases with increasing Fe content from 180 K for x = 0 to 110 K for x = 0.2 with a small increase for x = 0.05. The spontaneous magnetization decreases with increasing iron content from 3.25 μ_B/mole for x = 0 to 1.3 μ_B/mole for x = 0.2.

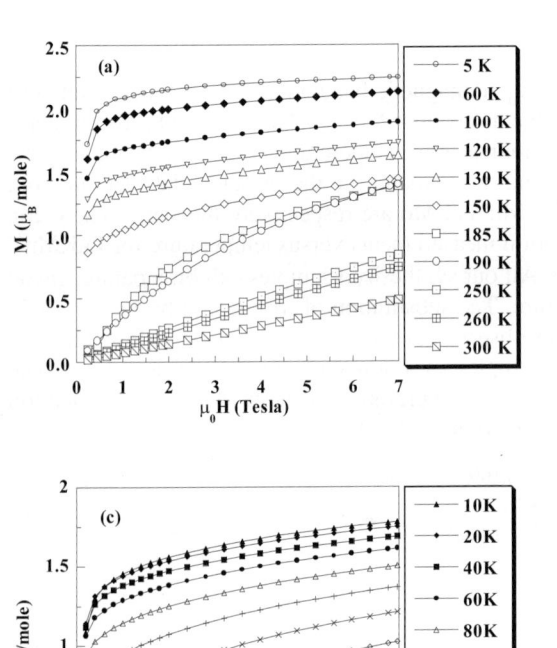
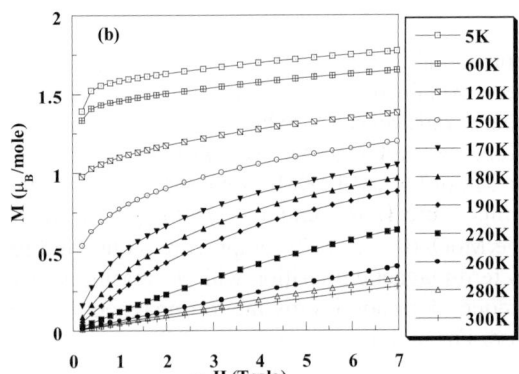
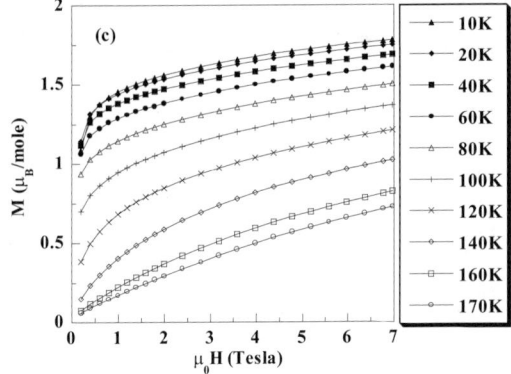

Fig. 4 Magnetization evolution versus magnetic applied field at different temperatures for LaKMn$_{1-x}$Fe$_x$MoO$_6$ compounds (x = 0.05 (a), 0.1 (b) and 0.2 (c)).

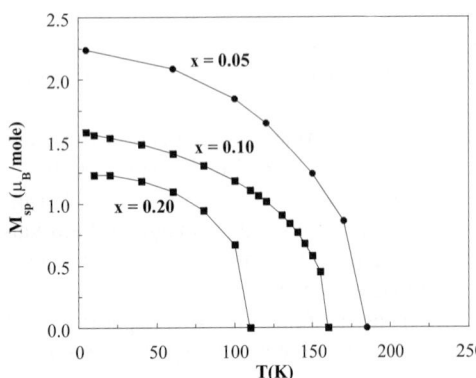

Fig. 5 The spontaneous magnetization (M$_{sp}$) as a function of temperature for LaKMn$_{1-x}$Fe$_x$MoO$_6$ samples (x = 0.05, 0.1 and 0.2).

Acknowledgements This study has been supported by the Tunisian Ministry of High Education, Scientific Research and Technology and by the CMCU collaboration (01 / F-1127).

References

[1] M. T. Anderson, K. B. Greenword, G. A. Taylor, and K. R. Peoppelmeier, Prog. Solid State Chem. **22**, 197 (1994).
[2] F. Galasso, F. C. Douglas, and R. Kasper, J. Chem. Phys. **44**, 1672 (1966).
[3] K. I. Kobayashi, T. Kimura, H. Sawada, K. Terakura, and Y. Tokura, Nature **395**, 677 (1998).
[4] K. I. Kobayashi, T. Kimura, H. Sawada, K. Terakura, and Y. Tokura, Phys. Rev. B **59**, 11159 (1999).

© 2004 WILEY-VCH Verlag GmbH & Co. KGaA, Weinheim

[5] F. K. Patterson, C. W. Moeller, and R. Wald, Inorg. Chem. **2**, 196 (1963).
[6] J. Longo and R. Wald, J. Am. Chem. Soc. **83**, 2816 (1961).
[7] S. Nakayama, T. Nakagawa, and S. Nomura, J. Phys. Soc. Jpn. **24**, 219 (1968); S. Nakayama, J. Phys. Soc. Jpn. **24**, 806 (1968).
[8] M. Itoh, I. Ohota, and Y. Inaguma, Mater. Sci. Eng. B **41**, 55 (1996).
[9] A. W. Sleigt and J. F. Weiher, J. Phys. Chem. Solids **33**, 679 (1972).
[10] S. Megdiche, M. Ellouze, A. Cheikh-rouhou, and R. Madar, J. Alloys Compd. **347**, 56 (2002).
[11] R. D. Shannon, Acta Crystallogr. B **25**, 925 (1969).

Ferromagnetic behaviour in PrKMnMoO$_6$ double perovskite oxide

S. Megdiche[*,1], **M. Ellouze**[1], **A. Cheikh-Rouhou**[1], and **R. Madar**[2]

[1] Laboratoire de Physique des Matériaux, Faculté des Sciences de Sfax, B.P. 802, 3018 Sfax, Tunisie
[2] Laboratoire des Matériaux et de Génie Physique, (UMR 5628 CNRS) ENSPG, B.P. 46, 38402 Saint Martin d'Hères, France

Received 31 August 2003, accepted 31 December 2003
Published online 25 March 2004

PACS 71.30.+h, 75.60.Ef, 81.20.Ev

PrKMnMoO$_6$ powder sample has been elaborated using the solid state technique at 1100 °C. Structural studies at room temperature show that our sample crystallizes in the orthorhombic structure with P222 space group. Magnetization measurements versus temperature of PrKMnMoO$_6$ sample show a ferromagnetic behaviour at T ≤ 80 K. The critical exponent γ, defined by $M_{sp} = M_{sp}(0)\,(1-(T/T_C))^\gamma$ and deduced from magnetization measurements versus magnetic applied field at several temperatures below T_C, is found to be 0.31.

© 2004 WILEY-VCH Verlag GmbH & Co. KGaA, Weinheim

1 Introduction

A$_2$FeMoO$_6$ and A$_2$FeReO$_6$ (A = Ca, Sr, Ba) double perovskite compounds have attracted recent attention because they may be half metals with high magnetic transition temperatures and have spin-dependent transport properties which may be useful in magnetic devices [1–3]. These compounds were discovered in the 1960s [4–6], and are members of the broad class of A$_2$B'B"O$_6$ double perovskites [7]. More generally, a survey of double perovskite A$_2$B'B"O$_6$ compounds shows that several of them are ferrimagnetic. Among them A$_2$FeMoO$_6$ (A = Ca, Sr, and Ba) are known to be ferrimagnetic with critical temperatures T_C above room temperature [8]. Sr$_2$MnMoO$_6$ and Sr$_2$CoMoO$_6$ do not show any trace of ferromagnetic transition down to 5 K [9]. However LaKMnMoO$_6$ exhibits a paramagnetic-ferromagnetic transition [10]. In this paper, we report the structural and magnetic properties of PrKMnMoO$_6$ double perovskite oxide.

2 Experimental

Polycrystalline PrKMnMoO$_6$ sample has been elaborated using the standard ceramic processing technique by mixing Pr$_6$O$_{11}$, K$_2$CO$_3$, MnO$_2$ and MoO$_3$ up to 99.9% purity in the desired proportions according to the reaction

$$1/6\,Pr_6O_{11} + 0.5\,K_2CO_3 + MnO_2 + MoO_3 \rightarrow PrKMnMoO_6 + \delta\,CO_2.$$

The precursors were mixed in an agate mortar and then sintered in air at about 950 °C for 72 hours with intermediate regrinding. A systematically annealing at high temperature is necessary to ensure a complete reaction. In fact the powder is pressed into pellets (of about 1mm thickness) and sintered at 1100 °C in air for 48 hours. Finally these pellets were rapidly quenched at room temperature in air. This step was

---

[*] Corresponding author: e-mail: smaiha@yahoo.fr, Phone: 00 216 74 274 923, Fax: 00 216 74 274 437

done in order to retain the structure achieved at high temperature. In fact the magnetic properties depend strongly on the quenching method.

Phase purity, homogeneity and cell dimensions were determined by powder X–ray diffraction at room temperature using a SIEMENS diffractometer with Fe radiation ($\lambda = 1.936$ Å). The cell parameters were obtained by least-squares calculations.

Magnetization measurements versus temperature were recorded by a vibrating sample magnetometer in the temperature range 50–300 K in an applied field of 500 Oe. Magnetization measurements versus magnetic applied field up to 7 T at different temperatures has been preformed using an extraction magnetometer.

3 Results and discussions

3.1 X-ray diffraction analysis

X-ray diffraction patterns at room temperature of $PrKMnMoO_6$ sample are shown in Fig. 1. Our sample is single phase and no impurity was detected. $PrKMnMoO_6$ sample crystallizes in the orthorhombic system with P222 space group, the same as $LaKMnMoO_6$ [10]. The lattice parameters are found to be: a = 10.4155 Å, b = 9.8541 Å, c = 13.145 Å. The unit cell volume is equal to 1349.14 Å^3. The unit cell volume of $LaKMnMoO_6$ is found to be V = 1360.32 Å [10], this result can be explained by the average ionic radius of La^{3+} which is larger than Pr^{3+} one ($r[Pr^{3+}] = 1.28$ Å and $r[La^{3+}] = 1.32$ Å) [11].

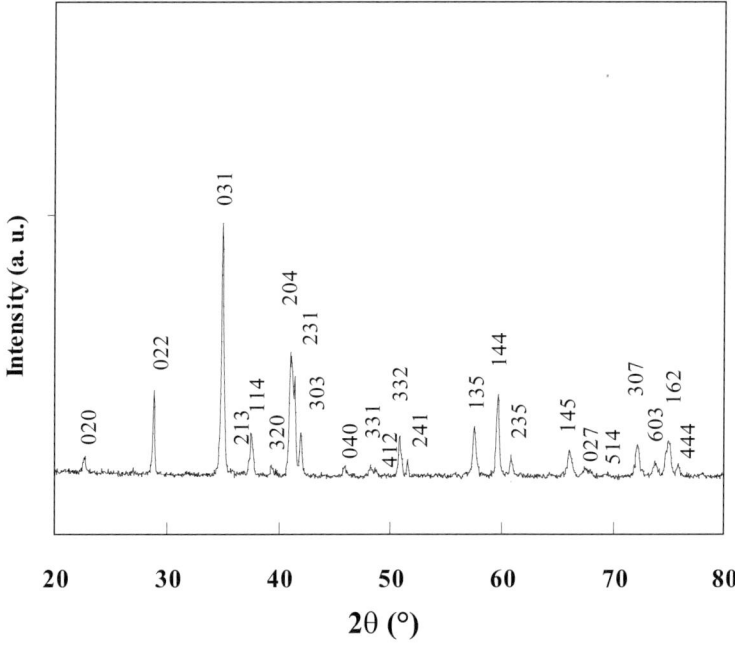

Fig. 1 The X-ray diffraction patterns at room temperature for $PrKMnMoO_6$ sample.

3.2 Magnetic properties

Figure 2 shows the temperature dependence of the magnetization for $PrKMnMoO_6$ double perovskite oxide. Our sample exhibits an ordered state at low temperatures. The transition temperature is found to be 80 K. Previous study [10] showed that $LaKMnMoO_6$ exhibits a paramagnetic to ferromagnetic transition with decreasing temperature. The Curie temperature T_C is found to be 180 K. This result can be probably explained by the average ionic radius of lanthanum which is larger than praseodymium one. It may be also explained by the praseodymium magnetic moment which leads to a weakness of the ferromagnetic component. Our result may indicate that the Curie temperature T_C depends strongly on the nature of the average ionic radius $\langle r_A \rangle$ of the A cation site.

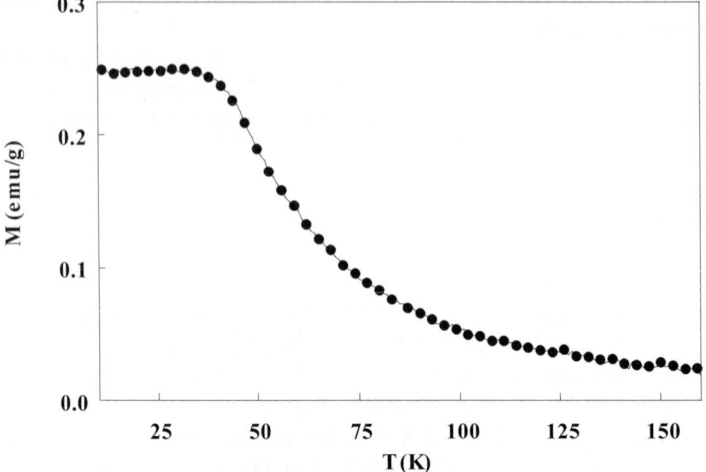

Fig. 2 Magnetization evolution versus temperature at 500 Oe for PrKMnMoO$_6$ sample.

In order to study the magnetic transition at low temperature, we performed magnetization measurements versus magnetic applied field up to 7 T at several temperatures for PrKMnMoO$_6$ sample (Fig. 3).

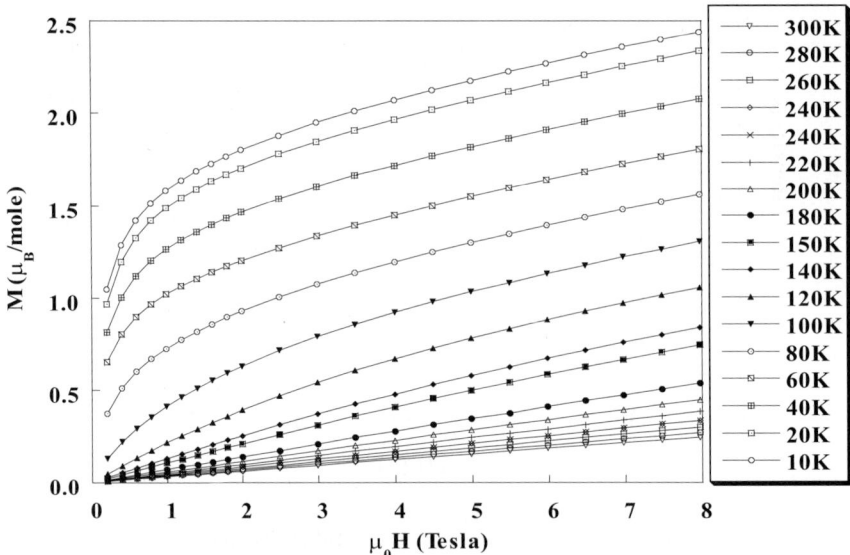

Fig. 3 Magnetization evolution versus magnetic applied field at different temperatures for PrKMnMoO$_6$ sample.

Magnetization, at low temperatures ($T < T_C$) increases sharply with magnetic applied field for $H < 1$ T but does not seem to saturate even for high magnetic applied field. Such result may indicate a spin canted state at low temperatures. The Curie temperature deduced from the Arrott curves (M^2 versus H/M) [13] (Fig. 4) is found to be 80 K.

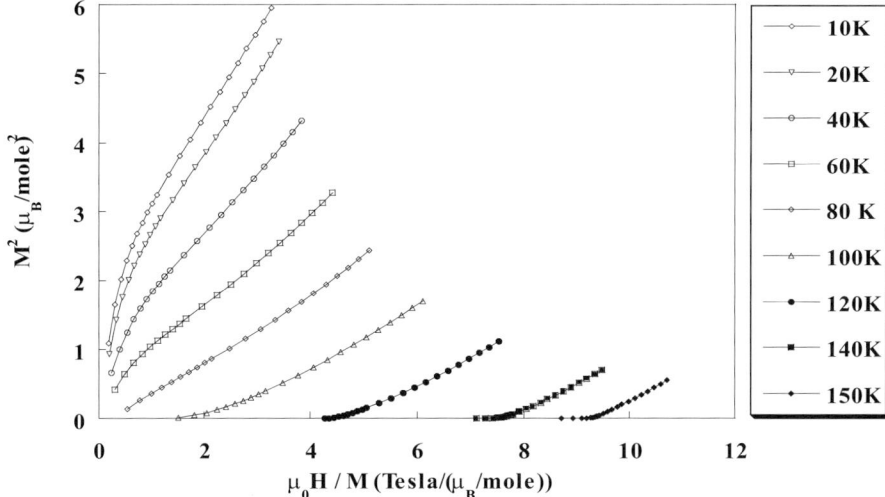

Fig. 4 Arrott curves for PrKMnMoO$_6$ sample.

The spontaneous magnetization of PrKMnMoO$_6$ is found to be 1.29 μ_B/mole at 10 K. This value is lower than that obtained in LaKMnMoO$_6$ (3.25 μ_B/mole at 10 K). The critical exponent γ defined by $M_{sp} = M_{sp}(0) \left(\dfrac{T_C - T}{T_C}\right)^{\gamma}$ is equal to 0.31 for PrKMnMoO$_6$ and 0.33 for LaKMnMoO$_6$. The difference may be explained by the praseodymium magnetic moment.

4 Conclusion

The structural and magnetic properties of PrKMnMoO$_6$ double perovskite sample have been investigated. Our synthesized sample crystallizes in the orthorhombic system with P222 space group and exhibits a paramagnetic to spin canted state transition with decreasing temperature. The transition temperature is found to be 80K, which is lower than that obtained in LaKMnMoO$_6$. The critical exponent γ deduced from spontaneous magnetization versus temperature is found to be 0.31 for PrKMnMoO$_6$.

Acknowledgements This study has been supported by the Tunisian Ministry of High Education, Scientific Research and Technology and by the CMCU collaboration (01 / F-1127).

References

[1] K. I. Kobayashi, T. Kimura, H. Sawada, K. Terakura, and Y. Tokura, Nature **395**, 677 (1998).
[2] K. I. Kobayashi, T. Kimura, H. Sawada, K. Terakura, and Y. Tokura, Phys. Rev B **59**, 11159 (1999).
[3] W. Prellier and V. Smolyaninova, J. Phys. C **12**, 965 (2000).
[4] F. K. Patterson, C. W. Moeller, and R. Wald, Inorg. Chem. **2**, 196 (1963).
[5] A. W. Sleigt and R. Ward, J. Am. Chem. Soc. **83**, 1088 (1961).
[6] T. Nakagawa, J. Phys. Soc. Jpn. **24**, 806 (1968).
[7] M. T. Anderson, K. B. Greenword, G. A. Taylor, and K. R. Peoppelmeier Prog. Solid State Chem. **22**, 197 (1993).
[8] F. Galasso, F. C. Douglas, and R. Kasper, J. Chem. Phys. **44**, 1672 (1966).
[9] Y. Moritomo, Sh. Xu, A. Machida, T. Akimoto, E. Nishibori, M. Takata, and M. Sakata, Phys. Rev. B **61**, 7827 (2000).
[10] S. Megdiche, M. Ellouze, A. Cheikh-rouhou, and R. Madar, J. Alloy. Comp. **347**, 56 (2002).
[11] R. D. Shannon, Acta crystal. B **25**, 925 (1969).
[12] A. W. Sleigt and J. F. Weiher, J. Phys. Chem. Solids **33**, 679 (1972).
[13] A. Arrott, Phys. Rev. **108**, 1394 (1957).

Study of the temperature dependence of magnetic permeability, selectivity and D.C. resistivity of LaKMnMoO$_6$

S. A. Saafan[*,1], **S. Magdiche**[3], **M. A. Elkestawy**[2], and **A. Cheikh-rouhou**[3]

[1] Physics Department, Faculty of Science, Tanta University, Tanta, Egypt
[2] Physics Department, Faculty of Education, Suez Canal University, Suez, Egypt
[3] Laboratoire de Physique des Matériaux, Faculté des Sciences de Sfax, B.P. 802, 3018 Sfax, Tunisia

Received 31 August 2003, accepted 31 December 2003
Published online 25 March 2004

PACS 71.30.+h, 75.30.Cr, 75.47.Pq

The magnetic permeability of LaKMnMoO$_6$ has been investigated using a resonance circuit. The temperature dependence of both permeability and selectivity was plotted in the temperature range 130–300 K approximately. The D.C. resistivity was also investigated and plotted as a function of temperature, the sample shows a semiconducting to metallic transition with decreasing temperature in the vicinity of T$_C$ – the Curie temperature – where a peak in the magnetic permeability plot has also occurred. The fairly high values of permeability in the whole temperature range show that LaKMnMoO$_6$ may be considered as a high-permeability material useful in many electronic applications.

© 2004 WILEY-VCH Verlag GmbH & Co. KGaA, Weinheim

1 Introduction

Our daily life and environment have become significantly dependent on materials with outstanding magnetic properties. Most of the progress achieved so far was due to the discovery of new materials. Further progress in these fields is possible if we succeed to develop materials with optimal property spectra where magnetic, electrical, mechanical, corrosive and thermal properties are simultaneously optimized for a certain application.

Recently, magnetic and electrical properties of double perovskite compounds with general formula A'A"B'B"O$_6$, where A' and A" are alkaline-earth or rare-earth elements and B' and B" are two transition metal elements, have been investigated [1–6]. Among the interesting features of those compounds are the colossal magnetoresistance and metal–insulator transition accompanied by charge and orbital ordering [2].

The polycrystalline LaKMnMoO$_6$, which belongs to this class of compounds, has been previously elaborated using the solid state reaction technique and pressed into pellets [1]. Some of the structural and magnetic properties have been studied [1]. The Curie temperature was reported to be 185K. In the present work, we present the study of the magnetic permeability, selectivity and D.C. resistivity of this compound.

2 Experimental

The previously prepared pellets [1], were ground thoroughly in an agate mortar. The resultant fine powder was repressed into a toroid-shaped sample, and sintered again at 1000 °C for 24 h. This toroid-shaped sample was used to study the magnetic permeability and the selectivity in the resonance circuit shown in Fig. 1 where the resonance frequency f is given by:

---

[*] Corresponding author: e-mail: samiasaafan@hotmail.com, Phone: + 2040 3312010 &+ 20106619326.

$$f = 1/\{2\pi[L(C+C')]^{1/2}\}; \tag{1}$$

where L is the inductance in the coil wound around the toroid-shaped sample, C is a variable capacitance connected in parallel with the coil and C' is a correction term added to account for any stray capacitances involved – but not seen – in the circuit (such as at the input of the oscilloscope). From equation (1), we can see that at each temperature value by changing C and plotting different values of $1/f^2$ as function of C, we can determine L from the slope of the resulting straight line and then calculate the magnetic permeability µ according to the relation:

$$L = (N^2A/D)\,\mu\,10^{-8}\text{ Henry} \tag{2}$$

where A is the cross sectional area of the toroid-shaped core, D is the average diameter, and N is the number of windings of the coil [7].

The selectivity S of a specific frequency is defined as the reciprocal of the quality factor Q: $S=1/Q=\Delta f/f$, where Δf is the band width $=(f_2 - f_1)$, and f_1 and f_2 are the frequencies corresponding to heights in the resonance spectrum equal to 0.7 of the height of the maximum resonance peak [8].

The D.C. resistance was roughly determined at first by a multimeter and was found to be of about 20Ω which makes it possible to use a two-probe method to study the resistivity of the sample. The measurements were carried out under vacuum to avoid the effect of humidity on the results.

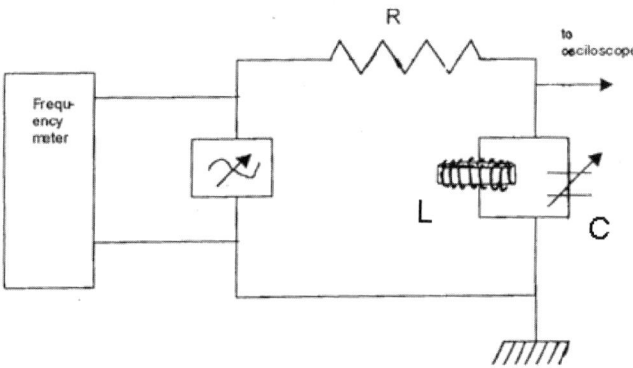

Fig. 1 The resonance circuit used to determine the permeability and the selectivity for the investigated composition.

3 Results and discussions

3.1 Magnetic permeability µ

Figure 2 shows the temperature dependence of the magnetic permeability of the double perovskite LaKMnMoO$_6$. The first interesting observation is the high values of the permeability in the whole temperature range. It is well known that recently, developing high-permeability magnetic materials is a desired goal in scientific researches [9], therefore it is quite fair to suggest that this composition should attract more interest since it may be very useful in many applications. The high values of permeability may be due to a vanishing anisotropy constant since it is known that µ is directly proportional to M^2_s/K_{eff} [8], where M_s is the saturated magnetization and K_{eff} is the effective anisotropy constant. Bearing in mind the values of magnetization previously reported for this compound [1], we should ascribe the very high values of permeability to a very small value of K_{eff}. Moreover, it may be useful to mention here that the permeability arises as the result of reversible displacements of the magnetic domain walls within the material when a high magnetocrystalline anisotropy exists and the rotation within a domain (domain rotation) contributes a little to permeability, but in the present composition the contribution to permeability seems to be mainly due to domain rotation as a consequence of the logic assumption of a vanishing anisotropy constant. Finally, it is important to mention that the permeability exhibits a peak at the Curie temperature T_c as expected [10], this finding enhances the accuracy of T_c determination in both the present work and the previously reported work [1], since it was determined by two independent and quite separate experiments. A similar increase in the permeability at T_C was reported by Belevtsev et al. [5] in $La_{0.5}Sr_{0.5}CoO_3$.

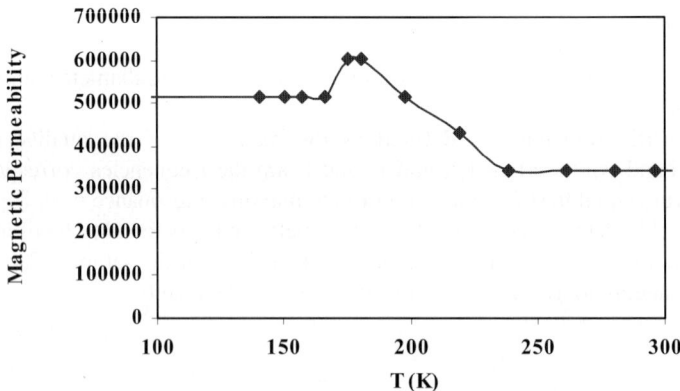

Fig. 2 Temperature dependence of permeability for the sample in the temperature range 140–303 K.

3.2 Selectivity

Fig. 3 shows the evolution of the selectivity versus temperature for a chosen frequency (1.6 MHz), it can be seen that the selectivity has at first a nearly constant value then it decreases upon cooling around T_C then it has a constant value again below 160 K.

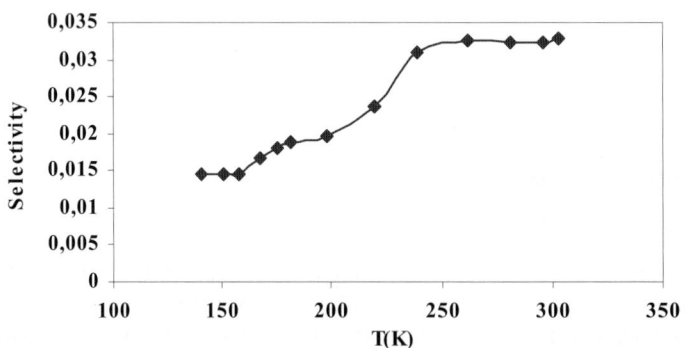

Fig. 3 Temperature dependence of selectivity at 1.6 MHz in the temperature range 140–303 K.

This can be interpreted by considering the definition of the selectivity; since it is the reciprocal of the quality factor Q: $S = 1/Q = R/X_L$; where X_L is the component inductance, R is not only the ohmic resistance of the coil but also including what is called the reflected resistance in the coil due to losses in the core. It will be discussed in the next section that above T_C the resistance of the sample increases with

© 2004 WILEY-VCH Verlag GmbH & Co. KGaA, Weinheim

decreasing temperature, therefore currents in the core will decrease, decreasing losses and consequently the reflected resistance in the coil decreases (notice that the ohmic resistance of the coil decreases also upon cooling), causing the decrease in R in the numerator. Whereas (below T_C) the resistance of the sample decreases (metallic behaviour) upon cooling resulting in a decrease in the selectivity also.

The selectivity for the chosen frequency has values in the range 0.015 to 0.033. It is known that the lower the selectivity value of a material the narrower will be the frequency bandwidth or the more useful a material will be to filter or separate out frequencies above or below a specific band [8]. The selectivity (often expressed by its reciprocal Q) becomes then a very important property usually categorized by the vendors according to frequency and inductance.

3.3 Electrical resistivity

Figure 4 shows the temperature dependence of the electrical resistivity. It is obviously seen that a significant transition occurs at T_C, where above T_C the sample exhibits a semiconduting behaviour while below T_C it exhibits a metallic one. This can be interpreted as follows: it is known that mixed valence solid solutions on the A-sublattice, or B ions in their less stable oxidation state can exhibit oxygen non-stoichiometry and/or metallic or semiconducting properties [12].

Moreover below T_C, increasing the magnetic order with decreasing temperature decreases carrier scattering and therefore results in a decrease in the resistivity [3]. For double perovskites both semiconducting and metallic behaviours were reported [2, 4]. Also, it may be useful to mention that Jonker and Van Santen in 1950 [12] had reported an insulator–metal-transition on cooling in $La_{1-x}S_xMnO_3$, and had interpreted this as due to the so called Zener double exchange where the electron retains a "memory" of its spin when it hops from one site to the next. Such hopping is therefore favored in the ferromagnetic state and this is why the composition exhibits metallic behaviour and ferromagnetism at the same time. A similar explanation may be valid for the present investigated composition.

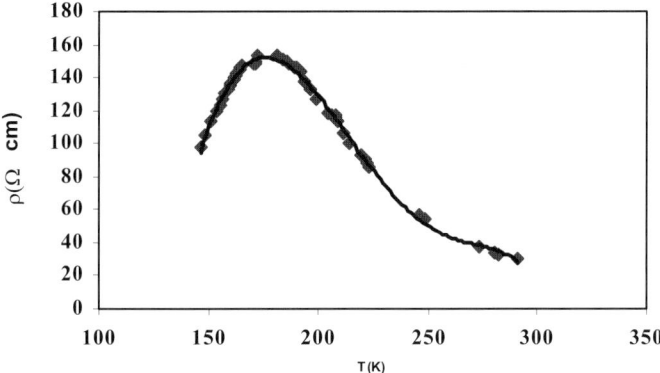

Fig. 4 Temperature dependence of electrical resistivity in the temperature range 140 K–303 K.

4 Conclusion

Both magnetic permeability and electrical resistivity measurements ensure the previously reported Curie temperature of the $LaKMnMoO_6$ double perovskite sample to be at about 180 K. A semiconductor to metallic transition is observed at T_C upon cooling.

The significantly high values of permeability and low values of selectivity should activate more investigations on this composition as a good candidate for electronic applications.

References

[1] S. Megdiche, M. Ellouse, A. Cheikh-rouhou, and R. Madar, J. Alloys Compd. **347**, 56 (2002).
[2] T. Nakajima, H. Kageyama, and Y. Ueda, Institute for Solid State Physics, University of Tokyo, Kashiwanoha, Kashiwa, Chiba, Japan, Internet Material **1–4**, 277 (2001).
[3] A. Samoilov, G. Beach, C. C. Fu, N.-C. Yeh, and R. P. Vasquez, J. Appl. Phys. **83**, 6998 (1998).
[4] R. D. Sanchez, D. Niebieskikwiat, A. Caneiro L. Morales, M. Vasquez-Mansilla, F. Rivadulla, and L. E. Hueso, J. Magn. Magn. Mater. **226–230**, 895 (2001).
[5] B. I. Belevtsev, A. Ya. Kirichenko, N. T. Cherpak, G.V. Golubnichaya, I. G.Maximchuck, A. B. Beznosov, V. B. Krasovitsky, P. P. Pal-Val, and I. N. Chukanova, J. Phys.: Condens. Matter **1**, 1 (2003).
[6] V. L. Mathe, K. K. Patankar, S. D. Lotke, P. B. Joshi, and S. A. Patil, Bull. Mater. Sci. **25**, 347 (2002).
[7] J. Smit and H. P. J. Wijn, "Ferrites" (Wiley and Sons Inc., New York, 1959).
[8] A. Goldman, "Modern Ferrite Technology" (Marcel Dekker Inc., New York, 1993).
[9] H. Kronmuller and J. M. D. Coey, "Materials: Science and Application; Magnetic Materials", Max-Plank Institut for metallforschung, Stuttgart, 92, Internet material.
[10] K. J. Standly,"Oxide Magnetic Materials" (Clarendon Press, Oxford, 1962).
[11] U. C. Berkeley, Encyclopedia of Crystal Structures, Materials Science **102** (1999), Internet material.
[12] R. Seshadri, Materials 218/UCSB: Class XVIII: Colossal magnetoresistance, Internet material.

Structural and magnetic study of the double-perovskites Ba$_2$(Fe, B')$_2$O$_6$ (B' = Mo, W and Re)

N. Rammeh[*,1,2], **K. G. Bramnik**[1], **H. Ehrenberg**[1], **C. Ritter**[3] **H. Fuess**[1], and **A. Cheikh-Rouhou**[2]

[1] Institute for Materials Sciences, University of Technology, 64287 Darmstadt, Germany.
[2] Laboratoire de Physique des Matériaux, Faculté des Sciences de Sfax, BP 802, 3018 Sfax, Tunisia
[3] Institute Laue Langevin, BP 156, 38042 Grenoble Cedex 9, France

Received 31 August 2003, accepted 31 December 2003
Published online 25 March 2004

PACS 61.66.Dk, 75.50.Bb, 75.60.Ej

Ceramics of Ba$_2$(Fe,B')$_2$O$_6$ double-perovskites have been prepared and studied for B' = Mo, W and Re. Rietveld analysis confirms that all samples crystallize in a cubic double-perovskite structure with Fm$\bar{3}$m space group. Magnetization measurements performed in the temperature range from 5 K to 350 K show a ferromagnetic behaviour for both materials Ba$_2$(Fe,Mo)$_2$O$_6$ and Ba$_2$(Fe,Re)$_2$O$_6$, with T_C = 335 K, 318 K respectively, and antiferromagnetic behaviour for Ba$_2$(Fe,W)$_2$O$_6$ with T_N = 20 K.

© 2004 WILEY-VCH Verlag GmbH & Co. KGaA, Weinheim

1 Introduction

The discovery of colossal magnetoresistance (CMR) in ternary manganese oxides with a perovskite-related structure has stimulated research on ferromagnetic metallic oxides. These materials are useful for the realization of magnetoelectronic devices and in particular for magnetic storage applications [1]. Physical properties of most of these compounds are very sensitive to even small structural changes. The double perovskites Ba$_2$FeReO$_6$, Ba$_2$FeMoO$_6$ and Ba$_3$Fe$_2$WO$_{8.42}$, synthesized for the first time almost 45 years ago [2, 3], were described as cubic with space group Fm$\bar{3}$m. More recently the Ba$_2$(Fe,W)$_2$O$_6$ structure was also reported as tetragonal with space group I4/m based on neutron powder diffraction data [4]. But the carefully performed high-resolution synchrotron diffraction analysis could confirm the initial cubic cell with a = 8.135 Å and with the presence of traces of Fe$_3$O$_4$ [5].

The disorder of Fe and Mo/Re/W on the B and B' sites of the double perovskite plays a dominant role for the structural and magnetic behaviour in these compounds. It results in the magnetic moment observed at saturation lower than the theoretical value [6]. The wide spectrum of the synthetic methods applied by different authors [5,7] hinders interpretation of variety of data concerning structure and magnetic properties of these compounds, because of the possible changing in the cation ordering and the stoichiometry.

Ba$_2$FeMoO$_6$ and Ba$_2$FeReO$_6$ were found to exhibit a similar magnetic behaviour, involving a ferrimagnetic coupling between iron and molybdenum or rhenium moments, with high Curie temperatures, which depends on the synthetic methods and the inhomogeneities involving Fe/(Mo,Re) site disorder [5, 7]. In contrast, there are discrepancies in the description of the magnetic properties of the tungsten-containing compound. A weak ferromagnetic behaviour was reported for the Ba$_3$Fe$_2$WO$_{8.42}$ [3], but recently antiferromagnetic ordering was reported for Ba$_2$(Fe,W)$_2$O$_6$ [5].

---

[*] Corresponding author: e-mail: nizar.rammeh@fss.rnu.tn, Phone: +00 49 6151 16 6003 999, Fax: +00 49 6151 16 6377

The present paper describes the detailed crystal and magnetic study of $Ba_2FeB'O_6$ with B' = Mo, W and Re.

2 Experimental

Polycrystalline samples of $Ba_2(Fe,B')_2O_6$ have been synthesized by the solid-state ceramic method: Firstly, the precursor oxides of composition $_{Ba2B'O5}$ (B' = Mo, W) were prepared by the reaction of stoichiometric amounts of $BaCO_3$ and MoO_3 or WO_3. The precursor $Ba_5Re_2O_{12}$ was synthesised by the following technique: $Ba(NO_3)_2$ was dissolved in distilled water, then a stoichiometric amount of NH_4ReO_4 was dissolved in the solution. After slow evaporation (at 100 °C), the dry residue was heated at 600 °C until gas elimination stopped, ground in an agate mortar and pressed into pellets. Annealing of the obtained products at 1000 °C for 24 hours in air resulted in pure $Ba_5Re_2O_{12}$. Secondly, the resultant oxides were mixed with the required quantities of $Fe_{0.95}O$, Fe_2O_3 (STREM Chemicals, 99.9%) and metallic Re (for B' = Re) to obtain the desired composition $Ba_2(Fe,B')_2O_6$ (B' = Mo, W and Re). Finally, pellets of these mixtures were heated in evacuated and sealed silica tubes (10–3 mbar) at 1223 K for 24 h with intermediate grinding. The formation of the $Ba_2(Fe,B')_2O_6$ phases was identified by X-ray powder diffraction using a STOE STADI P diffractometer with MoKα1 radiation. The crystal structure of Ba_2FeMoO_6 [7] was used as the starting model for structure refinement with the Rietveld program FULLPROF [8]. DC magnetization measurements on $Ba_2(Fe,B')_2O_6$ were performed using a Quantum Design superconducting Quantum Interference Device (SQUID) magnetometer. Magnetization versus temperature curves have been measured between 5 and 350 K in field cooled mode (FC), and field scans up to H = ± 5.5 T were recorded for different temperatures.

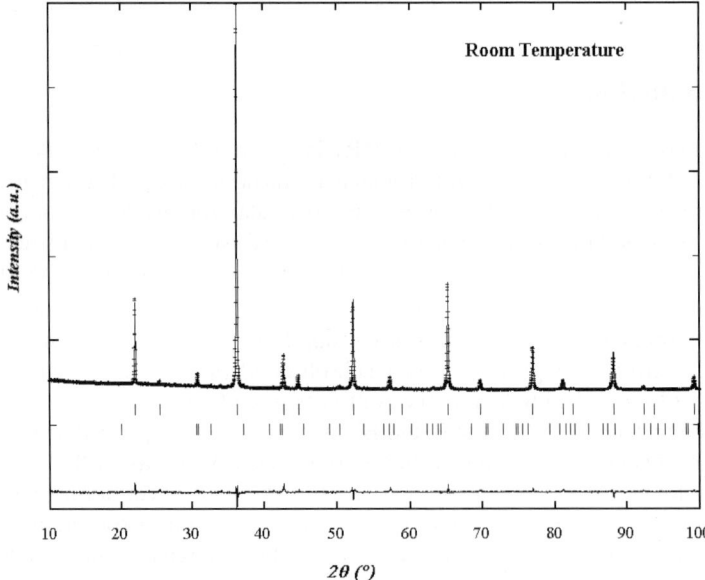

Fig. 1 Calculated, observed, and difference X-ray diffraction patterns for $Ba_2(Fe,W)_2O_6$. Additional lines of reflections belong to $BaWO_4$.

3 Results and discussion

3.1 Structural characterization

The X-ray powder diffraction patterns show the presence of $Ba_2(Fe,Re)_2O_6$ and $Ba_2(Fe,B')_2O_6$ with B' = Mo, W as the dominant phase with about 3% of $BaB'O_4$ as admixture, in accordance with previous reports [9]. These patterns could be completely indexed on the base F-centered cubic cell (space group $Fm\bar{3}m$ (No. 225)) and are in an agreement with previous reports [2, 4].

For the refinements of $Ba_2(Fe,Mo)_2O_6$, $Ba_2(Fe,Re)_2O_6$ and $Ba_2(Fe,W)_2O_6$ structures we first assumed an ordered distribution of Fe and B' (B' = Mo, Re and W) in the B-subcell of the perovskite structure. According to the symmetry analysis performed by Woodward [10], the ordering of the B cations in the $a = 2a_{per}$ (a_{per} is the unit cell parameter of an ideal cubic perovskite) face-centered unit cell corresponds to the absence of a tilting distortion of the perovskite framework ($a^0a^0a^0$ tilt system in Glazer's notations [11], space group $Fm\bar{3}m$). The agreement between observed and calculated X-ray powder diffraction pattern (e.g. $Ba_2(Fe,W)_2O_6$) is shown in Fig. 1. The bond distances of Fe–O in the FeO_6 octahedra are comparable with the distance of 2.08 Å calculated from the ionic radii according the Nguyen [7]. Similarly, the mean distances of W–O, Mo–O and Re–O in the $B'O_6$ octahedron are comparable with values of Shannon's radii [12].

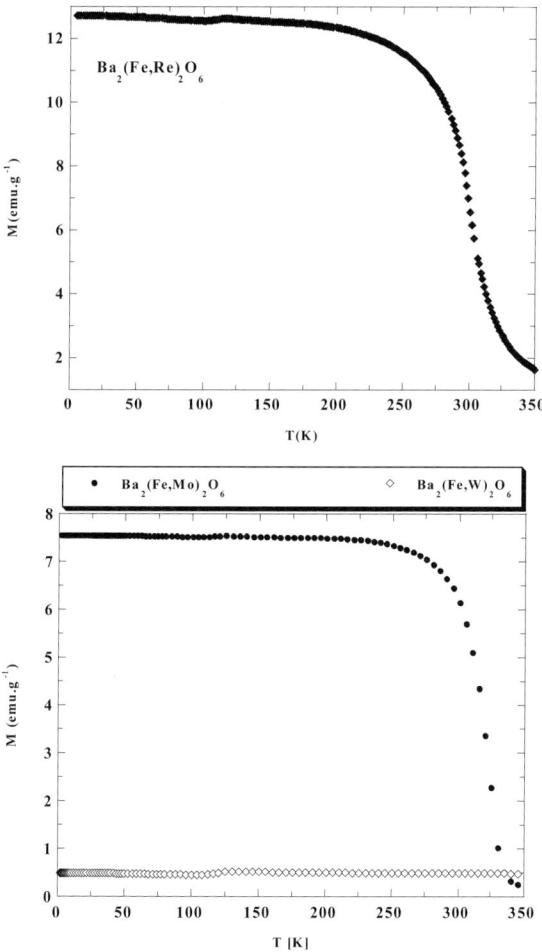

Fig. 2 Magnetization versus temperature curves at 0.2 T for $Ba_2(Fe,Re)_2O_6$ and 0.05 T for $Ba_2(Fe,W)_2O_6$ and $Ba_2(Fe,Mo)_2O_6$

3.2 Magnetic properties

The evolution of magnetization versus temperature with an applied field of 0.05 T for the $Ba_2FeB'O_6$ (B' = Mo and W) and 0.2 T for the $Ba_2(Fe,Re)_2O_6$ system is plotted in Fig. 2. The samples corresponding to

B' = Mo and Re exhibit a ferromagnetic to paramagnetic transition at T_C = 335 K and 318 K respectively, while the sample with B' = W with Fe^{2+} in the high–spin state (S = 2), as established by Mössbauer spectroscopy [5], and with the diamagnetic W^{6+} ion exhibits antiferromagnetic behaviour.

In order to study the magnetic behaviour at low temperatures, magnetization measurements versus applied magnetic field at several temperatures were carried out. The field dependence of magnetization was measured at different temperatures over the magnetic field range of –6T ≤ H ≤ 6T. The magnetic study at T = 10 K for $Ba_2(Fe,W)_2O_6$ reveals the existence of a weak ferromagnetic component which may be ascribed to the presence of a minor impurity of Fe_3O_4 as detected by synchrotron diffraction [5]. This assumption was supported by the neutron powder diffraction study performed, which confirms the antiferromagnetic ordering of $Ba_2(Fe,W)O_6$ with T_N = 20 K (Fig. 3).

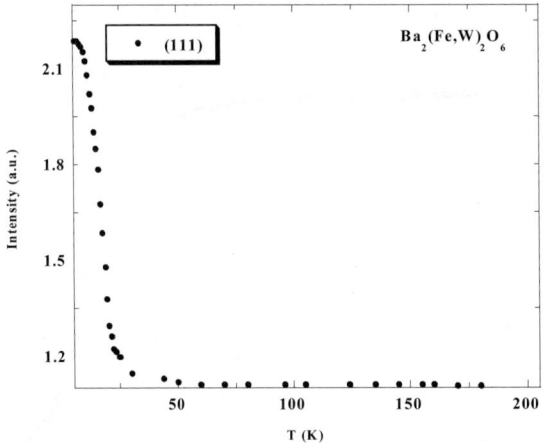

Fig. 3 Thermal evolution of the integrated intensity for the magnetic (111) reflexion for $Ba_2(Fe,W)_2O_6$.

The field dependence of magnetization for the other compounds (B' = Mo and Re) at different temperatures are shown in Figs. 4 and 5, respectively. The magnetization M at low temperature increases sharply with the applied magnetic field, and spontaneous magnetization increases with decreasing temperature. Fig. 6 shows the hysteresis loops measured at T = 10, 100 and 250 K for $Ba_2(Fe,Re)_2O_6$.

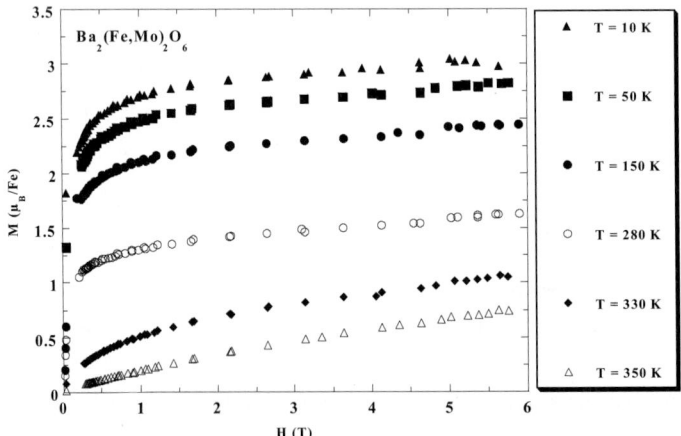

Fig. 4 Evolution of magnetization versus applied magnetic field for $Ba_2(Fe,Mo)_2O_6$.

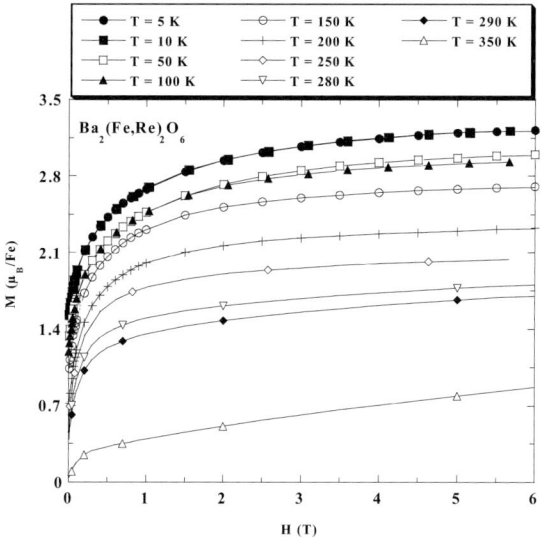

Fig. 5 Evolution of magnetization versus applied magnetic field for $Ba_2(Fe,Re)_2O_6$.

The remnant magnetization and coercive field falls rapidly as the temperature is increased. The spontaneous magnetization $M_{sp}(T)$ shows a Brillouin-type behaviour as plotted in Fig. 7. It confirms the ferrimagnetic or ferromagnetic behaviour for both samples B' = Mo and Re where the critical exponent γ defined as $M_{sp} = M_{sp(0)}(1-T/T_C)^\gamma$ is equal to 0.33 and 0.32, respectively. The value of the magnetization per iron ion at 6 T for Ba_2FeReO_6 is $3.02\mu_B$, which is close to the expected value based on Fe^{3+} moment ($3\ \mu_B$) for a ferrimagnetic state. In the same way, the high-field magnetization value of $3\mu_B$ was measured for $Ba_2(Fe,Mo)_2O_6$. This value is smaller than the theoretical value for a ferrimagnetic state ($M_{th} = 2(S_{Fe} - S_{Mo}) = 4\mu_B$). The difference may be explained by canting of the spin direction. Thus, these values are consistent with the ferrimagnetic type of ordering like for several other double-perovskite compounds [13, 14].

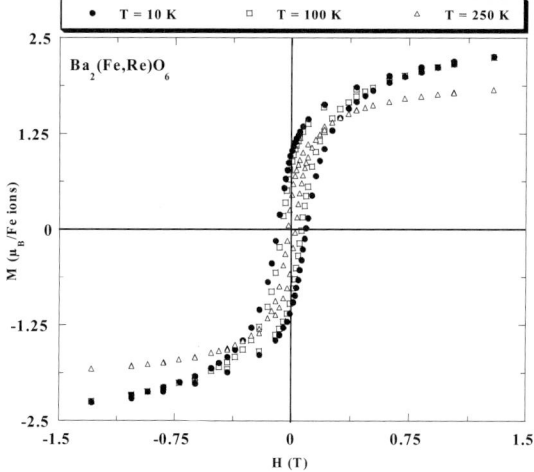

Fig. 6 Hysteresis loops of $Ba_2(Fe,Re)_2O_6$.

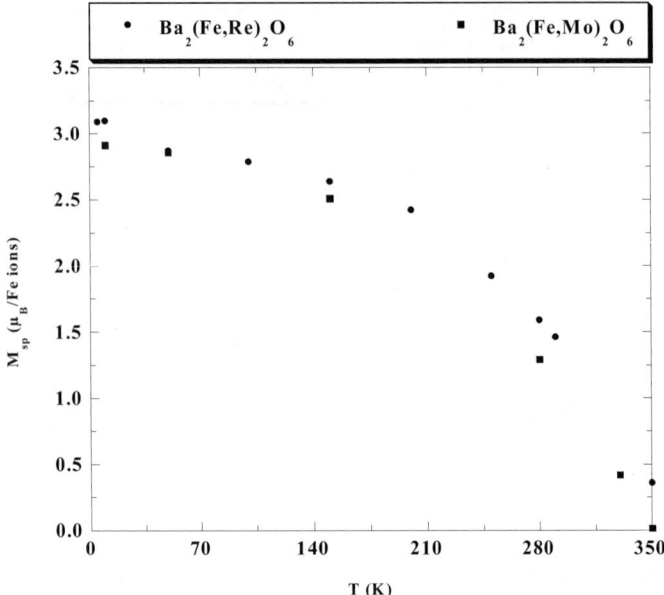

Fig. 7 Temperature dependence of the spontaneous magnetization of $Ba_2(Fe,Mo)_2O_6$ and $Ba_2(Fe,Re)_2O_6$.

Acknowledgements One of us (N. Rammeh) thanks the Deutscher Akademischer Austausch-Dienst (DAAD) for support in enabling the visit to Darmstadt (Germany). The authors are grateful for the support by a WTZ cooperation program of the Bundeministerium für Bildung und Forschung. The constant support from the Fonds der Chemischen Industrie is gratefully acknowledged. This study has been supported by the Tunisian Ministry of Scientific Research and Technology.

References

[1] N. Mott, "Metal–Insulator Transitions." (Taylor & Francis, London, 1990).
[2] F. K. Patterson, C. W. Moeller, and R. Ward, Inorg. Chem. **2**, 196 (1962).
[3] G. Matzen and P. Poix, J. Solid State Chem. **33**, 341 (1980).
[4] A. K. Azad, S.-G. Eriksson, A. Mellergard, S. A. Ivanov, J. Eriksen, and H. Rundlöf, Mater. Res. Bull. **37**, 1797 (2002).
[5] N. Rammeh, K. G. Bramnik, H. Ehrenberg, B. Stahl, H. Fuess, and A. Cheikh-Rouhou, J. Alloys Compd. **363**, 24 (2004).
[6] A. S. Ogale, S. B. Ogale, R. Ramesh, and T. Venkatesan, Appl. Phys. Lett. **75**, 573 (1999).
[7] N. Nguyen, F. Sriti, C. Martin, F. Bourée, J. M. Grenèche, A. Ducouret, F. Studer, and B. Raveau, J. Phys.: Condens. Matter **14**, 12629 (2002).
[8] T. Roisnel and J. Rodriguez-Carvajal, FULLPROF, version LLB-LCSIM, May 2000, France.
[9] T. Nakamura and Y. Gohshi, Chem. Lett. 1171 (1975).
[10] P. M. Woodward, Acta Cryst. B **53**, 32 (1997).
[11] A. M. Glazer, Acta Cryst. B **28**, 3384 (1972).
[12] R. D. Shannon, Acta Cryst. A **32**, 751 (1976).
[13] B. García-Landa, C. Ritter, M. R. Ibarra, J. Blasco, P. A. Algarabel, R. Mahendiran, and J. García, Solid State Commun. **110**, 435 (1999).
[14] J. Longo and R. Ward, J. Am. Chem. Soc. **83**, 2816 (1961).

Effect of praseodymium doping on the structural and magnetic properties of $La_{1.2-x}Pr_xSr_{1.8}Mn_2O_7$ bilayer manganese oxides

M. Triki, S. Zouari, N. Chniba, and A. Cheikhrouhou*

Laboratoire de Physique des Matériaux, Faculté des Sciences de Sfax, B. P. 802, 3018 Sfax, Tunisie

Received 31 August 2003, accepted 31 December 2003
Published online 25 March 2004

PACS 61.66.Dk, 67.72.Ww, 74.25.Ha, 75.30.Kz, 75.60.Ej

Structural and magnetic properties of $La_{1.2-x}Pr_xSr_{1.8}Mn_2O_7$ ($0 \leq x \leq 1.2$) bilayers manganese oxides have been investigated. Structural characterizations showed that all our synthesized samples crystallize in the tetragonal structure with I4/mmm space group. Magnetization measurements versus temperature showed that all of our samples exhibit a paramagnetic to ferromagnetic transition with decreasing temperature. The Curie temperature T_c decreases with increasing praseodymium content. It is found to decrease from 315 K for $x = 0$ to 265 K for $x = 1.2$.

© 2004 WILEY-VCH Verlag GmbH & Co. KGaA, Weinheim

1 Introduction

Ruddlesden–Popper manganese oxides with general formula $R_{2-x}M_{1+x}Mn_2O_7$ where R is a trivalent rare earth element and M is a divalent element have attracted extensive research during last few years due to the colossal magnetoresistance effect that they exhibit [1–6].

Magnetic and electrical properties in the ABO_3 type perovskite manganites, $R_{1-x}M_xMnO_3$, where R is a trivalent rare-earth element and M is a divalent or monovalent element are generally understood using the double exchange interaction mechanism [7, 8] which involves the ferromagnetic coupling between Mn^{3+} ($t_{2g}^3 e_g^1$) and Mn^{4+} (t_{2g}^3) spins [8–10].

Recently, the colossal magnetoresistance effect has been observed in the bilayered oxide $R_{2-x}M_{1+x}Mn_2O_7$ (n = 2 in the Ruddlesden–Popper series phase $A_{n+1}B_nO_{3n+1}$) [11–13]. However the properties are very different compared to the ABO_3 type oxides. With moving from perovskite-type ($n = \infty$) to the bilayered (n = 2) in the Ruddlesden–Popper phases, we obtain a two dimensional character and a reduction in the number of nearest neighbor Mn cations around a particular transition metal site which produces an anisotropic reduction in the width of the energy band derived from the 3d Mn orbitals. Such reduction modifies the physical properties of these materials [14–22].

In this paper, we report the study of the praseodymium doping effect on the structural and magnetic properties in $La_{1.2-x}Pr_xSr_{1.8}Mn_2O_7$ powder samples.

2 Experimental

$La_{1.2-x}Pr_xSr_{1.8}Mn_2O_7$ ($0 \leq x \leq 1.2$) powder samples were prepared using the conventional solid state reaction by mixing Pr_6O_{11}, La_2O_3, MnO_2 and $SrCO_3$ up to 99.9 % purity in the desired proportions.

The starting materials were intimately mixed in an agate mortar and heated then in air at 1000 °C for 48 hours. A systematically annealing at high temperature is necessary to ensure a complete reaction. In fact the powders are pressed into pellets (of about 1 mm thickness) and sintered at 1400 °C in air for 48 hours with intermediate regrinding and re-pelletizing. Finally these pellets were rapidly quenched to room temperature in air. This step was taken in order to freeze the structure at the annealed temperature.

---

* Corresponding author: e-mail: Abdel.Cheikhrouhou@fss.rnu.tn, Phone: +21674276400, Fax: +21674274437

Phase purity, homogeneity, and cell dimensions were determined by powder X-ray diffraction at room temperature (diffractometer using CuK$_\alpha$ radiation). Unit cell dimensions were obtained by least-squares calculations.

Magnetization measurements versus temperature were recorded using a vibrating sample magnetometer (VSM) in the temperature range 10 K–300 K and a magnetic applied field of 0.05 T. Magnetization measurements versus magnetic applied field up to 6 T were performed using an extraction magnetometer.

3 Results and discussion

3.1 X-ray diffraction analysis

X-ray diffraction patterns at room temperature for La$_{1.2-x}$Pr$_x$Sr$_{1.8}$Mn$_2$O$_7$ (x = 0; 0.6 and 1.2) are shown in Fig. 1. All our samples crystallize in the tetragonal structure with I4/mmm space group.

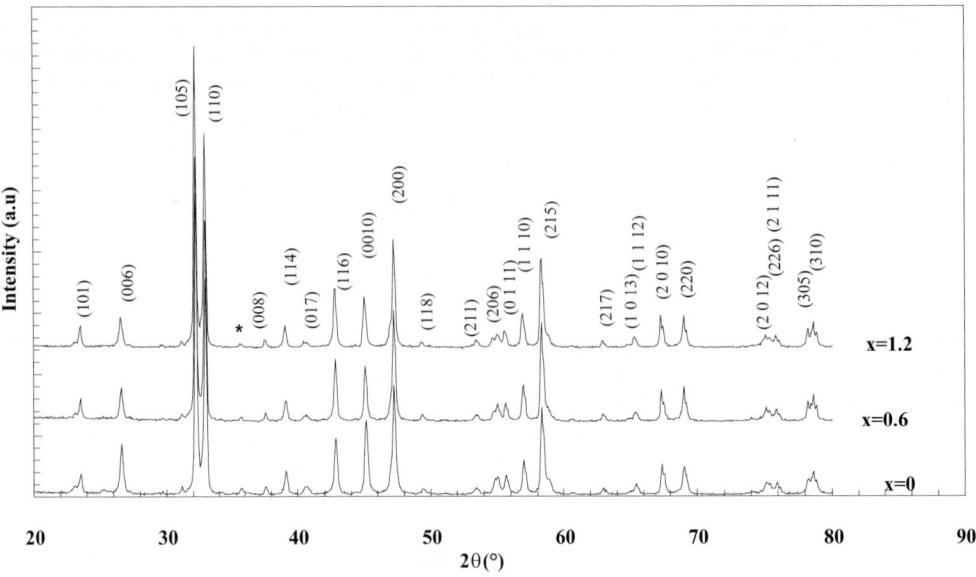

Fig. 1 X-ray diffraction patterns for La$_{1.2-x}$Pr$_x$Sr$_{1.8}$Mn$_2$O$_7$ samples.

As shown in Fig. 2, with increasing praseodymium content the a lattice parameter decreases and the c lattice parameters increases leading to a decrease in the unit cell volume (Fig. 3). This effect can be explained by the average ionic radius of lanthanum which is larger than the praseodymium one. Our crystallographic results are in agreement with previous work [12].

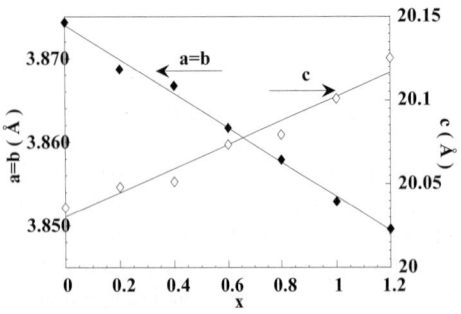

Fig. 2 Evolution of the lattice parameters versus x for La$_{1.2-x}$Pr$_x$Sr$_{1.8}$Mn$_2$O$_7$.

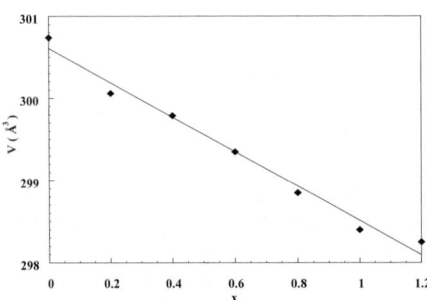

Fig. 3 Unit cell volume evolution versus x for La$_{1.2-x}$Pr$_x$Sr$_{1.8}$Mn$_2$O$_7$ samples.

3.2 Magnetic properties

We report in Fig. 4 the temperature dependence of the magnetization in a magnetic applied field of 0.05 T for $La_{1.2-x}Pr_xSr_{1.8}Mn_2O_7$ ($0 \leq x \leq 1.2$). Our synthesized samples exhibit a paramagnetic to ferromagnetic transition with decreasing temperature. The Curie temperature T_c decreases with increasing praseodymium content from 315 K for $x = 0$ to 265 K for $x = 1.2$ (Fig. 5). Mitchell et al. reported that the Jahn–Teller distortion of the MnO_6 octahedral in the quasi two dimensional $La_{1.2}Sr_{1.8}Mn_2O_7$ material enhances as charges are delocalized [23]. This suggests that the substitution of La^{3+} with large ionic radius by Pr^{3+} with small one will increase the internal chemical pressure within the compound and may lead to a decrease in T_c. Therefore, the variation of the local bond length around the Mn ions with increasing Pr amount may control the magnetic properties [24]. The T_c decrease with increasing Pr^{3+} content can be also explained by the antiferromagnetic coupling between praseodymium and manganese magnetic moments.

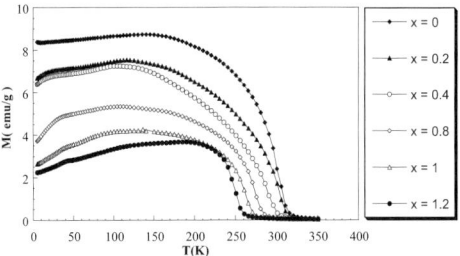
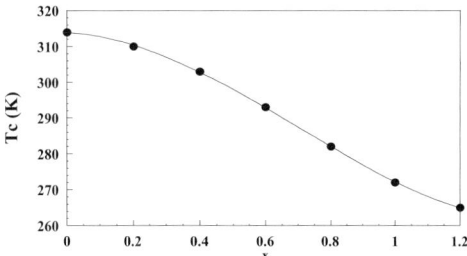

Fig. 4 Temperature dependence of magnetization for $La_{1.2-x}Pr_xSr_{1.8}Mn_2O_7$ measured at a field of 0.05 T.

Fig. 5 Curie temperature as a function of x for $La_{1.2-x}Pr_xSr_{1.8}Mn_2O_7$ samples.

Previous studies on $RE_{1.2}Sr_{1.8}Mn_2O_7$ (RE = La, Pr, Nd) [12] showed that the Curie temperature T_c in $La_{1.2}Sr_{1.8}Mn_2O_7$ is found to be 150 K. The difference may be explained by the elaborating method. We report in Fig. 6 the evolution of the magnetization versus magnetic applied field up to 6 T at several temperatures for $La_{0.2}PrSr_{1.8}Mn_2O_7$. The same behavior has been observed in all our synthesized samples. The magnetization does not saturate even for magnetic applied field up to 6 T. Such results may indicate that our samples exhibit a spin canted state at low temperatures. The maximum value of the magnetic moment at 5 K and in a magnetic applied field of 6 T is found to be $3\mu_B / f.u$, however the theoretical value of the magnetization is expected to be $6.4\mu_B / f.u$ for full spin alignment.

To study the effect of praseodymium in our samples, we plot in Fig. 7 the evolution of the magnetization versus magnetic applied field at 5 K for $x = 0$; 0.2 and 1.2. Compared to the parent compound, the magnetization in the praseodymium doped samples is smaller for low field values and larger for high field values. This result may be explained by the antiferromagnetic coupling between praseodymium and manganese moments at low field that becomes ferromagnetic with increasing magnetic applied field. Such results have been observed in the praseodymium perovskite manganese oxides $Pr_{0.7}Sr_{0.3}MnO_3$ compared to $La_{0.7}Sr_{0.3}MnO_3$ [25].

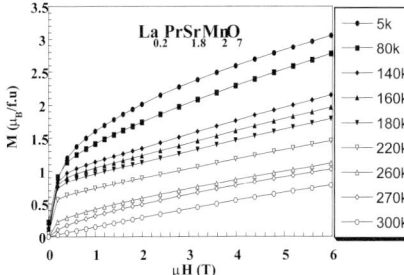
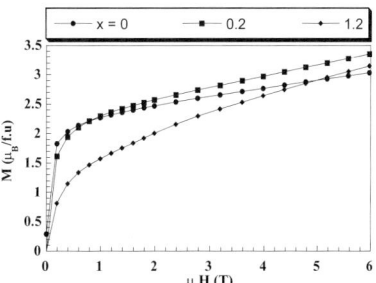

Fig. 6 Field dependence of magnetization for $La_{0.2}PrSr_{1.8}Mn_2O_7$ sample.

Fig. 7 Field dependence of magnetization at 5 K for $La_{1.2-x}Pr_xSr_{1.8}Mn_2O_7$ samples ($x = 0$; 0.2 and 1.2).

4 Conclusion We elaborate and investigate the structural and magnetic properties of $La_{1.2-x}Pr_xSr_{1.8}Mn_2O_7$ powder samples in the composition range $0 \leq x \leq 1.2$. Our synthesized samples crystallize in the tetragonal structure with I4/mmm space group. All synthesized samples exhibit a paramagnetic to ferromagnetic transition with the decreasing temperature. The Curie temperature decreases with increasing praseodymium content. Magnetization measurements versus magnetic applied field show that our samples exhibit a spin canted state at low temperatures.

Acknowledgements This work has been supported by the Tunisian Ministry of High Education, Scientific Research and Technology. Magnetization measurements have been performed in Laboratoire de Magnétisme Louis Neel, Grenoble.

References

[1] C. N. R. Rao, A. K. Cheetham, and R. Mahesh, Chem. Mater. **8**, 2421 (1996).
[2] R. Von Helmolt, J. Wecher, B. Holzapfel, L. Schultz, and K. Samwer, Phys. Rev. Lett. **71**, 2331 (1994).
[3] P. M. Levy, Science. **256**, 972 (1992).
[4] Y. Tokura, A. Urushibara, Y. Moritomo, T. Arima, A. Asamitsu, G. Kido, and N. Furukawa, J. Phys. Soc. Jpn. **63**, 3931 (1994).
[5] S. Jin, T. H. Tiefel, M. Mc. Cormack, R. A. Fastncht, R. Ramesh, and L. H.Chen, Science **264**, 413 (1994).
[6] Y. Moritomo, A. Asamitsu, H. Kuwahara, and Y. Tokura, Nature **380**, 141 (1996).
[7] P.-G. de Gennes, Phys. Rev. **118**, 141 (1960).
[8] C. Zener, Phys. Rev. **82**, 403 (1951).
[9] J. B. Goodenough, Phys. Rev. **100**, 564 (1955).
[10] P. W. Anderson and H. Hasegawa, Phys. Rev. **100**, 675 (1955).
[11] P. D. Battle, S. J. Blundell, M. A. Green, W. Hayes, M. Honold, A. K. Klehe, N. S. Laskey, J. E. Millburn, L. Murphy, M. J. Rosseinsky, N. A. Samarin, J. Singleton, N. E. Sluchanko, S. P. Sullivan, and J. F. Vente, J. Phys.: Condens. Matter **8**, L427 (1996).
[12] R. Seshardi, C. Martin, A. Maignan, M. Hervieu, B. Raveau, and C. N. R. Rao, J. Mater. Chem. **6**, 1585 (1996).
[13] S. N. Ruddlesden and P. Popper, Acta Crystallogr. **11**, 541 (1958).
[14] P. D. Battle, M. A. Green, N. S. Laskey, J. E. Millburn, P. G. Radaelli, M. J. Rosseinsky, S. P. Sullivan, and J. F. Vente, Phys. Rev. B **54**, 15967 (1996).
[15] R. Von Helmolt, J. Wecker, K. Samwer, and K. Barner, J. Magn. Magn. Mater. **151**, 411 (1995).
[16] T. Kimura, Y. Tomioka, H. Kuwahara, A. Asamitsu, M. Tamura, and Y. Tokura, Science **274**, 1698 (1996).
[17] R. M. Kusters, J. Singleton, D. A. Keen, R. McGreevy, and W. Hayes, Physica B **155**, 362 (1989).
[18] A. Ramirez, J. Phys.: Condens. Matter **6**, 8171 (1997).
[19] A. J. Millis, P. B. Littlewood, and B. I. Shraiman, Phys. Rev. Lett. **74**, 5144 (1995).
[20] G. H. Jonker and J. H. Santan, Physica **16**, 337 (1950).
[21] J. M. De Teresa, M. R. Ibara, P. A. Algarabel, C. Ritter, C. Marquina, J. Blasco, J. Garcia, A. del Moral, and Z. Arnold. Nature **386**, 256 (1997).
[22] K. H. Kim, J. Y. Gu, H. S. Choi, G. W. Park, and T. W. Noh, Phys. Rev. Lett. **77**, 1877 (1996).
[23] J. F. Mitchell, D. N. Argyriou, J. D. Jorgensen, D. G. Hinks, C. D. Potter, and S. D. Bader, Phys. Rev. B **55**, 63 (1997).
[24] R. S. Liu, C. H. Shen, S. F. Hu, J. G. Lin, and C. Y. Huang, J. Magn. Magn. Mater. **209**, 113 (2000).
[25] W. Boujelben, A. Cheikh-Rouhou, J. Pierre, and J. C. Joubert, J. Alloys Compd. **314**, 15 (2001).

Synthesis and characterization of SmNiO$_3$ thin films

N. Ihzaz[1,3], **S. Pignard**[1], **J. Kreisel**[1,*], **H. Vincent**[1], **J. Marcus**[2], **J. Dhahri**[3], and **M. Oumezzine**[3]

[1] Laboratoire des Matériaux et du Génie Physique, ENS de Physique de Grenoble, BP 46, 38402 Saint-Martin d'Hères Cedex, France
[2] Laboratoire d'Etudes des Propriétés Electroniques des Solides, BP 166-F-38042 Grenoble Cedex, France
[3] Laboratoire de Physico-Chimie des Matériaux, Département de Physique, Faculté des Sciences de Monastir, 5019 Monastir, Tunisia

Received 31 August 2003, accepted 31 December 2003
Published online 25 March 2004

PACS 61.10.Nz, 68.55.–a, 71.30.+h, 78.30.Hv, 81.15.Gh

Thin films of SmNiO$_3$ were grown on LaAlO$_3$ (012) substrates using the injection-MOCVD process, a method which has been in the past successfully applied to the growth of high-quality epitaxial oxide thin films. Our investigation of as-deposited films by X-ray diffraction allows concluding on the presence of a pure SmNiO$_3$ phase with an epitaxial growth on the LaAlO$_3$ substrate. It is interesting to note, that the growth of SmNiO$_3$ films on LaAlO$_3$ (012) leads to a non-negligible in-plane compressive stress near the interface, which possibly explains the stabilization of the SmNiO$_3$ phase, otherwise difficult to obtain. Further to the latter results we present and discuss temperature-dependent measurements of the electric properties : an electrical transition from insulting to metallic state is observed at 520 K with a minimum resistivity of 1.3 mΩ.cm. Finally, in order to further investigate the structure of the obtained films, we have performed a first-time Raman scattering study of SmNiO$_3$, the results are discussed in comparison to literature results obtained on NdNiO$_3$.

© 2004 WILEY-VCH Verlag GmbH & Co. KGaA, Weinheim

1 Introduction

Recently, perovskite-type nickelates RNiO$_3$ (R: rare earth elements) and its hole-doped analog $R_{1-x}A_x$NiO$_3$ (A: alkaline earth ions), have been proposed as appropriate systems for the investigation of fundamental aspects related to magneto-transport phenomena [1–6]. SmNiO$_3$ belongs to the family of orthorhombically-distorted perovskites RNiO$_3$ with GdFeO$_3$ structure (space groupe *Pbnm*) which undergo metal-to-insulator (MI) transitions at 130 K, 200 K, 400 K and 460 K for R = Pr, Nd, Sm and Eu, respectively [2]. It is, however, a much more complex material than indicated by its simple chemical composition, since it shows an unusual magnetic ordering with a propagation vector (1/2,0,1/2) belonging to an orthorhombic unit cell revealing the coexistence of ferromagnetic and antiferromagnetic interactions [7]. A specific feature that makes SmNiO$_3$ particularly interesting to investigate is the separation by 200 °C of magnetic and MI transitions (unlike the other members where they coincide); thus, both transitions can be studied independently. Unfortunately, SmNiO$_3$ is difficult to synthesize. In order to stabilize the less stable Ni^{3+} oxidization state of Nickel, ceramic samples have to be prepared at very high oxygen pressure explaining that investigations on this material are rather scarce. In order to circumvent the latter difficulties and motivated by the work of Novojilov *et al.* [8] the aim of the present work is to synthesize SmNiO$_3$ thin films by a peculiar CVD technique.

---

[*] Corresponding author: e-mail: jens.kreisel@inpg.fr

2 Experiment

Using "injection metal-organic chemical vapor deposition process" MOCVD [9, 10] we have prepared epitaxial thin films of SmNiO$_3$ on LaAlO$_3$ (012) single crystals. The advantages of the latter process are namely a rapid growth rate, the potential for a large area coverage, and easy adjust of film composition and morphology. Tris (tetramethylheptanedionato) samarium Sm(thd)$_3$ and tris (tetramethylheptanedionato) manganese Ni(thd)$_2$ precursors dissolved in 1.2-dimethoxyethane have been used for the preparation of the liquid solution. Droplets of few microliters are injected in an evaporation zone heated at 230 °C in order to sublimate the solid precursors. These vapors are then carried into the reaction zone where the classical CVD reaction occurs on the LaAlO$_3$ substrate heated at 680 °C in an oxygen partial pressure of 10 mbar. Optimized conditions ensuring the synthesis are given in Table 1. A short *in situ* annealing at the same temperature for 30 minutes in a pure oxygen atmosphere has been performed. The thickness of the film is about 1000 Å.

Table 1 Growth conditions.

Parameter	Value
Sm(thd)$_3$ dilution	2×10^{-2} mol/l
Ni(thd)$_2$ dilution	2×10^{-2} mol/l
Substrate temperature	680 K
Number of injections	1000
Volume of droplet	3.7 µl
Injection frequency	1 Hz
Total pressure	10 mbar
O$_2$ flow	600 cm^3/min

3 Results and discussion

Figure 1 shows X ray diffraction (XRD) θ–2θ scan of a SmNiO$_3$ thin film deposited on LaAlO$_3$ (012) (hereafter noted LAO). No second phase precipitates is visible suggesting a high quality of the growth. LAO has a rhombohedral structure which is usually described in the hexagonal system with $a = 5.36$ Å and $c = 13.11$ Å; in this system, the (012) plane (called *R*-plane in the rhombohedral system) defines a cubic symmetry with $a = 3.791$ Å. The reflections obtained for the SNO phase can be indexed 00*l* and coincide with 00*l* peaks of LAO described in the pseudo-cubic symmetry . The diffraction maxima of SNO peaks are strongly overlapped with those of the substrate indicating a strong texture of the film. The out-of-plane lattice parameter of SNO which can be deduced from the scan is $a = 3.821$ Å, indicating an in-plane compressive strain as the mean-lattice cubic parameter of SmNiO$_3$ is $a = 3.796$ Å at room temperature [11]. Measurements achieved with a texture diffractometer lead to the conclusion that the growth is epitaxial with a cube-on-cube coincidence. Unfortunately, the XRD pattern presented in Fig. 1 does not allow evidencing a non-cubic distortion of SmNiO$_3$ and, of course, even less the kind of potential distortion. We will address the latter point below in the Raman scattering section.

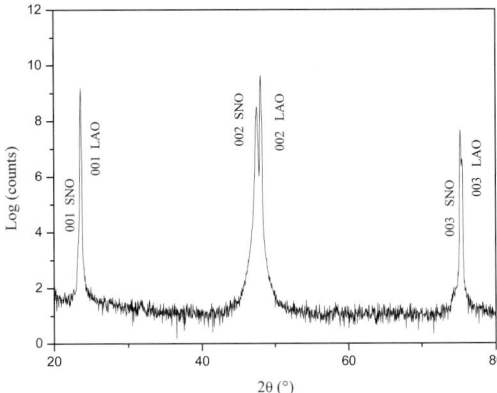

Fig. 1 X-ray diffraction pattern of an as-deposited SmNiO$_3$ film.

The evolution of the resistivity vs. temperature at zero field was measured by using a four probe method in the range of 300 K to 600 K (see Fig. 2.). An electrical transition from insulator to metal is observed around T_{IM} = 520 K ; note that this transition occurs at a much higher temperature than the magnetic transition (T_N = 220 K for bulk SNO). It is interesting to note, that the here-investigated SNO film presents a significantly higher T_{IM} than the 400 K observed for polycrystalline bulk samples [11]. As a matter of fact, the growth of SNO films on LAO leads to a non-negligible in-plane compressive stress near the interface and this might well be a possible reason for the above result. Several studies on the relation between strain and physical properties have been reported for perovskite-type oxides such as rare earth manganites [12, 13] revealing the relationship between strain relaxes, transport properties and transport behavior of bulk solid. However, further studies on SNO films with various thickness and on other substrates (leading to tensile strain) have to be carried out to confirm this hypothesis and to check the role of on the electric properties of SmNiO$_3$. Finally, let us notice the low value of the resistivity at T_{IM} compared to the literature, i.e. ρ_{IM} = 1.3 mΩ.cm, which is a further signature of the good crystalline quality of the sample.

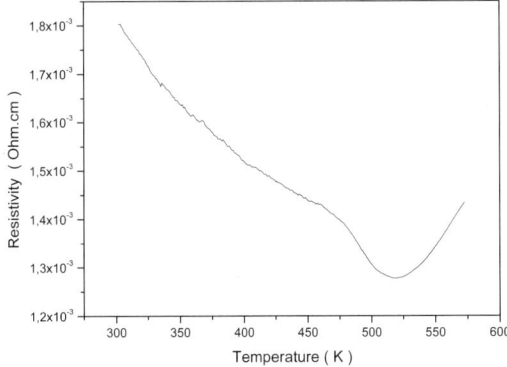

Fig. 2 Resistance vs. temperature at zero field for a SmNiO$_3$ film.

Let us now come back to the above-mentioned question concerning the actual structure of the SNO film. Raman spectroscopy is known to be a versatile technique for the investigation of, even slight, structural distortions in perovskite-type thin films. In this context we should remind that, for symmetry reasons, any first-order Raman scattering is forbidden for the Pm-3m cubic perovskite-prototype structure. As a consequence, the simple observation of a Raman spectrum gives direct evidence for a violation the cubic-structure-related selection rules. Figure 3 presents a room temperature Raman spectrum of the SNO film thus indicating unambiguously the presence of a non-cubic distortion. Although Raman results on bulk SNO are yet unavailable in the literature, it is instructive to compare the results of Fig. 1 with literature results on Raman scattering of the related NdNiO$_3$ [14]. As a matter of fact, the here-reported Raman signature of the SNO-film is very similar to that observed for a NdNiO$_3$ film in its low-temperature insulator regime owing a monoclinic $P2_1/n$ space group. As a consequence, considering that Raman scattering is a fingerprint technique, we propose that our deposited SNO films crystallize in a non-cubic structure presenting a monoclinic distortion. Unfortunately, in absence of any literature results on Raman scattering on bulk SNO samples we cannot discuss the strain-state of our film which could have been potentially investigated via a band position change of the film with respect to the bulk.

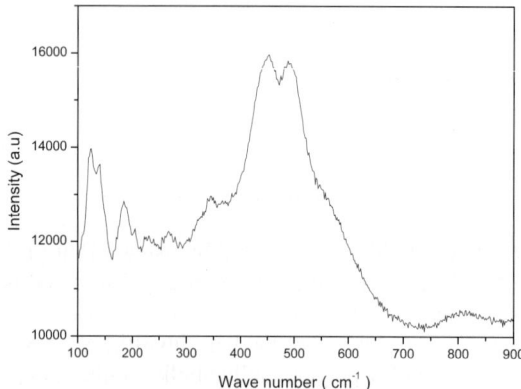

Fig. 3 Raman spectrum at RT for a SmNiO$_3$ thin film.

4 Conclusion

We have succeeded in preparing epitaxial SmNiO$_3$ thin films by injection-MOCVD on LaAlO$_3$ (012) single-crystal substrates. The prepared films are under in-plane compressive stress which stabilizes the formation of the SNO phase; no second phase precipitates is observed suggesting a good crystalline quality of the obtained films. The observed electrical properties show a high-temperature shift of the insalator–metal transition. The latter changes with respect to the bulk are possibly due to the non-negligible in-plane compressive stress near the interface but further investigations are needed in order to confirm this hypothesis. Finally, an investigation with Raman spectroscopy suggests a monoclinic structural distortion of the SNO films.

References

[1] P. Lacorre, J. B. Torrance, J. Pannetier, A. I. Nazzal, P. W. Wang, and T. C. Huang, J. Solid State Chem. **91**, 225 (1991).
[2] J. B. Torrance, P. Lacorre, A. I. Nazzal, E. J. Ansaldo, and Ch. Niedermayer, Phys. Rev. B **45**, 8209 (1992).
[3] J. B. Goodenough and P. Raccah, J. Appl. Phys. **36**, 1031 (1965).
[4] J. L. Garcia-Munoz, J. Rodriguez-Carvajal, P. Lacorre, and J. B. Torrance, Phys. Rev. B **46**, 4414 (1992).
[5] J. A. Alonso, M. J. Martinez-Lope, and M. A. Hidalgo, J. Solid State Chem. **116**, 146 (1995).
[6] J. L. Garcia-Munoz, M. Suaaidi, M. J. Martinez-Lope, and J. A. Alonso, Phys. Rev. B **52**, 13 563 (1995).
[7] J. L. Garcia-Munoz, Rodriguez-Carvajal, and P. Lacorre, Europhys. Lett. **20**, 241 (1992).
[8] M. A. Novojilov, O. Yu. Gorbenko, I. E.Graboy, and A. R. Kaul, Appl. Phys. Lett. **76**, 2041 (2000).
[9] J. P. Sénateur, F. Weiss, O. Thomas, R. Madar, and A. Abrutis, French patent No. 93/08838 PCT No. FR94/00858.
[10] J. P. Sénateur, F. Felten, S. Pignard, F. Weiss, A. Abrutis, V. Bigelyte, A. Teiserskis, Z. Saltyte, and B. Vengalis, J. Alloys Compd. **251**, 288 (1997).
[11] J. Pérez-Cacho, J. Blasco, J. Garcia, M. Castro, and J. Stankiewicz. J. Phys.: Condens. Matter. **11**, 405 (1999).
[12] R. A. Rao, Q. Gan, C. B. Eom, R. J. Cava, Y. Suzuki, J. J. Krajewwski, S. C.Gausepohl, and M. Lee, Appl. Phys. Lett. **70**, 3035 (1995).
[13] Y. Suzuki, H. Y. Hwang, S.-W. Cheong, and R. B. Van Dover, Appl. Phys. Letts. **71**, 140 (1997).
[14] M. Zaghrioui, A. Bulou, P. Lacorre, and P. Laffez. Phys. Rev. B **64**, 081102 (2001).

High field induced spin reorientations in $Ho_{0.24}Y_{2.76}Fe_5O_{12}$

A. Bouguerra[1,3], **S. Khène**[2], and **G. Fillion**[*,3]

[1] I.S.E.T, Centre Universitaire de Tebessa, 12002 Tebessa, Algeria
[2] Département de Physique, Université d'Annaba, BP-12 El-Hadjar, Annaba, Algeria
[3] Laboratoire Louis Néel, CNRS, BP166, F-38042 Grenoble Cedex 9, France

Received 31 August 2003, accepted 31 December 2003
Published online 18 March 2004

PACS 75.30.Kz, 75.50.Gg

Some precise magnetization measurements have been performed on $Ho_{0.24}Y_{2.76}Fe_5O_{12}$ single crystals under static magnetic fields up to 16 Tesla and in the temperature range (2–30 K). As previously observed in pulsed magnetic fields, the change from the spontaneous ferrimagnetic structure in zero magnetic field to the fully ferromagnetic one in high field, takes place through several intermediate phases separated by step-like magnetization jumps. Whole H–T phase diagrams have been determined for the three main cubic axes and also when the sample can rotate freely. We show that the Ising model, commonly used to account for the large magnetocrystalline local anisotropy of the Ho^{3+} moments, is in worst agreement with our static measurement than with the previous pulsed fields experiments and then a more realistic model is needed and will be discussed.

1 Introduction

In the ferrimagnetic holmium-yttrium iron garnet series $Ho_xY_{3-x}Fe_5O_{12}$ ($0 \leq x \leq 3$), for $x < \approx 0.8$, the resultant holmium moment is smaller than the iron one and there is no compensation point [1]. In these garnets, the exchange interactions between the irons are so strong that, for fields $< \approx 50$ T, the resulting iron moment can be considered as only one sublattice which behave like the YIG ($x = 0$) magnetization with practically no magnetic anisotropy. The Ho–Ho interactions are so small that the behaviour of the holmium moments depends only on the exchange interaction with the iron, their own magnetic anisotropy and the external applied field H. At sufficiently high temperature ($T > \approx 50$ K), when H is increased, the reversal of the Ho sublattices is achieved first by demagnetization, then by passing through zero values when H cancels the mean iron exchange molecular field H_{mol}, and finally, by remagnetization up to saturation [2]. Below some critical temperature T^*, depending on the concentration x, the reversal process can involve the occurrence of several intermediate canted phases differing from each other by spin reorientations. These "orientational" phase transitions correspond to step-like magnetization jumps, as previously observed by various pulsed fields experiments [3–7]. It was already noticed that the number of these jumps depends on the orientation of the external field with respect to the crystallographic axes. As an interpretation, a realignment of the magnetic structure under the field was called. The Ho^{3+} ions were considered in an extremely anisotropic Ising approximation with the Ho moments fixed to the $\langle 100 \rangle$, $\langle 010 \rangle$, and $\langle 001 \rangle$ directions of the cubic cell, forming an umbrella of axis $\langle 111 \rangle$ [8]. In this paper, in order to investigate the degree of reliability of this quasi-Ising model and to improve the determination of the involved parameters, we have undertaken precise magnetization measurements of these transitions in $Ho_{0.24}Y_{2.76}Fe_5O$ single crystals under high static magnetic field. Within the framework of the effective

---

[*] Corresponding author: e-mail: fillion@grenoble.cnrs.fr, Phone: +33 476 887 904, Fax: +33 476 881 191

spin Hamiltonian model, we have performed theoretical calculations of magnetization curves and H–T phase diagrams, either in the above Ising limit or in the isotropic case. The results will be discussed in comparison with the experimental data.

2 Experimental and results

The single crystals of $Ho_{0.24}Y_{2.76}Fe_5O_{12}$ were grown by standard PbO/PbF_2 flux method [1] and samples of about 0.3 g were put in an almost spherical shape of about 5 mm in diameter and oriented along the cubic $\langle 100 \rangle$, $\langle 110 \rangle$, and $\langle 111 \rangle$ directions by the X-ray Laüe technique within an error of 1°. The chemical composition was checked using scanning microscope (JSM-840A), electronic microprobe, and chemical analysis. The magnetization was recorded using an extraction technique in a 16 T superconducting magnet at a rate of one measure per minute. All the magnetization results are reported in Bohr magneton by $Ho_{0.24}Y_{2.76}Fe_5O_{12}$ formula unit and the magnetic field H is the applied one. As a check of calibration, the magnetization of a YIG single crystal was found to be 5.016 μ_B at 2 K.

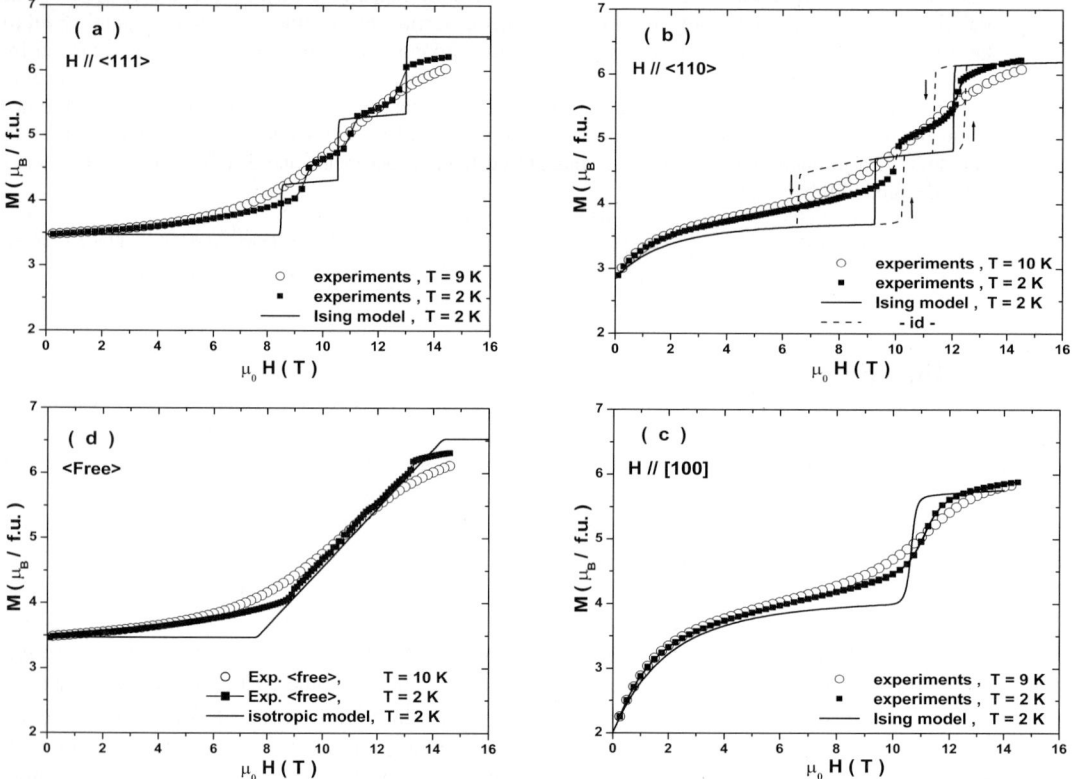

Fig. 1 Magnetization curves of $Ho_{0.24}Y_{2.76}Fe_5O_{12}$: a) $H//\langle 111 \rangle$, b) $H//\langle 110 \rangle$, c) $H//\langle 100 \rangle$, d) free rotating sample.

In Fig. 1a, 1b and 1c, the experimental magnetization curves reported at $T = 2$ K show rather sharp jumps, the number of which being in accordance with the theoretical predictions of the Ising model, i.e. 3, 2 and 1 respectively for the direction $\langle 111 \rangle$, $\langle 110 \rangle$ and $\langle 100 \rangle$ of the applied magnetic field H. When the sample is let to rotate freely (Fig. 1d), the curve obtained at 2 K presents an almost linear variation between two transitions fields, delimiting a single canted phase, as expected for the simple isotropic two sublattice model. In all cases, the jumps are smeared out when the temperature is increased and at 9 or 10 K there is no transition at all, but only an inflexion point around ~ 10.8 ÷ 11 T is observed. This is in agreement with the existence, in all these canted phase diagrams of a critical temperature T^* above which no other canted phase exist than the initial one, i.e.: the iron moment, after rotating towards the field, remain always parallel to H. It is important to notice that there is no significant hysteresis in all our curves, within the relative experimental errors of 0.1 % on the magnetization. The value of the spontaneous magnetization is M_o ($H//\langle 111 \rangle$) = 3.47 μ_B/f.u. which correspond to a mean Ho^{3+} moment of 6.42 μ_B/f.u. projected along the $\langle 111 \rangle$ easy axis.

The critical fields are reported on Fig. 2, together with those obtained by varying the temperature at constant applied magnetic fields (isofield) which are more suitable to determine the almost vertical lines of the diagrams. The error on determination of the transition points is about 0.1 T in field and 0.2 K in temperature. The critical temperature is the same for the diagrams along $\langle 111 \rangle$, $\langle 110 \rangle$ and for the free case: $T^* = 8.5 \pm 0.2$ K.

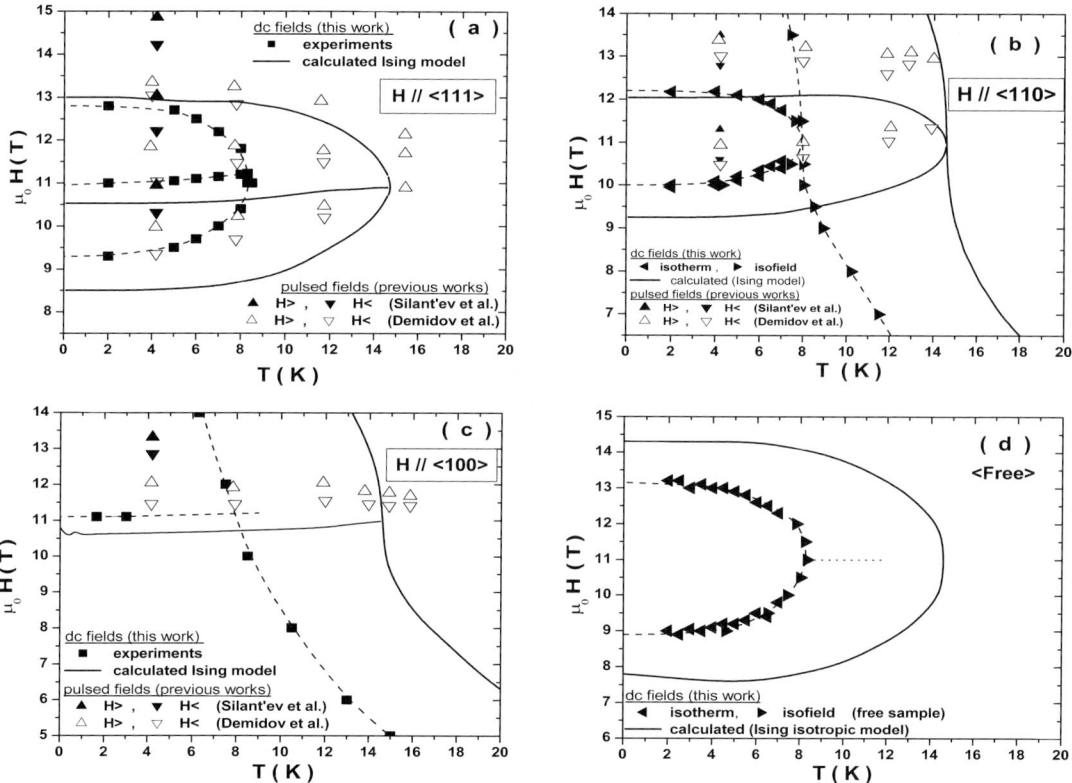

Fig. 2 Phase diagrams of $Ho_{0.24}Y_{2.76}Fe_5O_{12}$: a) $H//\langle 111 \rangle$, b) $H//\langle 110 \rangle$, c) $H//\langle 100 \rangle$, d) free rotating sample.

3 Analysis and discussion

Following Alben [9], the free energy F of one molecule of $Ho_xY_{3-x}Fe_5O_{12}$ is:

$$F = -\mathbf{M}_{Fe}\mathbf{H} - \frac{x}{6}k_BT\sum_{q=1}^{6}\ln 2\mathrm{ch}\left(\frac{\Delta_q}{k_BT}\right) \quad \text{where} \quad \Delta_q = \left| -\mu_B\mathbf{H}\,\tilde{\mathbf{g}}_q + \frac{M_{Fe}}{M_{Fe}}\tilde{\mathbf{G}}_q \right|,$$

$\tilde{g}$ is the paramagnetic tensor, $\tilde{G}_q = \mu_B M_{Fe}\tilde{n}\tilde{g}_q$ is the exchange tensor and $\tilde{n}$ is the molecular field coefficients tensor. The sum is over the six unequivalent holmium sites. These sites having D_2 symmetry, $\tilde{g}$ and $\tilde{G}$ are both diagonal in the local system of coordinates, and constitute the relevant parameters of anisotropy. For fixed T and H, the equilibrium configurations are given by numerical minimization of F

with respect to spherical angles of the iron sublattice magnetization vector M_{Fe}. The iron anisotropy is neglected and then, a variation in direction of M_{Fe} is accompanied by a variation of the resultant of the holmium moments M_{Ho} in such a way to have no torque on M_{Fe}. The calculations have been made with an isotropic exchange field $H_{mol} = -n\,M_{Fe}$; in opposite direction to M_{Fe}. In order to closely reproduce the observed spontaneous magnetisation, the following values have been used: $g_{xx} = g_{yy} = 0$, $g_{zz} = 22.5$ for the Ising case, and $g_{xx} = g_{yy} = g_{zz} = 13$ for the isotropic case. In both case, $M_{Fe} = 5\ \mu_B$ and $n = 2.2\ T/\mu_B$ ($H_{mol} = 11$ T), corresponding to a theoretical slope $\chi = 1/n$ in the isotropic canted phase very close to the experimental one, as seen on Fig. 1d. In fact, when the sample is let to rotate freely, the observed diagram is almost the same than for the isotropic model since the crystal can follow the rare earth resultant moment in such a manner that the total magnetization is always collinear to the applied field. The results are presented together with the experiments in Figs. 1 and 2. At 2 K, the agreement is qualitatively good as well for the critical fields as for the magnetization jumps, but the striking feature is that, in all cases, the transitions do not vanish below a temperature $T^* = 14.6$ K, which is much higher than the experimental one. The limits of stability of the different phases have been calculated in order to check the first order character of the transitions and to have an estimation of the maximum possible associated hysteresis. For $\langle 111 \rangle$ and $\langle 110 \rangle$ directions, this hysteresis is rather large, as shown in Fig. 1b as an example, and remains up to T^*. For $\langle 100 \rangle$, there is no hysteresis beyond $T = 0.75$ K and therefore the horizontal line is of second order, with continuous rotation of the moments. These features are well reflected by the observed hysteresis in pulsed fields experiments which seem surprisingly in better agreement with the calculations, but they are done in a much shorter time scale than our static ones and this is known to enhance hysteresis effects as well to deal with negative magnetocaloric effect due to the demagnetization of the holmium sublattices before the transitions, leading to apparent higher transition temperatures.

Several features can make the Ising model inadequate: i) this model does not account for the double umbrella found by neutron diffraction experiments on $Ho_3Fe_5O_{12}$ [11, 12], confirmed on diluted $Ho_xY_{3-x}Fe_5O_{12}$ by NMR experiments [13]; ii) the value which has to be taken for g_{zz} to fit the M_o corresponds to an holmium moment of 11.25 μ_B greater than the 10 μ_B maximum free ion value; iii) the model implies a quasi-doublet ground state approximation which, in the case of non Kramers ion like Ho^{3+} is valid for high fields whereas in our case, the external field is acting against the exchange field.

In conclusion, with a more complete CEF calculation, taking in account the non linear variation of the levels as well as the anisotropy of the exchange interactions, we should be able to fit our static magnetization results with a better accuracy together with other experiments like the neutrons diffraction.

Acknowledgements We warmly thank Dr A. Markosyan for providing us the samples, Prof. A. Zvezdin, Dr V. Nekvasil and Dr P. Haen for helpful discussions. This work was supported by French-Algerian training program (BFA#2002652).

References

[1] A. K. Gapeev, R. Z. Levitin, A. S. Markosyan, B. V. Mill', and T. M. Perakalina, Sov. Phys. JETP **40**(1), 117 (1975).
[2] R. Z. Levitin, B. K. Ponomarev, and Yu. F. Popov, Sov. Phys. JETP **32**, 1056 (1971).
[3] V. G. Demidov and R. Z. Levitin, Sov. Phys. JETP **45**(3), 581(1977).
[4] V. I. Silant'ev, A. I. Popov, R. Z. Levitin, and A. K. Zvezdin, Sov. Phys. JETP **51**(2), 323 (1980).
[5] G. A. Babushkin, A. K. Zvezdin, R. Z. Levitin, A. I. Popov, and V. I. Silant'ev, Sov. Phys. JETP **53**(5), 1015 (1981).
[6] A. S. Lagutin, Sov. Phys. JETP **72**(1), 189 (1991).
[7] A. S. Lagutin, Physica B, **201**, 63 (1994).
[8] A. K. Zvezdin, A. A. Mukhin, and A. I. Popov, Sov. Phys. JETP **45**(3), 573(1977)
[9] R. Alben, Phys. Rev. B **2**(7), 2767 (1970), and references therein.
[11] A. Herpin, W. Koehler, and P. Meriel, C. R. Acad. Sc. Paris **251**, 1350 (1960)
[12] M. Guillot, F. Tchéou, A. Marchand, and P. Feldmann, Z. Phys. B – Condensed Matter **56**, 29 (1984).
[13] J. Englich, H. Lütgemeier, M. W. Pieper, V. Nekvasil, and P. Novák, Solid State Commun. **56**(9), 825 (1985).

Optimized lithography and etching processes for a magnetic oxide micro-device

R. Soulimane[*,1], **M. Koubaa**[1], **A. M. Haghiri-Gosnet**[1], **B. Mercey**[2], **W. Prellier**[2], **Ph. Lecoeur**[2], **G. Poullain**[2], and **R. Bouregba**[2]

[1] Institut d'Electronique Fondamentale – IEF, CNRS UMR 8622, Université Paris Sud, Bâtiment 220, 91405 Orsay Cedex, France
[2] Laboratoire de Cristallographie et de Sciences des Matériaux, CRISMAT-ISMRA, UMR6508, 6 Boulevard du Maréchal Juin, 14050 Caen Cedex, France

Received 31 August 2003, accepted 31 December 2003
Published online 25 March 2004

PACS 85.40.–e, 85.40.Hp, 85.50.–n

A new process based on thick novolak UV resists is proposed for elaborating a Mott field effect transistor (FET) based on "ferroelectric/manganite" heterostructures. The whole process of this micro-device, which is a succession of photo-lithography, dry etching (IBE and RIE) and metal deposition steps, has been optimized and devices with vertical gate lengths as small as 5 µm have been successfully patterned. Special care was taken to prevent interference phenomena during UV lithography.

© 2004 WILEY-VCH Verlag GmbH & Co. KGaA, Weinheim

1 Introduction

Mixed-valent manganites $R_{1-x}A_x MnO_3$, in which some trivalent rare-earth ions R^{3+} are substituted by divalent alkaline-earth metal ions A^{2+}, exhibit near T_C a huge decrease of resistivity under high magnetic field, called the colossal magnetoresistance (CMR) effect [1]. With a carrier transfer process mainly governed by a double-exchange (DE) mechanism [2], the magneto-transport properties of these manganites are fully correlated to their structural lattice distortions. Thus, epitaxial strained manganite thin films or heterostructures offer the opportunity to deliberately manipulate the magneto-transport properties. It has been demonstrated that these properties are strongly affected either by the strain due to the substrate lattice mismatch [3] or by the film thickness [4]. This high sensitivity upon structural deformations can be judiciously used to produce large electro-resistance effects, in innovant micro-devices based on bi-layers "piezo-electric oxide/manganese oxide" [5]. In these Mott field effect transistors (MottFETs), the piezo-electric deformation can distort the manganite lattice and thus affect the metal–insulator transition. Kawaï et al. [6] have first constructed such a macro-device and reported an enhancement of Tc of about 8K in $La_{0.82}Sr_{0.18}MnO_3$ (200 nm) thin film by applying a 3 V-gate-voltage upon the $Pb(Zr_{0.52}Ti_{0.48})O_3$ gate (with a nominal thickness of 400 nm). The gate length of these first MottFETs was about 200 µm. To quantify such deformation at a micrometer scale, we propose to realize similar devices with gate lengths ranging from 5 µm to 50 µm in two types of manganite thin films, i.e. $La_{2/3}Sr_{1/3}MnO_3$ (LSMO) and $Pr_{0.5}Ca_{0.5}MnO_3$ (PCMO), covered by a $Pb(Zr_{0.65}Ti_{0.35})O_3$ (PZT).

This paper will focus on the optimisation of both optical lithography and dry etching processes at a 5 µm scale. Traditionally, dry etching processes in manganites are based on Al lifted masks. The removal of Al needs an immersion in concentrated basic solutions such as NaOH that strongly affect the magnetic

---

[*] Corresponding author: e-mail: ritha.soulimane@ief.u-psud.fr , Phone: +33 1 69 15 65 78, Fax: +33 1 69 15 41 11

properties. Thus, a new dry etching process based on thick resist masks which can be easily removed in acetone is proposed. The last passivization step for isolation of each device will be also detailed.

2 Experimental details

The pulsed laser deposition (PLD) technique working with an excimer laser (λ = 248 nm at a frequency of 3 Hz) [7] was used for the growth of manganite thin films on (100) oriented $SrTiO_3$ (STO) substrates. PCMO (110 nm) thin films have been deposited at 700 °C under high oxygen pressure (300 mTorr). The LSMO (40 nm) thin films were grown using low pressure PLD assisted by ozone (630 °C, 5.× 10^{-3} mbar). In a second step, the multi-targets sputtering technique (560 °C, 10–3 mbar) was used *ex-situ* for the growth of PZT (140 nm) on top of the manganite.

All the photolithography steps were performed using a simple-face Karl Suss MJB4 aligner working with a Hg lamp (λ = 405 nm). The Shipley S1818 (1.8 μm) positive resist and the AZ5214 (1.4 μm) negative resist were respectively used to get positive resist masks and for the last lift-off step (metallic pads). Ion beam etching (IBE) of both PZT and manganites was performed under an Ar atmosphere with an acceleration voltage of 500 V associated to a high current density of 0,5 mA/cm^2. The substrate holder is both rotating and tilted at 45°. Reactive ion etching (RIE) of the insulating SiO_2 film was achieved in a reactive plasma using SF_6/CHF_3 (1:5). With a r.f. power of 20 W and a pressure of 20 mTorr, the RIE rate of SiO_2 was found to be 14 nm/min.

3 Technological process steps

Figure 1 gives a geometrical scheme of the device. The device is designed in a four contacts geometry. The distance between the two voltage pads is named L and the lateral size of the PZT pattern is called the gate length (L_g), similarly to semiconductor transistors. In this study, L_g is varied from 50 μm to 5 μm and the ratio L/L_g is fixed at 1.5 for all transistors. Note that the volume of manganite located under the PZT gate (= $L_g \times t_{manganite}$) should be strongly distorted under the application of the gate voltage (V_G). Thus, the ratio L/L_g is the relevant parameter to compare devices with different L_g and to quantify the electro-resistance effects due to the PZT deformation.

The technological process of the device needs five principle steps. First, a layer of Pt (20 nm) is sputtered on the PZT film and Al alignment marks are patterned using a lift-off process. The PZT gates are defined during the first step of the process by IBE over thick S1818 resist masks (Fig. 2). Conventional IBE processes for manganites are based on metallic Al masks, that should be removed in basic solutions (NaOH for example). Such basic solutions are known to damage the manganite: an immersion in NaOH during 30 secondes produces a 10% decrease in the saturation moment at 300 K of a 40 nm-thin LSMO film. Here, thick resist masks are simply removed in acetone. So this resist process prevents the thin manganite to be altered.

Moreover, this step 1 is the most critical one, because the IBE should stop exactly at the PZT/manganite interface. The IBE etching rate of the PZT layer has thus to be precisely controlled and tuned.

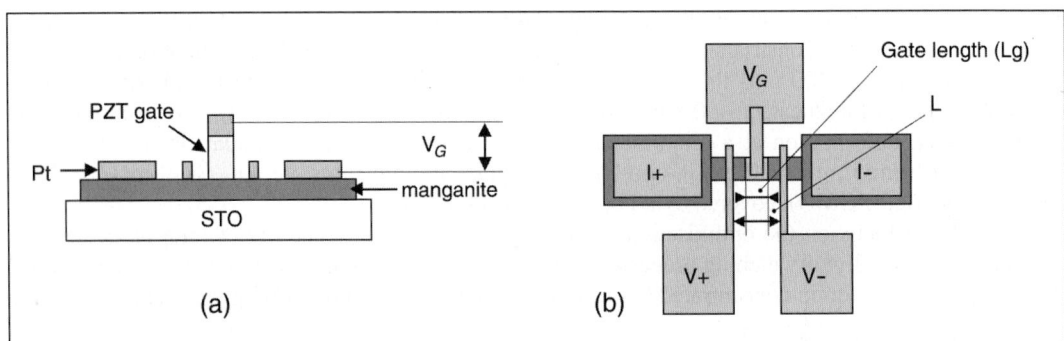

Fig. 1 Geometrical descriptive scheme of the device a) a cross-sectional view of the hetero-structure b) top view that shows the gate and the four contacts (I+,I–,V+ and V–).

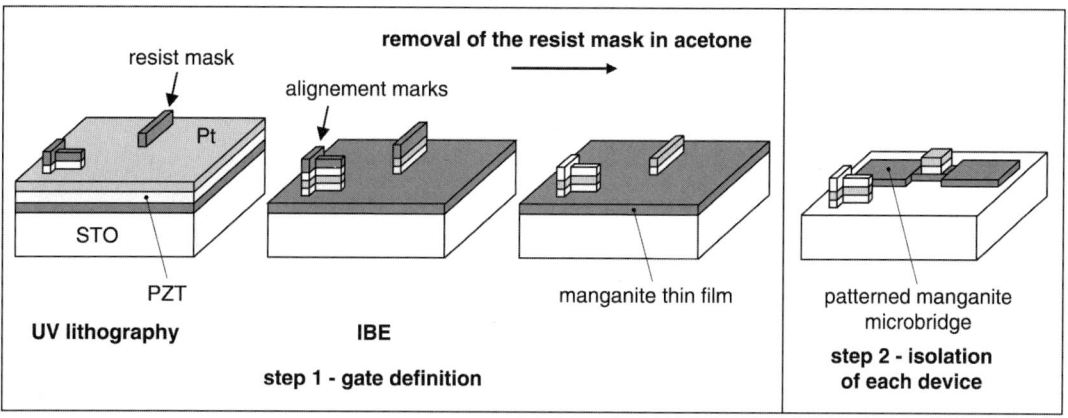

Fig. 2 The two first steps to generate the PZT gate and the manganite micro-bridge, using IBE over resist masks.

Table 1 IBE rates (nm/min) for 500 V and 0.5 mA/cm^2.

PZT	LSMO	PCMO	resist
15	17	11	30

Despite our very strong IBE parameters (500 V and 0.5 mA/cm^2), both oxide films appear to be very hard to etch with etching rates ranging between 11 and 17 nm/min (see Table 1). The end of the etching can be determined using optical reflectivity on the manganite surface since PZT and manganite exhibit different optical indexes. With such low rates, the over-etching is limited to 2 nm.

The use of a transparent substrate (STO) produces interference phenomena, which significantly affect the resist edge profiles. Figures 3b and d show typical gate resist patterns with wavy corners. The tip of the 5 μm –wide gate is particularly affected. To destroy such interferences during UV lithography, a black absorbing paper was placed under the substrate STO. The resulting resist masks exhibit smooth edges, as it can be observed on both optical top views (see Figs. 3a and c) and tilted scanning electron microscopy (SEM) image (see Fig. 4a). Moreover, to suppress any defect arriving from each gate tip, the gate was designed as a rectangle instead of a square. Each tips of this rectangle is then cut during the IBE step 2.

The second step of the process consists of the isolation of each device by etching the micro-bridge (Fig. 2). The good verticality of the etched patterns is illustrated on the Fig. 4b, which show a small alignment mark. The problem of re-deposition during IBE is suppressed by tilting and rotating the sample.

Fig. 3 Optical microscopy photographs of photo resist (S1818) patterns of a 25 μm-wide gate (a, b) and of a 5 μm-wide gate (c, d), with a black paper located under the substrate for (a, c) and without for (b, d).

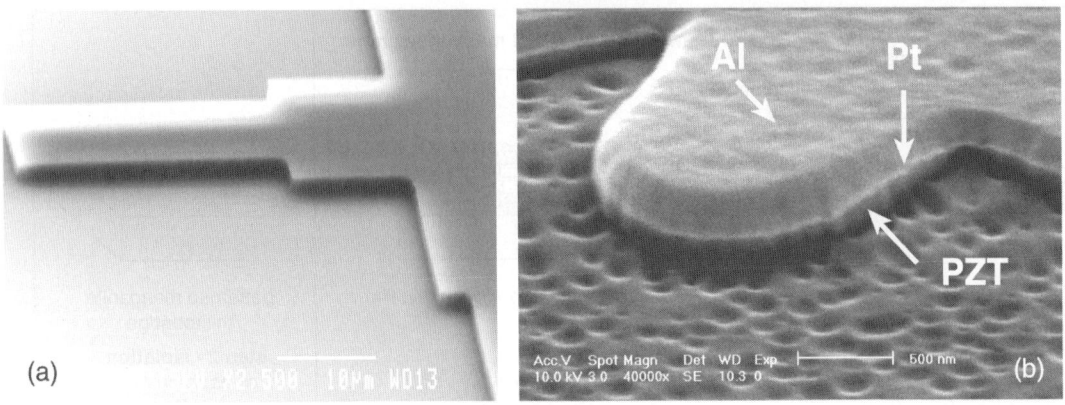

Fig. 4 SEM images of (a) a photoresist (S1818) alignment mark pattern and (b) PZT profiles after IBE etching observed on one alignment mark tip.

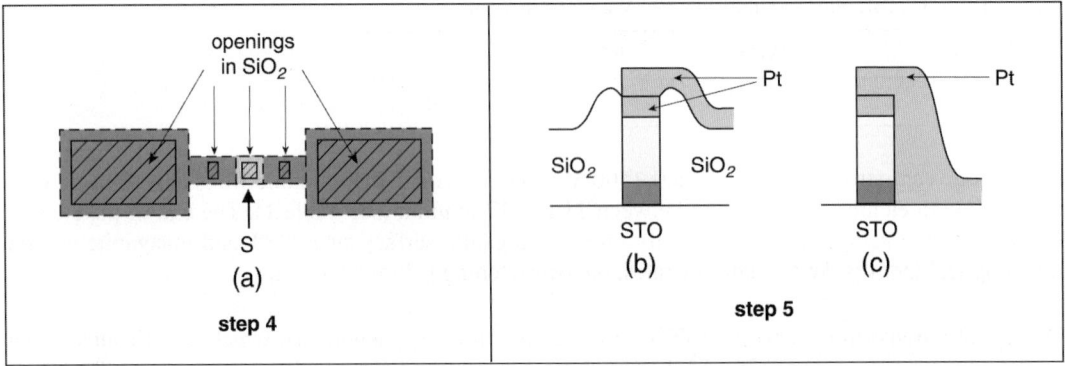

Fig. 5 Passivization of the device with SiO_2. a) Openings in the SiO_2 layer for Pt contacts pads, b) desired final cross section of the gate along the "S" direction (Fig. 5a)), c) what should be obtained in the absence of SiO_2 layer (contact between Pt gate and manganite bridge).

In the third step, a thin SiO_2 layer is deposed for the passivization of the devices. Openings in SiO_2 are etched for the last lifted Pt contacts levels (Figs. 5a and b). The passivization step is of great importance to avoid parasitic contact between gate and bridge (as shown in Fig. 5c).

In this paper, the design and the whole technological process of a MottFET based on bi-layers "piezo-electric oxide/ manganese oxide" has been described. The problem of interferences during UV lithography has been solved to get smooth and vertical resist patterns. A IBE process, with a 45° tilted substrate-holder and based on these thick resist masks, was used to produce both vertical PZT and manganite patterns. This etching process based on resist masks avoids the thin manganite to be damaged.

Acknowledgements We thank A. Charrier for graphics and illustrations.

References

[1] R. von Helmolt, J. Weker, B. Holzapfel, L. Shultz, and K. Samwer, Phys. Rev. Lett. **71**, 2331 (1993).
[2] C. Zener, Phys. Rev. **81**, 440 (1951).
[3] D. Dale and A. Fleet, App. Phys. Lett. **82**, 3725 (2003).
[4] W. Prellier, C. Simon, and A. M. Haghiri-Gosnet, Phys. Rev. B **62**, R16337 (2000).
[5] T. We and S. B. Ogale, Phys. Rev. Lett. **86**, 5998 (2001).
[6] H. Tabata and T. Kawai, Ieice. Trans. Electron **E80-C**, 918 (1997).
[7] W. Prellier and A. M. Haghiri-Gosnet, App. Phys. Lett. **77**, 1023 (2000).

© 2004 WILEY-VCH Verlag GmbH & Co. KGaA, Weinheim

Magnetic and structural properties of intermetallic compounds $Nd_{2-x}R_xFe_{17}$ (R = Sm, Gd)

M. S. Ben Kraiem[1,*], **M. Ellouze**[1], **A. Cheikh-Rouhou**[1], and **Ph. L'Héritier**[2]

[1] Laboratoire de Physique des Matériaux, FSS, B. P. 802–3018, Sfax, Tunisie
[2] Laboratoire des Matériaux et du Génie Physique (UMR 5628 CNRS) ENSPG, B. P. 46, 38402 Saint Martin d'Hères, France

Received 31 August 2003, accepted 31 December 2003
Published online 23 April 2004

PACS 61.10.Nz, 61.66.Dk, 75.30.Cr, 75.30.Kz, 75.50.Bb

The effects of Nd substitutions on the structural and magnetic properties in $Nd_{2-x}R_xFe_{17}$ compounds (R= Sm, Gd; x = 0.0–2.0) have been investigated. Nd substitutions lead to a decrease of the lattice parameters in $Nd_{2-x}R_xFe_{17}$ compounds with increasing Sm and Gd content. For both series, we observe a decrease of the saturated magnetization and an increase of the Curie temperature T_c with increasing x. The Nd substitution does not change the easy-plane anisotropy.

© 2004 WILEY-VCH Verlag GmbH & Co. KGaA, Weinheim

1 Introduction

In recent years, many investigations on the structure and magnetic properties of interstitial rare-earth iron alloys with hexagonal and rhombohedral 2:17-type structure have been reported. Large enhancements in the Curie temperature T_c of R_2Fe_{17} can be realized by addition of interstitial atoms, especially with nitrogen [1, 2] or carbon [3]. However, $Sm_2Fe_{17}M_y$ (M = N and C) compounds exhibit an easy c-axis anisotropy at room temperature [4]. An increase in Curie temperature of R_2Fe_{17} and a modification of the magnetocrystalline anisotropy can also be achieved by the substitution of Fe with other elements, such as Ga, Al and Si [5–11].

In this paper, we report on the structural and magnetic properties of substituted $Nd_{2-x}R_xFe_{17}$ alloys (R= Sm, Gd).

2 Experiment

$Nd_{2-x}R_xFe_{17}$ were prepared from starting materials up to 99.9% purity by induction melting. After melting, the polycrystalline specimens were sealed in silica tubes under argon atmosphere, annealed at 1273 K for 10 days and then quenched in water. This step was taken in order to freeze the structure at the annealed temperature. Phase purity was checked by X-ray diffraction (XRD) using a SIEMENS diffractometer with iron radiation (λ = 1.936 Å). The saturated magnetization (μ_s) has been determined using an extracting-sample magnetometer with a magnetic applied field up to 8 T. The Curie temperature were obtained from thermomagnetic analysis (TMA) using a home-made Faraday-type balance. The easy magnetization direction (EMD) was identified from X-ray diffraction pattern of the magnetically aligned samples. The aligning was done by mixing fine particles with epoxy resin and allowing them to harden in a magnetic field of 2 Tesla.

---

[*] Corresponding author: e-mail: ben_essalah@yahoo.fr, Phone: ++ 216 98 656 912, Fax: ++ 216 74 274 437.

3 Results and discussion

X-ray diffraction analysis show that all annealed $Nd_{2-x}R_xFe_{17}$ (R = Sm, Gd; x = 0.0–2.0) samples are single phase and crystallize in the rhombohedral Th_2Zn_{17}-type structure. We plot in Fig. 1 the X-ray diffraction patterns for Nd_2Fe_{17}, $Nd_{0.5}Sm_{1.5}Fe_{17}$ and $Nd_{0.5}Gd_{1.5}Fe_{17}$.

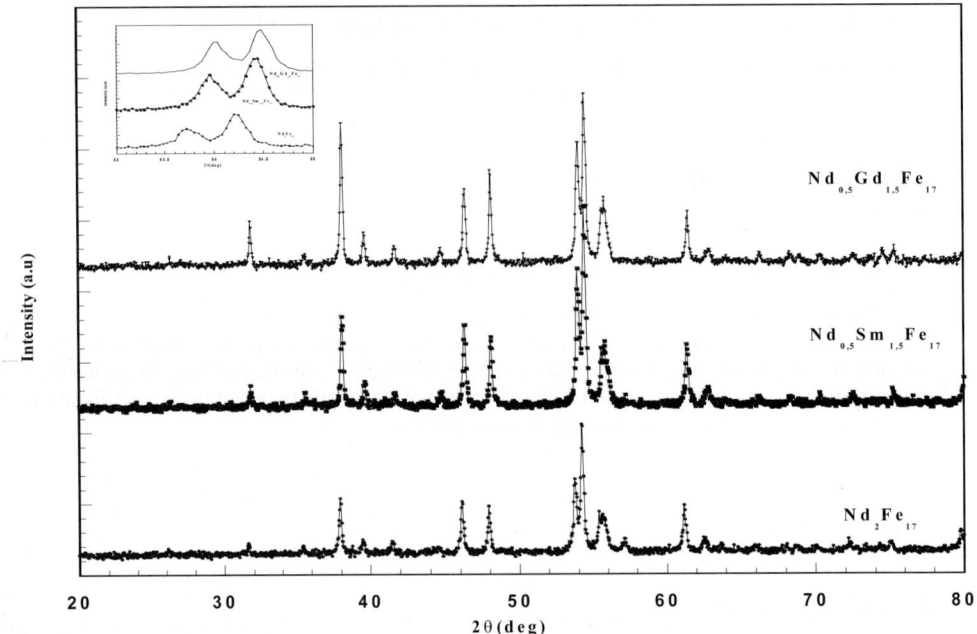

Fig. 1 X-ray diffraction patterns of Nd_2Fe_{17}, $Nd_{0.5}Sm_{1.5}Fe_{17}$ and $Nd_{0.5}Gd_{1.5}Fe_{17}$.

We plot in Fig. 2 the dependence of the lattice parameters a and c versus x in the $Nd_{2-x}R_xFe_{17}$ compounds (R = Sm, Gd). Numerical values are listed in Table 1. For both series, a and c lattice parameters and the unit cell volume decrease with increasing Sm or Gd content.

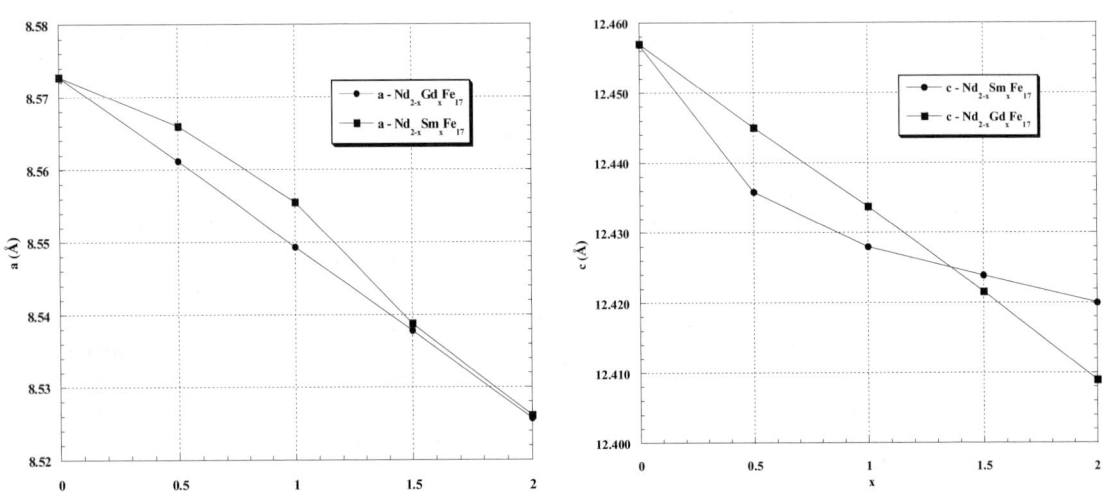

Fig. 2 Dependence of the lattice parameters a and c versus x in $Nd_{2-x}R_xFe_{17}$ compounds (R = Sm, Gd).

Table 1 Structural and magnetic data of $Nd_{2-x}R_xFe_{17}$ compounds.

$Nd_{2-x}Sm_xFe_{17}$	a (Å)	c (Å)	v (Å³)	T_C (K)	μ_s (μ_B/mol)
x = 0	8.572(8)	12.45(7)	792.84	330.3	28.50
x = 0.5	8.566(0)	12.43(6)	790.24	347.0	27.90
x = 1	8.555(5)	12.42(8)	787.81	370.6	27.10
x = 1.5	8.538(8)	12.42(4)	784.48	386.5	26.21
x = 2	8.526(1)	12.420(0)	781.90	399.0	25.32
$Nd_{2-x}Gd_xFe_{17}$	a (Å)	c (Å)	v (Å³)	T_C (K)	μ_s (μ_B/mol)
x = 0	8.572(8)	12.45(7)	792.84	330.3	28.50
x = 0.5	8.561(2)	12.44(5)	789.94	370.0	27.75
x = 1	8.549(3)	12.43(4)	787.04	400.0	24.10
x = 1.5	8.537(8)	12.42(2)	784.15	430.0	21.51
x = 2	8.525(7)	12.40(9)	781.14	460.0	17.60

Our result can be explained by the average ionic radius of Sm^{3+} and Gd^{3+} which are smaller than Nd^{3+} one.

Thermomagnetic analyses show that all our synthesized samples exhibit a ferromagnetic to paramagnetic transition with increasing temperature. Figure 3 shows the magnetization evolution versus temperature of both substituted series.

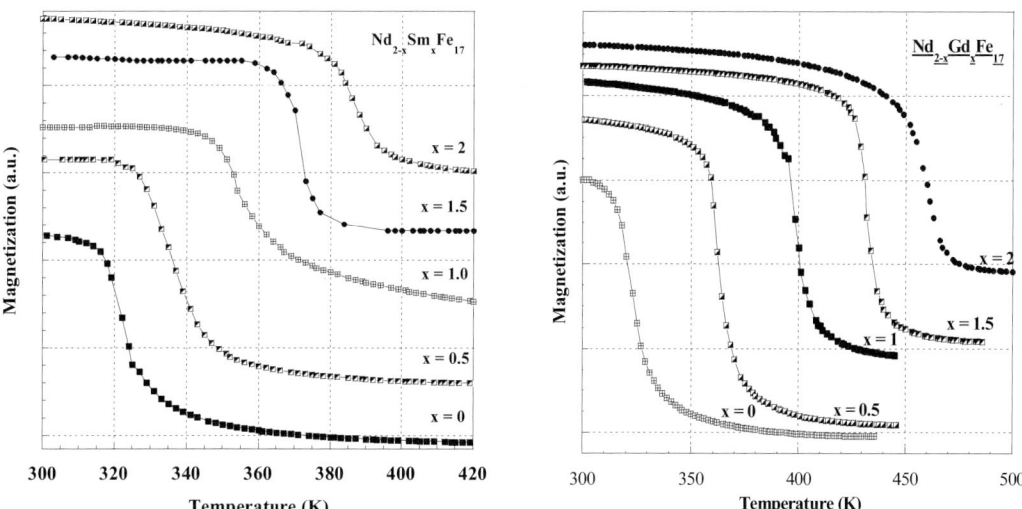

Fig. 3 Magnetization versus temperature for $Nd_{2-x}R_xFe_{17}$ (with R = Sm, Gd) samples.

The Curie temperature of the parent compound Nd_2Fe_{17} is foud to be 330.3 K. We plot in Fig. 4 the Curie temperature evolution versus x for both series. Neodymium substitution by samarium and gadolinium leads to an increase of the Curie temperature T_c in both series. It is found to be 399 K in Sm_2Fe_{17} and 460 K in Gd_2Fe_{17}. The Curie temperature enhancement with the neodymium substitution is more important in $Nd_{2-x}Gd_xFe_{17}$ than in $Nd_{2-x}Sm_xFe_{17}$.

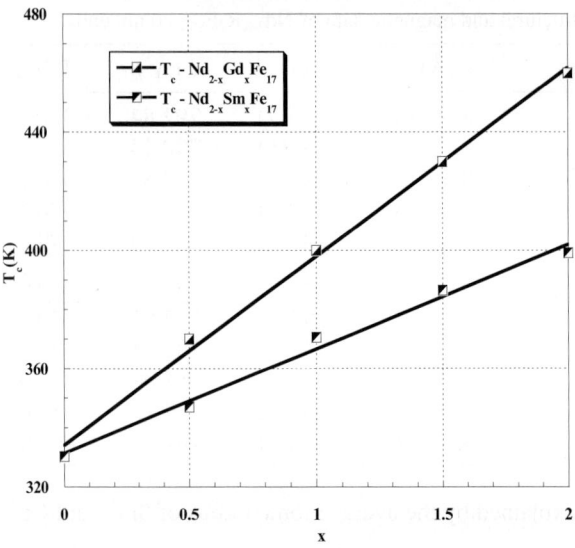

Fig. 4 Curie temperature T_C evolution versus x in $Nd_{2-x}Gd_xFe_{17}$ compounds.

Magnetization measurements as a function of the magnetic applied field up to 7 T have been performed at room temperature for all Sm or Gd contents. The evolutions of the saturated magnetization (μ_S) versus Sm or Gd contents are plotted in Fig. 5. The saturated magnetization was obtained by fitting the experimental data of M (H) versus H. The saturated magnetization is found to decrease with increasing Sm and Gd content.

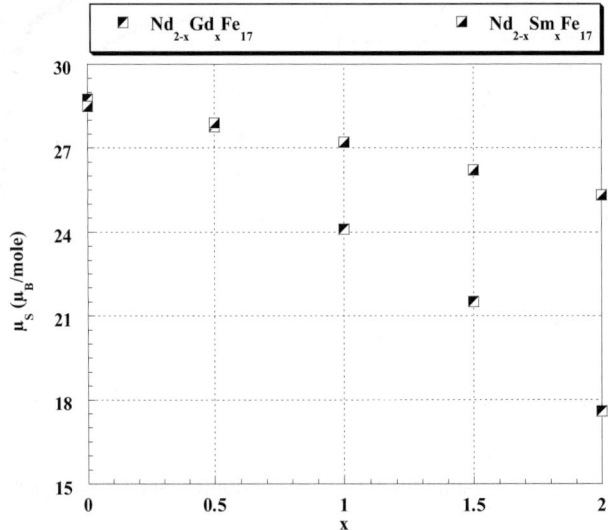

Fig. 5 The dependence of the saturated magnetization μ_S at 300 K with Sm and Gd content in $Nd_{2-x}R_xFe_{17}$ (R = Sm, Gd) samples.

The magnetic data can be explained by a two-sublattice mean field model [12]. Sm and Gd are expected to behave as a light rare earth with the total moment coupled parallel to the 3d moment. However, samarium behaves sometimes in such compounds as a heavy rare earth with a moment antiparallely coupled to the 3d moment. This can explain the above variation of the saturated magnetic moment.

Within the mean field approximation, the Curie temperature T_c is proportional to the square of the magnetic moment and the exchange integral J. The increase of the T_c in the system corresponds to a competition between positive and negative exchange interactions between Fe–Fe atoms [13–15] as a result of the increase in the interatomic Fe–Fe distance [16]. For Gd substituted samples, the Curie temperature increase is accompanied by a contraction of the unit-cell volume. This indicated that the composition dependence of the Curie temperature could not be explained on the basis of the variation of the unit-cell volume only. Substitution of Gd leads to an increase in all bond lengths and consequently decreases strongly the negative interactions, although the positive interactions are also weakned. However, the total Fe–Fe exchange interaction may become stronger, leading to an increase in the Curie temperature.

We present in Fig. 6 the XRD patterns for the magnetically aligned polycrystalline powder. XRD patterns indicate that for all $Nd_{2-x}R_xFe_{17}$ samples EMD is basal-plane direction.

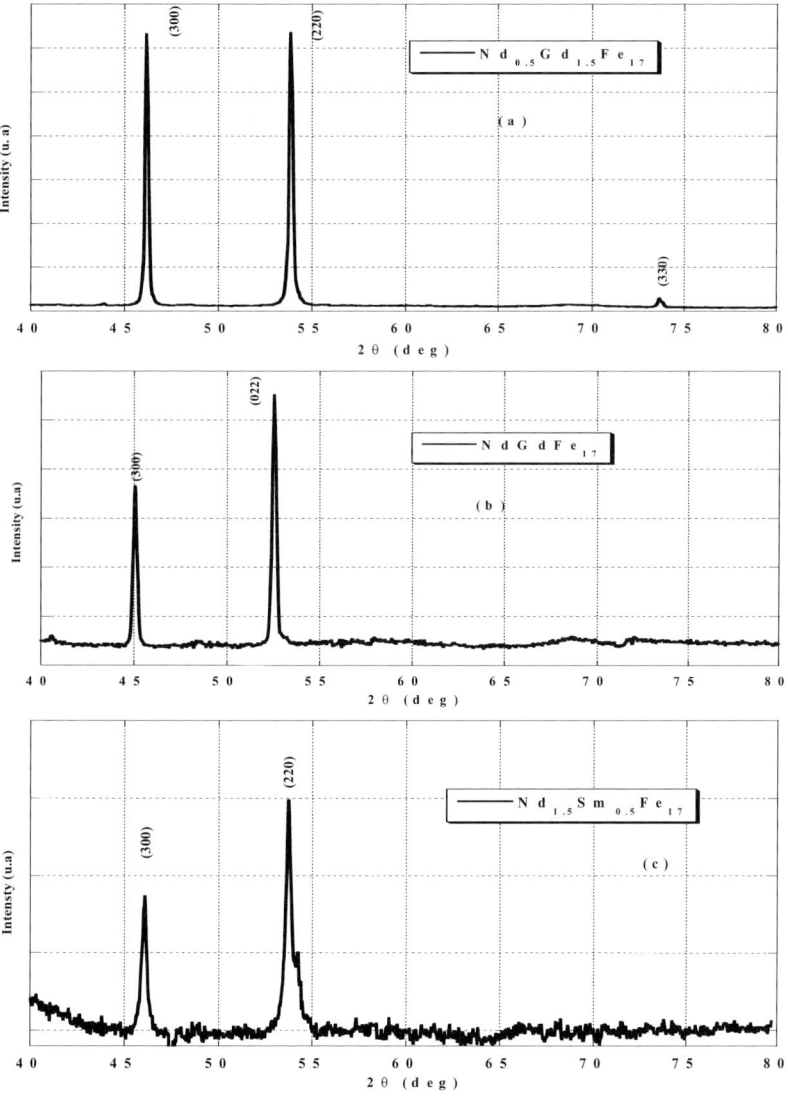

Fig. 6 XRD patterns of magnetically aligned $Nd_{1.5}Sm_{0.5}Fe_{17}$ (c), $NdGdFe_{17}$ (b) and $Nd_{0.5}Gd_{1.5}Fe_{17}$ (a).

In general, the overall magnetocrystalline anisotropy of a rare earth-transition metals intermetallic is the sum of 4f-sublattice and 3d-sublattice anisotropy. In the $Nd_{2-x}R_xFe_{17}$ compounds, the 3d-sublattice has uniaxial anisotropy [17]. The contribution of Sm and Gd are planar in the R_2Fe_{17} compounds. Therefore, the substitution of Sm and Gd enhances the planar anisotropy. In the $Nd_{2-x}R_xFe_{17}$ compound, the planar anisotropy of 4f-sublattice and the uniaxial anisotropy of 3d-sublattice apparently cancel.

4 Conclusion

The effect of Sm or Gd on the structural and magnetic properties of $Nd_{2-x}R_xFe_{17}$ (R = Sm, Gd) compounds has been studied. Substitution of Nd by Sm or Gd does not change the structure of Nd_2Fe_{17}. All compounds crystallize in the Th_2Zn_{17}-type structure. The substitution of Nd by Sm or Gd causes a decrease in the lattice parameters, a and c, and in the unit cell volume. The Curie temperature of $Nd_{2-x}R_xFe_{17}$ (R = Gd, Sm) compounds increases monotonically with increasing Sm or Gd content. The saturated magnetization of the $Nd_{2-x}R_xFe_{17}$ compounds decrease with increasing Sm or Gd concentration. The XRD patterns indicate that for all $Nd_{2-x}R_xFe_{17}$ samples, EMD is basal-plane direction.

Acknowledgements This study has been supported by the Tunisian Ministry of High Education, Scientific Research and Technology and by the CMCU collaboration (01/F-1127).

References

[1] J. M. D. Coey and H. Sun, J. Magn. Magn. Mater. **87**, L251 (1990).
[2] K. H. J. Buschow, R. Coehoorn, D. B. de Mooij, K. de Waard, and T. H. Jacobs, J. Magn. Magn. Mater. **92**, L35 (1990).
[3] J. M. D. Coey, H. Sun, Y. Otani, and D. P. F. Hurley, J. Magn. Magn. Mater. **98**, 76 (1991).
[4] H. Sun, J. M. D. Coey, Y. Otani, and D. F. F. Hurley, J. Phys.: Condens. Matter **2**, 6465 (1990).
[5] B. G. Shen, F. W. Wang, L. S. Kong, L. S. Kong, and L. Cao, J. Phys.: Condens. Matter **2**, L685 (1990.
[6] Z. Wang and R. A. Dunlap, J. Phys.: Condens. Matter **5**, 2407 (1993).
[7] B. G. Shen, L. S Kong, F. W. Wang, and L. Cao, Appl. Phys. Lett. **63**, 2288 (1993).
[8] T. H. Jacobs, K. H. J. Buschow, G. F. Zhou, X. Li, and F. R. de Boer, J. Magn. Magn. Mater. **116**, 220 (1992).
[9] E. E. Alp, A. M. Umarji, S. K. Malik, G. K. Snenoy, M. Q. Hung, E. B. Bltich, and W. E. Wallace, J. Magn. Magn. Mater. **68**, 305 (1987).
[10] Z. Hu, W. B. Yelon, S. Mishra, Gary J. Long, O. A. Pringle, D. P. Middleton, K. H. J. Buschow, and F. Grandjean, J. Appl. Phys. **76**, 443 (1994).
[11] D. P. Middleton and K. H. J. Bushow, J. Alloys Compd. **206**, L1 (1994).
[12] Cz. Kapusta, P. C. Riedi, G. J. Tomka, and R. Mycielski, J. Appl. Phys. **81**, 4563 (1997).
[13] F. Grandjean, Gary J. Long, S. Mishra, O. A. Pringle, O. Isnard, S. Miragle, and Fruchart, Hyperfine Interactions **95**, 277 (1995).
[14] I. A. Al-Omari, S. S. Jaswal, E. W. Singleton, D. J. Sellmyer, Y. Zhang, and G. C. Hadjipanayis, J. Magn. Magn. Mater. **151**, 145 (1995).
[15] I. A. Al-Omari, S. S. Jaswal, A. S. Fernando, and D. J. Sellmyer, J. Appl. Phys. **76**, 6159 (1994).
[16] Q. Liu, J. Liang, G. Rao, W. Tang, J. Sun, X. Chen, and B. Shen, Appl. Phys. Lett. **71**, 1869 (1997).

Chemical hydrogenation effects on $R_2Fe_{14}B$ compounds (with R = Ce, Nd and Gd)

M. S. Ben Kraiem[1,*], **M. Ellouze**[1], **A. Cheikh-Rouhou**[1], and **Ph. L'Héritier**[2]

[1] Laboratoire de Physique des Matériaux, FSS, B. P. 802 – 3018, Sfax, Tunisie
[2] Laboratoire des Matériaux et du Génie Physique (UMR 5628 CNRS) ENSPG, B. P. 46, 38402 Saint Martin d'Hères, France

Received 31 August 2003, accepted 31 December 2003
Published online 23 April 2004

PACS 61.10.Nz, 61.66.Dk, 75.30.Cr, 75.30.Kz, 75.50.Bb

The structural and magnetic properties of $R_2Fe_{14}BH_x$ compounds with R = Ce, Nd and Gd have been investigated by X-ray diffraction (XRD) and magnetic measurements. XRD patterns show that all samples are single phase and crystallize in the quadratic structure. The hydrogen insertion increases the lattice constants a, c and the unit cell volume. The lattice parameters values of the hydrides obtained by chemical method are very close to those elaborated by the classical method. Thermomagnetic analysis shows that the Curie temperature of the hydrides are larger than that of the parent compounds. Hydrogen insertion leads to an increase of the saturated magnetization at room temperature.

© 2004 WILEY-VCH Verlag GmbH & Co. KGaA, Weinheim

1 Introduction

After many years of a great activity in the 1960s and the 1970s, a new start in the permanent magnet research field came from the discovery of the $R_2Fe_{14}B$ intermetallic compounds [1, 2]. Due to its large economical consequences in the huge market of permanent magnets, a great deal of basic research was focused on $Nd_2Fe_{14}B$, which exhibits superior high-performance permanent-magnet properties over the earlier Sm–Co materials. This compound has a tetragonal crystal structure belonging to the $P4_2/mnm$ space group. As described before, most of the new potential materials for permanent magnets are ternary alloys of rare earth, iron and metalloid atoms, such as C or B. These compounds usually offer several advantages in comparison with the original binary compounds: higher magnetization and higher Curie temperature. The increase in Curie temperature is also partly understood in terms of the Néel–Slater model, which correlates the exchange interaction with the interatomic $3d_{metal}$–$3d_{metal}$ distances. The insertion of light elements within the structure leads to larger (Fe–Fe) distances as already proposed [3–5]. As the new hard magnet materials are usually based on the iron-rich compounds such as $Nd_2Fe_{14}B$ (82% Fe), the effects of the interstitial elements on the magnetic properties are now rather well understood. For example, in $R_2Fe_{14}X$ with X = B or C, the interstitial element increases the number of charges within the plane surrounding the rare earth. In this paper, we report a chemical method of hydrogen atoms insertion in $R_2Fe_{14}B$ compounds with R = Ce, Nd and Gd and we compare the structural and magnetic properties of hydrides obtained by chemical and classical method (under high pressure and heat treatment).

2 Experimental details

$R_2Fe_{14}B$ samples were prepared from elements (R = Ce, Nd and Gd) using induction melting. Obtained alloys have undergone thermal homogenization proceedings. Annealing samples have been realized in a

---

[*] Corresponding author: e-mail: ben_essalah@yahoo.fr, Phone: ++ 216 98 656 912, Fax: ++ 216 74 274 437.

bulb of quartz tube at T = 900 °C for 10 days. Finally the synthesized samples are rapidly quenched to room temperature in water. Our hydride samples ($R_2Fe_{14}BH_x$) were obtained by a chemical method at room temperature, using an aqueous solution of $NaBH_4$ according to the following reaction:

$$M + BH_4^- + 3H_2O \rightarrow M-H + H_2BO_3^- + 3{,}5H_2$$

where M and M–H are respectively the alloy and its hydride. After 24 h of reaction blends, the obtained compound was rinsed in alcohol then dried under air. Phase purity was checked by X-ray diffraction with iron radiation (λ = 1.9360 Å). Curie temperatures were obtained from thermomagnetic analysis using a home made Faraday-type balance. The saturated magnetization (M_s) was measured, at 300 K, using a vibrating sample magnetometer with an applied field up to 8 T.

3 Results and discussion

X-ray diffraction patterns show that all the hydrides conserve the quadratic structure of $Nd_2Fe_{14}B$ with $P4_2/mnm$ space group. The hydrogen insertion within the crystal lattice induces a significant increase of the unit cell volume of about 1.43 % for $Ce_2Fe_{14}B$, 2.73 % for $Nd_2Fe_{14}B$ and 1.14 % for $Gd_2Fe_{14}B$. This feature is quite general when hydrogen goes into intermetallic compounds and has already been observed in other series of magnetic materials such as R_2Fe_{17} [6–8], $R_2Fe_{17-x}Si_x$ [9] and $R_2Fe_{16}Ti$ [10] $RFe_{11}Ti$ [11]. We list in Table 1 the crystallographic and magnetic properties of the alloys and its hydrides.
The unit cell volume of our hydrides are slightly smaller than those reported by l'Héritier et al. [12]. This behaviour may be probably due to the hydrogen content.

Table 1 Crystallographic and magnetic data of $R_2Fe_{14}B$ compounds before and after hydrogen insertion.

Samples	a (Å)	c (Å)	c/a	V (Å^3)	$\Delta V/V$ (%)	T_c (K)	$\Delta T_c/T_c$ (%)	M_s (emu/g)
$Ce_2Fe_{14}B$	8.7573	12.1113	1.3830	804.38	-	441	-	139.7
$Ce_2Fe_{14}BH_x^*$	8.8020	12.1625	1.3817	816.05	1.43	535	17.6	152.5
$Ce_2Fe_{14}BH_{3.8}^1$	8.7881	12.2624	1.3953	820.15	1.92	546	19.2	-
$Nd_2Fe_{14}B$	8.8049	12.2056	1.3862	819.48	-	588	-	110.9
$Nd_2Fe_{14}BH_x^*$	8.8980	12.2915	1.3813	842.79	2.73	667	11.8	118.3
$Nd_2Fe_{14}BH_{4.3}^1$	8.9166	12.3656	1.3868	851.42	3.75	667	11.8	-
$Gd_2Fe_{14}B$	8.7841	12.0799	1.3752	807.21	-	666	-	76.9
$Gd_2Fe_{14}BH_{x^*}$	8.8120	12.1424	1.3779	816.55	1.14	711	6.3	84.5
$Gd_2Fe_{14}BH_{3.5}^1$	8.8687	12.2041	1.3760	831.30	1.77	712	6.4	-

* Hydrides elaborated by chemical method.
1 Hydrides elaborated by classical method [12].

Thermomagnetic analysis shows that all our hydrides exhibit a paramagnetic to ferromagnetic transition with decreasing temperature. Our synthesized hydrides present a higher Curie temperature compared to their parent alloys (Fig. 1). Such effect was already observed in the hydrides elaborated by classical method. In our hydrides, the Curie temperature values are similar to those obtained by gas–solid reaction. This effect confirm that the hydrogen content inserted by the chemical method is similar to that obtained by the classical method. The T_c increase can be attributed to the greater average Fe–Fe interatomic distances [13–15], which reinforces the positive character of the exchange interactions.

Magnetization (M) measurements versus magnetic applied field up to 8 T at 300 K were performed to determine the saturated magnetization (M_s) of the samples before and after hydrogen insertion. We plot in Fig. 2 the magnetization evolution versus applied field at 300 K. The saturated magnetization has been obtained by polynomial extrapolation of M versus $1/H^2$. The other major change upon hydrogen insertion is an increase of the saturated magnetization at room temperature as reported in Table 1. The highest value obtained is for the $Ce_2Fe_{14}BH_x$ sample. The increase of M_s is about 9.9 % for $Gd_2Fe_{14}B$, 9.2 % for $Ce_2Fe_{14}B$ and 6.7 % for $Nd_2Fe_{14}B$.

© 2004 WILEY-VCH Verlag GmbH & Co. KGaA, Weinheim

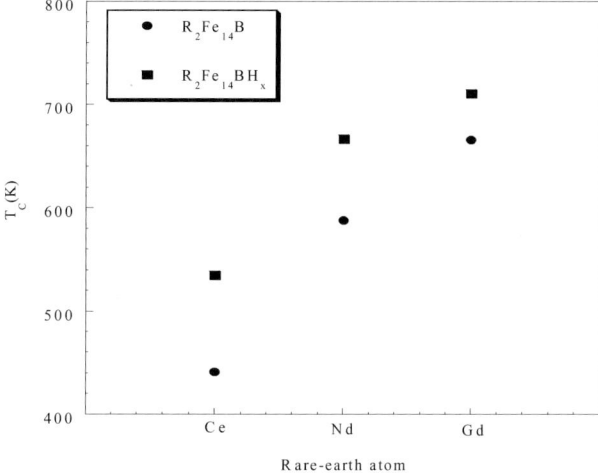

Fig. 1 Evolution of the Curie temperature before and after hydrogen insertion in $R_2Fe_{14}B$.

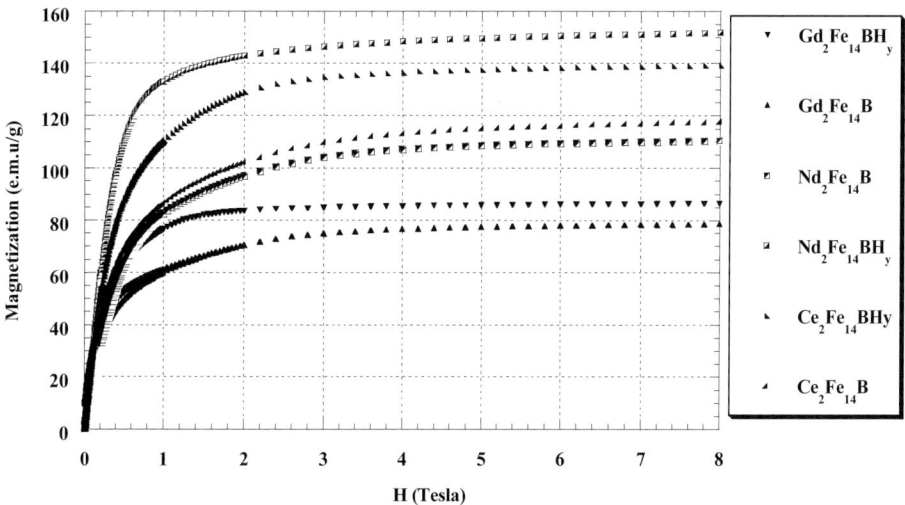

Fig. 2 Magnetization versus applied field at 300 K of $R_2Fe_{14}B$ with R = Ce, Nd and Gd before and after hydrogen insertion.

The insertion of hydrogen in the structure increases the distance between magnetic atoms and consequently the positive interaction between Fe–Fe and R–Fe atoms [16, 17]. In $Ce_2Fe_{14}B$, there are two ferromagnetic atoms Ce and Fe, which are parallely coupled in this system, this is why the saturated magnetization in this sample is higher than in the other samples. However, in the $Gd_2Fe_{14}B$ sample the magnetic moments are coupled antiparallely, which reduced the value of the magnetization compared to the other samples.

4 Conclusion

We have successfully inserted hydrogen atoms by chemical method in $R_2Fe_{14}B$ compounds with R = Nd, Ce and Gd. The hydrogen insertion leads to an increase of the lattice parameters, a and c, and the unit cell volume. The Curie temperatures, as well as the saturated magnetization, increase also after hydrogen insertion and the obtained values are higher than those obtained before hydrogen insertion and almost similar to those obtained by solid–gas reaction.

Acknowledgements This study has been supported by the Tunisian Ministry of High Education, Scientific Research and Technology and by the CMCU collaboration (01/F-1127).

References

[1] M. Sagawa, S. Fujimura, N. Togawa, H. Yamamoto, and Y. Matsuura, J. Appl. Phys. **55**, 2083 (1984).
[2] J. J. Croat, J. F. Herbest, R. W. Lee, and F. E. Pinkerton, Appl. Phys. Lett. **44**, 148 (1984).
[3] C. Christodoulou and T. Takeshia, J. Alloys Compd. **198**, 1 (1993).
[4] O. Isnard, S. Miraglia, J. L. Soubeyroux, D. Fruchart, and A. Stergion, J. Less-Common. Met. **162**, 273 (1990).
[5] O. Isnard, A. Sippel, M. Loewenhaupt, and R Bewley, J. Phys.: Condens. Matter **13**, 3533 (2001).
[6] H. Sun, J. M. D. Coey, Y. Otani, and D. P. F. Hurley, J. Phys.: Condens. Matter. **2**, 6465 (1990).
[7] J. M. D. Coey and H. Sun, J. Magn. Magn. Mater. **87**, L251 (1990).
[8] D. Fruchart, O. Isnard, S. Miraglia, and J.-L. Soubeyroux, J. Alloys Compd. **231**, 188 (1995).
[9] M. Artigas, D. Fruchart, C. Gasdeblay, O. Isnard, and S. Miraglia, J. Alloys Compd. **291**, 282 (1999)
[10] M. Ellouze, Ph. l'Héritier, A. Cheikh-Rouhou, and J.C. Joubert, J. Alloys Compd. **322**, 211 (2001)
[11] O. Isnard, S. Miraglia, M. Guillot, and D. Fruchart, J. Alloys Compd. **275–277**, 637 (1998)
[12] Ph. L'Héritier, R. Fruchart, D. Fruchart, S. Mmiraglia, and P. Wolfers, Z. Phys. Chem. Neue Folge **163**, 647 (1989).
[13] F. E. Pinkerton and W. R. Dunham, Bull. Am. Phys. Soc. **29**, 322 (1984).
[14] M. Sagawa, S. Fujimura, N. Togawa, H. Yamaoto, and Y. Matsuura, J. Appl. Phys. **55**, 2067 (1984).
[15] R. Kamal and Y. Andersson, Phys. Rev. B **32**, 1756 (1985).
[16] M. I. Bartashevich and A. V. Andreev, Physica B **162**, 52 (1990).
[17] N. M. Hong, Physica B **226**, 391 (1996).

Magnetic and structural properties of $Sm_{1.5}Gd_{0.5}Fe_{17-x}Co_xN_y$ with $0 \leq x \leq 3$ and $y \approx 3$

M. S. Ben Kraiem[1,*], **M. Ellouze**[1], **A. Cheikh-Rouhou**[1], and **Ph. L'Héritier**[2]

[1] Laboratoire de Physique des Matériaux, FSS, B. P. 802 – 3018, Sfax – Tunisie
[2] Laboratoire des Matériaux et du Génie Physique (UMR 5628 CNRS) ENSPG, B. P. 46, 38402 Saint Martin d'Hères, France

Received 31 December 2003, accepted 31 December 2003
Published online 23 April 2004

PACS 61.10.Nz, 61.66.Dk, 75.30.Cr, 75.30.Kz, 75.50.Bb

The effect of nitrogen insertion on the structural and magnetic properties of $(Sm_{1.5}Gd_{0.5})Fe_{17-x}Co_x$ compounds has been investigated. X-ray powder diffraction analysis shows that all the nitrides are single phase and crystallize in the Th_2Zn_{17}-type structure. The lattice parameters of the $(Sm/Gd)_2Fe_{17-x}Co_xN_y$ compounds increase with increasing x. Magnetization studies indicate that all the samples are ferromagnetic at room temperature. Nitrogen insertion leads to an increase in the Curie temperature T_c from 520 K for $Sm_{1.5}Gd_{0.5}Fe_{16}Co$ to 880 K for $Sm_{1.5}Gd_{0.5}Fe_{16}CoN_{2.9}$. Moreover, the nitrogen insertion increases the saturated magnetization M_s. M_s increases from 107.67 emu/g for x = 0 to 149.92 emu/g for y = 3. X-ray diffraction measurements on magnetically aligned powder samples of $Sm_{1.5}Gd_{0.5}Fe_{17-x}Co_xN_y$ reveal a change in the easy magnetization direction from planar to conical after nitrogen insertion.

© 2004 WILEY-VCH Verlag GmbH & Co. KGaA, Weinheim

1 Introduction

The R_2Fe_{17} intermetallic compounds (R = rare earth) crystallize in the rhombohedral Th_2Zn_{17}-type structure for the light R elements and in the hexagonal Th_2Ni_{17}-type structure for R heavier than Dy [1, 2]. The magnetic properties of the compounds have been investigated in great detail [3, 4]. Although the R_2Fe_{17} compounds are the most Fe rich of all binary rare-earth-Fe intermetallics. Owing to their low Curie temperatures and their nonuniaxial magnetocrystalline anisotropy, the R_2Fe_{17} compounds have not been used as permanent magnets. In 1990, Coey and Sun [5] improved successfully the magnetic properties of the R_2Fe_{17} compounds by introducing nitrogen to form interstitial nitrides $R_2Fe_{17}N_y$ with $y \approx 2.3$. The nitrogen insertion in R_2Fe_{17} compounds leads to a change in the anisotropy of Sm_2Fe_{17} from planar to uniaxial over the whole temperature range up to the Curie temperature. These outstanding intrinsic magnetic properties make $Sm_2Fe_{17}N_y$ a most promising candidate for permanent magnets. Therefore, the $R_2Fe_{17}N_y$ nitrides have recently attracted considerable interest in both fundamental and applied research. In the various compounds, the enhancement of Curie temperature doesn't result from the only unit-cell expansion as has been proposed earlier. In the present work, the crystal structure and magnetic properties of $Sm_{1.5}Gd_{0.5}Fe_{17-x}Co_xN_y$ compounds are reported. The effects of nitrogen insertion on the lattice parameters, Curie temperature, saturated magnetization and easy magnetization direction have been studied.

2 Experimental

$Sm_{1.5}Gd_{0.5}Fe_{17-x}Co_x$ powder samples were prepared from starting materials up to 99.9% purity by induction melting. After melting, the polycrystalline specimens were sealed in silica tubes under argon atmos-

---

[*] Corresponding author: e-mail: ben_essalah@yahoo.fr, Phone: ++ 216 98 656912, Fax: ++216 74 274 437.

© 2004 WILEY-VCH Verlag GmbH & Co. KGaA, Weinheim

phere, annealed at 1273 K for 10 days and then quenched to room temperature in water. Phase purity was checked by X-ray diffraction (XRD) using a SIEMENS diffractometer with iron radiation ($\lambda = 1.936$ Å). The saturated magnetization (μ_s) has been determined using an extracting-sample magnetometer with a magnetic applied field up to 8 T. The Curie temperature values were obtained from thermomagnetic analysis (TMA) using a home – made Faraday-type balance. The insertion of nitrogen in our cobalt substituted samples has been carried out under a static nitrogen gas under a pressure of 13 MPa at 723 K for 22 hours. The nitrogen content was checked by the weight increase in our samples. The nitrogen amount is found to be 2.9 ± 0.1 atoms.

3 Structural and magnetic results

3.1 Structural properties

X-ray diffraction analysis show that $Sm_{1.5}Gd_{0.5}Fe_{17-x}Co_xN_y$ samples with $x \leq 3$ are single 2:17 phase and crystallize in the rhombohedral Th_2Zn_{17}-type structure. The nitrogen insertion in the $Sm_{1.5}Gd_{0.5}Fe_{17-x}Co_x$ compounds leads to an expansion of the unit cell volume. The unit cell volume increases linearly after nitrogen insertion, as shown in Fig. 1. This expansion is found to be about 3 to 6.3 % for all Cobalt contents. Table 1 summarizes the lattice constants a and c, and the unit cell volumes v.

Table 1 Structural and magnetic data of $Sm_{1.5}Gd_{0.5}Fe_{17-x}Co_xN_y$ compounds.

Samples	a (Å)	c (Å)	v (Å^3)	y(N/u.f.)	T_c (K)	M_s (emu/g)
$Sm_{1.5}Gd_{0.5}Fe_{17}$	8.537(2)	12.41(3)	783.53		423	84.01
$Sm_{1.5}Gd_{0.5}Fe_{17}N_y$	8.389(3)	12.64(1)	807.65	3	707	107.62
$Sm_{1.5}Gd_{0.5}Fe_{16}Co$	8.533(3)	12.43(6)	784.22		523	116.31
$Sm_{1.5}Gd_{0.5}Fe_{16}CoN_y$	8.608(9)	12.71(8)	816.20	2.9	881	127.56
$Sm_{1.5}Gd_{0.5}Fe_{16}Co_2$	8.529(6)	12.45(6)	784.81		623	133.80
$Sm_{1.5}Gd_{0.5}Fe_{16}Co_2N_y$	8.646(1)	12.73(8)	824.64	2.9	1067	142.34
$Sm_{1.5}Gd_{0.5}Fe_{16}Co_3$	8.524(7)	12.48(5)	785.74		723	142.78
$Sm_{1.5}Gd_{0.5}Fe_{16}Co_3N_y$	8.715(8)	12.75(8)	839.30	2.8	–	149.97

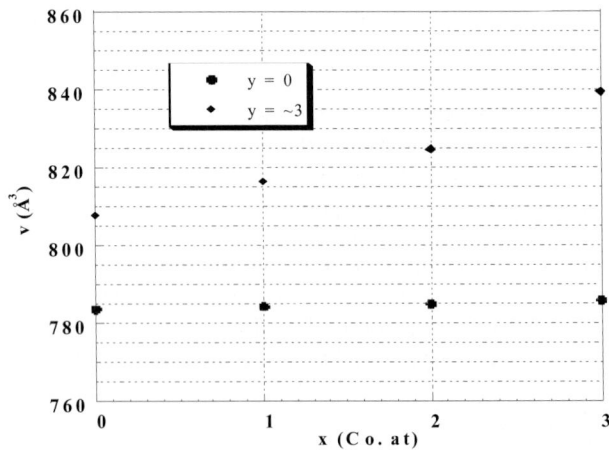

Fig. 1 Unit cell volume evolution versus cobalt content before and after nitrogen insertion.

3.2 Magnetic properties

Thermomagnetic analysis of $Sm_{1.5}Gd_{0.5}Fe_{17-x}Co_xN_y$ compounds show a typical ferromagnetic behaviour at room temperature. We plot in Fig. 2 the magnetization evolution versus temperature for both $Sm_{1.5}Gd_{0.5}Fe_{17}N_3$ and $Sm_{1.5}Gd_{0.5}Fe_{16}CoN_{2.9}$ samples. The nitrogen insertion leads to a strong increase in the Curie temperature T_c. The Curie temperature of R_2Fe_{17} compounds is determined by the combination of three types of exchange interactions: R–R, Fe–Fe and R–Fe [6, 7]. The enhancement of the Curie temperature in the interstitial compounds has been pointed out to be mainly due to the volume effect [8, 9]. Coey et al. [7] report about Curie temperature a large enhancement of the Fe–Fe interactions and a slight decrease of the R–Fe interactions in the interstitial compounds compared to their R_2Fe_{17} counterparts.

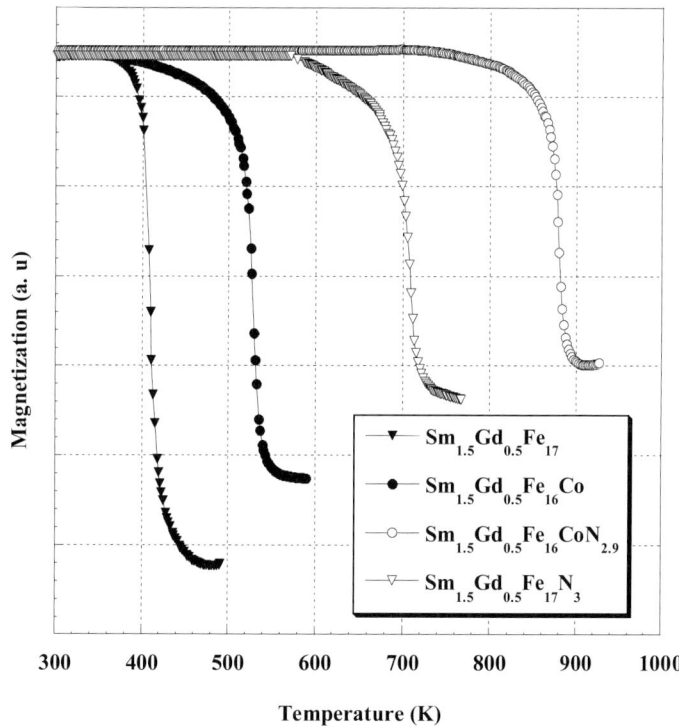

Fig. 2 Magnetization evolution versus temperature of $Sm_{1.5}Gd_{0.5}Fe_{17}N_3$ and $Sm_{1.5}Gd_{0.5}Fe_{16}CoN_{2.9}$.

The magnetization curves as a function of applied magnetic field M (H) of $Sm_{1.5}Gd_{0.5}Fe_{17-x}Co_xN_y$ compounds with different Co concentrations are plotted in Fig. 3a. The saturated magnetization (M_s) was obtained by fitting the experimental data of M versus $1/H^2$ using the polynomial law. The saturated magnetization at 300 K for $Sm_{1.5}Gd_{0.5}Fe_{17-x}Co_x$ and its nitrides are listed in Table 1. The values of the saturated magnetization before and after nitrogen insertion are plotted as a function of Co content in Fig. 3b. It can be seen that, as well as for the alloys, the saturated magnetization increases with increasing Co content, but the values of the nitride samples are higher than those obtained for the alloys ($Sm_{1.5}Gd_{0.5}Fe_{17-x}Co_x$). For example, at 300 K, M_s = 116.31 emu/g for $Sm_{1.5}Gd_{0.5}Fe_{16}Co$ and M_s = 127.57 emu/g for $Sm_{1.5}Gd_{0.5}Fe_{16}CoN_{2.9}$.

The X-ray diffraction patterns at room temperature of magnetically aligned powder samples of $Sm_{1.5}Gd_{0.5}Fe_{17-x}Co_xN_y$ with $x \leq 3$ are shown in Fig. 4. The X-ray diffraction studies on the aligned powder samples provide information concerning the magnetic anisotropy. For $Sm_{1.5}Gd_{0.5}Fe_{17-x}Co_xN_y$ compounds, the easy magnetization direction changes from the basal plane to conical.

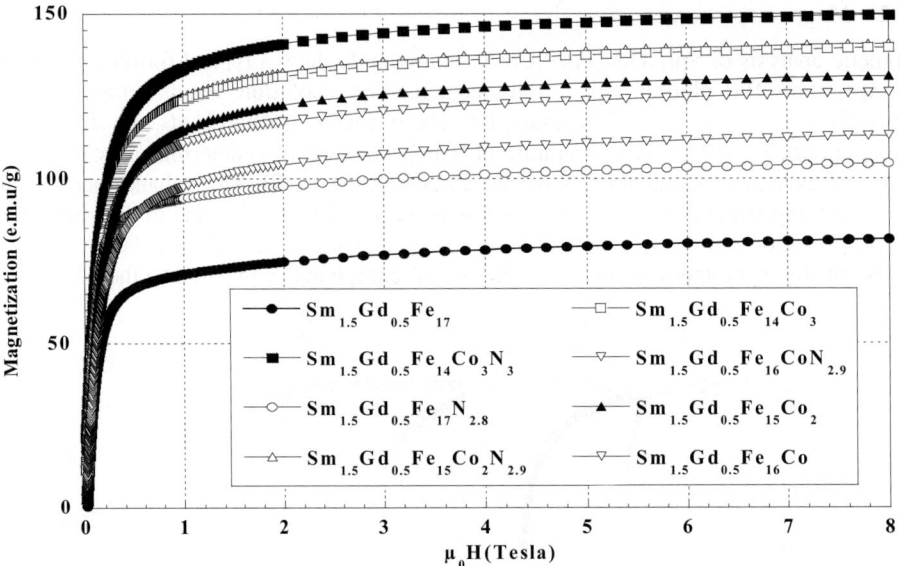

Fig. 3 a) Field dependence of the magnetization at room temperature of $Sm_{1.5}Gd_{0.5}Fe_{17-x}Co_xN_y$ before and after nitrogen insertion.

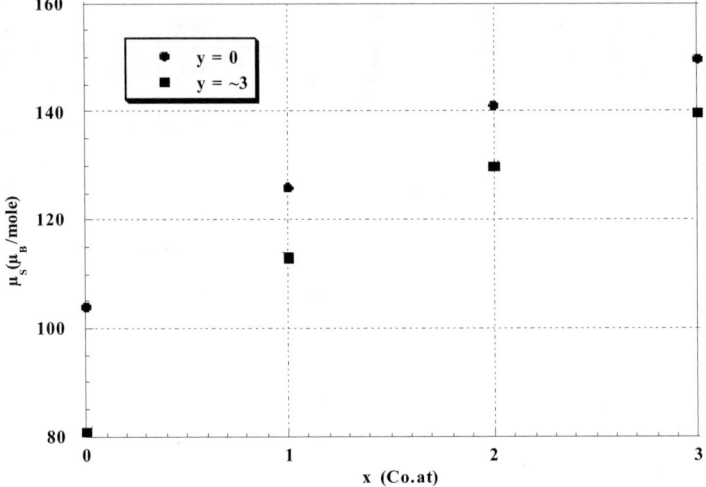

Fig. 3 b) The influence of cobalt content on the saturated magnetization at 300 K before and after nitrogen insertion.

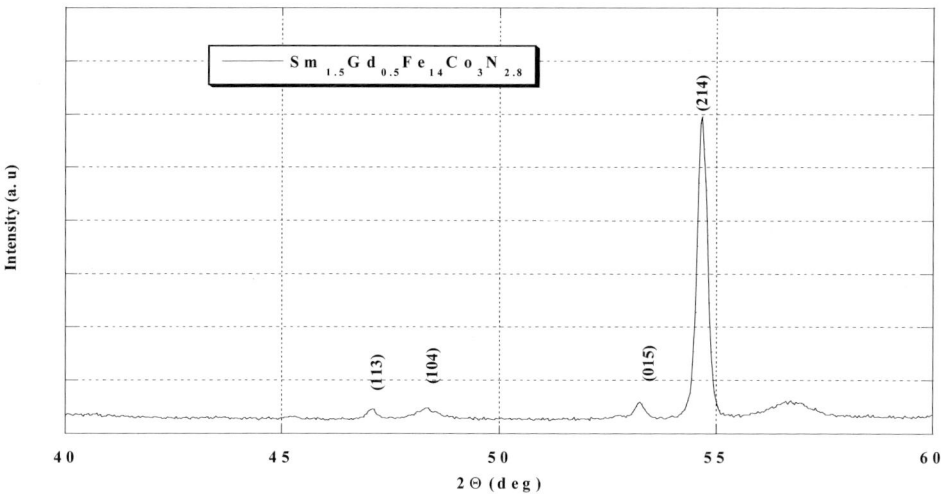

Fig. 4 X-ray diffraction patterns of magnetically aligned powder of $Sm_{1.5}Gd_{0.5}Fe_{14}Co_3N_{2.9}$.

4 Conclusion

We investigate the effect of the cobalt substitution and the nitrogen insertion on the structural and magnetic properties of $Sm_{1.5}Gd_{0.5}Fe_{17-x}Co_xN_y$ intermetallic alloys. The Th_2Zn_{17}-type structure of Sm_2Fe_{17} is retained. The nitrogen insertion leads to an increase of the unit-cell volume. The Curie temperature and the saturated magnetisation increase after nitrogen insertion. Finally, the easy direction of magnetization for nitrides samples with $x \geq 1$ is conical.

Acknowledgements This study has been supported by the Tunisian Ministry of high Education, Scientific research and technology and by the CMCU collaboration (01/F-1127).

References

[1] K. H. J. Buschow, J. Less-Common Met. **11**, 204 (1966).
[2] K. Hany, Acta Crystallogr. B **29**, 2502 (1973).
[3] K. H. J. Buschow, Rep. Prog. Phys. **40**, 1179 (1977).
[4] W. E. Wallace, Prog. Solid State Chem. **16**, 127 (1985).
[5] M. D. Coey and H. Sun, J. Magn. Magn. Mater. **87**, L251 (1990).
[6] J. M. D. Coey, Hong Sun, and Y. Otani, Proc. 6th Int. Symp. Magn. Anis. Coerc. RE-TM Alloys, Pittsburgh, PA, 1990, ed. S.G. Sankar (Carnegie Mellon Univ., Pittsburgh, PA, 1990), p. 36.
J. M. D. Coey, Hong Sun, and D. P. F. Hurely, J. Magn. Magn. Mater. **87**, L251 (1990).
[7] K. H. J. Buschow, Rep. Prog. Phys. **54**, 1123 (1991).
[8] D. Fruchart, O. Isnard, S. Mireglia, and J-L. Soubeyroux, J. Alloys Compd. **231**, 188 (1995).
[9] Q.-N. Qi, H. Sun, R. Skomski, and M. D. Coey, Phys. Rev. B **45**, 12278 (1992).

The crystallographic and magnetic properties of $Nd_{2-x}Gd_xFe_{16}Co$ solid solution and its nitrides

M. S. Ben Kraiem[1,*], **M. Ellouze**[1], **A. Cheikh-Rouhou**[1], and **Ph. L'Héritier**[2]

[1] Laboratoire de Physique des Matériaux, FSS, B. P. 802–3018, Sfax, Tunisie
[2] Laboratoire des Matériaux et du Génie Physique (UMR 5628 CNRS) ENSPG, B. P. 46, 38402 Saint Martin d'Hères, France

Received 31 August 2003, accepted 31 December 2003
Published online 23 April 2004

PACS 61.10.Nz, 61.66.Dk, 75.30.Cr, 75.30.Kz, 75.50.Bb

We elaborated and studied the physical properties of $Nd_{2-x}Gd_xFe_{16}Co$ alloys and its nitrides $Nd_{2-x}Gd_xFe_{16}CoN_y$. By means of X-ray powder diffraction analysis, it is shown for $0 \leq x \leq 2$ that all our synthesized compounds crystallize in the Th_2Zn_{17} structure. Nitrogen insertion leads to an increase of the unit cell volume. Curie temperature T_c increases with increasing gadolinium content in the alloys. Nitrogen insertion leads to an increase of T_c. The Curie temperature is found to increase from 441 K for $x = 0$ to 573 K for $x = 2$ in the alloys and from 773 K for $x = 0$ to 823 K for $x = 2$ in the nitrides.

© 2004 WILEY-VCH Verlag GmbH & Co. KGaA, Weinheim

1 Introduction

Carbides and nitrides of R_2Fe_{17} (R is rare earth) intermetallic alloys exhibit high potential permanent magnet properties. The insertion of light element (C,N) in these intermetallic alloys markedly reinforces the intrinsic properties such as Curie temperature, magnetization and mgnetocrystalline anisotropy [1].

During last years, studies on $R_2Fe_{17}N_x$ compounds have been reported by Coey and Sun [2] and it has been shown that the nitrogen content is important for improving magnetic properties. Previous neutron-powder diffraction has shown that the nitrogen atoms in $Nd_2Fe_{17}N_x$ occupy both 9e and 18g sites [3]. In this paper we report the structural and magnetic properties of $Nd_{2-x}Gd_xFe_{16}CoN_y$ powder samples.

2 Experimental

$Nd_{2-x}Gd_xFe_{16}Co$ compounds were prepared from starting materials of at least 99.9% purity by induction melting. After melting, the polycrystalline specimens were sealed in silica tubes under argon atmosphere and annealed at 1273 K for 10 days and then quenched in water. Phase purity was checked by X-ray diffraction (XRD) using SIEMENS diffractometer with iron radiation ($\lambda = 1.936$ Å). The saturated magnetization (μ_s) has been determined using a vibrating sample magnetometer (VSM) in a magnetic applied field up to 6 T. The Curie temperatures were obtained from thermomagnetic analysis (TMA) using a home – made Faraday-type balance. The insertion of nitrogen in our gadolinium substituted samples has been carried out under a static nitrogen gas under a pressure of 15 MPa at 733 K for 24 hours. The nitrogen content was checked by the weight increase in our samples. The nitrogen amount is found to be 2.9 ± 0.1 atoms.

---

[*] Corresponding author: e-mail: ben_essalah@yahoo.fr, Phone: ++ 216 98 656 912, Fax: ++216 74 274 437.

3 Results and discussion

X-ray powder diffraction study at room temperature shows that all our investigated alloys and their nitrides are single phase with very small amount of α-Fe as secondary phase in the nitrides. The $Nd_{2-x}Gd_xFe_{16}Co$ alloys and its nitrides crystallize in the rhombohedral Th_2Zn_{17}-type structure. We plot in Fig. 1 the unit cell volume evolution as a function of Gd content for $Nd_{2-x}Gd_xFe_{16}Co$ alloys and its nitrides.

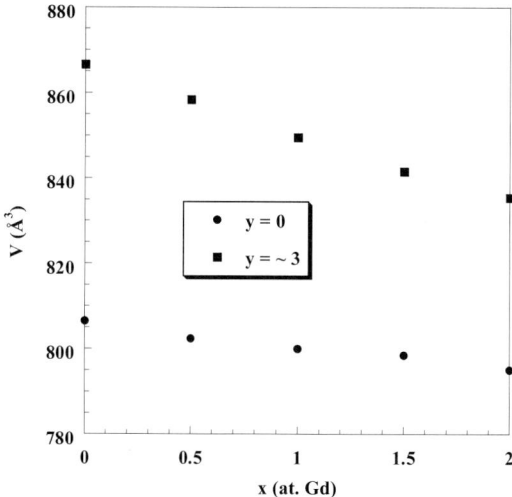

Fig. 1 The unit cell volume evolution as a function of gadolinium content in the $Nd_{2-x}Gd_xFe_{16}Co$ alloys and its nitrides.

The unit cell volume of the alloys decreases linearly with increasing Gd content, it is found to decrease from 806 Å^3 for x = 0 to 795 Å^3 for x = 2. Nitrogen insertion leads to an expansion of the unit cell volume, this expansion depends on the Gd content. The expansion of the unit-cell volume ΔV/V decreases with increasing Gd content from 6.94% for x = 0 to 5.7% for x = 2. We report in Table 1 the crystallographic data, Curie temperatures, and saturated magnetization of the synthesized alloys and its nitrides.

Table 1 Structural and magnetic parameters of $Nd_{2-x}Gd_xFe_{16}CoN_y$ compounds

Samples	a (Å)	c (Å)	V (Å^3)	y	ΔV/V(%)	T$_C$ (K)	μ$_S$ (μ$_B$/mol) (300 K)
$Nd_2Fe_{16}Co$	8.594(6)	12.46(9)	806	–		441	33.20
$Nd_2Fe_{16}CoN_y$	8.980(0)	12.40(9)	866	3	6.94	773	38.39
$Nd_{1.5}Gd_{0.5}Fe_{16}Co$	8.548(0)	12.47(6)	802	–		504	31.46
$Nd_{1.5}Gd_{0.5}Fe_{16}CoN_y$	8.932(7)	12.42(2)	858	2.9	6.52	795	35.83
$NdGdFe_{16}Co$	8.514(8)	12.49(4)	800	–		527	29.55
$NdGdFe_{16}CoN_y$	8.869(7)	12.47(4)	849	2.8	6.13	803	33.97
$Nd_{0.5}Gd_{1.5}Fe_{16}Co$	8.501(0)	12.50(5)	798	–		555	28.07
$Nd_{0.5}Gd_{1.5}Fe_{16}CoN_y$	8.795(4)	12.56(1)	844	2.9	5.8	813	33.34
$Gd_2Fe_{16}Co$	8.497(0)	12.52(0)	795	–		573	26.66
$Gd_2Fe_{16}CoN_y$	8.737(4)	12.63(7)	840	2.8	5.7	823	31.98

Thermomagnetic analysis of $Nd_{2-x}Gd_xFe_{16}CoN_y$ compounds show a typical ferromagnetic behavior at room temperature. We plot in Fig. 2 the temperature dependence of the magnetization for $Nd_{2-x}Gd_xFe_{16}Co$ (x = 0.5; 1) and its nitrides. Nitrogen insertion leads to an increase of the Curie temperature T_c.

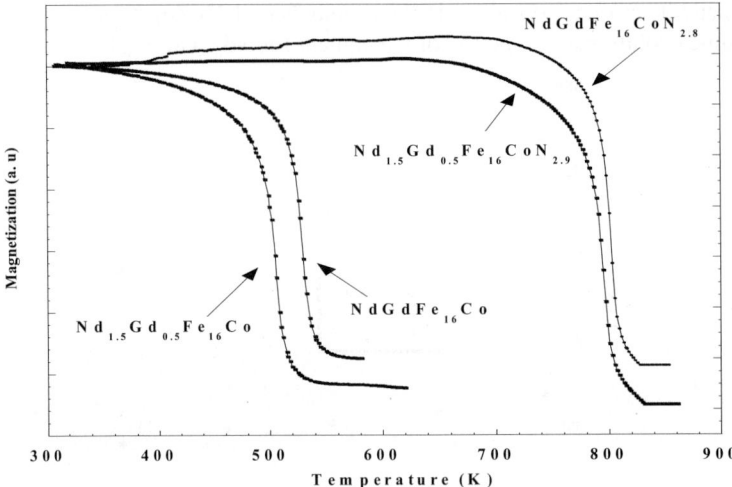

Fig. 2 Magnetization evolution versus temperature of $Nd_{1.5}Gd_{0.5}Fe_{16}Co$ and $NdGdFe_{16}CoN_{2.8}$ alloys and its nitrides.

We plot in Fig. 3 the T_c evolution versus Gd content for the alloys and its nitrides. The Curie temperature increases monotonically with increasing Gd content in both alloys and its nitrides. For the alloys, T_c is found to increase from 441 K for x = 0 to 573 K for x = 2 and from 773 K for x = 0 to 823 K for x = 2 in the nitrides. However the T_c enhancement with Gd amount is more important in the alloys than in the nitrides, in fact $\Delta T_c/T_c$ decreases with increasing Gd content from 43% for x = 0 to 30% for x = 2 after nitrogen insertion. Usually, the low Curie temperature in R_2Fe_{17} is suggested to exist owing to an antiferromagnetic interaction due to the short interatomic distance between Fe–Fe atoms on some sites. It has been demonstrated that in many cases the enhancement in curie temperature of rare-earth iron compounds correspond to an increase of the Fe–Fe exchange interactions as a result of the increasing interatomic Fe–Fe distance, owing to the addition of interstitial or substitutional atoms [4]. Therefore, the nitrogen insertion in the $Nd_{2-x}Gd_xFe_{16}Co$ leads an increase of Curie temperature (Fig. 3).

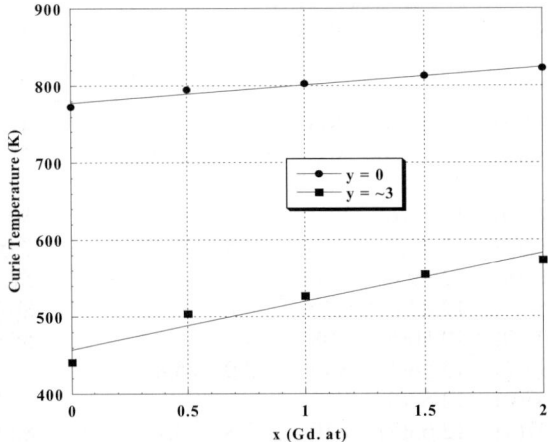

Fig. 3 The influence of Gd concentration on Curie temperature before and after nitridation.

The magnetization curves as a function of magnetic applied field up to 6 T at room temperature (300 K) of $Nd_{2-x}Gd_xFe_{16}Co$ alloys and its nitrides are plotted in Figs. 4a and 4b, respectively.

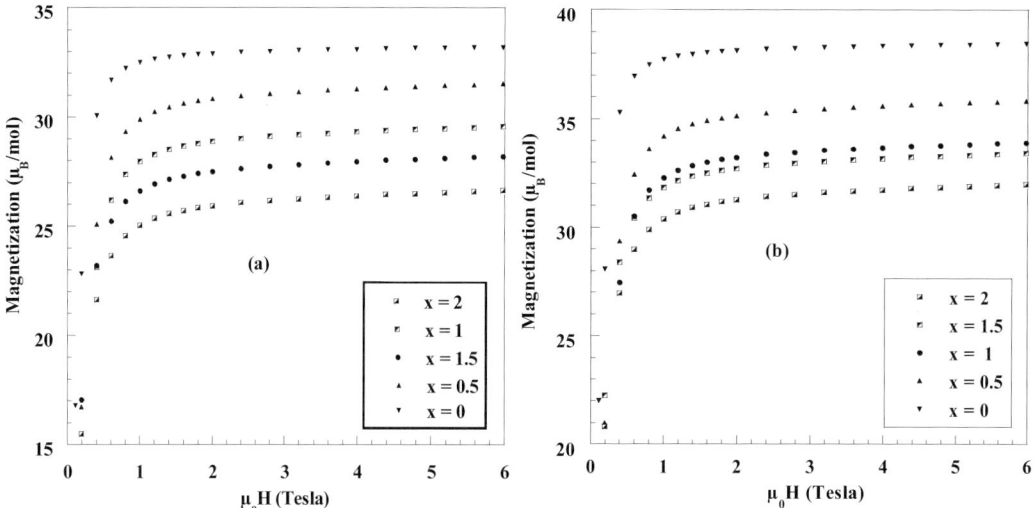

Fig. 4 Field dependence of the magnetization at room temperature of $Nd_{2-x}Gd_xFe_{16}Co$ (a) and $Nd_{2-x}Gd_xFe_{16}CoN_y$ (b).

The saturated magnetization (μ_s) was obtained by fitting the experimental data of M (1/ H^2) using the polynomial law. We plot in Fig. 5 the evolution of the saturated magnetization versus gadolinium content for the alloys and its nitrides.

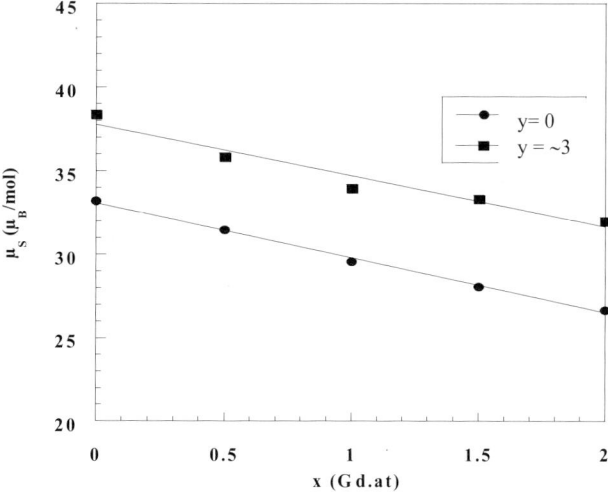

Fig. 5 The influence of Gd concentration on the saturated magnetization at 300 K before and after nitrogen insertion.

The saturated magnetization in both cases decreases monotonically with increasing Gd content, however, μ_s values for the nitrides are higher than those obtained for the $Nd_{2-x}Gd_xFe_{16}Co$ alloys (at 300 K, μ_s = 31.46 μ_B/mol for $Nd_{1.5}Gd_{0.5}Fe_{16}Co$ and μ_s = 35.83μ_B/mol for $Nd_{1.5}Gd_{0.5}Fe_{16}CoN_{2.9}$). Neutron diffraction studies [5] have shown that the nitrogen atoms occupy the 9e interstitial site and the magnetic moment of iron (Fe) increases due to a reduction in the Fe–N overlap caused by the increase in the average Fe–Fe distance. Furthermore, the magnetic moment of the Fe atoms in the 18f site, which is the nearest to the 9e-N atoms is the lowest owing to the hybridizations between the Fe-3d and N-2p states. On the contrary, the Fe-6c atoms, which are the farthest from the 9e-N atoms, have the highest magnetic moment. The largest reduction of the magnetization due to the strongest hybridization effects is found, either experimentally [6] and theoretically [7, 8], in the 18h and 18f sites.

4 Conclusion

The structural and magnetic properties of $Nd_{2-x}Gd_xFe_{16}Co$ alloys and its nitrides have been studied. All our samples crystallize in the Th_2Zn_{17}-type structure. Nitrogen insertion in $Nd_{2-x}Gd_xFe_{16}Co$ leads to an increase in the unit-cell volume. The Curie temperature of $Nd_{2-x}Gd_xFe_{16}CoN_y$ compounds increases linearly with increasing Gd content. The saturated magnetization μ_s increases after nitrogen insertion. μ_s in both alloys and nitrides decreases monotonically with increasing Gd content.

Acknowledgements This study has been supported by the Tunisian Ministry of High Education, Scientific Research and Technology and by the CMCU collaboration (01/F-1127).

References

[1] D. Fruchart, O. Isnard, S. Miraglia, and J. L.Soubeyroux, J. Alloys Compd. **231**, 188 (1995).
[2] J. M. D. Coey and H. Sun, J. Magn. Magn. Mater. **87**, L251 (1990).
[3] S. S. Jaswal, W. B. Yelon, G. C. Hadjipanayis, Y. Z. Wang, and D. J. Sellmyer, Phys. Rev. Lett. **67**, 644 (1991).
[4] B. G. Shen, Z. H. Cheng, H. Y. Gong, B. Liang, Q. W. Yan, F. W. Wang, J. X. Zhang, S. Y. Zhang, and H. Q. Guo, J. Alloys Compd. **226**, 51 (1995).
[5] S. S. Jaswal, W. B. Yelon, G. C. Hadjipanayis, Y. Z. Wang, and D. J. Sellmyer, Phys. Rev. Lett. **67**, 644 (1991).
T. Kajitani, Y. Morii, S. Funahashi, H. Kato, Y. Nakagawa, and K. Hiraya, J. Appl. Phys. **73**, 6032 (1993).
[6] O. Isnard and D. Fruchart, J. Alloys Compd. **205**, 1 (1994).
[7] R. Coehoorn, Phys. Rev. B **39**, 13072 (1989).
[8] T. Beuerle, P. Braum, and M. Faehnle, J. Magn. Magn. Mater. **94**, L11 (1991).

Structural and magnetic studies of 1% Ho:$Gd_{0.99-x}Lu_x$ alloys

I. A. Al-Omari[*,1], **A. Rais**[1], **M. S. Lataifeh**[2], and **A. A. Yousif**[1]

[1] Department of Physics, Sultan Qaboos University, P. O. Box 36, PC 123, Muscat, Sultanate of Oman
[2] Department of Physics, Mu'tahUniversity, P. O .Box 7, Al-Karak 61710, Jordan

Received 31 August 2003, accepted 31 December 2003
Published online 1 April 2004

PACS 61.66.Dk, 75.20 En, 75.30 Cr

We present a study of the magnetic and structural properties of 1% Ho:$Gd_{0.99-x}Lu_x$ alloys. X-ray diffraction patterns for 1% Ho:$Gd_{0.99-x}Lu_x$ (x = 0.01, 0.39, and 0.59) show a single hexagonal type phase structure and the lattice parameters are found to depend on the Lutetium concentration. Magnetic measurements were made at temperatures between 100 K and 850 K using a vibrating sample magnetometer with a maximum field of 13.5 kOe. All the samples under investigation show a ferromagnetic behavior up to Curie temperature. The Curie temperature (T_c) and the saturation magnetization (M_s) are found to decrease with increasing the Lutetium concentration. The saturation magnetization is found to increase with decreasing the temperature (T) for all samples. The magnetic moment of 1% Ho:$Gd_{0.98}Lu_{0.01}$ at 0 K was estimated, to be (7.05 ± 0.05) μ_B/f.u., by extrapolating the M_s versus $T^{3/2}$ to T = 0 K. The results clearly indicate the formation of solid solution alloys.

© 2004 WILEY-VCH Verlag GmbH & Co. KGaA, Weinheim

1 Introduction

The rare earth compounds have attracted many researchers in the last few years, that is because they play an important role in many technological applications such as permanent magnets, magnetic recording and memory devices. They also serve as an excellent example for the theory of ferromagnetic and ferrimagnetic alloys. NMR study of the magnetic anisotropy and field dependence of the hyperfine interactions of ^{165}Ho in ferromagnetically ordered single crystals of holmium with gadolinium have already been studied using NMR [1, 2]. The results reveal that the crystal field quenching of the holmium magnetic moment is negligible. A pilot NMR study of Ho in yttrium (1% Ho:Y) has been reported by Lataifeh [3]. In this study, they measured the field dependence of the magnetic dipole hyperfine splitting and found that Ho in the polycrystalline specimen of 1% Ho:Y acts as a free ion and their results were in reasonable agreement with the theoretical predictions based on Touborg's set of crystal field parameters. So, we expect that the magnetic moment for any rare earth ion embedded in the solid to be equal to its free ion value.

The lanthanide metals have some form of hexagonal structure at room temperature (RT). The heavy rare earth metals have a simple h.c.p structure [4]. In the alloys to be investigated the crystal structure is the same as for the constituent elements if they have the same structure. The distribution of the ions in the hexagonal lattice sites is random, because of the chemical similarity of heavy rare earth ions making the alloy. In all rare earth metals, gadolinium is exceptional as it has no orbital moment, has weak magnetic anisotropy and so behaves as a perfect host. The large molecular field set up by the isotropic gadolinium host makes it possible to rotate the holmium moment into magnetically hard directions with a small applied field. The reason for using low holmium percentage is that it has been used as a probe for NMR measurements. On the other hand Lu is a non-magnetic rare earth which acts as a solvent.

---

[*] Corresponding author: e-mail: ialomari@yahoo.com

Hanyu et al. [5] studied Gd–Lu multilayers using resonant photoemission and found that there is a new satellite structure above the 4f main peak for pure Gd, but this structure disappeared in Gd–Lu films, which is due to the intervening effect of Lu between Gd layers. Eccleston *et al.* [6, 7] studied the magnetic structure of $Gd_{76}Lu_{24}$ alloys in an applied magnetic field using ultrasound measurements at different temperatures and applied magnetic fields. This study added a new detail to the phase diagram and revealed the existence of Ferro-I in addition to helical antiferromagnetic and canted ferromagnetic phases. Recent studies by Venturini et al. [8] and Ijjaali et al. [9] of $(Gd–Lu)Ge_{2-x}$ ($0.33 < x < 0.5$) single crystal compounds, using X-ray diffraction and Rietveld refinement method, have shown that the samples display additional diffraction intensities suggesting ordering of the Ge vacancies. In theses studies, the superstructure lines were indexed by considering propagation vectors with respect to the ortho-hexagonal subcell of the AlB_2 structure.

In this work, we will study the effect of Lu substitution on the structural and magnetic properties of the alloy 1% Ho:$Gd_{0.99-x}Lu_x$ alloys where $x = 0.01$, 0.39 and 0.59. This is part of our project to study rare earth alloys using different techniques such as XRD, VSM, and NMR.

2 Experimental

The bulk samples of 1% Ho:$Gd_{0.99-x}Lu_x$ alloys ($x = 0$, 0.01, 0.39, and 0.59) were prepared from at least 99.9% pure elements by arc melting under a flowing argon atmosphere. Each sample was melted four to five times to insure homogeneity. Phase characterization of the samples was achieved by X-ray diffraction on powder samples with a Philips diffractometer, PW1820, by using Cu-K_α radiation. Magnetic measurements were performed by using a Vibrating Sample Magnetometer (VSM) at temperatures between 100 K and 850 K and in magnetic fields up to 13.5 kOe.

3 Results and discussion

Powder X-ray diffraction (XRD) patterns for 1% Ho:$Gd_{0.99-x}Lu_x$ alloys where $x = 0.01$, 0.39 and 0.59 have been recorded. Figure 1 shows the XRD pattern for a representative sample 1% Ho:$Gd_{0.40}Lu_{0.59}$. The other two samples show the same behavior with peaks' shift. The patterns show several sharp peaks with indices corresponding to a single hexagonal type phase structure. The lattice parameters a and c are listed in Table 1, and were found to decrease with increasing the lutetium concentration in the alloy. This decrease can be attributed to the smaller Lu atomic radius of (2.25 Å) compared with the Gd atomic radius of (2.54 Å). The lattice parameters for $x = 0.01$ are close to the lattice parameters of $a = 3.629$ Å and $c = 5.760$ Å for pure Gd [10].

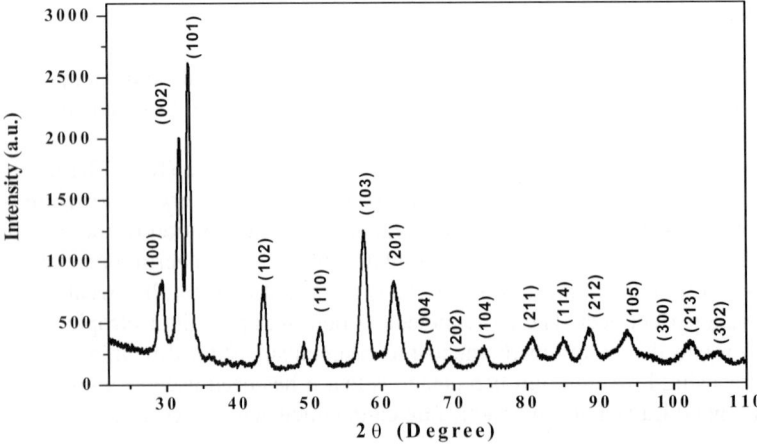

Fig. 1 X-ray diffraction patterns for 1% Ho: $Gd_{0.4}Lu_{0.59}$.

Table 1 Dependence of the lattice parameters a and c and the unit cell volume (V) of 1% Ho:$Gd_{0.98-x}Lu_x$ alloys on the Lu concentration (x).

x	a (Å)	c (Å)	V (Å^3)
0.01	3.626	5.756	65.538
0.39	3.584	5.682	63.205
0.59	3.560	5.640	61.901

Magnetization measurements on 1% Ho:$Gd_{0.99-x}Lu_x$, where x = 0.01, 0.39, and 0.59, have been done by using a VSM. Figure 2 shows the magnetization (M) versus the temperature (T) for the three different alloys. The magnetic curves indicate that the alloy follow a ferromagnetic behavior. The effect of lutetium, which is not magnetic because it has a full 4f shell and act as a solvent, is seen clearly in the figure. When the Lu content increases the spontaneous magnetization decreases. It was also observed that the Curie temperature T_c follow the same rule, T_c decreases with the increase in Lu concentration. The values of T_c were obtained by using the method of intersecting tangents for M versus T curves and found to be 330 K, 250 K, and 230 K for x = 0.01, 0.39, and 0.59, respectively. The decrease in M and T_c in the alloys as Lu content increases can be attributed to the reduction of the exchange field interaction between the spins of the magnetic ions (Gd and Ho). This reduction is attributed to the increase of interatomic spacing. The exchange field decreases in a similar manner as coulomb field does. A second order effect for the decrease of M and T_c may be attributed to the decrease of magnetic dipole–dipole interaction, since this interaction is inversely proportional to the cubic inter-atomic spacing [11].

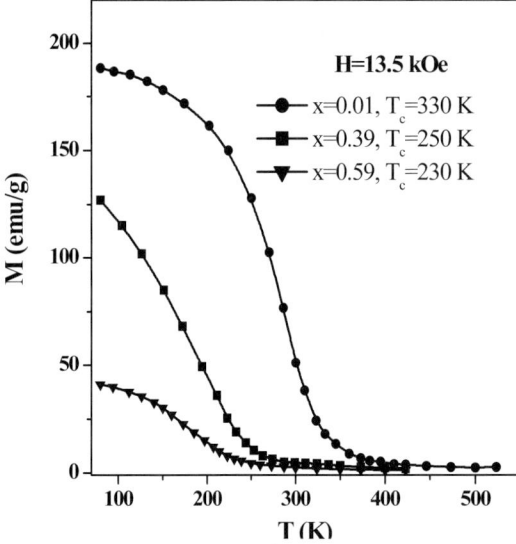

Fig. 2 Magnetization (M) vs temperature (T) at field of 13.5 kOe of 1% Ho:$Gd_{0.99-x}Lu_x$ alloys.

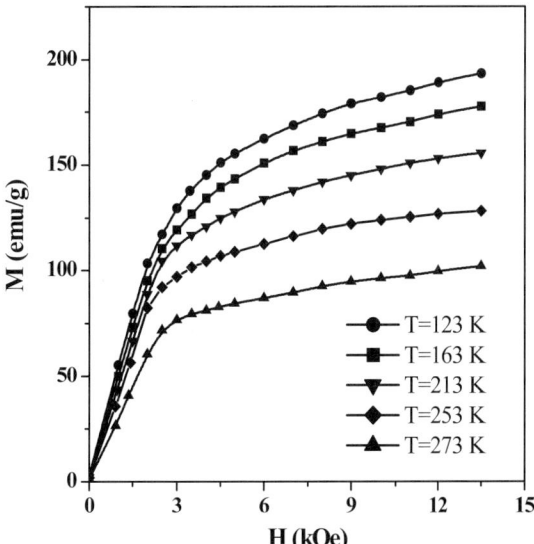

Fig. 3 Initial magnetization curves for 1% Ho:$Gd_{0.98}Lu_{0.01}$ alloy at different temperatures. The effect of small percent of holmium is pronounced on T_c and on M. The Curie temperature for pure gadolinium metal is 289 K, while it is 330 K for 1% Ho:$Gd_{0.98}Lu_{0.01}$ and that is because the Ho^{3+} has a magnetic moment (10 μ_B) greater than that of Gd (7 μ_B) and it has a smaller radius than that of gadolinium. These two effects will increase the exchange interaction responsible for the magnetic ordering.

Magnetic isothermal curves have been done on these alloys; representative results are shown in Fig. 3 for 1% Ho:Gd$_{0.98}$Lu$_{0.01}$ at different temperatures. Results for the other two samples show similar behavior. It is clear that the magnetization is reduced by the increase in temperature because of the increase in thermal agitation of the ions. In addition, the magnetic dipoles become disordered which lead to low magnetization. The initial magnetic susceptibility χ_{in} decreases with the increase of temperature because the response of the magnetic moments decreases with the increase in temperature.

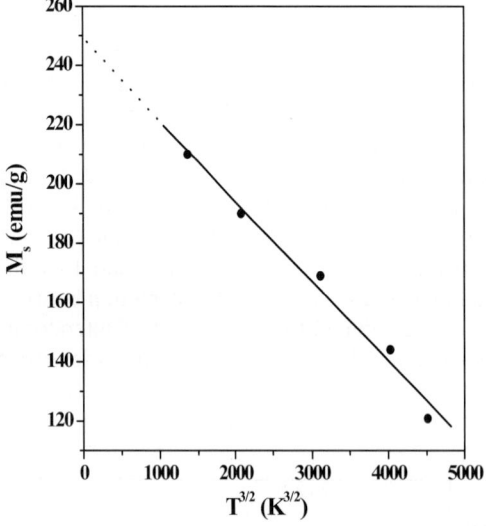

Fig. 4 Saturation magnetization (M$_s$) vs T$^{3/2}$ of 1% Ho:Gd$_{0.98}$Lu$_{0.01}$ alloy. The solid line represent the linear fitting based on Bloch T$^{3/2}$ law.

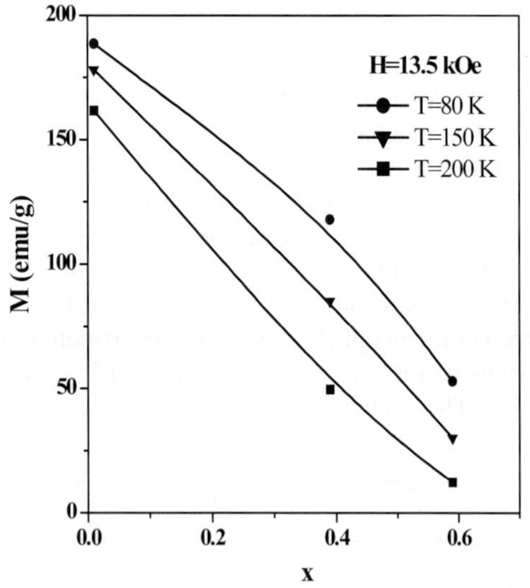

Fig. 5 Magnetization (M) vs Lu concentration (x) at field of 13.5 kOe, of 1% Ho:Gd$_{0.99-x}$Lu$_x$ alloy.

The saturation magnetization M$_s$ at different temperatures can be estimated by making a plot of M versus 1/H for high magnetic field values and extrapolate M to 1/H = 0. The saturation magnetization is found to increase with decreasing the temperature (T) for all samples. The dependence of the saturation magnetization on the temperature for 1% Ho:Gd$_{0.98}$Lu$_{0.01}$ alloy is shown in Fig. 4, M$_s$ as a function of T$^{3/2}$. The solid line represent the linear fitting based on Bloch T$^{3/2}$ law. We can see also from this figure the effect of temperature on magnetization. The magnetic moment of 1% Ho:Gd$_{0.98}$Lu$_{0.01}$ at 0 K was estimated, to be (7.05 ± 0.05) μ$_B$/f.u., by extrapolating the M$_s$ versus T$^{3/2}$ to T = 0 K. This value is in agreement with previous value for pure Gd, 7 μ$_B$/atom [11]. Due to the limitations of the maximum applied magnetic field (13.5 kOe) of our VSM, it is unreliable to determine the saturation magnetization for the samples with x = 0.39 and x = 0.59. However, the trend of the increase in the magnetization with decreasing the temperature is the same as for x = 0.01. Figure 5 shows the magnetization at an applied magnetic field of 13.5 kOe as a function of Lu concentration (x) at different temperatures, 80 K, 150 K and 200 K (below T$_c$). From this figure we can see that M decreases almost linearly with the increase of Lu concentration for all temperatures.

4 Conclusion

In this study we found that 1% Ho:Gd$_{0.98-x}$Lu$_x$ alloys have the hexagonal structure with lattice parameters depending on the Lu concentration. All the samples under investigation show ferromagnetic behavior up to Curie temperature. The saturation magnetization and the Curie temperature were found to decrease with increasing the Lu concentration. The above results clearly indicate the formation of solid solution alloys.

Acknowledgments We would like to thank Sultan Qaboos University for the support provided to perform this study and Dr. David Bunbury from the University of Manchester for supplying the samples.

References

[1] M. S. Lataifeh and Ph. D. Thesis, The Hyperfine Interactions of ^{165}Ho in Magnetically Ordered Single Crystals, 1989, pp. 81–118.
[2] C. Carboni, M. A. H. McCausland, D. St. P. Bunbury, B. L. Reid, M. Lataifeh, and J. S. Abell, J. Magn. Magn. Mater. **104–107**, 1513 (1992).
[3] M. S. Lataifeh, Mu'tah, J. Research and Studies **9**, No. 1, 113 (1994).
[4] M. A. H. McCausland and I. S. Mackenzie, Nuclear Magnetic Resonance, in: Rare Earth Metals (Taylor and Francis, London, 1980).
[5] T. Hanyu, H. Ishii, S. Hashimoto, T. Yokoyama, K. Jokura, H. Sato, and T. Miyahara, J. Electron. Spectrosc. Relat. Phenom. **78**, 67 (1996).
[6] R. S. Eccleston, A. R. Griffiths, M. L. Vrtis, G. J. Melntyre, D. Fort, and S. B. Palmer, Physica B **174**, 33 (1991).
[7] R. S. Eccleston and S. B. Palmer, J. Magn. Magn. Mater. **104–107**, 1527 (1992).
[8] G. Venturini, I. Ijjaali, and B. Malaman, J. Alloys Compd. **284**, 262 (1999).
[9] I. Ijjaali, G. Venturini, and B. Malaman, J. Alloys Compd. **284**, 237 (1999).
[10] Klemm, Bommer, Z. Anorg. Chem. **231**, 150 (1937).
[11] B. D. Cullity, Introduction to Magnetic Materials (Addison-Wesley Publishing Company, Reading (Mass.) 1972).

Structure and hyperfine parameters of nanocrystalline $R_{1-s}(Fe, M)_{5+2s}$

A. Nandra, S. Sab, E. Dorolti, L. Bessais, and C. Djéga-Mariadassou[*]

LCMTR, CNRS UPR 209, 2-8, rue Henri Dunant, 94320 Thiais, France

Received 31 August 2003, accepted 31 December 2003
Published online 18 March 2004

PACS 75.50.Bb, 75.50.Tt, 76.80.+y

Structure and hyperfine parameters of nanocrystalline metastable $SmFe_{9-y}M_y$ (M = Si, Ga) and $RTiFe_{9-x}Co_x$ (R = Pr, Sm) alloys, precursors of equilibrium $R\bar{3}m$ $Sm_2(Fe,M)_{17}$ and $I4/mmm$ $RTi(Fe,Co)_{11}$ phases have been investigated. The Rietveld analysis leads to the stoichiometry 1/9 and 1/10 for the precursor of the 2/17 and 1/12 alloys. Silicon, gallium and cobalt are located in $3g$ sites, titanium in $6l$ site of the $P6/mmm$ structure. The hyperfine parameters were assigned according to the relationship between the Wigner-Seitz cell volume of each iron site and their isomer shift. The recurrent sequences $\delta 2e > \delta 3g > \delta 6l$ and $H_{HF}2e > H_{HF}6l > H_{HF}3g$ are observed.

© 2004 WILEY-VCH Verlag GmbH & Co. KGaA, Weinheim

1 Introduction

The comprehension of the structure transformation from the nanocrystalline precursor to the equilibrium phases 2/17 and 1/12 is still open to discussion. Both precursors are relevant of the $P6/mmm$ space-group and generally known as $TbCu_7$, derived from $CaCu_5$ [1] while equilibrium structure $R\bar{3}m$ and $I4/mmm$ constitute super-lattices of the hexagonal sub-cell. However, we have recently demonstrated by Rietveld analysis coupled to Mössbauer spectroscopy that the precursor of $Sm(Fe,Ti)_{12}$ is $Sm(Fe,Ti)_{10}$ [2] and $SmFe_9$ is the precursor of Sm_2Fe_{17} [3].

The objective of this paper is the understanding of the specific precursors of Sm_2Fe_{17} and $RFe_{11}Ti$ (R = Sm, Pr), partially substituted respectively by Si, Ga and Co. The structural characteristics of high energy ball-milled and subsequently annealed alloys will be reported on various series $SmFe_{9-y}M_y$ (M = Si, Ga) and $RFe_{9-x}Co_xTi$ (R = Sm, Pr). On the analogy of $Sm_2Fe_{17-x}M_x$ (M = Si, Ga) and $RTiFe_{11-x}Co_x$ (R = Sm, Pr) alloys, for which the interesting magnetic properties are restricted to $x = 2$, we have limited the study to $y = 1$ for $Sm(Fe,M)_9$ and x =2 for $RFe_{9-x}Co_xTi$ alloys. Finally, an accurate and generalized interpretation of the Mössbauer spectra will then be presented on the basis of the structure information.

2 Experiment

We have prepared $RFe_{9-x}Co_xTi$ (R=Pr,Sm) and $SmFe_{9-y}M_y$ (M = Si, Ga) alloys by ball-milling inductively melted preloys with composition checked by (ICP-AES) for concentration x equal to 0.5, 1, 1.5, 2, and $y = 0.25, 0.5, 0.75, 1$. The resulting powders were subsequently annealed at 700–750 °C for 30 min. The X-ray spectra refinement was performed with the FULLPROF computing code based on the Rietveld technique. The Mössbauer spectra were collected using a conventional spectrometer with a source of ^{57}Co in Rh matrix.

3 Results and discussion

The problem concerning the $P6/mmm$ hexagonal phase is effectively its stoichiometry domain R_xM_{1-x}. The structure is derived from RM_5, it might lay theoretically from x = 0.16 (RM_5) down to $x = 0.07$ prelude to the 1:12 structure. Due to the nanoscale of the hexagonal phases preventing from a pertinent X-ray dispersive analysis, the phase composition had to be deduced from a Rietveld refinement.

---

[*] Corresponding author: e-mail: Catherine.Djega@glvt-cnrs.fr

We have explained the disordered $P6/mmm$ structure according to a generalization of the model developed previously by Givord et al. [4] on Co super-stoichiometric SmCo$_{9+\epsilon}$ alloys ($\epsilon = 0.22$). Let be $(1-s)$ the occupation rate of R (site $1a$ in 0,0,0 position), with decreasing R content, $2s$ Fe in $2e$ site (site of Fe dumbbells in $(0,0,Z)$ position) will be distributed on both sides of the empty Sm position. It results that only $2(1-3s)$ atoms remain located in the $2c$ special position $(1/3, 2/3, 0)$, and due to the presence of the $2e$ Fe dumbbells, $6s$ Fe are shifted towards the c axis in the more general $6l$ position $X, 2X, 0, (X < 1/3)$. The $3g$ site $1/2, 0, 1/2$, is totally occupied by the Fe atoms.

The Rietveld refinement of the XRD patterns is founded upon the structure model described above and on the fact that the transition between the out-of-equilibrium precursors and their equilibrium derivatives $R\bar{3}m$ or $I4/mmm$ appear without any noticeable displacement of the atom position [2].

The structure approach of the SmFe$_{9-y}$M$_y$ (M = Si, Ga) alloys is based on the M atom distribution correlation between the atomic positions of the $R\bar{3}m$ structure and those of the hexagonal $P6/mmm$. According to the Si and Ga location established previously [5, 6], up to $x = 2$, the Ga and Si atoms occupy preferentially the $18h$ site of the $R\bar{3}m$ structure in agreement with the results found on Nd$_2$Fe$_{17-x}$M$_x$, (M = Al, Ga, Si) for x lower than 2.2 [7].

The $6c$ and $18f$ positions of the $R\bar{3}m$ structure transform respectively into $2e$ and $6l$. The $R\bar{3}m$ $9d$ atoms constitute one third of the $3g$ $P6/mmm$ site. The remaining two thirds result from the $18h$ position. It comes that Si and Ga location is consistent with the exclusive occupation of the $3g$ site.

The same approach has been used for the RTiFe$_{9-x}$Co$_x$ (R = Pr, Sm) series. For the tetragonal $I4/mmm$ the $8i$ site generates the $2e$ site and one half of the $6l$ site in the $P6/mmm$ RM$_5$ motif. The whole $8f$ site constitutes two thirds of the $3g$ site. The remaining third of the $3g$ family and half of the $6l$ site result from the $8j$ site. Due to the close values of the Ti and Fe atomic diffusion factors, the X-ray analysis cannot solve the Ti preference. However, as it will be shown below by Mössbauer spectroscopy, the Ti occupation of the $2e$ site is excluded. The Rietveld refinement will be performed with Ti located in $6l$ site. According to the structural relationship between $I4/mmm$ and $P6/mmm$, and previous results showing that in 1/12 structure, cobalt is located in $8f$ site [8] like in YFe$_{11-x}$Co$_x$Ti compounds [9], we have performed the Rietveld refinement with cobalt distributed over the $3g$ position (Fig. 1).

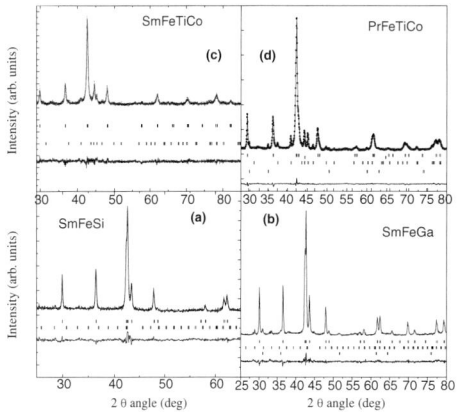 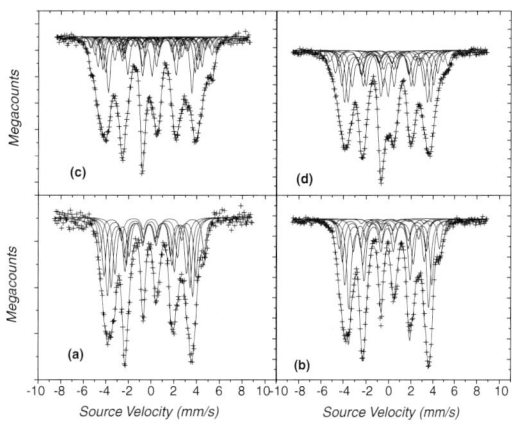

Fig. 1 Rietveld analysis for (a) SmFe$_{8.5}$Si$_{0.5}$, (b) SmFe$_{8.5}$Ga$_{0.5}$, SmFe$_{8.5}$Co$_{0.5}$Ti (c), and (d) PrFe$_{8.5}$Co$_{0.5}$Ti.

Fig. 2 The 293 K Mössbauer spectra of (a) SmFe$_{8.5}$Si$_{0.5}$, (b) SmFe$_{8.5}$Ga$_{0.5}$, (c) SmFe$_{8.5}$Co$_{0.5}$Ti, and (d) PrFe$_{8.5}$Co$_{0.5}$Ti.

The clue of the Mössbauer (Fig. 2) analysis that we have carried out derives from two criteria with intensities of random powders and Lamb-Mössbauer factors equal for all sites: (1) The possible distinction of the various iron sites, specified by their hyperfine parameter sets δ, H_{HF}, 2ϵ, results from a first order perturbation of a given central atom by its neighbors. The smallest the number of magnetic sites required fitting the spectra with line-width in the range 0.30 mm/s, the most pertinent is the solution. The efficient near-neighbor environment was limited to a coordination sphere of radius equal to 2.8 Å. The atomic neighboring of a given atom is calculated by means of the binomial law on the basis of the atom distribution given by the X-ray analysis. Only abundances higher than 1.8% were taken into account. (2) The assignment of the hyperfine parameter set of a given sextet to its right crystallographic site obeys the relationship between the isomer shift and the Wigner-Seitz cell (WSC) volumes.

Titanium occupies statistically the $6l$ site. On the one hand titanium radius is not consistent with the Rietveld fitted value $z\{2e\}$ on the other hand in the last step of the fit when all parameters are free, the fitted abundance of the $2e$ site versus cobalt content is found equal to that obtained with iron occupying the whole $2e$ site.

For a given x (or y), the isomer shift sequence observed is $\delta 2e > \delta 3g > \delta 6l$, as assumed from WSC volume correlation. The mean isomer shift of the $2e$ and $6l$ atoms upon Si, Ga and Co substitution increases while for the $3g$, it remains constant. This can be understood in terms of the preferential occupation of the substituting element in the $3g$ sites. The $6l$ and $2e$ atoms with higher probability to have Si, Ga and Co neighbor (six $3g$ neighbors) undergo a reduction of s electron density at the nucleus. The $3g$ site with only four $3g$ neighbors is less affected.

The following sequence of hyperfine field is observed whatever x (or y) in all series $H_{HF}2e > H_{HF}6l > H_{HF}3g$. The $2e$ site that derives from the $6c$ site and $8i$ sites of the equilibrium 2/17 and 1/12 phases is logically the highest. It is characterized clearly by the highest number of Fe neighbors. For the 1/10 alloys this sequence matches closely the number of (Fe,Co) near neighbors of each site. However, for a given y content (1/9 series), $6l$ and $3g$ sites have the same number of near neighbors. The sequence concerning these two last sites, which results from the isomer shift assignment, corroborates the sequence observed in the 2/17 system. The observed hyperfine field results from the competition between two terms: a positive conduction electrons term and a negative core electrons polarization term. In the case of Ga and Co substitution, the negative term increases more rapidly than the positive one. The opposite situation occurs for Si substitution.

4 Conclusion High energy ball milling and subsequent annealing of nanocrystalline $P6/mmm$ out of equilibrium precursor of the $R\bar{3}m$ $Sm_2(Fe,M)_{17}$ (M = Ga, Si) and $I4/mmm$ $RTi(Fe,Co)_{11}$ (R = Sm, Pr) obey respectively the stoichiometry 1/9 and 1/10 in the space-group $P6/mmm$. Silicon, gallium and cobalt are located in $3g$ sites, titanium in $6l$ site.

The Mössbauer spectra simulation has been carried out using the binomial distribution law, with the assignment of the hyperfine parameter set to the right crystallographic site based upon the relationship between the isomer shift δ and the Wigner-Seitz cell volume. The structure description is corroborated by the coherency of the hyperfine parameter evolution in all series with the following sequences for isomer shifts and hyperfine fields $\delta 2e > \delta 3g > \delta 6l$ and $H_{HF}2e > H_{HF}6l > H_{HF}3g$.

References

[1] J. Shield and B. E. Meacham, J. Appl. Phys. **87**, 2055 (2000).
[2] L. Bessais and C. Djéga-Mariadassou, Phys. Rev. B **63**, 054412 (2001).
[3] C. Djéga-Mariadassou and L. Bessais, J. Magn. Magn. Mater. **81**, 210 (2000).
[4] D. Givord, J. Laforest, J. Schweizer, and F. Tasset, J. Appl. Phys. **50**, 2008 (1979).
[5] C. Djéga-Mariadassou, L. Bessais, A. Nandra, J. M. Grenèche, and E. Burzo, Phys. Rev. B **65**, 014419 (2001).
[6] N. Mattern, L. Cao, A. Handstein, A. Teresiak, B. Gebel, M. Wolf, and K. H. Muller, phys. stat. sol. (a) **164**, 164 (1997).
[7] W. B. Yelon, Z. Hu, W. J. James, and G. K. Marasinghe, J. Appl. Phys. **79**, 5939 (1996).
[8] L. Bessais, S. Sab, C. Djéga-Mariadassou, and J. M. Grenèche , Phys. Rev. B **66**, 054430 (2002).
[9] Z. W. Li, X. Z. Zhou, and A. H. Morrish, J. Appl. Phys. **69**, 5602 (1991).

Magnetostriction and thermal expansion of polymer-bonded $Nd_4Fe_{77.5}B_{18.5}$ nano-composite

M. R. Alinejad[*,1], **A. Amirabadizadeh**[1], **N. Tajabor**[1], **F. Pourarian**[2], and **H. Kanekiyo**[3]

[1] Dept. of Physics, Faculty of Sciences, Ferdowsi University of Mashhad, Mashhad, Iran
[2] Carnegie-Mellon Inst., Carnegie-Mellon University, Pittsburgh PA15219, USA
[3] Sumitomo Special Metals Co. Ltd., Osaka 618-0013, Japan

Received 31 August 2003, accepted 31 December 2003
Published online 25 March 2004

PACS 75.50.Tt, 75.50.Ww, 75.80.+q

Magnetostriction and thermal expansion of polymer-bonded $Nd_4Fe_{77.5}B_{18.5}$ nano-composites are measured using strain gauge method. The XRD patterns and SEM pictures of the sample are used for microstructural studies which indicate of three nanometer-grains of α-Fe, Fe_3B and 2-14-1 phases embedded in an amorphous matrix. The overall behavior of the thermal expansion is similar to the 2-14-1 single phase alloy. The magnetostriction results show huge magnitudes in order of 10^{-4} in the presence of relatively weak external magnetic fields. Magnetostriction mainly originates from single (domain) particle interactions of Nd-sublattice in $Nd_2Fe_{14}B$ phase.

© 2004 WILEY-VCH Verlag GmbH & Co. KGaA, Weinheim

1 Introduction

Magnetostriction of composites received a large attention from physicists and engineers during the recent years [1–4]. Depending on the composition and microstructure, some of composites show large magnetostriction in the presence of not so large external fields. Currently, multi-phase Nd–Fe–B based magnets are recognized as first candidate for the preparation of magnets with largest energy product [5]. A typical composite of this type is formed by different percentages of $Nd_2Fe_{14}B$ as hard phase, α-Fe and Fe_3B as soft phases, some Nd-rich paramagnetic phases and an amorphous matrix. Each one of the soft and hard magnetic phases improves remanence and coercivity of the final magnet, respectively [6, 7]. The maximum energy product, $(BH)_{max}$, is obtained in a nano-structured magnet were grains are single domain. To our knowledge, although a great attention is paid to the different physical properties, as yet their magnetoelastic properties have not received sufficient attention. Hence, the thermal expansion and magnetostriction of a three-phase $Nd_4Fe_{77.5}B_{18.5}$ nano-composite magnet are studied.

Based on exchange-spring model of nano-composites Nd–Fe–B magnets, and assuming that the grains of soft phase are surrounded by magnetic hard phases, strongest inter-grain exchange interaction should be obtained for the case that the average grain size is smaller than a parameter b_{cm}, where:

$$b_{cm} = 2\pi \sqrt{\frac{J}{2K_1}} \tag{1}$$

J is the parameter of exchange interaction energy of the soft magnetic grains, and K_1 is the first constant of magnetocrystalline anisotropy of the hard magnetic phase. Substituting the reported values of J and K_1 for Fe_3B and $Nd_2Fe_{14}B$ phases in Eq. (1), we have $b_{cm} \approx 8$ nm, as a beneficial size of the soft phase [6].

---

[*] Corresponding author: e-mail: mralinejad@science1.um.ac.ir, Phone: +98 511 840 58 16, Fax: +98 511 843 80 32.

The $Nd_4Fe_{77.5}B_{18.5}$ sample was prepared by melt-spun method under conditions of v = 5 m/s as wheel velocity, and p =1.3 Pa as pressure of the chamber. The Average grain size is about 30 nm which is grater than the beneficial value of b_{cm} ($\approx$ 8 nm) [8–10].

2 Experimental method

The flakes of $Nd_4Fe_{77.5}B_{18.5}$ composition with about 200 μm in thickness were prepared via the melt-spinning method according to the ref. [1]. Existing phases in samples were analyzed by XRD method with Mo-K_α radiation. Primary flakes were grounded into powders in a mortar then mixed with about 5% weight of resin. Cylindrical samples with about 1 cm diameter were prepared using 0.5 GPa compressing pressure over the mixture. To improve the mechanical strength, the cylindrical samples have been annealed at 150 °C for two hours. The mechanical strength of the samples was good so that creep effects wrer not observed during strain measurements.

The microstructure of samples were analyzed using scanning electron microscopy (SEM). The grain size distribution was estimated by classification of at least 70 particles in SEM pictures based on their surface which has been measured with about 2 μm accuracy in crossed dimensions. Disk shape samples with about 6 mm in diameter and 2 mm thickness have been cut parallel to the cross section of cylinders for elasticity measurements. Magnetostriction and thermal expansion of the samples are measured between 77 and 300 K and in the presence of up to 1.5 T external fields using the familiar strain gauge method. Gradient of temperature hold below +1 °C/min during all of the measurements. Considering large magnetic anisotropy, the demagnetizing of the samples is too hard for perpendicular measurements, and so, the magnetostriction is only measured parallel to the applied field which is practically important quantity. The low-field ac-susceptibility of the sample is measured using a commercial ac-susceptometer.

3 Results and discussion

The XRD patterns of $Nd_4Fe_{77.5}B_{18.5}$ melt-spun sample have shown dispersed peaks because of the nano-size microstructure of the sample. These patterns indicate that the samples were composed of different percents of three α-Fe, Fe_3B and $Nd_2Fe_{14}B$ magnetic phases in addition to an amorphous matrix. Obviously, the 2-14-1 phase was formed directly in melt-spun alloy which is related to the selected conditions [11]. Because of the relatively small size of 2-14-1 grains, a quantitative analysis of peak lines of 2-14-1 phase in XRD patterns is not possible. Previous results showed that the grain size of 2-14-1 phase is about 20–30 nm [8]. The estimated size of α-Fe grains from XRD patterns is about 30 nm using Scherrer's approximation, which is consistent with the literature [6]. Since the grain size of α-Fe phase is greater than the critical value of b_{cm} in Eq. (1), domain walls can penetrate into the volume of grains that affect the magnetization and magnetostriction of the sample. SEM pictures, that are not shown here, show that the average size of particles in pressed sample is about 4 μm.

Figure 1 shows the linear thermal expansion and its coefficient for the resin bonded sample. A simple comparison between the coefficients of thermal expansion of magnetostrictive $Nd_2Fe_{14}B$ phase ($\sim 5.2 \times 10^{-6}$/°C) and the mechanically hard matrix ($\sim 10 \times 10^{-6}$/°C for α-Fe) indicates a negative pressure on the magnetostrictive phase which is created during melt-spun quenching. Following these considerations we can estimate the magnitude of this quantity. The introduced strain due to the difference between the coefficient of thermal expansion ($\Delta\alpha$) of constituent phases of the sample can be estimated by the following relation:

$$\varepsilon \approx \Delta\alpha \times \Delta T . \qquad (2)$$

Substituting $\Delta\alpha \sim 4.8 \times 10^{-6}$/°C and $\Delta T \approx 223$ K for current sample in this relation, we have $\varepsilon \sim 1.07 \times 10^{-3}$. The observed strain is $\varepsilon_{exp} \sim 1.3 \times 10^{-3}$ (Fig. 1). Therefore the average coefficient of linear thermal expansion is $\alpha \approx 1 \times 10^{-6}$ in the absence of internal stress. As shown in the inset of Fig. 1, the coefficient of thermal expansion deviates from linearity above about 125 K. In fact, the experimentally determined curve of the ac-susceptibility shows a peak at 125 K, at which the thermal expansion exhibits an anomalous behaviour. This behaviour is analogous with the spin reorientation observed at the same temperature in $Nd_2Fe_{14}B$ alloy [12]. Therefore, the anomalous behavior of thermal expansion can be attributed to the magnetovolume effects of thermal variations of the 2-14-1 phase magnetization.

The isotherm curves of magnetostriction of the original sample at typical temperatures are shown in Fig. 2. Apparently, the colossal magnetostrictive strains (in order of 10^{-4}) appear at the presence of relatively weak external fields (below 1 T). In single phase $Nd_2Fe_{14}B$ alloys with giant magnetocrystalline anisotropy, magnetostrictive strains of this order are observed at the presence of intense magnetic fields (about 17 T) [13]. However, in multi-phase composites in which the soft magnetic and mechanical phases present, giant magnetoelastic effects have been observed at the presence of relatively weak external fields [1, 3]. Here, soft magnetic phases (α-Fe and Fe_3B) enforces the effective magnetic induction field on magnetostrictive phase, and also mechanically flexible phases (amorphous phases) provide a suitable medium to transfer the introduced strains out of the body of sample. Phenomenalogicaly, overall behavior of experimental curves in Fig. 2 are comparable with following exponential function:

$$\lambda(H) = \lambda_s + A_1 \exp\left(\frac{-H}{H_1}\right), \qquad (3)$$

in which, H_1 and A_1 are adjustable parameters, and A_1/H_1 ratio represents slope of the magnetostriction curves at $H = 0$ limit (initial magnetostriction). Similar behavior has been observed in the isotherm curves of magnetostriction of as cast $Nd_2Fe_{14}B$ alloy at the presence of external fields up to $H = 15$ T [13]. This similarity is appeared because the effective magnetic induction field on 2-14-1 phase increases by the presence of α-Fe and Fe_3B soft phases in this nano-composite sample. Of course, saturation behavior of magnetostriction in Fig. 2 can be partly due to the balance between magnetostrictive and elastic forces, that latter appears in matrix against each deformation. Above similarity shows that $Nd_2Fe_{14}B$ crystallites are mainly responsible for the observed magnetostriction. Hence, the observed behavior should be consistent with the single particle theories [14]. To test this, the initial and saturation magnetostrictions were deduced from fitting of the experimental curves in Fig. 2 with Eq. (3). The results are shown in Fig. 3 as a function of temperature. Anomalous behavior of the curves in Fig. 3 about 125 K originated from the spin reorientation phenomena in $Nd_2Fe_{14}B$ phase which was also observed in the susceptibility results. In Callen's single particle theory, temperature dependence of the saturation magnetostriction described by the following scaling equation in terms of the reduced magnetization of Nd-sublattice, m, [15]:

$$\lambda_s(T) = \lambda_2(0) \, m^3 + \lambda_4(0) \, m^{10} + \lambda_6(0) \, m^{21}. \qquad (4)$$

The magnetostrictive coefficients, λ_i, are adjustable parameters. Using magnetization results in ref. [16], the curve of saturation magnetostriction in Fig. 3b is compared with Eq. (4). Relatively a good consistency is obtained through this comparison, neglecting some differences in the vicinity of the spin reorientation and room temperatures where the magnetocrystalline anisotropy of Nd-sublattice is weakened.

Therefore, from the overall similarity of the isofield curves of magnetostriction of our sample with the curves of 2-14-1 alloy at the presence of up to 18 T external fields once again confirm that soft phases firstly enhance the internal magnetic field, and secondly elastically transport the introduced strains into magnetostrictive 2-14-1 phase. More investigations show that, the hardening of matrix phase after annealing reduces its elasticity and so the curves of magnetostriction seems much different than the one obtained for 2-14-1 alloy.

4 Concluding remarks

We have studied the magnetostriction and thermal expansion of three-phase polymer-bonded $Nd_2Fe_{77.5}B_{18.5}$ magnet. Colossal magnetostriction ($\approx 10^{-4}$) was observed at the presence of relatively small external fields. Magnetostriction of the sample is originated mostly from the single particle effects of Nd-sublattice in $Nd_2Fe_{14}B$ phase. To use this magnet in magnetomechanical transformers, some of the important aspects such as: porosity, mechanical strength, magnetic anisotropy and coercivity of the composit have to be accurately determined.

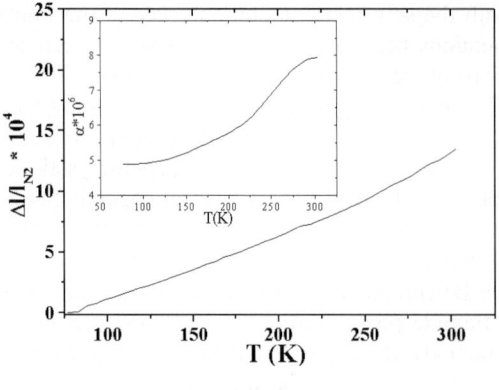

Fig. 1 Thermal expansion of polymer bonded $Nd_4Fe_{77.5}B_{18.5}$ nano-composite. Inset: coefficients of thermal expansion deduced via calculation of the derivation of the thermal expansion curves.

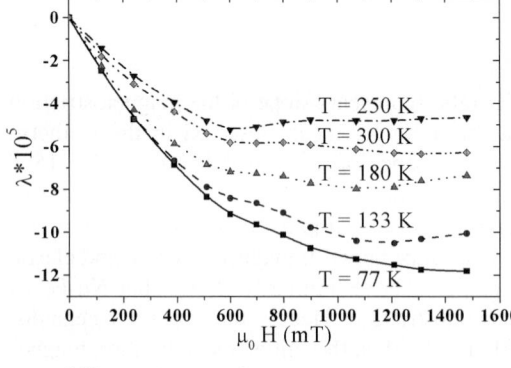

Fig. 2 Isotherms of magnetostriction of polymer bonded $Nd_4Fe_{77.5}B_{18.5}$ nano-composite as a function of applied field.

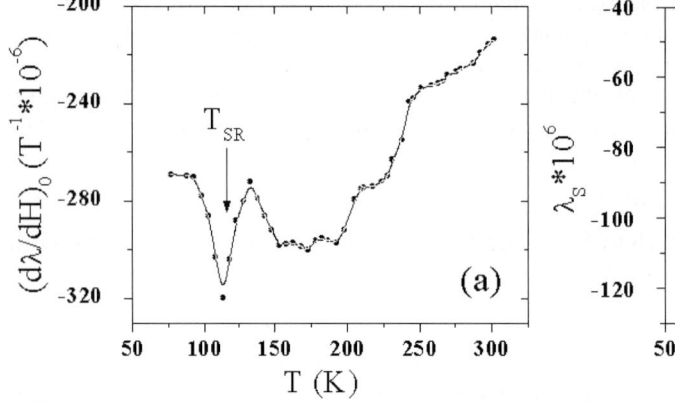
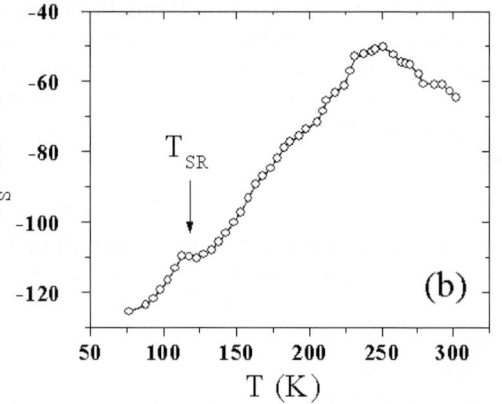

Fig. 3 Temperature dependence of (a) initial magnetostriction and (b) saturation magnetostriction of $Nd_4Fe_{77.5}B_{18.5}$ sample which are deduced from the analysis of experimental curves of Fig. 2.

References

[1] Y. Chen and J. E. Snyder, Appl. Phys. Lett. **74**, 1159 (1999).
[2] C. W. Nan and G. J. Weng, Phys. Rev. B **60**, 6723 (1999).
[3] F. E. Pinkerton, J. F. Herbst, and M. S. Meyer, J. Appl. Phys. **87**, 8653 (2000).
[4] C. W. Nan, Y. Huang, and G. J. Weng, J. Appl. Phys. **88**, 339 (2000).
[5] G. C. Hadjipanayis, J. Magn. Magn. Mater. **200**, 373 (1999).
[6] Q. F. Xiao, T. Zhao, Z. D. Zhang, E. Brück, K. H. J. Buschow, and F. R. de Boer, J. Magn. Magn. Mater. **223**, 215 (2001).
[7] E. F. Kneller and R. Hawing, IEEE Trans. Magnet. **27**, 3588 (1991).

[8] D. H. Ping, K. Hono, H. Kanekiyo, and S. Hirosawa, J. Appl. Phys. **85**, 2448 (1999).
[9] W. C. Chang, D. M. Hsing, B. M. Ma, and C. O. Bounds, IEEE Trans. Magn. **32**, 4425 (1996).
[10] L. H. Lewis, A. R. Moodenbaugh, D. O. Welch, and V. Panchanathan, J. Phys. D: Appl. Phys. **34**, 744 (2001).
[11] H. Kanekiyo and S. Hirosawa, J. Appl. Phys. **83**, 6265 (1998).
[12] N. Tajabor, M. R. Alinejad, and F. Pourarian, Physica B **32**, 163 (2002).
[13] P. A. Algarabel, A. del Moral, M. R. Ibarra, and C. Marquina, J. Magn. Magn. Mater. **114**, 161 (1992).
[14] N. Tajabor, M. R. Alinejad, and F. Pourarian, J. of Scientia Iranica **10**, 1 (2003).
[15] E. R. Callen and H. B. Callen, Phys. Rev. A **139**, 455 (1965).
[16] H. Onoedera, A. Fujita, H. Yamamoto, M. Sagawa, and S. Hirosawa, J. Magn. Magn. Mater. **68**, 6 (1987).

A comparative study on magnetostrictive strain in $GdCo_5$ and $Gd_{0.9}Pr_{0.1}Co_5$

A. Amirabadizadeh[*,1,2], **N. Tajabor**[2], **M. R. Alinejad**[2], and **F. Pourarian**[3]

[1] Department of Physics, Faculty of Science, University of Birjand, Birjand, Iran
[2] Department of Physics, Faculty of Science, Ferdowsi University of Mashad, Mashad, Iran
[3] Carnegie Mellon Institute, Carnegie Mellon University, Pittsburg, Pensylvania, 15219, USA

Received 31 August 2003, accepted 31 December 2003
Published online 18 March 2004

PACS 75.30.Sg, 75.80.+q

Results of the magnetostriction measurements on exchange coupled $GdCo_5$-type alloys, (with insignificant magnetic anisotropy of L-quenched Gd^{+3} ions) show lower values compared to $Gd_{0.9}Pr_{0.1}Co_5$ compound. In fact, partial substitution of Pr for Gd creates an orbital moment for the rare earth sublattice and therefore the magnetocrystalline anisotropy (MA) increases. The temperature dependence of MA introduces the spin reorientation transition in $Gd_{0.9}Pr_{0.1}Co_5$ compound. Simple model calculations indicate that a negative contribution in the first order anisotropy constant (K_1) appears after Pr substitution. The sign of K_1 changes from negative to positive at ~160 K with increasing temperature. The latter introduces a large increase in the anisotropic magnetostriction.

© 2004 WILEY-VCH Verlag GmbH & Co. KGaA, Weinheim

1 Introduction

The $RECo_5$ (RE is a rare earth element) compounds crystallize in the $CaCu_5$-type hexagonal structure. Many experimental and theoretical works have been devoted to the study of magnetic properties of $RECo_5$ compounds [1–14]. However the understanding of the interaction of localized and itinerant magnetism is far from being completely solved. The magnetic properties of these compounds are often discussed within two-sublattice model [2]. An example is the magnetocrystalline anisotropy (MA) of the rare earth sublattice and cobalt sublattice in $RECo_5$ series. MA energy (MAE), defined as the change of the ground state energy of a magnet upon rotation of the magnetization direction with respect to the crystal axes. In $RECo_5$ compounds, MAE arises from two major sources: (i) the spin–orbital interaction of the itinerant states (mainly the, Co, 3d states) which couples the magnetization direction to the anisotropic crystal environment and (ii) the interaction of the localized partially filled, RE, 4f shell with the crystal field [3]. The MAE is estimated by using anisotropy constants, K_i, where i = 1, 2, or higher are the first, second, and higher order anisotropy constants, respectively. Rotation of magnetization vector with respect to the crystal axes is known as spin reorientation transition, which occurs at certain temperatures (T_{SR}). Magnetostriction measurements are sensitive tool for the investigation of spin reorientation, which is related to the anisotropy K_1 and K_2 constants [4]. In the $GdCo_5$ compound, MAE is introduced from the effect of Co sublattice, where Gd half-filled 4f shell and do not interact with the crystal field [3]. However, for $PrCo_5$, Pr and Co sublattices contribute to the MAE and at T_{SR}~105 K the spin reorientation takes place from a cone to c-axis direction of the compound's hexagonal structure [5]. In this article, we present the magnetoelastic properties for the partial replacement of Pr for Gd in the $GdCo_5$ compound.

---
[*] Corresponding author: e-mail: amirabadi2000@yahoo.com, Phone: +98 511 8438033, Fax: +98511 8405816

2 Experimental technique and results

The samples preparation method is given in another work [6]. Measurements of magnetostriction were made using strain gauge method [7] in an applied field up to 1.5 T, and temperature range from 77 to 320 K. The strains were measured with field parallel and then perpendicular to the gauge. These strains are denoted by λ_l (longitudinal) and λ_t (transverse). We defined anisotropic magnetostriction and volume magnetostriction as $\lambda_a = \lambda_l - \lambda_t$ and $\omega = \lambda_l + 2\lambda_t$ respectively [7]. The samples used for measurements were in the form of disk, 6 mm in diameter and 3 mm thickness. No significant difference was observed between the strains measured in the plane and perpendicular to the plane of the disc of the sample, suggesting the absence of any preferred orientation effects.

For annealed $Gd_{1-x}Pr_xCo_5$ (x = 0 and 0.1) samples, X-ray diffraction patterns revealed that single phase samples were obtained in the composition range of x = 0 and 0.1 and were successfully indexed with the $CaCu_5$-type structure. The lattice parameters are given elsewhere [6].

Figure 1 shows the longitudinal magnetostriction, λ_l, as a function of temperature. For $Gd_{0.9}Pr_{0.1}Co_5$, the longitudinal magnetostriction is almost constant up to ~170 K and then decreases up to room temperature. The results for $GdCo_5$ is low compared to those obtained for the composition with x = 0.1 and they remain constant for all temperature ranges. Figure 2 shows the volume magnetostriction, ω, as a function of temperature. An anomalous change in the magnitude of volume magnetostriction occurs between 170 K and 280 K, from -50×10^{-6} to -250×10^{-6}, while no effect was observed for $GdCo_5$. Figure 3 shows results of anisotropic magnetostriction, λ_a, for two samples. For $Gd_{0.9}Pr_{0.1}Co_5$ compound, λ_a decreases beyond 170 K with increasing temperature.

3 Discussion

It is well known that Gd is located at the middle of rare earth elements with 7 electrons in 4f shell and it is L-quench. In this condition the magnetic anisotropy of Gd in $GdCo_5$ compound is negligible [5]. The easy direction of this compound was determined by the anisotropy of Co sublattice [1]. From low temperature to Curie temperature for $GdCo_5$ compound has not reported any spin reorientation transition and the c-axis easy direction does not change [5], and the first anisotropy constant, K_1, is almost constant throughout the experimental temperature range [8] (see also Fig. 4). The magnitude of the magnetostrictive strains for $GdCo_5$ is low compared to the other $RECo_5$ compounds. A similar magnetostrain property was also observed in the $RECo_2$ [9] and $RECo_3$ [10] series for RE = Gd.

The replacement of 10% Pr for Gd in $GdCo_5$ compound, modifies the Gd site symmetry and introduces an interaction of a partially filled localized, RE, 4f shell with the electric crystal field. In the Pr containing compound, the magnetocrystalline anisotropy energy and orbital moment anisotropy increase [3] and consequently the value of the magnetostriction becomes large compared to that observed in $GdCo_5$ compound.

The anomalies on magnetostrictive strain curves can be interpreted as follows. With partial Pr substitution for Gd, the compound is considered to be a two sublattices system, the RE sublattice and Co sublattice. The former is dominant at lower temperatures and the latter being dominant at higher temperatures [5]. At certain temperature ranges, the RE (4f) sublattice (planar anisotropy) and Co (3d) sublattice (uniaxial easy axis anisotropy) compete with each other, consequently spin reorientation transition (SR) takes place from a cone structure to the c-axis of the hexagonal structure. SR begins at the temperature where K_1, first magnetocrystalline anisotropy constant, changes sign. However it may extend over a wide temperature range according to the sign of K_2 (second order magnetocrystalline anisotropy constant) [11]. At spin reorientation a sharp change in the volume magnetostriction (ω) occurs, from -50×10^{-6} to -250×10^{-6}, as shown in Fig. 2. A similar drastic decrease in the anisotropic magnetostriction at SR (see Fig. 3) was also observed [12].

4 Modeling

Simple model calculations for the system anisotropy indicate that a negative contribution in the first order anisotropy constant (K_1) appears upon Pr substitution. The sign of K_1 changes from negative to positive at ~160 K an increasing with temperature.

Based on the two sublattices model [13], the first anisotropy constant for PrCo$_5$ compound can be calculated from

$$K_1 = (K_1^{Co} + K_1^R) + 2D^2/J'(1+2S/J')$$

where

$$D \equiv (K_1^{Co} - K_1^R M_{Co}/M_R), \quad J' = JM^2 M_{Co}/M_R \quad \text{and} \quad S = K_1^{Co} + K_1^R M_{Co}^2/M_R^2$$

and M_i is the saturation magnetization of each sublattice and J is the exchange integral. Now using value of K_1^{Co}, $K_1^R = 0.1 K_1^{Pr}$, M_{Co}, $M_R = M_{Pr}$ and J from references [13] and [14] and using above equation the calculated K_1 for $Gd_{0.9}Pr_{0.1}Co_5$ compound as a function of temperature is shown in Fig. 4. Obviously, a negative contribution in the first order anisotropy constant (K_1) appears after Pr substitution. The sign of K_1 changes from negative to positive at ~160 K with increasing temperature. The change of sign of K_1 occurs around temperatures close to that where anomalies on magnetostrctive strain curve have started. SR in $Gd_{0.9}Pr_{0.1}Co_5$ compound was confirmed by the low field ac-susceptibility measurements [6], where a wide peak appeared on this curve and it coincides with the magnetostrictive temperature anomalies.

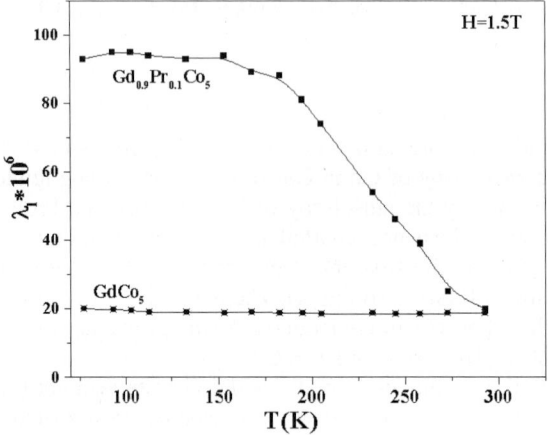

Fig. 1 Longitudinal magnetostriction versus temperature for $Gd_{1-x}Pr_xCo_5$ (x = 0 and 0.1) compounds.

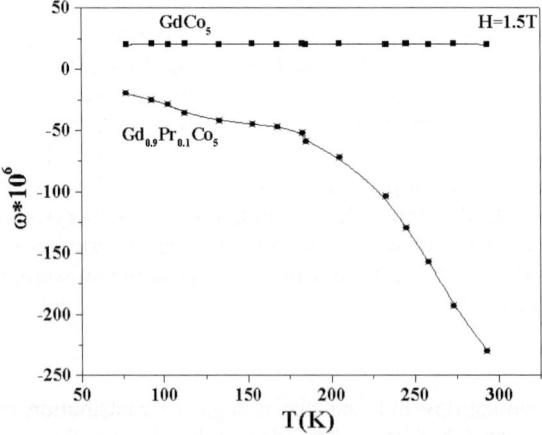

Fig. 2 Volume magnetostriction versus temperature for $Gd_{1-x}Pr_xCo_5$ (x = 0 and 0.1) compounds.

© 2004 WILEY-VCH Verlag GmbH & Co. KGaA, Weinheim

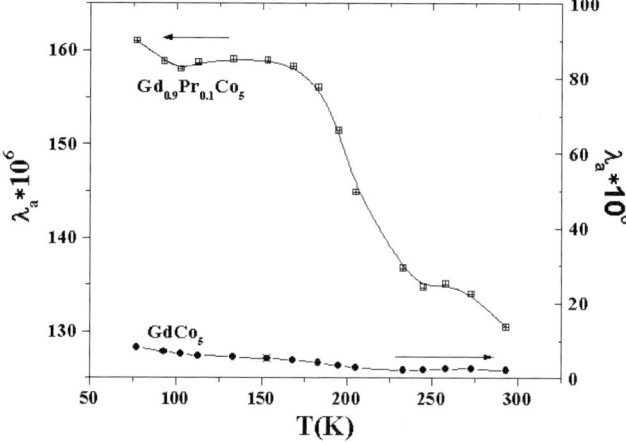

Fig. 3 Anisotropic magnetostriction versus temperature for $Gd_{1-x}Pr_xCo_5$ (x = 0 and 0.1) compounds.

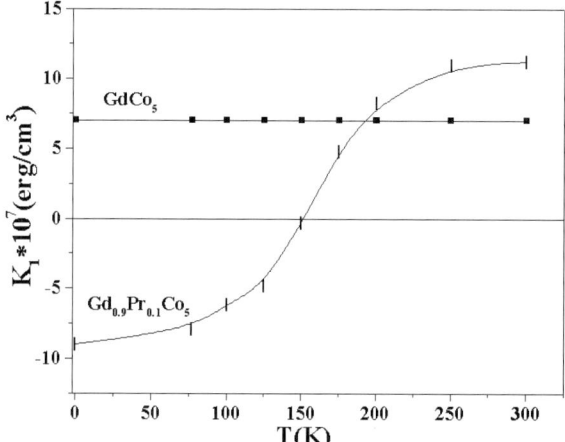

Fig. 4 First anisotropy constant (K_1) versus temperature (data for $GdCo_5$ compound are given from Ref. [8]).

References

[1] R.L. Streever, Phys. Rev. B **19**, 2704 (1979).
[2] K. Hammler, M. Liebs, T. Beaerle, P. Uble, and M. Fahnle. J. Magn. Magn. Mater. **140–144**, 851 (1995).
[3] L. Steinbeck, M. Richter, and H. Eschring. Phys. Rev. B **63**, 184431 (2001).
[4] F. Pourarian, M.V. Satyanargana, and W.E. Wallace, J. Magn. Magn. Mater. **25**, 113 (1981).
[5] A.V. Andreev, "Handbook of Magnetic Materials", edited by K.H.J. Buschhow, Vol. 8 (Elsevier, 1995), p. 86.
[6] A. Amirabadizadeh, N, Tajabor, M.R. Alinejad, and F. Pourarian, submitted to "International conference of superconductors and magnetic materials" MSM 03, Tunisia, Manistir (2003).
[7] N. Tajabor, M. R. Alinejad, and F. Pourarian, Physica B **321**, 63 (2002).
[8] B. Szpunar and P. A. Lindgard, J. Phys. F **9**(3), 56 (1979).
[9] F. Pourarian, Phys. Lett. **72A**, 175 (1979).
[10] F. Pourarain and N. Tajabor, phys. stat. sol. (a) **61**, 537 (1980).
[11] Bo-Ping Hu, Hong-Shuo Li, J.P. Gavigan, and J.M.D. Ceoy. J. Phys.: Condens. Matter **1**, 755 (1989).
[12] A.V. Andreev and S.M. Zadvorkin, Physica B **225**, 237 (1996).
[13] F. Bolzoni and M. F. Pirini, J. Appl. Phys. **66**(5), 2315 (1990).
[14] E.A. Nesbitt and J.H. Wernick, "Rare Earth Permanent Magnets" (Academic Press, 1973).

Structure and electrical resistivity of $Gd_{1-x}Pr_xCo_5$ compounds

A. Amirabadizadeh[*,1,2], **N. Tajabor**[1], **M. R. Alinejad**[1], and **F. Pourarian**[3]

[1] Department of Physics, Faculty of Science, University of Birjand, Birjand, Iran
[2] Department of Physics, Faculty of Science, Ferdowsi University of Mashhad, Iran
[3] Carnegie Mellon Research Institute, Carnegie Mellon University, Pittsburgh, Pennsylvania, 15219, USA

Received 31 August 2003, accepted 31 December 2003
Published online 18 March 2004

PACS 72.15.–v, 75.30.Fv, 75.50.Ww

The effect of partial substitution of Pr for Gd in $Gd_{1-x}Pr_xCo_5$ ($x = 0$, 0.1 and 0.3) on structural and resistivity properties are investigated. For $GdCo_5$, the temperature dependence of resistivity (from 80 K to 300 K) shows a negative curvature towards the temperature axis as one goes from low to high temperatures. For samples with $x = 0.1$ and 0.3 composition electrical resistivity curves show anomalies at certain temperatures. It is suggested that the anomalies of the resistivity are attributed to the triggered spin reorientation transition temperature. These anomalies were also observed in the temperature dependence of ac-susceptibility curves, which confirm the observed spin reorientation transition.

© 2004 WILEY-VCH Verlag GmbH & Co. KGaA, Weinheim

1 Introduction

Intermetallic compounds obtained by combining rare earth metals (R) with 3d-transition metal (T) form an important class of materials that find applications in permanent magnets, and magneto-optic recording [1]. The RCo_5 phases have an outstanding position among the 4f–3d intermetallic compounds. First, the discovery of a large magnetocrystalline anisotropy in these compounds led to the production of generation of powerful permanent magnet based on $SmCo_5$, widely used till now. Second, a variety of magnetic properties, in combination with other elements made RCo_5 very apt model system for studying magnetism in rare-earth-3d-metal alloys [2]. The RCo_5 compounds crystallize in the $CaCu_5$-type hexagonal structure. The cobalt atoms occupy two types of sites, (2c) and (3g), having different types and numbers of neighbouring atoms. The Co NMR studies show that Co atoms at (2c) sites have a large positive anisotropy contribution (favoring an easy c-axis alignment of the Co moment). The Co atoms at (3g) sites are found have smaller negative contribution (favoring an alignment of the moment s in the basal plan) [3]. Among these series $GdCo_5$ compound is ferrimagnetic while $PrCo_5$ is ferromagnetic at room temperature. $GdCo_5$ as an uniaxial anisotropy (c-axis) from low temperature up to its Curie temperature [2], but $PrCo_5$ shows a spin reorientation transition from cone to c-axis at a temperature $T_{SR} = 105$ K. The behavior of the electrical resistivity of magnetic materials is sensitive to type of magnetic structure and spin reorientation phase transition [2]. In this research we have investigated the effect of partial substitution of Pr for Gd in $GdCo_5$ compound on the structure and electrical resistivity behaviour.

2 Experimental technique

The polycrystalline samples were prepared by placing a proper amount of the constituents Gd (99.9%) and Co (99.99%) in a water-cooled boat using RF induction melting. A continuous flow of titanium-

---

[*] Corresponding author: e-mail: amirabadi2000@yahoo.com, Phone: +98 511 8405816, Fax: +98 511 8438032

getterd argon was maintained during heating. As-cast samples were wrapped in a tantalum foil, sealed in quartz tubes, filled with ~1/3 atmosphere of argon gas and annealed at 950 °C for six days.

Fairly detailed and accurate electrical resistivity measurements have been performed over 77 to 300 K using the four probe method. Pressure connections were used to provide electrical contacts. The dimensions of the sample were close to $1 \times 1 \times 6$ mm^3. During all measurements, the temperature of the sample was measured and controlled within ±0.1 K accuracy.

Temperature dependence of the low-field magnetic ac-susceptibility, $\chi_{ac}(T)$, was measured between 100 and 300 K using a modified commercial mutual inductance susceptometer at 333.3 Hz with ac-magnetic field 50 mA^{-1} peak values. The measurement was performed in small temperature intervals in order to obtain the critical behavior.

3 Results

For annealed X-ray diffraction (XRD) patterns of Gd$_{1-x}$Pr$_x$Co$_5$ ($x = 0$, 0.1 and 0.3) samples reveled a single phase which was successfully indexed with the CaCu$_5$-type structure. Refined lattice parameters a and c and unit cell volume are listed in Table 1. For $x = 0$ the values of lattice parameters are in good agreement with the data of literature [4].

In Fig. 1 we have shown the temperature dependence of electrical resistivity of GdCo$_5$ compound for a wide range of temperature (from 100 to 300 K). It shows a negative curvature towards the temperature axis as one goes from low to high temperatures. Figure 2 shows the behavior of electrical resistivity for Gd$_{0.9}$Pr$_{0.1}$Co$_5$ and Gd$_{0.7}$Pr$_{0.3}$Co$_5$ compounds. Obviously, for the case of $x = 0.1$, there are anomalies between 170 and 280 K. In this case electrical resistivity increase the almost linearly with temperature, except in the temperature range 170–280 K at which the slope of the resistivity decreases remarkably. For the case of $x = 0.3$, the anomaly occurs at ~150 K. The slope of the resistivity at 150 K decreases, before and after this temperature the electrical resistivity increases with temperature linearly.

The temperature dependence of ac-susceptibility (χ_{ac}) of compounds are shown in Fig. 3. For GdCo$_5$ compound there is no anomaly in our experimental temperature range. But, for $x = 0.1$ and 0.3 cases, an anomalous behavior is noticeable. For Gd$_{0.9}$Pr$_{0.1}$Co$_5$, the anomaly appears as a broad peak from ~170 to 280 K, while for the $x = 0.3$ case, there is an almost sharp peak about 150 K. The anomalies of ac-susceptibility occur at the same temperatures as was observed for the electrical resistivity anomalies.

4 Discussion

In the high temperature region (T > 70 K), the temperature dependence of electrical resistivity can be written in a general form [5]:

$$\rho(T) = B + CT - DT^3 , \qquad (1)$$

where B is a constant, C is a function of temperature and Fermi energy, and D is dependent on the density of states function (N), the first derivative of the density of state function (dN/dε), and the second derivative of the density of states function $\left(\dfrac{d^2 N}{d\varepsilon^2}\right)$ in the form of

Table 1 Lattice parameters of Gd$_{1-x}$Pr$_x$Co$_5$.

x	a(Å)	c(Å)	v(Å)^3
0	4.974	3.974	84.631
0.1	4.980	3.971	84.636
0.3	4.985	3.969	84.643

$$D \propto \left\{ 3\left(\frac{1}{N}\frac{dN}{d\varepsilon}\right)^2 - \frac{1}{N}\frac{d^2N}{d\varepsilon^2} \right\}. \qquad (2)$$

When the electrical resistivity data of the GdCo$_5$ compound in the temperature range 80–300 K were fitted according to Eq. (1), a good fit was obtained, as shown by the solid curve in Fig. 1. This is a characteristic of a weak ferromagnetic or ferromagnetic behavior [5].

From the shape of the resistivity curvature, one can discuss the correlation between the D coefficient (in Eq. (1)) and the electronic structure of this compound. A negative curvature (D is positive) implies that the right hand side of Eq. (2) must be positive. This may be possible if the Fermi level of this compound is near an inflection point of $N(\varepsilon)$ (density of states) [5]. The electronic structure calculation for GdCo$_5$ compound in Ref. [6] has been confirmed that the position of Fermi level is near such point.

For the case of $x = 0.1$ and 0.3 mainly, Gd$_{0.9}$Pr$_{0.1}$Co$_5$ and Gd$_{0.7}$Pr$_{0.3}$Co$_5$, the behavior of electrical resistivity (ρ) is similar to of the ferromagnetic materials like NdCo$_5$ [7] and Pr$_2$(Co$_{1-x}$Fe$_x$)$_{17}$ [8], in which both these ferromagnetic materials show spin reorientation transition at temperatures where electrical resistivity has anomalies. Therefore in addition to the usual resistivity due to impurities, phonon and spin disorder scattering, the electrical resistivity depends also on the angle θ between the saturation magnetization M_s and the c-axis. When this angle changes near a spin reorientation transition temperature (T_{SR}) a corresponding anomaly appears in ρ and $d\rho/dT$. It has been suggested that the anomalies of the resistivity attributed to the triggered spin reorientation transition temperature. The ac-susceptibility measurements confirmed our suggest, where for Gd$_{0.9}$Pr$_{0.1}$Co$_5$ sample a wide peak shows that a spin reorientation transition takes place from $T_{SR1} = 170$ K to $T_{SR2} = 280$ K (from cone to c-axis). For Gd$_{0.7}$Pr$_{0.3}$Co$_5$ sample, the spin reorientation occurs at a narrow temperature, about $T_{SR} = 150$ K.

5 Conclusion

Obviously, the electrical resistivity measurements of ferrimagnetic materials like GdCo$_5$ compound, is third order function of temperature, The negative curvature (D is positive) reflected from fluctuation of band structure of this compound near the Fermi level. Also with partial Pr substitution for Gd in GdCo$_5$ compound, it is suggested that the Pr content alloys become ferromagnetic with conical structure. With increasing temperature spin reorientation transition occurs and easy direction of them changes from cone to c-axis of hexagonal structure. For 10% Pr substitution for Gd the spin reorientation take place in wide temperature range from 170 to ~280 K, while in the case of 30% Pr substitution for Gd spin reorientation occurs in a nearly narrow temperature range (~150 K). It seems that with increasing of Pr, the spin reorientation transition temperature (T_{SR}) decreases, and suggests it will reach to $T_{SR,PrCo_5} = 105$ K with increasing of Pr concentration.

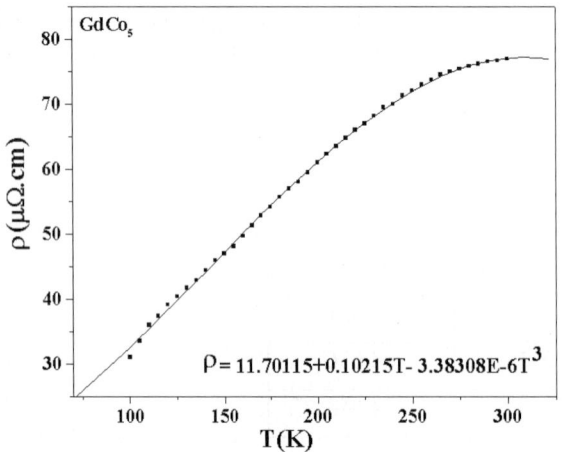

Fig. 1 Electrical resistivity of GdCo$_5$ compound versus temperature. (solid line shows the computer fitting of the curve and the relation is given in the inset).

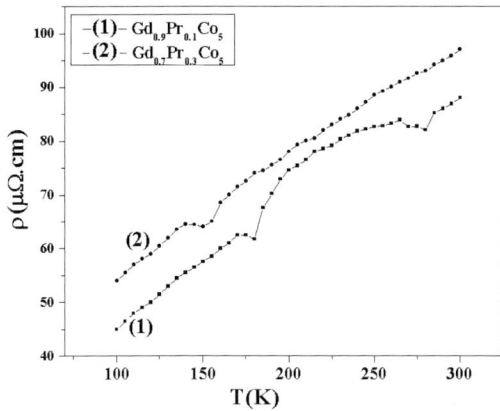
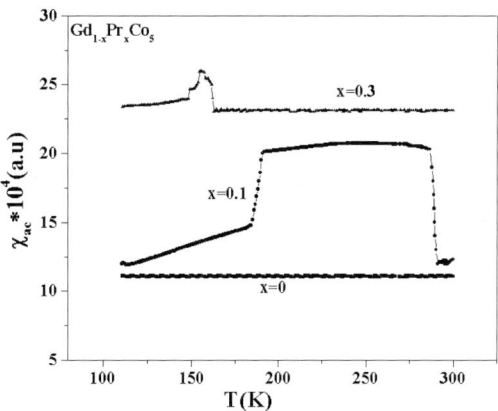

Fig. 2 Electrical resistivity of $Gd_{0.9}Pr_{0.1}Co_5$ and $Gd_{0.7}Pr_{0.3}Co_5$ compounds versus temperature.

Fig. 3 Temperature dependence of ac-susceptibility of $Gd_{1-x}Pr_xCo_5$ for $x = 0$, 0.1, and 0.3.

References

[1] A. Kowalczyk, J. Magn. Magn. Mater. **171**, 113 (1987).
[2] A. V. Andreev, "Handbook of Magnetic Materials", edited by K. H. J. Buschhow (Elsevier, 1995), Vol. 8.
[3] R. L. Streever, Phys. Rev. B **19**, 2704 (1979).
[4] C. V. Thung, N. H. Duc, M. M. Tun, N. P. Thuy, E. Bruck, B. E. Brommer, and J. J. M. France, J. Magn. Magn. Mater. **177–181**, 819 (1998).
[5] A. Kowalczyk and A. Jezierrski, J. Magn. Magn. Mater. **188**, 361 (1998).
[6] J. P. Rueff. R. M. Galera, Ch. Giorgetti, E. Dartgge, Ch. Brouder, and M. Alouani, Phys. Rev. B. **58**, 12271 (1998).
[7] J. B. Sousa, J. M. Moreia, A. Del Moral, P. Algarabel, and R. Ibarra, J. Phys.: Condens. Matter. **2**, 3897 (1995).
[8] J. B. Sousa, J. F. D. Montenegro, J. M. Moreira, and M. E. Braga, J. Phys. F: Met. Phys. **12**, 351 (1982).

Grain size effect on magnetic properties of Fe–28Cr–15Co permanent magnets as a function of Mo content

Z. Ahmed* and **A. ul Haq**

Metallurgy Division, Dr. A. Q. Khan Research Labs. P. O. Box 502, Rawalpindi, Pakistan

Received 31 August 2003, accepted 31 December 2003
Published online 23 April 2004

PACS 61.10.Nz, 68.37.Lp, 75.30.Cr, 75.50.Bb

The Fe–28Cr–15Co–(1–4)Mo magnetic alloys were studied using X-ray diffraction, optical microscopy, transmission electron microscopy and magnetometery techniques. Permanent magnets were produced by employing thermo-magnetic aging process. The Fe–28Cr–15Co–3.5Mo alloys produced the optimum magnetic properties as coercive force = 840 Oe (66.83 kA/m), remanence = 11 kG (1.1 T), saturation magnetization = 12.5 kG (1.25 T) and energy product = 5.4 MGOe (43 kJ/m^3). Molybdenum content above 3.5 wt% produced cracking and brittleness in the alloys due to the formation of unwanted σ phase. The results indicates that molybdenum content from 1 to 3.5 wt% extends the ferromagnetic α grain structure at annealed state and affects the particle size, amount and shape anisotropy of the strongly magnetic α_1 phase that in turn improves the magnetic properties of the ternary Fe–28Cr–15Co alloys.

© 2004 WILEY-VCH Verlag GmbH & Co. KGaA, Weinheim

1 Introduction

The Fe–Cr–Co based permanent magnets have found applications in hysteresis motors, aircraft magnetos, printers, stereo cartridges and are most suitable for small electromechanical devices, which are some time difficult to fabricate with Alnico or ferrite permanent magnets [1, 2]. Recent studies have shown that Fe–Cr–Co films can be used for magnetic recording applications [3]. Magnetic properties in the Fe–Cr–Co alloy system are due to spinodal decomposition during which α phase transforms into two spinodal phases known as α_1 and α_2.

In order to expand the applications of the Fe–Cr–Co alloys, enhancement of coercive force is essential. One possibility is the minor addition of one or two ferromagnetic forming elements such as Al, Ti, V, Mo to the ternary Fe–Cr–Co alloys [4]. The other possibility is the optimization of processing conditions [5]. It is known that addition of Mo increases the coercive force due to anisotropic spinodal decomposition along <100> or <111> directions [6]. Most of the studies have been conducted on Fe–Cr–Co–Mo alloys either by developing single crystals [7], <100> columnar grain structure [8] or by producing thin sheets of 0.5 mm thickness [6]. But it is expensive to produce single crystals or columnar grain structure due to the involvement of long heat treatment cycle. On the other hand sheet or wire shape magnets restrict the applications. However, few literature references are available on the magnetic properties of Fe–Cr–Co–Mo bulk permanent magnets. In the present work Fe–28Cr–15Co–(1–4)Mo bulk magnets were produced by thermo-magnetic aging process and their microstructural and magnetic properties were investigated.

2 Experimental procedures

The composition of the studied alloys is presented in Table 1. The alloy preparation is described elsewhere [9]. It was observed that the addition of Mo above 3.5 wt% produced undesirable sigma phase that

---

* Corresponding author: e-mail: anwar@comsats.net.pk

makes the alloy brittle in the cast state and could not be investigated further. Square samples of 10x10 mm dimensions were cut from the hot rolled bar and were used in the present study. The heat treatment of the alloys was composed of solution annealing for 30 minutes at 1250 °C followed by rapid quenching, for homogenization and development of α phase structure. Thermo-magnetic treatment for 40 minutes at 640 °C were performed for the development of permanent magnetic properties, referred to TMT-1. The samples were heated again at 615 °C for 4 hours (i.e., TMT-2) under magnetic field intensity of 2.5 kOe. Finally during the step-aging treatment, the samples were heated at 610 °C for 3 hours, cooled at the rate of 4 °C per hour to 500 °C and hold for 10 hours at 490 °C in order to increase the magnetic properties.

Microstructural examination was carried out on optical microscope, scanning and transmission electron microscope. The average grain size of the alloys in solution annealed state were measured by means of linear intercept method, using image processing and analysis system attached with optical microscope. Magnetic properties were measured at room temperature using ±10 kOe maximum applied field.

Table 1 Chemical composition of the studied alloys (wt%) with Fe balance.

Alloy designation	Cr	Co	Mo	C	S	N
A	28.10	15.13	0.00	0.002	0.010	0.0174
B	28.05	15.10	1.12	0.002	0.010	0.0173
C	28.01	15.10	2.01	0.001	0.012	0.0173
D	28.11	15.00	3.51	0.001	0.001	0.0175

3 Results and discussion

Figure 1 compares the microstructures of the alloys A and D after solution annealed state. These micrographs show coarse grains of bcc α phase along with nonmetallic inclusions as confirmed by X-ray diffraction studies. The black spots in the microstructure were identified as non-metallic Cr-rich inclusions in confirmation with the literature [10]. Figure 2 represents the mean α grain size of the alloys after solution treatment. It is evident from Fig. 2 that alloy A has α grain size of 540 µm which increases with the increase of Mo content and reaches to maximum of 980 µm in alloy D. The microstructural studies revealed that Mo extends the ferromagnetic α phase structure domain and it also affects the spinodal structure as depicted in the Fig. 4.

Figure 3 shows the changes in H_c and B_r values of the solution treated alloys. It shows that H_c and B_r values are low as H_c = 6.3 Oe and B_r = 0.15 kG in the alloy A that increases with the increase of Mo content and reaches to maximum values of H_c = 15 Oe and B_r = 0.33 kG. The better magnetic properties in alloy D is attributed to the formation of favorable α phase structure at annealed state.

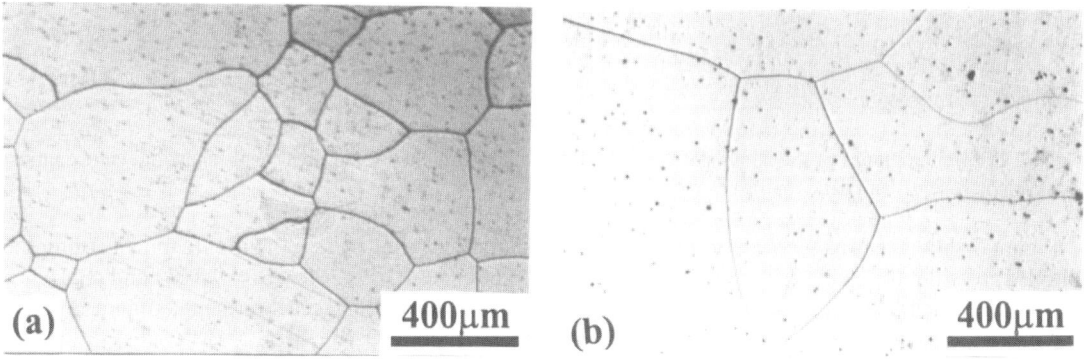

Fig. 1 Optical micrographs of the alloys showing coarsening of α grain (a) A-alloy and (b) D-alloy.

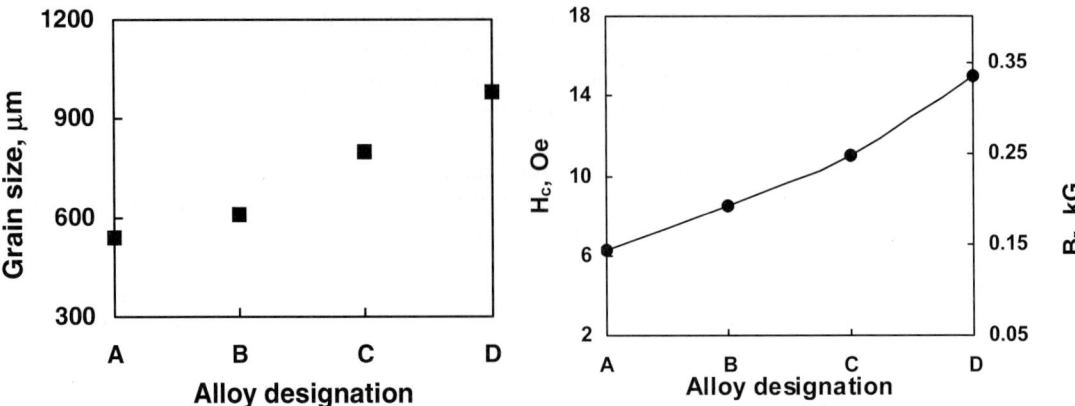

Fig. 2 Mean α grain size of the studied alloys obtained after treatment.

Fig. 3 Changes in H_c and B_r of the studied alloys after solution treatment.

Figure 4 compares the TEM micrographs of alloys A and D after the step-aging treatment. The micrographs shows that the α phase detected on annealed state have been spinodally decomposed into $α_1$ and $α_2$ phases. The white rods like particles in the micrographs were detected as FeCo-rich $α_1$ phase (ferromagnetic) embedded in dark Cr-rich $α_2$ phase (paramagnetic). The size of $α_1$ particles measured in alloy A was 15 ± 2 nm and in alloy D was 22 ± 2 nm. The volume fractions of $α_1$ particles in alloy A is detected as 19% and in alloy D as 37%. The ferromagnetic $α_1$ particles seem to be aligned and elongated in the applied field direction as marked with the arrow symbol. It was observed that the severity of alignment and elongation of the $α_1$ particles is better in alloy D as compared to alloy A. This suggests that good magnetic properties in alloy D are expected, better shape anisotropy of the magnetically rich $α_1$ particles.

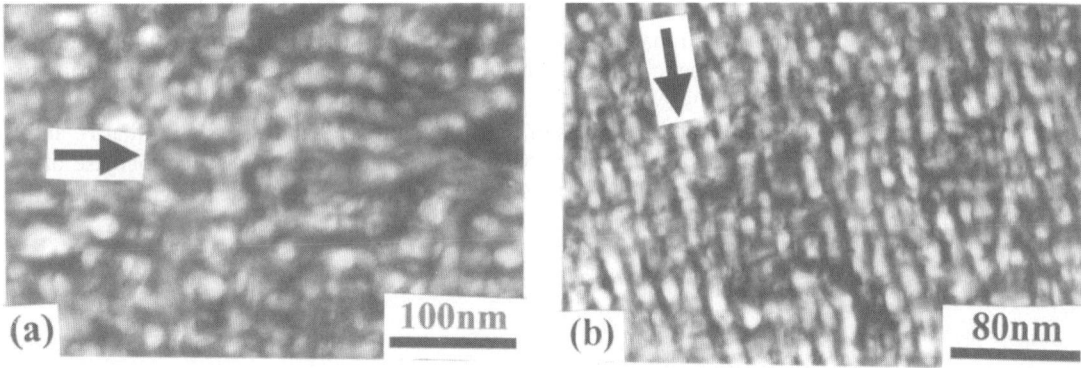

Fig. 4 TEM micrographs of the alloys A (a) and D (b) obtained after step-aging treatments showing white contrast (FeCo-rich phase $α_1$) and dark contrast (Cr-rich phase $α_2$).

Figure 5 shows the magnetic properties of the studied alloys after the step-aging treatment. Here, the alloy A gives the following magnetic properties as H_c = 600 Oe, B_r = 6.7 kG, B_s = 9.3 kG and $(BH)_{max}$ = 2.2 MGOe which increases with the increase of molybdenum content and reaches the maximum H_c = 840 Oe, B_r = 11 kG, B_s = 12.4 kG and $(BH)_{max}$ = 5.4 MGOe for the alloy D.

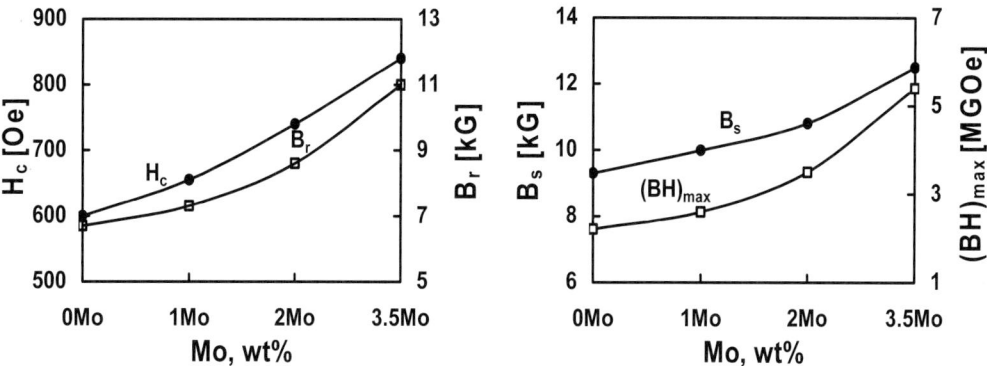

Fig. 5 Magnetic properties of the alloys obtained after the step-aging treatment.

It is concluded that magnetic properties of the ternary Fe–28Cr–15Co alloys can be improved by expanding the ferromagnetic α phase structure at annealed state and by developing shape anisotropy of spinodal phases. For example, Mo extends the α grain size at annealed state and increases the particle size and amount of the magnetically rich $α_1$ particles which in turn is responsible to the increase of the alloy magnetic properties. It is also observed that the excess amount of Mo addition to the ternary Fe–28Cr–15Co alloys produces cracking and brittleness due the formation of undesirable sigma phase. The studied anisotropic Fe–28Cr–15Co–3.5Mo permanent magnets have potential to replace the expensive Cunico, Cunife, Vicalloy, Alnico and ferrites magnets in terms of good ductility, better corrosion resistance, better working temperature, and cheaper raw materials along with excellent magnetic properties.

4 Conclusion

The microstructural and magnetic properties of Fe–28Cr–15Co–(1–4)Mo based permanent magnets have been investigated and compared with the ternary Fe–28Cr–15Co magnets. Magnetic properties of the ternary Fe–28Cr–15Co alloys can be improved with the addition of Mo. Molybdenum addition up to 3.5wt% is the optimum level for obtaining optimum magnetic properties. Moreover Mo addition above 3.5wt% produced brittleness and cracking. The better magnetic properties in Fe–28Cr–15Co–3.5Mo alloys are correlated to the formation of favorable microstructures with the optimum heat treatment conditions.

References

[1] N. Ikuta, M. Okada, M. Homma, and T. Minowa, J. Appl. Phys. **54**, 5400 (1983).
[2] S. Yin and G.Y. Chin, IEEE Trans. Magn. **MAG-23**, 3187 (1987).
[3] H.C. Chang, Y.H. Chang, and S.Y. Yao, Jpn. J. Appl. Phys. **37**, 151 (1998).
[4] Z. Ahmad, A. Ul Haq, S. W. Husain, and T. Abbas, Physica B **321**, 54 (2002).
[5] S. Sugimoto, J. Honda, Y. Ohtani, M. Okada, and M. Homma, IEEE Trans. Magn. MAG-**23**, 3193 (1987).
[6] S. Sugimoto, H. Satoh, M. Okada, and M. Homma, Mater. Trans. **32**, 557 (1991).
[7] N. Ikuta, M. Okada, M. Homma, and T. Minowa, J. Appl. Phys. Lett. **54**, 5400 (1983).
[8] M. Homma, E. Horikoshi, T. Minowa, and M. Okada, Appl. Phys. Lett. **37**, 92 (1980).
[9] Z. Ahmad, A. Ul Haq, S. W. Husain, and T. Abbas, Physica B **321**, 96 (2002).
[10] V. A. Sein, A. V. Andreyeva, and I. M. Milyayev, Phys. Met. Metallogr. **54** (6), 175 (1982).

Giant magnetoresistance and microstructure of FeCo–Al$_2$O$_3$ nanogranular films

Changzheng Wang, Zhenghong Guo, Yonghua Rong[*], and T. Y. Hsu (Xu Zuyao)

School of Material Science and Engineering, Shanghai Jiaotong University, Shanghai, 200030, P. R.China

Received 31 August 2003, accepted 31 December 2003
Published online 25 March 2004

PACS 75.47.De, 75.50.Tt, 75.70.Cn

The giant magnetoresistance (GMR) and microstructure of FeCo–Al$_2$O$_3$ films sputtered at various substrate temperatures (T_s) were investigated. The results indicate that the GMR of as-sputtered FeCo(41 vol.%)–Al$_2$O$_3$ granular films reaches the peak value of 6.9%, while for FeCo–Al$_2$O$_3$ granular films sputtered at 473 K, the dependence of GMR on volume fraction (f_v) displays 6% peak value with lower content about 35 vol.% FeCo. Based on the features of GMR of FeCo–Al$_2$O$_3$ granular films sputtered at different f_v and T_s, an optimum factor $A(D_e/L_c)$ is introduced to explain qualitatively the influence of spin-dependent interface scattering(D) and tunneling barrier's thickness(L) related to f_v and Ts on GMR. The phase separation of hcp α-Co from bcc α-Fe (Co) was revealed by analytical electron microscopy and its effect on GMR for different f_v and T_s was discussed.

© 2004 WILEY-VCH Verlag GmbH & Co. KGaA, Weinheim

1 Introduction   Since the discovery of the giant magnetoresistance (GMR) in granular films [1, 2], this phenomenon has been extensively studied due to its potential application in information industry. The GMR effect originates from spin-dependent scattering of conducting electrons at the interfaces between the ferromagnetic granules and the non-magnetic matrix, and within the ferromagnetic granules. Therefore, the GMR effect occurred easily in a system of ferromagnetic granule – non-ferromagnetic metal matrix in which the electronic mean free path and spin diffusion length are relatively long. In other words, the GMR cannot appear particularly in a material with high electrical resistivity. However, a pronounced GMR can be observed in some insulating granular films [3–5] and is believed to be based on the mechanism of spin-dependent tunnelling, therefore, the giant magnetoresistance in this case can be also called tunnelling magnetoresistance (TMR). The GMR has a closely relationship with chemical composition [6], measuring temperature [7], annealing conditions [8] or the substrate temperature. In this paper, we will investigate the effect of the composition and substrate temperature accompanying with microstructure on the GMR in FeCo–Al$_2$O$_3$ granular films.

2 Experimental procedure   The FeCo–Al$_2$O$_3$ granular films were sputtered respectively on KCl or glass substrate at 300 K and 473 K with a spc350 multi-target magnetron controlled sputtering system. FeCo target (the weight ratio of Fe and Co is 1:1) and Al$_2$O$_3$ target (99.9% purity) were separately installed on two independently controlled R.F. cathodes. The thickness of films for transmission electron microscope (TEM) and GMR measurement are respectively as about 50 nm and 300 nm. TEM observations were carried out on JEM-100CX and Philips Tecnai F20 with electron energy-loss spectrometer (EELS) respectively. The GMR effect was measured by conventional four-probe method.

---

[*] Corresponding author: e-mail: yhrong@sjtu.edu.cn, Phone: +86-21-6293-2558, Fax: +86-21-6293-2435

3 Results and discussion

3.1 Giant magnetoresistance Figure 1 shows the dependence of GMR on the FeCo volume fraction (f_v) in granular films sputtered at room temperature (RT, 300 K) and 473 K, respectively. As can be seen from Fig. 1, the granular films sputtered respectively at 300 K and 473 K possess the same change tendency of GMR as a function of FeCo volume fraction, namely, when $f_v > 0.52$, which is precisely the percolation threshold f_c for granular system, their GMR vanish. When f_v decreases from 0.52, GMR gradually increase, then up to peak values (at 300 K 6.9% for $f_v = f_{vm} = 41\%$, at 473 K 6.0% for $f_v = 36\%$), while GMR diminish with the further decrease of FeCo volume fraction. Such GMR behaviours of FeCo–Al$_2$O$_3$ granular films will be qualitatively explained as follows. It is obvious that the large GMR appears with f_v less than f_c (52%). For $f_v > f_c$, all the FeCo particles coalesce and gradually form a connecting network with multidomain structure. The electric transport in this region is not realized by tunnelling but carried by metallic conductance. As a result, GMR vanishes. Within the region $f_v < f_c$, the GMR increases with decreasing f_v from f_c, and then is up to a peak value. As f_v further decreases, the GMR drops. In the above region, the electric transport is governed by tunnel current between FeCo particles separated by Al$_2$O$_3$ matrix as tunnel barriers [9]. When f_v decreases from the f_c, the connecting network is broken and then large multidomain FeCo granules are separated into small single-domain or superparamagnetic granules. When these magnetic particles are magnetized under applied field, a larger GMR appears due to the spin-dependent tunnelling effect. With the further decrease of f_v from f_{vm}, FeCo particles are few and so far apart, i.e. the thickness of Al$_2$O$_3$ tunnel barriers become so large that it is larger than the spin diffusion length, therefore, the spin-flip tunnelling process may occur, giving rise to a reduction of GMR.

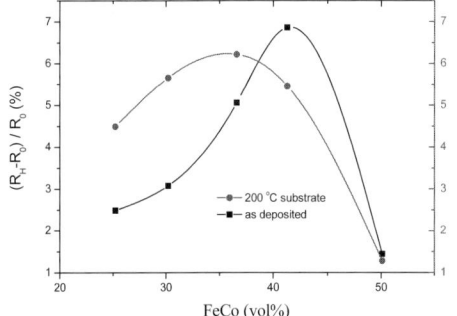

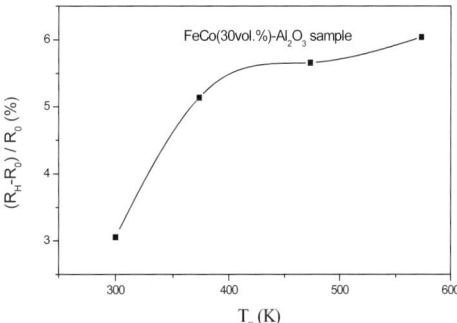

Fig. 1 Curve of GMR vs. FeCo volume fraction.

Fig. 2 The dependence of GMR on substrate temperature at RT and 473 K for FeCo(30 vol.%)–Al$_2$O$_3$ samples.

Yang [3] et al introduced the parameter L_c as a critical tunnel barrier's thickness related to the spin diffusion length based on their experiment of Fe–SiO$_2$ with GMR = 2.7% at 28 vol.%Fe. Their conclusion is that when the tunnel barrier's thickness L is smaller than L_c, the GMR is proportional to L, and above L_c, the GMR reduces due to the spin-flip process. However, their theory can be only used to explain part of phenomenon in our FeCo–Al$_2$O$_3$ granular films, i.e. the granular films sputtered at RT with $f_{vm} = 0.41$ at peak value corresponding to L_c. When the film with the same volume fraction is sputtered at 473 K, the tunnel barrier's thickness increases and is larger than L_c due to the coalescence of particles, which results in the drop of GMR. But their theory cannot be employed to explained the case that f_v is larger or smaller than f_{vm} (0.41). For example, for the granular films sputtered at RT with f_v smaller than f_{vm}, their tunnel barrier's thickness L is lager than L_c, and thus their L will become larger when the granular films with the same volume fraction is sputtered at higher temperature 473 K, obviously, GMR should drop based on their theory, however, our experiments show the opposite results. Therefore, their theory should be modified. We introduce a parameter A_{op} called as optimum factor, being proportional to D_c/L_c corresponding to maximum GMR for a ferromagnetic metal–insulator granular film, where D_c is a critical size of spin-dependent interface scattering and stands for maximum interface scattering, that is, when the mean size D of particle is smaller or larger than D_c, their spin-dependent interface scattering will become small, the former results from the currents passing by smaller particles more easily, while the later does the decrease of interface-to-volume ratio for lager particles [10]. For a given granular film with f_v at some T_s, its $A(D/L)$ is a constant [11], however, D/L can change when substrate temperature T_s or annealing time varies, which leads to increase or decrease of GMR. For example, for a FeCo–Al$_2$O$_3$ granular films with f_v smaller than f_{vm}, its $A(D/L)$ is smaller than A_{op} (D_c/L_c) since $D < D_c$ and $L > L_c$, when T_s is elevated from RT, if the enhancement of interface scattering(D) is larger than the increase of tunnelling barrier's thickness (L), in the case the change of T_s will improve GMR since its $A(D/L)$ increases and approaches A_{op} (D_c/L_c), as shown in Fig. 1, otherwise, will decrease GMR. The above idea is confirmed by another experiment. Fig. 2 shows the variation of GMR with substrate temperature T_s for FoCo (30 vol.%)–Al$_2$O$_3$ granular films. It can be found from Fig. 2 that GMR increases with increasing T_s from RT to 573 K, indicating that the enhancement of spin-dependent interface scattering is predominant with increasing T_s from RT to 573 K although the increase of tunnelling barrier's thickness, at the same time, is unfavourable. If T_s continue to be elevated over some temperature, it is conceivable that the multidomain formation and conjunction of particles will result in the drop of GMR until vanish. For a given granular film with $f_{vm} < f_v < f_c$ at RT, its $A(D/L)$ is larger than A_{op} (D_c/L_c) since $D > D_c$ and $L < L_c$. When T_s is elevated from RT, D always increases and is unfavourable for GMR; therefore, GMR will depend on the change of L. If the change of L makes $A(D/L)$ be far apart A_{op} (D_c/L_c), GMR will drop, otherwise, will increase GMR. The former can be verified by our experiment, as shown in Fig. 1. The drop of GMR for FeCo(41vol.%)–Al$_2$O$_3$ can result mainly from the easy conjunction and network formation of particles ($L \to 0$) when T_s is elevated up to 473 K from RT, since the extent between f_{vm}(41vol.% FeCo) and f_c (52%) is narrower. The later can be verified by experiment in Fe(42 vol.%)–SiO$_2$ granular films of Yang et al. [3]. The increase of GMR can result mainly from the increase of tunnelling barrier's thickness when T_s is elevated up to 473 K from RT, i.e. L approaches L_c since the extent between f_{vm}(26 vol.% Fe) and f_c (52%) is wider, implying the fact that granules do not easily form the connecting network.

Fig. 3 The microstructure of FeCo(41vol.%)–Al$_2$O$_3$ film. (a) Bright field image of α-Fe (Co) granules and their selected area diffraction pattern, (b) HRTEM image of amorphous clusters at RT sputtered film and its Fourier transformed pattern, (c) iron EELS map and (d) cobalt EELS map of 823 K sputtered granular films

3.2 Microstructure The microstructure of granular films affects the GMR effect significantly, hence it is necessary to study the relationship between microstructure and the GMR effect. Our previous TEM observation [12] indicates that in FeCo (41vol.%)–Al$_2$O$_3$ granular film sputtered at RT there are a little amount of α-Fe(Co) granules with bcc structure (Fig. 3(a)) and FeCo amorphous clusters dispersing in Al$_2$O$_3$ matrix (Fig. 3(b))and all granules and clusters contain both Fe and Co atoms.

In Co–Fe binary phase diagram there is a miscibility gap in wide composition range, namely, there is phase separation. However, this phenomenon was not observed in RT sputtered FeCo–Al$_2$O$_3$ film. The reason may be the difficulty of Fe and Co diffusion at RT. For the sake, the FeCo–Al$_2$O$_3$ film with same composition sputtered at 823 K will be investigated. Fig. 3(c), (d) are EELS maps of Fe and Co for 823 K sputtered FeCo–Al$_2$O$_3$ granular film respectively and show that Co and Fe particles (bright contrast) form individually in Al$_2$O$_3$ matrix. By electron diffraction Co particles are identified as hcp α-Co and Fe particles as bcc α-Fe, exhibiting the phase separation from α-Fe(Co) particle. By the comprehensive analysis of RT and 823 K sputtered films, the evolution sequence of the two-stage phase separations for the FeCo–Al$_2$O$_3$ granular film is revealed, namely, the amorphous phase or cluster containing both Fe and Co atoms will firstly transform to the super-saturation bcc α-Fe(Co) particles by amorphous crystallization, promoting the phase separation between α-Fe (Co) and Al$_2$O$_3$, then the super-saturation α-Fe (Co) gradually precipitates α-Co particles to further produce the phase separation between α-Co and α-Fe. It is worthy to point out that the microstructure of TEM specimen is somewhat different from that of GMR sample with the same composition and sputtered temperature. For example, in GMR sample of FeCo(41 vol.%)–Al$_2$O$_3$ sputtered at room temperature with the highest GMR effect, amorphous clusters can partially crystallize as α-Fe (Co), while majority of α-Fe(Co) particles undergo probably the phase separation due to the sputtered time of the GMR sample much greater than that of the TEM specimen. Such a conjecture on the microstructure of the above GMR sample may be verified by the experiment of FeCo(30 vol.%)–Al$_2$O$_3$ sputtered at 573 K temperature, in which the phase separation should occur or finish. Since FeCo(30 vol.%)–Al$_2$O$_3$ granular film sputtered at 573 K temperature exhibits 6.0% GMR and approaches 6.9% GMR of FeCo(41 vol.%)–Al$_2$O$_3$ sputtered at room temperature, it is reasonably believed that their microstructures should be similar. Moreover, it is worthy to emphasize that the phase separation enhance the spin-dependent interface scattering due to the increase of interface and the delay of particle growth, however, it accelerates the conjunction of particles due to *in-situ* precipitation, the later leads to the drop of GMR, specially for granular films with higher volume fraction, such as for FeCo–Al$_2$O$_3$ granular films with f_v equal to or larger than f_{vm}.

4 Conclusions (a) The GMR of as-sputtered FeCo(41 vol.%)–Al$_2$O$_3$ granular films reaches the peak value of 6.9%, while for FeCo–Al$_2$O$_3$ granular films sputtered on 473 K, the 6% peak value with 35 vol.% FeCo; (b) An optimum factor $A(D_c/L_c)$ is introduced to explain qualitatively the influence of spin-dependent interface scattering(D) and tunnelling barrier's thickness(L) related to f_v and T_s on GMR. For granular films with $f_v < f_{vm}$, GMR increases with increasing T_s from TR to 573 K since the enhancement of spin-dependent interface scattering is predominant, while for $f_{vm} < f_v < f_c$, GMR decreases with increasing T_s from RT to 473 K since the coalescence and conjunction of part of particles ($L \to 0$) is predominant; (c) Since the phase separation of hcp α-Co from bcc α-Fe(Co) enhance the spin-dependent interface scattering and delays the particle growth for FeCo–Al$_2$O$_3$ granular films with f_v smaller than f_{vm}, GMR will improve when T_s is elevated below some temperature. However, it also accelerates the conjunction of particles due to *in-situ* precipitation and leads to the drop of GMR for FeCo–Al$_2$O$_3$ granular films with f_v equal to or larger than f_{vm}.

Acknowledgements The present work is financially supported by the National Nature Science Foundation of China under Grant No. 50071033.

References

[1] A. E. Berkowitz, J. R. Mitekell, M. J. Corey et al., Phys. Rev. Lett. **68**, 3745 (1992).
[2] J. Q. Xiao, J. S. Jiang, and C. L. Chien, Phys. Rev. Lett. **68**, 3749 (1992).
[3] W. Yang, Z. S. Jiang, Y. W. Du et al., Solid State Commun. **104**, 479 (1997).
[4] Y. Hayakawa, N. Hasegawa, A. Makino et al., J. Magn. Magn. Mater. **154**, 175 (1996).
[5] T. Furubayashi and I. Nakatani, J. Appl. Phys. **79**, 6258 (1996).
[6] S. Mitani, H. Fujimori, and S. Ohnuma, J. Magn. Magn. Mater. **165**, 141 (1997).
[7] T. Zhu and Y. J. Wang, Phys. Rev. B **60**, 11918 (1999).
[8] Jae-geun Ha, S. Mitani, H. Fujimori et al., J. Magn. Magn. Mater. **198/199**, 21 (1999).
[9] M. Ohnuma, K. Hono, H. Fujimori, et al., Mater. Sci. Forum **307**, 171 (1999).
[10] Rong Yang, Wei Zhang, and W. J. Song, J. Appl. Phys. **84**, 2044 (1998).
[11] J. S. Helman and B. Abeles, Phys. Rev. Lett. **37**, 1429 (1976).
[12] N. Zhou, C. Wang, Z. Guo, Y. Rong, and T. Y. Hsu, Mater. Lett. **57**, 2168 (2003).

Structural, magnetic and magneto-transport properties of thermally evaporated Fe/Cu multilayers

K. Bouziane[*]**, M. Al-Busaidi, A. Gismelseed,** and **A. Al-Rawas**

Physics Department, College of Science, Sultan Qabos University, P. O. Box 36, Postal Code 123, Al-Khodh, Muscat, Sultanate of Oman

Received 31 August 2003, accepted 31 December 2003
Published online 25 March 2004

PACS 61.10.Nz, 68.37.Lp, 75.47.De, 75.50. Bb, 75.70.Cn, 76.80.+y

Structural, magnetic and magneto-transport properties of thermally evaporated Fe/Cu multilayers (MLs) have been investigated. Although multilayered structure has been successfuly obtained, a substantial interfacial roughness ranging from 0.6 nm to 1.2 nm has been determined. All Fe/Cu MLs were polycrystalline with an average grain size of about 10 nm. Fe was bcc and textured (110) whereas Cu was fcc (111). Transmission electron microscopy analysis showed that the fcc Cu layer was rather textured (110) and (100) at least in the first stage of growth of the Fe/Cu MLs. Conversion electron Mössbauer (CEMS) measurements indicated the existence of three phases. Two of them were magnetic with a dominant bcc Fe phase, followed by fcc Fe phase. The third phase was superparamagnetic. The CEMS results were explained in terms of the partial diffusion of Fe into Cu with three different zones. The small magnetoresistance (MR < 0.2%) was correlated to Fe clusters located at Fe–Cu interfaces.

1 Introduction

Recent developments in the synthesis of thin films by various techniques have made possible to prepare well-defined layered structures. Metallic MLs, composed of a ferromagnetic metal (such as Fe) alternating with a non-magnetic layer (such as Cr, Au, Ag or Cu) exhibit magnetoresistance (MR: change of resistivity under magnetic field) and indirect exchange coupling (J) [1]. Additionally, Fe–Cu system reveals further new physical features. At equilibrium, the bcc phase is the more stable. The fcc phase appears only at 1185 K [2]. However, the fcc metastable Fe phase was observed in mechanically alloyed FeCu below 1183 K [3–5]. Moreover, it has been shown that a very thin Fe layer, below a critical thickness (less than 1.5 nm), adopts the fcc phase when epitaxially grown on Cu(100) [6]. Although the solid solubility between Fe and Cu is negligible, a different degree of mutual diffusion has been obtained in mechanically alloyed FeCu and multilayers [3–5]. To the best of our knowledge, most of Fe/Cu multilayered structures were prepared by sputtering method or molecular-beam-epitaxy.

In this work we present the structural, magnetic and electronic properties of thermally evaporated Fe/Cu multilayers on glass substrates. We show that MR can be observed in Fe/Cu multilayers with relatively large interfacial roughness. The crystalline phase of Fe and the exchange coupling between Fe layers are also investigated and discussed here in.

2 Experimental details

Fe/Cu multilayers have been prepared by thermal evaporation with a base pressure below 10^{-6} mbar. The purity of either Fe or Cu, used as element sources, was 99.99%. All multilayers were deposited on 5.0 nm thick layer of Fe onto a glass substrate at room temperature. Fe/Cu MLs consist of 25 repeats of [Fe(1.5 nm)/Cu(t_{Cu})] with copper thickness between 0.5 and 4.0 nm. Other series of different bilayer repeats [Fe(1.5 nm)/Cu(t_{Cu})] were synthesized in order to optimise the highest magnetoresistive effect. 25 repeats constituted the optimal conditions. The rates of deposition of Fe and

---

[*] Corresponding author: e-mail: bouzi@squ.edu.om, Phone: +968 515 489, Fax: +968 514 228

Cu layers were determined from grazing X-ray diffraction (GXRD) analysis and by calibrating the crystal thickness monitor placed near the substrate position. The interfacial roughness Fe/Cu were calculated using a simulation programme with the computational code XREALM based on the recursive method [7] of GXRD spectra. The crystalline structure and the surface morphlogy were analysed using high X-ray diffraction (HXRD) with monochromated Cu $K_{\alpha1}$ radiation, and transmission electron microscopy (TEM) at room temperature (RT). DC-MR was measured at RT by the standard four-point probe in an in-plane arrangement with a field up to 10 kOe applied in plane and perpendicular to the sensing current. In-plane magnetic properties were recorded in a standard vibrating sample magnetometer (VSM-ADE 1660) in a field up to 6 kOe. CEMS with ^{57}Co(Rh) source was used to probe locally Fe at RT. The hyperfine parameters were determined by fitting CEMS spectra using the least-squares method.

3 Results and discussion The presence of the first-order Bragg peak in GXRD scans (not shown here) for most of the samples (except for MLs with $t_{Cu} \leq 1.0$ nm) was a good signature of a multilayered structure. The absence of higher-order peaks and the Kisseig fringes might be connected to relatively high interfacial roughness. The rms values of interfacial roughness (σ_{rms}) determined from the simulation of GXRD spectra are reported in Table 1. σ_{rms} ranged from 0.6 to 1.2 nm upon the thickness of Cu.

Table 1 The calculated root-mean-square roughness (σ_{rms}) from GXRD simulations for selected [Fe(1.5nm)/Cu(tCu)]$_{\times 25}$ MLs. All parameters are expressed in nm.

t_{Cu}	σ_{rms}-Cu/Fe	σ_{rms}-Fe/Cu
1.4	0.9	1.0
2.5	0.6	0.7
2.7	0.9	1.4
4.0	0.8	0.9

This roughness is relatively high as compared to that of similar Fe/Cu MLs deposited by energetic methods such as sputtering [8]. Furthermore, the roughness was dissimilar in alternate Fe/Cu (Cu grown on Fe) and Cu/Fe (Fe grown on Cu). The roughness of the Cu/Fe interface was slightly smaller than that of Fe/Cu interface. This behaviour was attributed, according to Lee et al. [8], to the phases that Fe and Cu might adopt under equilibrium conditions and to the lattice mismatch between Fe and Cu. fcc Fe can grow epitaxially on fcc Cu. Therefore, negligible lattice mismatch is expected in this case and hence Fe grows smoothly on Cu. Conversely, Cu adopts the fcc phase only, and a substantial distortion at the interface Fe/Cu is expected when fcc Cu grows on bcc Fe. This results in larger Fe/Cu interfacial roughness.

All Fe/Cu MLs were polycrystalline with an average grain size of ~ 10 nm (calculated from Debye-Scherrer law: $L = 0.9\lambda/[\Delta(2\theta)\cos\theta]$). The position of the Bragg peak in HXRD spectra at around $2\theta = 44.5°$ was associated to bcc Fe(110) and fcc Cu(111) textures. In fact, The position of the Bragg peak shifts to smaller angles as the thickness of Cu increases (Fig. 1).

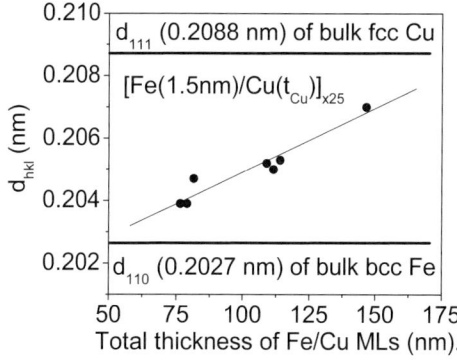

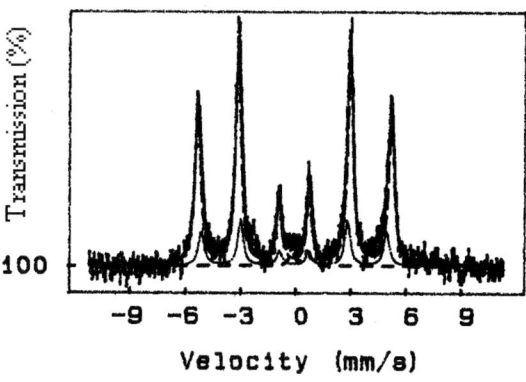

Fig. 1 Interplanar distance vs. total thickness of [Fe(1.5 nm)/Cu(t_{Cu})]$_{\times 25}$ MLs.

Fig. 2 Typical room-temperature CEMS spectrum of [Fe(1.5 nm)/Cu(4.0 nm)]$_{\times 25}$ multilayer (ML).

The lattice parameter of thin copper layers is presumably constrained (mismatch between Fe and Cu of ~ 11%) to iron bcc structure, growing along the bcc (110) plane (d = 2.027 Å). As the thickness of the Cu layer increases, relaxation of copper layers towards fcc Cu(111) plane (d = 2.088 Å) may explain this feature. No satellite peaks have been observed at larger angles. This may be due to the low contrast between the scattering factors of Fe and Cu, and to the presence of appreciable interfacial roughness.

In the above discussion, we have considered that Fe adopted a unique phase, namely bcc (110). This assumption may be incorrect for various reasons. The XRD technique gives a macroscopic analysis of the structure of films. Phase/texture in the film may be detected depending on their relative percentage per area and scattering factor [9]. CEMS has been used to probe the local structure of Fe. A representative room-temperature CEMS pattern of [Fe(1.5 nm)/Cu(4.0 nm)]$_{x25}$ MLs is shown in Fig. 2.

The corresponding CEMS parameters determined from the fit of the spectrum are displayed in Table 2.

Table 2 Mössbauer parameters of [Fe(1.5nm)/Cu(4.0nm)]$_{x25}$ ML.

IS (mm/s)	QS (mm/s)	LW (mm/s)	M S (mm/s)	A (%)
0.33 (0.07)	0.85 (0.09)	0.35 (0.00)	–	2.8 (1.0)
0.03 (0.02)	0.01 (0.01)	0.35 (0.00)	32.87 (0.01)	79.5 (1.2)
0.02 (0.04)	–0.01 (0.03)	0.35 (0.00)	31.36 (0.06)	17.7 (2.4)

The best fit of the spectrum (solid and dashed lines) was found to be a superposition of two sextet (A and B) components (having different hyperfine fields) and one doublet. It is expected that Fe-CEMS signals come from a pure bulk-like Fe region and an interface region between Fe and Cu layers [10]. The sextet component A is consistent with Fe bulk-like α-Fe corresponding to Fe atoms away from the interface region (with Fe nearest neighbours only). This component represents 79.5% per area of the total phase corresponding to bcc Fe(110) in accordance with XRD measurements. The other sextet component B, weighing 17.7%, was attributed to the CEMS signals of the Fe–Cu interface site (as it has smaller magnetic splitting than pure bulk-like Fe splitting). In this case, Fe atoms have Fe and some Cu atoms as the nearest neighbours. This component B was associated with fcc Fe. The doublet, weighing 2.8%, was attributed to the electric quadrupole interaction in a nonmagnetic or supeparamagnetic (SPM) phase. This doublet can be associated with granules of Fe embedded in Cu matrix at the interfaces. It is worth noting that the parameters determined from the fit of CEMS spectra, for MLs with various Cu thickness, were essentially the same. To identify the existence of textured Cu(100) favorable for fcc Fe phase [6], three samples of composition: (S_1) = Fe(5.0 nm), (S2) = Fe(5.0 nm)/Cu(2.5 nm) and (S3) = Fe(5.0 nm)/Cu(1.5 nm)/Fe(1.5 nm) were prepared for TEM investigation. Figure 3 shows the TEM elctrons diffraction patterns of the three samples S_1, S_2 and S_3. The ring configuration patterns suggest that the three samples are polycrystalline. For S_1, the strongest ring corresponds to bcc Fe(110), as expected. This ring persists in TEM spectra of S_2 and S_3 respectevely. The preferred orientations of Cu were found to be fcc(100) and fcc(110). This is not in desageement with XRD results and may suggest that the structure and textures are the same within the thickness of Fe/Cu multilayers.

Now we attempt to identify the spin dependent scattering sites. This can be analyzed in terms of the relationship between *MR* and the total magnetization *M* (defining the antiferromagnetic (AF) exchange coupling between adjacent Fe layers). Indeed, for coherent magnetization rotation of MLs with AF coupling, the resistance change ΔR [= $R(0)–R(H_s)$, H_s being the saturation field] is expected to be proportional to $\cos^2\theta \propto (M/M_s)^2$, where θ is the angle between magnetizations of adjacent Fe layers. ΔR of [Fe(1.5 nm)/Cu(t_{Cu})]$_{x25}$ as a function of $(M/M_s)^2$ is displayed in Fig. 4. ΔR of all samples is principally negligible up to the region near saturation (M ~ Ms). Most of the magnetoresistive effect comes from the saturation region. Regardless of the thickness of Cu, the coupling between Fe layers is essentially ferromagnetic as the M-H loops have nearly square shape (inset of Fig. 4). This also explains the absence of oscillation of MR as a function of Cu thickness. Moreover, the magnitude of MR varies slightly with Cu thickness (ranging from 0.05 to 0.2%). Therefore, we can infer that MR effect is not arising from spin-valve due to AF configuration [1].

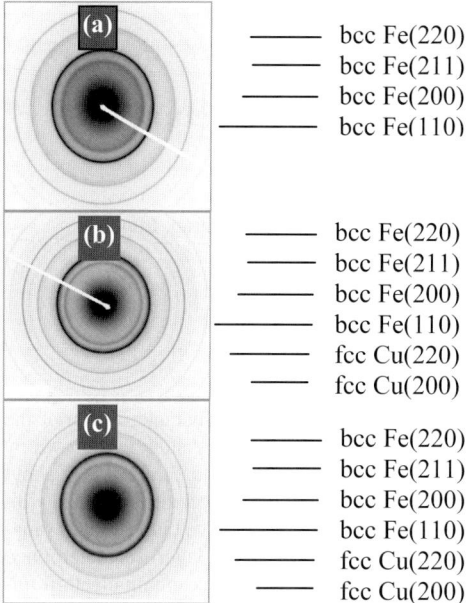
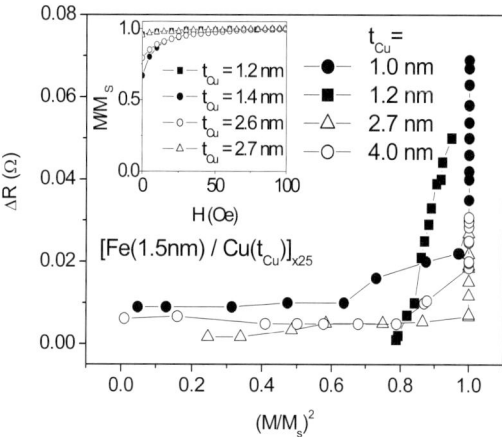

Fig. 3 TEM electrons diffraction patterns of samples (a) S_1, (b) S_2 and (c) S_3.

Fig. 4 Change of resistance MR versus $(M/M_s)^2$ of Fe/Cu MLs for different t_{Cu}. Inset is the M–H loops.

We can reasonably admit the existence of Fe granules embedded in Cu matrix at the Fe–Cu interfaces. We may suppose that those Fe granules correspond to the superparamagnetic phase detected by CEMS analyis. Their total magnetic moment contribute negligeably to the macroscopic magnetization of MLs. However, the rondom orientation of their magnetic moments can contribute significantly to spin dependent scattering of electrons. The alignements of theses moments by applying a magnetic field gives rise to *MR*.

4 Conclusion Thermally evaporated Fe/Cu MLs have been investigated. High interfacial roughness of the order 0.6–1.2 nm was observed in these MLs. We found that Fe has two crystalline phases: bcc and fcc both magnetic. A third dilute nonmagnetic phase of Fe was attributed to superparamagnetic state of Fe. The former was associated with the bulk of Fe layers and the latter with the Fe interfacial layers. This was attributed to granules of Fe at Fe–Cu interfaces. We think that these granules are the spin-dependent scattering centers of electrons at the origin of limited MR obesved in our Fe/Cu MLs.

References

[1] See for instance: "Magnetic Multilayers and Giant Magnetoresistance: Fundamentals and Applications", Hartman ed. (Springer-Verlag, Berlin, 2000).
[2] A. Clarke, P. J. Rous, M. Arnott, G. Jennings, and R. F. Willis, Surf. Sci. **192**, L843 (1987).
[3] U. Gonser, C. J. Meechan, A. H. Muir, and H. Wiedersich, J. Magn. Magn. Mater. **22**, 2373 (1963).
[4] K. Sumiyama, Y. Yoshitake, and Y. Nakamura, Acta Metall. **33**, 1785 (1985).
[5] J. Z. Jiang, Q. A. Pankhurst, C. E. Johnson et al., J. Phys.: Cond. Matter **6**, L227–L232 (1994).
[6] D. Q. Li, M. Frietag, J. Pearson, Z. Q. Qiu, and S. D. Bader, Phys. Rev. Lett. **72**, (1994) 3112.
[7] L. G. Parratt, Phys. Rev. **95**, 359(1954).
[8] D. W. Lee, D. H. Ryan, Z. Altounian, and A. Kuprin, Phys. Rev. B **59**, 7001–7009 (1999).
[9] W. F. Egelhoff, Jr. and M. T. Kief, IEEE Trans. Magn. **28**, No. 5, 2742–2744 (1992).
[10] H. M. Van Noort, F. J. A. den Broeder, and H. J. G. Draaisma, J. Magn. Magn. Mater. **51**, 273 (1985).

Co surface modification by bias sputtering in Cu/Co(V$_b$)/NiO/Si(100) magnetic multilayer structures

A. Z. Moshfegh[*] **and P. Sangpour**

Department of Physics, Sharif University of Technology, P. O. Box 11365-9161, Tehran, Iran

Received 31 August 2003, accepted 31 December 2003
Published online 25 March 2004

PACS 68.37.Hk, 68.37.Ps, 75.55.+f, 75.70.–i, 75.70.Cn, 81.15.Cd

To investigate the Ta/Co/Cu/Co/NiO/Si(100) spin valve structure, fabrication and characterization of the Cu/Co/NiO/Si(100) system was studied for further understanding the structure. The system was grown by employing combinative DC sputtering–evaporation technique. Nickel oxide with a thickness of about 30 nm was deposited on Si(100) substrate using thermal evaporation technique. The cobalt film, then, with a thickness of about 3 nm was grown by DC sputtering under various applied negative bias voltages ranging from 0 to – 80 V. The optimum bias voltage (V$_b$ = –60 V) for the growth of Co layer was determined by atomic force microscopy (AFM), four-point probe sheet measurement (R$_s$) and scanning electron microscopy (SEM). Following the Co deposition at the optimum condition, the Cu layer with a thickness of about 2 nm was deposited on the Co(V$_b$)/NiO/Si(100) structure by using DC magnetron sputtering technique. The Cu/Co(V$_b$)/NiO/Si(100) structure was examined by AFM and R$_s$ measurements. Our data analysis indicates that Cu possess a proper surface for the growth of next Co layer in the Co/Cu/Co active GMR region.

© 2004 WILEY-VCH Verlag GmbH & Co. KGaA, Weinheim

1 Introduction

In recent years, giant magneto resistance (GMR) material with high sensitivity has been studied to develop high-density magnetic storage system [1–3]. Among many magnetic multilayers, spin valve multilayers are promising for use in magnetoresistive read head, sensors and MRAM[1] devices [4–6].

Since the discovery of unidirectional exchange anisotropy in antiferromagnetic/ferromagnetic exchange coupled system, numerous investigations of this phenomenon have been carried out [7, 8].

Besides its fundamental interest, exchange anisotropy has received an increasing attention for ability to pin a ferromagnetic layer in spin valve structures. However, FeMn films have some drawbacks such as poor corrosion resistance and a relatively low blocking temperature [9–11]. Therefore, some other antiferromagnetic materials were studied. Among various antiferromagnetic materials used to pin the magnetization of an adjacent ferromagnetic layer, NiO is a possible candidate, which offers the advantage of no current shunting, excellent thermal stability, and strong resistance to corrosion with high blocking temperature [11, 12]. Considering these advantages, we have used NiO as a pinning layer in our investigation.

Among many multilayers investigated, the largest GMR effect has been observed in Co/Cu based structures [13]. So, in this study the Co surface modification by bias sputtering in Cu/Co(V$_b$)/NiO/Si(100) magnetic multilayers has been discussed specially during the growth of Co layer under different applied negative bias voltages.

---

[*] Corresponding author: e-mail: moshfegh@sharif.edu, Phone: +98 21 6164516, Fax: +98 21 6012983
[1] Magnetic Random Access Memory

2 Experimental The substrates used for this experiment were n-type Si(100) wafers with resistivity 5–8 Ω-cm and the dimension of 5×11 mm^2. After a standard RCA cleaning procedure and a short time dip in a diluted HF solution, the wafers were dried in high purity N2 (99.999%) environment then loaded in to the vacuum chamber. The chamber was evacuated to a base pressure of about 4×10^{-7} Torr prior to any deposition. We have used a combinative sputtering-evaporation technique to deposit the proposed magnetic multilayers. A detail arrangement of the deposition system can be found elsewhere [14]. A rotating quartz crystal oscillator is placed very near to the substrate to monitor the thickness of the desired material deposition. To deposit nickel oxide thin film, first a high purity NiO powder as a starting material was pressed and baked over night at 1400 °C in an atmospheric oven. This process yielded a green solid disk suitable for evaporation as also reported recently [15]. Al_2O_3 coated W was used as a working boat to prevent reaction between NiO and W. Before deposition of each layer, a pre-evaporation and pre-sputtering process was performed for about 3 and 10 min, respectively. Nickel oxide (NiO) with a thickness of about 30 nm was grown on Si(100) substrate using thermal evaporation method. Following the NiO deposition, cobalt layer with a thickness of about 3 nm and copper film then with a thickness of about 2 nm were grown without breaking the vacuum by utilizing conventional DC diode sputtering and DC magnetron sputtering methods, respectively. The cobalt layer was deposited on NiO surface at various negative bias voltages ranging from 0 to –80 V using DC sputtering technique under the similar deposition conditions. The other growth parameters for the fabrication of the $Cu/Co(V_b)/NiO/Si(100)$ structure are summarized in Table 1.

The optimum negative voltage (V_b = –60 V) was determined based on atomic force microscopy (AFM) and scanning electron microscopy (SEM) as well as sheet resistance(R_s) measurements.

The deposited Cu, Co and NiO were characterized by analytical techniques in order to determine the property and quality of each layer in the Cu/Co/NiO/Si(100) structure.

Table 1 The growth parameters of the $Cu/Co(Vb)/NiO/Si(100)$ structure.

Parameter	NiO	Co	Cu
Ar pressure (mtorr)	–	70	5
Applied power (W)	355	40.5	50
Bias voltage (V)	0	–60	0
Deposition rate (nm/sec)	0.03	0.01	1.3
Thickness (nm)	30	3	2

3 Results and discussion We have investigated the fabrication and characterization of both Co(Vb)/NiO/Si(100) and Cu/Co/NiO/Si(100) structures. The deposited films are analyzed using AFM and Rs measurements. The obtained results are described in the following section.

Figure 1 shows the electrical resistivity of cobalt thin films with thickness of about 3 nm as function of applied negative bias voltage at high Ar pressure (70 mtorr). It was found that the films deposited at zero bias voltage (anode potential) exhibit high resistivity, but, as the substrate bias voltage increases, the resistivity of the films decreased in a range from –20 to –60 V and then increased again. A similar behaviour was also obtained for the bias sputtered Ta films by our group and other investiagtors [16–18]. It is noted that the change in resistivity by ion bombardment have been previously explained by variation in impurity content [17] and lattice structure [18–20]. The increase in resistivity observed at higher bias voltages is believed due to stresses introduced in Co film by higher energy ion bombardment during the Co deposition [19]. In this study, the optimum bias voltage was determined by using AFM and R_s measurement. To correlate Co electrical property with its surface topography; we have also utilized AFM method to study the layer topography.

The aim of this investigation was to determine the optimum bias voltage at which Co layer possess the lowest resistivity with a good surface topography. Figure 2 illustrates the AFM micrograph of the Co/NiO/Si(100) thin film deposited at the optimum bias voltages (V_b = –60 V). It was observed that the Co surface topography formed at the optimum applied negative bias voltage (V_b = 60 V) exhibits smooth surface as compared with the surface of samples deposited at the other bias voltages. To further understand the properties of the deposited films, we have measured surface roughness of Co layer by using computer program. Figure 3 shows the variation of Co surface roughness as a function of applied negative bias voltages. According to our SEM and AFM observations, a minimum roughness obtained at the $V_b = -60$ V. This is consistent with our resistivity measurements of the same samples as shown earlier (Fig. 1). In addition to AFM and R_s measurements, SEM technique was also used to study the morphology of the surface resulting in lower electrical resistivity and higher density as also seen earlier [14] and improving surface roughness of the deposited Co/NiO/Si(100) structure. Figure 4 shows SEM micrographs of the representative samples deposited at different applied negative bias voltage of –20, –40 and –60 V as compared with the unbiased Co/NiO/Si(100) thin film. It is evident that as bias voltage increases, the voids density and their average size on the surface decrease indicating an improvement in the Co surface morphology. The Co sputtered at the optimum bias voltage resulted in smooth and relatively defect free structure. Therefore, according to our AFM, R_s and SEM data analysis, the Co films deposited at the optimum bias voltage (–60 V) can be used as an excellent under layer to grow Cu thin film in the Cu/Co/NiO/Si(100) system.

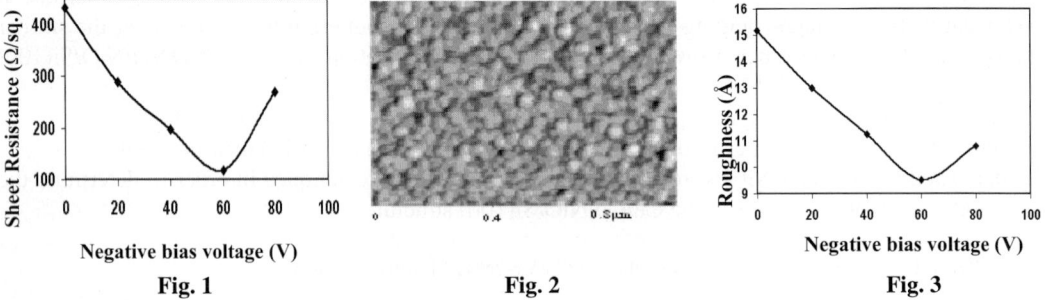

Fig. 1   Fig. 2   Fig. 3

Fig. 1   Sheet resistance of the Co/NiO/Si(100) thin films at different negative bias voltages.

Fig. 2   2D-AFM surface topography of the Co/NiO/Si(100) thin film at the optimum bias voltage (–60 V) during the growth of Co layer.

Fig. 3   Variation of the Co surface roughness in the Co/NiO/Si(100) structure at different negative bias voltages.

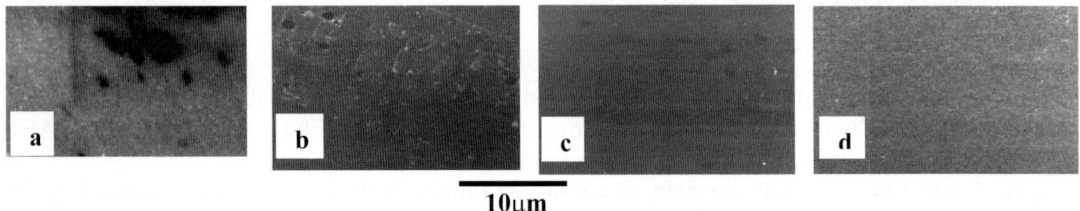

10μm

Fig. 4   SEM surface micrographs of the Co/NiO/Si(100) structure at various bias voltages: (a) 0, (b) –20, (c) –40 and (d) –60 V.

Following the Co deposition under the optimum conditions, Cu layer with a thickness of about 2 nm was grown by employing DC magnetron sputtering in Ar pressure (5 m/Torr). The electrical property of the deposited Cu/Co(V_b)/NiO/Si(100) thin films was measured by using four-point probe sheet resistance method at the room temperature. The average sheet resistance was measured about 131 Ω/□. To study the Cu surface roughness in the present structure, AFM method was used to measure its roughness. Based on the obtained data, the surface roughness of Cu was about 0.56 nm. Figure 5 illustrates the AFM micrograph of the Cu surface deposited on the Co(V_b)/NiO/Si(100) structure. According to our AFM analysis, the Cu was grown on the top of Co(V_b) have a least roughness with smooth topography, and based on our experimental observations, the Cu deposited under the optimum condition is suitable for the growth of next Co layer to form an active GMR region (Co/Cu/Co) for the spin valve systems. Fabrication and characterization of the Ta/Co/Cu/Co/NiO/Si(100) spin valve structure is under investigation.

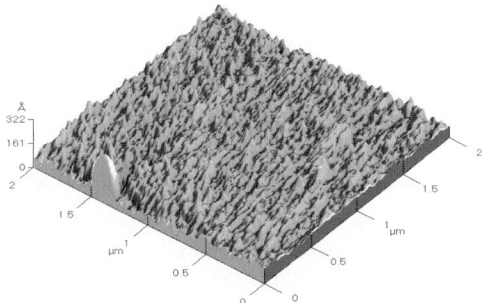

Fig. 5 3D-AFM surface micrograph of the Cu layer in the Cu/Co(V_b)/NiO/Si(100) thin film.

4 Conclusions The growth of both Co(V_b)/NiO/Si(100) and the Cu/Co(V_b)/NiO/Si(100) magnetic structures were studied by using combinative sputtering-evaporation technique. Cobalt thin films were initially deposited on NiO surface at various negative bias voltages ranging from 0 to –80 V. Different analytical methods including SEM, R_s, AFM and surface roughness computations were utilized to investigate properties of the deposited structures. An optimum applied negative bias voltage (–60 V) was determined for the growth of Co layer in the Co(V_b)/NiO/Si(100) structure showing the lowest electrical resistivity with the modified surface topography. Based on our AFM, R_s and surface roughness data, Cu thin film formed under our experimental conditions on the top of the Co(V_b) layer exhibits a good interlayer in the Co/Cu/Co active GMR structure, that can be used to fabricate the Ta/Co/Cu/Co/NiO/Si(100) spin valve system.

Acknowledgements The authors would like to thank the Research Council of Sharif University of Technology for financial support of the project, the useful discussion with Dr. Kavei, Dr. Iraji, Dr. Rahimitabar and Mr. Jafari for AFM analysis as well as the assistances of Mr. Gholami for R_s and Mr. Basirpour for SEM micrographs are greatly acknowledged.

References

[1] B. Dai, J. N. Coi, and W. Lai, J. Magn. Magn. Mater. **27**, 19 (2003).
[2] H. W. Jiang, M. H. Li, and G. H. Yu, J. Magn. Magn. Mater. **242–245**, 341 (2002).
[3] M. Menyhard, G. Zsolt, P. J. Chen. C. J. Powell, R. D. McMichael, and W. F. Egelholff, Jr., Appl. Surf. Sci. **180**, 315 (2001).
[4] J. Pelegri, J. B. Eje, D. Ramirez, and P. P. Freitas, Sens. Actuators A: Phys. **25**, 132 (2003).
[5] D. G. Hwang, C. M. Park, and S. S. Lee, J. Magn. Magn. Mater. **166**, 265 (1998).
[6] H. Chingtong, F. Liu, K. Stoeu, Y. Chen, X. Shi, and C. Qian, J. Magn. Magn. Mater. **239**, 106 (2002).
[7] M. Cartier, S. Auffret, Y. Samson, P. Bayle-Guillemaud, and B. Dieny, J. Magn. Magn. Mater. **223**, 63 (2001).
[8] B. Dieny, J. Magn. Magn. Mat. **136**, 335 (1994).
[9] B. Dieny, V. S. Speriosu, S. Metin, S. S, Parkin, B. A. Gurney, P. Baumgart, and D. R. Wilhit, J. Appl. Phys. **69**, 4774 (1991).
[10] T. Lim, C. Tsang, R. E. Fontana, and J. K. Howard, IEEE Trans. Magn. **31**, 2585 (1995).
[11] H. Yu., C. L. Chai, H. C. Zhao, F. W. Zhu, and J. M. Xiao, J. Magn. Magn. Mater. **224**, 61 (2001).
[12] W. Gouda and K. Shiiki, J. Magn. Magn. Mater. **205**, 136 (1999).
[13] S. S. Parkin, R. Bhadra, and K. P. Roche, Phys. Rev. Lett. **66**, 2152 (1991).
[14] A. Z. Moshfegh and O. Akhavan, J. Phys. D: Appl. Phys. **34**, (2001) 2103.
[15] I. Porqueras and E. Brtran, Thin Solid Films **398–399**, 41 (2001).
[16] A. Z. Moshfegh and O. Akhavan, Thin Solid Films **370**, 10 (2000).
[17] L. I. Maissel and P. M. Shaible, J. Appl. Phys. **35**, 237 (1965).
[18] P. N. Baker, Thin Solid Films **14**, 3 (1972).
[19] P. Catania, J. P. Doyle, and J. J. Cuomo, J. Vac. Sci. Technol. A **10**, 3318 (1992).
[20] W. D. Westwood, and F. C. Livemore, Thin Solid Films **5**, 407 (1970).

Growth process and characterization of magnetic semiconductors based on GeMn alloy films

N. Pinto[*,1], **L. Morresi**[1], **R. Murri**[1], **F. D'Orazio**[2], **F. Lucari**[2], **M. Passacantando**[2], and **P. Picozzi**[2]

[1] INFM – Dipartimento di Fisica, Università di Camerino, 62032 Camerino (MC), Italy
[2] INFM – Dipartimento di Fisica, Università di L'Aquila, 67010 L'Aquila, Italy

Received 31 August 2003, accepted 31 December 2003
Published online 18 March 2004

PACS 61.10.Nz, 61.14.Hg, 75.50.Pp, 75.70.Ak, 81.15.Hi

The growth mechanism of thin $Ge_{1-x}Mn_x$/Ge(100) diluted magnetic semiconductor films have been studied by reflection high energy electron diffraction (RHEED) technique and correlated to the structural and magnetic properties of the films provided by X-ray diffraction (XRD) and magneto-optical Kerr effect (MOKE), respectively. The RHEED analysis evidenced a transition from a bi-dimensional to a three-dimensional growth mechanism at deposition temperature, T_G, lower than 433 K while XRD characterization showed a polycrystalline structure with Ge grain size depending on T_G. At low T_G (343 K) all the $Ge_{1-x}Mn_x$ films behaved superparamagnetically, while at T_G = 433 K hysteresis loops were observed, with a maximum Curie temperature of ≈ 250 K, for $0.027 < x < 0.044$.

© 2004 WILEY-VCH Verlag GmbH & Co. KGaA, Weinheim

1 Introduction Diluted magnetic semiconductors (DMS) are semiconductors containing typically less than 5% of transition metal in their crystal lattice. In the recent past, several DMS systems have been investigated mainly those of the III–V [1–3] and II–VI [4, 5] groups but much less work has been done on IV-group DMSs, where a Curie room temperature has been predicted for Ge [6], C [7]. In particular, Mn_xGe_{1-x} alloy exhibits promising magnetic and electronic transport properties such as semiconductor-like resistivity and a p-type character [6]. As already demonstrated in Mn doped III–V DMS films, the low solid solubility of Mn in Ge requires the use both of non-equilibrium growth techniques such as molecular beam epitaxy (MBE) and very low growth temperatures in order to reduce phase separation and/or formation of Mn rich precipitates having composition and magnetic properties of difficult assessment. In fact, several Mn rich phases, ferromagnetic at room temperature, have been studied and reported in literature generally at T_G > 580 K [8, 9]. The aim of the present work is the investigation of the growth mechanism of the $Ge_{1-x}Mn_x$/Ge(100) films and the exploration of their structural and magnetic properties.

2 Experimental details Thin $Ge_{1-x}Mn_x$ films were grown by MBE on semi-insulating Ge(100) wafers, using Ge and Mn Knudsen cells. The deposition process was monitored in-situ by residual gas analyser (RGA) and X-ray photoelectron spectroscopy (XPS) techniques. The Ge wafers were cleaned by a wet etching [10], while the surface oxide was thermally desorbed at about 673 K. Before $Ge_{1-x}Mn_x$ evaporation at a fixed rate of 0.019 nm/s, a 10 nm thick Ge buffer layer was grown, at 673 K. The list of the grown samples is shown in table 1, while details of the growth procedure are reported elsewhere [11, 12]. A 10 keV RHEED technique was used to investigate the growth mechanism of the GeMn alloys, during film evaporation. For XRD and MOKE analysis, the films were capped by a 4 nm thick Ge layer,

---

[*] Corresponding author: e-mail: nicola.pinto@unicam.it, Phone: +39 0737 402528, Fax: +39 0737 402853

to prevent air oxidation. The crystal structure of the $Ge_{1-x}Mn_x$ films was analysed by XRD θ–2θ scans, using a Cu K_α source (λ =1.5406 Å), acquiring the spectra in grazing incidence configuration, at a fixed angle of 2°.

	T_G (K)	x	h (nm)	T_C (K)	G. size (nm)
Mn 8	343	0.042	68	-	12
Mn 23	343	0.026	68	-	-
Mn 24	433	0.027	40	250	27
Mn 25	433	0.053	40	250	-
Mn 26	433	0.044	40	50	68
Mn 28	433	0.044	20	-	-
Mn 37	433	0.072	20	-	-

The magnetic properties were investigated by MOKE measurements using the polar configuration, in the temperature range from 12 to 300 K. s-polarized radiation (λ = 2.0 µm) was used, incident at an angle of 45° on the sample surface. Details have been published elsewhere [11, 12].

3 Results
The growth mechanism of $Ge_{1-x}Mn_x$ films, deposited at different T_G, was studied by RHEED. Each experiment was carried out without epitaxy interruption. Figure 1 shows a sequence of RHEED patterns picked up at different stages of the growth process for the Mn23 film, evaporated at T_G = 343 K. Figure 1a shows a progressive conversion from a streaky pattern (Fig. 1a) to a spotty one, after about 12÷28 MLs (1 ML ≅ 0.142 nm) of thickness. These elongated spots marks the nucleation of large bi-dimensional (2D) islands on an underlying flat $Ge_{1-x}Mn_x$ layer as in Fig. 1b. The coexistence of the continuous layer with the 2D islands was observed up to about 160 MLs. Above this thickness, the GeMn islands progressively transforms from 2D to three-dimensional (3D) as in Fig. 1c. Around 320 ÷ 413 MLs, the 3D Bragg peaks split in four spots as in Fig. 1d. Similar results were observed both along the [100] direction and for the Mn8 alloy (RHEED patterns not reported in the present work).

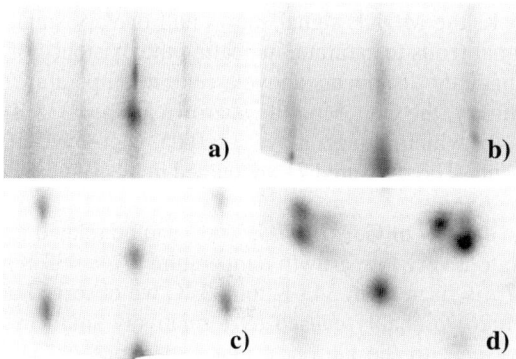

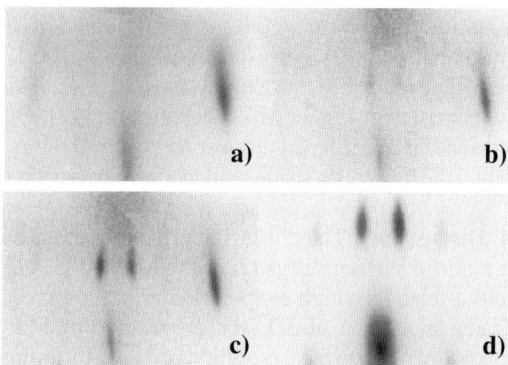

Fig. 1 RHEED patterns along the [110] direction for the Mn23 film at several stages of the growth process: a) 2×1 Ge(100) surface reconstruction. GeMn surface: b) 12 MLs; c) 320 MLs; d) 413 MLs.

Fig. 2 RHEED patterns along the [110] direction for the Mn28 film: a) 12 MLs; b) 67 MLs; c) 143 MLs. d) RHEED pattern of the film Mn37 after the growth of 143 MLs.

$Ge_{1-x}Mn_x$ alloys grown at T_G = 433 K and with a Mn concentration of x = 0.044 and x = 0.072, present different features in the RHEED patterns. Figure 2 shows the diffraction images for the Mn28 film. At the early stage of the evaporation process the 2×1 Ge(100) reconstruction changed into 1×1 elongated spots, continuously rotating in two opposite directions up to about 12 MLs as in Fig. 2a. These spots mark the nucleation of large 2D islands and the development of a very rough surface. At about 70 MLs, additional spots appeared, while the previous ones showed an evolution towards a 3D morphology as in

Fig. 2b. Continuing the epitaxy, the RHEED pattern did not change significantly indicating a stabilization of the surface morphology as in Fig. 2c. Similar results were observed for the sample Mn37, where at ≈ 15 MLs a complete transition from a 2D to a 3D growth mode with the final morphology as in Fig. 2d. XRD measurements were carried out on several $Ge_{1-x}Mn_x$ samples of both series. Figure 3, shows the XRD spectra for two $Ge_{1-x}Mn_x$ alloys. The diffraction peaks are relative to the (111), (220) and (311) planes of polycrystalline Ge in the cubic phase (JCPDS card no. 4-545). Other phases, if present, were not evidenced. In general, the distribution of peak heights remain almost the same, for the whole range of the x values, while peak half-width lowers with increasing T_G. According to the Scherrer equation, with a constant of 0.9, the crystallite size can be estimated from the half width of the (111) peak. The grain size values are quoted in the Table 1.

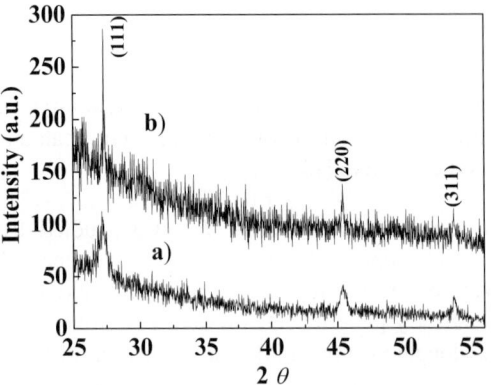

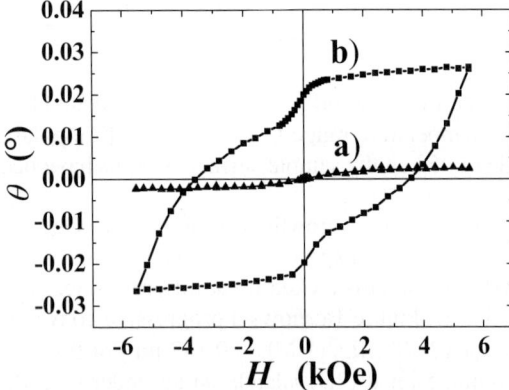

Fig. 3 X-ray diffraction of the samples Mn8 (a) and Mn26 (b).

Fig. 4 MOKE signal at 13 K, for two $Ge_{1-x}Mn_x$ films: a) Mn8; b) Mn28.

Depending on T_G all the $Ge_{1-x}Mn_x$ films present two distinct magnetic behaviors [11–13]. Figure 4 shows two typical MOKE curves for $Ge_{1-x}Mn_x$ DMS films belonging to the two series. For any Mn concentration in the range $0.02 < x < 0.08$ and at $T_G = 343$ K, the MOKE signal shows null or very small hysteresis [11, 12]. Nevertheless, below ≈ 100 K, the signal tends to saturate, indicating the existence of ferromagnetic (FM) properties as in Fig. 4a. This FM character disappears above a certain temperature, which rises for higher Mn concentration [12]. On the contrary, the $Ge_{1-x}Mn_x$ alloys grown at $T_G = 433$ K present large hysteresis loops, for $x < ≈ 0.04$ as in Fig. 4b, which reduce above $x ≅ 0.05$. Moreover, at any x the hysteresis shape evidence two magnetic components. The measured T_C are quoted in the Table 1.

4 Discussion The clustering effect observed during the epitaxy of $Ge_{1-x}Mn_x/Ge(100)$, can be related to a reduced surface diffusion of the Mn and Ge atoms, due to a very low growth temperature ($T_G < 433$ K) and a relatively high evaporation rate. However, when T_G is risen from 343 K to 433 K, we observed a faster nucleation of 3D islands (results for Mn8 and Mn26) probably caused by a relatively small increase of the surface mobility of the Mn atoms with respect to the Ge ones. Our experience on Mn evaporation onto Ge(100) suggests that, in the case of the $Ge_{1-x}Mn_x$ films, the morphology is controlled by the Mn atoms and the T_G value. Another feature in the RHEED pattern was the appearance of additional spots as in the Fig. 1d, suggesting the formation of GeMn 3D islands with a larger lattice parameter. Either a local increase in the Mn concentration of the alloy with respect to the surrounding GeMn matrix [6] or the nucleation of crystallographic defects could explain the observed feature. However, XRD measurements did not allow to evidence any phase different from that of the Ge. Previous works carried out by scanning tunneling microscopy (STM) and atomic force microscopy (AFM) showed a 3D island morphology, in agreement with the RHEED study [11, 14]. Moreover, the size of these islands measured by STM and AFM [12] is similar to that of the Ge grains calculated from the XRD spectra. Finally, the larger grain size of the film Mn26, with respect to that of the sample Mn8, suggests a higher long range order of the grains and then a better crystalline quality.

The huge difference in the magnetic behaviours of GeMn films of the two series could be related to their morphological and structural features. In fact, the absence of hysteresis at $T_G = 343$ K as in Fig. 4a, is probably due to the low magnetic anisotropy constant and to a reduced extension of the magnetic grains [12] as confirmed by XRD analysis. On the contrary, rising T_G (433 K) the improvement of the crystalline quality of the films may cause the appearance of the FM order with a relatively high T_C value. However, we have to consider the possibility that nano-sized FM precipitates made of an unknown Mn-rich GeMn phase could have nucleated in the $Ge_{1-x}Mn_x$ matrix. These have been found by Park et al. at $T_G \geq 343$ K [6]. Results of XRD analysis seem to exclude the presence of these Mn rich nano-clusters. Finally, the particular shape of the hysteresis curve is probably due to the 3D islands causing a difference in the coercivity from the film and its surface [15]. However, we cannot exclude another explanation about the origin of the magnetic behaviours exhibited by our samples. The use of techniques such as transmission electron microscopy (TEM), microprobe analysis of energy dispersive spectroscopy (EDS), etc. carried out on films cross-sections will give deeper insight of the GeMn system.

5 Conclusion In this work we reported on the growth mechanism, structural and magnetic properties of thin $Ge_{1-x}Mn_x$ DMS films. RHEED experiments revealed a transition from a 2D to a 3D growth mechanism with features depending on x and T_G. The XRD analysis evidenced a polycrystalline structure for all the GeMn films, with grains having a larger size for the sample grown at $T_G = 433$ K. From the MOKE characterization we detected two distinct magnetic behaviours, depending on T_G. At $T_G = 343$ K all the films were superparamagnetic while at $T_G = 433$ K magnetic hysteresis was observed. The highest measured T_C value was ≈ 250 K for an Mn concentration ranging from $x = 0.027$ to $x = 0.044$.

Acknowledgements This work has been supported by a grant from the Italian Institute for the Physics of Matter (INFM), project PAIS "GEMASE".

References

[1] G.A. Prinz, Science **282**, 1660 (1998).
[2] J.M. Kikkawa and D.D. Awschalom, Nature **397**, 139 (1999).
[3] H. Ohno, Science **281**, 951 (1998).
[4] G. Schmidt and L. W. Molenkamp, Physica E **9**, 202 (2001).
[5] J.K. Furdyna, J. Appl. Phys. **64**, R29 (1988).
[6] Y.D. Park, A.T. Hambicki, S.C. Erwin, C.S. Hellberg, J.M. Sullivan, J.E. Mattson, T.F. Ambrose, A. Wilson, G. Spanos, and B.T. Jonker, Science **295**, 651 (2002).
[7] T. Dietl, H. Hono, and F. Matsukura, Phys. Rev. B **63**, 195205 (2001).
[8] E. Abe, F. Matsukura, H. Yasuda, Y. Ohno, and H. Ohno, Physica E **7**, 981 (2000).
[9] Y.D. Park, A. Wilson, A.T. Hambicki, J.E. Mattson, T. Ambrose, G. Spanos, and B.T. Jonker, Appl. Phys. Lett. **78**, 2739 (2001).
[10] T. Akane, H. Okumura, J. Tanaka, and S. Matsumoto, Thin Solid Films **294**, 153 (1997).
[11] N. Pinto, L. Morresi, R. Gunnella, R. Murri, F. D'Orazio, F. Lucari, S. Santucci, P. Picozzi, M. Passacantando, and A. Verna, J. Mater. Sci. – Mater. Electron. **14**, 337 (2003).
[12] F. D'Orazio, F. Lucari, S. Santucci, P. Picozzi, A. Verna, M. Passacantando, N. Pinto, L. Morresi, R. Gunnella, and R. Murri, J. Magn. Magn. Mater. **262**, 158 (2003).
[13] F. D'Orazio, F. Lucari, N. Pinto, L. Morresi, and R. Murri, J. Magn. Magn. Mater. (in press).
[14] P. Castrucci, N. Pinto, L. Morresi, R. Gunnella, R. Murri, M. Scarselli, and M. De Crescenzi, J. Magn. Magn. Mater. (in press).
[15] M. Tanaka, K. Saito, and T. Nishinaga, Appl. Phys. Lett. **74**, 64 (1999).

Magnetoresistance effect and magnetoanisotropy of Co/Cu multilayered films prepared by electron beam evaporation

Y. Ueda[*], H. Adachi, W. Takakura, C. L. S. Rizal, and S. Chikazawa

Muroran Institute of Technology, 27-1 Mizumoto-cho, Muroran 050-8585, Japan

Received 31 August 2003, accepted 31 December 2003
Published online 1 April 2004

PACS 75.47.De, 75.50.Cc, 75.70.–i, 81.15.Ef

We attempted to prepare Co/Cu ferromagnetic layer films with uniaxial magnetic anisotropy by oblique incidence angle electron beam method. The purpose of the present note is to show the effect of the magnetic orientation in the ferromagnetic layer on the magnetoresistance of both GMR and AMR. The induced uniaxial magnetic anisotropy was observed in the all multilayer films formed by varying the oblique incidence angle of evaporation direction and the easy axis of the anisotropy is along the perpendicular direction (x-direction) of the incidence of evaporation. The sample produced near the oblique incidence angle of 45° shows the remarkable uniaxial magnetic anisotropy. The MR ratio of anisotropic sample is less than that of isotropic sample. In the weak magnetic field, the difference for the magnetic field dependence of MR is clearly observed with depending on the orientation of magnetization, that is, it is corresponding to the shape of the magnetization curves.

© 2004 WILEY-VCH Verlag GmbH & Co. KGaA, Weinheim

1 Introduction Research on the magnetoresistance effect has attracted attention in terms of physics and the application. The mechanism of giant magnetoresistance effect in multilayers [1, 2] composed of ferromagnetic and nonmagnetic layer was interpreted by considering the spin-dependent scattering of conduction electrons. It seemed that the scattering probability depends on the relationship between the spin direction of the conduction electron in nonmagnetic layer and the direction of magnetization in magnetic layers [3]. On the other hand the magnetic resistance effect with respected to the direction of magnetization observed depending on external magnetic field in the ferromagnetic materials has already known as the anisotropic magnetoresistance (AMR) for a long time. The origin of AMR is considered to be the orbital angular momentum of magnetic ions and the Lorentz force acting on conduction electrons [4]. This mechanism is essentially different from that of GMR. The state of orientation of magnetization in the ferromagnetic layer is considered to be the essential factor for the origin of the both GMR and AMR effects. Over the past few years, a considerable number of studies have been reported on the effect of magnetic field on MR [5, 6], however, there are very few reports how the state of the magnetic orientation in the film plane effects on the GMR and AMR. The magnetic orientation in the ferromagnetic layers of the multilayer films showing the GMR is random i.e. isotropic magnetic arrangement.

In this paper, we attempted to prepare Co/Cu ferromagnetic layer films with uniaxial magnetic anisotropy by oblique incidence angle electron beam method [7]. The purpose of the present note is to show the effect of the magnetic orientation in the ferromagnetic layer on the magnetoresistance of both GMR and AMR.

2 Sample preparation For the films deposited by the perpendicular incidence angle of 0°, the measured results of magnetization curves shows that the magnetization is randomly oriented, that is, the magnetism of the film is magnetically isotropic, as shown in Fig. 1(a) [7]. The induced uniaxial magnetic

---

[*] Corresponding author: e-mail: ueda@mmm.muroran-it.ac.jp, Phone: +81 143 46 5524, Fax: +81 143 46 5540

anisotropy was observed in the all multilayer films formed by varying evaporation direction of the oblique incidence angle, and the easy axis of the anisotropy is along the perpendicular direction (x-direction) of the incidence of evaporation. The sample produced near the oblique incidence angle of 45° shows minimum hysteresis loss for parallel direction (y-direction) corresponding to the remarkable uniaxial magnetic anisotropy.

Figure 1(b) shows the magnetization curves measured along to the parallel and perpendicular directions of the incidence of evaporation for the Co/Cu multilayer films prepared by evaporation method of oblique incidence angle at 45°. The remarkable uniaxial anisotropy was observed in these multilayer films deposited by the above method.

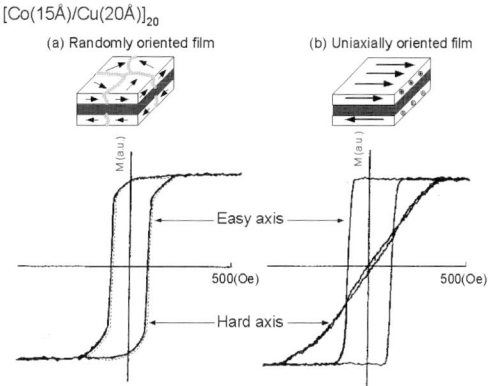

Fig. 1 Magnetism of a randomly oriented film and a uniaxially oriented film; (a) randomly oriented film; (b) uniaxially oriented film produced by the oblique incidence evaporation method.

3 Experiments Co/Cu multilayers with various layer thickness ratios of Cu to Co were prepared by electron beam evaporation on the glass substrate of 13 mm × 13 mm dimension. The pressure during deposition was 10^{-4} Pa, and the deposition rate was 1 Å/s. The average layer thickness of Co was within the range of 10–20 Å, and that of Cu was 15–30 Å. The stacking number of the set of Co and Cu layers in the multilayer film was 20. The incidence angle θ of the evaporation direction on the plane of the substrate was maintained in the range of θ = 0°–75°. The electrical resistivity and magnetization curves were measured by four probe method and vibrating sample magnetometer (VSM), respectively. The magnetic field dependence of MR was examined by varying the relative direction between the field *H* and current *I*. The MR ratio along to the transverse and longitude direction of measuring current was observed by measuring in the two cases for the magnetoisotropic film. For the films with uniaxial anisotropy, the MR ratio was measured in the four cases by varying the relation the relative direction between the field *H* and current *I* as shown in Fig. 2.

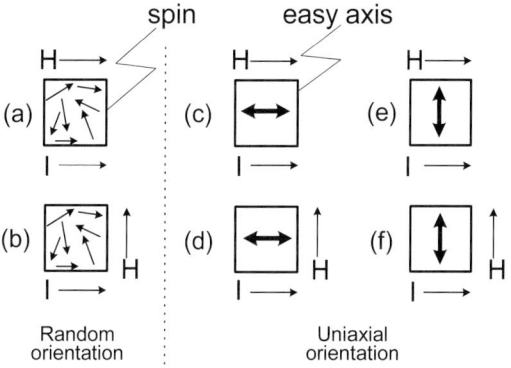

Fig. 2 Measurement method of magnetoresistance.

4 Results and discussion Figure 3(a) shows the magnetic field dependence of the MR for the magnetoisotropic film. Figures 3(b) and (c) shows those of the oriented films with uniaxially magnetic anisotropy [5]. The difference of field dependence was slightly observed by changing the direction of applied field against the measuring current, which difference is shown in the upper side of Fig. 3. As the tendency of the field dependence in the lower side of Figs. 3(a)–(c) shows the negative field dependence, and it seems to be GMR effect. The AMR (anisotropic MR) shown in the upper side of the Figs. 3(a)–(c) has a similar tendency each other. That is, it seems to be that the orientation of magnetization in the film has an effect on the GMR. The MR ratio observed by measuring current perpendicular to magnetic field is larger than that of measuring current parallel to field for Fig. 3(a) of isotropic sample. The MR ratio of magnetic field parallel to magnetic easy direction is larger than that of magnetic field parallel to magnetic hard direction independing on the direction of measuring current for the oriented sample.

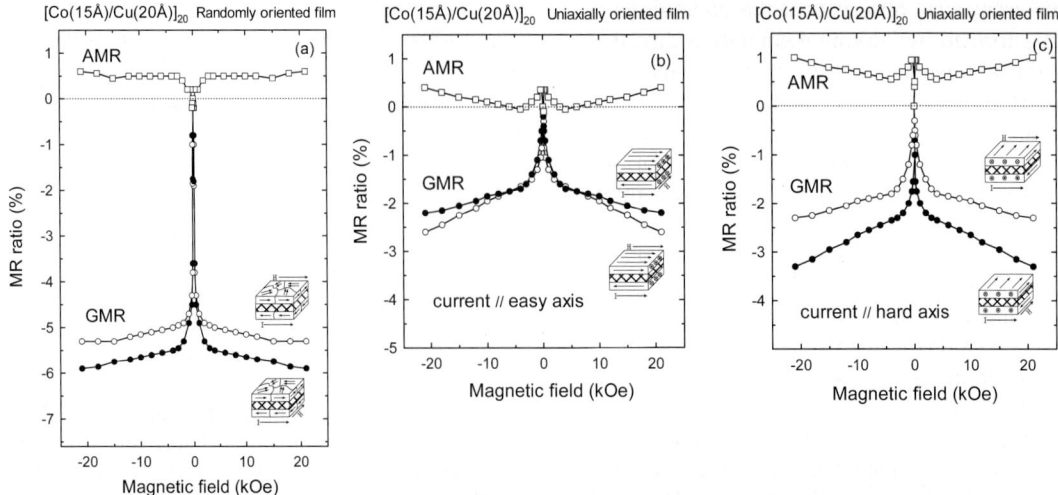

Fig. 3 Magnetic field dependence of the MR ratio for the randomly oriented and uniaxially oriented film of [Co (15 Å) /Cu (20 Å)]$_{20}$ multilayers; (a) randomly oriented film; (b) and (c) uniaxially oriented film produced by the oblique incidence evaporation; (b) measuring current // magnetic easy direction; (c) measuring current // magnetic hard direction.

Figure 4 shows Cu thickness dependence of the MR ratio at the magnetic field of 1 kOe and 21 kOe, for the films with the Co layer thickness of 15 Å observed in the six measuring cases, as shown in Fig. 2, where the MR ratio of sample with the magnetic isotropy shows the average of two values measured by

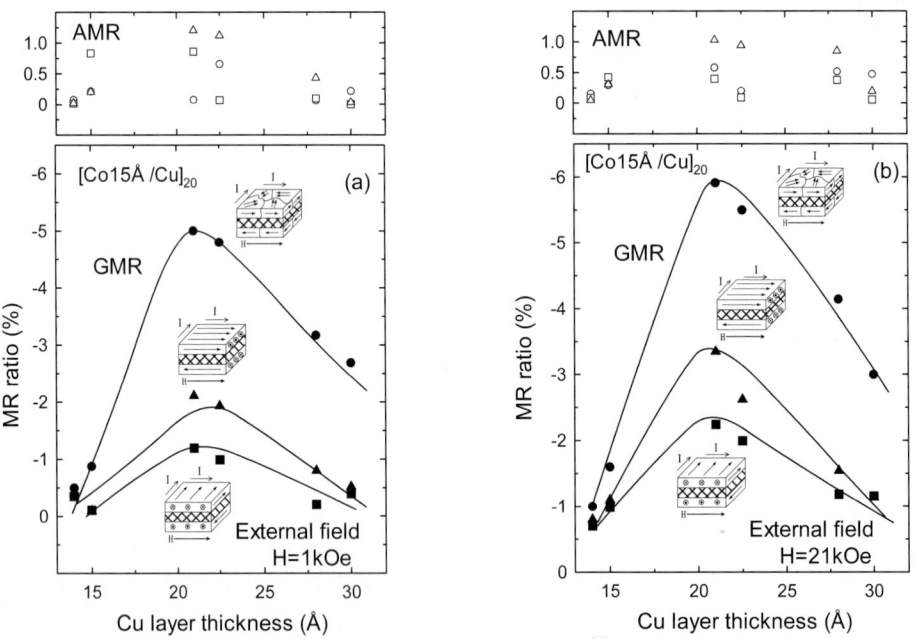

Fig. 4 Cu layer thickness dependence of the MR ratio for the Co thickness of 15 Å:
● ○ randomly oriented film; ▲ △ field // easy axis; ■ □ field // hard axis (a) at external field of 1 kOe, (b) at external field of 21 kOe

changing the orientation of measuring current at the field direction of the applied field. That of uniaxial anisotropy samples shows the averages of the two values measured by changing the direction of measuring current at the fixed applied field along to the magnetic easy and hard axis. The MR ratio of both the magnetoisotropic and anisotropic sample shows the maximum value at the Cu layer thickness of about 20 Å, and the value anisotropic (uniaxially oriented) sample is less than that of isotropic (randomly oriented) sample.

The MR ratio observed at the applied magnetic field along the hard axis shows the value less than that of easy axis. The reason for the showing the smaller MR ratio in the anisotropic sample is seemed to be as follows. The number of antiparallel alignment of the magnetic spin between the ferromagnetic layers adjacent to non-magnetic layers for the anisotropic (i.e., oriented) sample is smaller than that of isotropic sample.

In the weak magnetic field, the difference for the magnetic field dependence of MR is clearly observed with depending on the orientation of magnetization, that is, it is corresponding to the shape of the magnetization curves, as shown in Fig. 5. The field dependence of MR is not depending on the direction of measuring current, but, it only has a tendency to depend on the direction of the applied field. The magnetic spin state near zero magnetization (the step part) for the uniaxially oriented sample shown in Fig. 5(b) is antiparallel, where electric resistance shows high value. Therefore, the orientation of the magnetic spin remarkable corresponds to the value of the MR ratio. However, the oriented magnetization with uniaxial magnetic anisotropy is not of necessary in order to produce the large magnetoresistance effect.

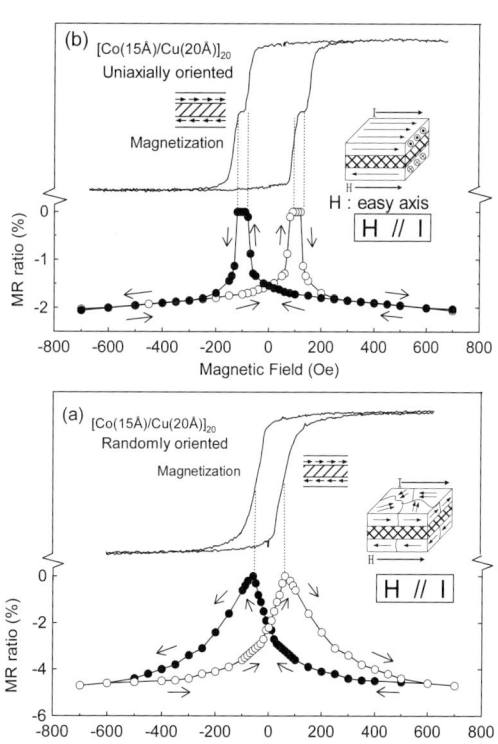

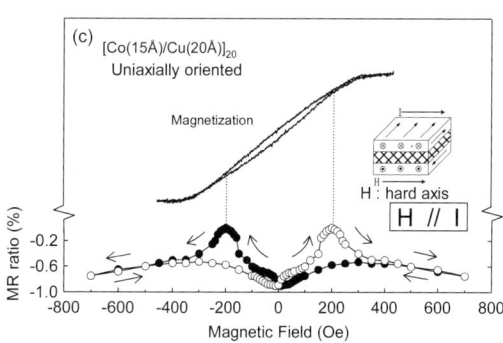

Fig. 5 Correspondence between magnetic field dependence of MR ratio and magnetization curves at weak field: (a) randomly oriented film, (b), (c) uniaxially oriented film.

References

[1] M. N. Baibich, J. M. Broto, A. Fert, Nguyen Van Dau, F. Petroff, P. Eitenne, G. Creuzet, A. Friederich, and J. Chazelas, Phys. Rev. Lett. **61**, 2472 (1988).
[2] S. S. Parkin, N. More, and P. Roche, Phys. Rev. Lett. **64**, 2304 (1990).
[3] J. Inoue, A. Oguri, and S. Maekawa, J. Phys. Soc. Jpn. **60**, 376 (1991).
[4] J. M. Ziman, "Principles of the Theory of Solids" 2nd ed. (University Press, Cambridge, 1972).
[5] G. Wen, H. Zhao, J. Zhao, and X. X. Zhang, Mater. Sci. Eng. C **16**, 81 (2001).
[6] Y. Ueda, T. Houga, H. Zaman, and A. Yamada, J. Solid State Chem. **147**, 274 (1999).
[7] W. Takakura, S. Ikeda, and Y. Ueda, Mater. Trans. **42**, 881 (2001).

Magnetic properties and magnetoresistance effect in Co/Au, Ag nano-structure films produced by pulse electrodeposition

C. L. S. Rizal[1], A. Yamada[2], Y. Hori[1], S. Ishida[1], M. Matsuda[1], and Y. Ueda*,[1]

[1] Muroran Institute of Technology, 27-1 Mizumoto-cho, Muroran 050-8585, Japan
[2] Tomakomai National College of Technology, 443 Nishikioka, Tomakomai 059-1275, Japan

Received 31 August 2003, accepted 31 December 2003
Published online 1 April 2004

PACS 75.47.De, 75.60.Ej, 75.75.+a

We have investigated the relationship between the magnetism and the magnetoresistance effect in the Co/Au, Ag multilayer films with layers produced in the atomic level by pulse electrodeposition method. The magnetoresistance effect is dependent on both the thickness of Co ferromagnetic layer and Ag, Au non-magnetic layers. The magnetization of these films shows the minimum value against the Ag and Au layer thickness. The Ag and Au layer thickness showing the maximum of MR ratio is not of necessary in agreement with the Ag and Au layer thickness showing the minimum of magnetization. Antiparallel alignment of magnetic spin is a necessary but not sufficient condition in order to generate the GMR of multilayer films. For the Co/Au multilayer films, the Au layer thickness showing the minimum of the magnetization shifts to higher side of the Au layer thickness.

© 2004 WILEY-VCH Verlag GmbH & Co. KGaA, Weinheim

1 Introduction The studies on the physical properties of the nano-ordered multilayer films have attracted much attention not only in the fundamental physics, but also in electronic engineering. The productions of multilayered films have been attempted actively by the several methods. Though the electrodeposition is not by no means easy to find a suitable condition for deposition, it has a significance as a useful means to prepare the nano-ordered thin films, even if the suitable deposition condition could be established successfully. Furthermore, it becomes a very useful method for the development of the research in nano-ordered multilayered films with the combination of ferromagnetic and non-magnetic layers, which shows the giant magnetoresistance effect [1–3]. It is possible to control the film composition, thickness of multilayer and grain size in an atomic order by regulating the pulse amplitude and width [4–7]. If the pulse width of deposition time to produce multilayers is reduced to very small values, the multilayer film with layers of two different atomic compositions eventually results in a binary alloy of solid solution type mixed in the atomic level.

In this present work, we have investigated the relationship between magnetism and magnetoresistance effect in the Co/Au, Ag films with layers produced in the atomic level by pulse electrodeposition method.

2 Experiments The electrolytic bath for Co/Ag multilayer film deposition was composed of $CoSO_4 \cdot 7H_2O$, AgI, KI and that for Co/Au multilayer film deposition was composed of $CoSO_4 \cdot 7H_2O$, $KAu(CN)_2$, $Na_3C_6H_5O_7 \cdot 2H_2O$, NaCl. The substrates were 150 Å copper thin films vapor deposited on glass plates. The multilayer films were deposited using a square pulse wave of current density range 0.1–25 mA/cm^2. The composition of the deposited films was determined by atomic absorption spectroscopy.

The ferromagnetic layer was Co-rich, which composition in Co/Ag and Co/Au multilayer films was 92 at%Co–8 at%Ag and 95 at%Co–5 at%Au, respectively. The MR ratio using a dc four probe method

---

* Corresponding author: e-mail: ueda@mmm.muroran-it.ac.jp, Phone: +81 143 46 5524, Fax: +81 143 46 5540

was measured under magnetic field up to 21 kOe. The magnetic properties were investigated using a VSM (Vibrating Sample Magnetometer).

3 Results and discussion Figure1(a) shows the Ag layer thickness dependence of MR ratio at the magnetic field of 21 kOe for the Co/Ag multilayer films. The MR ratio shows the maximum value against the Ag layer thickness, and the layer thickness of Ag shifts slightly to higher thickness of Ag layer with increasing Co layer thickness. The experimental results, shown, are the average of more than two measurements. The similar tendency is observed in the Co/Au multilayer films as shown in Fig. 2(a).

Figure1(b) shows the Co layer thickness dependence of MR ratio rearranged from these results of Ag dependence of Fig. 1(a). The additional experimental data have been included in the figure. The MR ratio shows the maximum value against the Co layer thickness, and the maximum MR ratio and the Co layer thickness showing the maximum MR ratio increases with increasing the Co and Ag layer thickness up to Ag layer thickness of 15 Å. The experimental results observed in Co/Au multilayer films have similar tendencies as shown in Fig. 2(b).

For the Ag layer thickness of 20 Å or higher than 15 Å, the value of the maximum MR ratio decreases and the Co layer thickness showing the maximum shifts very slightly toward the lower side of Co thickness. The reason for shifting toward lower side of Co thickness showing the maximum of MR for the

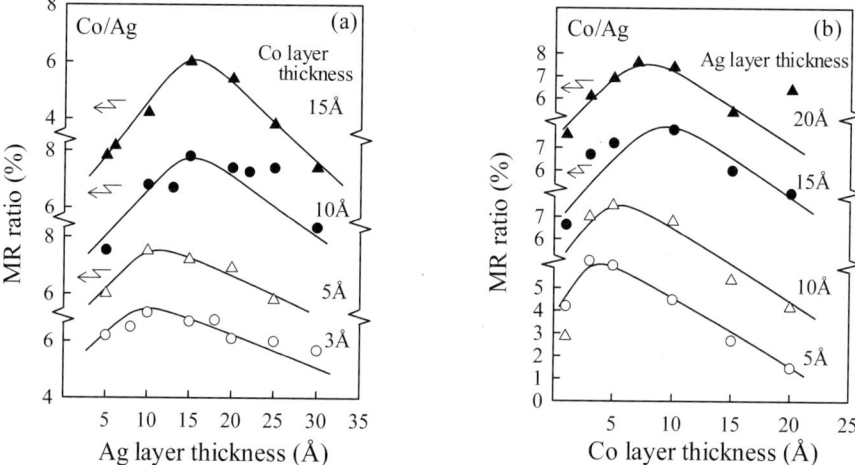

Fig. 1 MR ratio of the Co/Ag multilayer films as a function of (a) Ag layer thickness and (b) Co layer thickness.

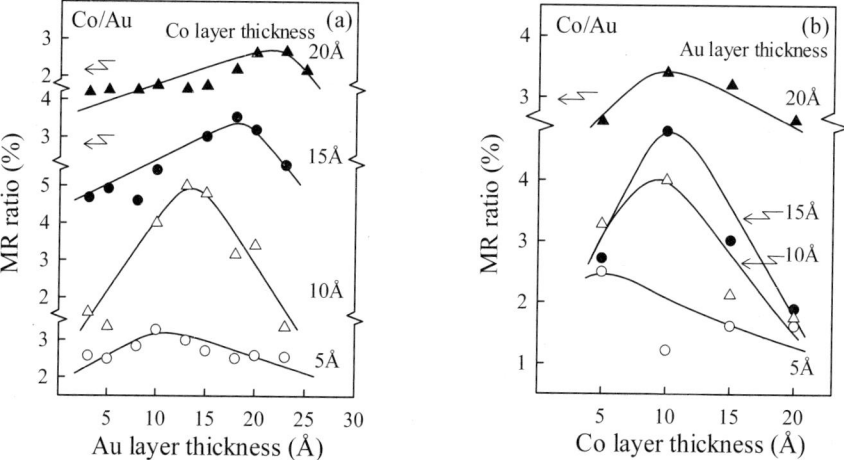

Fig. 2 MR ratio of the Co/Au multilayer films as a function of (a) Au layer thickness and (b) Co layer thickness.

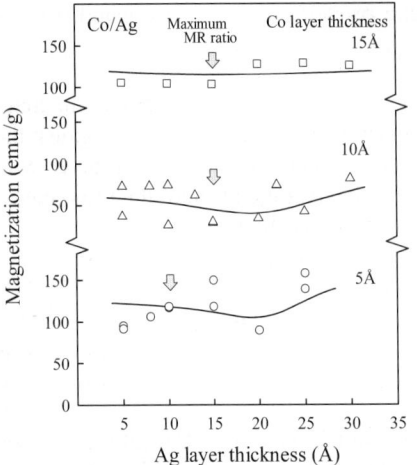

Fig. 3 The Ag layer thickness dependence of the magnetization per Co weight for the Co/Ag multilayer films at an applied magnetic field of 1 kOe.

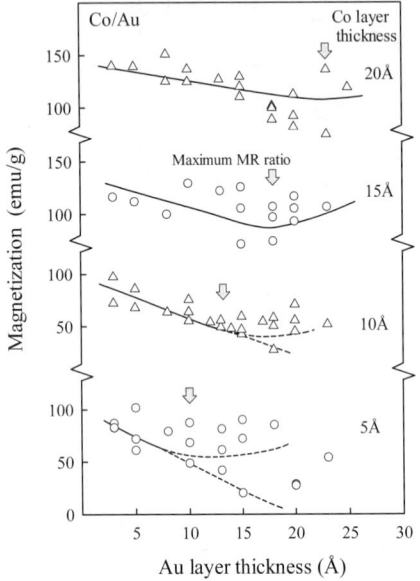

Fig. 4 The Au layer thickness dependence of the magnetization per Co weight for the Co/Au multilayer films at an applied magnetic

higher thickness of non-magnetic layer (Ag, Au: 20 Å) is seemed to be as follows. Since the antiferromagnetic coupling interaction, i.e., the interaction of antiparallel alignment of the spin between the ferromagnetic layers adjacent to non-magnetic layer reduces for the higher thickness of non-magnetic layer, the weak ferromagnetic coupling interaction corresponding with the lower thickness of Co layer in the ferromagnetic layer satisfies the condition in order to produce the antiparallel alignment of the spin.

The reason for the shifting of lower side in Co layer thickness showing the maximum of MR ratio for the lower thickness of the non-magnetic layer (Ag, Au: 5 Å) is seemed to be the following.

a) Since the ferromagnetic layer adjacent to non-magnetic layer becomes to be continuous for the lower thickness in an atomic scale of non-magnetic layer and higher thickness of ferromagnetic Co layer, the MR ratio decreases due to increase in the ferromagnetic region not showing antiparallel alignment of the spin.

b) On the other hand, as the ferromagnetic layer adjacent to non-magnetic layer becomes to be discontinuous for the lower Co and non-magnetic layer thickness in an atomic scale, the MR ratio increases due to the increase in the region of antiparallel alignment. Therefore the layer thickness shifts toward lower side of Co layer thickness.

Figures 3 and 4 show the magnetization of Co per weight of Co contained in the Co-rich ferromagnetic layers plotted as a function of Ag and Au layer thickness for the constant layer thickness of Co at the magnetic field of 1 kOe. If the magnetization of Co ferromagnetic layer is independent on the non-magnetic layer, the magnetization of Co per weight of Co should be same value even the variation of the Ag or Au thickness. The width of decrease in the magnetization against the layer thickness of Au for the higher layer thickness of Co 20 Å shows small value as expected, since there is weak interaction between the ferromagnetic layers as shown in Fig. 4. The magnetization has a tendency to show the minimum with decreasing the Co layer thickness. The minimum of magnetization is near the non-magnetic layer thickness of 20 Å for the Co thickness of 10 Å in Co/Ag films and Co thickness of 15 Å in Co/Au films. The magnetization becomes to decrease remarkably with decrease in Co layer thickness, and the layer thickness showing the minimum shifts toward the lower side of Au layer thickness for the Co/Au films. The fluctuation of magnetization in the region of larger Au layer thickness for the thinner Co films is larger than those of higher thickness of Co. The reason of the remarkable decrease in the magnetization with decrease in the Co layer thickness is seemed to be that the magnetism of the Co layer liable to be influenced greatly by the surrounding Au atom, as Co layer thickness decrease further. The layer thickness showing the decrease of the magnetization is in good agreement with the non-magnetic layer thickness showing the maximum of MR ratio, for the [Co10Å/Ag], and [Co15Å/Au] films. (This tendency is further described later in Fig. 5).

The non-magnetic layer thickness showing the minimum of the magnetization has a tendency to be not of necessary in agreement with the larger thickness showing the maximum of the MR ratio with decrease in Co layer thickness. Where, the arrow mark shown in Fig. 3 and Fig. 4 corresponds with the thickness showing the maximum of MR ratio in Fig. 1(a) and Fig. 2(a). Such a tendency is almost in agreement with our previous result measured for the Co/Cu multilayer films [7]. The reason seems to be that the origin to give the large MR ratio is different with the origin for the decrease in the magnetization. The MR ratio strongly depends on the antiparallel alignment of the magnetic spin between the Co ferromagnetic layers, however, the magnitude of the magnetization is apt to be influenced greatly by the crystal structure of Co, particle size and the surrounding effect due to the presence of Au atoms in the neighbourhood.

Figures 5(a) and (b) shows the correlation of the magnetization and the MR ratio for the [Co10Å/Ag]$_{50}$ and [Co15Å/Au]$_{50}$ multilayer films, that is, it seems to be that the magnetization is closely related to the MR ratio for these Co/Au and Co/Ag multilayer films. The results shown in this figure suggests that the antiparallel alignment of the magnetic spin between the adjacent ferromagnetic layers is very important in order to produce the large MR ratio.

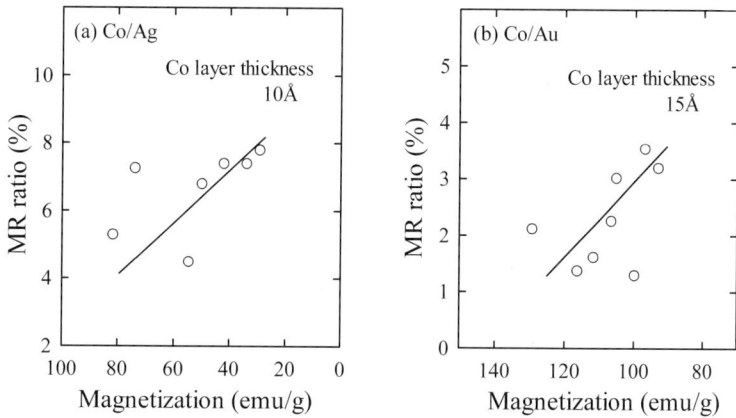

Fig. 5 The relationship between the magnetization (in Fig. 3 and Fig. 4) and the MR ratio (in Fig. 1 and Fig. 2) for (a) the [Co10Å/Ag] and (b) the [Co15Å/Au] films.

References

[1] M. Alper, K. Attenborough, R. Hart, S. J. Lance, D. S. Lashmore, C. Younes, and W. Schwarzacher, Appl. Phys. Lett. **63**, 2144 (1993).
[2] L. Piraux, J. M. George, J. F. Despres, C. Leroy, E. Ferain, R. Legras, K. Ounadjela, and A. Fert, Appl. Phys. Lett. **65**, 2484 (1994).
[3] Y. Ueda, N. Hataya, and H. Zaman, J. Magn. Magn. Mater. **156**, 350 (1996).
[4] W. Schwarzacher and D. S. Lashmore, IEEE Trans. Magn. **32**, 3133 (1996).
[5] Y. Ueda, T. Houga, H. Zaman, and A. Yamada, J. Solid State Chem. **147**, 274 (1999).
[6] L. Péter, A. Cziráki, L. Pogány, Z. Kupay, I. Bakonyi, M. Uhlemann, M. Herrich, B. Arnold, T. Bauer, and K. Wetzig, J. Electrochem. Soc. **148**, C168 (2001).
[7] A. Yamada, T. Houga, and Y. Ueda, J. Magn. Magn. Mater. **239**, 272 (2002).

Magnetism of Ni overlayers on Fe(111)

Naseem T. Shawagfeh[1] and **Jamil M. Khalifeh**[*,2]

[1] Department of Basic Sciences, Faculty of Engineering Technology, Al-Balqa' Applied University, Amman-11134, Jordan
[2] Department of Physics, University of Jordan, Amman-11942, Jordan

Received 31 August 2003, accepted 31 December 2003
Published online 25 March 2004

PACS 71.15.Ap, 73.20.At, 75.70.Cn

The magnetism and electronic structure of Ni/Fe(111) overlayer systems are explored within the *ab-initio* tight-binding linear muffin-tin orbital method in the atomic sphere approximation (TB-LMTO ASA). Surface formation energies and layer-resolved magnetic moments are calculated for Fe slabs with increased number of monolayers (MLs) and can be represented by at least 17 MLs.

© 2004 WILEY-VCH Verlag GmbH & Co. KGaA, Weinheim

1 Introduction Recent advancements in epitaxial growth of metastable phases of solids at their surfaces opened the door to deal with properties that are different from that of their bulk. Co(hcp) and Ni(fcc) in bcc structure were stabilized on Fe(bcc), therefore, experimental and theoretical studies were devoted to explore properties of these structures at their surfaces and interfaces[1–7]. For open surfaces, as in the case of (111), it is necessary to assure that slab thickness is enough to represent the bulk environment at the central layer(s). Theoretical studies for the Fe(111) surface [8], has indicated that coordination number and atomic arrangement affect the surface magnetism.

In our study, magnetism of the un-relaxed Fe(111) and Ni overlayer on Fe(111) surfaces are examined through slab calculations with different number of layers. The TB-LMTO-ASA method [10] is used through series of calculations for surface formation energy (E^{sf}), partial density of states (PDOS), and magnetic moments μ for the Fe(111) and Ni/Fe(111) surfaces.

2 The computational details Our calculations are preformed using the TB-LMTO-ASA within the LDA of the density-functional theory [9]. The convergence tolerance of energy and magnetic moment is less than 0.01 *mRy* and 0.02 μ_B, respectively. The bcc lattice constants for Fe(5.366 a.u.) and Ni(5.262 a.u.) are calculated using Murnaghan's equation of state[10]; their corresponding magnetic moments are (2.28 μ_B) and (0.49 μ_B) for Fe and Ni, respectively. However, the magnetic moment of bcc-Ni at the lattice constant of bcc-Fe is found to be (0.53 μ_B). The PDOS for Fe and Ni using our calculated parameters are shown in Fig. 1.

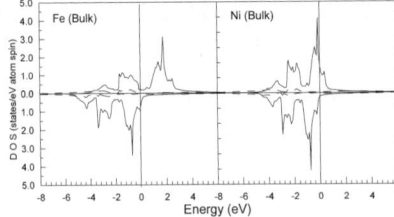

Fig. 1 The PDOS graphs for Fe and Ni in bcc structure, solid (dashed) line indicates the *d* (s and p) states and the upper (lower) half of the graph represents the minority (majority) spins.

---

[*] Corresponding author: e-mail: jkalifa@ju.edu.jo

Ni/Fe(111) surface is represented by a slab with Fe theoretical lattice spacing. The (111) surface atom has three nearest neighbours at the second layer and one at the fourth layer.

3 Results and discussion

3.1 The Fe (111) surface We have chosen E^{sf} and the value of the magnetic moment at the central layer to be the indicators for the best choice of slab thickness to represent the clean Fe(111) surface. The results in Table 1 show that the minimum thickness is 17 layers, where E^{sf} and the magnetic moments μ and $\bar{\mu}$ almost converged to a constant values; $\bar{\mu}$ is the average magnetic moment of Fe atoms in one slab. This thickness is larger than any one adopted in other calculations. Among the (111), (011), and (001) surfaces, the (111) surface is more open than the others – the atoms in the tope three layers are exposed to vacuum – and the interlayer distance is the least, so the interactions will extend through larger number of layers, and that justifies the large number of layers needed to represent the (111) clean surface.

Table 1 The formation energies (E^{sf}), surface (S), central(C), and the magnetic moment per Fe atom ($\bar{\mu}$) for slabs with different number of Fe layers.

Fe layers (ML)	E_{tot}/E_1	E^{sf} mRd/(1x1)cell	Magnetic moment (μ_B)		
			Fe(S)	Fe(C)	$\bar{\mu} = \mu$ /Fe atom
5	5.0006	353.5	3.06	2.51	2.686
11	11.0017	316.9	2.99	2.45	2.543
15	15.0025	305.1	2.97	2.31	2.451
17	17.0028	301.4	3.00	2.29	2.423
19	19.0032	300.0	2.99	2.29	2.420

The surface energy (E^{sf}) in Table 1 is obtained from the slab total energy as:

$$E^{sf} = 0.5\,(E_{slab} - nE_{bulk}), \tag{1}$$

where E_{slab} is the total energy of the slab containing n atom and E_{bulk} is the total energy of one atom in the bulk. These energies are calculated by the same structural parameters in order to overcome accuracy problems. Gay et al. [11] have suggested a method for calculating E^{sf} by graphical representation of the (n, E_{slab}) relation [12, 13], i.e.,

$$E_{slab} = nE_{bulk} + 2E^{sf}, \tag{2}$$

E_{bulk} can be calculated from the slop of (n, E_{slab}) relation. The obtained value is consistent with the – average-energy corresponding to adding one bulk atomic layer to a slab of n MLs,

$$\Delta E(n) = nE_{slab}(n) - E_{slab}(n-1) \tag{3}$$

The average surface energy $\bar{E}^{sf}$ of the slabs under consideration is obtained as half of the intercept with the ordinate. The final step of finding suitable slab thickness is to calculate E^{sf} – for all slabs – using eq. (1). The results are recorded in Table 1, where we have concluded that the central layer has reached the bulk environment for slabs of 17 and 19 MLs, while we cannot say the same thing for slabs with 15 MLs or less. Same conclusion can be drown by comparing the Fe bulk PDOS (Fig. 1) and that of the central layers of the considered slabs (Fig. 2), where we found that 17 (and 19) MLs is sufficient to obtain a good representation for the geometry of the surface and that of the bulk.

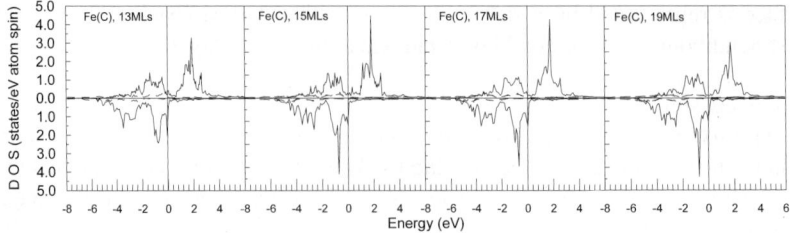

Fig. 2 The PDOS for Fe(C) in Fe(111) slabs consists of 13, 15, 17, and 19 MLs. [The caption is as Fig. 1].

Table 2 displays the magnetic structure of Fe(111) slabs; from this table we can draw the following:
* The surface magnetic moment is 2.99–3.00 μ_B approaches that of Fe(001) surface, 3.04 μ_B [14].
* The two subsurface layers are surface like since they are exposed to vacuum, which causes large enhancement to their magnetic moments as compared to corresponding layers in Fe(001) surface[14]. Moreover, their magnetic moments almost have equal values for the slabs with more than 7MLs.
* Due to openness of (111) surface, bulk like character starts from the central layer of 7 MLs slab, and for other slabs of greater thickness it starts from the (S-3) layer.

Table 2 Magnetic structure of Fe(111) slabs.

Fe MLs	S.	S-1	S-2	S-3	S-4	S-5	S-6	S-7	S-8	S-9	$\bar{\mu}$ / Fe atom
1	3.58										3.580
3	2.89	2.27									2.683
5	3.06	2.40	2.51								2.686
7	2.96	2.48	2.52	2.19							2.587
9	3.03	2.50	2.49	2.37	2.48						2.584
11	2.99	2.50	2.53	2.30	2.44	2.45					2.543
13	3.03	2.49	2.49	2.33	2.43	2.43	2.35				2.519
15	2.97	2.47	2.49	2.25	2.37	2.37	2.31	2.31			2.451
17	3.00	2.46	2.47	2.27	2.39	2.34	2.24	2.28	2.29		2.423
19	2.99	2.48	2.49	2.27	2.38	2.38	2.31	2.28	2.26	2.29	2.420

The first of the above comments is related to the coordination number, since the (111) and (001) surfaces have four nearest neighbours. The second comment is related to the atomic arrangement in the (111) orientation, which results in exposing the two subsurface atoms to vacuum and to have same coordination number. The last comment is related to the first, since the atom in (S-3) layer is one of the nearest neighbours for the surface atom and shares some of the *sp* electrons to induce more spin-bond *d* electrons at the surface. The magnetic moment of the unsupported Fe ML – 90% of the isolated atom value – displays the weakness of inter-atomic potential in the *x*–*y* plane because of the large distances between atoms. As a result, the bonding to the sub-surface layer is expected to increase and cause large relaxations for the top layers, which was proven experimentally and theoretically [8].

3.2 The Ni/Fe(111) surface By mixing the effects of Ni–Fe orbital interaction, coordination number, and atomic arrangement at the surface in (111) orientation one can expect a substantial enhancement for the interfacial Fe and Ni magnetic moments. This is reported in Table 3.

Table 3 Magnetic structure for the Ni/Fe(111) system.

Fe MLs	Ni Ni(S)	Fe Layers S-1	S-2	S-3	S-4	S-5	S-6	S-7	S-8	S-9	S-10	$\bar{\mu}$/Fe atom
15	0.93	2.76	2.47	2.42	2.43	2.36	2.31	2.31	2.34			2.431
17	0.90	2.74	2.46	2.39	2.40	2.37	2.32	2.25	2.28	2.30		2.396
19	0.92	2.76	2.47	2.41	2.42	2.38	2.33	2.30	2.31	2.27	2.24	2.396

These enhancements for Fe and Ni are about 72% and 21%, respectively, over that of their bcc bulk values. The Ni–Fe interaction can be seen from the magnetic moment enhancement of the Fe(S-1), Fe(S-3) in Fe(111) and Ni/Fe(111) surfaces, while Fe(S-2), Fe(S-5), .., have negligible changes.

Figure 3 displays PDOS of Fe(C), Fe(S-1), and Ni(S) in a 19 MLs of Fe slab. The effect of Ni–Fe interaction at the surface changes the DOS of both metals, all majority spin states are shifted down below the E^f, and DOS of the minority at E^f has increased to higher values.

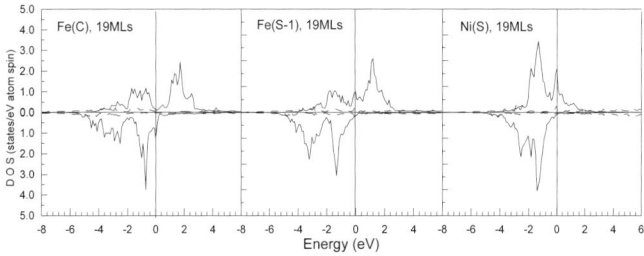

Fig. 3 PDOS for Fe(C), Fe(S-1), and Ni(S) in Ni/Fe(111) slab of 19 MLs of Fe. [The caption is as Fig. 1].

By comparing the interfacial Ni and Fe magnetic moments for Ni/Fe(001) [14] and Ni/Fe(111) systems, we notice that Ni(S) magnetic moment in Ni/Fe(111) system has a minor increase (~0.02 μ_B), while that of Fe(S-1) has a larger increase (~0.15 μ_B). This shows that Fe is more sensitive to the environment change than Ni. Since Fe(bcc) is a weak ferromagnet, small portion of the majority states are located above E^f – while Ni(bcc) is a strong ferromagnet, all of the majority electrons are located below E^f, see Fig. 1, then any change in the environment will affect Fe atoms more than Ni atoms at their corresponding positions.

4 Conclusion Finally, our main results as the following: The bulk of Fe in (111) orientations can be represented by at least 17 MLs. Fe(111) surface layer magnetic moment (~3.00) approaches that of the (001) surface (~3.04) as both have the same coordination number. Interfacial Ni and Fe moments in Ni/Fe(111) system are enhanced by 72% and 21%, respectively, compared to their bcc bulk magnetic moments.

Acknowledgements We highly appreciate fruitful discussions with Claude Demangeat and Salvador Aguilar. Partial financial support by the Deanship of Scientific Research at the University of Jordan is also acknowledged.

References

[1] A. J. Freeman, H. Krakauer, S. Ohnishi, Ding-Sheng Wang, M. Weinert, and E. Wimmer, J. Magn. Magn. Mater. 38, 269–272 (1983);
J. I. Lee, Soon C. Hong, and A. J. Freeman, Phys. Rev. B **47**, 810 (1993).
[2] Z. Q. Wang, S. H. Li, F. Jona, and P. M. Marcus, Solid State Commun. **61**, 673 (1987);
S. Ohnishi, A. J. Freeman, and M. Weinert, Phys. Rev. B **28**, 6741 (1983);
B. Heinrich, S. T. Purcell, J. R. Dutcher, K. B. Urquhart, J. F. Cochran, and A. S. Arrott, Phys. Rev. B **38**, 12879 (1988).
[3] V. L. Moruzzi, P. M. Marcus, K. Schwarz, and P. Mohan, Phys. Rev. B **34**, 1784 (1986);

V. L. Moruzzi, Phys. Rev. Lett. **57**, 2211 (1986);
V. L. Moruzzi, P. M. Marcus, and K. Schwarz, Phys. Rev. B **38**, 1613 (1988).
[4] B. Heinrich, J. F. Cochran, A. S. Arrott, S. T. Purcell, K. B. Urquhart, J. R. Dutcher, and W. F. Egelhoff, Jr., Appl. Phys. A **49**, 437–490 (1989);
N. B. Brookes, A. Clarke, and P. D. Johnson, Phys. Rev. B **46**, 237 (1992).
[5] A. A. Ostroukhov, V. M. Folka, and V. T. Tomilenko, J. Magn. Magn. Mater. **147**, 205–207 (1995).
[6] L. Duò, R. Bertacco, G. Isella, F. Ciccacci, and M. Richter, Phys. Rev. B **61**, 15294 (2000).
[7] J. Sokolov, F. Jona, and P. M. Marcus, Phys. Rev. B **33**, 1397 (1986);
Ruqian Wu and A. J. Freeman, Phys. Rev. B **47**, 3904 (1993);
Ruqian Wu, A. J. Freeman, and G. B. Olson, Phys. Rev. B **47**, 6855 (1993).
[8] A. M. N. Niklasson, B. Johansson, and H. L. Skriver, Phys. Rev. B **59**, 6373 (1999).
[9] O. K. Anderson and O. Jepsen, Phys. Rev. Lett. **53**, 2571 (1984);
O. K. Anderson, O. Jepsen, and Z. Pawalowska, Phys. Rev. B **34**, 5253 (1986);
P. Hohenberg, and W. Kohn, Phys. Rev. **136**, B864 (1964);
W. Kohn, and L. J. Sham, Phys. Rev. **140**, A1133 (1965);
U. von Berth and L. Hedin, J. Phys. C **5**, 1629 (1972).
[10] F. D. Murnaghan, Proc. Nat. Acad. Sci. U.S.A. **30**, 244 (1944).
[11] J. G. Gay, J. R. Smith, R. Richter, F. J. Arlinghaus, and R. H. Wagoner, J. Vac. Sci. Technol. A **2**, 931 (1984).
[12] J. C. Boettger, Phys. Rev. B **49**, 16798 (1994).
[13] V. Fiorentini and M. Methfessel, J. Phys.: Condens. Matter **8**, 6525 (1996).
[14] N. T. Shawagfeh and J. M. Khalifeh, Physica B **321**, Issues 1–4, 222–229 (2002).

Stability and magnetism of Ni–Fe alloyed overlayers on Fe(001)

Naseem T. Shawagfeh[1] and **Jamil M. Khalifeh**[*,2]

[1] Department of Basic Science, Faculty of Engineering Technology, Al-Balqa' Applied University, Amman-11134, Jordan.
[2] Department of Physics, University of Jordan, Amman-11942, Jordan.

Received 31 August 2003, accepted 31 December 2003
Published online 25 March 2004

PACS 71.15.Ap,75.70.Cn

The surface ordered alloy formation of Ni/Fe(001) overlayer system is explored, using the tight-binding linear muffin-tin orbital (TB-LMTO) method. The alloy formation for one- and two- surface layer of Ni–Fe on Fe(001) are considered by calculating the total and surface formation energies for different configurations of Ni overlayers on Fe(001). The preferred ground states for different configurations are determined and Ni (alloyed with Fe) is found to remain at the Fe surface. The magnetic profile for these configurations is found comparable to the experimental predictions.

© 2004 WILEY-VCH Verlag GmbH & Co. KGaA, Weinheim

1 Introduction Nickel overlayers on Fe(001) systems represent an interesting case among the 3d metal overlayers on Fe(001). The coupling between Ni and Fe in the magnetic ground state is ferromagnetic, and the enhancement of the magnetic moments of both metals is reported by theory and experiment [1–5]. An extensive experimental study of Fe and Ni(001) surfaces and the overlayer system (Ni/Fe) show some of the structural changes which depend on the number of Ni layers grown on Fe substrate [5, 6]. While previous theoretical studies are restricted to one- and two- pure Ni overlayers on Fe(001), the inter-mixing of Ni– Fe at the interface is confirmed by experiments.

The effects of magnetism on stability and alloy formation of the overlayer were under experimental and theoretical investigations [7–11]. These studies have included non-magnetic (Ag, Cu) overlayers on a magnetic (Fe) substrate. They concluded that coordination and magnetic structure have the main contributions to stability of the alloyed overlayer systems. The structural (magnetic) part stabilizes the system by increasing (decreasing) the coordination number.

In this study we report an analysis for total energy and magnetic structure of Ni/Fe(001) systems in different combinations, which includes Ni overlayer, Ni interlayer diffusion, and Ni– Fe alloy formations for one and two MLs on Fe(001). In order to identify the equilibrium configuration of Ni overlayer on Fe substrate; we have examined the formation energy gain by altering the order, the number of Ni atoms, and the number of Ni layers. The stability of the surface alloy formation is examined using a convention of energy calculation reported elsewhere [7, 8].

2 Method and calculation details Our calculations are preformed using the TB-LMTO-ASA method [12]. The convergence tolerance of energy and magnetic moment is less than 0.01 mRy and 0.02 μ_B, respectively. The bcc lattice constants for Fe(5.366 a.u.) and Ni(5.262 a.u.) are calculated using Murnaghan's equation of state[13]; their corresponding magnetic moments are (2.28 μ_B) and (0.49 μ_B) for Fe and Ni, respectively. However, the magnetic moment of bcc-Ni at the lattice constant of bcc-Fe is found to be (0.53 μ_B). The stability of slab surface is examined by calculating the total energy of the slab in

---

[*] Corresponding author: e-mail: jkalifa@sci.ju.edu.jo

different configurations of surface Ni and Fe, and by comparing their surface formation energy, defined as the required energy to create a unit area of that surface [14], which can be expressed as:

$$E^{sf} = \frac{1}{2A}[E_{slab} - N_{Ni}E_{Ni}^{bulk} - N_{Fe}E_{Fe}^{bulk}] \quad (1)$$

where A is the area occupied by one unit cell on the surface of the slab, E_{slab} is the total energy of the slab, N_{Ni} (N_{Fe}) is the number of Ni (Fe) atoms in the unit cell, E_{Ni}^{bulk} (E_{Fe}^{bulk}) is the energy of one Ni (Fe) atom in the bcc bulk at the equilibrium lattice constant.

3 Results and discussion

3.1 Stability with a complete surface layer substitution Table 1 reports the results of energy calculations for systems under consideration. By ordering these systems, with the same number and kind of atoms, according to their surface formation energies, we can address the following remarks:

* Substitution of Ni subsurface layer by Fe surface layer reduces the surface energy.
[energy gain (32.9 *mRyd*) towards stability of the Ni/Fe/Fe(001) system, row 2 and 3.]
* Substitution of Fe surface layer by Ni–Fe mixed subsurface layer reduces the surface energy.
[energy gain 16.9 *mRyd* towards stability of the Ni$_{50}$Fe$_{50}$/Fe/Fe(001), row 5 and 6.]

Then we can deduce that Ni, unlike the 3d transition-metal overlayers on noble metal substrates [7], has the trend to be at the surface or close to it.

Table 1 Surface formation energies for different configurations of Ni and Fe overlayers. Fe(001) ≡ 9 ML.

System		E^{sf} mRyd / (1x1) cell	E_{tot} Ryd/(1x1) cell
1.	Fe/Fe/Fe(001)	209.8	– 33068.11187
2.	Fe/Ni/Fe(001)	200.0	– 34059.75081
3.	Ni/Fe/Fe(001)	183.6	– 34059.78376
4.	Ni/Ni/Fe(001)	176.0	– 35051.41848
5.	Ni$_{50}$Fe$_{50}$/Fe/Fe(001)	196.0	– 33563.94919
6.	Fe/Ni$_{50}$Fe$_{50}$/Fe(001)	204.5	– 33563.93231
7.	Ni$_{50}$Fe$_{50}$/Ni/Fe(001)	182.7	– 34555.59526
8.	Ni/Ni$_{50}$Fe$_{50}$/Fe(001)	179.7	– 34555.60129
9.	2Ni$_{50}$Fe$_{50}$/Fe(001)	188.9	– 34059.77305

3.2 Stability of one and two Ni-Fe alloyed surface layers In Table 2, we have reported all of the above systems in which their surface combinations may produce one or two Ni–Fe alloyed overlayers on Fe(001). The last column contains the total energy differences for systems "A" and "B". The system "A" ("B") is considered to be more stable if $\Delta E < 0$ ($\Delta E > 0$). These energies are calculated as follows [7, 14, 15]:

$$\Delta E = [E_{2Ni_{50}Fe_{50}/Fe(001)} - E_{Ni/Fe/Fe(001)}] \quad (2\text{-a})$$

$$\Delta E = [E_{2Ni_{50}Fe_{50}/Fe(001)} - E_{Fe/Ni/Fe(001)}] \quad (2\text{-b})$$

$$\Delta E = [E_{2Ni_{50}Fe_{50}/Fe(001)} - 0.5[E_{Ni/Fe/Fe(001)} + E_{Fe/Ni/Fe(001)}]] \quad (2\text{-c})$$

$$\Delta E = [E_{2Ni_{50}Fe_{50}/Fe(001)} - 0.5[E_{Ni/Ni/Fe(001)} + E_{Fe/Fe/Fe(001)}]] \quad (2\text{-d})$$

for the two alloyed MLs systems, and

$$\Delta E = [E_{Ni/Ni_{50}Fe_{50}/Fe(001)} - 0.5[E_{Ni/Fe/Fe(001)} + E_{Ni/Ni/Fe(001)}]] \quad (3\text{-a})$$

$$\Delta E = [E_{Ni_{50}Fe_{50}/Fe/Fe(001)} - 0.5[E_{Ni/Fe/Fe(001)} + E_{Fe/Fe/Fe(001)}]] \quad (3\text{-b})$$

$$\Delta E = [E_{Ni_{50}Fe_{50}/Fe(001)} - 0.5[E_{Ni/Fe(001)} + E_{Fe/Fe(001)}]] \quad (3\text{-c})$$

for the one alloyed ML system. The results of energy differences indicate the following:
– The stability of Ni overlayer system Ni/Fe/Fe(001) is higher than the two-alloyed MLs system ($\Delta E = 10.7$ $mRyd$, see Eq. (2-a)). While the later is highly preferable over the buried Ni layer system Fe/Ni/Fe(001) ($\Delta E = -22.2$ $mRyd$, see Eq. (2–b)).
– The two-alloyed MLs system is preferable over the systems that might be formed from the phase separation of Fe/Ni/Fe(001) and Ni/Fe/Fe(001) systems ($\Delta E = -5.8$ $mRyd$, see Eq. (2-c)), or Ni/Ni/Fe(001) and Fe/Fe/Fe(001) systems ($\Delta E = -7.9$ $mRyd$, see Eq. (2-d)).
– The one-alloyed ML system $Ni_{50}Fe_{50}$/Fe(001) is more stable than the buried one-alloyed ML systems Fe/$Ni_{50}Fe_{50}$/Fe(001) ($\Delta E = -16.9$ $mRyd$, see Eq. (3-a)), and the Ni/$Ni_{50}Fe_{50}$/Fe(001) system is preferable over the $Ni_{50}Fe_{50}$/Ni/Fe(001) system ($\Delta E = 6$ $mRyd$, see Eq. (3-c)), i.e., the more Ni atoms present at the surface the more preferable it is.

Table 2 Energy differences of the Ni–Fe surface alloyed systems.

	System "A"	Thickness		System(s) "B"	$mRyd$ / (1x1)cell
1.	$2Ni_{50}Fe_{50}$/Fe(001)	13ML	Eq. (2-a)	Ni/Fe/Fe(001)	10.7
			Eq. (2-b)	Fe/Ni/Fe(001)	–22.2
			Eq. (2-c)	Fe/Ni/Fe(001) / Ni/Fe/Fe(001)	–5.8
			Eq. (2-d)	Ni/Ni/Fe(001) / Fe/Fe/Fe(001)	–7.9
2.	$Ni_{50}Fe_{50}$/Fe/Fe(001)	13ML	Eq. (3-a)	Fe/$Ni_{50}Fe_{50}$/Fe(001)	–16.9
3.	$Ni_{50}Fe_{50}$/Fe/Fe(001)	13ML	Eq. (3-b)	Fe/Fe/Fe(001) / Ni/Fe/Fe(001)	–1.4
4.	$Ni_{50}Fe_{50}$/Ni/Fe(001)	13ML	Eq. (3-c)	Ni/$Ni_{50}Fe_{50}$/Fe(001)	6.0
5.	$Ni_{50}Fe_{50}$/Ni/Fe(001)	13ML	Eq. (3-d)	Ni/Ni/Fe(001) / Fe/Ni/Fe(001)	–10.6
6.	Ni/Fe/Fe(001)	13ML	Eq. (4)	Ni/Ni/Fe(001) / Fe/Fe/Fe(001)	–18.6

The last conclusion is also obtained from the cases displayed in row 3, where the energy differences show that alloying is still favoured over the system with half of Fe surface covered by Ni and the other half covered by Fe ($\Delta E = -1.4$ $mRyd$, see Eq. (3-b)).

Finally, the last row shows that Ni ML on Fe is favoured over the system with half of Fe surface covered by 2 ML of Ni and the other half covered by Fe, i.e. phase separation of two ML height ($\Delta E = -18.6$ $mRyd$, see Eq. (4)). Asada *et al.* have concluded the same result ($\Delta E \approx -16$ $mRyd$) [15].

$$\Delta E = [E_{Ni/Fe/Fe(001)} - 0.5 \, [E_{Ni/Ni/Fe(001)} + E_{Fe/Fe/Fe(001)}]] \tag{4}$$

3.3 Magnetic structure of the Ni–Fe surface alloyed system Table 3 displays the magnetic moments of one- and two-surface alloyed systems. We see that the number and kind of the nearest neighbours affect the values of the magnetic moments of Ni and Fe. We can also conclude that within the same Ni–Fe alloyed layer the increase in the magnetic moment of Ni corresponds to a decrease in that of Fe, and vice versa. This behaviour can be attributed to the Ni–Fe interaction among the atoms in the surface and the subsurface layer.

The magnetism in iron surface atoms in $Ni_{50}Fe_{50}$/Fe/Fe(001) is less than that of $Ni_{50}Fe_{50}$/$Ni_{50}Fe_{50}$/Fe(001) owing to the presence of more Fe atoms in the subsurface layer of the former system. On the contrary, Ni magnetic moment is larger in $Ni_{50}Fe_{50}$/Fe/Fe(001) as compared to $Ni_{50}Fe_{50}$/$Ni_{50}Fe_{50}$/Fe(001) due to the induced moment by Fe.

Despite the discrepancies in experimental results [6], which give 0.59–0.89 μ_B range for the Ni magnetic moments at the 1–3 MLs region, we obtain an average of 0.72 μ_B for $Ni_{50}Fe_{50}$ /$Ni_{50}Fe_{50}$/Fe(001) system, and 0.74 μ_B for Ni/$Ni_{50}Fe_{50}$/Fe(001) system.

Table 3 Magnetic structure of Ni/$Ni_{50}Fe_{50}$/Fe(001), $Ni_{50}Fe_{50}$/Fe/Fe(001) and $Ni_{50}Fe_{50}$/$Ni_{50}Fe_{50}$/Fe(001) systems in unit of μ_B.

	Ni/$Ni_{50}Fe_{50}$/Fe(001)			$Ni_{50}Fe_{50}$/Fe/Fe(001)			$Ni_{50}Fe_{50}$/$Ni_{50}Fe_{50}$/Fe(001)		
S	Ni	0.82	0.82	Fe, Ni	3.01	0.894	Fe, Ni	3.09	0.79
S-1	Ni, Fe	0.58	2.59	Fe	2.39	2.39	Ni, Fe	0.65	2.41
S-2	Fe	2.54	2.54	Fe	2.40	2.44	Fe	2.56	2.60
S-3	Fe	2.40	2.38	Fe	2.34	2.34	Fe	2.38	2.38
S-4	Fe	2.34	2.34	Fe	2.31	2.28	Fe	2.34	2.34
S-5	Fe	2.29	2.31	Fe	2.27	2.27	Fe	2.29	2.29
C	Fe	2.28	2.28	Fe	2.25	2.27	Fe	2.27	2.28

4 Conclusions In this study we deduce that, unlike the 3d transition-metal overlayers on noble metal substrates, Ni overlayer has the trend to be at the surface or close to it. The buried alloyed layers are less preferable over the surface alloyed layers, but it is more preferable over surface phase separation. The magnetism of Ni and Fe is strongly affected by their surface alloying. The average magnetic moment for the considered systems is within the range of experimental findings (0.59–0.89 μ_B).

Acknowledgements We highly appreciate fruitful discussions with Claude Demangeat and Salvador Aguilar. Partial financial support by the Deanship of Scientific Research at the University of Jordan is also acknowledged.

References

[1] J. I. Lee, Soon C. Hong, and A. J. Freeman, Phys. Rev. B **47**, 810 (1993).
[2] N. B. Brookes, A. Clarke, and P. D. Johnson, Phys. Rev. B **46**, 237 (1992).
[3] N. T. Shawagfeh and J. M. Khalifeh, Physica B **321**, No. 1–4, 222–229 (2002).
[4] B. Heinrich, S. T. Purcell, J. R. Dutcher, K. B. Urquhart, J. F. Cochran, and A. S. Arrott, Phys. Rev. B **38**, 12879 (1988).
[5] Tao Lin, M. M. Schwickert, M. A. Tomaz, H. Chen, and G. R. Harp, Phys. Rev. B **59**, 13911 (1999).
[6] Y. Kamada and M. Matsui, J. Phys. Soc. Jpn. **66**, 658 (1997);
A. V. Mijiritskii, P. J. M. Smulders, V. Ya. Chumanov, O. C. Rogojanu, M. A. James, and D. O. Boerma, Phys. Rev. B **58**, 8960 (1998).
[7] S. Blügel, Appl. Phys. A **63**, 595 (1996);
G. Bilmayer, Ph. Kurz, and S. Blügel, Phys. Rev. B **62**, 4726 (2000).
[8] Wondong Kim, Wookje Kim, S. J. Oh, J. Seo, J. S. Kim, H. G. Min, and S. C. Hong, Phys. Rev. B **57**, 8823 (1998).
[9] O. Elmouhssine, G. Moraïtis, J. C. Parlebas, C. Demangeat, P. Schieffer, M. C. Hanf, C. Krembel, and G. Gewinner, J. Appl. Phys. **83**, 7013 (1998).
[10] M. Taguchi, O. Elmouhssine, C. Demangeat, and J. C. Parlebas, Phys. Rev. B **60**, 6273 (1999).
[11] A. Noguera, S. Bouarab, A. Mokrani, and C. Demangeat, Dreyssé, J. Magn. Magn. Mater. **156**, 21 (1996).
[12] O. K. Anderson and O. Jepsen, Phys. Rev. Lett. **53**, 2571 (1984);
O. K. Anderson, O. Jepsen, and Z. Pawalowska, Phys. Rev. B **34**, 5253 (1986);
P. Hohenberg and W. Kohn, Phys. Rev. **136**, B864 (1964);
W. Kohn, and L. J. Sham, Phys. Rev. **140**, A1133 (1965);
U. von Berth and L. Hedin, J. Phys. C **5**, 1629(1972).
[13] F. D. Murnaghan, Proc. Nat. Acad. Sci. U.S.A. **30**, 244(1944).
[14] H. K. Skriver and N. M. Rosengaard, Phys. Rev. B **46**, 7157(1992);
M. J. Mehl and D. A. Papaconstantopoulos, Phys. Rev. B **54**, 4519(1996).
[15] T. Asada, G. Bihlmayer, S. Handschuh, S. Heinze, Ph. Kurz, and S. Blügel, J. Phys.: Condens. Matter **11**, 9347 (1999).

The effect of Bi mole ratio on phase formation in $Bi_xY_{3-x}Fe_5O_{12}$ nanoparticles

J. Amighian[*], A. Hasanpour, and M. Mozaffari

Department of Physics, Faculty of Science, University of Isfahan, Isfahan, Iran

Received 31 August 2003, accepted 31 December 2003
Published online 18 March 2004

PACS 75.50.Tt, 81.20.Ev, 81.40.Rs

A series of Bi substituted yttrium iron garnet (Bi-YIG) nanoparticles with nominal formula of $Bi_xY_{3-x}Fe_5O_{12}$, in which x varied in the steps of 0.0, 0.5, 1.0 and 1.5, were prepared via mechanochemical processing (MCP). A milling time of 5 hours was found to be suitable for all the samples. Minimum annealing temperature for completion of the phase formation was determined for each sample. The optimum annealing temperature of each sample is much lower than the annealing temperature for the completion of the related phases in the conventional ceramic technique. The average sizes of the particles were determined by Scherrer's formula and were in the range of 30 nm to 40 nm. The effect of Bi^{3+} substitution for Y^{3+} on lattice parameters and magnetic properties of pure YIG has been investigated.

© 2004 WILEY-VCH Verlag GmbH & Co. KGaA, Weinheim

1 Introduction

Magnetic fine particles are very attractive objects of magnetic research. This is because they are single magnetic domain and accordingly their magnetic properties and their mutual interaction can be studied without magnetic domain effects [1]. In addition from an industrial viewpoint, they are applicable to media for high-density magnetic or magneto-optical information storage systems. Among the magnetic materials, polycrystalline yttrium iron garnets and substituted ones have received a great deal of attraction in laser, microwave devices and magneto-optics. Bi-substituted garnets are suitable materials for use in magneto-optical discs and also magneto-optical display devices [2].

A variety of processing techniques has been tried to prepare single-phase substituted garnet fine powders [3]. These include coprecipitation [4], mist pyrolysis [5], sol–gel [6], hydrothermal [7], alkoxide [8] and homogeneous precipitation [9]. Recently mechanochemical processing (MCP) has found wide application in producing nanosized magnetic particles [10, 11]. In this work we have used MCP to prepare $Bi_xY_{3-x}Fe_5O_{12}$ nanosized particles and to see the effect of different x values on preparation, crystal structure and magnetic parameters.

2 Experimental

Polycrystalline substituted yttrium iron garnets of composition $Bi_xY_{3-x}Fe_5O_{12}$ with $0 \leq x \leq 1.5$ were prepared by mechanical alloying and subsequent heat treatments. The raw materials were Fe_2O_3, Y_2O_3 and Bi_2O_3 all from Merck Company. The oxides were first weighed in different mole ratios and then milled in a Spex 8000D Mixer/Mill unit in different milling times. All the compositions were corrected for extra pick up on iron from ball milling using iron deficient compositions. A programmable electrical furnace (HT17/4 Nabertherm Co.) was used for subsequent heat treatment. The as milled powders were annealed in air at different temperature for 3 hours to achieve a single-phase composition.

XRD investigation of the as milled powders was performed to see any new phase formation during mechanical alloying. The results show that subsequent heat treatments are necessary to obtain single

---

[*] Corresponding author: e-mail: jamighian@sci.ui.ac.ir, Phone: +98-311-7392419, Fax: +98-311-7932419

phase $Bi_xY_{3-x}Fe_5O_{12}$. The formation of garnet phase was investigated by means of a Philips X'Pert diffractometer, using Cu-Kα radiation. Using XRD patterns and Scherrer formula, the lattice parameters and mean particle size of all samples were calculated. Magnetizations of the samples were measured at room temperature by a vibrating sample magnetometer.

3 Results and discussion XRD spectra of the as-milled powders and powders annealed up to 700 °C show that the garnet phase has not been formed for any of the $Bi_xY_{3-x}Fe_5O_{12}$ compositions. But for the powders annealed at higher temperatures the garnet phase starts to form and completes finally at 775 °C, 800 °C, 850 °C and 925 °C for x = 1.5, 1.0, 0.5 and 0.0, respectively. This shows that the minimum annealing temperature for each sample is depended on the Bi mole ratio in $Bi_xY_{3-x}Fe_5O_{12}$. This can be due to the low melting temperature of Bi_2O_3 (817 °C) compared with the value corresponding to Y_2O_3 (2420 °C). The values of the lattice parameter determined from X-ray data for x = 0 to x = 1.5 are plotted as a function of x in Fig. 1. The lattice constant exhibits a very slow linear increase with increasing. This is understandable, because Y^{3+} with the smaller ionic radius (1.016 Å) is being replaced by the lager Bi^{3+} ionic radius (1.132 Å).

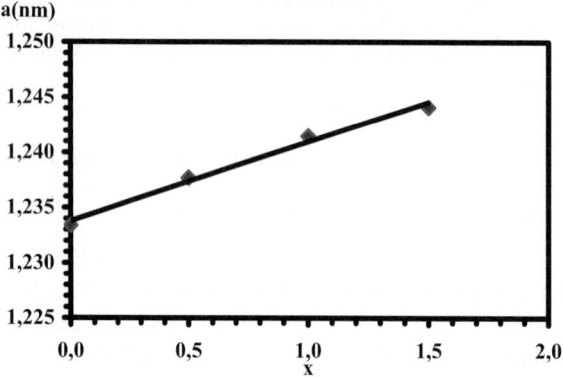

Fig. 1 Variation of lattice constant as a function of x.

The average size of the particles for each single-phase composition was determined by Scherrer's formula. The measured values are in the range of 30 nm to 40 nm.

The magnetic hysteresis of one of the substituted bismuth garnets (x = 0.5) annealed at 850 °C is shown in Fig. 2. The saturation magnetizations at room temperature were obtained from the hysteresis curve for each sample. The dependence of the saturation magnetization on the Bi content is plotted in

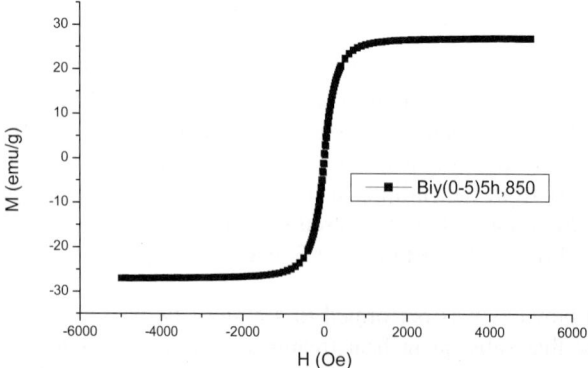

Fig. 2 The magnetic hysteresis of a typical Bi-substituted garnet (x = 0.5) annealed at 850 °C.

Fig. 3. As can be seen there is a linear decrease of saturation magnetization as x increases. This is due to the fact that by increasing Bi^{3+}, the densities of the samples will increase. This is obvious because the effect of Bi^{3+} atomic weight on the density of Bi-substituted garnets is more dominant than its effect of the ionic radius. This in turn leads to lower saturation magnetization (σ_s) as x increases.

The values of measured σ_s in Fig. 3 also show that the values are smaller than the values associated with the values obtained in conventional ceramic technique [12]. The decrease is due to the surface effect in nanoparticles (core-shell model), which has been discussed elsewhere [13].

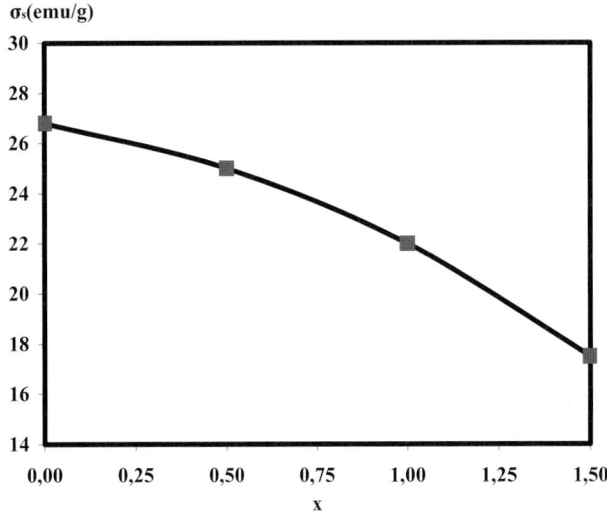

Fig. 3 The variation of saturation magnetization with respect to x.

Acknowledgements The authors wish to thank Dr. Grace Lin from National Taiwan University for VSM measurements and also Mrs. Sarrami from Isfahan University of Technology for XRD measurements.

References

[1] S. Tsketomi, Phys. Rev. E **57**, 3073 (1998).
[2] N. Kawai, E.Komuro, T. Namikawa, and Y. Yamazaki, IEEE **Mag-30**, No. 6, 4446 (1994).
[3] M. H. Han and Y. S. Ahn, J. Mater. Sci. **31**, 4233 (1996).
[4] P. Vaqueiro, M. A. Lopez-quintela, and J. Rivas, J. Mater. Chem. **7**, 501 (1997).
[5] K. Matsumoto, Y. Yamanobe, S. Sasaki, and T. Fujii, J. Appl. Phys. **70**, 5912 (1999).
[6] P. Vaqueiro, M.P. Crosnier-Lopez, and M.A. Lopez-Quintela, J. Solid State. Chem. **126**, 161 (1996).
[7] Y.S. Cho, V.L. Butfivk, and V.R.W. Amarakoon, J. Amer. Ceram. Soc. **80**, 1605 (1997).
[8] S. Taketomi, K. Kawasaki, Y. Ozaki, S. Yuasa, Y. Otani, and H. Miyajima, J. Amer. Ceram. Soc. **77**, 1787 (1994).
[9] C. S. Kuroda, T. Y. Kim, K. Yoshida, T. Namikawa, and Y. Yamazadi, Electrochimica Acta **44**, 3921 (1999).
[10] D. J. Fatemi, V. G. Harris, and V. M. Browing, J. Appl. Phys. **83**, 6867 (1998).
[11] K. J. D. Mackenzie, J. Temuujin, T. S. Jadambaa, M. E. Smith, and P. Angerer, J. Mater. Sci. **35**, 5529 (2000).
[12] S. Geller, H. J. Williams, G. P. Espinosa, R. C. Sherwood, and M. A. Gilleo, Appl. Phys. Lett. **3**, 21 (1963).
[13] M. Muroi, R. Street, P. G. McCormick, and J. Amighian, Phys. Rev. B **63**, 184414 (2001).

The coupling between antiferromagnetic transition and martensitic transformation in γ-MnFe based alloys

Jihua Zhang[*], Wenyi Peng, Ping Lu, and T.Y. Hsu (Xu Zuyao)

School of Materials Science and Engineering, Shanghais Jiao Tong University, Shanghai 200030, China

Received 31 August 2003, accepted 31 December 2003
Published online 18 March 2004

PACS 62.40.+i, 64.70.Kb, 72.15.Eb, 75.50.Ee

The internal friction (Q^{-1}), elastic modulus (E) and electrical resistance (R) of γ-MnFe based alloys were measured as the function of temperature. It was found that in the alloys with Mn-content (x) ranging from 0.4 to 0.6 in atomic fraction, the R–T curves turn up and there is no obvious anomaly in Q^{-1} and E curves, as generally resulted from only the contribution of antiferromagnetic transition. However, when $0.7 < x < 1.0$, the coupling between this transition and fcc ↔ fct martensitic transformation makes the R–T curves turn down and the anomalies in the dependences of Q^{-1} and E on temperature become greater. With increasing Mn-content, the characteristic temperatures for both antiferromagnetic transition and martensitic transformation, T_N and M_s, move closer to each other, revealing that a stronger coupling occurs between them. In the alloys with $x > 0.8$, a Q^{-1} peak appears when E starts changing abnormally at the temperature corresponding to T_N as is defined by the R–T measurement.

© 2004 WILEY-VCH Verlag GmbH & Co. KGaA, Weinheim

1 Introduction

Mn-rich solid solution alloys quenched from high temperature γ-phase are known to become antiferromagnetically ordered below the Néel temperature T_N being followed by distortion of their original *fcc* lattice. When the lattice distortion is larger than 5×10^{-3}, the antiferromagnetic transition may induce *fcc→fct* martensitic transformation, that is known as strain relaxation mechanism [1]. For example, In the Mn–Cu alloys, antiferromagnetic transition temperature, T_N, is almost consistent to martensitic transformation, M_s. But in binary Mn–Ni alloys with the composition of Mn > 80 at%, M_s is far off T_N [2] and the antiferromagnetic transition does not induce the following martensitic transformation. In low-Mn Mn–Fe alloys martensitic transformation changes its type from γ→α' into γ→ε with increasing manganese content, experimental results in Fe–Mn–Si showed that increasing the manganese content makes the Néel point rise gradually and the martensitic transformation will be restrained completely in the end [3]. For Mn-rich Mn–Fe alloys the change of the characteristic temperatures of the antiferromagnetic transition and *fcc→fct* martensitic transformation are in relation with the composition of alloys [4]. But the relation between both in a wide concentration is still not clear. The purpose of the paper is to investigate the coupling relation between antiferromagnetic transition and *fcc→fct* martinsitic transformation by means of measuring electric resistance, internal fraction and modulus with changing temperature and to discover the influence of antiferromagnetic transition on following martensitic transformation in Mn-rich Mn–Fe alloys.

---

[*] Corresponding author: e-mail: jihua@sjtu.edu.cn, Phone: +86 21 6293 2408, Fax: +86 21 6402 0422

2 Experimental methods

The alloys were prepared from electrolytic iron and manganese by medium-frequency induction melting in argon atmosphere. When the Mn-content exceeds 70 at%, the addition of 5 at% Cu makes it possible to stabilize the γ phase. Ingots were homogenized at 960 °C, and then quenched into water. No second phase particle was observed in the alloys prepared. The plates with 0.5×2×10 mm were used for electrical resistance determination by using KEITHLEY 2182 and 2400 instruments, the internal friction and modulus of the specimens were measured with Dynamic Mechanical Analyser (DMA). The manganese content of alloys was determined by Energy Dispersive Spectrometer system in a JEOL JSM-820 scanning microscope and the results are listed in Table 1.

Table 1 EDS-measured Manganese content of the Mn_x–Fe_{1-x}(Cu) alloys.

Sample No.	#1	#2	#3	#4	#5	#6
Mn (at%)	45.9	53.1	61.4	71.3	80.8	86.4

3 Results and discussion

3.1 Change of resistance vs. temperature

The resistance is shown in Fig. 1 as a function of temperature. The turning point, which is shown by an arrow, is assigned to be the Néel temperature [5]. It can be seen that when manganese content is lower than 62 at%, the R–T curves turn up and when 70–90 at%, the R–T curves turn down.

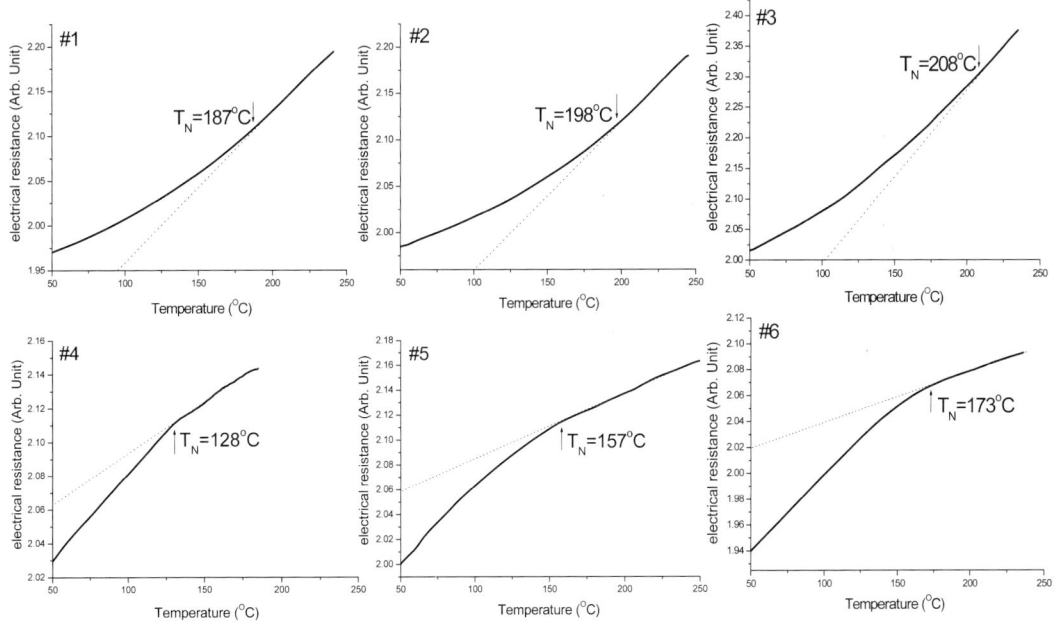

Fig. 1 The curves of R ($\times 10^{-2}$ Ω) – T (°C) in Mn_x–Fe_{1-x}(Cu) alloys
(#1) Sample, x = 0.459; (#2) Sample, x = 0.531; (#3) Sample, x = 0.614;
(#4) Sample, x = 0.713; (#5) Sample, x = 0.808; (#6) Sample, x = 0.864.

Generally, the change of R–T curves during antiferromagnetic transition shows to turn up [6]. Kunio et al. [7] plotted the relation of the antiferromagnetic transition, T_N, and the martensitic transformation, M_s, on Mn content in Mn-base metastable γ-phase alloys. It was seen that when Mn content is lower than 70 at%, M_s are far off T_N and when Mn content is larger than 70 at%, the both move closer to each other, so the different changes of the electric resistance may be because of the coupling between antiferromagnetic transition and martensitic transformation in the alloys. Because a martensitic transformation takes place in following, that results in the decreasing of antiferromagnetic structural resistance.

3.2 Internal friction and modulus

The results of the internal friction $(Q^{-1} = \tan\delta)$ and modulus are shown in Fig. 2. In the figure the modulus is shown with dot lines and the internal friction with solid lines. There is an abnormal change of modulus at the temperature corresponding to T_N, which is consistent to the turning point of R–T curves. But the internal friction of antiferromagnetic transition is low because of its short relaxation time. It is seen that when Mn content < 70 at%, the change of the modulus and the peak of internal friction are not obvious, as exemplified by the "#3" in the Fig. 2, but when Mn > 70 at%, the modulus decreases obviously at T_N. That results from the coupling between antiferromagnetic transition and martensitic transformation such as Mn–Cu alloy. The more rapid the decreasing of modulus, the closer to each other the temperatures between Ms and T_N, the higher Mn content is. A stronger coupling occurs between them.. In Fig. 2 it is seen that the modulus of #5 and #6 alloys decreases only up to ~75% and that of #4 alloy can be up to 59%. Because the M_s temperature is nearly consistent with T_N in #5 and #6 alloys, when martensitic transformation is finished, the modulus of martensite is higher and softening is over so that the modulus is only decreased ~75%. When there is more different between M_s and T_N, modulus can continuously decrease up to temperature of martensitic transformation, M_s, so modulus can be continuously decreased up to 59% as shown for #4 alloy in Fig. 2.

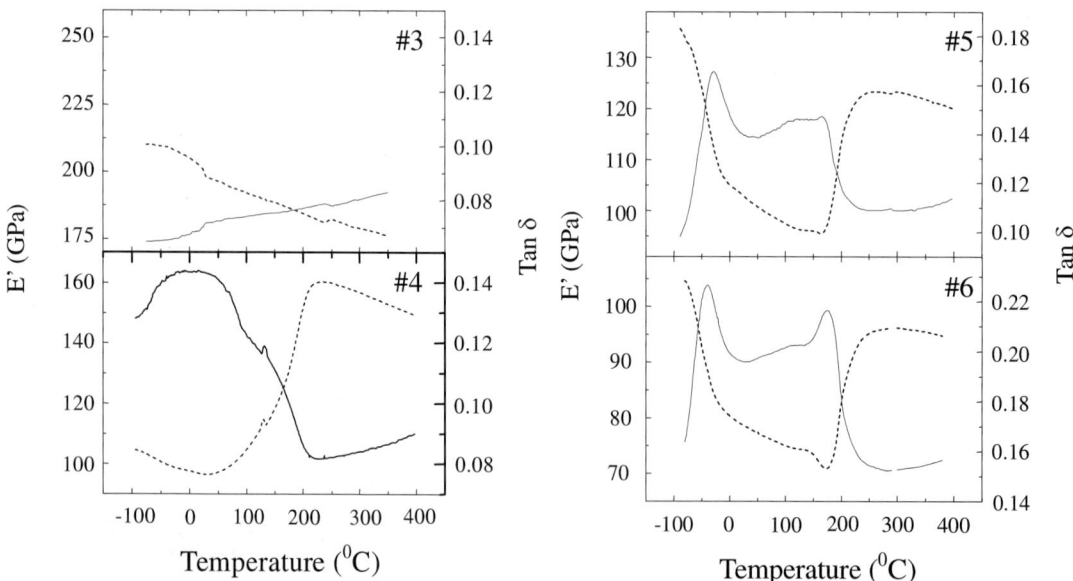

Fig. 2 Variation of internal friction and elastic modulus with temperature in Mn_x–Fe_{1-x}(Cu) alloys. (#3) Sample, x = 0.614; (#4) Sample, x = 0.713; (#5) Sample, x = 0.808; (#6) Sample , x = 0.864. f = 0.5 Hz, amplitude = 0.3 mm; solid line (—): tanδ; dotted line (·····): E'.

In fact, there are three internal friction peaks in the process: first is that of antiferromagnetic transition which is lower at T_N temperature, second is that formed by martensitic transformation and third is twin internal friction which is at about 0 °C. When M_s moves closer to T_N, a stronger coupling occurs between antiferromagnetic transition and martensitic transformation, the internal friction peak of antiferromagnetic transition and the peak of martensitic transformation are overlapped, the overlapped peak is a separated twin peak, for example the #5 and 6# alloys in Fig. 2. With decreasing Mn-content, characteristic temperatures for M_s and T_N are separated each other, the peak of martensitic transformation moves closer to twin peak, resulting in widening of twin internal friction peak, as shown for #4 alloy in Fig. 2.

Nosova and Vintaikin [7] have investigated two-way shape memory effect in a 80%Mn–19%Cu–1%Ni, the results show that there is a reversible change of the shape with a very small hysteresis for the alloy. Actually, the alloy can show a hysteresis-free nature of shape memory effect, because T_N is consistent with M_s. Above experimental results show that when the coupling between antiferromagnetic transition and martensitic transformation occurs, M_s moves closer to T_N. Antiferromagnetic transition can induce the following martensitic transformation and its reversal transformation and martensitic transformation, M_s, is consistent to its reversible temperature, A_s, which may result in a hysteresis-free process of the transformation and the shape memory effect.

4 Conclusions

4.1 There is the coupling between the antiferromagnetic transition and the *fcc→fct* martensitic transformation in Mn–Fe alloy when Mn content is greater than 70 at%.

4.2 The coupling results in turning down the R–T curve through T_N and obvious anomalies of Q^{-1} and E at T_N temperature.

4.3 The coupling makes the reversal temperature of *fct* martensitic transformation be consistent to T_N and M_s, which may predict a hysteresis-free shape memory effect in the alloys.

Acknowledgements The present work was financially supported by the National Natural Science Foundation of China (Grant No. 50171041).

References

[1] J.A. Hedley, Met. Sci. J. **2**, 129–137 (1968).
[2] T.J. Hicks, A.R. Pepper, and J.H. Smith, J. Phys. C (Proc. Phys. Soc.) **1**, 1683–1689 (1968).
[3] S.C. Chen, C.Y. Chung, and T.Y. Hsu(Xu Zuyao), Mater. Sci. Eng. A **264**, 262–268 (1999).
[4] Y. Endoh and Y. Ishikawa, J. Phys. Soc. Jpn. **30**, 1614–1627 (1971).
[5] H.M. Deng, C.Y. Chung, J.H. Zhang, and S.C. Chen, Scr. Mater. **44**, 87–90 (2001).
[6] Zhang Yansheng, Acta Metall. Sin. **22**, A470–475 (1986), in Chinese.
[7] G. Nosova and E. Vintaikin, Scr. Mater. **40**, 347–351 (1999).

The longitudinally driven giant magneto-impedance effect of a Co-based amorphous ribbon

Jianchao Zheng, Chengyuan Dong, Shipu Chen*, and T. Y. Hsu (Xu Zuyao)

School of Materials Science and Engineering, Shanghai Jiao Tong University, Shanghai 200030, China

Received 31 August 2003, accepted 31 December 2003
Published online 18 March 2004

PACS 75.30.Gw, 75.50.Kj, 75.60.Ej, 75.70.Kw, 85.70.Kh

A new kind of GMI effect, the Longitudinally Driven Giant Magneto-Impedance (LDGMI), was recently proposed by Yang *et al.* (Chinese Science Bulletin **43**, 1051–1053 (1998), in Chinese). The main concept is that an alternating magnetic field generated by a solenoid is applied along a wire or the length of a thin film or ribbon, and not circumferentially to or transversally to the sample axis (TDGMI) as usually done. By taking such a driving mode, some specific behaviors observed in the LDGMI effect of the melt-spun $Co_{66}Fe_4Ni_1Si_{15}B_{14}$ amorphous ribbons are reported in the present paper. The results show great improvement in GMI ratio and its sensitivity upon the external magnetic field, and seemingly bring forth promising potential of GMI materials for application to sensors. The related mechanism is also discussed briefly.

© 2004 WILEY-VCH Verlag GmbH & Co. KGaA, Weinheim

1 Introduction Giant magneto-impedance (GMI) effect, discovered in 1992 [1], has attracted more and more attention because of its expected applications to sensors and magnetic recording [2]. Nowadays, it is expected that GMI effect results from the skin effect [3]. Dong *et al.* [4] further pointed out that the applied magnetic field changes the effective permeability (μ_{eff}) in a ferromagnet and, thus, its skin penetration depth and impedance. When measuring the GMI effect in ribbons or films, the so-called four point method is usually adopted where the driven alternating magnetic field is applied along the transverse direction of the sample. Therefore, in this case one can call it Transversally Driven Giant Magneto-Impedance (TDGMI) effect. On the other hand, the use of a longitudinally driven giant magneto-impedance (LDGMI) effect has been proposed recently. In that case the GMI ratio ($\Delta Z/Z$) of the ferromagnet is much more greater than that obtained by the TDGMI effect and the measurement is also simplified [5, 6]. However, the LDGMI effect has not been studied in detail till now so that many phenomena associated with it remain to be explored. By taking such a driving mode, some specific behaviors observed from the melt-spun $Co_{66}Fe_4Ni_1Si_{15}B_{14}$ amorphous ribbons are reported in the present paper. The results have shown great improvement in the GMI ratio and its sensitivity upon the external magnetic field, and seemingly brought forth promising potential of GMI materials for application to sensors.

2 Experimental procedures The amorphous ribbon with composition $Co_{66}Fe_4Ni_1Si_{15}B_{14}$ was supplied by Antai Ltd., with dimensions of about 4.5 mm width and 30 μm thickness. The amorphous structure of the as-prepared ribbons was verified by X-ray diffraction using Cu-K_α radiation. For measuring the LDGMI effect, a ribbon about 40 mm long was placed in a small solenoid (20 mm long), its length being parallel to the solenoid axis. Such a "complex" impedance component was situated at the center of a pair of Helmholtz coils producing a DC external magnetic field H_{ext}. The angle between H_{ext} and the solenoid axis, θ, has been varied from 0° to 90°. An AC current with frequency of 100 kHz was used, producing an AC longitudinal driving field to the sample located inside the solenoid. The dependences of the mag-

---

* Corresponding author: e-mail: spchen@sjtu.edu.cn, Phone: +86 21-6293 2134, Fax: +86 21 6293 2435.

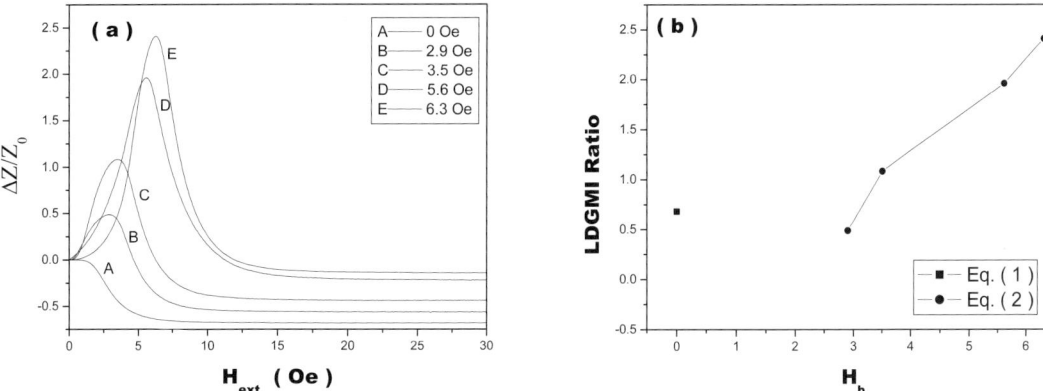

Fig. 1 (a) The dependence of $\Delta Z/Z_0$ on applied magnetic field H_{ext} with various bias fields, and (b) the dependence of *LDGMI Ratio* on the bias field H_b.

neto-impedance on the static external field H_{ext} were acquired and recorded by using a home-made computer-controlled system. The dependence of the *LDGMI Ratio*, $\Delta Z/Z_0 = [\,Z(H_{ext})\,/\,Z(H_{ext}=0)\,] - 1$, on the applied H_{ext} can be recorded directly. The static bias magnetic field $\boldsymbol{H}_b$, applied in the opposite direction of $\boldsymbol{H}_{ext}$, was provided by a pair of permanent magnets with adjustable distance. H_b was measured by a gaussmeter.

3 Results and discussion Figure 1(a) shows the dependence of $\Delta Z/Z_0$ on the applied magnetic field H_{ext} with different bias fields $H_b = 0$, 2.9, 3.5, 5.6 and 6.3 Oe values. It was found that the applied bias field dramatically influences the shape of $\Delta Z/Z_0 \sim H_{ext}$ curve. In the case of $H_b = 0$, $\Delta Z/Z_0$ decreases monotonically when $H_{ext} < 5$ Oe and then tends to be constant, while if $H_b \neq 0$, a positive peak appears at different H_{ext} values depending on applied H_b. The peak $\Delta Z/Z_0$ value increases and its position shifts to higher H_{ext} with increasing bias field. In order to evaluate the dependence of the LDGMI effect on H_b in the present case, we define the

$$LDGMI\ Ratio = 1 - [\,Z(H_{ext}=30\ \text{Oe})\,/\,Z(H_{ext}=0)\,]\ \text{when}\ H_b = 0 \qquad (1)$$

and

$$LDGMI\ Ratio = [\,Z(H_{ext}=H_p)\,/\,Z(H_{ext}=0)\,] - 1\ \text{for}\ H_b \neq 0\ . \qquad (2)$$

Here, H_p indicates the H_{ext} field at which the LDGMI effect shows a positive peak.

As shown in Fig. 1(b), the *LDGMI Ratio* increases with the bias field when $H_b > 2.9$ Oe. In particular, the *LDGMI Ratio* can reach 240% when $H_b = 6.3$ Oe is applied, showing a distinctive improvement in GMI effect compared with that for zero bias field. It is believed that this improvement results from the change of the reference point for H_{ext}. If $H_b = 0$, the data collecting system takes 0 Oe as origin and when $H_b \neq 0$, the reference point practically moves to $-H_b$ so that the peak of *LDGMI Ratio* shifts to $H_{ext} \approx H_b$. The above results seem promising. For example, the bias field could be used to switch and easily control the sensitive range of a magnetic sensor.

The θ angle between the applied magnetic field $\boldsymbol{H}_{ext}$ and the solenoid axis also strongly influences the LDGMI effect of amorphous ribbons. The dependence of $\Delta Z/Z_0$ on H_{ext} is shown in Fig. 2(a). The shape of the $\Delta Z/Z_0 \sim H_{ext}$ curve changes markedly when θ is switched from 0° to 90°. Compared with the case for the solenoid axis is parallel to $\boldsymbol{H}_{ext}$, the $\Delta Z/Z_0 \sim H_{ext}$ curve looks flatter when $\boldsymbol{H}_{ext}$ and $\boldsymbol{H}_b$ are perpendicular to each other so that $\Delta Z/Z_0$ does not saturate even when $\boldsymbol{H}_{ext}$ reaches 30 Oe.

Besides, it can be seen in Fig. 2(b) that the dependence of LDGMI effect on θ for > 80° behaves dramatically different than at smaller angle θ. The *LDGMI Ratio* decreases slowly with θ in general, but when θ is higher than 80° it suddenly falls down much quickly. The influence of θ on the *LDGMI Ratio* might be useful for designing an angle sensor, especially if θ is dedicatedly set at a region close to 90°.

© 2004 WILEY-VCH Verlag GmbH & Co. KGaA, Weinheim

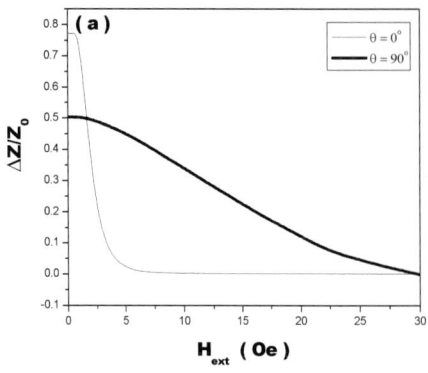
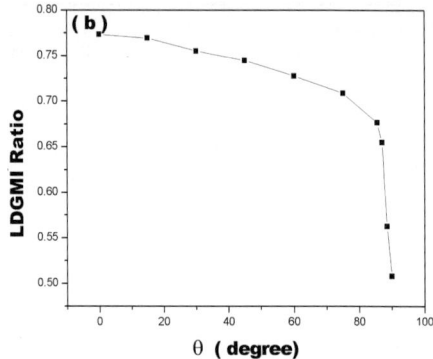

Fig. 2 (a) The dependence of $\Delta Z/Z_0$ on the applied magnetic field H_{ext} with $\theta = 0°$ and $90°$, and (b) the dependence of *LDGMI Ratio* on θ.

Figure 2 tells that θ is a critical factor for the LDGMI effect. In order to explain the influence of H_{ext}, a simple model of the LDGMI effect for amorphous ribbons is proposed, starting from the sketch shown in Fig. 3. The solenoid in the model is supposed to be infinitely long so that the magnetic field in the solenoid expresses as [7]

$$H = ni \qquad (3)$$

where n is the turn number per unit length of the solenoid and i the AC current. Therefore, the inductance of the equivalent component can be expressed as

$$L = \mu_{eff} N_{eff} S_{eff} n \qquad (4)$$

where μ_{eff} is the effective permeability, N_{eff} the effective turn number of the solenoid and S_{eff} the effective area of the "complex" component. For simplicity, another assumption made is that the impedance of the component is only governed by the inductance, i.e.

$$Z = \omega L = \mu_{eff} N_{eff} S_{eff} n \omega \qquad (5)$$

where ω is the angular frequency of the alternating current. In Eq. (5) N_{eff}, S_{eff}, n and ω are not affected by the applied static magnetic field H_{ext}, and, therefore, the impedance change is mainly due to the variation of μ_{eff} with H_{ext}. For a better understanding, it is necessary to consider the magnetization process of amorphous ribbons. First of all, a longitudinal magnetic anisotropy is produced from solidification in the as-prepared amorphous ribbons as was reported by Wang *et al.* [5]. As shown in Fig. 4(a), when $\theta = 0°$ the external static magnetic field induced magnetization process is dominated by domain-wall movement and the alternating magnetization by domain-wall oscillation. The second process is governed by the initial domain-wall position controlled by the first process. In aletrnating magnetization, the domain-wall movement produces a change of the magnetic energy, resulting in a decrease of μ_{eff} and Z. When the domain-wall movement completes, μ_{eff} and Z do not change with H_{ext} so that the *LDGMI Ratio* ~ H_{ext} curve becomes flat. In contrast, if $\theta = 90°$ as shown in Fig. 4(b), the static magnetization process is dominated by magnetization rotation instead of domain-wall movement. The alternating magnetization process is still due to domain-wall oscillation that is again controlled by the external magnetic field magnetization. It is well known that magnetization rotation is not as easy as domain-wall movement, so the wall energy change in the alternating magnetization proc-

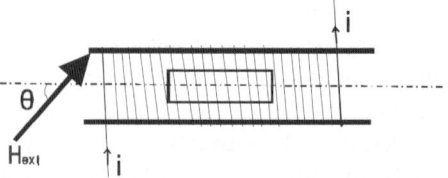

Fig. 3 A simple model of LDGMI effect for amorphous ribbon

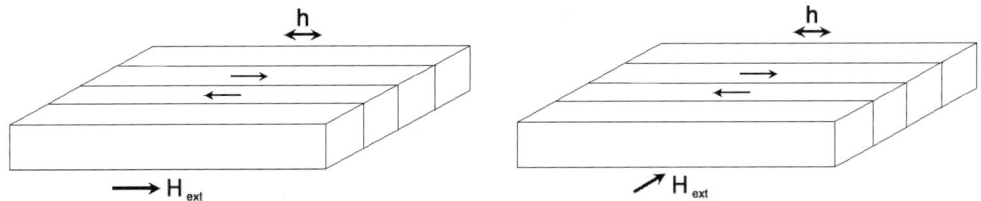

Fig. 4 The sketch of LDGMI effect in an amorphous ribbon with longitudinal magnetic anisotropy: (a) $\theta = 0°$, and (b) $\theta = 90°$. **h** indicate the longitudinal driven magnitic field.

ess varies much more slightly with the applied static magnetic field, resulting in a small change of μ_{eff} and Z with H_{ext}. Therefore, the *LDGMI Ratio* ~ H_{ext} curve with $\theta = 90°$ is more flatter than that obtained for $\theta = 0°$, as seen in Fig. 2(a). It is worthy to notice that the *LDGMI Ratio* ~ H_{ext} curve for $\theta = 90°$ does not saturate even when H_{ext} = 30 Oe but it does for $\theta = 0°$. Since the *LDGMI Raito* is defined by Eq. (1), it is obvious that the *LDGMI Ratio* for $\theta = 90°$ is smaller than that for $\theta = 0°$.

4 Conclusions

(1) A bias field was applied to improve the LDGMI effect when H_b is larger than 2.9 Oe. This may be attributed to the change of the reference point by the bias field.
(2) The angle between the applied magnetic field and the solenoid axis influences the curve shape of the LDGMI effect and the *LDGMI Ratio* decreases with θ, in particular, more dramatically when θ becomes closer to 90°.
(3) A simple model has been proposed to interpret the observed result, *i. e.* the dependence of the LDGMI effect on θ in amorphous ribbons.

Acknowledgments The authors are indebted to Prof. X. L. Yang and Mr. L. P. Liu, East China Normal University, for their assistance in LDGMI effect measurement. The financial support of the Emerson Electric Co., USA for the present work is greatly appreciated.

References

[1] K. Mohri, T. Kohzawa, K. Kawashima, and L. V. Panina , IEEE Trans. Magn. **28**, 3150 (1992).
[2] K. Mohri, K. Kawashima, T. Kohzawa, and H. Yoshida, IEEE Trans. Mag. **29**, 1245 (1993).
[3] L. V. Panina and K. Mohri, J. Magn. Soc. Jpn. **18**, 245 (1994).
[4] C. Y. Dong, S. P. Chen, and T. Y. Hsu (Xu Zuyao), J. Magn. Magn. Mater. **263**, 78 (2003).
[5] Z. C. Wang, F. F. Gong, X. L. Yang, and D. P. Yang, J. Appl. Phys. **87**, 4819 (2000).
[6] Z. J. Zhao, F. Bendjaballah, X. L. Yang , L. Zeng, G. Chen, J. X. Yang, S. M. Qian, and D. P. Yang, J. Magn. Magn. Mater. **246**, 62 (2002).
[7] J. D. Kraus and D. A. Fleisch, Electromagnetics with Application (McGraw-Hill Press, New York 1999).

Influence of Nb-addition on LDGMI effect in CoFeSiB amorphous ribbons

Chengyuan Dong, Shipu Chen[*], **and T.Y. Hsu (Xu Zuyao)**

School of Materials Science and Engineering, Shanghai Jiao Tong University, Shanghai 200030, China

Received 31 August 2003, accepted 31 December 2003
Published online 18 March 2004

PACS 75.30.Gw, 75.50.Kj, 75.60.Ej, 75.70.Kw

The longitudinally driven giant magneto-impedance (LDGMI) effect in the melt-spun $Co_{70.5-x}Fe_{4.5}Nb_xSi_{15}B_{10}$ ($x = 0, 1, 2, 3$ and 4) amorphous ribbons was systematically investigated. The results show that the *LDGMI Ratio*, $[Z(0)/Z(50)] -1$, increases with increasing Nb-content, especially in the high-frequency regime of driving current. For instance, the GMI ratio varies from 179% ($x = 0$) to 212% ($x = 4$) when driven by an alternating current with frequency of 0.1 MHz and from 48% ($x = 0$) to 64% ($x = 4$) by a 3.0 MHz current respectively. The reason of the observed improvement in LDGMI effect is presumably attributed to the enhanced glass-forming ability of Co-based alloys resulted from Nb-addition.

© 2004 WILEY-VCH Verlag GmbH & Co. KGaA, Weinheim

1 Introduction

Giant magneto-impedance (GMI) effect, a phenomenon that the impedance of ferromagnet can be dramatically changed by the applied static magnetic field, has shown great application potential in the fields of sensor and magnetic recording [1, 2]. Nowadays, the GMI effect being investigated covers various geometries including wire [3], ribbon [4] and film [5]. Conventionally the GMI effect in a ribbon or film is driven in the transverse direction [6], but recently a longitudinally driven giant magneto-impedance (LDGMI) effect has been proposed in which the GMI ratio ($\Delta Z/Z$) of ferromagnet is much larger than that in transversely driven GMI effect and the measurement has also been simplified [7, 8].

Up to now, the GMI effect has been studied in almost all available soft magnetic materials, especially in amorphous alloys. Referred to the FINEMET concept [9] that was successfully applied to the Fe-based soft magnetic alloys, a new series of novel GMI alloys was designed by adding Nb into a Co–Fe–Si–B alloy. The crystallization behavior of melt-spun amorphous ribbons upon annealing and their LDGMI effect were investigated in order to see whether niobium-addition could improve the GMI effect of Co-based alloys in the as-prepared amorphous state.

2 Alloy design and experimental procedures

In a systematical study, Makino et al. [10] found that a composition of $Co_{70.5}Fe_{4.5}Si_{15}B_{10}$ (in at%) gave rise to the best soft magnetic property among the Co–Fe–Si–B amorphous alloys. Therefore, it was taken as the initial composition for the new alloys in the present work. The nominal composition of a series of Co-based alloys was designed as $Co_{70.5-x}Fe_{4.5}Nb_xSi_{15}B_{10}$ (in at%) by adding certain amounts ($x = 0, 1, 2, 3$ and 4) of niobium to replace cobalt. Makino et al. [11] did also propose a similar composition design in which, however, the metalloid constituents were slightly different, *i.e.* 10 at% Si, 15%B and less than 2%Nb contained, and no magnetic property was reported. The alloy ribbons with composition $Co_{70.5-x}Fe_{4.5}Nb_xSi_{15}B_{10}$ were prepared by using single-roller melt-spinning technique, in dimensions about 50 mm wide and 0.03 mm thick. The amorphous structure of the as-prepared ribbons was examined by both X-ray diffraction using Cu-K$_\alpha$

---

[*] Corresponding author: e-mail: spchen@sjtu.edu.cn, Phone: +86 21 6293 2134, Fax: +86 21 6293 2435

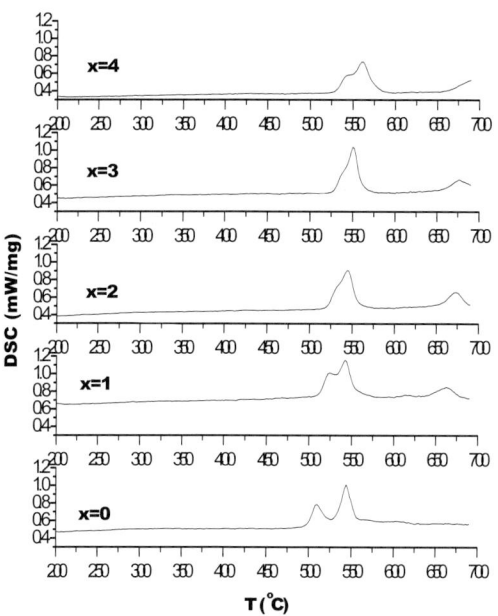

Fig. 1 DSC curves of $Co_{70.5-x}Fe_{4.5}Nb_xSi_{15}B_{10}$ amorphous ribbons.

radiation and transmission electron microscopy (TEM). Crystallization temperature T_{start} and latent heat of crystallization ΔH_x were measured by differential scanning calorimetry (DSC) at a scanning rate of 10 K/min and the onset temperature of exothermic peak related to crystallization was defined as crystallization temperature. For measuring LDGMI effect, a ribbon (1.5 cm long) is placed in a small solenoid (2 cm long), with its length parallel to the solenoid axis. In such a configuration, an equivalent impedance component is composed. The solenoid was situated at the center of a pair of Helmholtz coils producing a DC external magnetic field parallel to the solenoid axis. An AC current was applied to the solenoid, which produces an AC longitudinal driving field for the sample inside the solenoid. The megneto-impedance spectra were acquired and recorded by using a home-made computer-controlled system. *LDGMI Ratio* was defined to characterize the GMI effect in different alloys, that is

$$\Delta Z/Z = [Z(0)/Z(50)] - 1 \qquad (1)$$

where $Z(0)$ and $Z(50)$ are the measured impedances when the external static magnetic field H_{ext} is 0 and 50 Oe respectively.

3 Results and discussion The DSC curves of $Co_{70.5-x}Fe_{4.5}Nb_xSi_{15}B_{10}$ amorphous alloys in a temperature range of 200–700 °C are shown in Fig. 1. It can be noticed that there exist two exothermic peaks in the alloy without Nb addition. These two peaks are supposed to indicate a two-stage crystallization upon heating [10]:

$$Am \rightarrow Am + hcp\ Co + Co_2Si$$
$$\rightarrow hcp\ Co + fcc\ Co + Co_2Si + Co_3B \qquad (2)$$

where Am represents amorphous phase. With increasing Nb-addition, the exothermic peaks shift to higher temperature. The first peak initially at 520 °C shifts faster than the second so that they combine into one in the alloys with high Nb addition. Fig. 2 shows the dependence of Nb addition on the crystallization start temperature, T_{start}, and the latent heat of crystallization, ΔH_x. As is seen in Fig. 2(a), the T_{start}

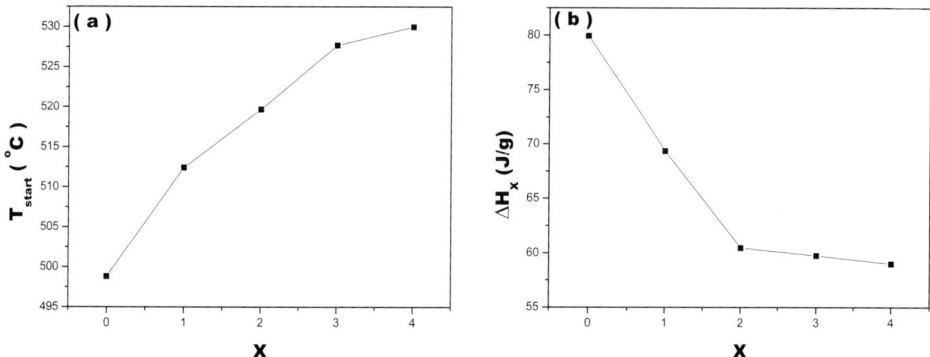

Fig. 2 The dependence of x on T_{start} (a) and ΔH_x (b) in $Co_{70.5-x}Fe_{4.5}Nb_xSi_{15}B_{10}$ ribbons.

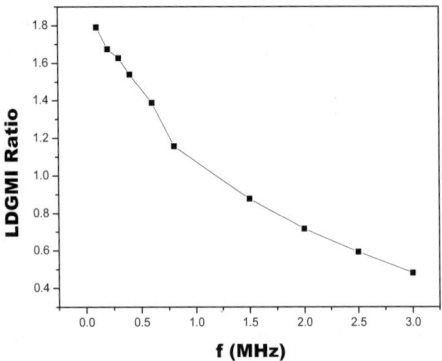

Fig. 3 The dependence of alternating current frequency on LDGMI ratio in as-prepared $Co_{70.5-x}Fe_{4.5}Nb_xSi_{15}B_{10}$ amorphous ribbons.

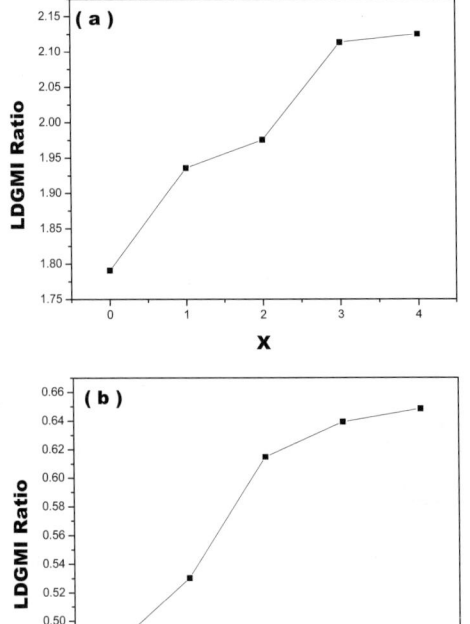

Fig. 4 The dependence of x on LDGMI ratio in $Co_{70.5-x}Fe_{4.5}Nb_xSi_{15}B_{10}$ amorphous ribbons at current frequencies 0.1 MHz (a) and 3.0 MHz (b).

increases from 498 °C ($x = 0$) to 531 °C ($x = 4$), which implies that Nb can enhance the stability of amorphous phase in Co-based alloys. Besides, ΔH_x decreases rapidly with small addition of Nb, but its change tends to be flat when $x > 2$.

Figure 3 is the dependence of alternating current frequency (f) on *LDGMI Ratio* in $Co_{70.5}Fe_{4.5}Si_{15}B_{10}$ ($x = 0$) amorphous ribbon. It was found that better GMI effect was improved from 48% to 179% as the frequency decreases from 3.0 to 0.1 MHz. Similar results were also observed in the alloys with various Nb addition. It is widely acknowledged that the magneto-impedance of a ferromagnet strongly depends on skin effect and the best GMI effect occurs when the penetration depth approximately equals the sample thickness [12], probably corresponding to the case in the present alloys when $f \leq 0.1$ MHz.

Figure 4 shows the remarkable improvement in the *LDGMI Ratio* of $Co_{70.5-x}Fe_{4.5}Nb_xSi_{15}B_{10}$ amorphous ribbons resulted from Nb addition both in low and high frequency regimes. By setting the alternating current frequency at 0.1 MHz [Fig. 4(a)], the LDGMI Ratio increases from 179% ($x = 0$) to 212% ($x = 4$) with 19% improved. It is seen that the effect of Nb on the improvement of LDGMI Ratio is relatively stronger at higher frequency. As is found in Fig. 4(b), when $f = 3.0$ MHz the LDGMI Ratio increases from 48% ($x = 0$) to 65% ($x = 4$), implying that a 35% enhancement is achieved.

The reason why Nb can improve the LDGMI effect in $Co_{70.5-x}Fe_{4.5}Nb_xSi_{15}B_{10}$ amorphous alloys is presumably attributed to an enhanced glass-forming ability of the alloys containing Nb which causes greater chemical disorder in liquid state. When the alloy solidifies, long-range diffusion may have been more effectively hindered and thus favorable to the formation of amorphous structure. Makino *et al.* [11] reported that the smaller the latent heat of crystallization, the greater is the glass-forming ability of the glass-forming CoFeSiB alloys. Fig. 2(b) also indicates a decrease of ΔH_x due to Nb addition. Therefore, it can be drawn into that Nb addition enhances the glass-forming ability of $Co_{70.5-x}Fe_{4.5}Nb_xSi_{15}B_{10}$ and then results in the improvement of LDGMI effect.

It has been well established that the magnetization process of a ferromagnet is predominately governed by domain-wall motion and magnetization rotation at low and high frequency regimes respectively [12, 13]. The GMI effect is sensitive to the volume fraction of amorphous phase, especially when magnetization rotation dominates. This may explain why the relative GMI improvement becomes obvious at high frequencies.

4 Conclusions A series of novel Co-based GMI alloys is proposed by substituting Co with Nb, based on a conventional $Co_{70.5}Fe_{4.5}Si_{15}B_{10}$ amorphous alloy.
(1) The addition of Nb remarkably improves the LDGMI effect of the conventional CoFeSiB based amorphous alloys, especially at high frequencies, presumably attributed to the enhanced glass-forming ability.
(2) The higher T_{start}, i.e. the stability at higher temperatures of Co-based amorphous ribbon can also be achieved by Nb addition.

Acknowledgements The authors are indebted to Prof. X. L. Yang and Mr. L. P. Liu, East China Normal University, for their assistance in LDGMI effect measurement. The financial support of the Emerson Electric Co., USA for the present work is greatly appreciated.

References

[1] K. Mohri, T. Kohzawa, K. Kawashima, H. Yoshida, and L. V. Panina, IEEE Trans. Magn. **28**, 3150 (1992).
[2] L. V. Panina and K. Mohri, J. Magn. Soc. Jpn. **18**, 245 (1994).
[3] K. Mohri, K. Kawashima, T. Kohzawa, and H. Yoshida, IEEE Trans. Mag. **29**, 1245 (1993).
[4] S. S. Yoon, S. C. Yu, G. H. Ryum, and C. G. Kim, J. Appl. Phys. **85**, 5432 (1999).
[5] T. Uchiyama, K. Mohri, M. Jimbom, and S. Tsunashima, J. Magn. Soc. Jpn. **19**, 481 (1995).
[6] D. Menard, M. Britel, P. Ciureanum, and A. Yelon, J. Appl. Phys. **84**, 2805 (1998).
[7] Z. C. Wang, F. F. Gong, X. L. Yang, L. Zeng, G. Chen, J. X. Yang, S. M. Qian, and D. P. Yang, J. Appl. Phys. **87**, 4819 (2000).
[8] Z. J. Zhao, F. Bendjaballah, X. L. Yang, and D. P. Yang, J. Magn. Magn. Mater. **246**, 62 (2002).
[9] Y. Yoshizawa, S. Oguma, and K. Yamaguchi, J. Appl. Phys. **64**, 6044 (1988) .
[10] A. Makino, A. Inoue, and T. Masumoto, Mater. Trans. JIM **31**, 884 (1990).
[11] A. Makino, A. Inoue, and T. Masumoto, Mater. Trans. JIM **31**, 891 (1990).
[12] L. V. Panina, K. Mohri, T. Uchiyama, and M. Noda, IEEE Trans. Magn. **31**, 1249 (1995).
[13] Chengyuan Dong, Shipu Chen, and T. Y. Hsu (Xu Zuyao), J. Magn. Magn. Mater. **263**, 78 (2003).

Influence of field direction on magnetization measurement for NbTi wire

Dali Mao*, Ling Jiang, and **Chengkang Chang**

School of Materials Science and Engineering, Shanghai Jiaotong University, 1954 Huashan road, Shanghai, 200030, P.R.China

Received 31 August 2003, accepted 31 December 2003
Published online 18 March 2004

PACS 74.25.Ha, 74.25.Op, 74.70.Ad

Upper critical field, $\mu_0 H_{c2}$ was important to the type-II superconductors and interested to study for the nature of transition points near the upper critical field. A standard NbTi wire was used to explore the influence of magnetic field direction on magnetization measurement. The external field was applied in the direction of perpendicular or parallel to the sample axis. It swept up to 12 Tesla with the temperature changing from 4.2 K to 9.5 K in a Vibrating Sample Magnetometer (VSM) system. Upper critical fields such as $\mu_0 H_{irr}(T)$ and $\mu_0 H_{c2}(T)$ were observed on the measured curves. These upper critical fields showed a linear relation dependent on temperatures.

© 2004 WILEY-VCH Verlag GmbH & Co. KGaA, Weinheim

1 Introduction Conventional type-II NbTi wire has been used in many aspects of magnetic field application [1]. Because of its nature of flux pinning, the measurement of magnetic properties near the upper critical field has been focused. The criteria for the definition of the upper critical field were studied [2]. Researchers also noticed the importance of the applied field direction on the superconducting behaviour. It was found that the angle between the applied magnetic field and sample axis affected the Ic measurement for the Bi-oxide tape [3–5]. VAMAS (Versailles Agreement on Advanced Materials and Standards) test results reported the importance of the standard measurement of magnetic properties for superconductors [6]. As the author's knowledge, there was no detail work about the effects of magnetic field direction on the magnetization measurement for the NbTi wire and it was worth studying on it. In this paper, we measured the magnetization properties for a standard NbTi sample under the magnetic field till to 12 Tesla by VSM. The influence of direction of magnetic field on magnetization measurement was studied.

2 Experimental The sample used for study was a standard NbTi wire fabricated by NBS (National Bureau of Standards) as a normal sample for VAMAS test. The wire was cut into 10 pieces in the length of 5 mm and bundled together by Stycast resin. Magnetization was measured in an VSM system made by Oxford company. The magnetic field was swept up to 12 Tesla at the rate of 0.5 T/min. In order to measure the local temperature more accurately, a Ru-oxide sensor was mounted to the sample. The magnetic field was applied as perpendicular or parallel direction to sample axis, respectively. The magnetization loop was obtained at 4.2 K, 5 K, 5.5 K, 6 K, 6.5 K, 7 K, 7.5 K, 8 K, 8.5 K, 9 K, and 9.5 K.

3 Results and discussion Figure 1 showed two typical magnetization curves measured in two directions at 4.2 K. The shape of the curve on perpendicular direction was different from that of parallel direction. In Fig. 1(a), it was a normal loop shape similar to other superconductors in magnetization meas-

---

* Corresponding author: e-mail: dlmao@sjtu.edu.cn, Tel.: ++86-21-62932582, Fax: ++86-21-62932587

urement, and $\mu_0 H_{c2}$ was around 10.5 T from the curve at 4.2 K, while $\mu_0 H_{c2}$ was around at 11 T for the parallel direction case in Fig. 1(b).

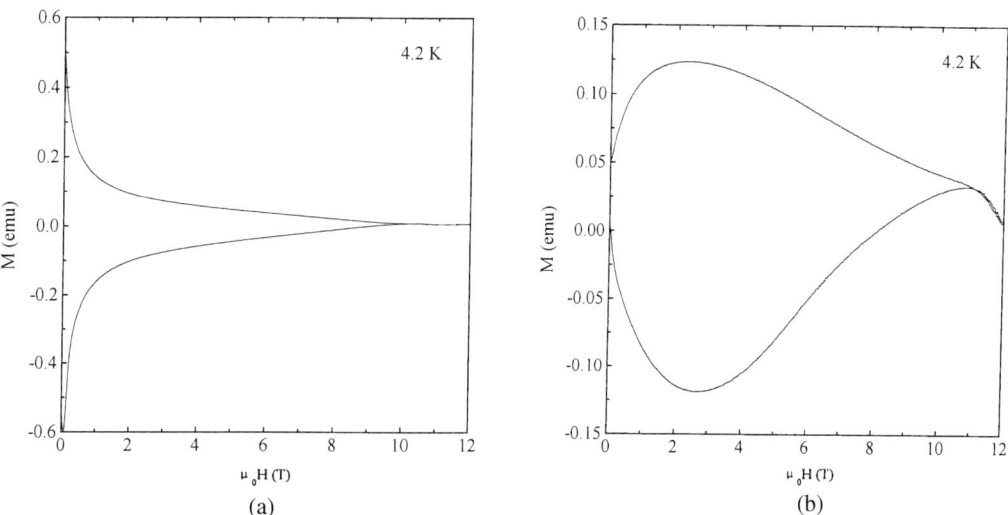

Fig. 1 Magnetization curve at 4.2 K for an NbTi sample under different directions. (a) Magnetic field perpendicular to the sample axis. (b) Magnetic field parallel to the sample axis.

The difference between the upper branch and lower branch of magnetization M–μ_0H loop in Fig. 1, i.e. $2\mu_0\Delta M$ vs. magnetic field $\mu_0 H$, was shown on a semi-logarithmic scale in Fig. 2. In Fig. 2(a), the magnetic moment decreased with the increase of magnetic field and showed similar tendencies at each temperature. In the case of parallel direction, the curve showed a different shape and there was a peak near the zero fields (Fig. 2(b)). The magnetic moment decreased with the increase of the magnetic field in the figure. If the curve was extrapolated to the x axis, the $\mu_0 H_{c2}$ at 4.2 K under parallel direction was about 11 T larger than that (10.5 T) under perpendicular direction.

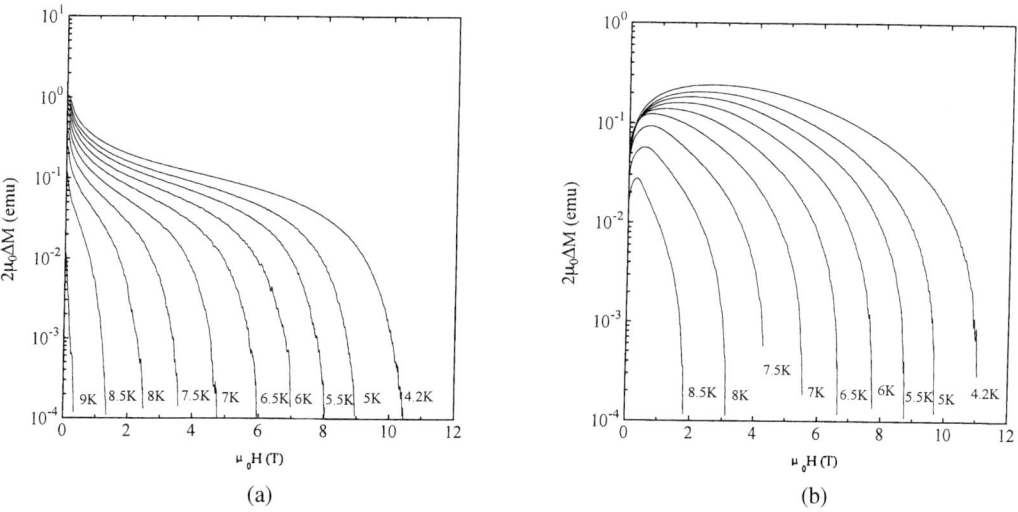

Fig. 2 $2\mu_0\Delta M$ vs. magnetic field $\mu_0 H$ curve. (a) Magnetic field perpendicular to the sample axis. (b) Magnetic field parallel to the sample axis.

It was interest to notice the turning points on the magnetization curve near the upper critical field. In the case of the perpendicular direction, three obvious transition points on the curve were noted at each temperature and the typical curve was shown in Figure 3. NbTi superconductor had a critical temperature about at 9 K and there was almost no hysteresis existing at this temperature. NbTi wire exhibited a linear relation with magnetic field as a normal conductor at 9 K. Below 9 K, the curve showed a hysteresis phenoma. At certain point, the magnetic moment did not change as a same route while the magnetic field increased or decreased. We defined this point as $\mu_0H_{irr}(T)$, i.e. an irreversible magnetic moment occurred while the magnetic field changed. μ_0H_{c2} was defined as the starting point that the magnetic moment overlapped on the line which was measured as a normal state. This point was preferable to define the transition as the first departure from superconductivity, rather than other critical fields.

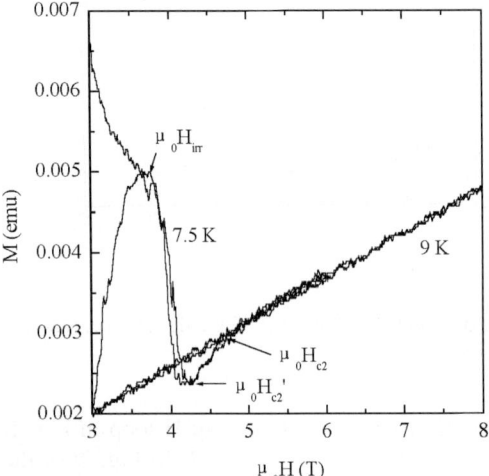

Fig. 3 Definition of turning point at perpendicular direction.

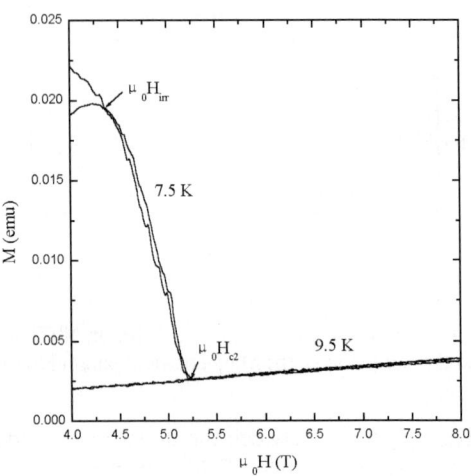

Fig. 4 Definition of turning point at parallel direction.

These two points were observed in both directions (see also in Fig. 4). Different from the case of parallel direction, another obvious turning point noted as μ_0H_{c2}' was observed in perpendicular direction case. At this point, the magnetic moment changed reversiblely but deviated the line of normal state. Among these 3 turning points, $\mu_0H_{irr} < \mu_0H_{c2}' < \mu_0H_{c2}$.

These upper critical fields μ_0H_{irr}, μ_0H_{c2} and μ_0H_{c2}' at different temperatures were plotted in Fig. 5, which all showed a linear dependence of temperature. The values under parallel direction were slightly higher than those under perpendicular direction.

The problem of definition has become more complex when the upper critical field μ_0H_{c2} was measured. The origin of the problem of definition lay in the basic heterogeneity of an optimized superconductor in which, for Nb-based alloys of very high dislocation density. It has been accepted that fluxoids penetrating into type II superconductors

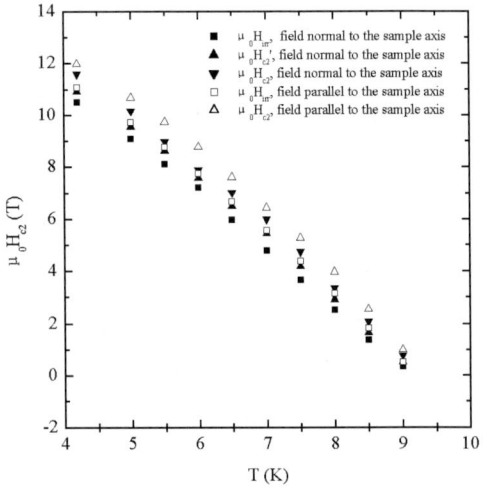

Fig. 5 Upper critical fields vs. temperature.

were pinned at structural imperfections, such as dislocations, precipitates, voids, grain boundaries and surfaces. Development of multifilamentary conductors of the high field superconductors such as NbTi has tended to proceed of fundamental studies of flux pinning interactions. The most common method of $\mu_0 H_{c2}$ determination was by a resistive measurement. In an optimized material, the resistive transition was inherently broad (0.5–0.7 T being typical),but additonal uncertainty came from the dependence of the position of the transition on the measuring critical current density for values were negligibly small (0.05–5 A/mm^2). Accroding to the critical state model, i.e. the $2\mu_0\Delta M$ was propotional to critical current I_c and effective diameter of the filament d_{eff}, the dependence of the position of the transition on magnetization mesurment was observed and $\mu_0 H_{c2}$ value measured in the experimental was appreciate for critical field while d_{eff} was not changed. The very small scale of the coherence length ξ made it difficult to characterize fully the defects which were responsible for the fundamental flux pinning interactions. The wire was fabricated by a standard process, the cold work made the imperfection along the sample longitude direction different from other direction. The nature of NbTi superconductor properties especially near critical field was determined by larger scale defects in the filamental microstructure and the pinning behavior should be different in different direction, and resulting in a different critical field.

4 Conclusion The magnetization of NbTi wire was measured with the applied magnetic field from 4.2 K to 9.5 K. Influence of field direction in perpendicular or parallel to the sample axis was checked. Upper critical field $\mu_0 H_{irr}$, $\mu_0 H_{c2}$ and $\mu_0 H_{c2}'$ were defined from magnetization vs. applied field curves in the case of perpendicular direction, while no $\mu_0 H_{c2}'$ was observed for the case of parallel direction. These upper critical fields showed a linear relation dependent on temperatures.

Acknowledgements The author appreciates very much the financial support for the STA Fellowship funding and also thanks Dr. H.Wada, Dr. K.Itoh a lot for their help of experiments and discussions at Tsukuba magnetic lab.

References

[1] David C. Larbalestier, Superconductor Materials Science: Metallurgy Fabrication and Applications, Chap. 3, edited by Simon Foner and Brian B. Schwartz, NATO Advanced Study Institutes Series, Series B, Physics, Vol. 68 (Plenum Press, New York, 1981), p. 133.
[2] M. Suenaga, A.K. Ghosh, Youwen Xu, and D.O. Welch, Phys. Rev. Lett. **66**(13), 1777 (1991).
[3] M. Dhalle, M.N. Cuthbert, L.F. Cohen et al., J. Superconductivity **8**(1), 37 (1995).
[4] H. Kumakura, K. Togano, and H. Maeda, Appl. Phys. Lett. **58**(24), 2830 (1991).
[5] Y. Murakami, K. Itoh, M. Yuyama, and H. Wada, ICMC'95 Columbus, Ohio, USA, July 17–21, 1995.
[6] K. Itoh, H. Wada, T. Ando et al., Advances in Cryogenic Engineering Materials, Vol. 36, edited by R.P. Reed and F.R. Fickett (Plenum Press, New York, 1990), p. 199.

Anomalous behavior of electrical resistivity in NdFe$_{11}$Ti

N. Tajabor[*,1], **A. Amirabadizadeh**[1,2], **M.R. Alinejad**[1], and **F. Pourarian**[3]

[1] Department of Physics, Faculty of Science, Ferdowsi University of Mashhad, Iran
[2] Department of Physics, Faculty of Science, University of Birjand, Birjand, Iran
[3] Carnegie Mellon Research Institute, Carnegie Mellon University, Pittsburgh, Pennsylvania, 15219, USA

Received 31 August 2003, accepted 31 December 2003
Published online 18 March 2004

PACS 72.15.–v, 75.30.Fv

The electrical resistivity, ϱ, and temperature derivative of resistivity ($d\varrho/dT$) of polycrystalline NdFe$_{11}$Ti are investigated. The resistivity exhibits an anomalous behavior at about 160 K and 220 K. These anomalies attributed to spin reorientation phenomena originate from two magnetic phase transition. A modified theoretical model for electrical resistivity, ϱ and $d\varrho/dT$, behavior in the spin reorientation temperature (T_{SR}) region is presented. The results suggest that the average magnetic moment at $T_{SR1} \sim 160$ K is aligned along a cone angle and at $T_{SR2} \sim 220$ K along c-axis. The temperature dependence of ac-susceptibility, $\chi_{ac}(T)$, is presented. The results show the existence of the spin reorientation related to a magnetic phase transition.

© 2004 WILEY-VCH Verlag GmbH & Co. KGaA, Weinheim

1 Introduction

REFe$_{11}$Ti crystallizes in the tetragonal structure of ThMn$_{12}$ type (I4/mmm space group, 2 formula units per unit cell) [1]. The ternary compound exhibit interesting magnetic properties. The anisotropy constant is extremely sensitive to temperature [2, 3]. As temperature increases, the easy direction of the magnetization rotates from a basal plane (or cone) to an axial direction at T_{SR}, the spin reorientation point. In most cases the total spin reorientation angle is 90°, but it can also be an incomplete reorientation; giving rise from a conical magnetic structure to c-axis. Such observations are usually obtained from a precise magnetization measurements performed on single crystals. The peak shown in the magnetization curve of NdFe$_{11}$Ti at 200 K [4] is thought to be due to a spin reorientation transition (SR) from conical to unixal anisotropy, but for a polycrystalline sample this temperature may be extended over a wide temperature range according to the sign of the second anisotropy constant (K_2) [4] (i.e. spin reorientation starts from a T_{SR1}, and completes at T_{SR2}). For polycrystalline samples, temperature-induced spin reorientation transition can be detected by performing an accurate measurement of electrical resistivity (ϱ) and the derivative of electrical resistivity with respect to temperature ($d\varrho/dT$) [5, 6]. In Section 2 we will present a theoretical model for the electrical resistivity in the spin reorientation temperature region. Earlier, a similar model was also used by Sausa et al. [5, 6]. However, they used their model for a SR transition from plane to c-axis, while this model will be modified here for our experimental results on NdFe$_{11}$Ti sample. In Section 3 we will explain the experimental results and will make a calculation to find the angle between magnetic moment and c-axis as a function of temperature.

---

[*] Corresponding author: e-mail: tajabor@ferdowsi.um.ac.ir, Phone: +98 511 8405816, Fax: +98 511 8438032

2 Theoretical model

We consider, in addition to the usual resistivity due to impurities, phonon and spin disorder scattering, the electrical resistivity to be dependent also on the angle θ between the saturation magnetization M_s and the c-axis. When this angle change near a spin reorientation transition temperature (T_{SR}) a corresponding anomaly appears in ρ and dρ/dT. Considering that ρ is an even function of θ (or sinθ) and, in general, a second-order parameter of ρ(θ) is sufficient to describe such dependence:

$$\rho(T) = \rho_n(T) + b(T)\sin^2\theta \tag{1}$$

where $\rho_n(T)$ is the resistivity contribution due to impurities, phonon and spin disorder scattering and $b(T) \propto M_s^2(T)$ [5]. For $T \ll T_c$ we can neglect the temperature dependence of M_s and taking b as a constant. From relation (1) we obtain

$$\frac{d\rho}{dT} = \frac{d\rho_n(T)}{dT} + 2b\cos\theta\sin\theta\frac{d\theta}{dT} . \tag{2}$$

Let us consider the expected behavior of $\frac{d\rho}{dT}$ near T_{SR1} and T_{SR2}. For $T < T_{SR1}$ we have θ = constant and $\frac{d\theta}{dT} = 0$, thus $\frac{d\rho}{dT} = \frac{d\rho_n}{dT}$. Just above T_{SR1} (i.e. $T = \varepsilon + T_{SR1}$) one can write

$$\cos\theta = \sqrt{C(T - T_{SR1}) + C'} \tag{3}$$

where C and C' are constants and for the rotation of spin from the basal plane to the c-axis, C' is zero. This leads to

$$\frac{d\rho}{dT} = \frac{d\rho_n}{dT} - bC \tag{4}$$

which shows that $\frac{d\rho}{dT}$ has a discontinuity at T_{SR1} given by:

$$\Delta\left(\frac{d\rho}{dT}\right)_{SR1} = -bC . \tag{5}$$

In a similar manner, for $T > T_{SR2}$ we have θ = 0 and $\frac{d\rho}{dT} = \frac{d\rho_n}{dT}$. Just below T_{SR2} (i.e. $T = T_{SR2} - \varepsilon$) we can write

$$\sin\theta = \sqrt{C''(T_{SR2} - T)} \tag{6}$$

where C'' is a constant. This leads to a discontinuity in $\frac{d\rho}{dT}$ at T_{SR2} given by:

$$\Delta\left(\frac{d\rho}{dT}\right)_{SR2} = -bC'' . \tag{7}$$

One can obtain $\Delta\rho_{ESR}$ (the effective resistivity of such polycrystalline sample at T_{SR}) from equations (5) and (7) simply by calculating the area under the peaks:

$$\Delta\rho_{ESRi} = \int_{peak} \Delta\left(\frac{d\rho}{dT}\right)dT , \; i = 1, 2 . \tag{8}$$

3 Experimental and results

Polycrystalline sample was prepared by melting the constituents Nd (99.95%), Fe and Ti (99.99%) in a water-cooled cupper boat using RF induction heating. A continuous flow of titanium-gettered argon was maintained during heating. As-cast sample was wrapped in a tantalum foil, sealed in quartz tubes, filled with ~ 1/3 atmosphere of argon gas, and annealed at ~ 1000 °C for seven days. Scanning electron microscope (SEM) and X-ray diffraction (XRD) analyses of the annealed sample show the compound was single phase. The lattice parameters obtained for the powder sample, a = 8.584 Å and c = 4.791 Å, are in good agreement with the literature [4].

Accurate resistivity measurements in zero field have been performed over a wide range of temperatures (77 – 300 K) on the sample. The resistivity study was carried out by the four probe technique using a microvoltmeter and a constant source. Pressure contacts were used to provide electrical contacts. The dimensions of the sample were close to 1×1×6 mm^3.

The temperature dependence of the low-field ac susceptibility $\chi_{ac}(T)$ was performed between 100 and 300 K using a modified commercial mutual inductance at 333.3 Hz with ac magnetic field 50 mA^{-1} peak value. The measurement was performed in small temperature intervals in order to obtain the critical behavior.

Figure 1(a) shows electrical resistivity (ρ) as a function of temperature for NdFe$_{11}$Ti compound. For temperatures below 160 K and above 220 K the electrical resistivity increases almost linearly with increasing temperature, but between ~160 K and 220 K, the slope decreases noticeably. This is due to the onset of a spin reorientation from a conical ferromagnetic structure to c-axis with the angle θ changing rapidly with temperature. Obviously, the spin reorientation transition is associated with the two minima observed in the dρ/dT near T_{SR1} and T_{SR2} (see Fig. 1(b)). To our knowledge, this is the first experimental observation of such localized critical features in the resistivity temperature derivative of NdFe$_{11}$Ti compound.

The temperature dependence of χ_{ac} for the measured compound is displayed in Fig. 2. The observed anomalies are associated with the spin reorientation phase transition. The reorientation starts at about 160 K, where slope of $\chi_{ac}(T)$ curve changes, and ends at about 220 K, namely from the onset of the spin reorientation to c-axis at 220 K.

Now, consider the special case of studied sample where the angle between M$_s$ and the tetragonal c-axis is 53 in degree about T = 160 K [7]. So, the value of C' can be obtained via Eq. (3) ($C' = 0.632$). Using relation (8) for experimentally obtained temperature dependence of $\frac{d\rho}{dT}$ in Fig. 1(b), we get the values of $\Delta\rho_{ESRi}$ (i = 1, 2) which are equal to 10.2 μΩ and 70 μΩ at 160 K and 220 K, respectively. Taking integral from both sides of Eqs. (5) and (7) and using Eqs. (3) and (6), the values of C and C'' are obtained. The calculated variation of θ as a function of temperature is shown in Fig. 3. This curve shows that the spin reorientation in the studied sample starts at temperature T_{SR1} ~ 160 K from a conical structure (with θ = 53°) and is completed at temperature T_{SR2} ~ 220 K along the tetragonal c-axis (θ = 0°).

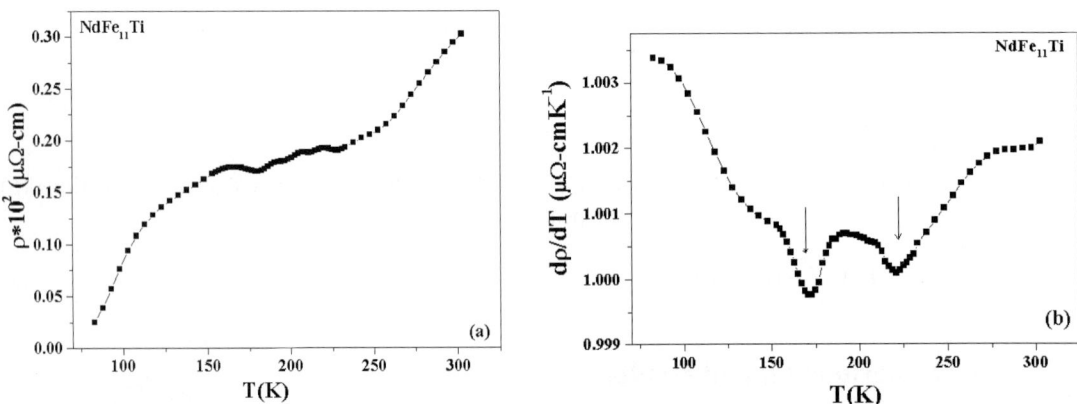

Fig. 1 (a) and (b) Electrical resistivity and temperature derivative of electrical resistivity versus temperature for NdFe$_{11}$Ti compound.

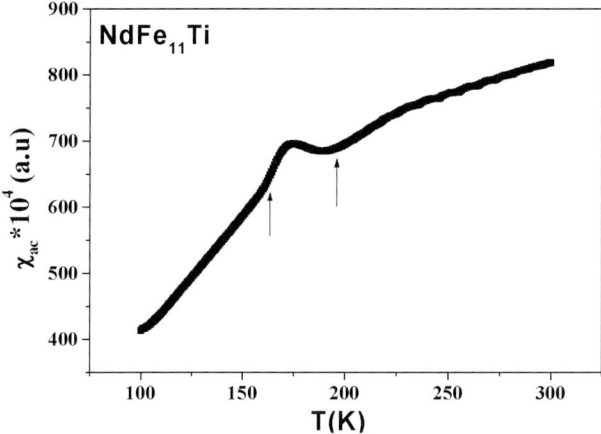

Fig. 2 ac susceptibility versus temperature for NdFe$_{11}$Ti compound.

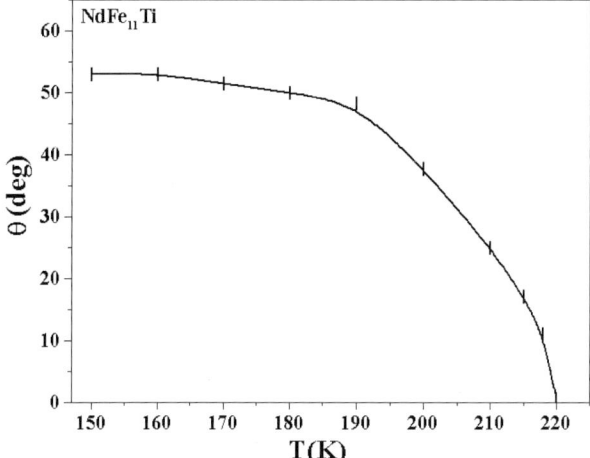

Fig. 3 Calculated temperature dependence of θ (the angle between M$_s$ and the c-axis).

References

[1] A.V. Andreev and S.M. Zadvorkin, Physica B **225**, 237 (1996).
[2] I.S. Tereshshina, S.A. Nikitin, T.I. Ivanova, and K.P. Skokov, J. Alloys Compd. **275–277**, 625 (1998).
[3] T.I. Ivanova, Yu.G. Pastushenkov, K.P. Skokov, I.V. Telegina, and I.A. Tskhadadze, J. Alloys Compd. **280**, 20 (1998).
[4] Bo-Ping Hu, Hong-Shuo Li, J.P. Gavigan, and J.M.D. Ceoy, J. Phys.: Condens. Matter **1**, 755 (1989).
[5] J.B. Sousa, J.M. Moreira, A. Del Moral, P. Algarabel, and R. Ibarra. J. Phys.: Condens. Matter **2**, 3897 (1990).
[6] J.B. Sousa, J.F.D. Montenegro, J.M. Moreira, and E.M. Braga, J. Phys. F: Met. Phys. **12**, 351 (1982).
[7] W. Liu, X.K. Sun, Z.D. Zhang, Q.F. Xiao, E. Bruck, and F.R. de Boer, J. Alloys Compd. **296**, 39 (2000).

Investigation of the garnet–perovskite transition in Nd doped YIG by means of magnetic disaccommodation

P. Hernández-Gómez[*,1], **C. De Francisco**[1], **C. Torres**[1], **J. Iñiguez**[2], **V. Raposo**[2], **J. M. Perdigao**[3], and **A. R. Ferreira**[3]

[1] Dpt. Electricidad y Electrónica, Universidad de Valladolid, Prado de la Magdalena s/n, 47071 Valladolid, Spain
[2] Dpt. Física Aplicada, Universidad de Salamanca, Pza. la Merced s/n, 37071 Salamanca, Spain
[3] Dpt. Engenharia Electrotécnica e Computadores, Universidade de Coimbra, Pinhal de Marrocos, 3030 Coimbra. Portugal

Received 31 August 2003, accepted 31 December 2003
Published online 18 March 2004

PACS 75.50.Gg, 75.60.Lr, 81.40.Rs

The garnet-perovskite transition is studied in this work with magnetic disaccommodation on polycrystalline Nd doped YIG samples $Y_{3-x}Nd_xFe_5O_{12}$ ($0 < x < 2.5$). A very different behaviour with the sintering atmosphere is observed. The results of magnetic disaccommodation for samples sintered in air reveal that the low temperature peak of YIG at 120 K disappear and another peak emerge at 300 K with $x = 2.0$, when the solubility limit is surpassed and the compound contains secondary perovskite phase and magnetite. For CO_2 sintered samples, the process at 120 K exhibit magnetic accommodation features (i.e. negative disaccommodation), and vanishes with increasing Nd substitution, together with the appearance of the 300 K peak associated with the garnet–perovskite transition which takes place with $x \approx 1.4$ doping. Thermal evolution of magnetic permeability and X-ray diffraction support the behaviour observed.

© 2004 WILEY-VCH Verlag GmbH & Co. KGaA, Weinheim

1 Introduction

The study of magnetic relaxations, which is technologically important to minimize losses, provides also information about the underlying mechanisms governing the dynamic behaviour of Bloch walls [1]. Measurement of magnetic disaccommodation is a sensitive way to study magnetic relaxations in ferrites. It consists in the time variation of the mobility of domain walls after a magnetic shock, and is shown by a temporal evolution of the magnetic permeability after a demagnetization stage. The origin of this relaxation phenomenon, which has been observed both in cubic spinel ferrites [1], hexaferrites [2] and garnets [3], has been attributed to either the rearrangement or the diffusion of anisotropic point defects (lattice vacancies, interstitials) within the Bloch walls [1], and the relaxation time which characterizes each relaxation process is strongly temperature-dependent. Yttrium iron garnets (YIG) are excellent ferrimagnetic materials for use in microwave devices due to their good properties – high electrical resistivity and narrow resonance linewidth – at high frequencies together with the possibility of tailor their properties by introducing dopants [4]. Usually the tailoring is made by substitution of cations in tetrahedral or octahedral sites. The substitution of part of yttrium ions in dodecahedral sites with a rare earth element has been studied so far [5–7], and it has been established that in the garnet lattice total substitution of Y by lighter 4f rare earth elements La, Ce, Pr and Nd is not possible, and that above the solubility limit a secondary phase perovskite forms together with the magnetic garnet. In this paper we investigate the garnet–perovskite transition in Nd doped YIG by using magnetic disaccommodation techniques.

---
[*] Corresponding author: e-mail: pabloher@ee.uva.es, Phone: +34 983 423895, Fax: +34 983 423225

2 Experimental setup

For this work, polycrystalline garnet samples with different Nd doping rate $Y_{3-x}Nd_xFe_5O_{12}$ (0 < x < 2.5) have been prepared by means of standard ceramic techniques. The mixtures were prepared with high purity starting oxides Fe_2O_3, Y_2O_3 and Nd_2O_3, mixed with the appropriate molar ratio in agate mortar for 1 h., pressed at 10000 Kg/cm^2 in cylindrical form (diameter: 5 mm, length: 15 mm), and sintered at temperatures in the 1380 °C < T < 1420 °C range in air or CO_2 sintering atmospheres. The samples were rapidly quenched to avoid phase annealing and to provide the presence of crystal vacancies.

Magnetic disaccommodation measurements were carried out with a computer aided system based on a LCR bridge [8], in the 77 K < T < 400 K temperature range. The results have been represented as isochronal curves, i.e, the relative variation of the initial permeability after sample demagnetization between an initial time t_1 = 2 s and different window times t_2 = 4, 8, 16, 32, 64 and 128 s in the form

$$\frac{\mu(t_1,T)-\mu(t_2,T)}{\mu(t_1,T)}(\%). \tag{1}$$

When the time window (t_2–t_1) is of the same order of magnitude that the relaxation time at a specified temperature, this curve exhibits a maximum. In this way, isochronal spectra disclose the different aftereffect processes in the temperature range tested.

3 Results and discussion

X-ray diffraction patterns of representative samples are shown in Fig. 1. Thermal evolution of magnetic permeability and isochronal disaccommodation spectra for air and CO_2 sintered samples are represented in Figs. 2a and 2b, and 3a and 3b, respectively. The garnet–perovskite transition can be observed in all the experimental results. XRD for air sintered samples (Figs. 1a to 1c) reveals the appearance of additional lines, related to $NdFeO_3$ distorted perovskite and magnetite, once reached the solubility limit, which in our case takes place with x = 2.0 doping rate. The calculated garnet lattice parameter *a* obtained for our samples, increase for 12.405 Å for x = 0.6 to 12.525 Å for x = 2 and 12.529 Å for x = 2.4, similar to the literature data [5,7], and which indicate the nearly lineal increase of unit cell with doping rate until the solubility limit is reached, which following [7] takes place with a substitution rate x = 1.9. The transition is also evident in the thermal evolution of magnetic permeability: in Fig. 2a we can observe that Nd doped YIG behave in three different manners depending that the doping rate lies in the range 0 < x < 1, 1 < x < 2 or x > 2. In the latter case the magnetic permeability falls down to a very low value, which is due to the formation of substantial amount of antiferromagnetic perovskite (see Fig. 1c). Finally, the isochronal disaccommodation spectra obtained for air sintered samples (Fig. 3a) show a low temperature disaccommodation process at around 120 K characteristic of yttrium magnetic garnets, called II peak in [3]. This peak is due to Fe^{3+}–Fe^{2+} electronic hopping relaxation mechanism in octahedral sites of garnet lattice, and its peak temperature, and hence the activation energy of the relaxation process, decrease with the Nd-induced increase of lattice unit cell. The garnet–perovskite transition is observed by the strong decrease of this relaxation process, which finally disappears, due to the small amount of garnet formed compared with perovskite and magnetite. Another disaccommodation peak at room temperature (300 K) also reveals that the solubility limit has been reached with x = 2.0. This relaxation process is observed due to the presence of magnetite in the sample and their origin is caused by ionic reorientation of anisotropic ferrous cations in octahedral sites of spinel lattice.

In CO_2 sintered YIG (Fig. 3b) we can observe accommodation processes (i.e. negative disaccommodation) at low temperatures. This fact is due to the existence of time dependent wall resonance effects in garnets, at frequencies close to that employed in our measurement device (1 kHz) [9], so that the time evolution of magnetic permeability can increase after demagnetization and hence accommodation can be observed. The peak also shifts to lower temperatures, and the 300 K peak characteristic of the presence of magnetite in the samples, and hence indicating that the solubility limit has been reached, appears with $x = 1.6$, a lower doping rate than air sintered samples. For higher substitution rates the low temperature garnet-related process disappears and it can be observed only the room temperature peak, due to the great amount of perovskite and magnetite formed. The thermal dependence of magnetic permeability (Fig. 2b) support this conclusion. We observe also three steps, depending that $x < 1.6$, $x = 1.6$ or $x > 1.6$, indicating, respectively, pure garnet formation, onset of secondary perovskite and magnetite formation, and substantial amount of perovskite and magnetite in the sample (and hence strong diminution of magnetic permeability). XRD obtained for the CO_2 sintered sample with $x = 1.6$ doping rate (Fig. 1c) confirms that the solubility limit has been surpassed, as the intensities of $NdFeO_3$ perovskite are comparable to that the corresponding of YIG (whose lattice parameter calculated is $a = 12.465$ Å). Taking into account the value for air atmosphere [7] and comparing the relative intensities of Figs. 1b and 1d, the most probable value for this limit in CO_2 is $x \approx 1.4$, but additional samples are needed to check it. We can conclude that the solubility limit in this system decreases with reducing atmospheres, probably due to a higher oxygen vacancy amount [9] that favours secondary phases formation. This fact also opens the possibility of increase the Nd doping rate by increasing oxygen partial pressure, for which further research is neccesary.

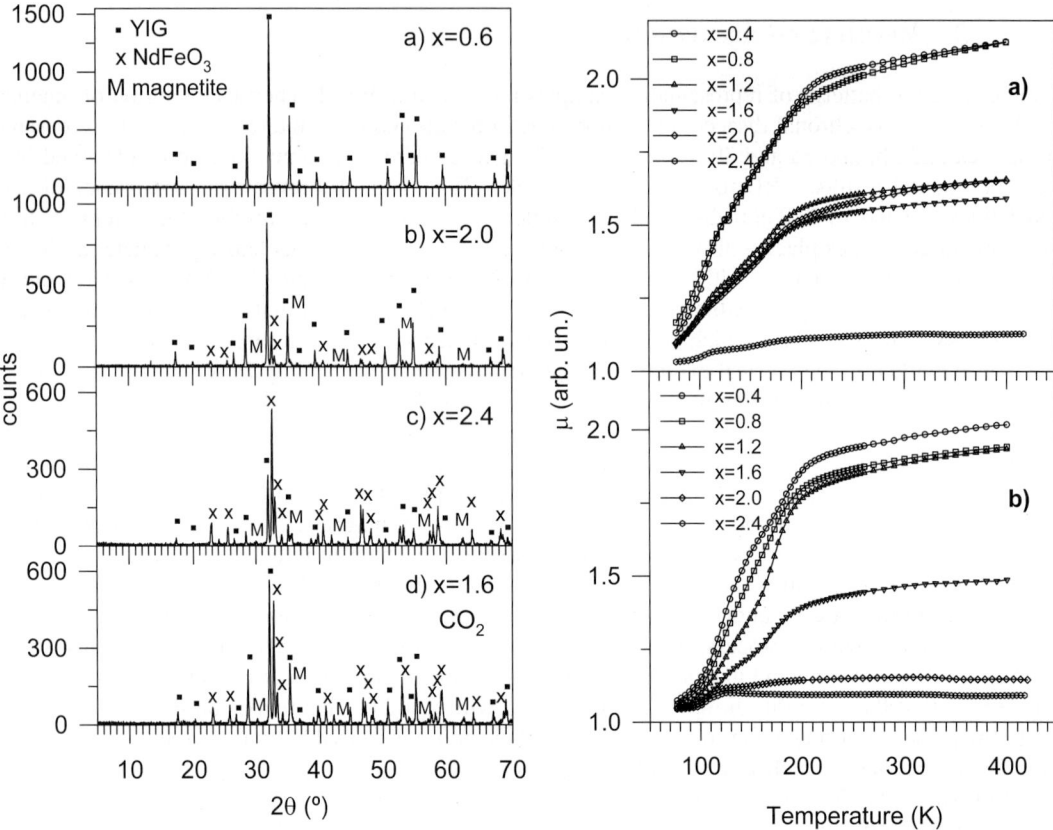

Fig. 1 XRD patterns of $Y_{3-x}Nd_xFe_5O_{12}$ polycrystalline samples a) $x = 0.6$, b) $x = 2.0$, c) $x = 2.4$ and d) $x = 1.6$ (sintered in CO_2).[■: YIG; x: $NdFeO_3$; M: magnetite].

Fig. 2 Thermal evolution of magnetic permeability of $Y_{3-x}Nd_xFe_5O_{12}$ polycrystalline samples sintered in a) air and b) CO_2.

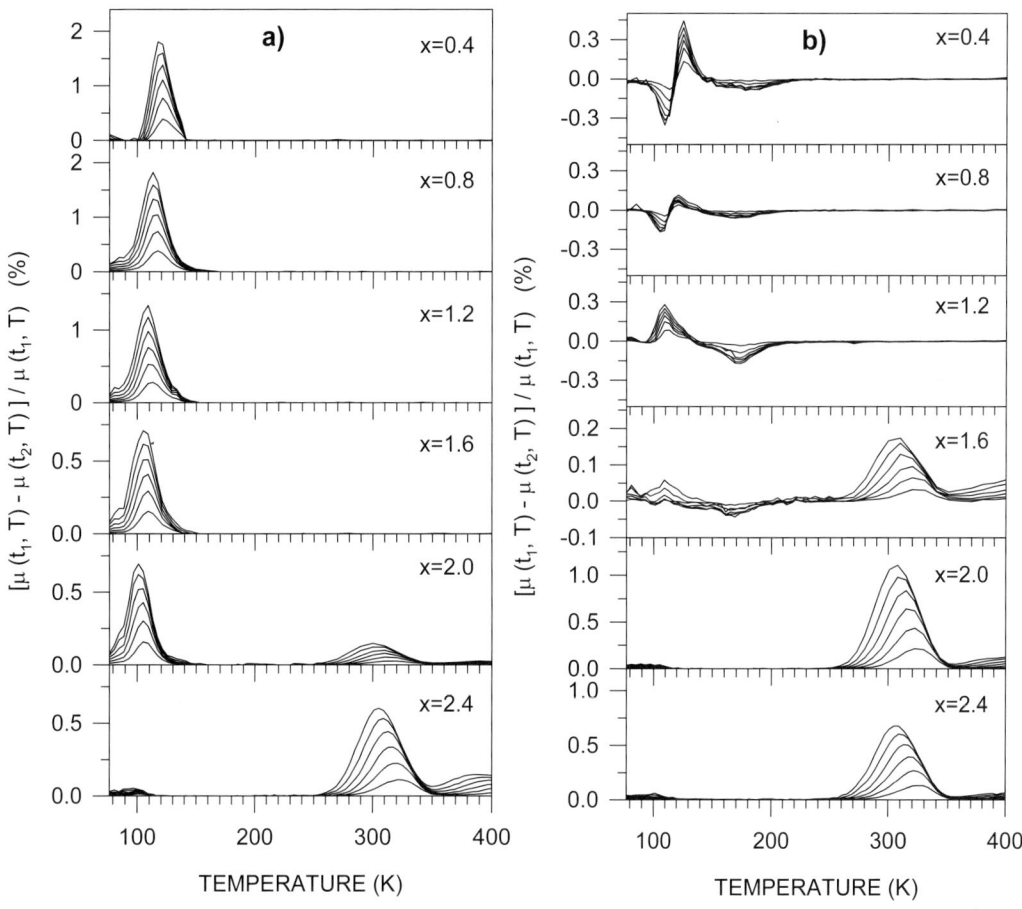

Fig. 3 Isochronal disaccommodation spectra of $Y_{3-x}Nd_xFe_5O_{12}$ polycrystalline samples sintered in a) air and b) CO_2 sintering atmospheres.

Acknowledgements This work has been partially supported by MCYT "Acciones Integradas Hispano-Lusas" and Junta de Castilla y León, project numbers HP2002-0014 and SA/010/03 resp.

References

[1] F. Walz, V.A.M. Brabers, S. Chikazumi, H. Kronmuller, and M.O. Rigo, phys. stat. sol. (b) **110**, 471Q (1982).
[2] P. Hernández-Gómez, P.G. Bercoff, C. de Francisco, J.M. Muñoz, O. Alejos, C. Torres, and H.R. Bertorello, J. Appl. Phys. **87**(9), 6250 (2000).
[3] L. Torres, F. Walz, J. Iñiguez, and H. Kronmüller, phys. stat. sol. (a) **159**, 485 (1997).
[4] M. Pardavi-Horvath, J. Magn. Magn. Mater. **215–216**, 171 (2000).
[5] J. Loriers and G. Villiers, Compt. Rend. Acad. Sci. **252**, 1590 (1961).
[6] H. Pascard, J. Magn. Magn. Mater. **15–18**, 1313 (1980).
[7] S. Geller, H.J. Williams, and R.C. Sherwood, Phys. Rev. **123**(5), 1692 (1961).
[8] C. de Francisco, J. Iñiguez, J.M. Muñoz, and J. Ayala, IEEE Trans. Mag. **23**, 1866 (1987).
[9] C. Torres, O. Alejos, J. M. Muñoz, P. Hernández-Gómez, and C. de Francisco, Phys. Rev. B **66**, 024410 (2002).

Magnetism of iron in face-centered cubic 4d metals

M. Elzain*, A. Al Rawas, A. Yousif, A. Gismelseed, A. Rais, I. Al Omari, and H. Widatallah

Physics Department, College of Science, Box 36, Sultan Qaboos University, Al Khod 123, Oman

Received 31 August 2003, accepted 31 December 2003
Published online 18 March 2004

PACS 75.30.–m, 75.50.Bb

The magnetic moments and hyperfine fields at iron sites embedded in Rh, Pd and Ag face centered cubic structures were calculated using the first principle discrete variational method (DVM) and the full-potential linear-augment plane wave (FP-LAPW) method. In DVM the systems were represented by, clusters of atoms, while in FP-LAPW supercells were used. The objectives of this work are to compare and contrast results from the two different computational methods in addition to comparison to experimental data. Large magnetic moments were obtained for iron in Pd, relatively smaller moments for iron in Ag and smaller moments for iron in Rh. Iron atoms were found to couple ferromagnetically to Pd atoms and antiferromagnetically to Rh. No moment is induced on the Ag atom.

1 Introduction Iron forms solid solutions as well as ordered alloys with the 4d elements rhodium and palladium, while it is completely immiscible in silver. The equiatomic FeRh composition has an ordered B2 (CsCl) structure. Above room temperature it undergoes a transition from antiferromagnetic (AFM) to ferromagnetic (FM) phase. Intense ball milling and plastic deformation result in a disordered FCC phase [1, 2]. The rhodium-rich FCC phase exhibits competing ferro- and antiferromagnetic local spin configurations behaving as a spin glass [3]. Introduction of Fe into Pd results in a giant magnetic moment, where a large atomic moment of Fe is maintained in addition to the induced ferromagnetic polarization of the surrounding Pd atoms ([4] and references therein). Various ordered PdFe alloys crystallize, depending on the concentrations of the two alloying components. On the hand, since Fe and Ag are immiscible, supersaturated out of equilibrium alloys can only be prepared by, vapor quenching, mechanical alloying or implantation [5]. At low temperatures AgFe alloys have a spin-glass phase. As the concentration of Fe varies, the alloys undergo transitions to paramagnetic or ferromagnetic phases with increasing temperature [6].

On the theoretical side the electronic and magnetic properties of Fe in the face-centered Rh, Pd and Ag were calculated using the density functional approach [4, 7–9]. The equiatomic BCC FeRh system received more attention because of its peculiar features. In both AFM and FM phases Fe maintains a large magnetic moment of order $3\mu_B$, whereas the moment on Rh is about $0.8\mu_B$ in the FM phase and zero in the AFM phase [10]. The local magnetic moment at Fe site in Pd, obtained using KKR-Green function calculation, is $3.47\mu_B$. The magnetic moment at Pd neighboring site was found to be of order $0.1\mu_B$ [4]. Nogueira and Petrilli [9] calculated the magnetic moments and hyperfine fields at Fe sites in Ag using the real-space linear muffin-tin orbital formalism, within atomic sphere approximation. The single Fe impurity moment of $3.07\mu_B$ was found to decrease as the Fe impurities start to interact. A magnetic hyperfine field of –10.9 T was obtained for the isolated Fe impurity.

The objective of the present work is to compare and contrast results of the calculation of the local magnetic moments and hyperfine field at Fe isolated impurities in Rh, Pd and Ag using two different

---

* Corresponding author: e-mail: elzain@squ.edu.om

approaches: full real-space approach and full k-space approach. In the real-space discrete variational method (DVM), the systems are represented by, small clusters of 55 atoms. The Kohn–Sham equation is solved using linear combinations of atomic orbitals (LCAO) as basis functions [11 and references therein]. In DVM the local contact hyperfine field is split into core and valence components. The valence component is directly calculated, whereas the core component is assumed to be proportional to local d-magnetic moment [11]. The constant of proportionality is taken here as -11 T/μ_B. In the full-potential linear-augmented-plane-waves formalism (FP-LAPW), the augmented-plane-waves + local oribitals are used as basis function. The Kohn–Sham equations are solved using the WIEN2k code of the Vienna Group [12]. Supercells of 32 atoms are used for the calculations of the local properties of Fe impurities in Rh, Pd and Ag.

The isomer shift at an Fe site in a system is proportional the difference in contact charge densities at Fe sites in α-Fe and in the system. The constant of proportionality is -0.25 mm/s per unit charge density in atomic units [11].

We found that the results for Fe in Pd and Ag using both approaches are in good agreement in all respects. However, different values were obtained for the local magnetic moment of Fe in Rh.

2 Results and discussions We have calculated the local properties at Fe site in α- and expanded γ-Fe (Table 1), in addition to local properties at Fe site in equiatomic BCC FeRh and FCC Rh, Pd and Ag (Table 2). The local moments obtained from DVM are the sums of d, s and p contributions, whereas the moments quoted for FP-LAPW are those within the muffin-tin sphere.

In Table 1 we note that the local moments in the BCC phase calculated through the two methods, are almost equal. In the case of the expanded γ-Fe, the values obtained using the two methods are in agreement for $a = 7.36$ and $a = 7.73$ au. However, for $a = 7.18$ au the DMV gives a relatively smaller value. We note that the DMV gives an AFM solution for $a = 7.0$ au, while a FM solution is obtained by FP-LAPW. This could be attributed to the number of atoms used in each calculation, where 55 atoms are used in DVM, while a one-atom primitive cell is used in FP-LAPW. Indeed the result obtained by, DVM using a 19-atom cluster gives a larger moment.

The isomer shifts, relative to α-Fe, are in reasonable agreement between the two methods. The contact hyperfine fields obtained using the DVM are in general large because of the large contribution resulting from the valence component. Apart from these larger DVM values, the trends of the hyperfine fields are satisfactorily in agreement.

The Fe magnetic moments for the BCC ordered equiatomic FeRh ferromagnetic alloy shown in Table 2 are identical and the hyperfine fields are combatable. The isomer shifts are zero and are in close agreement with experiment [2]. The same is true for Fe in Pd and to a reasonable degree for Fe in Ag. However, the results for Fe in Rh are in disparity exhibiting the trends discussed above for the expanded γ-Fe with $a = 7.18$ au (the lattice constant of Rh). We note that Fe magnetic moment calculated using DVM is in agreement with the experimental value quoted in Ref. [3].

When comparing the results shown in Tables 1 and 2, we observe that the values of the local moment for Fe in expanded γ-Fe with $a = 7.73$ au and in Ag are close, while the hyperfine fields are very different. The difference between the isomer shifts is negligible. This indicates that the d–d interaction is week in this expanded γ-Fe as well as in Ag. On other hand the s–d interaction is strong and in the case of the expanded γ-Fe it leads to a negative contribution to the hyperfine field and positive contribution for Fe in Ag. The total charge density is slightly affected.

The electronic d–d interactions of Fe in Rh and Fe in Pd are strong leading to changes in the magnetic moments and to charge transfer. We propose that the d electrons of Pd interact more with Fe majority electron and less with the minority electrons leading to the observed large increase in the Fe magnetic moment accompanied with a loss of about 0.2 d-electrons from Fe to Pd. This also induces positive moments at the Pd sites. On the other hand the Rh d-electrons interact comparably with Fe majority and the minority electrons giving the same d-moment as that for expanded γ-Fe. The observed increase in Fe local moment is due to sp contribution. Indeed, the components of local moments at Fe site in Rh are $\mu_d = 2.04\mu_B$ and $\mu_{sp} = 0.20\mu_B$, whereas the corresponding moments for expanded γ-Fe are $\mu_d = 2.00\mu_B$ and

$\mu_{sp} = -0.02\mu_B$, respectively. Consequently, negative moments are induced on Rh atoms since the Fe minority d-states extend more into the Rh sites. The hyperfine field is reduced because the induced moments at Rh and Pd sites are small and not enough to generate negative contributions to the hyperfine field through s–d interaction. The isomer shifts do not change very much except for Fe in Rh as compared to the corresponding values of expanded γ-Fe with Rh lattice constant.

Table 1 The local magnetic moment (μ_{tot}), the isomer shift (IS) relative to α-Fe and the hyperfine field (B_{hf}) at Fe site BCC and expanded FCC with lattice constants (a) corresponding to those of Rh, Pd and Ag obtained using method (1) DVM and (2) FP-LAPW formalisms.

	BCC		FCC					
a (au)	5.417		7.18		7.36		7.73	
method	1	2	1	2	1	2	1	2
μ_{tot} (μ_B)	2.30	2.27	1.98	2.73	2.75	2.86	2.88	3.01
IS (mm/s)	0.0	0.0	0.16	0.15	0.24	0.23	0.39	0.34
B_{hf} (T)	−39	−32	−35	−39	−41	−38	−42	−37

Table 2 The local magnetic moment (μ_{tot}), the isomer shift (IS) relative to α-Fe and the hyperfine field (B_{hf}) at Fe site in BCC ferromagnetic equiatomic FeRh and in FCC Rh, Pd and Ag obtained using method (1) DVM and (2) FP-LAPW formalisms.

	BCC		FCC					
system	FeRh	FeRh	RhFe		PdFe		AgFe	
method	1	2	1	2	1	2	1	2
μ_{tot} (μ_B)	3.19	3.19	2.22	2.94	3.40	3.43	2.93	3.07
IS (mm/s)	0.0	0.0	0.02	0.12	0.23	0.23	0.49	0.42
B_{hf} (T)	−21	−17	−19	−14	−18	−18	−16	−11

3 Conclusion We have calculated the local magnetic moments, hyperfine fields and isomer shifts at Fe sites in Rh, Pd and Ag using the first principle DV and FP-LAPW methods. The results in general agree with experiment. It was found that the two methods give close results for Fe in Pd and Ag and different results for Fe in Rh. For the latter the magnetic moment calculated using DVM agrees more with experimental data. When comparing results with the corresponding expanded γ-Fe results, we found that in the case of Fe in Ag the local moments and isomer shifts are similar to those in the expanded γ-Fe with lattice constant of Ag, whereas the hyperfine fields differ. No magnetic moment is induced on the Ag atoms. On the hand, the magnetic moments at Fe in Rh and Pd were found to increase relative to that of the corresponding expanded γ-Fe. The Pd atoms couple ferromagnetically to Fe acquiring positive moment while the Rh atoms couple antiferromagnetically and acquire negative moment. These observations were explained in terms of the differences in the strength of the electronic d–d and s–d interactions.

Acknowledgements This work is supported by the Sultan Qaboos University research grant number IG/SCI/PHYS/02/01.

References

[1] V. Kuncser, M. Rosenberg, G. Principi, U. Russo, A. Hernando, E. Navarro, and G. Filoti, J. Alloys and Compounds **308**, 21 (2000); M Rosenberg, V Kumar, O Crisan, A. Hernando, E. Navarro, and G. Filot, J. Magn. Magn. Mater. **177–181**, 135 (1998).
[2] L. S. Peng and G. Collins, Mater. Res. Soc. Symp. Proc. **481**, 631 (1998).
[3] V. P. Parfenova, N. N. Delyagin, A. L. Erzinkyan, and S. I. Reyman, phys. stat. sol. (b) **214**, R1 (1999).
V. P. Parfenova, A. L. Erzinkyan, N. N. Delyagin, and S. I. Reyman, phys. stat. sol. (b) **228**, 731 (2001).

[4] V. A. Gubanov, A. I. Liechtenstein, and A. V. Postnikov, Magnetism and Electronic Structure of Crystals, Springer Series in Solid-State Sciences 98 (Springer-Verlag, 1992), p. 125.
[5] M. A. Morales, E. C. Passamani, and E. Baggio-Saitovitch, Phys. Rev. B **66**, 144422 (2002).
[6] V. Manns, B. Scholz, W. Keune, K. P. Schletz, M. Braun, and E. F. Wassermann, J. de Physique Colloque C **8** suppl. 12, 1149 (1988)
[7] H. Moon, W. Kim, S. Oh, J. Park, J. G. Park, E. Cho, J. Lee, and H. Ri, J. Korean Phys. Soc. **36**, 49 (2000)
[8] Y. Shi, D. Qian, G. Dong, and D. Wang, Phys. Rev. B **65**, 172410 (2002).
J. Khalifeh and B. Hamad, Physica B **321**, 230 (2002)
[9] R. Nogueira and H. Petrilli, Phys. Rev. B **60**, 4120 (1999)
[10] S. Lounis, M. Benakki and C. Demangeat, Phys. Rev. B **67**, 94432 (2003)
[11] M E Elzain, D E Ellis and D Guenzberger Phys. Rev. B **34**, 1430 (1986).
[12] P. Blaha, K. Schwarz, G. K. H. Madsen, D. Kvasnicka, and J. Luitz, WIEN2k, An Augmented Plane Wave + Local Orbitals Program for Calculating Crystal Properties (Karlheinz Schwarz, Tech. Universitat Wien, Austria), 2001. ISBN 3-9501031-1-2.

Coercive properties of epitaxial magnetic garnet films after heat treatment in reducing atmosphere

G. Vértesy[*,1] **and I. Tomáš**[2]

[1] Research Institute for Technical Physics and Materials Science, Hungarian Academy of Sciences, 1525 Budapest, P.O.B. 49, Hungary
[2] Institute of Physics, Academy of Sciences of the Czech Republic, Na Slovance 2, 18221 Praha, Czech Republic

Received 31 August 2003, accepted 31 December 2003
Published online 18 March 2004

PACS 75.50.Bb, 75.60.–d, 75.70.Ak, 81.40.Rs

The domain wall pinning field of epitaxial magnetic rare earth garnet film was modified in certain temperature range by annealing in a reducing atmosphere at temperature as low as 600 °C. The result is analyzed on the assumption of existence of two mutually independent wall-pinning sets of material defects. The heat treatment had no influence on the *anisotropy* of the sample and on the characteristic *periods* of the defect distributions in any of the sets, while at the certain temperature range (between 100 and 300 K) the pinning *strength* of the defects in one set was increased. The result supports the assumed model of coercivity and shows that by an appropriate processing of the sample some of the wall-pinning sets can be modified independently on the others.

© 2004 WILEY-VCH Verlag GmbH & Co. KGaA, Weinheim

1 Introduction

Epitaxial magnetic garnet films belong to the most perfect single-crystalline materials, as their growth by liquid-phase epitaxy (LPE) technology and the control of their magnetic properties are very well established methods. These films have several application possibilities and they play an important role in the study of basic magnetization processes.

The coercivity is commonly considered to be one of the most important properties in applied magnetism as it is connected with the fundamental magnetic parameters of the material such as magnetization, anisotropy and exchange constant. On the other hand, it also characterizes the structural nonuniformities (defects) of the material via their interaction with the magnetization vector and with the domain walls. The domain walls act as probes of the defects, and the value of the measured domain wall pinning field contains information on the distribution and quality of the defects. The knowledge of the temperature dependence of coercive properties is very important, because it mediates an insight into the properties of the wall-pinning centers, and makes it possible to identify the pinning centers with the known defects.

Value of the domain wall pinning field, H_{cp}, in highly anisotropic, epitaxially grown magnetic rare-earth garnet films increases very rapidly when temperature, T, is decreased. This steep temperature dependence of H_{cp} was analyzed previously and it was found that $H_{cp}(T)$ can be described by a piece-wise exponential curve [1]. The quantitative analysis revealed the shape of the

$$H_{cp}(T) = H_A(T) \, \alpha \, f(\delta(T)/d) , \qquad (1)$$

to be caused by the exponential temperature dependence of the anisotropy field, $H_A(T) = H_{A0}\exp(-T/T_0)$, where H_{A0} and T_0 are constants, and modified by functions $\alpha_i f(\delta/d_i)$, describing efficiency of the wall-defect interactions, usually at least for two different sets ($i = 1, 2$) of defects, see [2] and [3]. Here α_i is

---

[*] Corresponding author: e-mail: vertesyg@mfa.kfki.hu

the relative strength (amplitude) and d_i the effective period of the i-th set of defects and $\delta = \pi(A/K)^{1/2}$ (A and K are the exchange and the uniaxial anisotropy constants, respectively) is the domain wall width. The typical characteristic periods d_1, d_2 of the two sets of defects were determined and the defects were identified in the first case with point nonuniformities caused by local fluctuation (α_1) of the anisotropy constant at sites of individual rare earth ions in the crystal lattice and in the other case with fluctuation of local stresses (and through the stress fluctuation (α_2) of the local anisotropy constant again) due to local variability of the material composition, see [4].

It was shown lately [3] that annealing up to 1200 ^{0}C of the garnet films in *air* atmosphere did *not* change the *pinning field*, $H_{cp}(T)$, of the samples in the whole investigated range of temperature (20 K ≤ T ≤ 300 K). At the same time, however, the $H_A(T)$ curves were observed to *decrease* as a whole more and more, as soon as the annealing temperature was more and more increased above 800 ^{0}C. The whole series of measurements was analyzed in [3] and the surprising parallel decrease of the $H_A(T)$-curves and the unchanged values and shape of the $H_{cp}(T)$-curve were explained by the contemporary changes of the efficiency of the wall-defect interactions, due to modifications of the α_i and d_i parameters of the sets of defects existing in the samples.

The aim of the present work is to study the influence of a similar annealing of a similar garnet sample, this time in a *reducing atmosphere*, however. It will be shown that in contrast to the experiment discussed in [3], annealing in such atmosphere at temperature as low as 600 °C *does modify* shape of the $H_{cp}(T)$-curve and is able almost to double its values at a certain temperature range. It will be shown also, that this annealing does *not* change the $H_A(T)$-curve at all, however. Conclusions made from these facts will be suggested.

2 Experimental results Results of measurements are presented, which were performed on a thin magnetic garnet film with the nominal composition $(YSmCa)_3(FeGe)_5O_{12}$ and thickness $h = 5.3$ µm. The film was grown by liquid phase epitaxy on (111)-oriented gadolinium gallium garnet substrate (GGG) from the traditional $PbO-B_2O_3$ melt–solution system [5]. This film has no compensation temperature. Its Néel temperature is 473 K. The sample has a small cubic anisotropy upon which a much larger uniaxial one with its easy axis of magnetization perpendicular to the sample plane is superimposed.

The saturation magnetization $4\pi M_s$, uniaxial anisotropy field H_A, zero field stripe domain period p_o and domain wall pinning field H_{cp} were experimentally determined. Two basic experimental techniques were used for the measurement of the magnetic parameters.

In the temperature range below 300 K the parameters were measured in a PAR vibrating sample magnetometer Model 155 with a helium cryostat. The magnetic contribution of the paramagnetic GGG substrate was always subtracted from the net measured magnetic moment. The saturation magnetization was determined from the major hysteresis loop measured with the external field applied along the easy axis. The absolute error in the measurement of the sample magnetic moment did not exceed ±0.5 mT (the relative error is lower than 1%), but the probable inaccuracy of the saturation magnetization values presented here is about 8%, which is predominantly due to the poorer precision in determination of the magnetic film volume. The value of the uniaxial anisotropy field, H_A, was determined from differences between the middle slopes of major loops, measured with the external field, H, oriented in the parallel and perpendicular directions with respect to the easy axis. It was assumed that the finite slope of the hysteresis loop for H parallel to the easy anisotropy axis is caused by the finite demagnetizing factor of the sample. The uniaxial anisotropy field was determined from the difference between slopes of the loops, $\Delta\chi$, close to the origin of coordinates, and from the m value of the saturation magnetic moment of the sample using the expression $H_A = m/\Delta\chi$. We estimate that the uncertainty in the magnitude of H_A did not exceed 3%. The domain wall pinning field, H_{cp}, was determined from the half-width of minor loops, which characterize the coercive properties of domain wall motion [6]. These loops were recorded after demagnetization of the sample by an AC magnetic field with its amplitude gradually decreasing to zero. This procedure was used to ensure that we always started from the domain structure with the lowest energy anhysteretic equilibrium period, p_o. The domain wall pinning field was determined with the accuracy of ±0.2 Oe.

In the temperature range between 80 and 500 K magnetooptical methods were used. The sample holder, isolated from its surroundings by vacuum, can be cooled by a stream of evaporated liquid nitrogen and can be heated by an electric current. Magnetizing coils generating AC and DC fields are built into the sample holder. The beam of the He–Ne laser used for measurements is able to pass through the system. The zero field stripe domain period, p_o, was measured using the first-order diffracted light from the domain pattern [7], with the accuracy of 2%. The saturation magnetization, $4\pi M_s$, was calculated using the measured value of zero field stripe domain period, the bubble domain collapse field, and the film thickness [8]. The uncertainty of this value is estimated to be 5%. The domain wall pinning field was measured using the stripe domain-wall oscillation method [9]; that is, the amplitude of an AC field oriented perpendicular to the sample plane was increased linearly from zero, and the field H_{cp} – when the domain walls started to move – was detected photo-electrically. The error of the measurement is ±0.05 Oe. The magnetooptical measurement technique of the uniaxial anisotropy field, H_A, involves application of a large DC field and a small AC field parallel and perpendicular to the sample plane, respectively [10]. Polarized light passing through the sample is modulated by the Faraday rotation of the sample and it is detected at the modulation frequency by a lock-in amplifier. The observed signal is proportional to the transverse susceptibility of the film and the anisotropy field is determined from the behavior of the susceptibility in the high field region. This measurement was done in a small-size electromagnet, placed on the optical bench, cooling or heating the sample was done by cold or warm nitrogen flow. Uncertainty of H_A is 6%.

The results of the two types of measurements (one performed in the vibrating sample magnetometer, the other using the magnetooptical method) were compared with each other. In a wide range of temperature (between 80 and 300 K), the above mentioned parameters were measured by both methods on the same sample. All the measured values were identical within 2% in the whole overlapping temperature range.

The thickness of the magnetic film was measured at room temperature and this value was used for the calculations at all temperatures. The linear thermal expansion coefficient for yttrium iron garnet is $\alpha_T = 10.35 \times 10^{-6}$/K [11], so that the changes in thickness, in the whole range of the temperature, due to the thermal expansion, are less than 1% and do not cause any observable error in the magnetic quantities.

The sample was annealed in 15% H_2 and 85% N_2 atmosphere at 600 ^{0}C temperature, for sixty minutes, and the magnetic parameters were compared for the as-grown and for the annealed state of the sample.

The temperature dependence of the saturation magnetization is shown in Fig. 1, for both as-grown and annealed cases. Figs. 2 and 3 show the temperature dependence of the uniaxial anisotropy field and of the domain wall pinning field, respectively, again for both as-grown and annealed cases.

3 Discussion

The applied heat treatment had no influence either on the saturation magnetization or on the uniaxial anisotropy field (see Figs. 1 and 2), but it modified the temperature dependence of the pinning field, Fig. 3. However, only one part of the $H_{cp}(T)$ curve was changed (mainly in the temperature range between 100 and 300 K), the other, namely the low temperature range, remained unchanged, as it is seen in Fig. 3. This influence can be analyzed on the base of domain wall pinning theory.

A number of approaches to the description of coercive properties of magnetic domain walls can be summarized [12] in a general expression (1) for the temperature dependence $H_{cp}(T)$, which is very close to that introduced as early as 1944 by Kersten [13]. A good quantitative insight into the temperature dependence of the efficiency function $\alpha f(\delta/d)$ is obtained by dividing the measured $H_{cp}(T)$ by the corresponding $H_A(T)$ values as shown in Fig. 4. If the shape of defects is described by a sinusoidal two-dimensional fluctuation of local values of the anisotropy constant, the efficiency function can be explicitly expressed [14] as $f(x) = \pi x^2/\sinh(\pi x)$ and its temperature dependence is

$$\frac{H_{cp}(T)}{H_A(T)} = \alpha\pi \frac{x_0^2 \exp(T/T_0)}{\sinh[\pi x_0 \exp(T/2T_0)]} ,\qquad(2)$$

where $x_0 = \pi\sqrt{2A/(d^2 H_{A0} M_S)} = $ const , (see also [2] and [3]). Using constants $H_{A0} = 14953$ Oe and $T_0 = 86.4$ K obtained from the exponential fit of the measured $H_A(T)$, two efficiency functions for two different sets of defects (one set governing the temperature range 1, the other set governing the range 2) were applied to fit the experimental data of each of the sample states presented in Fig. 4 (see the solid lines through the data). The relative amplitudes, α_i, and the effective periods, d_i, of the two sets of defects corresponding to the best fit are, in the case of the as-grown sample $\alpha^{ag}_1 = 0.035$, $d^{ag}_1 = 33$ nm (range 1), and $\alpha^{ag}_2 = 0.0037$, $d^{ag}_2 = 200$ nm (range 2), and in the case of the annealed sample $\alpha^{ann}_1 = 0.035$, $d^{ann}_1 = 33$ nm (range 1), and $\alpha^{ann}_2 = 0.0052$, $d^{ann}_2 = 200$ nm (range 2). We estimate the uncertainty of these values, obtained from the curve fitting, to be about 10% even though the numerical results from the computation of each *single* curve speak about even better precision. The very god fit of the experimental points with the theoretical curves shows that the proposed model [2] is well applicable to the classification of the $H_{cp}(T)$ dependence in rare earth magnetic garnets, earlier described as piece-wise exponential curves [1, 15].

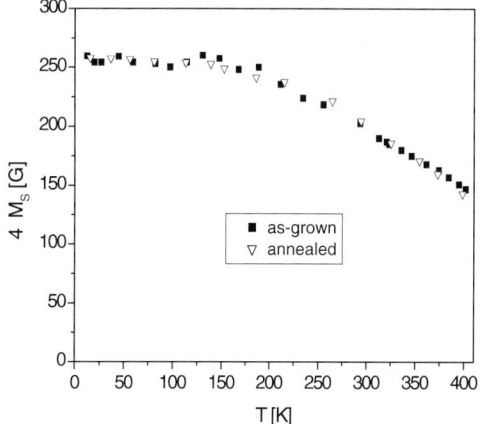

Fig. 1 Temperature dependence of saturation magnetization, $4\pi M_s$, for the as-grown and for the annealed sample.

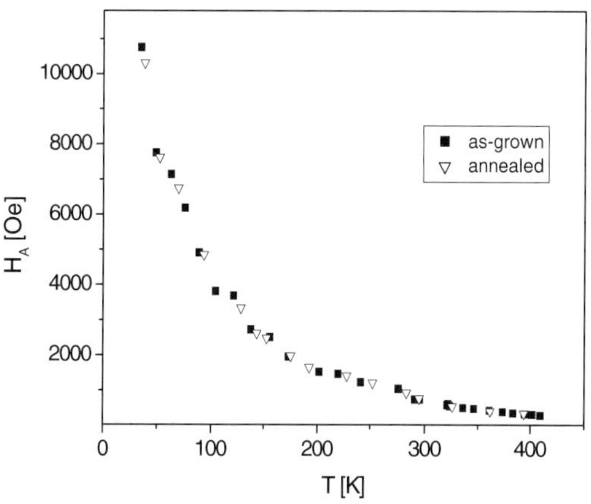

Fig. 2 Temperature dependence of uniaxial anisotropy field, H_A, for the as-grown and for the annealed sample.

The results showed that the applied annealing did not change any property of the set 1, i.e. of those defects, which were predominantly efficient in the low temperature range 1 ($\alpha^{ag}_1 = \alpha^{ann}_1 = 0.035$, $d^{ag}_1 = d^{ann}_1 = 33$ nm). This set 1 is probably connected with the positions of the individual rare-earth ions and with their local anisotropy influence, and the applied annealing temperature of 600 ^{0}C was not able to change here anything, even when the reducing atmosphere was applied. In contrast to this, the temperature range 2, where the governing defects are probably connected with variation of local chemical composition, was evidently vulnerable to the aggressive reducing properties of hydrogen in the applied atmosphere. The period d_2 of this set 2 remained also unchanged after the annealing, but the relative amplitude α_2 was increased through the annealing by about 40%, which lead to almost doubling of the H_{cp} values at their maximum change (see Figs. 3 and/or 4).

As a hypothesis explaining such behavior we suggest, that stability of oxygen (which was probably the ion the most attacked by the aggressive reducing atmosphere) certainly selectively depended on the local fluctuation of chemical composition of the garnet thin film. This stability was then disturbed by the atmosphere at the *selected* positions with the same effective period d_2 as was the period of the compositional variation [4], identified as the set 2 of defects. This disturbance was then reflected in a modification (in the observed case in an *increase*) of amplitude of the local stress and anisotropy fluctuation, α_2, resulting in modification of the H_{cp} values at the temperature range 2, where the set 2 was responsible for the domain wall pinning.

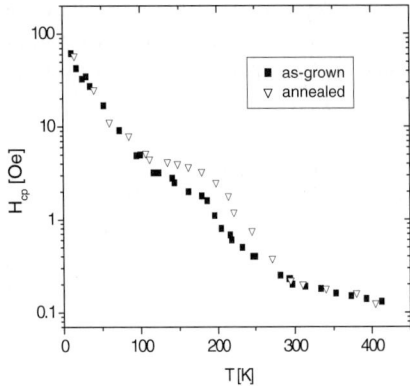 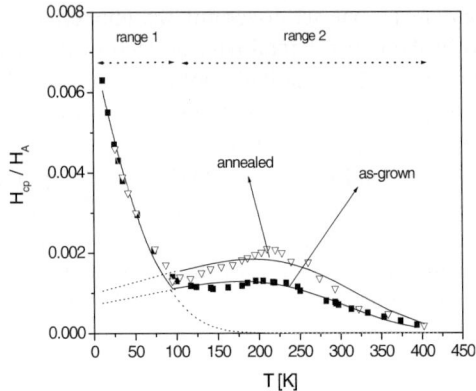

Fig. 3 Temperature dependence of domain wall pinning field, H_{cp}, for the as-grown and for the annealed sample.

Fig. 4 The ratio $H_{cp}/H_A(T)$ as a function of temperature for the as-grown and for the annealed sample. The fitting curves are identical in the temperature range 1 but differ from one another in range 2.

4 Conclusion Rare earth magnetic garnet epitaxial film was annealed at temperature 600 °C in a reducing atmosphere of 15% H_2, 85% N_2. Annealing in air atmosphere [3] of a similar sample changed neither coercive nor anisotropy properties of the material. In contrast to the case [3], the reducing atmosphere was able to modify selectively the temperature dependence of the domain wall pinning field $H_{cp}(T)$.

The observed results were interpreted on the grounds of the coercivity model of multiple contributions of different sets of defects [1, 2]. The analysis strengthened reliability of the model, it confirmed the earlier observed values of the typical defect distributions periods and relative strengths, and it supported probability of correct identification of the origin of the defects sets 1 and 2.

It was found possible to modify the domain wall pinning field of epitaxial rare earth magnetic garnet films by appropriate processing in a selected temperature range and to leave it at the same time unchanged for other temperatures.

Acknowledgements The authors are grateful to Dr. B. Keszei for supplying the sample. The financial support by Hungarian Scientific Research Fund through project T-035264 and by the Grant Agency of the Czech Republic, project No.101/02/0236, and by project K1010104 is appreciated.

References

[1] G. Vértesy and I. Tomáš, J. Appl. Phys. **77**, 6426 (1995).
[2] J. Kadlecová, K. Metlov, I. Tomáš, and G. Vértesy, J. Phys. IV **8**, 307 (1998).
[3] G. Vértesy and I. Tomáš, J. Phys. D: Appl. Phys. (accepted for publication).
[4] M. Pardavi-Horváth, A. Cziráki, I. Fellegvári, G. Vértesy, J. Vandlik, and B. Keszei, IEEE Trans Magn. MAG-**20**, 1123 (1984).
[5] P. Görnert, R. Hergt, E. Sinn, M. Wendt, B. Keszei, and J. Vandlik, J. Cryst. Growth **87**, 331 (1988).
[6] G. Vértesy, L. Půst, I. Tomáš, J. Pačes, J. Phys. D: Appl. Phys. **24**, 1482 (1991).
[7] R.D. Henry, IEEE Trans. Magn. **MAG-13**, 1527 (1977).
[8] D.C. Fowlis and J.A. Copeland, AIP Conf. Proc. **5**, 240 (1972).
[9] J.A. Seitchik, W.D. Doyle, and G.K. Goldberg, J. Appl. Phys. **42**, 1272 (1971).
[10] R.M. Josephs, AIP Conf. Proc. **10**, 286 (1973).
[11] S. Geller, G.P. Espinoza, and P.B. Candrall, J. Appl. Crystallogr. **2**, 86 (1969).
[12] K.L. Metlov, Thesis, Donetsk State University, 1995, Donetsk, Ukraine.
[13] M. Kersten, Grundlagen einer Theorie der ferromagnetischen Hysterese und der Koerzitivkraft, 1944, Hirzel, Leipzig.
[14] M.A. Shamsutdinov, Fiz. Tverd. Tela **33**, 3336 (1991).
[15] G. Vértesy, I. Tomáš, L. Půst, J. Pačes, J. Appl. Phys. **71**, 3462 (1992).

Mössbauer studies of the mechanically alloyed Cu–30 at.% Fe

I. A. Al-Omari[*]

Department of Physics, P.O. Box 36, Sultan Qaboos University, PC 123, Muscat, Sultanate of Oman

Received 31 August 2003, accepted 31 December 2003
Published online 18 March 2004

PACS 61.10.Nz, 75.50.Bb, 75.50.Ww, 75.60.Ej, 76.80.+y

We report Mössbauer and structural studies of the mechanically alloyed Cu–30 at% Fe for milling times between 0 and 18 hours. Samples are prepared by mechanical alloying of Fe and Cu powders. X-ray diffraction and Mössbauer spectroscopy are used to study the magnetic and structural properties of the alloyed samples as a function of milling time. The results of X-ray diffraction patterns show the coexistence of bcc α-Fe and fcc Cu phases. After extended milling times, X-ray diffraction show an increase in the peaks' width, line broadening, indicating a reduction in the particles size. The position of the diffraction peaks as a function of milling time suggest the appearance of new phases. Mössbauer spectroscopy is used to monitor the phase transformation as a function of milling time. It is found that for small milling times, less than 3 hours, the spectrum is consist of six absorption lines corresponding to α-Fe. For milling times higher than 3 hours, new absorption lines start to appear which identified as γ-Fe phase and fcc Fe/Cu solid solution phase. The percentage of each of the three phases is calculated from the intensity of each subspectra after the fitting. α-Fe phase is found to decrease from 100% at milling times of 3 hours or less to 0% at milling time of 18 hours, while γ-Fe phase and fcc Fe/Cu solid solution phase are found to increase from 0% to 34%, for γ-Fe, and from 0% to 66% for Fe/Cu solid solution, for milling times of 3 hours to 18 hours.

© 2004 WILEY-VCH Verlag GmbH & Co. KGaA, Weinheim

1 Introduction

Fe–Cu alloys have interesting properties from scientific point of view and for different technological applications and have been studied by many researchers even they are almost immiscible. Different techniques have been used to prepare Fe/Cu alloys such as ball milling [1–5], rapid quenching [6], sputtering [7], evaporation [8, 9], ion-beam mixing [10]. Fe/Cu alloys were found to form the bcc structure for samples with high Fe concentration, while they form the fcc structure for samples with low Fe concentration. The magnetic properties depend strongly on the structure, where fcc-phase is paramagnetic while bcc-phase is magnetic [7, 11, 12]. Uenishi et al. [1] have shown that the magnetic moment for the iron atom is a round 2.2 μ_B for Fe contents higher than 50%, but it falls to zero for contents less than 20%. Chien et al. [12] have shown that the Curie temperature is very sensitive to the composition.

Macri et al. [13] studied the structural properties of $Fe_{50}Cu_{50}$ alloys prepared by mechanical alloying and fond that the mechanical alloying process initially reduces the crystallite size of Fe and Cu and the Fe atoms incorporate to the copper matrix to form the fcc phase. They also studied the magnetic properties for samples milled for 16 hours using Mössbauer spectroscopy. In this study, they fitted the spectrum with one broadened sextet (line widths = 1.09 mm/s and hyperfine magnetic field = 218 kOe), one non-magnetic doublet (quadrupole splitting = 0.95 mm/s and isomer shift 0.13 mm/s), and one singlet (isomer shift = –0.04 mm/s). The magnetic sextet, the doublet, and the singlet are related to the Fe/Cu solid solution, Cu-rich solid solution, and γ–iron, respectively. Mössbauer study by Majumdar et al. [14] for $Fe_{100-x}Cu_x$ (where x = 10, 20, 25, and 50) samples prepared by mechanical alloying showed the presence of γ-iron in all the samples in addition to the presence of fcc Cu-rich Fe/Cu solid solution and

---

[*] Corresponding author: e-mail: ialomari@yahoo.com, Phone: +968-515463, Fax: +968-514228

Fe$_3$O$_4$. They also found that excess milling time leads to the precipitation of bcc iron from the fcc iron phase.

The lattice constant for fcc γ-iron (3.661 Å) [13] is almost equal to that of fcc Fe/Cu solid solution (3.660 Å) [1] and it is very difficult to distinguish between them form the X-ray diffraction pattern but they can be easily resolved in the Mössbauer spectra. In this study, we present a detailed study of structural and magnetic properties of the mechanically alloyed Cu–30 at%Fe at different milling times between 0 and 18 hours.

2 Experimental methods The Cu–30 at% Fe samples were prepared from bcc α-Fe with 30 μm diameter and fcc Cu powder in the form of platelets with 30–80 μm by 1–2 μm in thickness. The powders were mixed to have a composition of Cu–30 at% Fe, then milled incrementally for 18 hours in Spex 8000 mixer/mill using three 12 g WC balls in a WC lined grinding chamber containing 18 g of powder. During milling, the vial's temperature was kept below 350 K by forced-air cooling. The structure of the samples were examined using X-ray diffraction with Philips diffractometer by using Cu-K$_\alpha$ radiation. Magnetic properties were measured by Mössbauer spectrometer at room temperature. The samples for Mössbauer studies were circular disks of diameter 2 cm prepared by sprinkling a thin layer of the finely powder on a piece of scotch tape. The γ-ray source was a 25 mCi Co^{57} in palladium matrix. Isomer shifts were measured relative to the centroid of the α-iron spectrum at room temperature, and α-iron spectrum was also used for calibration. The spectra were fitted using home made fitting routines.

3 Results and discussion Figure 1 shows the X-ray diffraction (XRD) pattern of the alloy system Cu–30 at%Fe at different milling times between 3 and 18 hours and the XRD diffraction patterns for pure Fe and pure Cu powders. The patterns for the samples with milling times 3, 6, 8, and 12 hours show the presence of both bcc and fcc phases. The bcc phase is related to α-Fe and the fcc phase is related to Cu or Cu/Fe solid solution. As the milling time increases the peaks start to broaden and overlap. In addition to that, the intensity of the bcc α–Fe peaks start to decrease as a function of milling time. It can be seen from Fig. 1 that the main peak of iron (110) decreases with increasing the milling time and it disappears completely after milling for 18 hours. The broadening in the peaks indicates a decrease in the particle and crystallite size. Using Scherrer's formula ($D = 0.9\lambda/W\cos\theta$, where D is the crystallite size, λ is the wavelength, W is the width at half maximum, and θ is the peak's angular position) for the (111) peak of the fcc phase we calculated the crystallite size as a function of milling time and found that it decrease from 364 Å at zero milling time to 137 Å at 18 hrs of milling time. The decrease in the peaks' intensity with increasing the milling time indicates that the iron atoms are incorporated into the fcc phase. We also found that the peaks' position of the fcc phase has shifted to lower angles with increasing the milling time indicating the formation of Fe/Cu solid solution. The lattice constant for fcc Cu is 3.620 Å while after milling for 18 hours the lattice constant of the fcc Fe/Cu solid solution phase is equal to 3.667 Å. The expansion of the Cu lattice constant can be attributed to the larger atomic radius of Fe (1.72 Å) compared to the atomic radius of Cu (1.57 Å) when Cu forms the Fe/Cu solid solution.

Figure 2 shows the room temperature Mössbauer spectra (dots) of the alloy system Cu–30 at % Fe at milling times of 3, 6, 8, 12 and 18 hours, and the fitting is represented by the solid curves. It is clear from this figure that the spectra of the samples milled for 3, 6, 8, and 12 hours show a magnetically split component, and a central paramagnetic lines start to appear at milling times of 6 hours and above. After milling for 18 hours the sample is completely paramagnetic. It can be seen from Fig. 2 that the intensity of the magnetic component decreases with increasing the milling time, while the intensity of the singlet (paramagnetic) lines increases with increasing the milling time. The spectra were fitted with one sextet for milling time of 3 hours, with one sextet and two singlets for milling times of 6, 8, and 12 hours, and with two singlets for milling time of 18 hours. The sextet has a magnetic hyperfine field of 332 kOe and a zero isomer shift compared to the centroid of α-Fe, this sextet is attributed to the bcc α-Fe phase. The two singlets are attributed to γ-Fe phase with isomer shift of (–0.06 ± 0.02) mm/s and to Fe/Cu solid solution with isomer shift of (0.27 ± 0.02) mm/s. These values of the isomer shift are in agreement with previous values of –0.04 mm/s for γ–Fe by Macri *et al.* [13] and –0.06 mm/s by Gonser *et al.* [16], and of 0.223 mm/s for Fe/Cu by Ron *et al.* [15] and 0.27 mm/s by Gonser *et al.* [16].

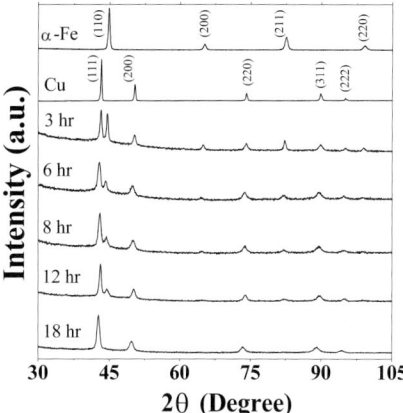

Fig. 1 X-ray diffraction patterns for Cu–30 at%Fe at different milling times, and XRD diffraction patterns for pure Fe and pure Cu powders.

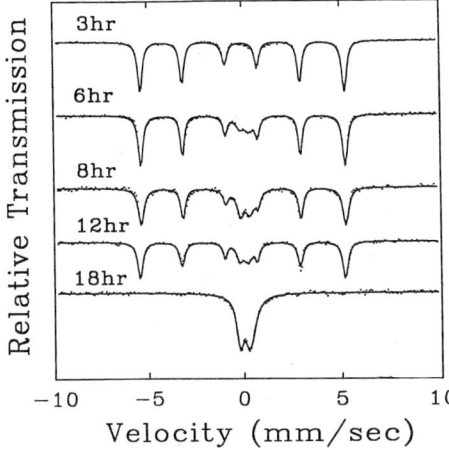

Fig. 2 Room temperature Mössbauer spectra of Cu–30 at%Fe at different milling times (the solid curves represent the fitting).

The intensity of each subspectra in the Mössbauer spectrum is related to the atomic percentage of Fe in each phase. From the fitting we determined the percentage of each phase for different milling times, as shown in Fig. 3. It can be seen from this figure that the percentage of the bcc α-Fe phase decreases with increasing the milling time and it reaches zero after 18 hours of milling time, which is consistent with our X-ray diffraction results. As the milling time increases the percentage of the fcc Fe/Cu solid solution phase start to increase in addition to the development of the fcc γ-Fe phase. After 18 hours of milling time the percentage of the fcc Fe/Cu solid solution is 66%. Taking the total percentage of 30 at%Fe in the sample, the solubility of Fe in Cu will be about 20%, which is higher than previous reported values.

4 Conclusions The magnetic and structural properties of Cu–30 at%Fe prepared by mechanical alloying have been investigated by XRD and Mössbauer spectroscopy. X-ray diffraction results show that both bcc and fcc phases coexist up to milling times of 12 hours, while only fcc phase exist after 18 hours of milling time. After extended milling times, X-ray diffraction show an increase in the peaks' width, line broadening, indicating a reduction in the particles size. Mössbauer spectroscopy determined that α-iron persist up to 12 hours of milling time with a mixture of Fe–Cu solid solution and γ-iron. The percentage of each of the three phases is calculated from the intensity of each subspectra after the fitting. α-

Fe phase is found to decrease from 100% at milling times of 3 hours or less to 0% at milling time of 18 hours, while γ-Fe phase and fcc Fe/Cu solid solution phase are found to increase from 0% to 34% for γ-Fe, and from 0% to 66% for Fe/Cu solid solution, for milling times of 3 hours to 18 hours, indicating that there is a 20% Fe solubility in Cu.

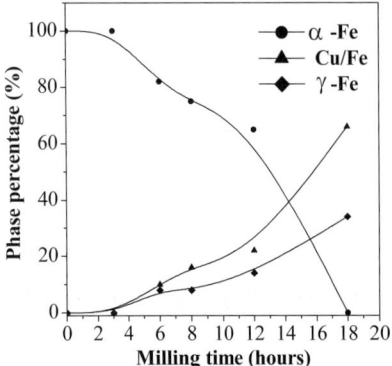

Fig. 3 The phase transformation behavior, determined from the subspectra of the Mössbauer data, of Cu–30 at%Fe as a function of milling time.

Aknowledgments We would like to thank Dr. R. Schalek for helpful assistant and discussion.

References

[1] K. Uenishi, K.F. Kobayashi, S. Nasu, H. Hatano, K.N. Ishihara, and P.H. Shingu, Z. Metallkd. **83**, 132 (1992).
[2] J. Eckert, J.C. Holzer, C.E. Krill, and W.L. Johnson, J. Mater. Res. **7**, 1980 (1992).
[3] J. Eckert, J.C. Holzer, and W.L. Johnson, J. Appl. Phys. **73**, 131 (1993).
[4] A.R. Yavari, P.J. Desre, and T. Benameur, Phys. Rev. Lett. **68**, 2235 (1992).
[5] G. Mazzone and M.V. Antisari, Phys. Rev. B **54**, 441 (1996).
[6] K. Sumiyama, T. Yoshitabe, and Y. Nakamura, J. Phys. Soc. Jpn. **53**, 2160 (1984).
[7] K. Sumiyama and Y. Nakamura, J. Magn. Magn. Mater. **35**, 219 (1983).
[8] C. Peng and D. Dai, J. Appl. Phys. **76**, 2986 (1994).
[9] W. Keune, A. Schatz, R.D. Ellerbrock, A. Fuest, Katrin Wilmers, and R.A. Brand, J. Appl. Phys. **79**, 4265 (1996).
[10] M. Murayama, K. Takahiro, S. Nagata, T. Konno, and S. Yamagucgi, Surf. Coating Technol. **83**, 74 (1986).
[11] Y. Ueda and N. Kikuchi, Jpn. J. Phys. **32**, 1779 (1993).
[12] C.L. Chien, S.H. Liou, D. Kofalt, Wu Yu, T. Egami, and T. R. McGuire, Phys. Rev. B **33**, 3247 (1986).
[13] P.P. Macri, P. Rose, R. Frattini, S. Enzo, G. Principi, W. X. Hu, and N. Cowlam, J. Appl. Phys. **76**, 4061 (1994).
[14] B. Majumdar, M. Manivel Raja, A. Navayanasamy, and K. Chattopadhyay, J. Alloys Compd. **248**, 192 (1997).
[15] M. Ron, A. Rosencwaig, H. Shechter, and A. Kidron, Phys. Lett. **22**, 44 (1966).
[16] U. Gonser, R.W. Grant, A.H. Muir, Jr., and H. Wiedersich, Acta Metall. **14**, 259 (1966).

Mössbauer and structural studies of $Fe_{0.7-x}V_xSi_{0.3}$ alloy system

I. A. Al-Omari[*,1] **and H. H. Hamdeh**[2]

[1] Department of Physics, P.O. Box 36, Sultan Qaboos University, PC 123, Muscat, Sultanate of Oman
[2] Department of Physics, Wichita State University, Wichita, KS 67260, USA

Received 31 August 2003, accepted 31 December 2003
Published online 18 March 2004

PACS 61.10.Nz, 75.50.Bb, 75.50.Ww, 75.60.Ej

X-ray diffraction patterns for $Fe_{0.7-x}V_xSi_{0.3}$ ($x = 0, 0.05, 0.1, 0.15, 0.2$, and 0.3) yield a single BCC-type phase for small values of x and a minor non-cubic phase starts to develop for large values of x. The lattice parameter is found increase with increasing vanadium concentration. Mössbauer experiments were made at room temperature and 20 K. Room temperature spectra show magnetic order for $x = 0, 0.05$, and 0.1, a central paramagnetic line of high intensity superimposed to the magnetically ordered component for $x = 0.15$, and a almost completely paramagnetic state for $x = 0.2$ and 0.3. A similar behavior prevails at 20 K, except that hyperfine splittings are larger and the magnetic ordering starts to show up for $x = 0.2$ and 0.3. In data analysis, each room-temperature and 20-K spectrum is fitted with a distribution of hyperfine sextets. The results clearly indicate the expected atomic disorder in the materials. The average magnetic hyperfine field is found to decrease with increasing x, reflecting the replacement of magnetic iron by nonmagnetic vanadium. The average isomer shift shows the same vanadium-concentration dependence.

© 2004 WILEY-VCH Verlag GmbH & Co. KGaA, Weinheim

1 Introduction The structural and magnetic properties of the intermetallic Fe–Si alloys depend on the Si concentration. $FeSi_2$ was found to form the tetragonal type structure, FeSi forms the B20 type structure, while Fe_3Si forms the DO_3 type cubic structure. Fe_3Si is a well ordered ferromagnetic alloy with four sites; 2-equivalent sites occupied by Fe and the other two sites are occupied by Fe and Si [1]. $Fe_{1-x}Si_x$ forms a continuous rage of solid solution with a bcc structure between $x = 0$ and 0.265 [2]. Mössbauer and NMR [3, 4] investigations for this system have yielded information about the hyperfine field and Si site occupation. These studies showed that the hyperfine field and the magnetic moment at the Fe sites strongly depend on the Si occupancy and the number of Fe nearest neighbors and they decrease with increasing the Si concentration.

Compounds based on Fe_3Al and Fe_3Si alloy systems are of great interest , because of their high temperature strength, excellent oxidation and corrosion resistance. Substitution of Fe by transition metal element affect the magnetic properties, the lattice parameter, and the structural ordering of these compounds[5–10]. Nielsen *et. al.* [11] Studied the $(Fe_{1-x}V_x)_3Al$ alloy and found that the magnetization and Curie temperature decrease with increasing the V concentration. Another study by Nielsen *et al.* [12] for Fe_2VAl showed that this alloy exhibit a magnetic state and a semiconductor like behavior. Al-Nawashi *et al.* [9] studied the effects of Mn substituion for Fe in the alloy system $Fe_{3-x}Mn_xSi$, where x between 0 and 0.5, using Mössbauer spectroscopy. This study reveals that the Mn atoms replace the Fe atoms in the B site of the DO_3 structure and resulting in an atomic disorder. They also found that the average hyperfine field decreases with increasing the Mn concentration. Previous magnetic study by Yoon and Booth [13] for $Fe_{3-x}Mn_xSi$ alloys have shown that this system exhibits ferromagnetic behavior at low Mn concentration ($x < 0.75$), while at higher concentrations, a complex magnetic behavior evolves and the re-

---

[*] Corresponding author: e-mail: ialomari@yahoo.com, Phone: +968-515463, Fax: +968-514228

placement of Fe by Mn causes a decrease in the magnetic moment and Curie temperature with increasing the Mn concentration.

In this work, the effects of Si substitution for Fe on the magnetic and structural properties of $Fe_{0.7-x}V_xSi_{0.3}$ alloys are studied.

2 Experimental methods The bulk samples of $Fe_{0.7-x}V_xSi_{0.3}$ Alloys, where x = 0, 0.05, 0.1, 0.15, 0.2, and 0.3, were prepared from at least 99.9% pure elements by arc melting under a flowing argon atmosphere. Each sample was melted four to five times to insure homogeneity. The structure of the samples were examined using X-ray diffraction with Philips diffractometer by using Cu-K_α radiation. Magnetic properties were measured by Mössbauer spectrometer in the temperature range 20–300 K. The samples for Mössbauer studies were circular disks of diameter 1.3 cm prepared by sprinkling a thin layer of the finely powdered alloys on a piece of scotch tape. The γ-ray source was a 100 mCi Co^{57} in a palladium matrix. Isomer shifts were measured relative to the centroid of the α-iron spectrum at room temperature, and α-iron spectrum was also used for calibration. The spectra are fitted with distributions of hyperfine fields program in which the fields are linearly correlated with the isomer shift for all samples with hyperfine field distribution.

3 Results and discussion Figure 1 shows the typical X-ray diffraction (XRD) pattern of the alloy system $Fe_{0.7-x}V_xSi_{0.3}$ for x = 0.0, and 0.2. All the samples studied have similar behavior with a shift in the peaks' position. The patterns for the samples with x = 0.0, 0.05, 0.1, and 0.15 are consistent with a single phase of the BCC type with lattice parameter a = 2.829 Å for x = 0. However, the patterns for the samples with x = 0.2 and x = 0.3 show additional weak peaks indicating the presence of a minor non-cubic phase. The indices for the various peaks are shown in the diffraction patterns. The lattice parameter is found to increase linearly with increasing the V concentration from 2.829 Å for x = 0 to 2.841 Å for x = 0.3. The increase in the lattice parameter and hence the unit cell volume with increasing the V concentration is due to the volume expansion because of the larger atomic radius of V (1.92 Å) compared with the atomic radius of Fe (1.72 Å). This behavior is in agreement with previous observations by Shobaki et al. [14] for $Fe_{0.7-x}V_xAl_{0.3}$ alloys. The linear increase in the lattice parameter for these alloys as we increase the V concentration suggest a simple dilution process.

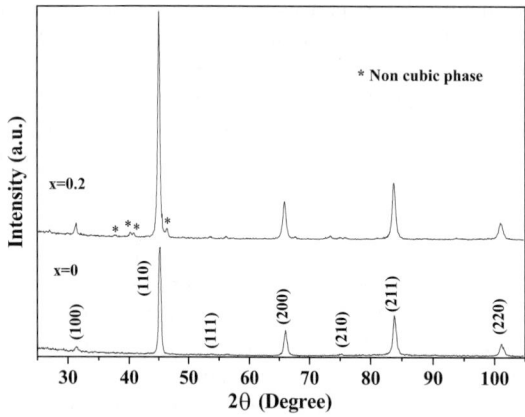

Fig. 1 Typical X-ray diffraction pattern for $Fe_{0.7-x}V_xSi_{0.3}$ alloys.

Figures 2 and 3 show the, room temperature and 20 K, Mössbauer spectra (dots) for the alloy system $Fe_{0.7-x}V_xSi_{0.3}$, and the fitting is represented by the solid curves. It is clear from these figures that the spectra of all the samples show a magnetically split component, and a central paramagnetic line that starts to appear at x = 0.1 at room temperature and at x = 0.15 at 20 K with a very low intensity. As x increases, the paramagnetic line becomes more intense. The spectra are fitted with a distribution of magnetic hyperfine fields P(H), and the results are also shown in Figs. 2 and 3. The average hyperfine field is found to decrease with increasing the

V concentration at room temperature and at 20 K, as shown in Fig. 4. This decrease is due to the replacement of magnetic Fe by non-magnetic V and can be attributed to the reduction of the interatomic exchange interaction, and hence that this reduction drops the Curie temperature and has the effect of depressing the saturation magnetization as well as the magnetic hyperfine field. Another reason for the decrease is when Fe replaced by V, the vanadium surrounding the iron will hybridize the iron's 3d, 4s, and 4p atomic orbits and broaden their energy levels into energy bands. The increase in the average hyperfine field with decreasing the temperature can be attributed to the increase in the interatomic exchange interaction and the enhancement of the magnetic moments with decreasing the temperature below Curie temperature.

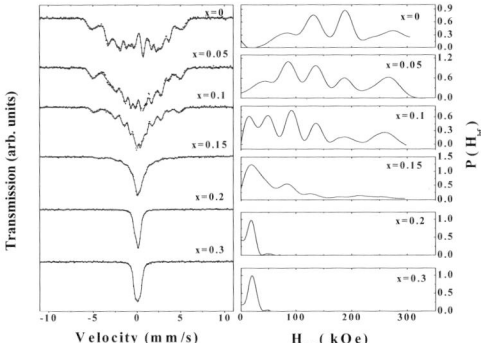

Fig. 2 Room temperature Mössbauer spectra of $Fe_{0.7-x}V_xSi_{0.3}$ alloys (the solid curves represent the fitting), and the probability hyperfine field distribution for the different alloys.

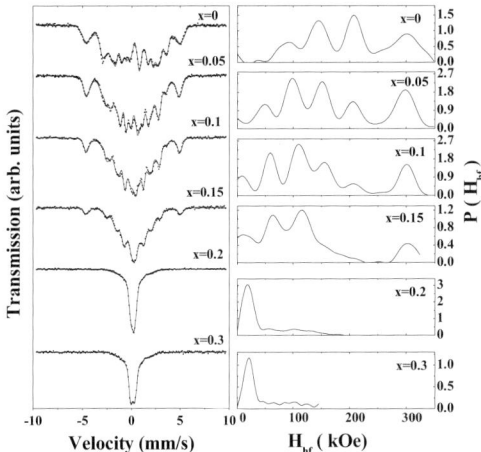

Fig. 3 Low temperature (20 K) Mössbauer spectra of $Fe_{0.7-x}V_xSi_{0.3}$ alloys (the solid curves represent the fitting), and the probability hyperfine field distribution for the different alloys.

The average isomer shift is found to be positive and to decrease with increasing the vanadium concentration from 0.30 mm/s for $Fe_{0.7}Si_{0.3}$ to 0.16 mm/s for $Fe_{0.4}Si_{0.3}V_{0.3}$ at room temperature, and from 0.39 mm/s for $Fe_{0.7}Si_{0.3}$ to 0.25 mm/s for $Fe_{0.4}Si_{0.3}V_{0.3}$ at 20 K. Using the average value of the magnetic hyperfine field for $Fe_{0.7}Si_{0.3}$ and the correlation parameters between the isomer shift and the hyperfine field from Ref. [15] for $Fe_{0.76}Si_{0.24}$, we calculated a value of 0.31 mm/s for the average isomer shift at room temperature for $Fe_{0.7}Si_{0.3}$. This indicates that the average isomer shift for $Fe_{0.7}Si_{0.3}$ is in good agreement with the calculated value. The positive isomer shift in Fe–Si alloys is due to the screening of the 4s electrons in Fe as some of the s, p electrons partially transfer from Si to the iron's 3d band. The decrease of the average isomer shift with increasing the V concentration can be attributed to the transfer of electrons from the V atoms to the Fe atoms during the substitution. This behavior is in agreement with previous observations for similar compounds [14, 16]. The isomer shift for all samples at 20 K is larger than that at room temperature by 0.09 mm/s. This difference is equal to the second order Doppler shift (δ_R). This value of 0.09 mm/s is close to the value of 0.11 mm/s for $(Fe_{0.67}V_{0.33})_3Al$ by Bara et al. [17]. The difference between these two values can be attributed to the difference in Debye temperature (θ_D) for Al and Si.

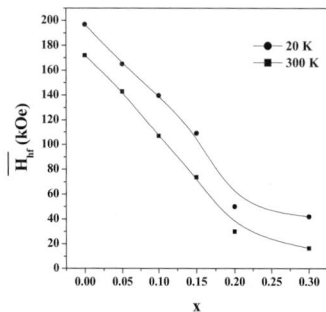

Fig. 4 Dependence of the average hyperfine field on the V concentration (x) for $Fe_{0.7-x}V_xSi_{0.3}$ alloys at room temperature and 20 K.

4 Conclusions The effect of V substitution for Fe in the alloy system $Fe_{0.7-x}V_xSi_{0.3}$ have been investigated by XRD and Mössbauer spectroscopy. A unit cell volume expansion is observed in these V substituted alloys. V substitution for Fe leads to a decrease in the average hyperfine field with increasing x at room temperature and 20 K, which is due to the decrease in the Fe moment and the decrease in T_c. The average isomer shift is found to decrease with increasing the vanadium concentration, which can be attributed to the magnetic dilution upon the substitution and the charge transfer from V atoms to Fe atoms.

Acknowledgments Al-Omari would like to thank Sultan Qaboos University for the support provided to perform this study.

References

[1] V.A. Niculescu, T.J. Burch, and J.I. Budnick, J. Magn. Magn. Mater. **39**, 223 (1983).
[2] M. Hansen, Constitution of Binary Alloys (McGraw-Hill, New York, 1958).
[3] M.B. Stearns, Phys. Rev. **129**, 1336 (1963).
[4] J.J. Budnick, S. Skalski, T.J. Burch, and J.H. Wernick, J. Appl. Phys. **38**, 1137 (1967).
[5] K. Ishikawa, R. Kainuma, I. Ohnuma, K. Aoki, and K. Ishida, Acta Materialia **50**, 2233 (2002).
[6] S. Suga, S. Imada, A. Yamasaki, S. Ueda, T. Muro, and Y. Saitho, J. Magn. Magn. Mater. **233**, 60 (2001).
[7] G.L.F. Fraga, P. Pureur, D.E. Brandao, Solid State Commun. **124**, 7 (2002).
[8] J.W. Dong, J. Lu, J.Q. Xie, L. C. Chen, R.D. James, S. Mckernan, and C.J. Palmstrom, Physica E **10**, 428 (2001).
[9] G.A. Al-Nawashi, S.H. Mahmood, A.D. Lehlooh, and A.S. Saleh, Physica B **321**, 167 (2002).
[10] M.R. Said, Y.A. Hamam, I. Abu-Aljarayesh, and S. Mahmood, J. Magn. Magn. Mater. **195**, 679 (1999).
[11] Y. Nishino, C. Kumada, and S. Asano, Scr. Mater. **36**, 461 (1997).
[12] Y. Nishino, M. Kato, S. Asano, K. Soda, M. Hayasaki, and U. Mizutani, Phys. Rev. Lett. **79**, 1909 (1997).
[13] S. Yoon and J. Booth, J. Phys. F: Metal Phys. **7**, 1079 (1997).
[14] J. Shobaki, I.A. Al-Omari, M.K. Hassan, B.A. Albiss, K.A. Azez, M-Ali Al-Akhras, H.H. Hamdeh, and S.H. Mahmood, J. Magn. Magn. Mater. **213**, 51 (2000).
[15] B. Flutz, Z-Q. Gao, H.H. Hamdeh, and S.A. Oliver, Phys. Rev. B **49**, 6312 (1994).
[16] S.M. Dubiel and W. Zinn, J. Magn. Magn. Mater. **45**, 298 (1984).
[17] J.J. Bara et al., J. Magn. Magn. Mater. **59**, 208 (1986).

Superconducting phase in κ-(BEDT-TTF)$_2$X compounds

R. Charguia[*,1], **A. Ben Ali**[1], **S. Charfi-Kaddour**[1], **R. Bennaceur**[1], and **M. Héritier**[2]

[1] LPMC, Département de Physique, Faculté des Sciences de Tunis, Campus Universitaire, 1060 Tunis, Tunisia
[2] Laboratoire de Physique des Solides, UMR CNRS-Paris XI, Bat. 510, 91405 Orsay, France

Received 31 August 2003, accepted 31 December 2003
Published online 1 April 2004

PACS 74.25.Jb, 74.70.Kn

We study within a mean field theory the superconducting phase of κ-(BEDT-TTF)$_2$X organic superconductors. These two dimensional compounds can be described by a two-band model which depends on two parameters t_1 and t_2. We have considered that the superconducting state is induced by spin fluctuations and the gap symmetry is of d_{xy} type. We have then determined the dependence of the energy gap on the ratio t_1/t_2, which gives us the behavior of the superconducting transition temperature. We show that t_1/t_2 may mimics either a chemical pressure or a hydrostatic one. Our results are in a good agreement with recent experiments suggesting a d_{xy} wave symmetry.

© 2004 WILEY-VCH Verlag GmbH & Co. KGaA, Weinheim

1 Introduction

The organic superconducting salts κ-(BEDT-TTF)$_2$X where the anion X = Cu(NCS)$_2$, Cu[N(CN)$_2$]C are the subject of intensive studies because of a variety of novel electronic phases [1]. These compounds are characterized by a layered structure of alternating sheets of conducting BEDT-TTF radical cations and insulating anions X. The κ type molecular arrangement of organic compounds yields to a rich phase diagram with a several ground states, in particular insulating and superconducting phases. The phase diagram of these compounds shows a very interesting feature, which consists in the proximity of an antiferromagnetic state and a superconducting one [2]. This proximity seems to be crucial for the stability of the superconducting stae.

Several experimental observations suggest that the superconducting state in these compounds has an unconventional pairing with a very anisotropic gap [3–7]. Moreover, recent studies on organic salts have confirmed a d_{xy} order parameter [8–10].

In this paper we focus on the symmetry of the superconducting order parameter. We will confirm, as it has been proposed in reference [10], that the gap symmetry is d_{xy}. In Section 2 we present our model based on a mean field theory. The results are discussed in Section 3. Section 4 is devoted to the concluding remarks.

2 The model

Band structure calculations and several theoretical studies have been carried out on the κ-(BEDT-TTF)$_2$X family. They have led to almost the same shape of the Fermi surface which is in a qualitative agreement with experimental data such as Seebek effect and Shubnikov de Haas experiments. The κ-(BEDT-TTF)$_2$X organic compounds have a large and structurally complex unit cell. Caulfield et al. [11] have obtained a simplified dispersion relation that reproduces band structure calculations and fits well

---

[*] Corresponding author: e-mail: raihen.charguia@fst.rnu.tn, Phone: +00 216 98 946 337

the Shubnikov de Haas data. We have transposed this band model for the κ-(BEDT-TTF)$_2$Cu[N(CN)$_2$]Br [10] where the corresponding conducting plane is (k_x,k_z). It is described by a two parameter (t_1 and t_2) 2D surface model as follows:

$$\varepsilon_k^{(1)} = 2\, t_1 (\cos k_z - \cos 0.7\pi) + 2\, t_2 \sqrt{(1+\cos k_x)(1+\cos k_z)}$$

$$\varepsilon_k^{(2)} = 2\, t_1 (\cos k_z - \cos 0.7\pi) - 2\, t_2 \sqrt{(1+\cos k_x)(1+\cos k_z)}$$

It has been shown that a superconducting state is stabilized in the band (2), which has $\varepsilon_k^{(2)}$ as dispersion relation. This superconductivity is due to spin fluctuations in the band (1) described by $\varepsilon_k^{(1)}$ relation [10]. The theoretical studies seem to further a d_{xy} symmetry for the superconducting order parameter [10]. The mean field Hamiltonian corresponding to superconducting state is:

$$H = \sum_{k,\sigma} \xi_S(k) c^+_{k,\sigma} c_{k,\sigma} - \sum_{k,k'} g_S(k,k') c^+_{k,\uparrow} c^+_{-k,\downarrow} c_{-k',\downarrow} c_{k',\uparrow}$$

where $\xi_S(k) = \varepsilon_k^{(2)}$, $c^+_{k,\sigma}$ ($c_{k,\sigma}$) is the electron creation (annihilation) operators, σ is the spin index and $g_S(k,k')$ is the superconductivity coupling constant. The latter depends on the gap symmetry as follows:

$$g_S(k, k') = g_{0S}\, f_S(k) f_S(k')$$

where $f_S(k)$ is a function giving the symmetry of the gap and $k = (k_x,k_z)$.

Recent experiments have provided evidence of a d_{xy} symmetry for the order parameter. We will then choose the following expression for $f_S(k)$ corresponding to this symmetry:

$$f_s(k) = \sin k_x \sin k_z .$$

To confirm the experimental observations, we will determine the superconducting gap $\Delta_S(k)$, which obeys to:

$$1 = g_{0s} \sum_k f_s^2(k) \frac{\mathrm{th}\left(\frac{E_k^s}{2kT}\right)}{2 E_k^s}, \tag{1}$$

$$E_k^s = \sqrt{\xi_s^2(k) + \Delta_s^2(k)} \quad \text{and} \quad \Delta_S(k) = \Delta_{0S}\, f_S(k).$$

In the next, we will study the dependence of the order parameter on the temperature.

3 Results and discussions

We have solved numerically the self-consistent equation (1) for different values of t_1/t_2 ratio. The coupling constant g_{0S} has been chosen to fit the experimentally determined transition temperature T_c. In Table 1, we have reported the numerical results concerning $\Delta_{0S}(T\rightarrow 0)$ which correspond to the value of Δ_{0S} at zero temperature. As it was suggested in Ref. [9], the superconductivity is madiated by the spin fluctuations which are present in the system. Since these spin fluctuations are suppressed progessively as the t_1/t_2 (or the pressure) increases, g_{0S} is necessarily a decreasing function of t_1/t_2 and T_c follows evidently the same behavior.

Table 1 Numerical results of the superconducting coupling constant g_{0S}, the critical temperatures T_c and the superconducting gap Δ_{0S} for different values of t_1/t_2.

t_1/t_2	g_{0S} (eV)	T_c(K)	$\Delta_{0S}(T\rightarrow 0)/k_B T_c$
0.3	0.375	13	5.603
0.4	0.271	11.6	4.603
0.5	0.201	10.4	3.561
0.7	0.118	3.6	2.389

As it is shown in the Table 1, the ratio $\Delta_{0S}(T\to 0)/k_B T_c$ depends on t_1/t_2. Contrary to a conventional s wave superconductors. This ratio is different from the BCS constant value 1.76, which clearly confirms the non conventional character of the superconductivity in this salts as found experimentally [8, 9] and as suggested theoretically [10].

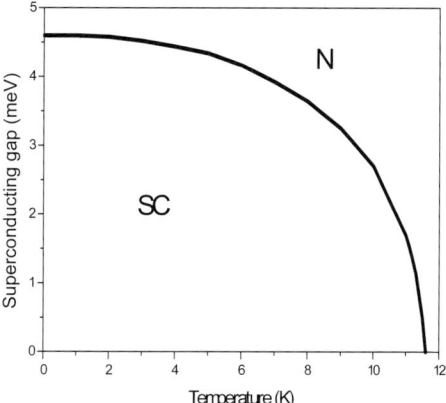

Fig. 1 Temperature dependence of superconducting gap with $g_{0S} = 0.271$ eV and $t_1/t_2 = 0.4$.

The temperature dependence of the superconducting order parameter Δ_{0S} is depicted in Fig. 1. In figure 2, we have presented the variation of the superconducting transition temperature with the ratio t_1/t_2. This variation is the same as that obtained by varying the hydrostatic pressure. We, then, suggest that the t_1/t_2 axis in Fig. 2 may be regarded as the pressure axis in κ-(BEDT-TTF)$_2$X compounds. Therefore, the effect of t_1/t_2 may mimics the effect of a hydrostatic pressure or a chemical one [10], since the pressure axis corresponds to the succession of κ-(BEDT-TTF)$_2$X salts.

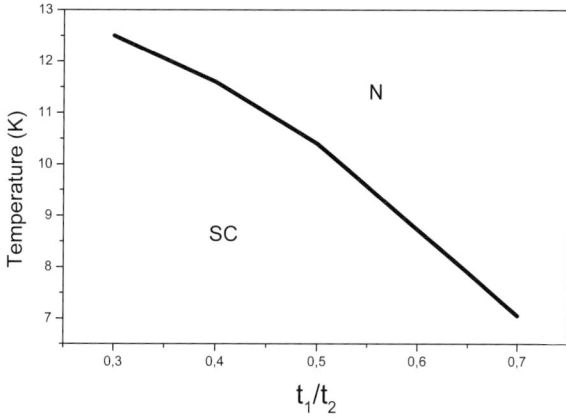

Fig. 2 Superconducting critical temperature as a function of t_1/t_2.

5 Conclusion

In this paper we have discussed the symmetry gap of the superconducting state in κ-(BEDT-TTF)$_2$X salts. Our results confirm the non conventional character of this gap. We have argued that the symmetry of the latter is d_{xy}. On the other hand, we have found that this gap depends substantially on the ratio t_1/t_2. We have shown that this ratio can be regarded as a hydrostatic or chemical pressure. Our calculations are in good agreement with the experimental data in the superconducting state, which corroborate recent theoretical studies.

Acknowledgements We are grateful to S. Haddad and R. Louati for stimulating discussions.

References

[1] T. Ishiguro, K. Yamaji, and G. S. Saito, Organic superconductors (Springer, Berlin, 1998);
J. M. Williams et al., Organic superconductors (Prentice Hall, New Jersey, 1992). For a review see, P. Wzietek et al., J. Phys. I France **6**, 2011 (1996).
[2] S. Lefèbvre et al., Phys. Rev. Lett. **85**, 5420 (2000);
C. Pasquier, private communication.
[3] D. Achikar et al., Phys. Rev. B **47**, 11595 (1993).
[4] H. Mayaffre et al., Phys. Rev. Lett. **75**, 4122 (1995).
[5] K. Kanoda et al., Phys. Rev. B **54**, 76 (1996).
[6] S. Friemel, Ph. D. thesis, Université Paris-sud, Orsay (1997) (unpublished).
[7] M. Pinterié et al., Phys. Rev. B **61**, 7033 (2000).
[8] T. Arai et al., Phys. Rev. B **63**, 104518 (2001).
[9] K. Izawa et al., Phys. Rev. Lett. **88**, 027002 (2002).
[10] R. Louati et al., Phys. Rev. B **62**, 5957 (2000).
[11] J. Caulfield et al., J. Phys.: Condens. Matter **6**, 2911 (1994).

Superconducting phase of fulleride family

D. Meddeb[1], **S. Charfi-Kaddour**[*,1], **R. Bennaceur**[1], and **M. Héritier**[2]

[1] LPMC, Département de Physique, Faculté des Sciences de Tunis, Campus Universitaire, 1060 Tunis, Tunisia
[2] Laboratoire de Physique des Solides, UMR CNRS-Paris XI, Bat. 510, 91405 Orsay, France

Received 31 August 2003, accepted 31 December 2003
Published online 8 April 2004

PACS 71.20.Tx, 74.25.Jb, 74.70.Wz

We have performed a study of the superconducting phase of the Fulleride compounds as a function of n, i.e. the band filling of the LUMO energy band. We discuss the role of strong electron–electron correlation, related to electron–vibron interaction and Jahn–Teller distorsion. Superconducting coupling is expected for odd values of n and is expected to be strongest for n = 3 in electron doped systems. We take into account the correlated nature of the electron Fermi liquid, which should exhibit a pseudogap phenomenon. We have calculated the critical temperature using a strong coupling approach following the McMillan approximation. We show that $T_c(n)$ decrease away from n = 3 on both sides of this maximum. Moreover, for even values of n, we obtain that the system is the neighbourhood of metal–insulator transition.

© 2004 WILEY-VCH Verlag GmbH & Co. KGaA, Weinheim

1 Introduction

It is well known that Fullerene forms with alkali metals definite chemical compounds A_nC_{60}, in which the alkali atom (A = K, Rb, Cs) is assumed to provide n electrons per C_{60} molecule. The physical properties of these compounds strongly depend on n. A fairly high T_c superconductivity is found for n = 3, strongly contrasting with the insulating ground states found for n = 2 or 4 [1]. The superconducting phase appears for a number of electrons (n) per C_{60} higher than 2.6 with a maximum value of T_c at n = 3. The phase diagram is therefore in agreement with known bulk superconductivity in alkali fullerides A_nC_{60}.

At first sight, the strong peak in $T_c(n)$ displayed at exactly n = 3 is not simple to understand. It seems to rule out a model of strongly correlated electrons, in which A_3C_{60} is a Mott insulator and superconducting fullerides are slightly doped Mott insulators, in which $T_c(n)$ should strongly increases when departing from n = 3, at variance with the data. Other arguments seem to favor a BCS-like explanation for the fullerene superconductivity. In a BCS model, one expect that the maximum critical temperature should correspond either to the highest density of states, or to the higher phonon energy, or to the strongest electron–electron attractive interaction. Obviously, the case n = 3 corresponds to half filling of the LUMO fullerene band, which does not correspond at all to the highest single particle density of states in a fcc lattice, which does not preserve the electron–hole symmetry. Since one does not expect any effect of n on the phonon energy, one reaches the conclusion that the maximum of $T_c(n)$ stems from the effective electron–electron attraction. Usually, this attraction is a balance of electron–phonon induced electron–electron coupling, in a B.C.S. scheme, and Coulomb repulsion, both of which should be independant of n. In the fullerene case, the degeneracy of the t1u LUMO triplet can be lifted by a dynamycal Jahn–Teller effect, which contributes to the effective electron–electron interaction. Indeed, recently, the photo-

---

[*] Corresponding author: e-mail: samia.kaddour@fst.rnu.tn, Phone: +00 216 98 925 884, Fax: +00 216 71 885 073

© 2004 WILEY-VCH Verlag GmbH & Co. KGaA, Weinheim

emission on C60 involving Jahn–Teller (JT) in the final state have been studied experimentally and theoratically showing that (JT) effect plays a important role in the fullerides [2].

2 The model

We describe the competition between the kinetic energy and the Coulomb repulsion, for electrons occupying the t_{1u} by the following Hamiltonian:

$$H = H_0 + H_U$$

with

$$H_0 = \sum_{i,m,\sigma} \varepsilon_{t1u} n_{im\sigma} + \sum_{<i,j>m,m',\sigma} t_{ijmm'} c^+_{im\sigma} c_{jm'\sigma},$$

$$H_U = U \sum_{i,m} n_{im\uparrow} n_{im\downarrow} + U' \sum_{i\sigma\sigma'} \sum_{m<m'} n_{im\sigma} n_{im'\sigma'}$$

where $c^+_{im\sigma}$ is a creation operation of an electron on the site j in orbital m with the spin σ, $t_{ijmm'}$ is the hoping integral between the C_{60} molecules and U (resp. U') is the on-site Coulomb interaction for equal (resp. Different) orbitals. We have neglected the exchange interaction compared to the intra-molecular term and we assume U = U'.

However, in the Fullerene case, an original effect should be taken into account in discussing the effective correlations between electrons. the degeneracy of the t_{1u} LUMO triplet can be lifted by a distortion of the C_{60} molecule. The result is a dynamical Jahn–Teller effect, which contributes to the effective electron–electron interaction. Besides, a strong coupling of the electrons with intramolecular modes of vibration of the C_{60} is an important source of electron correlation The three-fold degeneracy of these states is lifted by any quadrupolar deformation that makes the cartesians axes inequivalent. The relevant modes are those having the H_g symmetry. This degeneracy lifting can induce strong electron correlations.

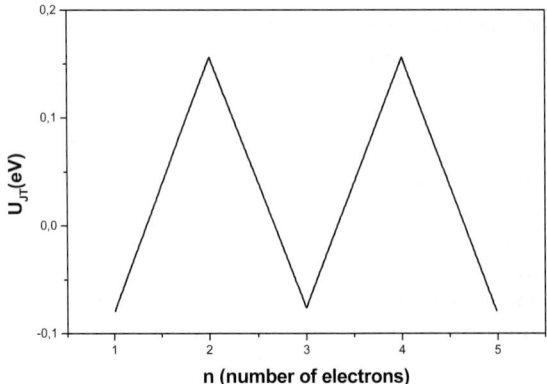

Fig. 1 Jahn–Teller distortion energy U_{JT} as a function of the filling of the t_{1u} band n.

The Jahn–Teller distortion and the Coulomb repulsion play a central role to explain this phase diagram. The dynamical Jahn–Teller effect was studied by Auerbach, Manini [2] and by Victoroff [3]. The Jahn–Teller distortion induces a correlation energy U_{JT} which we define in the following way :

$$U_{JT} = E(n+1) + E(n-1) - 2 E(n)$$

where E(n) is the energy of a molecule occupied by n electrons in the limit of vanishing t_{ij}. A positive value of U_{JT} was obtained for even values of n (n = 2,4) where n is the number of electrons per C_{60}. In this case, U_{JT} and the intramolecular Coulomb repulsion add their effect to localize the electrons. However, for odd values, U_{JT} is negative unducing a tendency to superconducting pairing. We have reported in Fig. 1, the variation of U_{JT} as a function of filling of the t_{1u} band.

Taking into account the Jahn–Teller effect, we can consider an effective Coulomb interaction term H_U^{eff} defined as follows :

$$H_U^{eff} = U_{eff} \sum_{i,m} n_{im\uparrow} n_{im\downarrow}$$

where

$$U_{eff} = U_{JT} + U$$

is the expected value of the effective Coulomb repulsion U given by the Jahn–Teller energy U_{JT} added to the intramolecular Coulomb repulsion U.

We have studied the superconducting region in the neighbourhood of n = 3 as a function of the carrier concentration using a calculating of the critical temperature within a Mc-Millan strong coupling approach [4] as follows:

$$T_c = \hbar\omega_0 \exp\left(\frac{1+\lambda}{\lambda - \mu^*(1+\lambda\eta)}\right), \quad \text{where} \quad \mu^* = \frac{U_{eff} N}{1 + U_{eff} N \, \ln\left(\frac{U_{eff} + B}{2\hbar\omega_0}\right)}$$

and V = NV where V is the pairing coupling via the electron–phonon interaction, $\hbar\omega_0$ is the highest phonon energy and N is the density of states at the Fermi level of the electron correlated system which is given by :

$$N = N_0 \left(1 - \left(\frac{U_{eff}}{U_c}\right)^2\right)$$

where U_c is the Coulomb interaction corresponding to the metal–insulator transition.

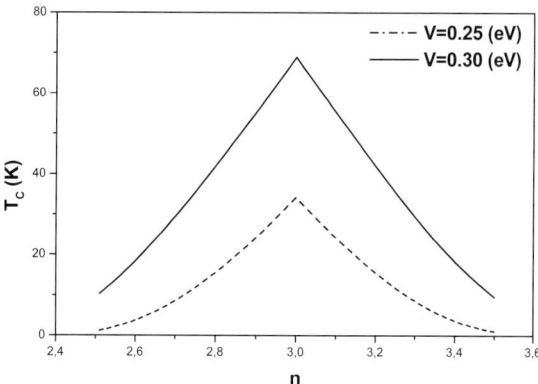

Fig. 2 Critical temperature for superconductivity as a function of the C_{60} valence n for two values of the coupling constant V and ω_0 = 0.1 eV (phonon energy) and B = 0.5 eV (bandwidth).

The critical temperature is displayed in Fig. 2. The behavior of the superconducting phase is a consequence of the variation of the effective repulsion U_{eff}. In fact, as n deviates from 3 on either side, U_{eff} increases rapidly, which causes an increase of μ^* and a reduction of the density of states at the Fermi level. Consequently, the critical temperature rapidly drops around n = 3. This phase diagram is in agreement with the critical temperature variation vs carrier concentration obtained experimentally [5]. It can be considered as a non trivial consequence of a many body effect, related to the strong electron–electron interactions.

In an other hand, we can suggest that for n = 2 and n = 4, the system has a maximum value of U_{eff} which can favor a metal–insulator transition in the neighbourhood of these two values.

3 Conclusion

In summary, we have derived a phase diagram for continous values of the number of electrons par C_{60}. We have shown that the superconducting critical temperature is maximum for n = 3. It decreases rapidly in both sides and vanishes near 2.5 and 3.5 electrons par C_{60} in a good agreement with experimental data.

Acknowledgements This work was supported by French–Tunisian cooperation CMCU project 01/F1303.

References

[1] A. F. Hebard et al., Nature **350**, 600 (1991);
    O. Gunnarsson, Rev. Mod. Phys. **69**, 575 (1997).
[2] A. Auerbach, N. Manini, and E. Tosatti, Phys. Rev. B **49**, 12998 (1994);
    N. Manini, E. Tosatti, and A. Auerbach, Phys. Rev. B **49**, 13008 (1994);
    N. Manini, Philos. Mag. B **81**, 793 (2001);
    N. Manini and Tosatti, Phys. Rev. Lett. **90**, 249601 (2003);
    Canton et al., Phys. Rev. Lett. **89**, 045502 (2002).
[3] W. Victoroff and M. Héritier, J. Phys. France I **6**, 2175 (1996);
    W. Victoroff, M. Héritier, and S. Charfi-Kaddour, Mol. Mater. **7**, 257 (1996).
[4] W. L. McMillan, Phys. Rev. **167**, 331 (1968).
[5] T. Yildirim, L. Barbedette, J. E. Fischer, C. L. Lin, J. Robert, P. Petit, and T. M. Palstra, Phys. Rev. Lett. **77**, 167 (1996).

Temperature dependence of vortex flux pinning in melt-textured superconductors

I. A. Al-Omari[*,1], **M. K. Hasan (Qaseer)**[2], **A. Rais**[1], and **K. A. Azez**[1]

[1] Department of Physics, P.O. Box 36, Sultan Qaboos University, PC 123, Muscat, Sultanate of Oman
[2] Department of Physics, Jordan University of Science and Technology, P.O. Box 3030 Irbid-22110, Jordan

Received 31 August 2003, accepted 31 December 2003
Published online 8 April 2004

PACS 74.25.Ha, 74.25.Qt, 74.62.Bf

Magnetic properties of melt-textured $YBa_2Cu_3O_7$ superconductor are studied at different temperatures from 77 K up to the critical temperature, $T_C = 91$ K, using vibrating sample magnetometer. Initial magnetization curve is measured from the demagnetized state in an applied magnetic field from 0 to 13.5 kOe, followed by measuring the full hysteresis loop. This procedure is repeated at different temperatures between 77 and 91 K. In all the measurements, the magnetic field was applied along the ab-plane of the sample which is perpendicular to the c-axis. Within the framework of the Bean critical state model, our results show that the maximum pinning forces which are represented by the magnetization remanence at which the internal field is zero are found to decrease with increasing temperature and they obey the empirical scale law $\Delta M(emu/cm^3) = 34 \times (1 - T/T_C)^{1/n}$, where $n \approx 2.9$.

© 2004 WILEY-VCH Verlag GmbH & Co. KGaA, Weinheim

1 Introduction

One of the most promising high-temperature superconductors that carry high critical currents are those who has been prepared using melt textured technique [1]. Samples of single domain and of several hundred grams were fabricated [2–4]. One advantage of this preparation technique is the production of the big size materials. This size cannot be achieved in the case of single crystals. Although these materials can carry high critical currents densities but their magnetic behavior is somewhat different from single crystals. Different pinning agents have been introduced such as oxygen vacancies or irradiations and different dependency of the critical current density has been observed. Also, fish-teal peak has been observed in Y–B–C–O single crystal and in Y–B–C–O melt textured.

Therefore, it is of considerable interest to study the magnetic behavior and the vortex flux pinning in these materials. In this study we will focus in the temperature dependence of the vortex flux pinning over the temperature range from 77 K to Tc. This range of temperature is the most important for superconductor application technology.. We will use the magnetization curves to find the maximum pinning forces within the framework of the Bean critical state model [5, 6].

2 Experiment results and discussion

Our sample is a melt-textured $YBa_2Cu_3O_7$ prepared at Argonne National Laboratory. The sample is disk shape of 5.12 mm in diameter and 4.1 mm thickness. The c-axis of the sample is along the perpendicular to the disk plane. The magnetization M is measured using a vibrating sample magnetometer (VSM). The sample was cooled below $T_c = 91$ K in zero fields and the temperature is kept constant. The external field

---

[*] Corresponding author: e-mail: ialomari@yahoo.com

(H_e) is applied along the ab-plane which is parallel to the plane of the disk. The field is raised from zero and cycled by ± 13.5 kOe and the magnetization M is recorded as a function of the applied field. This method is repeated at different temperatures.

Figure 1 shows a typical initial and magnetization curves for $YBa_2Cu_3O_7$ sample at 80 K. Measurements at different temperatures showed similar behavior, with different hysteresis widths at H = 0. Typical example of the corrected magnetization curves is shown in Fig. 2 for $YBa_2Cu_3O_7$ sample at 80 K. In Figure 2 we plot M versus the internal field H. The data has been corrected for the demagnetization fields. The internal field $H = H_e - DM$, where D is the demagnetization factor. By assuming that the sample has a portable spheroid shape, we estimated the demagnetization factor along the ab-plane in which the external field is applied as equal to D = 3.8.

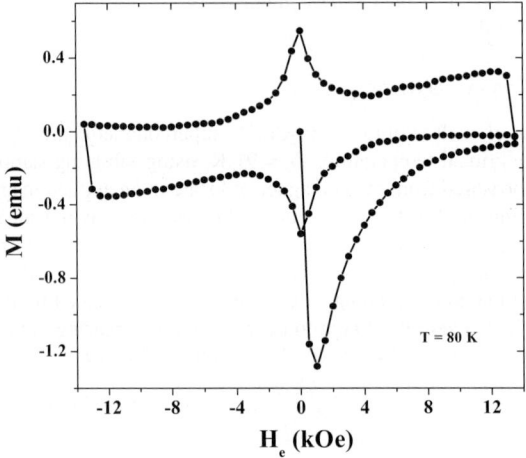

Fig. 1 Magnetization (M) versus the external magnetic field (H_e) at T = 80 K. The external magnetic field applied along ab-plane.

The applicability of the Bean critical state models has been checked at the 77 K. At this temperature we found that by applying a field of 8 kOe the magnetization remanence is saturated and the data represents the major loop. In this experiment the magnetic field exceeds the 8 kOe and cycled by ± 13.5 kOe. Therefore, the critical state has been established and the magnetization is directly proportional to the critical currents and to the pinning forces. Moreover, as we can see from Fig. 2, the magnetization remanence has its maximum value at H = 0. The hysteresis width at H = 0 is directly proportional to the maximum pinning forces. We get the average value of the magnetization remenace of the two points at H = 0 and represent it by ΔM. Accordingly ΔM is directly proportional to the maximum pinning.

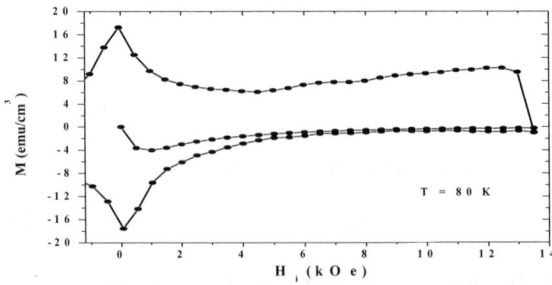

Fig. 2 Magnetization (M) versus the internal magnetic field (H_i) at T = 80 K. The external magnetic field applied along ab-plane.

In order to get the temperature dependence of ΔM we repeat the experiment at different temperatures by cycling the field by ±15.3 kOe. It is worth mentioning here that if the magnetization is saturated at 77 K by 8 kOe it should saturate at lower fields at higher temperatures.

In Fig. 3 we plot ΔM versus the reduced temperature T/T_c, where T_c = 91 K. The figure shows that ΔM decreases with increasing the temperature as expected. The data has been fitted by $\Delta M(emu/cm^3) = 34 \times (1 - T/T_C)^{1/n}$, where n ≈ 2.9. As can be seen from the figure as T approaches T_c, the temperature has more influence in the pinning forces. In fact ΔM represents the average pinning acting on the vortices. The value of the exponent term n ≈ 2.9 is a little bit higher than what was found in similar samples where n ≈ 2 [7].

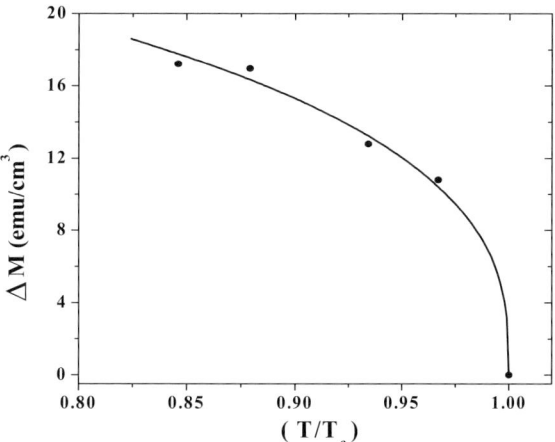

Fig. 3 ΔM versus the reduced temperature (T/T_c). The solid line represent fitting of $\Delta M(emu/cm^3) = 34 \times (1 - T/T_C)^{1/n}$, where n ≈ 2.9.

3 Conclusion

Magnetization measurements were done in a melt-textured $YBa_2Cu_3O_7$ sample by applying the magnetic field along the ab-plane which is perpendicular to the c-axis. Within the framework of the Bean critical state model ΔM at H = 0 represented as a measurable quantity to the maximum pinning forces, we found that these pinning forces decrease with increasing temperature. We also found that ΔM follows the empirical law $\Delta M(emu/cm^3) = 34 \times (1 - T/T_C)^{1/n}$, where n ≈ 2.9.

Acknowledgement We are very grateful to H. Claus from Material Science Division, Argonne National Laboratory, Illinois, USA.

References

[1] M. Murakami, M. Morita, K. Doi, and M. Miyamoto, Jpn. J. Appl. Phys. **28**, 1189 (1989).
[2] P. Gauter Picard, E. Beauynon, X. Chaud, A. Sulpice, and R. Tournier, Physica C **308**, 161 (1998).
[3] M. Ullrich, H. Walter, A. Leenders, and H.C. Freyhardt, Physica C **311**, 86 (1999).
[4] A. E. Garrrillo, P. Rodriguez, T. Puig, A. Palau, X. Obradors, H. Zheng, U. Welp, L. Chen, B. W. Veal, H. Claus, and G.W. Crabtree, Physica C **372**, 1119 (2002).
[5] C.P. Bean, Phys. Rev. Lett. **8**, 250 (1962).
[6] C.P. Bean, Phys. Rev. Mod. **36**, 31 (1964).
[7] Karapetrov, V. Cambel, W.K. Kwok, G.W. Crabtree, H. Zheng, and B. W. Veal, Physica B **284**, 2065 (2000).

Possible coexistence of antiferromagnetism, spin–glass, and superconductivity in ScFe$_4$Al$_8$, and YFe$_4$Al$_8$ single crystals

V. M. Dmitriev[1,3], **J. Stępień–Damm**[2], **W. Suski**[*,2,3], **E. Talik**[4], and **N. N. Prentslau**[1]

[1] National Academy of Sciences of Ukraine, B. I. Verkin Physico–Technical Institute of Low Temperatures, pr. Lenina 47, 61103 Kharkov, Ukraine
[2] Polish Academy of Sciences, W. Trzebiatowski Institute of Low Temperature and Structure Research, P.O. Box 1410, 50–950 Wrocław 2, Poland
[3] International Laboratory of High Magnetic Fields and Low Temperatures, P.O. Box 4714, 50–985 Wrocław 47, Poland
[4] Silesian University, A. Chełkowski Institute of Physics, ul. Uniwersytecka 4, 40–007 Katowice, Poland

Received 31 August 2003, accepted 31 December 2003
Published online 1 April 2004

PACS 72.15.Eb, 75.50.Ee, 74.70.Dd, 75.50.Lk

ScFe$_4$Al$_8$ and YFe$_4$Al$_8$ crystallise in the tetragonal structure of the ThMn$_{12}$-type. Whilst there exists antiferromagnetic ordering in the Fe sublattice below about 100–200 K in these compounds, they can be considered the reference materials for the RFe$_4$Al$_8$ where R = magnetic lanthanide, U or Np. Our recent magnetic measurements in a low magnetic field of 50 Oe have revealed another pronounced anomaly in the temperature dependence of magnetic susceptibility at lower temperature (LT) than the Néel point. There is a clear difference in the temperature dependence of the magnetic susceptibility measured in zero field cooled (ZFC) and field cooled mode (FC). Thus a deviation from stoichiometry, spin-reorientation transition (SRT) and spin–glass (SG) state are discussed as possible reasons for the LT anomalies. The magnetic susceptibility of unoriented single crystals of ScFe$_4$Al$_8$ and YFe$_4$Al$_8$ does not follow either the Curie–Weiss or the modified Curie–Weiss law. At T ≈ 6 K both single crystals are partially transferred to the superconducting state.

© 2004 WILEY-VCH Verlag GmbH & Co. KGaA, Weinheim

1 Introduction

A "fractional" superconductivity which exits in a fraction of compound's volume only, has recently been reported for some ThMn$_{12}$-type ternaries of rare earths at low temperature (for references see [1]). For the other hand, the compounds of nonmagnetic rare earths exhibiting, in principle, antiferromagnetic character, show complex magnetic properties at low temperature (see e.g. [2]).

RFe$_4$Al$_8$ intermetallics crystallise in the tetragonal, ThMn$_{12}$ structure (*I4/mmm*), which contains four different lattice positions. In an ideal material all R (Sc,Y) atoms are located on site 2*a*, whereas sites 8*i* and 8*j* are solely occupied by Al atoms and site 8*f* exclusively by Fe atoms.

The results of magnetic, Mössbauer effect and neutron diffraction experiments collected for polycrystalline ScFe$_4$Al$_8$ and polycrystalline as well as single crystals of YFe$_4$Al$_8$ have been collected in introduction to Ref. [2]. The most important information is that given in [3, 4]. There is mentioned that magnetization and magnetic susceptibility of YFe$_4$Al$_8$ show another anomaly except of the Néel point at 60 K. However, this observation did not receive proper consideration. Neutron diffraction (ND) experiment performed on a non-stoichiometric single crystal [4] shows that a small excess of the Fe atoms substitute for Al at the *j* position. The difference in the stoichiometry decreases the Néel point and the magnetic

---
[*] Corresponding author: e-mail: suski@int.pan.wroc.pl, Phone: +48 71 3435 021, Fax: +48 71 3441 029

ordering changes from the cycloid modulation found in stoichiometric compound to an amplitude modulated wave with a much shorter period. The anomaly at 60 K does not appear clearly in the temperature dependence of magnitude of the magnetic satellites (ND) [4]. In [1] the temperature dependence of the electrical resistivity, $\rho(T)$, of $ScFe_4Al_8$ and YFe_4Al_8 single crystals show in the 25–55 K region a peak in the $\rho(T)$ plot, which is damped by a low magnetic field of 50 Oe. Therefore, in the above mentioned temperature range the effect of the negative magnetoresistivity is observed. For the Sc compound this effect is more pronounced that for the Y single crystal. The experimental data on surface resistance and heat capacity of polycrystalline aluminides suggest "partial" superconductivity at temperature close to 6 K [5]. For the yttrium compound the existence of superconductivity was also proved by the Andreev–reflection in the point contact measurements [6]. The results of examination of the heat transport and preliminary magnetic measurements have been presented in [2]. The contradictory information inclined us to check low field magnetic susceptibility in zero field (ZFC) and field cooled (FC) mode, and superconductivity on unoriented single crystal of these materials.

2 Experiment

The details of obtaining polycrystalline and single crystal samples are described in Ref. [2]. X–ray investigations were carried out at T = 15(1) and 13(1) K for the Sc and Y compounds, respectively, and at 293 K for both samples. The results are presented in [2]. The magnetic susceptibility of the ternary compounds was examined in the temperature range 1.9–300 K, in a magnetic field of 50 Oe for zero field cooled (ZFC) samples, in a magnetic field of 100 Oe and 200 Oe for field cooled (FC) in 50 Oe and 100 Oe, respectively, for the Sc sample (Fig. 1a). For the Y sample FC run was performed in 100 Oe after cooling in 50 Oe (Fig. 1b). The magnetization was measured at 1.9 K in a magnetic field up to 50 kOe (Fig. 2). Both samples were investigated using a SQUID magnetometer. The surface resistivity, and the bulk resistivity were examined as described in Refs. [1, 5].

3 Results and discussion

In Fig. 1a the temperature dependence of the magnetic susceptibility of $ScFe_4Al_8$ is presented. The susceptibility of the sample cooled in a zero magnetic field (ZFC) and measured in a magnetic field of 50 Oe is taken from Ref. [2]. $\chi_{ZFC}(T)$ shows two maxima at 120 K (T_N) and more pronounced peak at 60 K, whereas the FC plots have different character. The low temperature maximum disappears and instead there is a strong increase of the magnetic susceptibility as temperature decreases. This increase is stronger in a lower cooling field. One can see that at about 100 K all runs reach the same value, while above this temperature the ZFC susceptibility is higher, and the FC susceptibilities are equal.

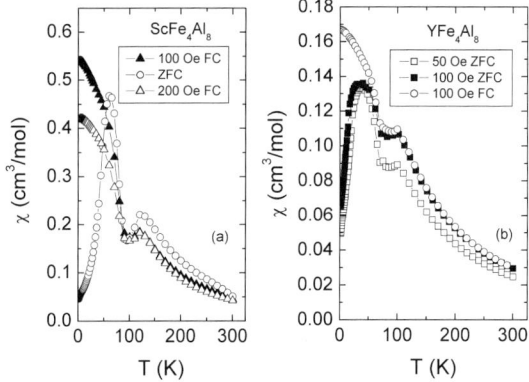

Fig. 1 The magnetic susceptibility versus temperature measured in various magnetic fields in ZFC [2] and FC modes. For the $ScFe_4Al_8$ (a) and YFe_4Al_8 (b).

The temperature dependence of the magnetic susceptibility of YFe_4Al_8 obtained in ZFC and FC mode is presented in Fig. 1b $\chi_{ZFC}(T)$ plot obtained in a magnetic field of 50 Oe is from Ref. [2] and shows two maxima at 100 K (T_N) and at 38 K, both less pronounced than for the Sc compound. The character of the ZFC curve obtained in a magnetic field of 100 Oe is very close to the 50 Oe one below about 50 K but above this temperature, in turn, is very close to the FC curve. In the last dependence, at low temperature the maximum disappears and the susceptibility is monotonously increasing with diminution of temperature. These results are different than those presented in Fig. 1b in Ref.[4]. The reason for observed disagreement could be slight difference in stoichiometry of both alloys.

The results presented in Figs. 1a and 1b are very similar (qualitatively) to those obtained for UFe_4Al_8 and $NpFe_4Al_8$ (see Fig. 2 in Ref. [7]) and considered an indication of the spin–glass transition (SG). Moreover, in the present experiments, as in [7] an increase of magnetic field results in a shift of the $\chi(T)$ plot in the direction of lower temperatures which corresponds to diminution of T_{SG}. All the mentioned observations suggest that at temperatures below $T_{max} = 60$ and 38 K for the Sc and Y compounds, respectively, the states of spin-glass and antiferromagnetic ordering coexist. However, the driving mechanism in presently investigated alloys could be different than in the actinide compounds because in the last compounds the magnetic actinide atoms might introduce additional magnetic interaction whereas in the rare earths compounds containing nonmagnetic elements the reason for the SG state is the location of Fe atoms in various crystallographic positions.

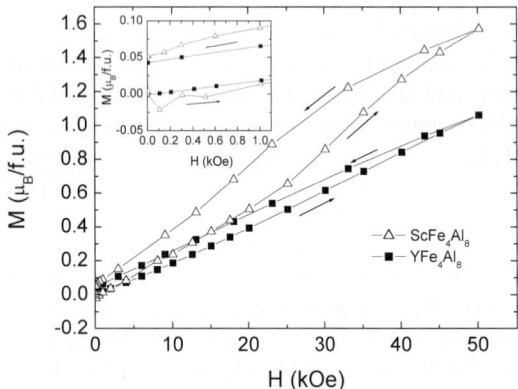

Fig. 2 The magnetization of $ScFe_4Al_8$ and YFe_4Al_8 versus magnetic field at T = 1.9 K.

Presented in Fig. 2 the field dependencies of the magnetization at T = 1.9 K for both samples are fairly close to linearity, however, with tiny hysteresis. Strong decrease of the magnetic susceptibility at low temperature seen for ZFC plots (Figs.1a, 1b) excludes a presence of ferromagnetic impurities as the reason for hysteresis but strongly suggests a field induced change of magnetic properties if a slight deviation from 1:4:8 stoichiometry is present. At very low fields (below 50 Oe, see inset in Fig. 2) a negative magnetization is observed for both samples. It can be the confirmation of a "partial" superconductivity (see Fig. 3a).

In Fig. 3a the temperature dependence of the electrical resistivity, $\rho(T)$, for $ScFe_4Al_8$ and YFe_4Al_8 single crystals in the vicinity of "partial" superconducting transition (T = 4–7 K) is shown. The electrical resistivity is measured along the direction of single crystal growth and zero magnetic field and in 50 Oe [1]. It is seen that the magnetic field of 50 Oe destroys superconductivity. Therefore, the remanent magnetic moment in Fig. 2 after switching–off the magnetic field can result from the frozen flux inside the lattice of "partial" superconductivity.

In Fig. 3b the temperature dependence of the inverse magnetic susceptibility for $ScFe_4Al_8$ and YFe_4Al_8 is presented. It is seen that both compounds do not follow either the Curie–Weiss law or the modified Curie–Weiss law. It might result from strong crystal field (CEF) acting on the Fe atom or considerable contribution of the Pauli paramagnetism.

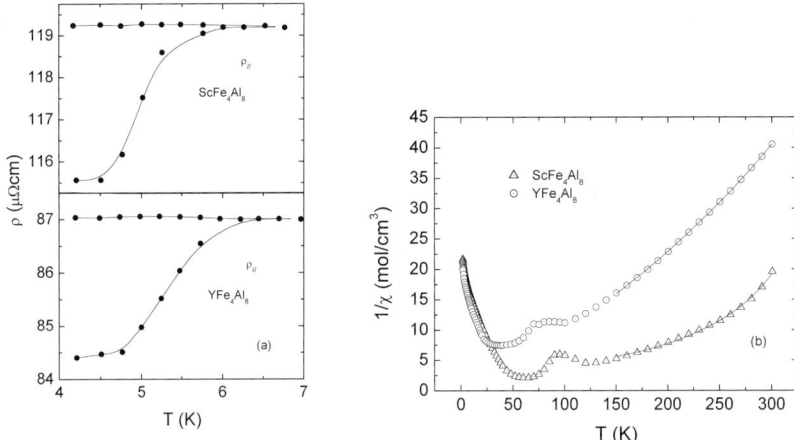

Fig. 3 a) The electrical resistivity, ρ, of ScFe$_4$Al$_8$ (upper panel) and YFe$_4$Al$_8$ (lower panel) versus temperature close to the superconducting transition measured along the direction of a crystal growth. The lower lines correspond to measurements in zero magnetic field whereas the upper lines were obtained in 50 Oe. b) The inverse magnetic susceptibility of ScFe$_4$Al$_8$ and YFe$_4$Al$_8$ measured in magnetic field of 50 Oe (ZFC).

4 Conclusions

It is clear that ScFe$_4$Al$_8$ and YFe$_4$Al$_8$ represent an interesting class of materials for further research in which the coexistence of the antiferromagnetism(AF), spin-glass(SG) and superconducting (SC) states is detected. Although, it is broadly accepted that the antiferromagnetism is related to Fe atoms in the *f* sites, but the reason for SG and SC states is obscure at present. It would be particularly interesting to grow single crystals in which the Fe atom is located exclusively in the *f* position with high probability and then to check if such a material exhibits a SG or SC states.

References

[1] V. M. Dmitriev, N. N. Prentslau, L. A. Ischenko, B. Ya. Kotur, W. Suski, E. Talik, A. V. Terekhov, and I. V. Zolochevskii, Low Temp. Phys. **29**, 1189 (2003).
[2] H. Misiorek, J. Stępień–Damm, W. Suski, E. Talik, B. Ya. Kotur, and V. M. Dmitriev, J. Alloys Comp. **263**, 81 (2004).
[3] J. A. Paixão, S. Langridge, S. Aa. Sørensen, B. Lebech, A. P. Gonçalves, G. H. Lander, P. J. Brown, P. Burlet, and E. Talik, Physica B **234–236**, 614 (1997).
[4] J. A. Paixão, M. Ramos Silva, J.-C. Waerenborgh, A. P. Gonçalves, G. H. Lander, P. J. Brown, M. Godinho, and P. Burlet, Phys. Rev. B **63**, 054410 (2001).
[5] A. M. Gurevich, V. M. Dmitriev, V. N. Eropkin, B. Ya. Kotur, N. N. Prentslau, W. Suski, A. T. Terekhov, and L. V. Shlyk, Low Temp. Phys. **27**, 1308 (2001).
[6] V. M. Dmitriev, L. F. Rybaltchenko, P. Wyder, A. G. M. Jansen, N. N. Prentslau, and W. Suski, Low Temp. Phys. **28**, 260 (2002).
[7] J. Gal, I. Yaar, D. Regev, S. Fredo, G. Shani, E. Arbaboff, W. Potzel, K. Aggarwal, J. A. Pereda, G. M. Kalvius, F. J. Litterst, W. Schäfer, and G. Will, Phys. Rev. B **42**, 8507 (1990).

High temperature superconductors as a two-dimensional electron gas

M. R. Mohammadizadeh and M. Akhavan[*]

Magnet Research Laboratory (MRL), Department of Physics, Sharif University of Technology, P.O. Box 11365-9161, Tehran, Iran

Received 31 August 2003, accepted 31 December 2003
Published online 5 April 2004

PACS 71.30.+h, 74.25.Fy, 74.72.Bk

Based on transport and magnetic measurements on $Gd(Ba_{2-x}Pr_x)Cu_3O_{7+\delta}$, and some other properties of high temperature superconductors (HTSC), we have extracted similarities between superconductors, two-dimensional electron gas (2D-EG) i.e., MOSFETs. These are based on properties such as superconductor–insulator transition in superconductors and metal–insulator transition (MIT) in 2D-EG with doping and magnetic field, localization in transport conduction, quantum unit of resistance at MIT, larger change in resistivity from critical doping to the insulating side in comparison with change from critical doping to the metallic side, and strongly electron–electron coupling. These similarities could lead to a deeper understanding of HTSC and 2D-EG systems.

© 2004 WILEY-VCH Verlag GmbH & Co. KGaA, Weinheim

1 Introduction A large number of experimental and theoretical investigations indicate that two-dimensionality of the normal and superconducting state is one of the key factors in high temperature superconductivity [1]. In 2D systems, it is not expected to be conducting in either weak or very strong interactions between carriers; with decreasing temperature the resistance is expected to grow logarithmically (weak localization) or exponentially (strong localization), becoming infinite as T approaches zero [2]. In recent years, however, systematic studies of the temperature dependence of the resistance in zero magnetic field in a variety of dilute, low-disordered 2D systems have suggested that metallic behavior ($d\rho/dT > 0$) is possible and it has been observed down to the lowest accessible temperatures at electron or hole densities above some critical density. Neither the metallic behavior nor its suppression by a magnetic field is currently understood [3].

In our previous reports, we have studied structural properties of polycrystalline $(Gd_{1-x}Pr_x)Ba_2Cu_3O_{7-\delta}$ HTSC [4], normal state resistivity [5], magnetoresistance [6], and structural properties of $Gd(Ba_{2-x}Pr_x)Cu_3O_{7+\delta}$ [7] compounds. In the various properties of the above called compounds, there are some 2D characteristics, which are very similar to the corresponding real 2D systems. In this paper, we will study the similarities of HTSC and real 2D compounds. This could lead us to a better understanding of superconductivity mechanism and maybe the MIT in 2D systems.

2 Experimental The $Gd(Ba_{2-x}Pr_x)Cu_3O_{7+\delta}$ single phase samples with $0.00 \leq x \leq 1.00$ were synthesized by the standard solid state reaction technique. Appropriate amounts of Gd_2O_3, Pr_6O_{11}, $BaCO_3$, and CuO powders with 99.9 % purity were mixed, ground, and calcined at 840 °C for 24 h in an air atmosphere. Calcination was repeated twice with intermediate grinding. Then powders were reground, pressed into pellets, and synthesized at 930 °C for 24 h in an oxygen atmosphere. The samples were cooled to 550 °C and retained under oxygen flow for 16 h. Finally, they were furnace cooled to room temperature. The

---

[*] Corresponding author: e-mail: akhavan@sharif.edu, Phone: +98 21 6164510, Fax: +98 21 6012983

oxygen content of samples has been determined by the iodometric titration technique with ±0.03 accuracy. An ac four-probe method with f = 33 Hz and 10 mA current was used for the conductivity measurements of the samples within the temperature range of 10 to 300 K. A Lake Shore-330 temperature controller with two Pt-100 resistors was used for measuring and controlling temperature to within ±10 mK. For the magnetoresistivity measurements a magnetic field of maximum value 20 kOe was used.

3 Results and discussion In HTSC, there is a superconductor–insulator transition (SIT) with carrier density doping. Figure 1 shows the variation of hole carrier density in the CuO_2 plane, Cu–O chain, and whole unit cell of $Gd(Ba_{2-x}Pr_x)Cu_3O_{7+\delta}$(GdBaPr-123) compound, which have been derived from Tokura et al. method [8]. Pr-doping decreases the hole content in the CuO_2 planes until x = 0.35, where there is a drop in the hole content of the unit cell, Cu–O chains, and CuO_2 planes. The inset figure also shows the variation of superconducting transition temperature (T_c) with Pr-doping (x), which indicates the SIT at x_c = 0.35. Resistivity vs. temperature curves of the samples in Fig. 2 show the SIT in x = 0.35, too.

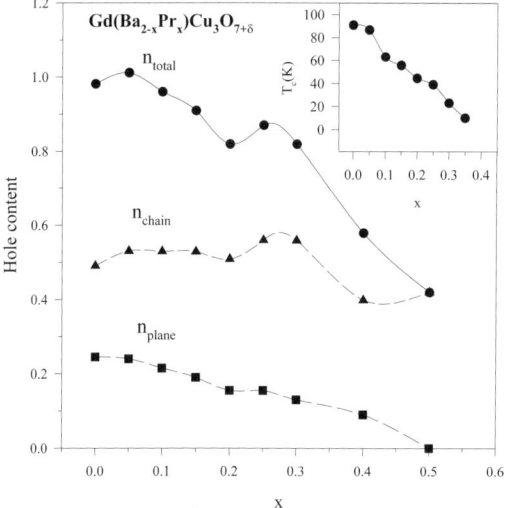 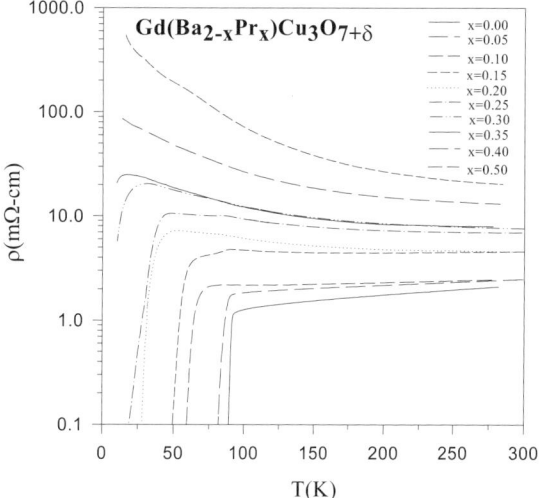

Fig. 1 Hole content in the CuO_2 plane (n_{plane}), Cu–O chain (n_{chain}), and unit cell (n_{total}) vs. different amounts of Pr-doping (x). The inset shows superconducting transition temperature (T_c) vs. different amounts of Pr-doping (x). The lines are guides to the eye.

Fig. 2 Resistivity vs. temperature for different amounts of Pr-doping (x).

A similar MIT behavior is observed in dilute 2D-EG systems [9]. For example, in Si-MOSFET, the resistance changes from insulating behavior at electron density equal 7.12×10^{10} cm^{-2} to metallic behavior at 13.7×10^{10} cm^{-2}, which are shown in Fig. 3 [10]. Thus, at low electron densities, the resistance grows monotonically as temperature decreases; at densities higher than the critical n_c, the resistance decreases as temperature decreases. This is the first similarity between these systems.

In the insulating part of resistivity vs. temperature of 2D-EG systems, it was found that $\rho(T)$ obeys the variable range hopping (VRH) between localized states under the influence of the Coulomb interaction [11]. The normal state resistivity of HTSC, have extensively been studied to find the dominant conduction mechanism [12]. For example, in $(Gd_{1-x}Pr_x)Ba_2Cu_3O_{7-\delta}$ compound the 3D-VRH and VRH with Coulomb interaction have been found as a preferable conduction mechanism for different amounts of Pr-doping [13]. In our recent report [5], we have extensively studied the GdBaPr-123 HTSC compounds. Between 3D-VRH, 2D-VRH, and Coulomb gap (CG) regimes, the 2D-VRH is preferable. As it is evident in this example and so many other HTSC (as have been given in the literature [5]), 2D-VRH between localized states is the preferable mechanism for dominant conduction process in normal state. Hence, in both HTSC and 2D-EG, the hopping between localized states is the dominant mechanism for conduction.

In $(Y_{1-x}Pr_x)Ba_2Cu_3O_{7-\delta}$ (YPr-123) thin films, there is a SIT with Pr-doping, when the sheet resistance per copper oxide bilayers is ~7 kΩ [14]. This is close to the quantum resistance for cooper pairs, $R_Q = h/(2e)^2 \sim 6.45$ kΩ. Such a cross over is also observed in oxygen-deficient and Zinc-doped Y-123 thin films [15]. The authors observe the transition at a value of resistance, which is independent of the combination of oxygen or Zn content that is used. This gives confidence that the observation is a universal phenomenon, when the sheet resistance per CuO_2 bilayer is close to R_Q. Moreover, in-plane resistivity for Zn-substituted single crystal of Y-123 and $(La_{2-x}Sr_x)CuO_4$ in the under-doped regime of hole densities, show the SIT near the universal resistance of 2D resistance [16].

The interesting point is that in low-disordered 2D systems at the critical density, where MIT happens, the resistivity is found to be nearly independent of temperature and of the order of the quantum unit of resistance, $h/e^2 \approx 25.6$ kΩ [9]. It should be noted that due to pair carriers in superconductors, the quantum unit of resistance is $h/(2e)^2$ while in 2D semiconductor systems is h/e^2. So, HTSC and 2D electron gas systems show a SIT (or maybe MIT in 2D-EG) at a universal quantum resistance $R_Q = h/q^2$ (q = 2e in superconductors and e in 2D-EG). Since, this transition has also been observed in bulk single crystals of HTSC [16], the R_Q may indeed indicate the 2D-behavior of HTSC.

In the 2D-EG systems for the same amounts of carrier density doping from critical density n_c towards insulating and metallic sides, the change of resistivity in insulating side is about two orders of magnitude larger than the change of resistivity in metallic side. For example in Si-MOSFET of Kravchenko et al. [17], the critical density is $n_c = 9.02 \times 10^{10}$ cm^{-2}, where the resistance is almost independent of temperature [18]. With a change of n_c to 13.7×10^{10} cm^{-2} (metallic side), which means $\Delta n = 4.68 \times 10^{10}$ cm^{-2}, the resistance changes from about 2 to 0.1 R_Q in the lowest temperature. While, for $\Delta n = 2.1 \times 10^{10}$ cm^{-2} toward the insulating side cause the resistance changes from about 2 to 10^3 R_Q in the lowest temperature. So, although in the metallic side Δn is twice larger than in the insulating side, but the $\Delta \rho$ in the insulating side is about two orders of magnitude larger than in the metallic side.

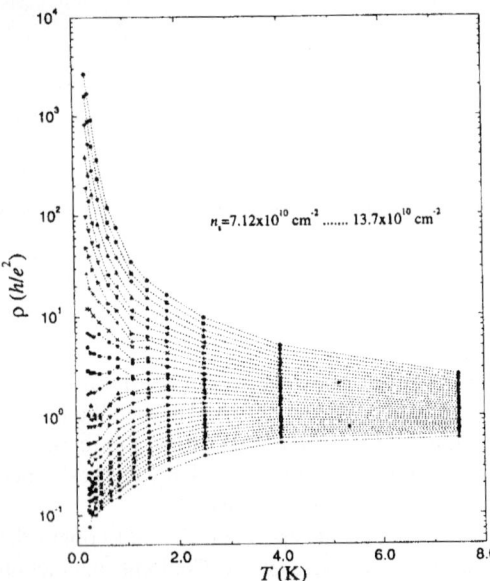

Fig. 3 Resistivity vs. temperature for different amounts of electronic density from 7.12×10^{10} cm^{-2} to 13.7×10^{10} cm^{-2} in Si-MOSFET (after Kravchenko et al. [10]).

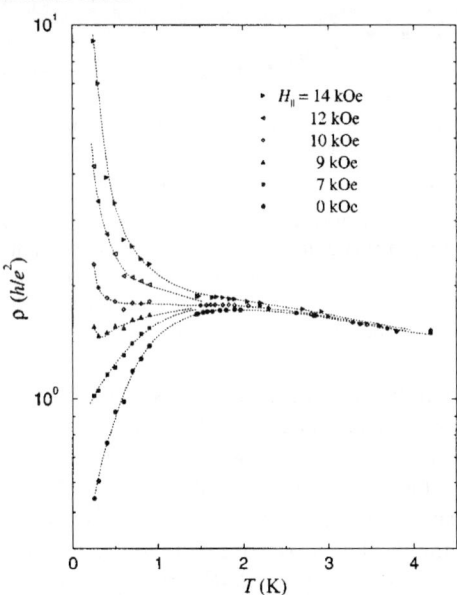

Fig. 4 Magnetoresistance vs. temperature plot of Si-MOSFET with 8.83×10^{10} cm^{-2} electronic density (after Simonian et al. [19]).

The critical x for SIT in GdBaPr-123 samples is 0.35. For $0.00 \leq x < 0.35$, resistivity changes from about 1 to 20 mΩcm in the normal state while, for $x \geq 0.35$ resistivity changes from about 10 to

10^3 mΩcm. In this compound, due to solubility limit of Pr at Ba site, the 123 structure does not form for x > 0.5 [7]. However, in the insulating side, the change in resistivity is two orders of magnitude larger than in the metallic side. Therefore, this is another similarity between the corresponding systems, which shows that changing the critical parameter (carrier density in HTSC and 2D-EG, and thickness in ultrathin films) has more effects on the insulating side with respect to the metallic side in resistivity of the samples.

The magnetoresistance of GdBaPr-123 samples under applied magnetic show a SIT [6]. Figure 4 shows the resistance vs. temperature for a Si-MOSFET with 8.83×10^{10} cm^{-2} electronic density [19]. The 2D-EG systems also show such a MIT under magnetic field. The zero-field curve exhibits typical behavior of metallic systems. In a magnetic filed of only 14 kOe (the upper curve), the metallic drop in the resistance is completely suppressed, so that the system is now strongly insulating over the entire temperature range. Therefore, HTSC and 2D-EG show a SIT/MIT under applied magnetic field. This also enhances the similarities of these systems.

In summary, we have explained the similar characters of HTSC and 2D-EG in: 1) MIT under variation of carriers density and applied magnetic field, 2) localization in normal state of HTSC and insulating phase of 2D-EG, 3) quantum unit of resistance at SIT in HTSC and MIT at 2D-EG systems, and 4) larger change of resistance from critical doping to insulating phase in comparison with change of resistance from critical doping to metallic side. Neither HTSC mechanism, nor MIT in 2D systems have been fully understood, but the presented evidences of similarities may attribute to a unique origin of completely 2D-character of HTSC, or 2D-character of HTSC under special conditions. In addition, these similarities may confirm the superconducting character of metallic phase in 2D-EG systems, which has also been proposed. However, working on the similarities between the mentioned systems, could help to explore the physics of the two complicated unknown HTSC and 2D systems.

Acknowledgements This work was supported in part by the Offices of Vice President for Research and Dean of Graduate Studies at Sharif University of Technology.

References

[1] P. W. Anderson and Z. Zou, Phys. Rev. Lett. **60**, 132 (1988).
[2] A. M. Goldman and N. Markovic, Phys. Today **51**, 39 (1998).
[3] Hilgenkamp and J. Mannhart, Rev. Mod. Phys. **74**, 485 (2002).
[4] M. R. Mohammadizadeh, H. Khosroabadi, and M. Akhavan, Physica B **321**, 301 (2002).
[5] M. R. Mohammadizadeh and M. Akhavan, Eur. Phys. J. B **33**, 381 (2003).
[6] M. R. Mohammadizadeh and M. Akhavan, Supercond. Sci. Technol. **16**, 538 (2003);
   M. R. Mohammadizadeh and M. Akhavan, Physica C **390**, 134 (2003).
[7] M. R. Mohammadizadeh and M. Akhavan, Phys. Rev. B **68**, 104516 (2003).
[8] Y. Tokura, J. B. Torrance, T. C. Huang, and A. I. Nazzal, Phys. Rev. B **38**, 7156 (1988).
[9] E. Abrahams, S. V. Kravchenko, and M. P. Sarachik, Rev. Mod. Phys. **73**, 251 (2001).
[10] S. V. Kravchenko and T. M. Klapwijk, Phys. Rev. Lett **84**, 2909 (2000); cond-mat/9909458.
[11] A. L. Efros and B. I. Shklovskii, J. Phys. C: Solid State Phys. **8**, L49 (1975).
[12] C. Quitmann, D. Andrich, C. Jarchow, M. Fleuster, B. Beschoten, G. Guntherodt, V. V. Moshchalkov, G. Mante, and R. Manzke, Phys. Rev. B **46**, 11813 (1992).
[13] Z. Yamani and M. Akhavan, Solid State Commun. **107**, 197 (1998).
[14] M. Covington and L. H. Greene, Phys. Rev. B **62**, 12440 (2000).
[15] D. J. C. Walker, A. P. Mackenzie, and J. R. Cooper, Phys. Rev. B **51**, 15653 (1995).
[16] Y. Fukuzumi, K. Mizuhashi, K. Takenaka, and S. Uchida, Phys. Rev. Lett. **76**, 684 (1996).
[17] J. E. Furneaux, S. V. Kravchenko, W. Mason, V. M. Pudalov, and M. D'Iorio, Surf. Sci. **361/362**, 949 (1996);
   S. V. Kravchenko, W. E. Mason, G. E. Bowker, J. E. Furneaux, V. M. Pudalov, and M. D'Iorio, Phys. Rev. B **51**, 7038 (1995).
[18] M. R. Mohammadizadeh and M. Akhavan, Supercond. Sci. Technol. **16**, 1216 (2003).
[19] D. Simonian, S. V. Kravchenko, M. P. Sarachik, and V. M. Pudalov, Phys. Rev. Lett. **79**, 2304 (1997).

Advances in doping MgB_2: tuning the Fermi level to the "shape resonance" by Sc substitution

S. Agrestini[*,1], **C. Metallo**[1], **M. Filippi**[1], **G. Campi**[1], **C. Sanipoli**[1], **A. Saccone**[2], **S. De Negri**[2], **M. Giovannini**[2], **A. Latini**[3], and **A. Bianconi**[1]

[1] Dipartimento di Fisica and Unità INFM, Università di Roma "La Sapienza", P. le Aldo Moro 2, 00185 Roma, Italy
[2] Dipartimento di Chimica e Chimica Industriale, Università di Genova, Via Dodecaneso 31, 16146 Genova, Italy
[3] Dipartimento di Chimica, Università di Roma "La Sapienza", P. le Aldo Moro 2, 00185 Roma, Italy

Received 31 August 2003, accepted 31 December 2003
Published online 8 April 2004

PACS 74.25.Jb, 74.62.–c, 74.70.Ad

We have synthesized superconducting $Mg_{1-x}Sc_xB_2$ ternary system in order to explore the influence of electron doping on the electronic properties. An accurate characterization of several samples by X-ray diffraction and conductance measurements has been performed. The T_c vanishes at the critical concentration x at which the Fermi level is tuned to the top of the σ band. Sc doping in $Mg_{1-x}Sc_xB_2$ allows to tune the Fermi level to the "shape resonance" of the σ boron superlattice with negligible changes of the lattice and we have found a large T_c amplification by approaching the "shape resonance" as predicted by our previous works.

1 Introduction

About two years ago a great excitement in the solid state physics community was raised by the surprising discovery of superconductivity in the binary boride, MgB_2, with a T_c of about 39 K, which is the highest known transition temperature for a non-copper-oxide bulk material [1]. This discovery stimulated intense experimental and theoretical investigations [2] for understanding the mechanism of superconductivity in this compound. In fact its critical temperature of about 39 K is close to or above the theoretical value predicted for a single band BCS theory, which indicates that it is a non-conventional superconductor. Unconventional high T_c superconductors are expected to be made of superlattices of superconducting wires or layers [3] where the T_c amplification is driven by a positive quantum interference effect (inter-channel pairing) between two different subbands giving a two gap superconductor. It was quickly shown that the case of MgB_2 is a particular realization of a superlattice of boron monolayers and the Fermi level is not far from a "shape resonance" defined as the energy at which the Fermi surface of the boron σ band exhibits a crossover from 2 to 3 dimensionality [4]. The two expected superconducting gaps have been observed by different techniques [5–7]: a large gap for the two quasi-two-dimensional boron σ bands and a small gap for the π bands showing the need of inter-channel pairing for a theoretical interpretation [8, 9] of both the normal and the superconducting properties. In particular the superconducting critical temperature is increased by a factor 1.5–2 in comparison with the expected value for a single-band model. For some authors the presence of the π bands with a second gap has been considered only a complication while for other authors the inter-channel pairing near a "shape resonance" is the key term for obtaining high T_c [4, 10].

---

[*] Corresponding author: e-mail: stefano.agrestini@roma1.infn.it, Phone: +39 06 4991 4391, Fax: +39 06 4957 697

Chemical substitution for Mg and B is one of the approaches used to get further insight into superconductivity of this system, because it can induce changes in the electronic properties. In fact the introduction of holes or electrons into MgB_2 through the replacement of divalent Mg by a metal atom, having lower (or higher) valence, changes the charge density in the boron layers and the Fermi level is shifted downward (or upward). Moreover the variation of the intercalated ions between the boron honeycomb monolayers modifies both the transversal band dispersion (by changing the spacing between the boron layers) and the electron–lattice interaction (by varying the micro-strain of the B–B distance) [11]. Therefore chemical substitution is an effective way to determinate the effect of each of these parameters on the electronic properties of the system. Consequently during the last two years several attempts of chemical substitution in MgB_2 were made [2]. However a complete or large substitution was obtained only in the case of Al replacing Mg [11–15] and C replacing B [16]. In all the other cases the substitution for Mg and B was minimal and often dubious [17]. Al is a trivalent metal with a much smaller atomic radius than that of Mg, thus the replacement of Mg by Al results not only in the addition of electrons but also in a compression of the MgB_2 lattice structure. On the contrary Sc and Mg have nearly the same atomic radius, therefore the synthesis and investigation of the $Mg_{1-x}Sc_xB_2$ section are very useful and of great relevance because it can allow to study the band filling effect on the superconducting properties with a minimum effect on the structure size. With this motivation, we have studied the $Mg_{1-x}Sc_xB_2$ section in order to explore the influence of charge density on the electronic properties. Until now MgB_2 have been considered insoluble with ScB_2 [17]. Here we show for the first time that by optimizing the preparation conditions a solid solution $Mg_{1-x}Sc_xB_2$ can be synthesized. In this paper we have reported the characterization of several $Mg_{1-x}Sc_xB_2$ samples of their structural and superconducting properties using X-ray diffraction and complex conductivity measurements. On the basis of the results shown we argue that the holes in the boron σ bands play an essential role for the occurrence of superconductivity in the diborides. Finally our findings point out the important contribution of the inter-band coupling around the "shape resonance" to obtain the unusual high T_c.

2 Experimental

Several samples of the $Mg_{1-x}Sc_xB_2$ system were synthesized by direct reaction method of the elemental magnesium and scandium (powder, 99.9 mass% nominal purity), boron (99.5 % pure <60 mesh powder). The starting materials were mixed in stoichiometric ratio and pressed into a pellet of 8 mm in diameter. The pellet was enclosed in tantalum crucible and sealed by arc welding under argon atmosphere. The Ta crucibles were then heated in a furnace Centorr M60 under high-pure Ar atmosphere for 14 hours in the temperature range between 1280 and 950 °C. The samples were characterized for their superconducting properties by the temperature dependence of the complex conductivity using the single-coil inductance method [18]. The diffraction patterns of $Mg_{1-x}Sc_xB_2$ samples were measured in the Bragg–Brentano θ–θ geometry by a vertical X'Pert Pro MPD diffractometer using a Cu $K_α$ radiation.

3 Results and discussion

Analysis of the X-ray powder diffraction patterns measured on the samples of the $Mg_{1-x}Sc_xB_2$ system at $T = 300$ K has shown that all the peaks can be indexed according to the hexagonal AlB_2 structure type (P6/mmm space group). Very little impurity phases of MgB_4 (<4%) and MgO (<1%) have been noticed in the Sc-doped samples. The Sc content dependence of the lattice parameters has been determined by Rietveld method and the GSAS program has been used for this fitting procedure. Special care has been taken to check the positions of the MgB_4 impurity peaks in order to have results independent from diffractometer calibration errors. The variations of the a and c-axes as a function of Sc content x are plotted in Fig. 1. By increasing Sc content x a linear elongation of a-axis is observed while the c-axis is nearly constant. From XRD data it has been found that for low Sc concentration ($0.03 \leq x < 0.13$) there is a miscibility gap where a MgB_2 nearly pure phase (phase 1) coexists with a $Mg_{1-x}Sc_xB_2$ phase (phase 2), which shows that the miscibility of ScB_2 in MgB_2 is only partial.

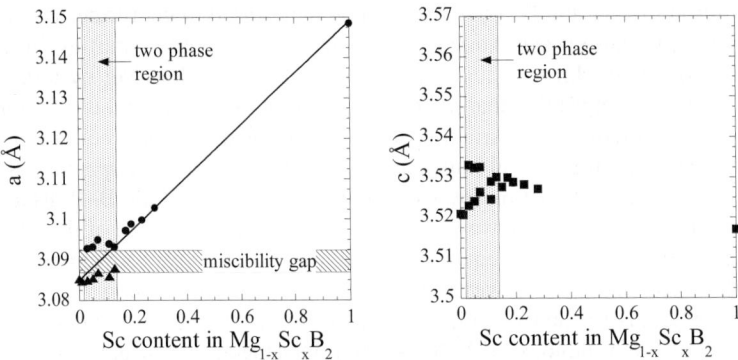

Fig. 1 Variation of a and c axes with Sc doping x in $Mg_{1-x}Sc_xB_2$ samples. For low Sc content has been observed the coexistence of two phases with different Sc concentration: the phase 1 (triangles) having an a-axis close to that of MgB_2 (3.085 Å); the phase 2 (circles) with an a-axis intermediate between that of MgB_2 and ScB_2 (3.148 Å).

The temperature dependence of f_0^2/f^2, probing radio-frequency conductivity, measured on $Mg_{1-x}Sc_xB_2$ system for several Sc contents is shown in Fig 2a. In the low Sc concentration range $0.03 \leq x < 0.13$ a double superconductive transition is observed due to the existence of two phases with different Sc content. With increasing the Sc content from $x = 0.13$ to $x = 0.28$ the superconductive transition shifts to lower temperatures. The superconducting properties appear to have a clear analogy with the diffraction measurements: both resistivity and diffraction measurements show the presence of a two-phase region at low Sc content, i.e. phase 1 and phase 2 from XRD and low- and high-T_c phases from the two steps-like superconductive transition. Consequently we could associate the high T_c phase with the phase 1, determined by diffraction, having an a-axis close to that of MgB_2, and the low T_c phase with the phase 2 having a larger a-axis. Therefore it has been possible to make an one to one correspondence between the lattice and the superconductive properties of the different phases. By assuming a linear dependence of a parameter on Scandium concentration, we have estimated the real Sc concentration in each phase. The same criterion was previously used by [19, 20] to estimate Carbon content in $Mg(B_{1-x}C_x)_2$ system.

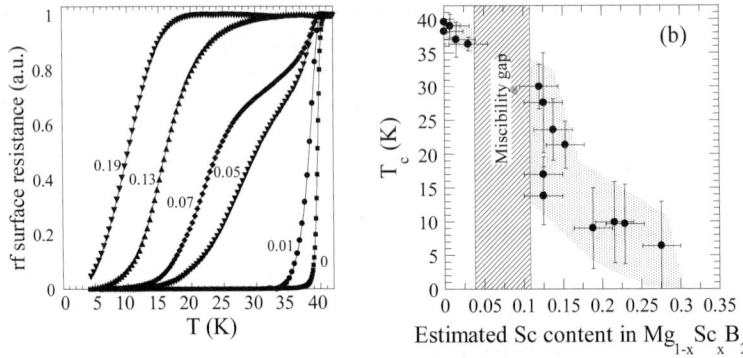

Fig. 2 (a) Radio frequency conductivity probed by the ratio f_0^2/f^2, where $f_0(T)$ and $f(T)$ are the resonance frequencies of the probing LC circuit measured without and with the sample respectively. The ratio f_0^2/f^2 is measured on $Mg_{1-x}Sc_xB_2$ samples with different values of nominal Scandium concentration. (b) Evolution of the critical temperature as a function of the estimated Sc content x.

The evolution of the critical temperature for the superconductivity in the $Mg_{1-x}Sc_xB_2$ has been reported in Fig. 2b as a function of the estimated Sc content x: the T_c shows various regimes by increasing x. A high T_c regime (39.5 K > T_c > 30 K) can be observed for $0 < x < 0.14$, where the critical temperature exhibits a gradual decrease with the increasing Sc content. This high T_c regime is partially hidden by the presence of the miscibility gap. A sharp drop of the T_c down to a low T_c superconductivity regime (15 K > T_c > 5 K) occurs at x ~ 0.15. The superconductivity disappears in the proximity of x = 0.30.

The decrease of the T_c in $Mg_{1-x}Sc_xB_2$ could be explained by considering the electronic band structure. The small changes in the lattice size between MgB_2 and ScB_2 compounds indicates that the effect of substituting Sc for the Mg can be believed primarily as a simple filling of available electronic states, with one electron donated per Sc, within the rigid band picture. Rigid band calculations [21] show that the density of states (DOS) at the Fermi level decreases suddenly at an electron concentration analogous to the one where the jump in the T_c is observed. Therefore it can be concluded that the sharp drop of the T_c in the low T_c regime is associated with the decrease of the DOS at the Fermi level. The critical electron doping of x = 0.25, at which the number of holes in the σ bands vanishes, is only slightly lower than the Sc concentration where the superconductivity is suppressed. Consequently, the observed loss of the superconductivity in the $Mg_{1-x}Sc_xB_2$ system can be explained as a result of the filling of the σ bands.

However the fall in the DOS should give rise to a T_c drop down to 10 mK in the range 0.15 < x < 0.28 while the present experimental results show a critical temperature in the range of 15–5 K. We can hypothesize that this unexpected T_c amplification of the order 100–1000 is given by a tuning of the Fermi level to the superlattice "shape resonance' for Sc substitution x = 0.15 [4, 10].

In summary we have investigated the $Mg_{1-x}Sc_xB_2$ system as a function of Sc content, in order to determinate the effect of charge density on the superconducting properties with minimal changes of the lattice structure size. T_c decreases by increasing the Sc doping and the superconductivity disappears near the critical doping value 0.25. Theoretical band calculations show that at these critical doping values the boron σ bands are filled. The shape resonance occurs at x = 0.15 in the $Mg_{1-x}Sc_xB_2$. The key role of T_c amplification due to a positive effect of the inter-channel pairing by tuning the Fermi level near the "shape resonance" of the superlattice of boron layers has been supposed on the basis of the present experimental results.

Acknowledgements This work was supported by MIUR in the frame of the project "Leghe e composti intermetallici: stabilità termodinamica, proprietà fisiche e reattività", by "Istituto Nazionale Fisica della Materia" (INFM), and by "Consiglio Nazionale delle Ricerche" (CNR) in the frame of the project "5% superconduttività".

References

[1] J. Nagamatsu, N. Nakagawa, T. Muranaka, Y. Zenitani, and J. Akimitsu, Nature **410**, 63 (2001).
[2] C. Buzea and T. Yamashita, Supercond. Sci. Technol. **14**, R115 (2001).
[3] A. Bianconi, A. Valletta, A. Perali, and N.L. Saini, Physica C **296**, 269 (1998).
[4] A. Bianconi et al., J. Phys.: Condens. Matter **13**, 7383 (2001).
[5] F. Giubileo et al., Phys. Rev. Lett. **87**, 177008 (2001).
[6] S. Tsuda et al., Phys. Rev. Lett. **87**, 177006 (2001).
[7] Y. Wang, T. Plackowski, and A. Junod, Physica C **355**, 179 (2001).
[8] A. Y. Liu, I. I. Mazin, and J. Kortus, Phys. Rev. Lett. **87**, 87005 (2001).
[9] H. J. Choi, D. Roundy, M. L. Cohen, and S. G. Louie, Nature **418**, 758 (2002).
[10] A. Bussmann-Holder and A. Bianconi, Phys. Rev. B **67**, 132509 (2003).
[11] S. Agrestini et al., J. Phys.: Condens. Matter **13**, 11689 (2001).
[12] A. Bianconi et al., Phys. Rev. B **65**, 174515 (2002).
[13] D. Di Castro et al., Europhys. Lett. **58**, 278 (2002).
[14] J. S. Sluski et al., Nature **410**, 343 (2001).
[15] J. Q. Li, L. Li, F. M. Liu, C. Dong, J. Y. Xiang, and Z. X. Zhao, Phys Rev. B **66**, 012511 (2002).
[16] A. Bharathi et al., Physica C **370**, 211 (2002).
[17] R. J. Cava, H. W. Zandbergen, and K. Inumaru, Physica C **385**, 8 (2003).
[18] D. Di Castro, N. L. Saini, A. Bianconi, and A. Lanzara, Physica C **332**, 405 (2000).
[19] M. Avdeev, J. D. Jorgensen, R. A. Ribeiro, S. L. Bud'ko, and P. C. Canfield, Physica C **387**, 301 (2003).
[20] S. M. Kazakov, J. Karpinski, J. Jun, P. Geiser, N. D. Zhigadlo, R. Puzniak, and A. V. Mironov, cond-mat/0304656.
[21] J. M. An and W. E. Pickett, Phys. Rev. Lett. **86**, 4366 (2001).

The role of spin diffusion quasiparticle in CMR/HTSC heterostructures

S. Soltan[*,1,2], **J. Albrecht**[1], **G. Cristani**[1], and **H.-U. Habermeier**[1]

[1] Max-Planck-Institut für Festkörperforschung, Heisenbergstr. 1, 70569 Stuttgart, Germany
[2] Physics Department, Faculty of Science, Helwan University, 11792-Cairo, Egypt

Received 31 August 2003, accepted 31 December 2003
Published online 25 March 2004

PACS 74.72.Bk, 75.30.Vn

Half metal–colossal magnetoresistive $La_{2/3}Ca_{1/3}MnO_3$ (HM–CMR) and $YBa_2Cu_3O_{7-\delta}$ high-T_c superconductors (HTSC) bilayer structures are grown on $SrTiO_3$ (100) single crystalline substrates using pulsed laser deposition. Magnetization and magneto-optical measurements are carried out that show the coexistence of ferromagnetism and superconductivity at low temperatures. Using the HM–CMR layer as an electrode for spin polarized electrons, we investigate spin polarized self injection into the HTSC layer. The experimental results are in good agreement with a theoretical model, that is developed with respect to spin polarized quasiparticle diffusion.

© 2004 WILEY-VCH Verlag GmbH & Co. KGaA, Weinheim

Recently, much attention has been paid to junctions consisting of colossal magnetoresistance material $La_{2/3}Ca_{1/3}MnO_3$ (LCMO) and the high-T_c superconductor $YBa_2Cu_3O_{7-\delta}$[1, 2, 3, 4, 5]. With these junctions information on the properties of high-T_c superconductors can be obtained, to find applications of new superconducting devices. The combination of ferromagnetic and superconducting materials could end up in devices like a spin-controlled transistor with high gain current and short switching times. Such devices, that exploit the spin of electrons rather than the charge are summarized as "spintronics" and are already in commercial use, e. g. in read-out heads of hard disks. It has been suggested by Si [6] that electron injection from hole-doped rare earth manganites into high-temperature superconductors can yield information on spin-charge separation in high-T_c superconductors. The spin-polarization, in addition to the ferromagnetic ordering nature, is one particular property of hole-doped rare earth manganite of the form $R_{1-x}A_xMnO_3$ (R = trivalent rare-earth ions, A = divalent alkaline-earth ions). In case of $La_{2/3}Ca_{1/3}MnO_3$ a nearly full spin polarization of the transport electrons can be found [7]. To carry out experiments on the spin diffusion from the manganite into the superconductor it is necessary to take the quality of the interface between both materials into account. Only a small difference occurs in the in-plane lattice parameters of $La_{2/3}Ca_{1/3}MnO_3$ and $YBa_2Cu_3O_{7-\delta}$, this leads together with a small surface roughness of the single layer films to the growth of LCMO/YBCO bilayers with high interface transparency and only few structural defects [1, 4].

Recently there have been a lot of experimental and theoretical efforts [3, 8, 9] to estimate the spin diffusion length, ξ_{FM}, in CMR/HTSC heterostructures, resulting in values 20 nm < ξ_{FM} < 100 nm but up to now no distinct answer could be given. In this paper, we show experimental data on ferromagnetic and superconducting bilayers that show that the spin diffusion length ξ_{FM} has to be much smaller than 30 nm at low temperatures. We have prepared and characterized high quality bilayer thin films consisting of a ferromagnetic half metal CMR layer and a high-T_c superconducting layer. The coexistence of ferromagnetism and superconductivity in these samples at low temperature is shown by magnetization measurements. By

---

[*] Corresponding author: e-mail: S.Soltan@fkf.mpg.de

means of quantitative magneto-optics the critical current density j_c has been measured for all bilayer structures, in order to see the effect of the magnetic layer on the superconductor.

Thin films of $La_{2/3}Ca_{1/3}MnO_3$ and $YBa_2Cu_3O_{7-\delta}$ are grown using pulsed laser deposition. As substrates we used 5×5 mm² $SrTiO_3$ (100) (STO) single crystals. The preparation conditions are identical for all of the bilayer structures. The substrate is kept at a constant temperature of 780 °C while the film growth, controlled by a far-infrared pyrometer. Afterwards the bilayer is in-situ annealed at 530 °C in oxygen pressure of 1.0 bar for 30-60 min. This procedure results in films of high crystalline quality and sharp film-substrate interfaces [1]. The temperature dependence of the magnetization $M(T)$ of these samples is recorded in a magnetic field parallel to the film plane using the Quantum Design MPMS superconducting quantum interface device (SQUID) magnetometer. The determination of the critical current density is additionally carried out by quantitative magneto-optics, the details of this technique can be found in Jooss et al. [10]. By XRD measurements it is found that for both layers strongly textured c-axis oriented growth occurs, detailed results will be published elsewhere. We want to remark, that the same strong texture of the film is also present if the LCMO is grown on top of the YBCO film. The XRD results agree with transmission electron microscopy (TEM) measurements [1], which have been carried out for LCMO/YBCO superlattices grown by the same PLD set-up. These investigations lead to the fact that the LCMO buffer layer and the YBCO layer on top are grown epitaxially with a high-quality interface. This enables us to study now the electronic properties of the sample in first order without concerning the defect structure of the interface.

Fig.1a depicts the normalized magnetization after zero-field cooling $M(T)/M_s(T)$; where $M_s(T)$ is the saturation magnetization in the diamagnetic state. The measurement is performed in an in-plane external magnetic field of $H_{ex} = 10$ Oe for sample with thicknesses of $d_{LCMO} = 50$ nm and $d_{YBCO} = 30$ nm. The experimental data clearly show the coexistence of superconductivity and ferromagnetism at low temperatures. We find an onset of the magnetic ordering at $T_{Curie} \approx 150$ K and a strong diamagnetic drop at $T_c \approx 75$ K that indicates the superconducting transition. The superconducting transition temperature is now determined systematically for bilayer samples with different thickness of the YBCO layer. These results are depicted in Fig. 1b as black squares. We find a strong suppression of the superconducting transition temperature for small YBCO thickness and a nearly full recovery of T_c for thicknesses larger than $d_{YBCO} = 30$ nm.

Fig.1c shows the critical current density j_c versus the thickness of the YBCO layer. The determination of the critical current density is performed by a quantitative magneto-optical technique, for more details see [12, 13]. The critical current density j_c is reduced in all bilayer structures (j_c= 0 - 1.25×10¹¹ A/m²) compared with unmodified single layers of YBCO with similar thicknesses ($j_c \approx$ 2-3×10¹¹A/m²) [10]. Though we do not want to focus on the absolute magnitude of the critical current density, the important feature in Fig.1c is the recovery of the critical current for YBCO thicknesses larger than $d = 30$ nm in these bilayer structures. The magnetization and the critical current density measurements show that the superconductivity is strongly influenced by the presence of the ferromagnetic layer. The superconductivity in the YBCO layer is strongly suppressed up to thickness of 20-30 nm, films with a larger thickness are able to recover superconductivity.

The pair breaking in CMR/HTSC heterostructures, as well as in the classical FM/SC superlattices, can not only due to the proximity effect [1, 4]. Also the quasiparticle injection (QPI) into the superconducting layer has to be taken into account [3, 14, 15]. This phenomenon has been very early investigated by Gray et al. [14]. The pair breaking can be written as:

$$\frac{\Delta(n_{qp})}{\Delta(0)} \approx 1 - \frac{2n_{qp}}{4N(0)\Delta(0)}, \qquad (1)$$

Where Δn_{qp} is the energy that is required to suppress the order parameter of the superconductor due to a finite quasiparticle density n_{qp}. $N(0)$ and $\Delta(0)$ are the density of state and the order parameter at $T = 0$ K, respectively. It is now assumed that the n_{qp} in our model describes the self-injection of QP electrons [3]

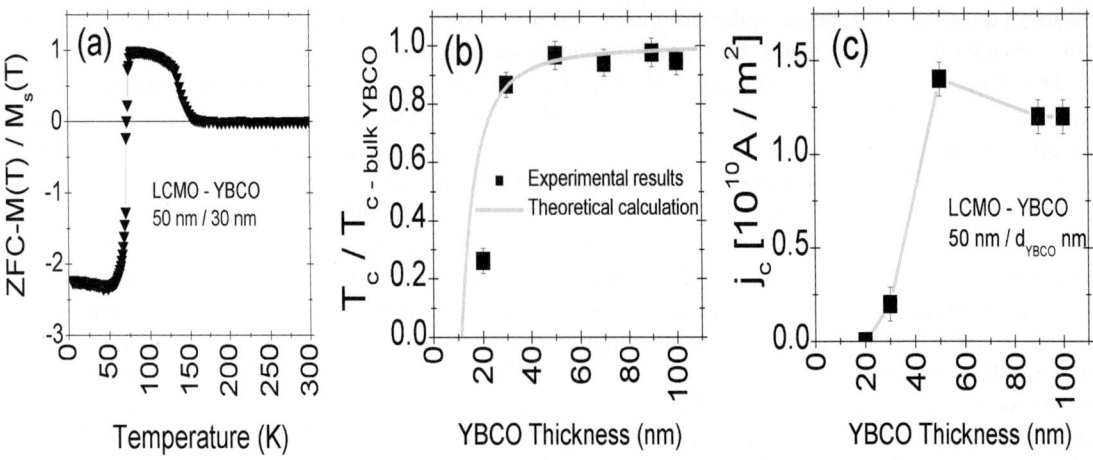

Fig. 1 (a) The normalized ZFC-M(T)/M$_s$(T) measurement after application of in-plane magnetic field $H_{ex} = 10$ Oe for a bilayer sample with $d_{LCMO} = 50$ nm and $d_{YBCO} = 30$ nm. It is found that the superconducting transition temperature is nearly recovered to the bulk value for thicknesses larger than $d_{YBCO} \approx 30$ nm. (b) Superconducting transition temperature for the samples with different thicknesses of the YBCO film (black squares). The solid grey line refers to the model that is described in the text. (c) The critical current density j_c determined by magneto-optics on the bilayer samples at a temperature of $T = 7$ K. Similar to the critical temperature an increase starting around a critical thickness of $d \approx 30$ nm occurs.

only along the c-axis direction, because of the high transparency of the interface. The high exchange splittng energy of $\Delta E_{ex} \approx 3$ eV of the magnetic layer leads to a energetically favorable injection of the QP into the superconductor. The temperature dependence of the quasiparticle injection can be written in the following form:

$$n_{qp}(T) \approx 4N(0)\Delta(0)\sqrt{\frac{\pi}{2}\frac{\Delta(T)k_B T}{\Delta(0)^2}}\frac{1}{e^{\frac{\Delta(T)}{k_B T}}}, \quad (2)$$

The estimation of the spin diffusion length ξ_{FM} can be derived after Ref.[14] in analogy to the classical FM/SC as:

$$\xi_{FM} \approx \sqrt{\ell_o v_F \tau_s}, \quad (3)$$

$\ell_o(T = 0 \text{ K}) \approx 20$ nm gives the mean free path of YBCO [3, 14], the spin diffusion relaxation time τ_s is:

$$\tau_s \approx 3.7 \frac{\hbar k_B T_c}{\Delta E_{ex} \Delta(T)}, \quad (4)$$

with

$$\Delta(T) \sim \Delta(0)\sqrt{1 - (\frac{T}{T_c})}, \quad (5)$$

and $\Delta(0) \approx 20$ meV for YBCO [3]. From equation 3, replacing the related parameter inside the equation, we end up with a relation where the temperature dependence and the length scale of the spin diffusion length ξ_{FM} is included. First we have to consider the spin concentration in the superconductor. It is assumed that spins in high-temperature superconductors can be described as unitary scatterers. From Zn doping in YBCO measurements it is known that a critical doping in the range of 2-10 % strongly reduces

T_c [16, 17, 18]. This critical concentration of spins is achieved at a distance of $d = \alpha \xi_{FM}$, with $\alpha \approx 3$. This d is now identified with the YBCO film thickness. This enables us to model the experimental data by:

$$d = \alpha \xi_{FM} \cong 3.7 \frac{\alpha m^* \hbar v_F^2}{\Delta(0) \Delta E_{ex} n_{qp}(0) e^2} \frac{\sqrt{T/n_{qp}(T)}}{\sqrt[4]{1-(T/T_c)}}, \quad (6)$$

with m^* and e the electron effective mass and charge, respectively. Fig.1b shows the calculated spin diffusion length ξ_{FM} related to equation 6. The nice fit using only one free parameter, $n_{qp}(T)$, of the theoretical calculation to the experimental data shows that a description of the results in terms of a spin diffusion process is adequate. Additionally, it is suggested that the spin diffusion length is about $\xi_{FM} \approx 10 \pm 1$ nm at low temperatures. This is in good agreement with results, that have been found by Holden et al. [5] from far-infrared ellipsometric measurements at superlattices, where a critical thickness for the YBCO of $d \approx 20$ nm is found. Note, that in the case of superlattices the spin diffusion QP penetrate from both sides into to the superconducting films. The accordance between the results shown in this paper and other groups using different experimental techniques for measurements of the effect of magnetic layers on superconducting films gives rise to a spin diffusion length ξ_{FM} from LCMO into YBCO in the order of 10 nm at low temperatures.

We have investigated experimentally and theoretically the role of the spin diffusion in bilayer systems consisting of thin films of ferromagnetic and superconducting materials. Magnetization measurements of epitaxially grown bilayers of LCMO and YBCO show a coexistence of ferromagnetism and superconductivity in these bilayer samples at low temperatures. Both magnetization measurements which are used to determine T_c and magneto-optical measurements for the j_c evaluation of the superconducting film show up a critical thickness for the YBCO film of about $d = 30$ nm, below which the superconductivity is strongly affected by the magnetic film. We presented a theoretical model which is able to describe the occurring suppression of the superconductivity in terms of a polarized spin diffusion process from the magnetic layer into the superconductor. a quantitative evaluation provides a value of the spin diffusion length in this system of $\xi_{FM} \approx 10$ nm at low temperatures.

References

[1] H. -U. Habermeier, G. Cristiani, R. K. Kremer, O. Lebedev, and G. Van Tendeloo, Physica C **364**, 298 (2001).
[2] A. M. Goldman, V. A. Vas'ko, P. A. Kraus, K. R. Nikolaev, and V. A. Larkin, J. Magn. Magn. Mater. **200**, 69 (1999).
[3] N. C. Yeh, R. P. Vasquez, C. C. Fu, A. V. Samoilov, Y. Li, and K. Vakili, Phys. Rev. B **60**, 10522 (1999).
[4] Z. Sefrioui, D. Arias, V. Pena, J. E. Villegas, M. Varela, P. Prieto, C. Leon, J. L. Martinez, and J. Santamaria, Phys. Rev. B **67**, 214511 (2003).
[5] T. Holden, H. -U. Habermeier, G. Cristiani, A. Golnik, A. Boris, A. Pimenov, J. Humlicek, O. Lebedev, G. Van Tendeloo, B. Keimer, and C. Bernhard, cond-mat./**0303284** and submitted to Phys. Rev. B.
[6] Q. Si, Phys. Rev. Lett. **78**, 1767 (1997).
[7] R. J. Soulen, Jr., J. M. Byers, M. S. Osofsky, B. Nadgorny, T. Ambrose, S.F. Cheng, P. R. Broussard, C. T. Tanaka, J. Nowak, J. S. Moodera, A. Barry, and J. M. D. Coey, Science **282**, 85(1998).
[8] J. Y. T. Wei, J. Superconduct. **15**, 67 (2002).
[9] Y. Gim, A. W. Kleinsasser, and J. B. Barner, J. of Appl. Phys. **90**, 4063 (2001).
[10] Ch. Jooss, J. Albrecht, H. Kuhn, S. Leonhardt, and H. Kronmüller, Rep. Prog. Phys. **65**, 651 (2002).
[11] Ch. Jooss, R. Warthmann, A. Forkl, and H. Kronmüller, Physica C **299**, 215 (1998).
[12] J. Albrecht, S. Soltan, and H.-U. Habermeier, EuroPhys. Lett. **63**, 881 (2003).
[13] S. Soltan, J. Albrecht, and H.-U. Habermeier, submitted to Phys. Rev. B.
[14] K. E. Gray, "Nonequilibrium Superconductivity, Phonons and Kapitza Boundaries", edited by K. E. Gray (Plenum, New York, 1981).
[15] P. M. Tedrow and R. Meservey, Phys. Rev. Lett. **26**, 192 (1971).
[16] P. J. Hirschfeld and N. D. Goldenfeld, Phys. Rev. B **48**, 4219 (1993).
[17] M. Franz, C. Kallin, A. J. Berlinsky, and M. I. Salkola, Phys. Rev. B **56**, 7882 (1997).
[18] C. Y. Yang, A. R. Moodenbaugh, Y. L. Wang, Youwen Xu, S. M. Heald, D. O. Welch, M. Suenaga, D. A. Fischer, and J. E. PennerHahn, Phys. Rev. B **42**, 2231 (1990).

Electrical resistivity of magnetic fluids

B. A. Al Shalabi and **H . M. El-Ghanem**[*]

Physics Department Jordan University of Science and Technology, Irbid, Jordan

Received 31 August 2003, accepted 31 December 2003
Published online 1 April 2004

PACS 72.80.Ga, 74.47.Pp, 75.50.Mm

Resistance (R) of Fe_3O_4-based magnetic fluids (MFs) of various liquid carriers, concentration of Fe_3O_4 particles (ε) and applied magnetic fields (H), have been estimated from the voltage–current measurements at room temperature. The resistivity was found to follow an empirical relation of the form $\rho = \rho_0 \varepsilon^{-n}$, where ρ_0 is a constant and the exponent n was found to depend on the viscosity of the liquid carrier. The voltage current relation was found to be linear up to 120 V. A simple model was proposed to explain the dependence of the resistance on the viscosity of the liquid carrier and on chain length.

© 2004 WILEY-VCH Verlag GmbH & Co. KGaA, Weinheim

1 Introduction

Magnetic fluids (MFs) are stable colloidal suspension of single-domain magnetic particles (~ 100A) in a suitable liquid carrier, such as kerosene, diester, Isopar-M, etc. [1–6]. MFs continue to generate interest because of present and wide potential industrial and technological applications as well as of academic importance [1–4]. Carrier liquids in general are occupying the intermediate stage between ion dielectics and aqueous electrolyte solutions in their resistivity. The Fe_3O_4 which is widely used as a magnetic dispersed phase has a relatively low electrical resistivity, which is several orders higher than that of metals. In MFs the magnetic particles are separated from each other by coating each particle with a surfactant layer with one molecule thick. Oleic acid is oftenly used for this purpose. Oleic acid is an organic molecule whose tail is a chain of 18 carbon atoms and whose head is a polar carboxyl group (COOH) [4]. The chemical affinity to the carrier liquid and values of the mobility of charge carriers and their concentrations of this surfactant are close to those of the carrier [2].

The distribution of particle sizes which exists in any real MF system plays a profound role in its magnetic, electrical, rheological properties [3, 7].

The results of numerical simulations as well as of actual experimental results under the effect of, magnetic fields or the dipole–dipole interactions, reveal that there exist chain – like structures of magnetic particles which play an important role in physical phenomena observed in the MFs such as magneto-mechanical, magneto-optical and magnetic properties of MFs [8–12].

Dababneh et al. [13] measurements of the relative resistivity of Fe_3O_4 magnetic fluids for $0.00526 < \varepsilon < 0.0526$ and their results showed that the relative resistivity is inversely proportional to the concentration of magnetic particles. Although the magnetic, magneto-optical, rheological properties of MFs have been studied extensively using different techniques [1–3, 12–14]. the electrical behavior of MFs have been reported to a much lesser extent.

In this work, measurements of the voltage–current (I–V) characteristics of several magnetic fluids over a wide range of concentrations and carriers at room temperature have been carried out with and

---

[*] Corresponding author: e-mail: hmel@just.edu.jo

without the application of an external magnetic field. The results of this work are believed to be helpful in understanding the electrical properties of MFs.

2 Experimental procedure

The original Fe_3O_4 magnetic fluids samples were purchased from ferrofluidics GmbH and different concentration of magnetic particles were varied by adding proper amounts of the corresponding liquid carriers. The volumic packing fraction (ε) for each sample was estimated using the following relation:

$$\varepsilon = \frac{\rho_s - \rho_c}{\rho_p - \rho_c} \quad (1)$$

where ρ_s, ρ_c and ρ_p are the sample, carrier and Fe_3O_4 densities, respectively.

Direct measurements of the current at various concentrations were made using a Metravo (ABB) multi-meter with 10 MΩ constant internal resistance which enabled measurements of currents as low as 0.1 nA. The potential difference across a square cell of gold-plated copper plates of 1 cm^2 area was measured using Philips multi-digital voltmeter with 10 µV accuracy.

The magneto-resistance was measured with the applied magnetic field direction parallel to the direction of the electric field. The intensity of the magnetic field was kept uniform through out the cell area..

The viscosity (η_0) of all used liquid carriers were measured via Oswald viscometer, the results are listed in Table 1.

The electrical measurements have been taken several times and repeated over a long period of time to insure reproducibility of the results. Prior to each measurement the sample homogeneity was insured by placing the sample for 10 minutes in a 3000 RPM centrifuge.

3 Results and discussion

Figures 1 and 2 show the (I–V) characteristics for Fe_3O_4 based magnetic fluids in kerosene and diester, respectively, for volumic packing fractions in the range $55 \times 10^{-4} < \varepsilon < 243 \times 10^{-4}$.

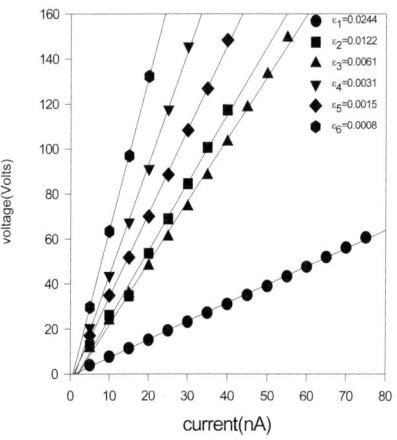 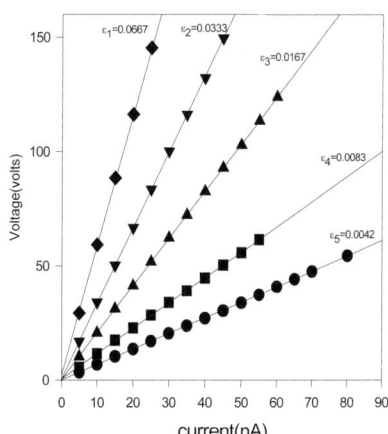

Fig. 1 The I–V curves for Kerosene based MFs. **Fig. 2** The I–V curves for Diester based MFs.

For each ε a linear relationship between voltage (V) and current (I) was observed, with all lines pass through the origin. Using the least squares method for a straight line, the slope (resistance) of each curve was obtained and plotted versus ε on a log–log scale as shown in Fig. 3 for the kerosene and in Fig. 4 for the diester based samples. Similar measurements and analysis were made on olive oil and toluene based samples from which the following empirical relation was derived:

$$R = R_0 \varepsilon^{-n} \tag{2}$$

where R_0 and n are constants determined from the graph.

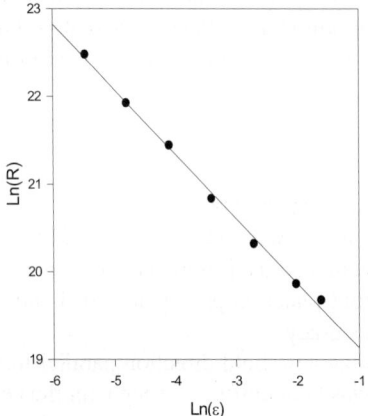

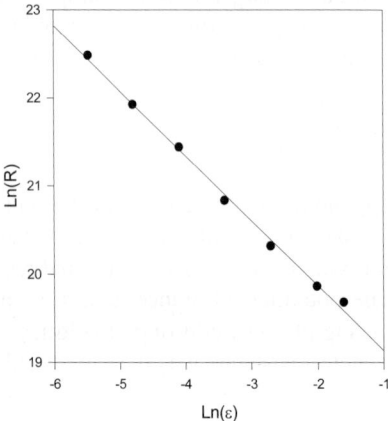

Fig. 3 Ln(ε) versus ln(R) for Kerosene based MFs.

Fig. 4 (Ln(ε) versus ln(R) for Diester based MFs.

The determined values for R_0 and n for each case was calculated from the graphs and the results are shown in Table 1, together with the values of viscosity (η_0).

Table 1 The viscosity (η_0), the exponent (n) and the constant R_0, as given in equation (2) for different MFs.

	Kerosene	Diester	Olive oil	Toluene
R_0(MΩ)	30.52	98.0	349	21.9
n	0.88	0.74	0.62	0.90
η_0	0.92	22.4	84	0.95

The magneto-resistance measurements for the Fe_3O_4 based MFs in kerosene was carried out for two packing fractions $\varepsilon_1 = 0.0061$ and $\varepsilon_2 = 0.001524$.

The results are shown in Fig. 5. It can be seen from the figure that the magneto-resistance decreases with increasing H and continues to do till it reaches asymptotic value for H > 800 Oe.

To examine the Ohmicity of the resistance of MFs, the (I–V) characteristic curves were measured for a selected concentration namely $\varepsilon = 0.0122$ for different cell lengths (L). The results are presented in Fig. 6. The inset shows that R varies linearly with L in accordance with Ohm's law in the voltage range 0 < V< 24 kV/m. From the sample geometry of the electrodes, the resisitivity of this sample was estimated to be equal to $\rho = 5.24 \times 10^8$ Ωm.

The results can be explained qualitatively based on the following assumptions:
- The magnetic particles are non-interacting spheres of uniform size with radius (a) and,
- The magnetic particles form linear chains of a uniform size with an average length L.

The number of these chains (N) is proportional to $[(a/d)^2 P]$, where d is the radius of the electrode, and P is the probability that a chain is being formed on the electrode parallel to H, so one writes

$$N \sim (a^2/d^2)P . \tag{3}$$

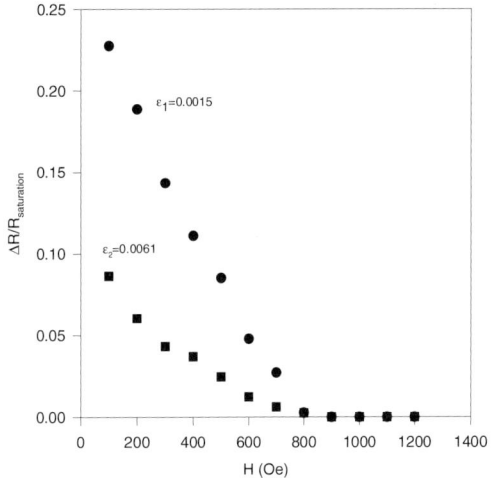
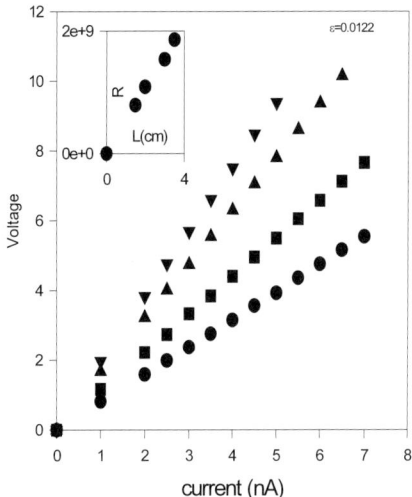

Fig. 5 Magneto-resistance vs. H for two Kerosene samples. **Fig. 6** The I–V curves for Kerosene MFs.

The resistance of one chain can be written as

$$r = (\rho_0 L)/(\pi a^2) \tag{4}$$

and as these chains are connected in parallel, the conductance (1/R) of the sample is given by;

$$\frac{1}{R} = \sum_{i=1}^{N} \left(\frac{1}{r}\right)_i = \frac{N\pi a^2}{\rho_0 L} \tag{5}$$

substituting Eq. (3), in Eq. (5) gives

$$R = \frac{\rho_0 L d^2}{\pi a^4 P} \,. \tag{6}$$

Equation (6) indicates that the resistance is inversely proportional to the probability values of forming chains (P). It worth mentioning that P is a function of ε, H, liquid carrier and size of the dispersed particles as well as of the inter particle interactions.

The dependence of R on ε, can be explained as follows:
since the chains attained a constant terminal velocity (v_t) equals the drift velocity (v_d), which is approximated by

$$v_d = \frac{Eq}{6\pi \eta L} \tag{7}$$

where E is the electric field and q is the charge on the chain.
Thus

$$R = \frac{6\pi L d \eta}{nq^2 A} \tag{8}$$

where A is the area of the electrode and η is the viscosity of MFs, which is given by [1]

$$\eta = \frac{\eta_0}{\left(1 - a\varepsilon - b\varepsilon^2\right)} \tag{9}$$

whence R can be rewritten as

$$R = R_0 \varepsilon^{-n} \quad (10)$$

where R_0 is a function of η_0, the carrier viscosity.

Figure 7 depicts the relation between the exponent (n) and (η_0) for all carrier liquids used, where it can be seen that a relation of the form

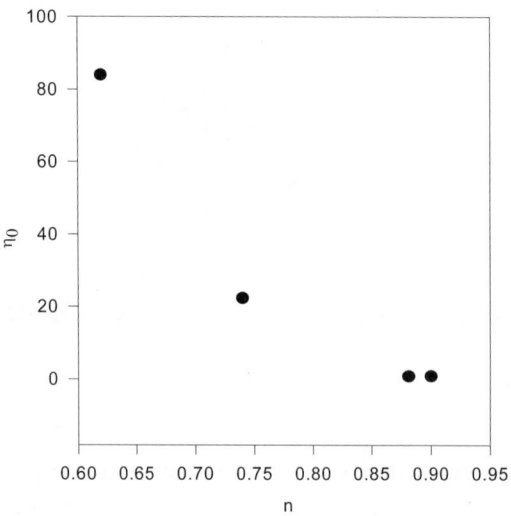

Fig. 7 The exponent n versus η_0 for all carrier fluids.

$$\eta_0^{1/2} = \alpha + \beta \, n^{-1} \quad (11)$$

holds and the values of α and β are -18.08 and 16.89, respectively.

Equations (10) and (11) can be used to estimate η_0 from electrical measurements and further more we conclude that source of electric conduction does not originate from ion trapping as mentioned in Ref. [13], but is due to electron hopping through chains or partially due to uncoated Fe_3O_4 particles and electrophoreses, which is due to the polarizing tails attached to the magnetic particles forming these chains.

4 Conclusion

The ohmic behaviour of the current voltage relation of all ferrofliuds used in this study, and irrespective of the type of the carrier liquid can be used to determine indirectly the viscosity of a small volume of the ferrofliud. The origin of electric conduction in the non-metallic fluids needs further investigation, especially with the application of an a.c current.

Acknowledgments The authors wish to thank Jordan University of Science & Technology for financially supporting this project

References

[1] R. E. Rosensweig, "Ferrohydrodynamics" (Cambridge Univ. Press, Cambridge, 1985).
[2] V. E. Fertman, Magnetic Fluids Guidebook: Properties and Applications (Hemisphere, New York, 1990).
[3] S. W. Charles and J. Popplewell, in: Ferromagnetic Materials, Vol. 2, Ed. E. P. Wohlfarth (North-Holland, New York, 1980), p. 509.
[4] R. E. Rosensweig, Sci. Am. **247**, 136 (1982).
[5] P. C. Scholten, J. Magn. Magn. Mater. **39**, 99 (1983); ibid. IEEE Trans. Magnetics **16**, 77 (1980).

[6] M. I. Shliomis, Sov. Phys. Usp. **17**, 153 (1974).
[7] N. Wiser, J. Magn. Magn. Mater. **159**, 119 (1996).
[8] A. Satoh, R. W. Chantrell, S. Kamiyama, and G. N. Coverdale, J. Colloid Interface Sci. **181**, 422 (1996).
[9] J. Cernak, P. Macko, and M. Kasparkova, J. Phys. D **24**, 1609 (1991).
[10] C. Johansson, M. Hanson, and P. Lundqvist, J. Magn. Magn. Mater. **157/158**, 599 (1996).
[11] S. Chikazumi, S. Taketomi, M. Ukita, M. Mizukami, H. Miyajima, M. Setogawa, and Y. Kurihara J. Magn. Magn. Mater. **65**, 245 (1987).
[12] C. F. Hayes and S. R. Hwang, J. Colloid Interface Sci. **60**, 443 (1977).
[13] M. S. Dababneh, N. Y. Ayoub, I. Odeh, and N. M. Laham, J. Magn. Magn. Mater. **125**, 34 (1993).
[14] E. Blums, A. Cebers, and M. M. Mairrov, "Magnetic Fluids" (Walter de Gruyter, NewYork, 1997).

The BM_5Se_9 phases (B = Al, Ga, Ge, Sb, Sn; M = V, Nb, Ta): superconductors or ferromagnets?

A. Leblanc-Soreau[*,1], **P. Molinié**[1], and **J. C. Jumas**[2]

[1] I.M.J.R.-CNRS, 2, rue de la Houssinière, BP 32229, 44322 Nantes Cedex 3, France
[2] LAMMI-CNRS, place E. Bataillon, 34095 Montpellier Cedex 5, France

Received 31 August 2003, accepted 31 December 2003
Published online 23 April 2004

PACS 74.70.Dd, 75.50.Cc, 76.30.Fc, 76.80.+y

The dissolution of some intermetallic A15 compounds in lamellar metallic diselenides (2H-NbSe$_2$, 2H-TaSe$_2$ and 1T-VSe$_2$) results in original phases. Syntheses performed in the Nb$_3$Sn/NbSe$_2$, Nb$_3$Ge/NbSe$_2$, Nb$_3$Sn/TaSe$_2$, V$_3$Ga/NbSe$_2$, Nb$_3$Sb/NbSe$_2$ V$_3$Ga/VSe$_2$ systems lead to the formation of the BM_5Se_9 phases (superconductors or ferromagnets). The T_c values vary from 17.5(2) K for SnNb$_5$Se$_9$ to 4.5(2) K for GeNb$_5$Se$_9$. For the ferromagnet GaV$_5$Se$_9$, the ESR study shows two vanadium sites with axial symmetries. Previous results in Mössbauer and Raman studies showing the existence of B-M entities, are in agreement with the Mössbauer ones obtained on SnNb$_5$Se$_9$ which could be obtained without any superconducting behaviour. The comparison could be done between the 3 superconductors GaNb$_5$Se$_9$, SnNb$_5$Se$_9$, SnNb$_{0.5}$Ta$_{4.5}$Se$_9$ and the 3 ferromagnets GaV$_5$Se$_9$, Sn$_{0.94}$Ga$_{0.05}$Nb$_{4.70}$V$_{0.30}$Se$_9$, GaV$_{0.15}$Ta$_{4.85}$Se$_9$, the presence of V probably modifies the superconducting exchanges and allows ferromagnetic couplings.

© 2004 WILEY-VCH Verlag GmbH & Co. KGaA, Weinheim

1 Introduction

In order to try to get multi-layer systems [1, 2], the inclusion of superconducting chains has been tested in the metallic diselenides {MSe$_2$ with M = (Nb, Ta, V)}. However the outcome derivatives are the result of the dissolution of small amount of intermetallic A15 compounds {A$_3$B with A = (Nb, V) and B = (Al, Ga, Ge, Sb, Sn)} in the host matrix 2H-NbSe$_2$, 2H-TaSe$_2$ and 1T-VSe$_2$. In the case of the two superconductors 2H-NbSe$_2$ and 2H-TaSe$_2$, electronic instabilities lead only to an incommensurate charge density waves distortion (ICDW) whereas in 1T-VSe$_2$ they lead to an ICDW followed by commensurate charge density waves (CCDW). Moreover 1T-VSe$_2$ is always non-stoichiometric and its formula is V$_{1+x}$Se$_2$ with x ≈ 0.005.

After a very precise study of the Nb$_3$Sn/NbSe$_2$ system [3, 4], the chemical process of the synthesis can be understood in order to study new systems. In all cases the chemical formula of the new obtained phases can be written as following B(M/A)$_5$Se$_9$ were B can be Al, Ga, Ge, Sb, Sn, A is Nb or V, and M stands for Nb, Ta, V. In addition, and to further explore the incidence of the MSe$_2$ network as that of the metal B of the starting A$_3$B (A15), the Nb$_3$Sn/VSe$_2$, V$_3$Ga/NbSe$_2$, V$_3$Ga/TaSe$_2$ systems are studied, and completed through the B (or B' ≠ B) metals used in front of the A15 phases. Some of the new phases are superconductors, the others being ferromagnets.

The ^{119}Sn Mössbauer measurements realized on the superconducting SnNb$_5$Se$_9$ [5] show 3 types of Sn sites,but SnNb$_5$Se$_9$ could also exist without any superconducting properties; its Mössbauer studies reveal a great modification: only one site Sn(II) exists.

2 Experimental

The magnetic properties of powder samples were carried out in the temperature interval from 2 to 300 K using a commercial SQUID magnetometer "MPMS-5" from Quantum Design. The powder samples were placed in medical caps. Powder sample temperature dependent ESR measurements

---

[*] Corresponding author: leblanc@cnrs-imn.fr, Fax: 33 2 40 37 39 95

were performed on a Bruker ER 200D-SRC operating at 9.418 GHz and equipped with a continuous nitrogen flow system ER 4111VT in the temperature range 100 K to 300 K. Each spectrum was simulated with the pole integration method [6], which provided the parameters of the possible g distributions, which compose the signal. ^{119}Sn Mössbauer spectra were recorded in the constant-acceleration mode on a ELSCINT-AME40 spectrometer (the processes are described in [5].

Before being studied according to these techniques all the samples were analysed by X-ray powder diffraction, density measurements, EDAX analysis, metallographic microscopy. The elemental analysis was obtained by using a scanning electron microscope equipped with an EDX system (Jeol JMS-35C sonde tracor).

3 Synthesis results

At the beginning only two systems Nb$_3$Sn/NbSe$_2$ and V$_3$Ga/VSe$_2$ were studied, this yielded two new phases [3], the superconductor SnNb$_5$Se$_9$ and the ferromagnet GaV$_5$Se$_9$ The chemical synthesis process of these phases was understood by the discovery of an original fact: a small amount of B metal must be added to the A$_3$B/MSe$_2$ mixture in order to lead to a stoichiometric formula (A,B,M)$_2$Se$_3$. In fact it was possible to explore many other systems with or without changing the B metal from those used in the pristine A15, so we can study the role of the extra metal B (or B') in the synthesis (Table 1). But A$_3$B/VSe$_2$ systems where B = Sn or combined with Sn for B ≠ Sn led to V$_x$Se$_y$ or Sn$_z$V$_x$Se$_y$. All syntheses confirm the complete dissolution of the pristine A15 phase in the MSe$_2$ host structure and the insertion of the extra B (or B') metal in the new phase (Table 1).

Table 1 the pristine mixtures, the formulas and the critical temperatures T$_C$ of the superconducting and ferromagnetic phases measured under a field strength of 100 G.

Superconducting systems	Resulting phases	Transition temperatures (T$_c$)
Nb$_3$Sn/NbSe$_2$ + Sn	SnNb$_5$Se$_9$	17.5(2) K
Nb$_3$Sn/NbSe$_2$ + Ga	Ga$_{0.84}$Sn$_{0.15}$Nb$_5$Se$_9$	18.0(2) K
V$_3$Ga/NbSe$_2$ + Ga	GaNb$_5$Se$_9$	12.0(2) K
Nb$_3$Sn/TaSe$_2$ + Sn	SnNb$_{0.5}$Ta$_{4.5}$Se$_9$	17.5(2) K
Nb$_3$Sb/NbSe$_2$ + Sb	SbNb$_5$Se$_9$	6.0(2) K
Nb$_3$Al/NbSe$_2$ + Al	AlNb$_5$Se$_9$	18.0(2) K
Nb$_3$Ge/NbSe$_2$ + Ge	GeNb$_5$Se$_9$	4.5(2) K
Nb$_3$Al/Ge/NbSe$_2$ + Al/Ge or Nb$_3$Al$_x$Ge$_{1-x}$/NbSe$_2$ + Ge/Al	Ge$_x$Al$_{1-x}$Nb$_5$Se$_9$	16.5(2) K
Ferromagnetic systems	Resulting Phases	Transition Temperatures (T$_c$)
V$_3$Ga/VSe$_2$ + Ga	GaV$_5$Se$_9$	21.0(5) K
V$_3$Ga/NbSe$_2$ + Sn	Sn$_{0.94}$Ga$_{0.05}$Nb$_{4.70}$V$_{0.30}$Se$_9$	35.7(5) K
V$_3$Ga/TaSe$_2$ + Ga	GaV$_{0.15}$Ta$_{4.85}$Se$_9$	11.1(5) K

Indexations of the X-ray powder patterns show that the unit cells are, excepted AlNb$_5$Se$_9$ [5], a substructure of the pristine cell of MSe$_2$ (hexagonal cell parameters a',c') with a ≅ 4a', b ≅ 4a'√3,c ≅ 9/2 c'.

4 Superconducting phases

The ^{119}Sn Mössbauer measurements realized on SnNb$_5$Se$_9$ [5] show 3 types of Sn: site 1 corresponds to Sn surrounded by Nb atoms, site 2 (symmetrical) and site 3 (asymmetrical) to Sn in an Se environment. At 90 K a reversible dynamic conversion between sites 2 and 3 occurs correlated to a possible CDW distortion. However when a non-stoichiometric NbSe$_2$ is used for the synthesis, a non-superconducting SnNb$_5$Se$_9$ is obtained, thus only one singlet of Sn(II) surrounded by Se is observed at 298 K, then below 80 K a quadrupolar splitting appears (Table 2).

Table 2 Mössbauer parameters of non-superconducting $SnNb_5Se_9$ (IS relative to $BaSnO_3$).

Temperature measurements	IS δ mm/s	Q.S. Δ mm/s	LW Γ mm/s	Contribution C %	χ^2
298 K	3.614(4)	–	0.91(1)	100.0	0.42
80 K	3.658(8)	0.36(2)	0.97(2)	96.5	0.64
4 K	3.68(1)	0.28(3)	1.00(2)	97.0	0.76
298K not relaxed	3.598(3)	–	0.92(1)	92.5	0.43

The susceptibility curves shown in Figs. 1, 2, 3 reveal the superconducting behaviours for powder samples of BM_5Se_9 phases (Table 1) (ZFC, 10 mT field strength).

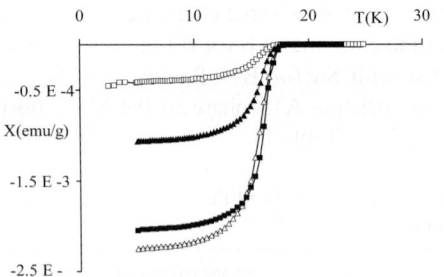

Fig. 1 $SnNb_5Se_9$: (□) Synthesized with 5 % in mass of Nb_3Sn; (▲) synthesized with 6 % in mass; (■) and (Δ) powder samples, quenched with two different speeds.

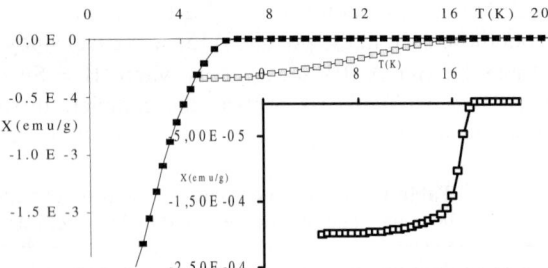

Fig. 2 $SbNb_5Se_9$ (■), $AlNb_5Se_9$ (□); in insert $Ga_{0.84}Sn_{0.15}Nb_5Se_9$.

5 Ferromagnetic phases

The ferromagnetic state of GaV_5Se_9 (powder samples) appears at 21 K (Fig. 4 and Table 1) with a strong anisotropy in the coercive field [3, 4]. The Curie constant value obtained from the high susceptibility curve is equal to 0.24 emu.K.mol^{-1}, which gives $\mu_{eff} = 0.62\mu_B$ per V atom. From the magnetization measurement the value of the total effective magnetic moment at 0K is equal to $0.7\mu_B$ ($0.14\mu_B$ per V).

The susceptibility curves shown in Figs. 5, 6 reveal a ferromagnetic behaviour observed for $GaV_{0.15}Ta_{4.85}Se_9$ and $Sn_{0.94}Ga_{0.05}Nb_{4.70}V_{0.30}Se_9$. In the latter case the ferromagnetic behaviour is more complicated with may be two transition temperatures.

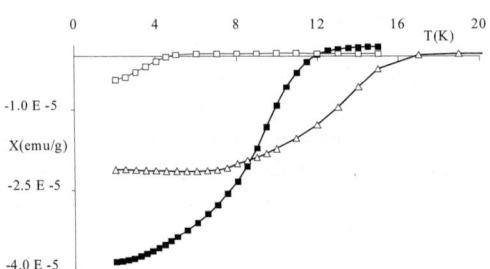

Fig. 3 $GaNb_5Se_9$(■), $GeNb_5Se_9$(□), $(Ge/Al)Nb_5Se_9$ (Δ).

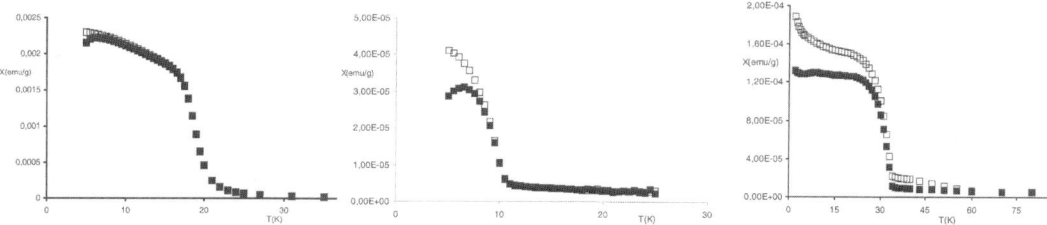

Fig. 4 GaV$_5$Se$_9$ ■ ZFC; □ FC.

Fig. 5 GaV$_{0.15}$Ta$_{4.85}$Se$_9$. ■ ZFC; □ FC.

Fig. 6 Sn$_{0.94}$Ga$_{0.05}$Nb$_{4.70}$V$_{0.30}$Se$_9$. ■ ZFC; □ FC.

6 ESR study of the ferromagnet GaV$_5$Se$_9$

The ESR spectra obtained for a powdered sample in the temperature range 100 K–300 K indicate that the Landé factor g and width (W) are almost temperature independent. In order to simulate the ESR spectrum (Fig. 8) we use two Zeeman distributions with axial symmetry and lorentzian shape. The simulation leads to the following parameters: the first one, noted D_1, with $g_{D1}// = 1.990(1)$, $g_{D1}\perp = 1.795(1)$, the width $W_{D1} = 150$ G and the relative intensity $I_{D1} \approx 0.6$; the second D_2 with $g_{D2}// = 1.970(1)$, $g_{D2}\perp = 1.910(1)$, $W_{D2} = 130$ G and $I_{D2} \approx 0.4$. The g isotropic values of two distributions are 1.929(8) for D_1 and 1.951(5) for D_2. It is very difficult to determine the oxidation state of the vanadium atoms in the paramagnetic domain (T > 22 K). The observed Curie constant is small compared to the spin only value of V^{IV} ($\mu_{eff} = 1.73\mu_B$; S = 1/2; 3d^1). However the ESR study reveals two different types of vanadium atoms and the difference in their populations agrees with the effective moment observed in the ferromagnetic state (the fifth of the effective moment observed in the paramagnetic state).

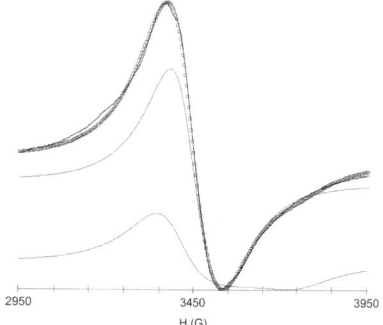

Fig. 7 Simulated ESR spectrum of powdered sample of GaV$_5$Se$_9$ at T = 105 K; the broad full lines are the experimental spectra, the light full lines are components representing the D_1 and D_2 distributions and the empty square stands for the simulated spectrum.

7 Discussion

Phases with vanadium are ferromagnets, but phases without any V are superconductors: for example, SnNb$_{0.5}$Ta$_{4.5}$Se$_9$ is superconductor but GaV$_{0.15}$Ta$_{4.85}$Se$_9$ is ferromagnetic, as Sn/GaNb$_5$Se$_9$ and (Sn,Ga)(Nb,V)$_5$Se$_9$. So when some vanadium atoms enter in the frame, they take fewer places than Nb, Ta atoms and electronic repulsions appear which destroy superconductivity state and replace them by ferromagnetic one. This could be compared with the role played by V atoms in Van Der Waals gaps in $V_{1+x}Se_2$.

The Mössbauer study of SnNb$_5$Se$_9$ [5] has shown the existence of 3 tin sites corresponding to 2 tin species, surrounded by niobium atoms (Sn site 1) and by selenium atoms (Sn II, sites 2 and 3); Raman studies on SnNb$_5$Se$_9$ and GaV$_5$Se$_9$ give evidence that the Se–Nb–Se or Se–V–Se slabs are kept and a partial preservation of some A15 structural fragments exists [7]. But from a non-perfect NbSe$_2$, SnNb$_5$Se$_9$ is obtained without any superconducting behaviour and its Mössbauer spectrum shows only the existence of one tin species (Sn II). We could imagine that the superconductivity properties could be due to the tin surrounded by niobium atoms in SnNb$_5$Se$_9$, as the evolution of the T_c values with the nature of B–M species.

The ESR results on GaV$_5$Se$_9$ show the existence of 2 sites for the vanadium atoms. So in all cases M in the BM$_5$Se$_9$ phases has one environment with Se and the second with B; so the superconducting behaviour could be depending on the existence of Sn–Nb, Ga–Ta, Ga–Nb, Sb–Nb species and the ferromagnetic state could be due to interactions between vanadium atoms in the "M_5Se_9" bulks. These results are in agreement with those obtained by Point-Contact-Probing for the $SnNb_5Se_9$ superconducting properties [8]. Point-Contact measurements of N–c–N and S–c–N types (N = normal metal, c = constriction, S = superconductor), show that two kinds of superconductors entities were observed for S–c–N case, thus $SnNb_5Se_9$ could be a mesoscopic superconductor.

References

[1] K. Kawaguchi and M. Sohma, Phys. Rev. B **46**, 14722 (1992).
[2] Th. Mühge, K. Westerhold, H. Zabel, N. N. Garifyanov, Yu. V. Gorynov, I. A. Garifullin, and G. G. Khaliullin, Phys. Rev. B **55**, 8945 (1997).
[3] P. Molinié, A. Leblanc-Soreau, E. Faulques, Z. Ouili, J. C. Jumas, and C. Ayache, in: Spectroscopy of superconducting materials, Ed. E. Faulques (ACS Symposiums series 730, Oxford University Press, Washington D. C., 1999), p. 21.
[4] P. Molinié and A. Leblanc, Phys. Chem. News, in press, ACMS-II 2001.
[5] A. Soreau-Leblanc, P. Molinié, and J. C. Jumas, Physica B **321**, 138 (2002).
[6] T. P. Nguyen, M. Giffard, and P. Molinié, J. Phys. Chem. **100**, 8340 (1994).
[7] A. Leblanc-Soreau, Z. Ouili, P. Molinié, and E. Faulques, Phys. Chem. New, to be published., Congr. CIS 2003.
[8] G. V. Kamarchuk, P. N. Chubov, A. Leblanc-Soreau, P. Molinié, and E. Faulques, Europhys. Conf. Abstracts, Vol. 26; to be published in Nato Science Series (Kluwer, 2004, Vol. 26A).

Conduction mechanism in Pr-doped GdBa$_2$Cu$_3$O$_7$

M. R. Mohammadizadeh and **M. Akhavan**[*]

Magnet Research Laboratory(MRL), Department of Physics, Sharif University of Technology,
P.O. Box 11365-9161, Tehran, Iran

Received 31 August 2003, accepted 31 December 2003
Published online 1 April 2004

PACS 71.30.+h, 74.25.Dw, 74.25.Fy, 74.72.Jt

The normal state resistivity of single phase polycrystalline Gd(Ba$_{2-x}$Pr$_x$)Cu$_3$O$_{7+\delta}$ samples with $0.0 \leq x \leq 0.50$ have been investigated. The two-dimensional variable range hopping is dominant in the normal state resistivity of the samples. The conduction shows that Pr-doping strongly localizes the carriers in normal state, and finally causes the suppression of superconductivity. Based on the resistivity measurements, the effect of Pr substitution in 123 structure of HTSC at Gd or Ba site is to increase the pseudogap temperature T_s, although, Pr at Ba site has a stronger effect on the suppression of T_s and superconductivity. Pr-doping not only reduces the carrier density and induces pseudogap, but also simultaneously increases T_s and suppresses the linearity behavior of resistivity vs. temperature in the normal state.

© 2004 WILEY-VCH Verlag GmbH & Co. KGaA, Weinheim

1 Introduction The normal state of the high temperature cuprate superconductors shows many unusual properties, which are far from the standard Fermi liquid behavior. An example is the so-called pseudogap behavior observed in the under-doped region near the Mott insulating phase, which is a frequency threshold for the strong excitation of spin and charge modes [1]. Among different models for describing the charge transport in materials, hopping conduction between localized states have been widely used for normal state of HTSC, semiconductors, and perovskites. In hopping conduction, the temperature dependence of resistivity is $\rho = \rho_0 (T/T_0)^{2p} \exp(T_0/T)^p$, where T_0 is a characteristic temperature, which will be discussed later, and $p = 1/3$ in two dimensions (2D), $p = 1/4$ in three dimensions (3D), and $p = 1/2$ in Coulomb gap (CG) regime [2]. The CG, 2D-VRH, and 3D-VRH have been reported for normal state resistivity of different HTSCs [3].

The pseudogap has been observed from different experiments. In addition, due to the extremely complex structural and electronic nature of the high T_c cuprates [4], and the somewhat controversial nature of much of the current experimental data, there are many theoretical attempts to explain the pseudogap behavior. A number of families of HTSC have also shown evidences of pseudogap [5]. In this paper, through resistivity measurements, we will investigate the effect of Pr at Ba and R sites in GdBa$_2$Cu$_3$O$_7$-based compound from the conduction mechanism and pseudogap points of view.

2 Experimental details The Gd(Ba$_{2-x}$Pr$_x$)Cu$_3$O$_{7+\delta}$ single phase polycrystalline samples with $0.00 \leq x \leq 0.50$ were synthesized by the standard solid state reaction technique. The amount of nominal dopings were confirmed by the Rietveld refinement. The Gd$_2$O$_3$, Pr$_6$O$_{11}$, BaCO$_3$, and CuO powders with 99.9 % purity were mixed, ground, and calcined at 840 °C for 24 h in an air atmosphere. Calcination was repeated twice with intermediate grinding. Then powders were reground, pressed into pellets, and synthesized at 930 °C for 24 h in an oxygen atmosphere. The samples were cooled to 550 °C and retained under oxygen flow for 16 h. Finally, they were furnace cooled to room temperature. An ac four-probe method with f = 33 Hz and 0.3 A/cm^2 current density was used for the conductivity measurements of the samples

---

[*] Corresponding author: e-mail: akhavan@sharif.edu, Phone: +98 21 6164510, Fax: +98 21 6012983

within the temperature range of 10 to 300 K. A Lake Shore-330 temperature controller with two Pt-100 resistors was used for measuring and controlling the temperature to within ±10 mK.

3 Results and discussion   In the samples with $x \leq 0.35$, the superconducting transition occurs, while for $x \geq 0.4$, there is no transition down to 10 K. Therefore, within our 0.05 steps in samples preparation, the critical doping for superconductor–insulator transition (SIT) is $x_c^{SIT} = 0.35$.

To test the VRH and CG regimes, we have fitted 2D-VRH, 3D-VRH, and CG regimes separately for all the samples (i.e. fixed-p method). These results are shown in Fig. 1. To recognize the best fit to the curves, the χ^2s for the fits are also evaluated. It is evident that the numerical results for the 2D-VRH is preferable than the others. However, it would be interesting to leave the exponent value p as a variable, and obtain the valid regime automatically (i.e. the variable-p method). For almost all the samples, the value of p tends to 1/3 with a reasonable fit's results, which corresponds to the 2D-VRH (Table 1). It is promising to see that this result is consistent with the above fixed-p method. The 2D behavior is the HTSC's well known character, which has been established for many years [6].

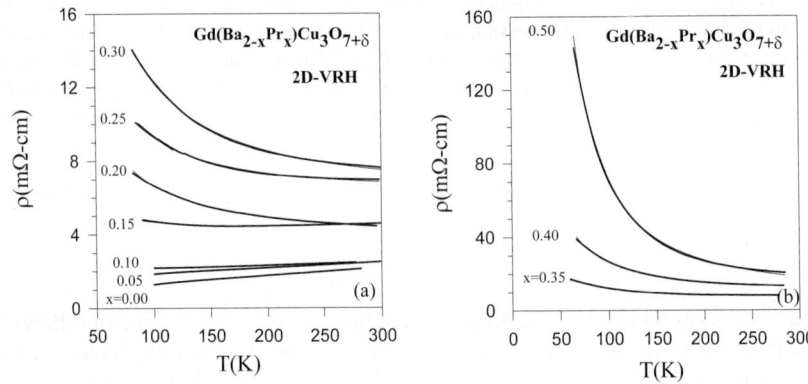

Fig. 1   The 2D-VRH fit (fixed-p method) for different amounts of Pr-doping (x). The dots appearing as a thick line are the experimental data, and the thin lines are the 2D-VRH regime predictions; (a) $0.00 \leq x \leq 0.30$ and (b) $0.35 \leq x \leq 0.50$.

Table 1   The resistivity fit with the variable-p method in hopping conduction for different amounts of Pr doping (x). The ρ_0, T_0, and p are the fit parameters and χ^2 shows the goodness of fit.

x	ρ_0(mΩcm)	T_0(K)	p	χ^2
0.00	0.5	98	0.39	0.9994
0.05	1.0	349	0.37	0.9988
0.10	1.2	676	0.36	0.9998
0.15	2.4	1144	0.38	0.9902
0.20	2.1	9274	0.30	0.9998
0.25	3.7	2083	0.39	0.9994
0.30	3.8	3700	0.37	0.9995
0.35	3.9	4036	0.36	0.9945
0.40	5.4	7404	0.36	0.9983
0.50	3.4	38493	0.32	0.9986

In the metallic samples ($d\rho/dT > 0$), T_s is defined as the deviation point of resistivity from linear behavior at temperature at which pseudogap (or spin gap) is opened [7], as is shown in Fig. 2. The resistance hump before superconducting transtion in $x = 0.10$, and 0.15 samples has been discussed elsewhere [8]. It is observed that the temperature T_s associated with the appearance of the pseudogap state, becomes equal to T_c around optimum doping, but is larger than T_c in the under-doped regime. This is consistent with HTSC phase diagram [9]. The x dependence of T_s for GdBaPr-123 and GdPr-123 compounds [3] are presented in Fig. 3. In both systems with increasing Pr-doping, T_s increases, but the change of T_s in GdPr-123 is closer to the linear behavior than in the GdBaPr-123 compound. This is in agreement with other $R_{1-x}Pr_x$-123 compounds [7]. In GdBaPr-123 system, T_s grows exponentially with x, and for a fixed amount of Pr-doping, its T_s is larger than the corresponding value for GdPr-123. This shows that the effect of Pr at Ba is stronger for changing the linear temperature dependence of resistivity and opening the pseudogap in comparison with the effect of Pr at R site. In comparison, T_c decreases with Pr doping x [8].

Based on the above results, it is noteworthy that the real position of Pr in the constructed structure is important not only for superconducting properties [8], but also for the normal state behaviors [10], as well as magnetic properties [11]. Therefore, undesired substitution of Pr at Ba site (mis-substitution) in $PrBa_2Cu_3O_{7-\delta}$ compound during the preparation process could have destructive effects on the exhibition of superconductivity. On the other hand, the very close neutron scattering length of Ba (0.507×10^{-12} cm), and Pr (0.458×10^{-12} cm) [12] have also caused in not distinguishing their real positions in the structure. So, determination of Pr real position in each 123 compound is a crucial step in explaining its influence.

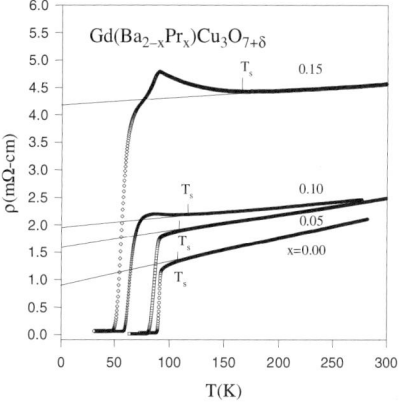
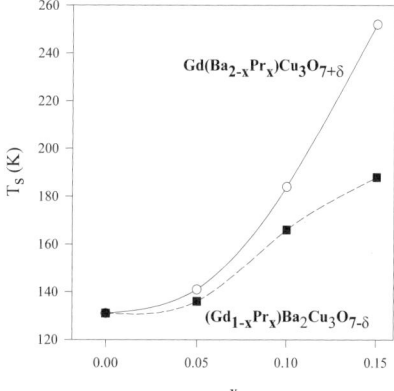

Fig. 2 Resistivity vs. temperature curves for $Gd(Ba_{2-x}Pr_x)Cu_3O_{7+\delta}$ with $0.00 \leq x < 0.20$. The deviation of each curve from linear behavior defines the pseudogap temperature T_s.

Fig. 3 Variation of pseudogap temperature vs. the amounts of Pr doping (x) for $Gd(Ba_{2-x}Pr_x)Cu_3O_{7+\delta}$ and $(Gd_{1-x}Pr_x)Ba_2Cu_3O_{7-\delta}$. The lines are guides to the eye.

The strong T_s–T_c dependence on oxygen deficiency and Pr position in the structure, suggest that there should be a correlation between superconducting gap – which is related to superconducting transition temperature – and pseudogap. This result has also been concluded for the c-axis oriented $NdBa_2Cu_3O_{7-\delta}$ thin films [13]. In addition, on the basis of a theory of dynamic stripe-induced superconductivity, T_s and T_c are correlated through the pseudogap, which induces a gap in the single-particle energies that persists into the superconducting state [14]. On the other hand, based on the similarity in the scattering rate spectrum of the under-doped and optimally doped Y-123, it has been concluded that the pseudogap in the normal state and the superconducting gap are closely related [15]. However, the direct measurements of electronic spectrum such as ARPES [16] have indicated the similarity between the pseudogap and superconducting gap. While, using the intrinsic tunneling spectroscopy [17], it has been found that the pseudogap is coexisting with the superconducting gap, indicating a different nature of the two phenomena, or it is known that disorder decreases T_c without affecting T_s. So, although there are crucial requirements to more and exact experiments to find the origin of pseudogap, but it seems that a correlation of T_s and T_c is to be expected.

In summary, we have studied the normal state resistivity versus temperature in $Gd(Ba_{2-x}Pr_x)Cu_3O_{7+\delta}$ system. The appropriate conduction mechanism in the normal state is 2D-VRH. This dominant mechanism has been obtained with the fixed-p and variable-p methods. Therefore, with Pr-doping, the normal state conduction is strongly affected. The localization of carriers in normal state with Pr-doping causes the suppression of superconductivity in Pr-doped systems. Finally, our results indicate evidences for strong correlations between normal and superconducting states in HTSCs. By comparing the T_s–x of $Gd(Ba_{2-x}Pr_x)Cu_3O_{7+\delta}$ and $(Gd_{1-x}Pr_x)Ba_2Cu_3O_{7-\delta}$ compounds, we have concluded that the destructive effect of Pr substitution at Ba site is more pronounced than Pr at R site for superconductivity suppression in Gd-123 based HTSC. In agreement with most other reports, a correlation of superconducting gap and the normal state pseudogap is evident from the T_s–T_c curves. It is also shown that with Pr-doping, not only the carrier density reduces, which induces the pseudogap, but also simultaneously increases T_s and suppresses linearity behavior of resistivity vs. temperature in the normal state. However, the knowledge of the real position of Pr in the structure is necessary to explain the suppression mechanism of superconductivity in HTSC. Further clarification of the pseudogap evidence is highly desired for a more general understanding of the physical origin of HTSC.

Acknowledgements This work was supported in part by the Offices of Vice President for Research and Dean of Graduate Studies at Sharif University of Technology.

References

[1] M. Imada, A. Fujimori, and Y. Tokura, Rev. Mod. Phys. **70**, 1039 (1998).
[2] B. I. Shklovskii and A. L. Efros, Electronic Properties of Doped Semiconductors, eds. M. Cardona, P. Fulde, and H.-J. Queisser (Springer-Verlag, Berlin, 1984).
[3] M. R. Mohammadizadeh, H. Khosroabadi, and M. Akhavan, Physica B **321**, 301 (2002).
[4] V. M. Loktev, Low Temp. Phys. **22**, 1 (1996).
[5] M. R. Mohammadizadeh and M. Akhavan, Physica B **336**, 410 (2003).
[6] P. B. Littlewood and C. M. Varma, Phys. Rev. B **45**, 12636 (1992).
[7] S. J. Liu and W. Guan, Phys. Rev. B **58**, 11716 (1998).
[8] M. R. Mohammadizadeh and M. Akhavan, Phys. Rev. B **68**, 104516 (2003).
[9] J. L. Tallon, J. R. Cooper, P. S. I. P. N. de Silva, G. V. M. Williams, and J. W. Loram, Phys. Rev. Lett. **75**, 4114 (1995).
[10] M. R. Mohammadizadeh and M. Akhavan, Eur. Phys. J. B **33**, 381 (2003).
[11] M. R. Mohammadizadeh and M. Akhavan, Supercond. Sci. Technol. **16**, 538 (2003); M. R. Mohammadizadeh and M. Akhavan, Physica C **390**, 134 (2003).
[12] V. F. Sears, Neutron News **3**, 26 (1992).
[13] W. H. Tang and J. Gao, J. Phys.: Condens. Matter **11**, 8555 (1999).
[14] A. Bussmann-Holder, A. R. Bishop, H. Buttner, T. Egami, R. Micnas, and K. A. Muller, J. Phys.: Condens. Matter **13**, L545 (2001).
[15] T. Timusk and B. Statt, Rep. Prog. Phys. **62**, 61 (1999).
[16] H. Ding, T. Yokaya, J. C. Campuzano, T. Takahashi, M. Randeria, M. R. Norman, T. Mochiku, K. Kadowaki, and J. Giapinzakis, Nature **382**, 51 (1996).
[17] V. M. Krasnov, A. Yurgens, D. Winkler, P. Delsing, and T. Claeson, Phys. Rev. Lett. **84**, 5860 (2000).

Role of Pr doping in Gd1113

M. Kariminezhad, H. Khosroabadi, and M. Akhavan*

Magnet Research Laboratory (MRL), Department of Physics, Sharif University of Technology, P.O. Box 11365-9161, Tehran, Iran

Received 31 August 2003, accepted 31 December 2003
Published online 8 April 2004

PACS 74.25.Fy, 74.25.Ha, 74.72.Jt

Tetragonal $Gd_{1-x}Pr_xBaSrCu_3O_{7-\delta}$ polycrystalline samples ($0 \leq x \leq 0.9$) have been prepared by the standard solid-state reaction and characterized by XRD. A hump on the $\rho(T)$ curve have been observed at about 80 K for large values of x. The normal state resistivity have been investigated with two and three dimensional variable range hopping (2&3D-VRH) and Coulomb gap (CG). For low concentration of Pr, 2D-VRH is the dominant mechanism, but with the increase of x, 3D-VRH is dominant. The magnetoresistance measurements of the samples have been analysed by the Ambegakor and Halperin (AH) phase slip model. The field dependences of the pinning energy and critical current density have been studied for different amounts of Pr-doping.

© 2004 WILEY-VCH Verlag GmbH & Co. KGaA, Weinheim

1 Introduction After the discovery of high temperature superconductivity, effects of substitutions on each of the cation sites in the $RBa_2Cu_3O_{7-\delta}$ (R123) phase have been extensively studied. Substitution of non-isovalent cations gives a mean of determining the effect of hole concentration on T_c, while substitution of isovalent cations provides a probe for determining which lattice sites are significant for superconductivity. It has been shown that the substitution of rare earth elements, even magnetic ions except Pr on the R site has very little effect on the superconducting properties. For a recent review see Ref. [1]. In contrast, the substitution of Sr on the Ba site decreases T_c substantially. These observations imply that the Ba site is more important than the Y site in superconductivity. Among all rare earth ions that form 123 structure, the only non-superconducting Pr123 has produced a lot of research on compounds of the type $R_{1-x}Pr_xBa_2Cu_3O_{7-\delta}$ (RPr123). Pr substitution has also been tried in the so called distorted $RBaSrCu_3O_{7-\delta}$ (R1113) structures, viz: RPr1113 [2]. A comparison study on the structural details and superconducting properties of RPr1113 and RPr123 by considering the revival of superconductivity by Sr substitution can be helpful in understanding the mechanism of suppression of superconductivity by Pr-doping.

2 Experimental details The $Gd_{1-x}Pr_xBaSrCu_3O_{7-\delta}$ (GdPr1113) samples with $0 \leq x \leq 0.9$ were synthesized by the standard solid-state reaction from Gd_2O_3, Pr_6O_{11}, $BaCO_3$, $SrCO_3$ and CuO powders with 99.9% purity. The preparation procedures employed in Ref. [3]. Oxygen content for all samples were determined by iodometric titration analysis. The XRD measurements were done by Philips PW-3710 powder diffractometer with CuK_α radiation ($\lambda = 1.54056$ Å) in room temperature with 0.02 step width and 0.5 second step time in the range of $0° < 2\theta < 120°$. The resistance and magnetoresistance of the samples were measured at I = 5 mA and f = 40 Hz by ac four-probe method within the temperature range of 10 to 300 K by a helium closed-cycle refrigerator. A Lake Shore-330 temperature controller with two Pt-100 resistor sensors was used for measuring and controlling the temperature to ±10 mK. For the magnetoresistivity measurements a magnetic field of maximum 20 kOe was used.

---

* Corresponding author: e-mail: akhavan@sharif.edu, Phone: +98 21 6164510, Fax: +98 21 6012983

3 Results and discussion

The XRD patterns show that all samples have tetragonal symmetry in spite of their oxygen content of about 6.91–7.06. Small impurity peaks, due to the presence of $BaPrO_3$ and Pr_2O_3 phases, are present. No evidence of Sr phases are found in the impurity phases. Impurity peaks begin at $x = 0.3$ and their intensity increase with the increase of x. In comparison, all GdPr123 samples with $0 \leq x \leq 1$ have orthorhombic symmetry but, with the increase of Pr content, orthorhombicity decreases. A small amount of $BaCuO_2$ and $BaPrO_3$ is observed in some of the rich Pr-content samples [3].

The Sr ion, being smaller than Ba in size, causes local lattice distortions, which may disturb the ordering in the basal plane [4]. Ionic radius of Sr^{2+} in comparison with Ba^{2+} is closer to the ionic radius of R^{3+}, and Sr has its electronegativity closer to R. These conditions may cause site mixing between Sr^{2+} and R^{3+}. This cation disorder may be another factor for the tetragonal structure. Another words, the trivalent R ions enter the Ba site and changes the oxygen sublattice. The long range ordered Cu–O chains are destroyed and tetragonality is induced.

Figure 1 shows the resistivity of $x = 0$ to $x = 0.6$ samples. With the increase of x, T_c decreases and the width of transition temperature (ΔT_c) increases. With the increase of the number of insulating parts in the grains, the homogeneity of the grains decreases, which leads to the larger ΔT_c. T_c of the Gd1113 sample is 72 K with $\Delta T_c = 10$ K in comparison with Gd123 which has $T_c = 92$ K and $\Delta T_c = 1$ K. Disorder introduced into the lattice by Sr substitution causes oxygen disorder . This disorder effects are reflected by the widths of ΔT_c [5]. Site mixing between Sr and Gd could be another factor for the lower T_c and large ΔT_c in Gd1113 sample. There is a superconducting insulator transition at $x_c^{SIT} = 0.7$. With the increase of x, normal state resistivity of the samples increases. A change is also observed in the rate of variation of resistivity at room temperature at $x = 0.55$, meaning the occurance of metal insulator transition at $x_c^{MIT} = 0.55$. In GdPr123 system the $x_c^{SIT} = 0.45$ and $x_c^{MIT} = 0.25$ [3]. The larger values of x_c^{SIT} and x_c^{MIT} in GdPr1113 compared to GdPr123 indicate the lower effect of Pr in suppressing the superconductivity, and hence revival of superconductivity by Sr substitution for Ba in Gd123. In the sample with $x = 0.6$ before transition, resistivity increases. This can be due to the disorder related to Pr substitution or the fluctuations related to phase transition [3].

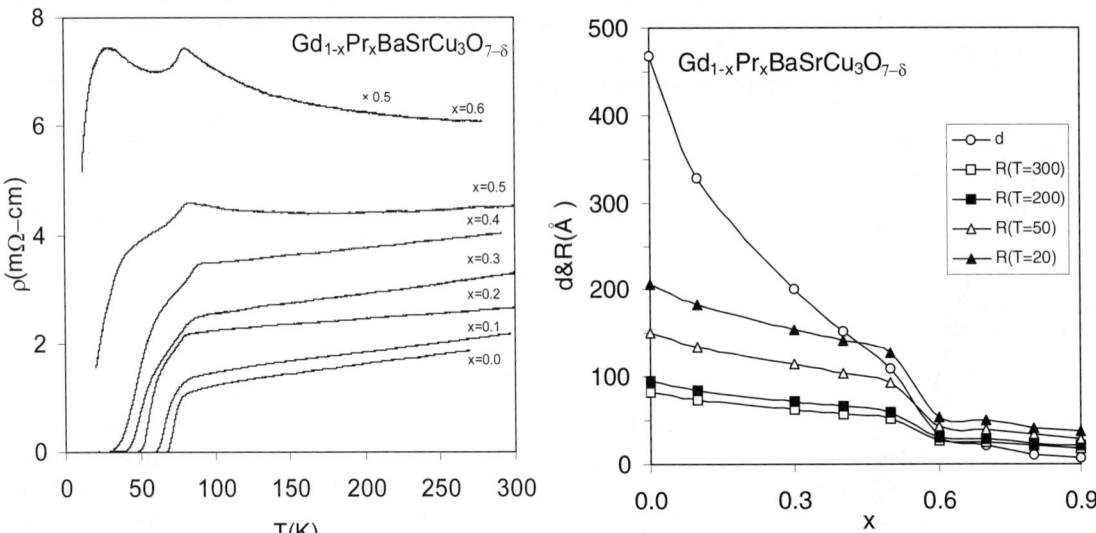

Fig. 1 Resistivity of GdPr1113 samples.   **Fig. 2** Localization length and hopping range.

An unusual hump on the $\rho(T)$ curve has been observed at about 80 K in GdPr1113 samples with large x values. With the increase of x, the hump becomes more clear. Even, in the $x = 0.7$ sample, which is an insulator, this hump presists at about 80 K. This may also exist in samples with $x > 0.7$, but because of their large resistivity it could not be observed. It should be noted that these humps are reproducible through different measurements, and are not due to any measurement errors. The impurity phases, having non-superconducting nature, cannot be the cause of the humps. Similar peaks on ρ(T) curves of YPr123[6], YPr1113[7], and $GdBa_{2-x}Pr_x123$ (GdBaPr123) compounds [8] have also been observed. It is noteworthy that in these reports humps are at about 80 K, same as the hump temperature for our samples. But, in our samples the hump is larger and more clear. The samples with x = 0.4, 0.5, 0.6, and 0.7 have been synthesized with different time-temperature calcination and sintering in different batches, and in all cases, the hump persists at about 80 K. ρ(T) curve of the sample were measured with different currents and application of magnetic field. These factors causes no change on the humps. In the Zou's superconducting Pr123 samples [9], one possibility for the larger c lattice parameter with respect to other 123 structures, is related to Ba atoms at Pr site (Ba_{Pr}). Narozhnyi et al. [10], based on the effective magnetic moment of Pr atom, have concluded that in the Zou's superconducting Pr123 sample there should be some Ba_{Pr}. We think that Ba_{Pr} could lead to superconductivity in some parts of the grains at about 80 K, which appears as a hump on ρ(T) curve. An evidence is that the temperature that hump on ρ(T) curve appears in our samples at about 80 K, has the same value as the superconducting transition temperature of Pr123 in Ref. [9]. The T_c^{onset} of the Gd1113 sample is 80 K, so the hump may be thought to be related to some Gd1113 regions, but we do not think so, because with the increase of Pr and decreasing Gd, the hump becomes more clear, and the intensity of impurity peaks related to Pr increases. These facts can be shown with increasing Pr: Ba_{Pr} increases, hence impurity peaks and the hump become more clear.

Table 1 Parameters obtained from the VRH fitting.

	p = 1/4			p = 1/3			p = 1/2		
x	T_0	$ρ_0$	$χ^2$	T_0	$ρ_0$	$χ^2$	T_0	$ρ_0$	$χ^2$
0.0	2.4247	0.13	0.9526	73.993	0.41	0.997	230.67	0.64	0.9984
0.1	22.313	0.35	0.9809	151.09	0.62	0.998	298.08	0.81	0.9982
0.3	255.85	1.14	0.9953	401.72	1.32	0.9993	409.16	1.41	0.9982
0.4	797.9	1.83	0.9995	698.67	1.89	0.9998	505.39	1.92	0.9971
0.5	2694.7	2.39	0.9996	1340.5	2.38	0.9998	662.56	2.34	0.997
0.6	7395	6.46	0.9999	2346.8	6.63	0.9996	828.88	6.68	0.9966
0.7	22961	9.45	0.9999	4742.2	10.81	0.9994	1181.7	11.90	0.9962
0.8	179397	11.95	0.9999	17288	22.16	0.9992	2188	37.24	0.9949
0.9	847315	9.28	0.9991	51179	31.75	0.9998	4181.3	95.41	0.9987

In order to investigate the normal state resistivity, we have fitted 2D-VRH, 3D-VRH, and CG regimes separately for all the samples with $0 \leq x \leq 0.9$, i.e. we chose p as a fix parameter. In this model, the temperature dependence of resistivity is $ρ(T) = ρ_0(T/T_0)^{2p}\exp(T_0/T)^p$ where p is related to the dimensionality of the hopping process. T_0 and hopping range (R) are defined as: $T_0^{2D} = 14/K_B N_{2D}(E_F)d^2$, $T_0^{3D} = 21/K_B N_{3D}(E_F)d^3$, $R_{2D} = [d/\{πN(E_F)K_B T\}]^{1/3}$, $R_{3D} = [3d/\{2πN(E_F)K_B T\}]^{1/4}$. For definition of the parameters see Ref. [11]. T_0 and $ρ_0$ derived from fitting the resistivity data in the temperature range with semilog plot of ρ(T) vs. T^{-p} is linear. The hump on the ρ(T) curve is not considered in the fitting. The fitting results are shown in Table 1. For x = 0 distinction between 2D-VRH and CG is very difficult. But for low concentration of Pr ($x < x_c^{MIT}$), 2D-VRH is the dominant mechanism; with the increase of x, 3D-VRH is dominant. Near the x_c^{MIT} distinction between 2D-VRH and 3D-VRH becomes difficult. For different amounts of Pr-doping, VRH gives very good agreement with experimental data. We can obtain the valid regime for conduction automatically, i.e. taking p as variable. For all the samples, the results are consistent with the fixed-p method. In Ref. [8], the 2D-VRH conduction mechanism, was reported for Gd123. In another report on GdPr123 samples, [3] the normal state resistivity for $x \geq 0.35$, follows a 3D-VRH mechanism, where as for x < 0.35 it shows a CG behavior. With increasing Pr-doping, T_0 increases, and the activation energy increases. Hence, the conductivity decreases. The derived localization length (d) and hopping range for different amount of Pr doping are shown in Fig. 2. In the GdPr1113 with the increase of x, d and R decrease. Therefore, Pr doping localizes the carriers in the normal state. Localization length is very large for x = 0 sample. This means that due to the very large d with respect to the distance of the neighboring atoms, the overlap of carriers' wave function is enough for the conduction to perform easily. For VRH mechanism, R should be larger than d. For small x, d is larger than R. For example, for x = 0.4, 2D-VRH occurres in temperature lower than 20 K, but for larger x, the temperature that VRH is dominant mechanism have the tendency to larger temperatures. In the GdBaPr123, for x = 0.5 even in room temperature, R is larger than d. Therefore, VRH is dominant conduction. But, in GdPr1113 for x = 0.5 even in 50 K R is smaller than d, therefore VRH is dominant mechanism only in low temperatures. We can conclude that Pr at Ba site causes more localization of carriers rather than at R site, or substitution of Sr at Ba site lowers the localization effect of Pr.

The dissipation part of $\rho(T)$ under magnetic field has been studied by the AH theory, as $\rho(T) = \rho_n[I_0(\gamma/2)]^{-2}$, $\gamma = U_0/K_BT = A/H (1-t)^q$ in the $I \ll I_c(T)$ limit. for description of this model and parameters see Ref. [12]. In this fitting process, we use ρ_n, $C(H) = A/H$, and q as free parameters. Critical current density at zero temperature $J_{cj}(0)$ is obtained from the AH theory. The value of $J_{cj}(0)$ for x = 0 sample varies from 40000 A/cm^2 at H = 0 to 2260 A/cm^2 at H = 15 kOe. For all the samples, the pinning energy

Table 2 Magnetic field dependence of the pinning energy.

x	0.0	0.1	0.2	0.3
α	0.865	0.804	0.751	0.485
β	0.099	0.124	0.24	0.253

and critical current density decrease with the increse of magnetic field. Also, with the increase of Pr-doping, $J_{cj}(0)$ decreases to 6300 A/cm^2 for x = 0.3 at H = 0. Magnetic field dependence of the pinning energy for H ≤ 2 kOe vary in the form $H^{-\alpha}$ and for H > 2 kOe, as $H^{-\beta}$. The values of α and β are listed in Table 2. With the increase of Pr-doping, the values of α and β become closer to each other. In low fields, for low level of Pr-doping, average value of α in GdPr123 is 0.2 and in GdPr1113 is 0.8. Therefore, the pinning energy in GdPr1113 compound is more sensitive to the magnetic field.

Acknowledgement The authors wish to thank M. Mirzadeh, P. Maleki, Z. Mokhtari. This work was supported in part by the Offices of Vice President for Research and Dean of Graduate Studies at Sharif University of Technology.

References

[1] M. Akhavan, Physica B **321**, 265 (2002).
[2] G. Cao, Y. Qian, X. Li, H. Wu, Z. Chen, and Y. Zhang, Physica C **248**, 92 (1995).
[3] Z. Yamani and M. Akhavan, Solid State Commun. **107**, 197 (1998).
[4] A.K. Ganguli and M.A. Subramanian, Mater. Res. Bull. **26**, 869 (1991).
[5] G. Hilscher, T. Holubar, G. Schaudy, J. Dumschat, M. Strecker, G. Wortmann, X.Z. Wang, B. Hellebrand, and D. Bäuerle, Physica C **224**, 330 (1994).
[6] J.L. Peng, P. Klavins, R.N. Shelton, H.B. Radousky, P.A. Hahn, and L. Bernardez, Phys. Rev. B **40**, 4517 (1989).
[7] G. Cao, Y. Qian, Z. Chen, X. Li, H. Wu, and Y. Zhang, Phys. Lett. A **196**, 263 (1994).
[8] M.R. Mohammadizadeh, and M. Akhavan, Phys. Rev. B **68**, 104516 (2003).
[9] Z. Zou, J. Ye, K. Oka, and Y. Nishihara, Phys. Rev. Lett. **80**, 1074 (1998).
[10] V.N. Narozhnyi and S.-L. Drechsler, Phys. Rev. Lett. **82**, 461 (1999).
[11] M.R. Mohammadizadeh and M. Akhavan, Eur. Phys. J. B **33**, 381 (2003).
[12] H. Shakeripour and M. Akhavan, Supercond. Sci. Technol. **14**, 234 (2001).

High pressure effects in YBCO and YSCO

H. Khosroabadi, B. Mossalla, and **M. Akhavan**[*]

Magnet Research Laboratory (MRL), Department of Physics, Sharif University of Technology, P.O. Box 11365-9161, Tehran, Iran

Received 31 August 2003, accepted 31 December 2003
Published online 8 April 2004

PACS 62.50.+p, 71.15.Ap, 74.72.Bk

Electronic structure calculations for two isofamilies, $YBa_2Cu_3O_7$ (YBCO) and $YSr_2Cu_3O_7$ (YSCO) high T_c cuprates, have been done for ambient and high pressures (–15 GPa < p < 15 GPa) by pseudopotential density functional theory in the local density approximation. Results show the transfer of hole charge in both CuO_2 planes and CuO chains for both systems. It is concluded that in spite of the larger dT_c/dp for YSCO in comparison with YBCO, the charge transfer in YSCO is lower than in YBCO. This apparent discrepancy has been explained against the pressure induced charge transfer (PICT) model.

© 2004 WILEY-VCH Verlag GmbH & Co. KGaA, Weinheim

1 Introduction In recent years, extensive studies have been done on high-pressure effects (HPE) in fullerids [1], organic superconductors [2], and layered high temperature superconductors (HTSCs) [3]. These studies have been focused mostly on the correlation between normal and superconducting states, different pressure derivatives of $T_c(dT_c/dp)$ in different families and compounds of HTSCs, and their anisotropic structural and transport dependences. Another advantage of investigating the HPE in HTSCs is to reveal higher T_c by correct structural or internal parameter changes by chemical pressure [4]. In addition to the externally applied pressure, the chemical pressure by substitution of a smaller ion has also been suggested to be useful for increasing T_c at ambient pressure [5]. To interpret the experimental data, models such as resonating valence bond [4], van-Hove singularity [6], extended Hubbard model [7], and pressure-induced charge transfer (PICT) [8] have been suggested. The PICT model is presented as redistribution of charge density, and as a result, the increase of the hole content in the CuO_2 planes in these compounds, which is considered as a suitable approach to explain dT_c/dp. Of course, there has also been other studies, which indicate the insufficiency of this model [9, 10].

For comparison between the experimental data and the models in the normal and superconducting states, one needs to determine the electronic parameters such as hole carrier content and coupling constant between them, $N(E_F)$, van-Hove singularity position, one site Coulomb interaction, hopping amplitudes, and vibrational and structural parameters (e.g., phonon frequency distribution, ionic position, bond lengths, or compressibility) at ambient pressure and their pressure derivatives. Electronic structure calculation is one of the suitable approaches to determine these parameters especially by electronic and ionic relaxation calculation, by ruling out the non-intrinsic effects observed in the experimental studies [3].

The results of many studies in this area show that the pressure dependence of T_c varies strongly, even by small changes of the component stoichiometry. In recent HPE study, a $dT_c/dp = 0.64$ K/GPa [10] has been derived for YBCO single crystals up to 6 GPa, but this value in some isostructural YBCO-related compounds is completely different from YBCO, such as 3 K/GPa [11], 5.5 K/GPa [12], and 7.4 K/GPa [13] for YSCO, Y124, and Pr123, respectively. In this paper we have compared the HPE on two HTSC isofamilies, YBCO and its chemically pressured YSCO by density functional theory in the range of –15

---
[*] Corresponding author: e-mail: akhavan@sharif.edu, Phone: +98 21 6164510, Fax: +98 21 6012983

to +15 Gpa, to clarify the differences in the electronic and structural parameter changes, and explain the great difference in dT_c/dp values in both systems.

2 Computational details The total energy calculations have been performed using ab initio pseudopotentials in the local density functional theory. We have used the VASP code [14], by using the projector augmented wave method. The lattice parameters for the tetragonal YSCO at ambient pressure have been taken from the experimental data [11]. For comparison with YBCO system, these parameters have been changed to the orthorhombic symmetry by the orthorhombicity factor $200*(a-b)/(a+b) = 1.32$. The change of lattice parameters by pressure for YBCO [15] has been considered for YSCO too. The atomic pseudopotential for the Y4d&5s, Ba5p&6s, Sr5s, Cu3d&4s, and O2s&2p orbitals have been employed from the VASP library files. The cut off energy and k point sampling are determined to be 800 eV and 50 k, respectively, for convergence of the total energy to better than 10^{-4} eV/(unit cell). Ion relaxation has been done to achieve forces of the order 1 mRy/Å for each pressure. Charge density calculations have been done in 40*40*120 mesh points in the unit cell. Employing the above criteria, we have calculated charge density, hole carriers in each plane, equilibrium ionic positions, and bond lengths for the pressure range of –15 GPa to +15 GPa for YBCO and YSCO.

3 Results and discussion The z coordination of the equilibrium ionic positions of Ba(Sr), Cu(2), O(4), O(3), and O(2) ions at ambient pressure, have been derived from ionic relaxation for YBCO as 0.18258, 0.35365, 0.15972, 0.3789, 0.3798, and for YSCO as 0.18093, 0.3453, 0.1639, 0.37283, 0.37432, respectively. At first, the equilibrium ionic positions were considered for each pressure, and they were relaxed to new equilibrium positions. The x and y coordinations have not been changed because of their zero forces due to symmetry. The data for ambient pressure are in good agreement with the experimental study [16]. Charge density, band structure, density of states, equilibrium ionic positions, equilibrium unit cell volume, and bulk modulus have been calculated. Comparison of the result for ambient pressure of the YBCO system with other works, shows that the pseudopotential VASP code is able and efficient for calculating the electronic properties of HTSC as good as other approaches such as LAPW or LMTO [17].

The chemical pressure by replacing Ba by Sr (the cell parameters of YSCO are smaller than those of YBCO and become equivalent at nearly 10 GPa pressure) causes non-scaling behavior for YSCO, i.e., the z values for YBCO except for $Z_{O(4)}$ are larger than the values for YSCO. The $Z_{O(4)}$ is increased for YSCO, so that the same value for the Cu(1)–O(4) bond lengths (~1.865 Å) for both systems are obtained. In contrary, the Cu(2)–O(4) bond length for YSCO is 0.20 Å smaller than the value for YBCO. Also, the thickness of each CuO_2 plane, $Z_{O(2)}$ & $Z_{O(3)}-Z_{Cu(2)}$, remain nearly constant for both systems.

The derived equilibrium positions indicate a scaling relation (expected linear relation) for Ba, Sr, and Cu(2) atoms for both systems by applied pressure, but not for the other atoms: O(4), O(2), and O(3). Determination of the non-scaling behavior in these compounds is important because it causes large redistribution of charge density in these systems, especially in the CuO and CuO_2 layers, which are important for the superconducting properties of the layered HTSCs. The Cu–O(4) bond length decreases from 1.90 to 1.80 Å in YSCO for the whole range of applied pressure. Of course, this value changes from 1.842 for p = –15 GPa to 1.859 for ambient pressure, and then decreases to 1.839 Å for +15 GPa. Such a small change relative to the scaling, shows the stiffness of the bond, in good agreement with the neutron scattering study [8]. It should be noted that the apical oxygen – the bridge between the superconducting CuO_2 layer and charge reservoir CuO chain – is important in the electronic and superconducting properties [18]. The apical oxygen can transfer holes to the CuO_2 planes as will be pointed out in the following.

Figures 1b and 2b show the distribution of charge density (calculated by integration for all 120 planes perpendicular to c-direction) versus c-direction, at ambient pressure in the half unit cell of YBCO and YSCO systems, respectively. The similarity between these figures is due to the same structure and chemical characteristics. The only considerable difference is the charge density in the SrO layer relative to the BaO layer because of the absence of the 5p orbital in Sr pseudopotential. The characteristics of these figures for similar calculations have been described in more details in Ref. [19]. Figures 1a, 1c, 2a, and

2c show the differences of the charge distribution in the c-direction for the positive and negative high pressures for YBCO and YSCO, respectively. These figures show the change of charge density distribution by applied pressure, which indicate the transfer of charge density in these systems. There are three dips and peaks in the distribution for both systems, whose values reduce in order of CuO_2, BaO (SrO), and CuO. The symmetry between these dips and peaks for the negative and positive pressures show the intrinsic effect of pressure in the charge density. Of course, their values for the negative pressure are higher than those for positive pressures. We can see that the dip and the peak in the CuO_2 planes for YBCO for negative pressures is higher relative to the positive pressures for YBCO, which is due to the different behavior of bond length for the positive and negative pressure in the system. These dip and peak positions are a little different for the YBCO to the YSCO system. These figures show the electric dipole of charges about BaO (SrO) and CuO_2 planes with opposite polarity, although there is no net polarization in the unit cell because of the mirror symmetry to Y plane. The movement of O(4) upward, and O(2) and O(3) downward with respect to the scaling atomic positions by applied pressure, have an important role in the electric dipole formation.

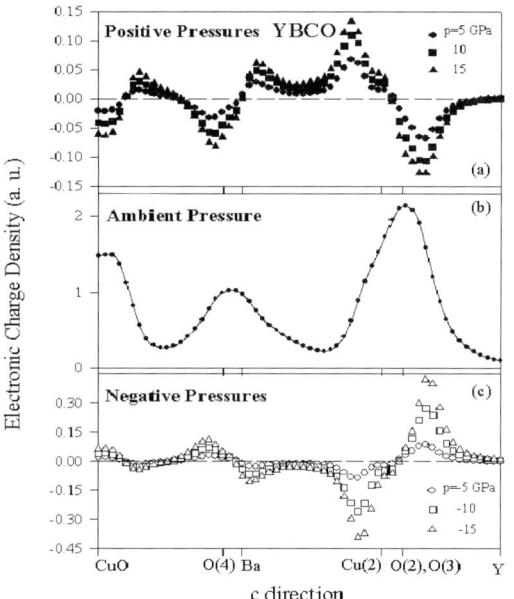
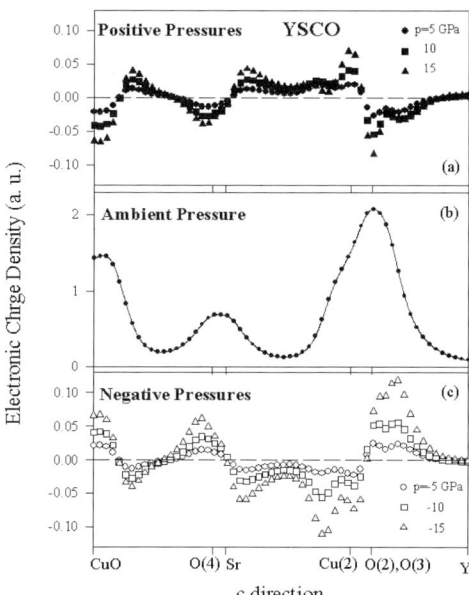

Fig. 1 Charge density distribution in the c direction YBCO for ambient pressure (b); differences with respect to ambient pressure for positive (a) and negative (c) pressures.

Fig. 2 Charge density distribution in the c direction of YBCO with respect to ambient pressure for positive (a) and negative (c) pressures.

The charge density in each layer has been calculated by integrating the charge density for the full width at half maximum (FWHM) range (Figs. 1b and 2b). The calculation shows that the hole content in both the CuO_2 plane and the CuO chain increase with applied pressure, in agreement with our previous study [15] and other reports [9]. The values are different for YBCO and YSCO as 0.022 and 0.026 hole/(GPa-unit cell) for the CuO chains and 0.026 and 0.015 hole/(GPa-unit cell) for the CuO_2 planes, respectively. The difference between these values and the pervious study [15] is due to the more accurate calculation in this study and a more suitable definition employed for the thickness of the planes (i.e. FWHM) in this work, so that the thickness of the layer in FWHM is about four times larger than the previous definition. Similar results can be obtained by considering these criteria. These values are high relative to the neutron experimental studies [8]. This is probably due to the fact that the calculated values come from all 36 bands of Cu3d–O2p hybridization, while the experimental data come mostly from the four bands intersecting the Fermi energy, which are important in the electronic properties. As it has been shown by the resistivity measurements [9] and the charge density picture in our study, it seems that there is no evidence for hole carrier charge transfer from the CuO reservoir chains to the conducting CuO_2 planes, in contrast with the bond valence sum picture [8].

The increase of hole content in the CuO_2 plane by applied pressure is lower for YSCO than YBCO. As mentioned by Cao et al. [11], their overdoped samples have $dT_c/dp = 3.0$ K/GPa, which is in contrast with the PICT model, since we expect a decrease of T_c with the increase of hole content by applied pressure in the overdoped regime. It is resulted that in spite of the lower charge transfer to the CuO_2 plane in YSCO relative to YBCO, the dT_c/dp for YSCO is larger than for YBCO.

Some reasons have been suggested for the discrepancy: 1) Change of hole content is not the only parameter for determining dT_c/dp, as concluded from other studies [10]. 2) The effect of Sr in the superconducting state in this system maybe different from Ba in spite of their similar chemical properties. 3) The most important effect is probably due to the different oxygen content in YSCO [11] and this calculation, because the oxygen content plays a crucial role in the dT_c/dp.

4 Conclusion Ionic relaxation shows the non-scaling behaviour for all the ions by chemical pressure in YSCO relative to YBCO. The $Z_{(O4)}$ increases in YSCO so that an equal Cu(1)–O(4) bound length have been found for both systems, while the z position of other ions decrease. A scaling relation for Ba, Sr and Cu(2) positions was found, while no scaling relation was found for O(4), O(2), and O(3). The difference of charge density redistribution by applied pressure indicates the increase of holes in the CuO_2 and CuO layers in both YBCO and YSCO systems. The values are different for YBCO and YSCO as 0.022 and 0.026 hole/(GPa-unit cell) for the CuO chains and 0.026 and 0.015 hole/(GPa-unit cell) for the CuO_2 planes, respectively. In spite of the larger dT_c/dp in YSCO, the change of hole content in the CuO_2 plane by applied pressure is smaller for YSCO than in YBCO. This discrepancy can be explained by the insufficiency of just the hole concentration for determination of dT_c/dp or the role of oxygen content in the dT_c/dp.

Acknowledgements The authors whish to thank G. Kresse and J. Furthmuller, the writers of the VASP code. This work was supported in part by the Offices of Vice President for Research and Dean of Graduate Studies at Sharif University of Technology.

References

[1] C.M. Brown, T. Takenobu, K. Kordatos, K. Prassides, Y. Iwasa, and K. Tanigiaki, Phys. Rev. B **59**, 4439 (1999).
[2] R. Kondo, S. Kagoshima, and M. Maesato, Synthetic Metals **133–134**, 137 (2003).
[3] J.S. Schilling and S. Klotz, in: D.M. Ginsberg (Ed.), Physical Properties of High Temperature Superconductors III (World Scientific, Singapore, 1992), pp. 59–177.
[4] G. Baskaran, Phys. Rev. Lett. **90**, 197007 (2003).
[5] F. Licci, A. Gauzzi, M. Marezio, G.P. Radelli, R. Masini, and C. Chaillout-Bougerol, Phys. Rev. B **58**, 15208 (1998).
[6] B.K. Agrawal and S. Agrawal, Phys. Rev. B **52**, 12556 (1995).
[7] E.V.L. de Mello and C. Acha, Phys. Rev. B **56**, 466 (1997).
[8] J.D. Jorgensen, S. Pei, P. Lightfoot, D.G. Hinks, B.W. Veal, B. Dabrouski, A.P. Poulikas, and R. Kleb, Physica C **171**, 93 (1990).
[9] K. Yoshida, A.I. Rykov, S. Tajima, and I. Terasaki, Phys. Rev. B **60**, R15035 (1999).
[10] S.W. Tozer, J.L. Koston, and E.M. McCarron, Phys. Rev. B **47**, 8089 (1993).
[11] Y. Cao, T.L. Hudson, Y.S. Wang, S.H. Xu, Y.Y. Xue, and C.W. Chu, Phys. Rev. B **58**, 11201 (1998).
[12] B. Bucher, J. Karpinski, E. Kaldis, and P. Wachter, Physica C **157**, 478 (1989).
[13] J. Ye, Z. Zou, A. Matsushita, K. Oka, Y. Nishihara, and T. Matsumoto, Phys. Rev. B **58**, R619 (1998).
[14] G. Kresse and J. Furthmuller, Comput. Mat. Sci. **6**, 15 (1996); Phys. Rev. B **54**, 11169 (1996).
[15] H. Khosroabadi, M.R. Mohammadizadeh, and M. Akhavan, Physica C **370**, 85 (2002).
[16] R.M. Hazen, in: D.M. Ginsberg (Ed.), Physical Properties of High Temperature Superconductors II (World Scientific, Singapore, 1998), Chap. 3.
[17] W.E. Pickett, Rev. Mod. Phys. **61**, 433 (1989).
[18] L.F. Finer, M. Grilli, and C.D. Castro, Phys. Rev. B **45**, 10647 (1992).
[19] H. Khosroabadi, M.R. Mohammadizadeh, and M. Akhavan, Physica B **321**, 360 (2002).

Hole carrier transfer by apical oxygen in YBCO

H. Khosroabadi and **M. Akhavan**[*]

Magnet Research Laboratory (MRL), Department of Physics, Sharif University of Technology,
P.O. Box 11365-9161, Tehran, Iran

Received 31 August 2003, accepted 31 December 2003
Published online 1 April 2004

PACS 71.75.Ap, 74.25.Jb, 74.72.Bk

The electronic structure of high temperature superconductor YBCO is calculated by pseudopotential density functional theory using VASP code for different values of $Z_{O(4)}$. Comparison of the results of equilibrium $Z_{O(4)}$ with other calculation techniques indicates the capability of pseudopotential VASP code in energy band structure calculations for HTSCs. Both charge distribution and band structure calculations indicate the transfer of hole carriers from the CuO chains to the CuO_2 planes with the increase of $Z_{O(4)}$. It is resulted that the increase of Cu(1)–O(4) bond lengths causes the creation of holes in the CuO_2 planes. The redistribution of charge density confirms the charge transfer phenomenon. Also, the increase of $Z_{O(4)}$ causes a decrease in E_F and $N(E_F)$. The hole transfer by moving O(4) towards the CuO_2 planes could be considered for optimization of HTSC properties.

© 2004 WILEY-VCH Verlag GmbH & Co. KGaA, Weinheim

1 Introduction Although, there is yet no comprehensive theory for explaining the properties of high temperature superconducting (HTSC) systems, it is generally accepted that the hole content in the CuO_2 plane has very important role in their superconducting properties. Chemical substitution [1], or increasing the oxygen content [2] in these compounds, in many cases, causes the creation of holes in the CuO_2 planes, and as a result, the observation of superconductivity. It is found that a universal relation exists between the superconducting transition temperature T_c and the hole content in the CuO_2 plane (n_{CuO2}) as $T_c/T_c^{max} = 1 - A(n_{CuO2} - n_{opt})^2$ [3]. This results that the increase of n_{CuO2} in the underdoped (overdoeped) HTSC can increase (decrease) T_c. The change of T_c by applied high pressure in HTSC is a similar phenomenon due to the charge transfer. In both models, the transfer of holes to the CuO_2 superconducting planes causes the observation of superconductivity or change of T_c, and as a result, the optimization of T_c.

There are some physical parameters, which affect the charge transfer phenomenon such as doping [4], change of oxygen content [4, 5], and change of lattice parameters and bond lengths by mechanical or chemical pressure [6]. Because of the collective effect of these parameters in optimization of T_c, determination of the role of each parameter in the charge transfer can help to optimize the HTSC properties. The main parameter in $RBa_2Cu_3O_7$ (RBCO, R is Y and rare earth with superconducting behavior) systems is the apical oxygen bond length [6, 7]; a universal relation between T_c and Cu(1)–O(4) bond length has been deduced in this system [8]. Electronic structure calculations is a suitable way to determine the role of apical oxygen, because in the calculations, in contrast to the experimental finding, there are no mixing effects. The local density functional theory (LDFT) method has been successfully applied to HTSC, e.g., YBCO, concerning various aspects of these materials such as structural and vibrational properties, phonon frequency, electronic charge density, and optical spectra [9]. In this study, the electronic structures of YBCO system such as energy band structure (BS), electronic charge density (ECD), and total density of state (TDOS) have been calculated in the LDFT for different values of $Z_{O(4)}$.

---

[*] Corresponding author: e-mail: akhavan@sharif.edu, Phone: +98 21 6164510, Fax: +98 21 6012983

2 Computational details The total energy calculations have been performed using the ab initio total energy-pseudopotential technique. We have used the Vienna Ab initio Simulation Package (VASP) code [10, 11] based on the LDFT by using projector augmented wave (PAW) method. The norm-conserving ultra-soft pseudopotentials for the O2s, O2p, Cu3d, Cu4s, Y4d, Y5s, Ba5p, and Ba6s orbitals have been employed from the VASP library files. It was concluded that 800 eV cut-off is sufficient for convergence of the total energy to better than 10^{-4} eV/(unit cell). The repeated calculations for two different cut-off energy values of 400 eV (VASP default) and 800 eV has no considerable effect on the calculated electronic properties such as EDS, BS and TDOS.

50 k points have been implemented for the Brillouin zone integration in the irreducible wedge, which correspond to the 9*9*3 grids in the scheme of Monkhorst and Pack. Lattice parameters for YBCO system were considered to be $a = 3.83$, $b = 3.88$, and $c = 11.68$ Å. Equilibrium ionic positions were obtained by minimizing the ionic force to less than 0.5×10^{-2} eV/Å. Charge density of the system was calculated for 40*40*120 mesh in the unit cell. The grid-points mesh for a second finer FFT (NGFX, Y, Z in the VASP) was employed to be 36, 36, and 108 to avoid wrap around errors in the VASP code. The calculation was repeated for different values of $Z_{O(4)}$ =0.1500, 0.1550, 0.1570, 0.1590, 0.15972, 0.1610, 0.1630, 0.1650, and 0.1700. The total energy in each calculation was converged to better than 10^{-4} eV/(unit-cell).

3 Results and discussion The equilibrium ionic position for the YBCO system in the c-direction was obtained to be 0.15972, 0.18258, 0.35365, 0.3789, and 0.3798 for the O(4), Ba, Cu(2), O(3), and O(2) ions, respectively. These values are in good agreement with the results of X-ray diffraction analysis [12]. By using the Z values, the BS, TDOS and ECD have been calculated. The BS is calculated for the high symmetry lines in the $k_z = 0.0$ and $k_z = 0.5$, and in the k_z directions in the first Brillouin zone. The 36 bands from the Cu3d–O2p hybridization are distributed in the range of –7.14 to 1.88 eV relative to the Fermi energy E_F including four bands intersecting the Fermi level. These are in agreement with other calculations by different methods and approximations [13]. The TDOS at E_F ($N(E_F)$) is derived to be 4.93 state/(eV-unit cell). The overall shape of TDOS and $N(E_F)$ are in good agreement with the more accurate calculations by other methods [13]. Comparing these results with other reported calculations for YBCO system confirm the capability of the pseudopotential approach by VASP code for calculation of the electronic structure of the complicated HTSCs crystal structure.

Figures 1a–1d show the four bands intersecting Fermi level in the high symmetry lines of the Brillouin zone in the k_z=0.0 plane for different $Z_{(O4)}$ values. These four bands which are formed from the hybridization of the Cu and O ions in the CuO chains and CuO$_2$ planes [15], have crucial effects on the electronic properties of the system. Some parts of these bands which are from the CuO chain are strongly affected and are disturbed with the O(4) shift, but the other parts do not change considerably. This indicates that O(4) interacts stronger with the CuO chains than the CuO$_2$ planes. This shows that Cu(1)–O(4) bond is a very strong bond (the Cu(1)–O(4) bond length is 1.866 Å, about 0.42 Å smaller than Cu(2)–O(4) bond length) relative to other Cu–O bonds. This point has also been confirmed by other studies [6].

Another important feature in Figs. 1a–1d is the creation of new states in the vicinity of S point of the 1st band, Y point of the 3rd band, and X point of the 4th band. In the 1st band related to the CuO chains, the hole carriers annihilate with the increase of $Z_{O(4)}$, while the hole carriers are created in the 2nd and 3rd bands related to the CuO$_2$ planes. This shows that the hole transfers from the CuO chains to the CuO$_2$ planes. The right shoulder of the 4th band with nearly 0.64 eV/0.02 c and the peak in the 2nd band with 0.57 eV/0.02 c are the most affected parts without any charge transfer. This result indicates that to determine the electronic properties of the RBCO family, it is important to consider the O(4) position in the system, or its position change by doping different elements, oxygen deficiency, or exerting high pressures. By the increase of $Z_{O(4)}$ from 0.1500 to 0.1700, the values of E_F and $N(E_F)$ change from –0.029 to –0.163 eV and 6.7 to 3.39 state/(eV-unit cell), respectively. This seems to be due to the lower interaction in the system by the larger distance of O(4) from the CuO chains. These changes in E_F and $N(E_F)$ again indicate the importance of the O(4) position in the electronic properties of the system.

To clarify the change of charge redistribution in the CuO and CuO$_2$ layers, we have plotted the distribution of ECD in the c-direction of YBCO unit cell in Fig. 2a. The results reveal large amount of charge density in the vicinity of CuO$_2$ planes and CuO chains, which indicate the importance of these layers in

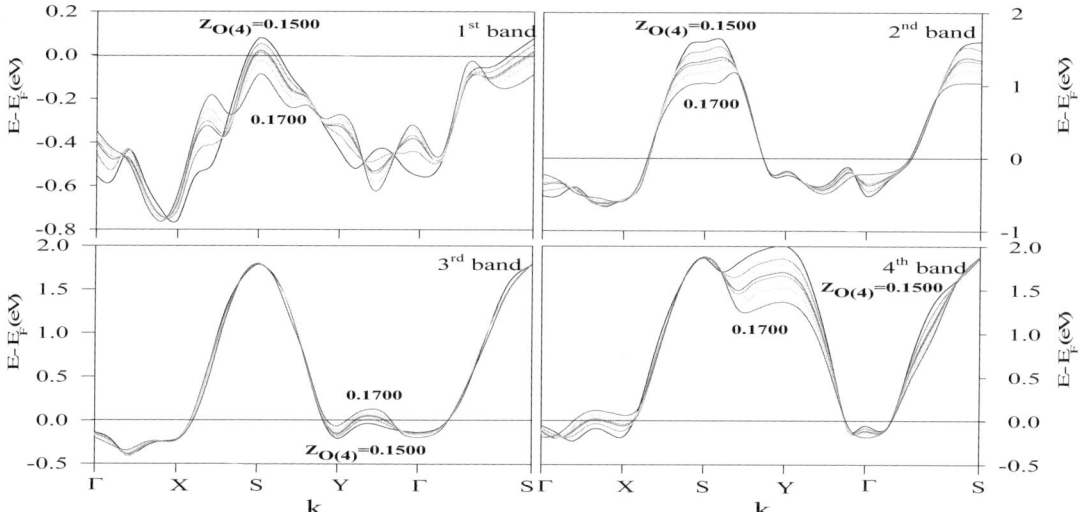

Fig. 1 The four calculated bands of YBCO intersecting the Fermi level for different value of $Z_{(O4)}$.

electronic and superconducting properties of the system, as also confirmed by experimental studies. The generally accepted idea is that the change of hole carrier content in the CuO$_2$ planes is strongly related to the superconducting properties, while the CuO chains act as charge reservoir for the CuO$_2$ planes.

The differences of ECD distribution for different values of $Z_{O(4)}$ with respect to the equilibrium ionic position of O(4) ($Z_{O(4)} = 0.15972$) (Figs. 2b and 2c) show a dominant charge redistribution in the vicinity of BaO layer, but a very small charge redistribution in the vicinity of CuO$_2$ and CuO layers. Due to the movement of O(4)$^{2-}$ ion, a large amount of charge redistributes in the vicinity of BaO layer. This movement of O(4)$^{2-}$ ion also redistributes the ECD in the CuO$_2$ planes and CuO chains; the electronic charge transfers from the CuO chains to the CuO$_2$ planes with the decrease of $Z_{O(4)}$. This is equivalent to the hole carrier transfer from the CuO chains to the CuO$_2$ planes by moving O(4) toward the CuO$_2$ planes. Since charge transfer affects the electronic and superconducting properties, it is concluded that there is a strong correlation of T$_c$ to Cu(1)–O(4) bond lengths in YBCO. From the above results, it is suggested that to obtain a maximum T$_c$ in the underdoped (overdoped) HTSC, we should obtain maximum (minimum) Cu(1)–O(4) bond length. To quantify the charge transfer, we have calculated the ECD for different planes. The full width at half maximum for each peak of the charge density plot, shown in Fig. 2a, has been considered as the thickness of the layer [14]. The changes in ECD are 1.09, 11.88, –1.01, and 0.32 electron/c for the CuO, BaO, CuO$_2$ and Y planes, respectively. The negative value for the CuO$_2$ planes indicates an increase of holes in these planes.

4 Conclusion From the ECD calculations of YBCO system for different values of $Z_{O(4)}$, it is resulted that moving of O(4) ion towards the CuO$_2$ planes causes a hole carrier charge transfer from the CuO chains to CuO$_2$ planes. The change of ECD values for the CuO, BaO, CuO$_2$ and Y planes have been determined to be 1.09, 11.88, –1.01, and 0.32 electron/c, respectively. It is found that the position of O(4) is very important in the shape of the four bands intersecting the Fermi level. The highly affected parts of the bands, related to the CuO chains, indicate a strong interaction between the O(4) and CuO chains. The BS calculation also shows the hole transfer from the CuO chains to CuO$_2$ plane with the increase of $Z_{O(4)}$. The highly affected bands near the Fermi level, and the changes of E$_F$ and N(E$_F$) indicate the high sensitivity of electronic structure of the system to the position of O(4).

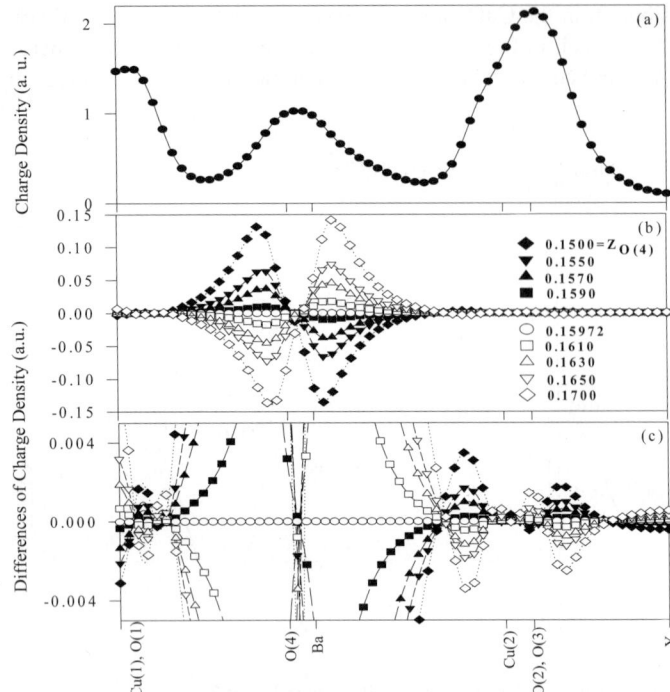

Fig. 2 Charge density distribution at equilibrium value of $Z_{O(4)}$ (a), and its changes for different values of $Z_{O(4)}$ in two different ranges: (b) and (c).

Acknowledgements The author whish to thank G. Kresse and J. Furthmuller, the writers of the VASP code, and Z. Nasrollahi for her assistance. This work was supported in part by the Offices of Vice President for Research and Dean of Graduate Studies at Sharif University of Technology.

References

[1] Y. F. Xiong, Y. S. Yao, L. F. Xu, F. Wu, D. Jin, and Z. X. Zhao, Solid State Commun. **107**, 509 (1998).
[2] R. J. Cava, A. W. Hewat, E. A. Hewat, B. Batlogg, M. Marezio, K. M. Rable, J. J. Krajewski, W. F. Peck, Jr., and L.W. Rupp, Jr., Physica C **165**, 419 (1990).
[3] J. L. Tallon, C. Bernhard, H. Shaked, R. L. Hitterman, and J. D. Jorgensen, Phys. Rev. B **51**, 12911 (1995).
[4] H. Khosroabadi, V. Daadmehr, and M. Akhavan, Mod. Phys. Lett. B **16**, 943 (2002).
[5] J. D. Jorgensen, B. W. Veal, A. P. Paulikas, L. J. Nowicki, G. W. Crabtree, H. Ciaus, and W. K. Kowk, Phys. Rev. B **41**, 1863 (1990).
[6] J. D. Jorgensen, S. Pei, P. Lightfoot, D. G. Hinks, B. W. Veal, B. Dabrowski, A. P. Paulikas, and R. Kleb, Physica C **171**, 93 (1990).
[7] L. F. Finer, M. Grilli, and C. D. Castro, Phys. Rev. B **45**, 10647 (1992).
[8] P. F. Miceli, J. M. Tarascon, L. H. Greene, P. Barboux, F. J. Rotella, and J. D. Jorgensen, Phys. Rev. B **37**, 593 (1988).
[9] R. Kouba, C. Ambrosch-Draxl, and B. Zangger, Phys. Rev. B **60**, 9321 (1999).
[10] G. Kresse and J. Furthmüller, Comput. Mater. Sci. **6**, 15 (1996).
[11] G. Kresse and J. Furthmüller, Phys. Rev. B **54**, 11169 (1996).
[12] R. M. Hazen, in: D. M. Ginsberg (Ed.), Physical Properties of High Temperature Superconductors II (World Scientific, Singapore, 1998), Chap. 3.
[13] W. E. Pickett, Rev. Mod. Phys. **61**, 433 (1989).
[14] H. Khosroabadi, B. Mossalla, and M. Akhavan, phys. stat. sol. (c) **1**, No. 7 (2004), this conference.
[15] H. Krakauer, W. E. Pickett, and R. E. Cohen, J. Supercond. **1**, 111 (1988).

Role of Pr/Ba disorder in Pr123 superconductor

H. Khosroabadi, M. Modarreszadeh, P. Taheri, and M. Akhavan[*]

Magnet Research Laboratory (MRL), Department of Physics, Sharif University of Technology, P.O. Box 11365-9161, Tehran, Iran

Received 31 August 2003, accepted 31 December 2003
Published online 1 April 2004

PACS 74.25.Fy, 74.72.Jt

Polycrystalline $(Pr_{1-x}Ba_x)Ba_2Cu_3O_{7-\delta}$ samples ($0.0 \leq x \leq 0.6$) is prepared by the standard solid state reaction technique. X-ray measurements confirm the formation of the 123 phase for $x \leq 0.3$, but for $0.4 \leq x \leq 0.6$, due to the solubility limit, the 123 phase is not a dominant phase. Resistivity measurements show the increasing of resistivity with the increase of Ba-doping up to about $x=0.08$, and then decreases in the 123 phase up to $x = 0.3$. There is no observation of superconducting state in our samples, but we observe the sharp decrease of resistivity for $x>0.1$, which is related to possible superconductivity in Pr123 system by Ba-doping. It is concluded that the Pr/Ba disorder has important effect in the transport electrical properties, specially occuring a superconducting state in Pr123. It is also suggested that extra Ba concentration in Pr123 system can create superconducting state such as in Ca-doped Pr123.

© 2004 WILEY-VCH Verlag GmbH & Co. KGaA, Weinheim

1 Introduction

Although, tremendous amount of research have been carried out for understanding the high temperature superconductivity (HTSC) mechanism, there is not yet any complete theory for describing it. Research on metal–insulator and superconductor–insulator transitions (MIT & SIT), magnetic transitions, and phase diagrams can help us to this understanding. One singularity in the rare earth-based HTSC is the suppression of superconductivity in $RBa_2Cu_3O_7$ (R123) family (R is Y and rare earth elements except Ce, Pm, and Tb) by Pr-doping in the Ba [1] or R [2] sites, and then, the Pr123 system is the only R123 orthorhombic structure with insulator behaviour. So, the role of Pr in the superconductor–insulator anomaly could help to understand the mechanism of superconductivity, at least in R123. Models such as pair breaking [3], hole filling [4], hole localization [5], and Pr/Ba disorder [6] have been considered for explaining this anomaly. Within these models, no single complete and convincing model exists for describing the experimental data on the Pr anomaly. Ref. [7] gives a more complete survey on these models for solving the Pr anomaly.

While, many researchers are seeking to find the answer to the insulating Pr123 through new models or the combination of the above models [8], some groups have reported superconductivity in thin films [9], single crystal [10], and powder and film samples [11] of Pr123. These observations have caused some confusion on the Pr anomaly, and renew the study of the Pr anomaly by the Pr/Ba disorder in RPr123. In some recent reports, the nominal substitution of Pr at Ba site has been discussed for the determination of the role of Pr/Ba disorder in the anomalous behaviour of Pr123. So, before attempting to solve the Pr problem, it should be settled whether Pr123 is really a superconductor or an insulator. Zou et al. is one of the groups who have reported superconductivity in Pr123 in their samples grown by travelling-solvent-floating-zone (TSFZ) method [12]. Of course, this observation is in contrast to many other reports indicating that Pr123 is indeed on insulator. Observation of superconductivity in $Pr_{0.5}Ca_{0.5}123$ [13] with similar atomic properties of Ba and Ca, and the magnetic study on $(Pr_{1-x}Ba_x)Ba_2Cu_3O_{7-\delta}$ [14] can suggest the Pr/Ba miss-substitution scenario for explaining the unusual properties of Pr123 system. To fully

---

[*] Corresponding author: e-mail: akhavan@ sharif.edu, Phone: +98 21 6164510, Fax: +98 21 6012983

clarify this matter, we investigate the structural and normal electrical transport properties of $(Pr_{1-x}Ba_x)Ba_2Cu_3O_{7-\delta}$ system.

2 Experiments Single-phase polycrystalline $(Pr_{1-x}Ba_x)Ba_2Cu_3O_{7-\delta}$ (PrBa123) samples with x = 0.00, 0.025, 0.05, 0.10, 0.15, 0.20, 0.30, 0.40, 0.50, and 0.60 stoichiometry have been prepared by the standard solid state reaction technique. Pr_6O_{11}, $BaCO_3$, and CuO powders with 99.9 % purity have been used for the samples. The calcination and sintering temperature were 840 °C and 930 °C, respectively, and the samples were oxygenated at 450 °C in oxygen atmosphere. The sintered samples were characterized by powder X-ray diffraction (XRD) using Cu K radiation in the range of 5°< 2θ < 120°, by 0.02 ° step. The resistivity measurements were carried out by the standard ac four-probe method using PAR 3490A multimeter. The temperature was measured and controlled to ±10 mK by a Lake Shore-330 temperature controller with two Pt resistor sensors. The temperature in the range of 300–30 K was provided by a helium closed-cycle refrigerator.

3 Results and discussion XRD measurements of the samples have been analyzed by comparing the pattern with the standard JCPDS cards and Rietveld refinement. It is resulted that for x ≤ 0.3, 123 phase is a dominant phase structure, while for x ≥ 0.4, the impurity phases, i.e. Pr_2BaCuO_5 and BaO_2 grow in the samples, and finally for x=0.6, the impurity phases dominate. The solubility limit of the system found to be x = 0.4 in the standard solid state reaction preparation procedure. The limit is due to the large differences between Pr and Ba ionic sizes (i.e. $R(Pr^{3+})$ = 0.99 Å and $R(Ba^{2+})$ = 1.35 Å), and their different valence states. This limit is smaller than the value for R $(Ba_{2-x}Pr_x)Cu_3O_{7-\delta}$(R (BaPr)123) [15] and PrCa123 [16] systems. This can be explained by the Madelung potential energy for R^{3+} at Ba^{2+} site [17], and the smaller difference between the ionic size of Pr and Ca (i.e. $R(Ca^{2+})$ = 1.00 Å) with respect to the difference of Pr and Ba, for PrCa123 system. For x ≤ 0.3, the samples have dominant 123 structure phase with small impurity phases below 2θ = 30 ° The intensity of impurity phases grows smoothly with the increase of x, but for x=0.4 there is a discontinuity in the formation of these phases. Rietveld refinement of the XRD patterns show that the occupation number N for Cu(1), Cu(2) and Ba ions are about one, and for Pr and Ba_{Pr} (Ba at Pr site) are nearly the same as the stoichiometry.

Figures 1a and 1b show the normalized resistivity of the samples with respect to the resistivity at room temperature in the range of 250 to 50 K. All of the samples have the insulating behaviour in this temperature range. The insulating behaviour of bulk Pr123 sample is in agreement with many measurements [7], but is in contrast to the reports by Zou et al. [10] and Blakstead et al. [11]. Zou et al. have reported the wide range of conductivity from insulator to superconductor for Pr123; two pieces of their sample have T_c=56.5 K and 81 K [18]. In our sample, the resistivity increases by Ba doping at Pr site for low doping values (x < 0.1) (Fig. 1a), and then, decreases sharply with the increase of x up to x = 0.3 (Fig. 1b). The non-formation of pure 123 phase for x ≥ 0.4, as pointed out by the XRD measurements, causes a great increase in the resistivity, as it is observed from the resistivity measurements. The observed behaviour of resistivity of the samples is due to the Ba-doping at R site, and not from other parameters such as oxygen content, since all of the samples were prepared under the same preparation and annealing procedures.

In the range of x = 0.0–0.3 (below the solubility limit), there are two distinct features in the resistivity versus x: Increase with x up to about x = 0.08, and then sharp decrease up to x = 0.30. For the lower values of x, the scattering of carriers is more dominant, while for larger values of x, the increase of hole content in the CuO_2 planes is the dominant factor. The resistivity versus x plot for different temperature values show the decrease of resistivity in Pr123 samples with the increase of x above 0.3. It is suggested that PrBa123 samples for the values of x larger than 0.3 may have superconducting transition if the 123 structure could be formed. Other experimental data on similar systems like $Pr_{0.5}Ca_{0.5}123$ [16] confirm this proposal. Comparing our results with PrCa123 system indicates an insulator–superconductor transition with doping the divalent Ba or Ca in the Pr site. For these reasons we suggest that Ba rich Pr123 samples could have a superconducting transition. This result can clarify the reported superconductivity in Pr123 by Zou et al. [10].

© 2004 WILEY-VCH Verlag GmbH & Co. KGaA, Weinheim

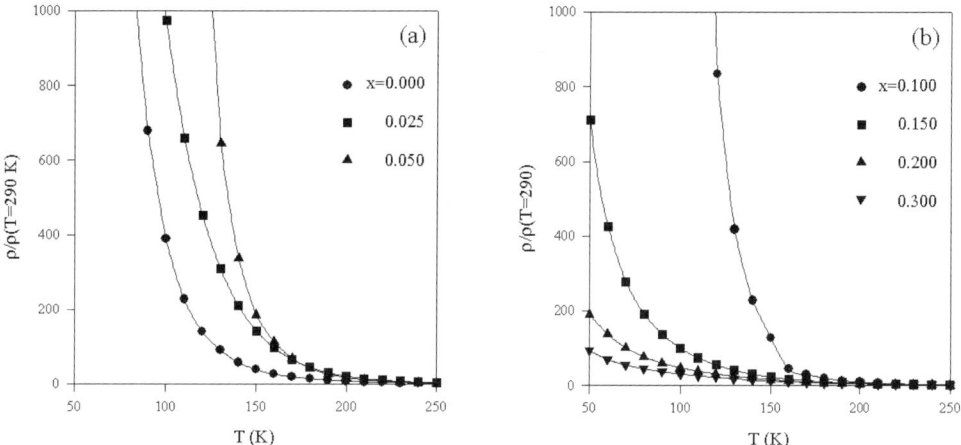

Fig. 1 Normalized resistivity of $(Pr_{1-x}Ba_x)123$ for x = 0.000, 0.025, and 0.050 (a), and x = 0.10–0.30 (b).

The cell parameters for the Pr123 superconductor reported by Zou et al. are larger than those for the Pr123 insulator sample grown by the flux method. By noting the difference between Ba and Pr ionic sizes, the difference between the lattice parameters could be understood by miss-substitution of Ba at Pr site in the Pr123 superconductor samples. The increase of the Cu (2)–Cu (2) distance in the samples also confirms this point. Another unusual characteristic of Pr123 superconducting samples is the large change of T_c under high pressure for these samples (i.e. dT_c/dp = 7.4 and 3.5 K/GPa for different samples). This result is compared with the measurement of T_c with applied pressure for the oxygen deficient Y123 samples [19], where the same overall behaviour is concluded for both systems [20]. This comparison and the same value of dT_c/dp for the two systems lead to two points: 1) the oxygen contents for Pr123 superconducting samples have not been determined directly, but they have been determined by comparison of their T_cs with that of Y123 samples. In our view, the lower T_c can be due to lower Ba content in the samples. 2) Due to the strong hybridization between Pr and oxygen, we should not expect the same behaviour and the same value of the dT_c/dp for the Pr123 and Y123 systems. High pressure experimental data on Pr-doping in NdPr123 [21] and YPr123 [22] systems show a negative dT_c/dp in contrast with the large positive value for Pr123 superconductor. The large dT_c/dp observed for Pr123 samples, can be explained if Ba is located at Pr site, because the decrease of distance between Ba and CuO_2 planes in high pressure can increase the hole content in the CuO_2 planes, and as a result of the pressure induced charge transfer, the T_c increases.

Another property of the samples, as shown by Narozhnyi and Drecshler [14], is the estimated effective magnetic moment and Curie constant of Zou et al.'s data which are 2.09 μ_B and 0.546 emu-K/mol, respectively, but with the Zou's correction, the magnetic moment is μ = 2.29μ_B [23]. These values are different from the value expected for Pr123 single crystals. To reach an agreement, it is suggested that about 50 percent of non magnetic Ba ion at the Pr sites or Ba rich samples can decrease these values. The difference between T_c of some parts of the Zou's samples prepared by the same TSFZ method, where the conducting properties of samples were smeared from superconductor to insulator behaviour [24] shows that the superconducting properties such as T_c are very sensitive to the details of the preparation procedure.

From our data and the above evidences, we can conclude the importance of the role of Ba in the Pr123 superconducting samples prepared by Zou et al. Finally, the Pr123 system with correct stoichiometry is indeed an insulator.

4 Conclusion From the analysis of electrical resistivity measurements of $(Pr_{1-x}Ba_x)Ba_2Cu_3O_{7-}$ system, it is shown that for low values of x, an increase of resistivity is due to the dominant scattering of carriers from lattice dislocations. For high values of x, but lower than the solubility limit, the increase of hole carriers in the CuO_2 planes causes the increase of conductivity in this system. It is suggested that Ba doping at Pr site can revive superconductivity in Pr123 systems if the solubility limit could be enhanced by some sample preparation techniques. Finally, we conclude that Pr123 superconducting samples observed by Zou et al. are due to Ba rich Pr123 samples; Pr123 with correct stoichiometry is indeed an insulator.

Acknowledgements This work was supported in part by the Offices of the Vice President for Research and Dean of Graduate Studies at Sharif University of Technology.

References

[1] V. E. Gasumyants, M. V. Elizarova, and R. Suryanarayanan, Phys. Rev. B **61**, 12404 (2000).
[2] Y. Xu and W. Guan, Phys. Rev. B **45**, 3176 (1992).
[3] C.-S. Jee, A. Kebede, D. Nicholas, J. E. Crow, T. Mihalisin, G. H. Myer, I. Perez, R. E. Salomon, and P. Schlottmann, Solid State Commun. **69**, 379 (1989).
[4] A. Kebede, C. S. Jee, J. Schwegler, J. E. Crow, T. Mihalisin, G. H. Myer, R. E. Salomon, P. Schlottmann, M. V. Kuric, S. H. Bloom, and R. P. Guertin, Phys. Rev. B **40**, 4453 (1989).
[5] H.-C. I. Kao, F. C. Yu, and W. Guan, Physica C **292**, 53 (1997);
    P. Wei, H. W. Ying, and Z. Q. Qi, Physica C **209**, 400 (1993).
[6] H. A. Blackstead and J. D. Dow, Phys. Rev. B **51**, 11830 (1995).
[7] M. Akhavan, Physica B **321**, 265 (2002).
[8] H. Khosroabadi, V. Daadmehr, and M. Akhavan, Mod. Phys. Lett. B **16**, 943 (2002).
[9] T. Usagawa, Y. Ishimaru, J. Wen, T. Utagawa, S. Koyama, and Y. Enomoto, Jpn. J. Appl. Phys. **36**, L1583 (1997).
[10] Z. Zou, K. Oka, T. Ito, and Y. Nishihara, Jpn. J. Appl. Phys. **36**, L18 (1997).
[11] H. A. Blackstead, J. D. Dow, D. B. Chrisey, J. S. Horwitz, M. A. Black, P. J. McGinn, A. E. Klunzinger, and D. B. Pulling, Phys. Rev. B **54**, 6122 (1996).
[12] K. Oka, Z. Zou, and J. Ye, Physica C **300**, 200 (1998).
[13] Y. S. Yao, Y. F. Xiong, D. Jin, J. W. Li, F. Wu, J. L. Luo, and Z. X. Zhao, Physica C **282–287**, 49 (1997).
[14] V. N. Narozhnyi and S.-L. Drechsler, Phys. Rev. Lett. **82**, 461 (1999).
[15] Y. T. Ren, Y. Y. Xue, Y. Y. Sun, and C. W. Chu, Physica C **213**, 224 (1993).
[16] Y. F. Xiong, Y. S. Yao, L. F. Xu, F. Wu, D. Jin, and Z. X. Zhao, Solid State Commun. **107**, 509 (1998).
[17] M. Muroi and R. Street, Physica C **314**, 172 (1999).
[18] J. Ye, Z. Zou, A. Matsushita, K. Oka, Y. Nishihara, and T. Matsumoto, Phys. Rev. B **58**, R619 (1998).
[19] S. Sadewasser, J. S. Schilling, A. P. Paulikas, and B. W. Veal, Phys. Rev. B **61**, 741 (2000).
[20] J. Ye, S. Sadewasser, J. S. Schilling, Z. Zou, A. Matsushita, and T. Matsumoto, Physica C **328**, 111 (1999).
[21] J. G. Lin, C. Y. Huang, and J. C. Ho, Physica C **341–348**, 625 (2000).
[22] J. J. Neumeier, M. B. Maple, and M. S. Torikachvili, Physica C **156**, 574 (1988).
[23] Z. Zou and Y. Nishihara, Phys. Rev. Lett. **82**, 462 (1999).
[24] Z. Zou, J. Ye, K. Oka, and Y. Nishihara, Phys. Rev. Lett. **80**, 1074 (1998).

Electrical and magnetic properties of Pr-doped Nd123

P. Maleki, H. Khosroabadi, and **M. Akhavan**[*]

Magnet Research Laboratory (MRL), Department of Physics, Sharif University of Technology,
P.O. Box 11365-9161, Tehran, Iran

Received 31 August 2003, accepted 31 December 2003
Published online 8 April 2004

PACS 74.25.Fy, 74.25.Ha, 74.72.Jt

Single-phase polycrystalline $NdBa_{2-x}Pr_xCu_3O_{7+\delta}$ ($0 \leq x \leq 0.5$) have been prepared by the standard solid-state reaction. The results of X-ray diffraction measurements confirm the formation of 123 single phase. A superconductor–insulator transition at $x_c^{SIT} = 0.3$ has been observed for the samples. The magnetoresistance of the samples have been investigated within the thermally activated flux creep model. The pinning energy decreases with the magnetic field application and Pr-doping. A scaling relation as power law dependence of pinning energy to the magnetic field has been obtained. The derived pinning energy shows that the Pr-doping plays the role of weak link. The normal state resistivity analysis shows that with the increase of x, there is a cross-over from 2D to 3D variable range hopping mechanism.

© 2004 WILEY-VCH Verlag GmbH & Co. KGaA, Weinheim

1 Introduction

One of the anomalies in high temperature superconductors (HTSC) is the superconductor–insulator transition with the increase of Pr-doping (x_c^{SIT}) at R site in $RBa_2Cu_3O_7$ (R123, R is Y and rare earth elements except Tb, Pm, and Ce) [1]. Because of the anomalous insulating behaviour of Pr in the R123 family, it seems that solving the Pr anomaly could help the understanding of the HTSC mechanism. The most important models presented for resolving this anomaly are: hole filling [2], hole localization [3], and Pr/Ba disorder [4]. In spite of the presented models, there is no consensus on which characteristics of Pr ion has the dominant effect on the unusual electrical and magnetic [5] behaviours of Pr123 system. One of these problems is the dependency of x_c to the rare earth ion size, which has been concluded from the experimental data. Recently, Akhavan [6] has reviewed the Pr anomaly in HTSC.

Although, many groups are studying the Pr anomaly, some have reported superconductivity in Pr123 [7, 8]. These observations are in disagreement with the many reports about the insulating behavior of Pr123. As suggested by Narozhny and Drechsler [9], it has been deduced that the Ba at Pr site causes the superconducting bulk Pr123. Though, the superconducting Pr123 has not been established, but it has also been extensively accepted that Pr123 is the only insulating R123 with orthorhombic structure, and the Pr/Ba disorder has been taken into account in recent studies [10]. Blackstead and Dow have claimed that the Pr at Ba site can disturb the superconductivity in unobserved superconductor Pr123, and deduced that the Pr at the real site R in R123 has the superconducting behavior like other members of the R123 family [8].

The Blackstead's claim on Pr/Ba disorder has caused a number of groups to renew the study of Pr problem and the different behaviors of Pr at R and Ba sites [11]. It seems that the $RBa_{2-x}Pr_xCu_3O_7$ (R(BaPr)123) mis-substitution can be helpful in investigating the Pr anomaly. Also, the comparison between electronic and magnetic properties of Pr-doped at the R and Ba sites is another good approach for this subject. In this work, we have studied the Nd(BaPr)123 system in the normal and superconducting states. The results have been compared with $Nd_{1-x}Pr_xBa_2Cu_3O_{7-\delta}$ (NdPr123) and heavy rare earth-based systems.

---

[*] Corresponding author: e-mail: akhavan@sharif.edu, Phone: +98 21 6164544, Fax: +98 21 6012983

2 Experimental   Single-phase polycrystalline $NdBa_{2-x}Pr_xCu_3O_{7+\delta}$ ($0 \leq x \leq 0.5$) has been prepared by standard solid-state reaction. The mixture of Nd_2O_3, Pr_6O_{11}, $BaCO_3$, and CuO powders with 99.9% purity in proper stoichiometry have been calcinated at 840 °C in air atmosphere for 24 h. The calcination was repeated twice with intermediate grinding. The compounds were then pressed into pellets and sintered in oxygen atmosphere at 930 °C for 36 h; annealing was done in 450 °C for 24 h to obtain fully oxygenated samples, and then cooled slowly to room temperature. The sintered samples were characterized by powder X-ray diffraction (XRD) using CuK_α radiation ($\lambda = 1.5406$ Å) in room temperature in the range of $5° \leq 2\theta \leq 120°$ with 0.02° step.

The resistivity measurements were carried out by the standard ac four-probe method using a PAR 124A lock-in amplifier with current of 5 mA. The temperature was measured and controlled to ±10 mK using a Lake Shore-330 temperature controller with a Pt resistor sensor. A helium closed-cycle refrigerator provided the temperature in the range of 300–20 K. For the magnetoresistivity measurements a dc magnetic field of maximum value 15 kOe was applied perpendicular to the current and sample's surface.

3 Results and discussion   The results of X-ray diffraction measurements confirm the formation of 123 phase as the dominant phase with no impurity phases. Due to the closeness of Ba^{2+} ionic size (1.34 Å) and that of the light rare earths (i.e. Nd^{3+}: 1.12 Å), the mis-substitution becomes more probable. This effect still appears as a 123 phase in XRD pattern. Therefore, preparation of single phase light rare earth-based HTSC is more difficult.

Figure 1 shows the samples resistivity in the range of 300 to 20 K. It is observed that T_c decreases and ΔT_c increases with the increase of Pr-doping value due to disorder. The normal resistivity of the samples increases with x increment, and for larger x values the insulating behaviour prevails. The slope of the resistivity of the normal state $(d\rho/dT)$ decreases with the increase of x. These facts indicate that the suppression of superconductivity with increase of Pr-doping may be due to hole filling by replacing Ba^{2+} for Pr^{3+}, or even localization of hole carriers in the CuO_2 plane in these samples.

The superconductor–insulator transition occurs at $x_c^{SIT} = 0.3$. This result is in agreement with other reports on the system [12]. It should be noted that the x_c^{SIT} reported for NdPr123 [13] has the same value as the Nd(BaPr)123 system. This is in spite of the reports that in R123 family, the Pr at Ba site suppresses superconductivity stronger than it does when Pr is located at R site [14]. In this system, x_c^{SIT} is smaller than the heavy rare earth such as Tm(BaPr)123 ($x_c^{SIT} = 0.45$) [15] and Gd(BaPr)123 [16]. This is similar to the relation of x_c in RPr123. It is concluded that the decrease of x_c in systems with larger ionic radius has similar behaviour in Pr substitution both at Ba and R sites.

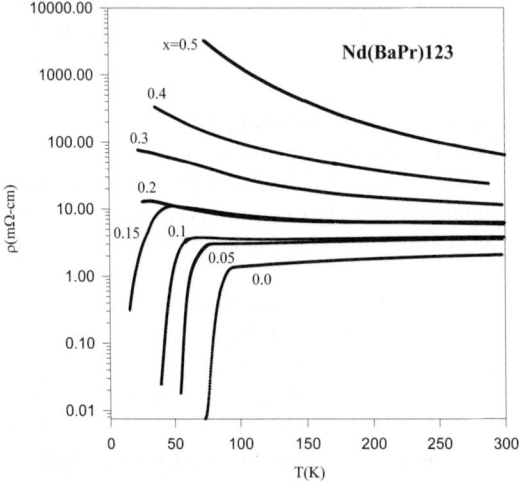

Fig. 1   Resistivity behaviour of $Nd(Ba_{2-x}Pr_x)Cu_3O_{7+\delta}$ compounds as a function of temperature for various values of x.

Figure 2 shows the typical magnetoresistance for the x = 0.05 sample in the range of 0–15 kOe. Inset to the Fig. 2 shows the Arrehenius plot of the magnetoresistance data. At low temperatures the linear part of the lnρ versus 1/T has been considered for thermally activated flux creep model (TAFC). The TAFC model presented for the unusual broadening of resistivity in the vicinity of T_c has been used for HTSC materials. From the figure, we obtain the average microscopic pinning energy by the $\rho = \rho_0 \exp(-U_0/k_B T)$, where k_B is the Boltzman constant and U_0 is the pinning energy. The derived U_0 values are smaller than the values reported for other R123 [17]. This is expected due to the larger transition temperature width observed for our samples. For the fixed magnetic field, pinning energy decreases with the increase of Pr-doping. The derived parameter U_0 versus magnetic field shows the decrease of pinning energy by applied magnetic field. A scaling relation has been found for U_0 versus H (e.g. $U_0 \sim H^{-\beta}$) for the three samples ($0 \leq x \leq 0.1$). The derived value for β is in the range of 0.40 to 0.89. The derived pinning energy shows that the Pr-doping plays the role of weak link.

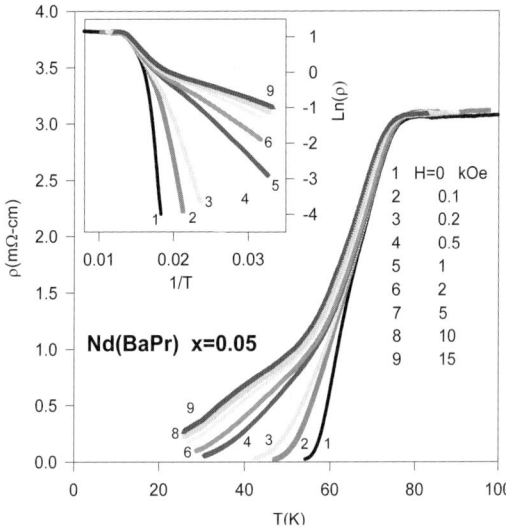

Fig. 2 Magnetoresistance of Nd(Ba$_{1.95}$Pr$_{0.05}$)Cu$_3$O$_{7-\delta}$ in the range of 0–15 kOe. The inset shows the Arrehenius plot of the data.

The normal state resistivity was fitted to $\rho = \rho_0 (T/T_0)^{2p} \exp(T_0/T)^p$, (p = 1/(1 + D); D is the dimensionality) variable range hopping (VRH) equation for p = 1/3, 1/4, and p = 1/2 Coulomb gap (CG). For the fitting, we have considered the linear part of $\ln(\rho/T^{2p})$ vs. $T^{-1/p}$. The fitting shows that the hopping conduction mechanism can be applied for the normal resistivity of the superconducting and insulating samples, as pointed out by other reports [18]. This analysis shows that there is a cross-over from 2D-VRH to 3D-VRH regime with the increase of x. The characterization temperature T_0 is found to increase with the increase of Pr for VRH regime. The localization length $d \sim (N(E_F)T_0)^{-1/D}$ in VRH regime decreases with the increase of x. It shows that Pr-doping localizes the carriers in the normal state, which causes the suppression of superconductivity in this system.

4 Conclusion We have determined that the superconductor–insulator transition in Nd(BaPr)123 which occurs at $x_c^{SIT} = 0.3$, is the same value as for the NdPr123 system. The derived pinning energy, obtained from the thermally activated flux creep model, decreases with the applied magnetic field and Pr-doping. A scaling relation for pinning energy to the magnetic field has been obtained as $U_0 \sim H^{-\beta}$. The derived pinning energy shows that the Pr-doping plays the role of weak link. Our analysis of the normal state resistivity shows that there is a cross-over from 2D-VRH to 3D-VRH regime with the increase of x.

Acknowledgements The authors would like to thank M. Kariminezhad, M. Mirzadeh, and Z. Mokhtari for assistance and useful discussions, and M.R. Mohammadizadeh for useful discussions. This work was supported in part by the Offices of Vice President for Research and Dean of Graduate Studies at Sharif University of Technology.

References

[1] Z. Yamani and M. Akhavan, Physica C **268**, 78 (1996).
[2] A. Kebede, C.S. Lee, J. Schagler, J.E. Crow, T. Milhalisin, G.H. Meger, R.F. Salomon, P. Schlottman, M.V. Kuric, S.H. Bloom, and R.P. Guertin Phys. Rev. B **40**, 4453 (1989).
[3] R. Fehrenbacher and T.M. Rice, Phys. Rev. Lett. **70**, 3471 (1993).
[4] H.A. Blackstead and J.D. Dow, Phys. Rev. B **51**, 11830 (1995).
[5] V.Stabub, L. Soderholm, R. Osborn, T.J. Goodwin, H.B. Radousky, and R.N. Shelton, J. Phys.: Condens. Matter **10**, 4637 (1998).
[6] M. Akhavan, Physica B **321**, 265 (2002).
[7] Z. Zou, J. Ye, K. Oka, and Y. Nishihara, Phys. Rev. Lett. **80**, 1074 (1998);
T. Usagawa, Y. Ishimaru, J. Wen, T. Utagawa, S. Koyama, and Y. Enomoto, Jpn. J. Appl. Phys. **36**, L1583 (1997).
[8] H.A. Blackstead, J.D. Dow, D.B. Chrisey, J.S. Horwitz, M.A. Black, P.J. McGinn, A.E. Klunzinger, and D.B. Pulling, Phys. Rev. B **54**, 6122 (1996).
[9] V.N. Narozhnyi and S.-L. Drechsler, Phys. Rev. Lett. **82**, 461 (1999).
[10] M.R. Mohammdizadeh, H. Khosroabadi, and M. Akhavan, Physica B **321**, 301(2002).
[11] V.E. Gasumyants, V. Elizarova, and R. Surynanaryanan, Phys. Rev. B **61**, 12404 (2000).
[12] L. Colonescu, J. Berthon, R. Suryanarayanan, and I. Zelenay, Physica C **291**, 85 (1997).
[13] L. Colonescu, F. Cairon, J. Berthon, I. Zelenay, and R. Suryanarayanan, Physica B **259–261**, 528 (1999).
[14] W.H. Tang and J. Gao, Physica C **315**, 59 (1999).
[15] Z. Mokhtari, H. Khosroabadi, and M. Akhavan, phys. stat. sol. (c) **1**, No. 7 (2004), this conference.
[16] M.R. Mohammadizadeh and M. Akhavan, Eur. Phys. J. B **33**, 381 (2003).
[17] H.S. Gamchi, G.J. Russell, and K.N.R. Taylor, Phys. Rev. B **50**, 12950 (1994).
[18] X.W. Cao, Z.H. Wang, and Y.J. Tang, Physica B **212**, 411 (1995).

Normal state conduction and TAFC in Gd(BaLn)123 (Ln = La, Nd)

M. Mirzadeh, H. Khosroabadi, and **M. Akhavan**[*]

Magnet Research Laboratory (MRL), Department of Physics, Sharif University of Technology,
P. O. Box 11365-9161, Tehran, Iran

Received 31 August 2003, accepted 31 December 2003
Published online 8 April 2004

PACS 74.25.Fy, 74.25.Ha, 74.72.Jt

We have studied the electrical and magnetic properties of single phase polycrystalline $Gd(Ba_{2-x}La_x)Cu_3O_{7+\delta}$ and $Gd(Ba_{2-x}Nd_x)Cu_3O_{7+\delta}$ ($0 \leq x \leq 0.5$) samples. The samples were prepared by the standard solid-state reaction technique. The electrical measurements show the decrease of T_c and increase of ΔT_c with the increase of Nd and La-doping. The magnetoresistance of samples has been studied within the thermally activated flux creep (TAFC) model. The pinning energy decreases with the increase of magnetic field and doping for both systems in a similar way. The decrease of pinning energy with applied magnetic field can be scaled with two power law relations for the magnetic field: smaller and larger than 1 kOe. In the $H \leq 1$ kOe region, the power coefficient increases with the increase of doping. The normal state resistivity of samples, investigated by the variable range hopping and Columb gap models, indicates that for the superconductor samples the CG model and for the insulator samples 2D-VRH are the dominant mechanisms for both compounds.

© 2004 WILEY-VCH Verlag GmbH & Co. KGaA, Weinheim

1 Introduction

It is well known that chemical substitution of Y in $YBa_2Cu_3O_7$ by rare earth elements (except for Ce, Tb, Pr) neither affects its superconducting properties nor significantly changes the corresponding superconducting transition temperature (T_c) of ~92 K. The high T_c cuprate superconductor $RBa_2Cu_3O_7$ (R = Y and rare earth except Ce, Pr and Tb) exhibits similar properties regardless of magnetic or nonmagnetic R ion [1]. T_c is dramatically affected by oxygen content, oxygen distribution, and oxygen symmetry. In addition, the oxygen content and the crystal symmetry of $RBa_2Cu_3O_7$ compounds may be changed by chemical substitution [2].

$PrBa_2Cu_3O_7$ (Pr123) is isostructure to other R123, but exhibits the semiconducting behaviour [3]. The origin of this anomalous behaviour might give important clues in the study of the superconducting mechanism for high T_c cuprates. Several possible mechanisms, such as hole filling, magnetic pair breaking, and hole localization [4] have been proposed to explain the T_c suppression in Pr123. Recently, it has been argued that the Pr at Ba site not R site destroys the superconductivity [5]. So, the occupancy of rare earth elements at Ba site in R123 has given rise to extensive research. It seems that the substitution of rare earth at R site or Ba site may play a different role on superconductivity.

In this paper, the system of Gd123 with La and Nd-doping is investigated for the effects of magnetic and nonmagnetic rare earth at Ba site on the superconductivity. Also, we study the magnetoresistance and normal state resistivity of the systems.

2 Experimental details

The samples of $Gd(Ba_{2-x}Ln_x)Cu_3O_7$, Gd(BaLn)123, (Ln = La and Nd) with $x = 0.0, 0.1, 0.2, 0.3, 0.4, 0.5$ were prepared by solid-state reaction. The mixture of the powders of Nd_2O_3, La_2O_3, $BaCO_3$, Gd_2O_3 and CuO with purity 99.9% in proper stoichiometry were ground and

---

[*] Corresponding author: e-mail: akhavan@sharif.edu, Phone: +98 21 6164510, Fax: +98 21 6012983

calcined in air at 840 °C for 24 h. The calcinations were repeated twice with grinding. The compounds were pressed into pellets and sintered in oxygen atmosphere at 930 °C for 24 h. Then, the samples were cooled to 550 °C and retained there for 4 h and finally cooled down to room temperature at a slow cooling rate of 1 °C/min. For the details of sample preparation procedures see Ref. [6]. The final samples were examined by X-ray diffraction with Cu K radiation ($\lambda = 1.5406$ Å) at room temperature. The oxygen content of the samples was determined by the iodometic titration technique. The resistivity was measured within the temperature range 10–300 K by ac four probe method with the applied current of 5 mA and frequency of 33Hz. A Lake Shore-330 temperature controller with a Pt-100 resistor was used for measuring and controlling the temperature to within ±10 mK. A magnetic field with a maximum value of 15 kOe was applied perpendicular to the current for the magnetoresistance measurements.

3 Results and discussion The XRD patterns show that all samples are isostructure with 123 phases. There is a structural transition from orthorhombic to tetragonal at $x = 0.2$ for the two series of compounds. This transition occurs at $x = 0.2$ in the Gd(BaPr)123 system [7]. Iodometric titration analysis show that the oxygen content of samples increases with the increase of La-doping from 6.98 for $x = 0$ to 7.17 for $x = 0.5$. This is expected because of the substitution of trivalent La and Nd ions for divalent Ba ion, and the charge neutrality requirement.

Figure 1 shows the resistivity versus temperature for different amounts of La-doping. The resistivity of the system with Nd-doping is higher than the La-doped system. It is shown that the doping of Nd is more effective than La in the increase of resistivity in Gd123 system. In comparison, the rise in resistivity of Gd123 with Pr-doping is more than for La and Nd-doping [7]. The resistivity of samples indicates that the transition temperature T_c decreases and the transition width ΔT_c increases with the increase of doping. The metallic behaviour in the normal state is remained up to $x = 0.3$ for La-doping, and to $x = 0.2$ for Nd-doping. There is an insulator like behaviour for $x \geq 0.4$ in both systems.

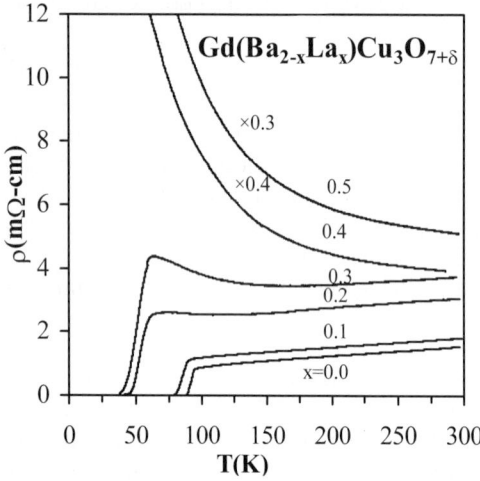

Fig. 1 Resistivity vs. temperature for different amounts of doping in Gd(Ba$_{2-x}$La$_x$)Cu$_3$O$_{7+\delta}$ system.

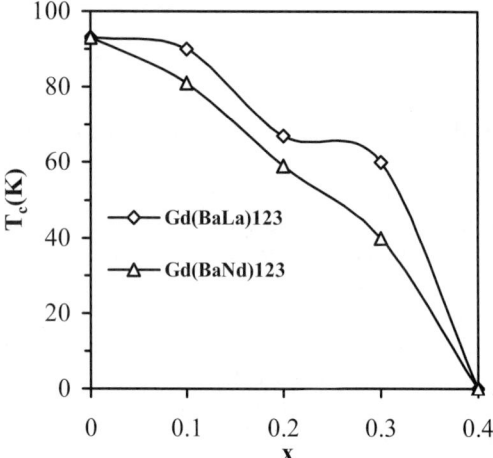

Fig. 2 Superconducting transition temperature T_c vs. La and Nd-doping. The lines are guides to the eye.

Figure 2 shows the dependence of T_c on the La and Nd content x. The suppression of superconductivity with La-doping can be due to the difference in the valences of La^{3+} and Ba^{2+}. The superconductivity of p-type doped superconductors depends on the holes predominantly on the oxygen sites in the CuO$_2$ planes, and the substitution of R^{3+} for Ba^{2+} would require donor electrons, then it cancels partial holes in the CuO$_2$ planes. So, T_c should decrease with the increase of R^{3+} content. There is a nearly linear dependence of T_c to Nd content, while the La-doped system shows a non-linear relation. The decrease of T_c with increase of doping x in the Gd(BaNd)123 is more than in Gd(BaLa)123. It means that there is more to the suppression of superconducting than just the hole filling mechanism in the Gd(BaNd)123 system.

The linear dependence of T_c to Nd content indicates that the Abrikosov and Gor'kov (AG) theory of pair breaking [8] due to magnetic moment of Nd ($\mu = 3.54\mu_B$) is applicable in the suppression of superconductivity.

Broadening of transition was observed in resistivity curves. Figure 3 shows a typical magnetic field dependence of resistivity for x = 0.1 in Gd(BaLa)123 system. For all samples, with increasing applied magnetic field, the normal state resistivity remain unchanged and the onset transition temperature remains almost unaffected, while the temperature of zero resistivity moves to considerably lower temperature.

We have examined the thermally activated flux creep model for interpreting the broadened resistivity data in this region. In the framework of this model, the broadened resistivity curves for the high T_c oxides can be expressed as $\rho(T,H) = \rho_0 \exp[-U(T,H)/k_BT]$ [9]. $U(T,H)$ is the pinning energy that depends on temperature and magnetic field. This model is only applicable if $U > k_BT$. For x = 0.3 at H > 1 kOe, this condition is not valid. Therefore, the flux creep model is not employed in this region. The values of U can be deduced from the slope of the plot of $\ln\rho$ vs. 1/T. These values are calculated from the linear data in the tail part of the plot. The derived pinning energy at different magnetic field and for different doping of La is presented in Fig. 4. The pinning energy decreases similarly with the increase of doping and magnetic field for the two series of compounds. This is in agreement with the results reported for GdPr-123 [10] and Gd(BaPr)123 [5].

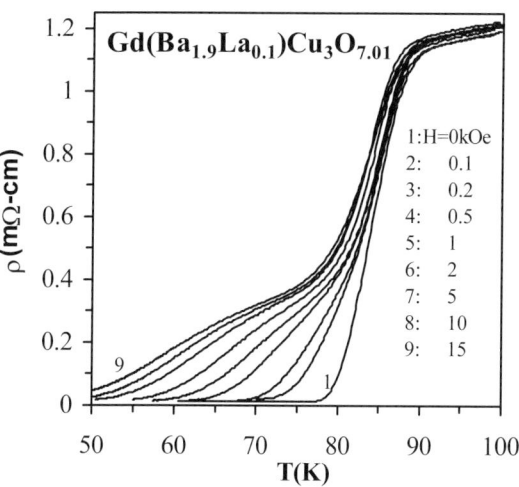
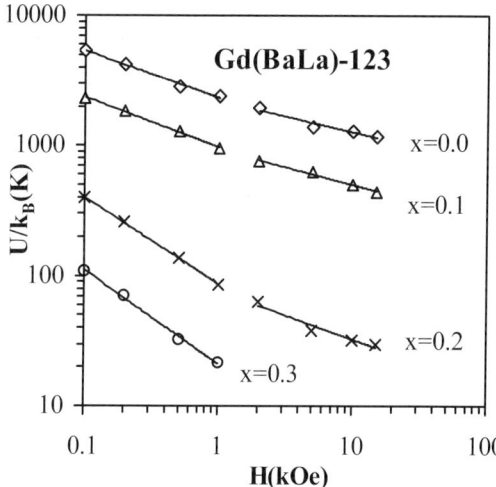

Fig. 3 Resistivity vs. temperature under applied magnetic field for Gd(Ba$_{1.9}$La$_{0.1}$)Cu$_3$O$_{7.01}$.

Fig. 4 Pinning energy versus applied magnetic field for different amount of La doping. The linear behaviour of the pinning energy is separated into two parts: less, and greater than 1 kOe.

The pinning energy versus applied magnetic field can be scaled by a power law relation ($U \sim H^{-\beta}$). The derived values of β and the goodness of fit χ^2 have been obtained. There is a break at about H = 1 kOe in the curves of U vs. H. Similar breaks have also been reported in other studies [7, 11]. The value of β for different amount of doping values varies in $H \leq 1$ kOe and H > 1 kOe regions from 0.36 to 0.72 and 0.23 to 0.37, respectively. In the entire range of magnetic field, the values of β increases with the increase of doping. This shows that the dependence of U to the magnetic field becomes weaker with the increase of doping. The pinning energy decreases with the increase of La, but the Nd-doping is more effective. In comparison, Pr doping at Ba site in Gd123 decreases the pinning energy more than La and Nd-doping. Therefore, in Gd123 system, La and Nd substitutions at Ba site affect like weak link, and cause the vortices to creep more easily. This is in agreement with the effect of Pr in the Gd(BaPr)123 [7] and GdPr123 [10].

We have investigated the normal state resistivity of $Gd(Ba_{2-x}La_x)Cu_3O_{7+\delta}$ and $Gd(Ba_{2-x}Nd_x)Cu_3O_{7+\delta}$ samples by the Coulomb gap (CG) and variable range hopping (VRH) methods [12]. We have fitted $\ln(\rho/T^{2p})$ vs. $1/T^p$ curve by a linear function. The results show that in the Gd(BaLa)123, for superconductor samples ($x \leq 0.3$), the CG model prevails the conduction mechanism, and for insulator samples (x=0.4, 0.5) 2D-VRH is dominant. The conduction mechanism in Gd(BaLa)123 system is same as Gd(BaNd)123. It means that the increase of La and Nd-doping change the conduction mechanism in a similar way.

4 Conclusion The electrical resistivity and magnetoresistance of Gd(BaLa)123 and Gd(BaNd)123 samples have been investigated. The observed transition temperature T_c decreases and ΔT_c increases similarly with the increase of La and Nd-doping. It seems that the AG magnetic pair breaking is more effective in suppression of superconductivity of Gd(BaNd)123 system. Broadening of the resistivity in the tail part has been investigated by the thermally activated flux creep model. The results show the pinning energy, with the increase of doping and magnetic field, decreases equally for both systems, but Nd-doping reduces the pinning energy stronger than La-doping. There is a break in the lnU–lnH curves, and the power coefficient of the relation of U(H) is different for $H \leq 1$ kOe and $H > 1$ kOe. The conduction mechanism varies from CG to 2D-VRH at x = 0.4 for both systems.

Acknowledgments We wish to thank M. Kariminezhad, P. Maleki and Z. Mokhtari for assistance and useful discussions. We also thank M. R. Mohammadizadeh for useful discussions. This work was supported in part by the Offices of Vice President for Research and Dean of Graduate Studies at Sharif University of Technology.

References

[1] M. K. Wu, J. R. Ashbarn, C. J. Torng, P. H. Hor, R. L. Meng, L. Gao, Z. J. Huang, Y. Q. Wang, and C. W. Chu, Phys. Rev. Lett. **58**, 9082 (1987).
[2] S. Li, E. A. Hayri, K. V. Ramianujachary, and M. Greenblatt, Phys. Rev. B **38**, 2450 (1988).
[3] A. Kebede, C. S. Jee, J. Schwegler, J. E. Crow, T. Mihalisin, G. H. Myer, R. E. Salemon, P. Sehlottman, K. V. Kuric, S. H. Bloom, and R. P. Guertin, Phys. Rev. B **40**, 4453 (1989).
[4] H. Khosroabadi, V. Daadmehr, and M. Akhavan, Mod. Phys. Lett. B **16**, 943 (2002).
[5] H. A. Blackstead and J. D. Dow, Appl. Phys. Lett. **70**, 1891 (1997).
[6] Z. Yamani and M. Akhavan, Supercond. Sci. Technol. **10**, 427 (1997).
[7] M. Mohammadizadeh and M. Akhavan, Supercond. Sci. Technol. **16**, 538 (2003).
[8] A. A. Abrikosov and L. P. Gor'kov, Sov. Phys. JETP **12**, 1243 (1961).
[9] M. Tinkham, Introduction to Superconductivity (McGraw Hill, New York, 1975).
[10] V. Daadmehr and M. Akhavan, phys. stat. sol. (a) **193**, 153 (2002).
[11] Z. H. Wang and X. W. Cao, Solid State Commun. **109**, 709 (1999).
[12] N. F. Mott and E. A. Davis, Electronic Processes in Non-crystalline Materials, 2nd ed. (Clarendon, Oxford, 1979).

Appearance of a new superconducting phase in Gd(Ba$_{2-x}$Pr$_x$)Cu$_3$O$_{7+\delta}$

M. R. Mohammadizadeh and **M. Akhavan**[*]

Magnet Research Laboratory (MRL), Department of Physics, Sharif University of Technology, P.O. Box 11365-9161, Tehran, Iran

Received 31 August 2003, accepted 31 December 2003
Published online 1 April 2004

PACS 74.25.Fy, 74.25.Ha, 74.62.Dh, 74.72.Jt

An unusual hump on the resistivity vs. temperature curve of Gd(Ba$_{2-x}$Pr$_x$)Cu$_3$O$_{7+\delta}$ samples have been observed for particular values of Pr doping. Investigation of the origin of this effect has led to an important result. Based on the Rietveld refinement of the XRD patterns, we have found that the Ba atom substitution at the rare earth site could lead to superconductivity in some parts of the grains at $T_p \sim$ 80–90 K, which appears as a hump on the $\rho(T)$ curve. Our result is in line with the previously proposed possibility of existence of superconductivity in Pr–123 due to the Ba atom substituted at Pr site.

© 2004 WILEY-VCH Verlag GmbH & Co. KGaA, Weinheim

1 Introduction The PrBa$_2$Cu$_3$O$_{7-\delta}$ compound in the orthorhombic phase, in contrast to other rare-earth-based R-123 compounds, is an insulator [1]. Different anomalous effects have been observed when Pr atom is substituted in R-123 and other HTSC. Although, many attempts have been made to explain the insulating behavior of Pr-123, some groups have reported observation of superconductivity in the powder, single crystal, polycrystalline, and thin films of this compound [2]. Therefore, the phenomenon of Pr-123 system has become more complicated: The question is under what conditions, if ever, the Pr-123 compound could become a superconductor. For a recent review on the role of Pr in HTSC see Ref. [3].

Among different theories for explaining the superconductor–insulator transition (SIT) in RPr-123 such as hole filling, pair breaking, and hybridization, the substitution of Pr at Ba site (Pr$_{Ba}$) (site-switching or mis-substitution effect) rather than at R site has been proposed for the insulating behavior of Pr-123 [4]. This proposal seems relevant due to the nearly equivalent positions of R and Ba sites in the center of the imperfect R-123 perovskite structure, and the fact that superconductivity is suppressed in R(Ba$_{2-x}$R'$_x$)Cu$_3$O$_{7+\delta}$ compounds [5].

Recently, we have prepared Gd(Ba$_{2-x}$Pr$_x$)Cu$_3$O$_{7+\delta}$ compound and have investigated its normal and superconducting properties [6] in comparison with (Gd$_{1-x}$Pr$_x$)Ba$_2$Cu$_3$O$_{7-\delta}$ system [7]. In the resistivity vs. temperature curve we have observed a hump above the superconducting transition temperature (T_c) for some particular Pr doping values. In this paper, we will explain the possible scenario to account for the resistance anomaly. Our final conclusion is important for resolving the ρ–T anomaly and the possibility of superconductivity in Pr-123 system.

2 Experimental The Gd(Ba$_{2-x}$Pr$_x$)Cu$_3$O$_{7+\delta}$ single phase samples with $0.00 \leq x \leq 0.50$ were synthesized by the standard solid state reaction technique. Appropriate amounts of Gd$_2$O$_3$, Pr$_6$O$_{11}$, BaCO$_3$, and CuO powders with 99.9 % purity were mixed, ground, and calcined at 840 °C for 24 h in an air atmosphere. Calcination was repeated twice with intermediate grinding. Then powders were reground, pressed into pellets, and synthesized at 930 °C for 24 h in an oxygen atmosphere. The samples were cooled to 550 °C

---

[*] Corresponding author: e-mail: akhavan@sharif.edu, Phone: +98 21 6164510, Fax: +98 21 6012983

and retained under oxygen flow for 16 h. Finally, they were furnace cooled to room temperature. The oxygen content of samples has been determined by the iodometric titration technique with ±0.03 accuracy.

The XRD measurements have been done by Philips PW-3710 powder diffractometer with Cu K_α radiation and $\lambda = 1.5406$ Å in room temperature with 0.02° step width and 0.5 second step time. The XRD results have been analyzed with DBW3.2S-PC-9207 package based on the Rietveld method [8]. An ac four-probe method with f = 33 Hz was used for the conductivity measurements of the samples within the temperature range of 10 to 300 K. A Lake Shore-330 temperature controller with two Pt-100 resistors was used for measuring and controlling the temperature to within ±10 mK.

3 Results and discussion

The resistivity of the samples have been measured at I =10 mA and presented in Fig. 1. With increasing x, the superconducting transition temperature decreases and the width of transition temperature (ΔT_c) as well as the normal state resistivity increase. With increasing the number of insulating parts in the grains (i.e. Pr substituted unit cells) the homogeneity of the grains decreases, which leads to larger ΔT_c.

There is a hump present on the curves above T_c at $T_p \sim 80$–90 K for x = 0.15, 0.20, 0.25, and 0.30 samples. This hump is different from S-shaped curvature in the resistivity vs. temperature curve, which happens and develops with Pr-doping [9]. The S-shaped curvature is characteristic of either oxygen depletion [10] or cation substitutions for chain Cu atoms [11]. Figure 2 shows the variation of T_c and hump temperature (T_p) vs. x. The hump on the resistivity curve is present only for some particular values of x. So, this cannot be due to some characteristic structural modification, but a special event has occurred only for some particular dopings. It should be emphasized that these humps are reproducible through different measurements, and are not due to any measurement errors. In addition, there is no impurity phase in the $0.15 \leq x \leq 0.30$ range, meaning that the impurity phase cannot be the origin of these humps.

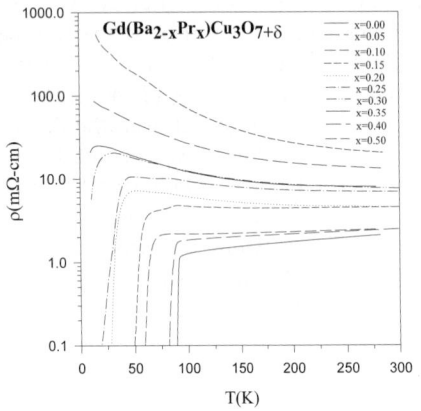
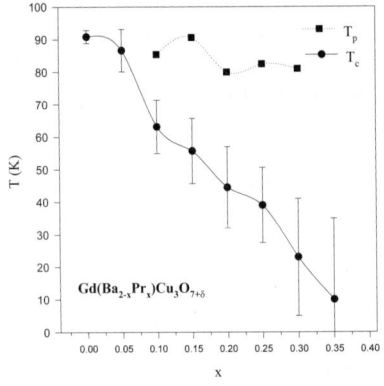

Fig. 1 Resistivity curve for Gd(Ba$_{2-x}$Pr$_x$)Cu$_3$O$_{7+\delta}$ samples in the range 10 K < T < 300 K for $0.0 \leq x \leq 0.5$.

Fig. 2 Variation of superconductivity transition temperature (T_c) and the temperature at which the resistivity hump occurs (T_p) vs. different amounts of Pr doping (x). The lines through the points are guide to the eye.

The x = 0.0, 0.1, and 0.2 samples were prepared in one batch, and the x = 0.05, 0.15, and 0.25 in another batch, and the x = 0.30, 0.40, and 0.50 in a third batch, with exactly the same fabrication procedures. As some of the samples in each batch show the hump on their ρ–T curves, it is evident that the preparation procedure could not have caused these humps.

Figure 3 shows the positions of O(2), O(3), and Cu(2) atoms in the unit cell for different x. From the $Z_{Cu(2)}$–x curve it is evident that for $0.1 \leq x \leq 0.25$, the position of Cu(2) atom is lower than the expected value from the curve; the $Z_{O(2)}$ and $Z_{O(3)}$ at x = 0.15 are obviously out of the expected range (Fig. 3). The decrease in the values of $Z_{O(2)}$, $Z_{O(3)}$, and $Z_{Cu(2)}$ can be understood if one large atom, larger than the typical R atomic size, is substituted at R site, which could push away the CuO_2 plane, i.e. elongation of the distances between the CuO_2 planes. The larger atom, probably with positive charge, but smaller than Gd^{3+}, pushes away the $O(2)^{2-}$ and $O(3)^{2-}$ anions. This reduction in the value of positions of O(2) and O(3) atoms could also reject the O(4) atom from occupying its expected equilibrium position. The Ba atom is the most probable candidate to fit this scenario, as it has the larger ionic radius ($r_{Ba^{2+}} > r_{Pr^{3+}} > r_{Gd^{3+}}$), and smaller ionic charge (+2).

To test the above proposal, we have repeated the Rietveld refinement with the allowing the of Ba atom to relocate at R site (Ba_R) for all values of x. The interesting result is that for the special values of $0.15 \leq x \leq 0.30$, the Ba atom occupies the R site (Fig. 4); the Ba_R mis-substitution has the highest value at x = 0.2. This domain of x is almost the same as the domain for which a hump is present on the ρ–T curves. It would be interesting to find a relation between the Ba_R and the ρ–T anomaly.

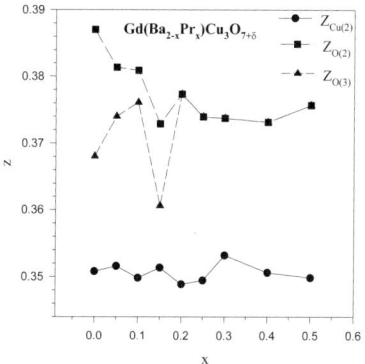 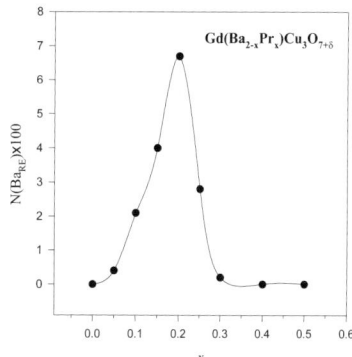

Fig. 3 Position of atoms in the CuO_2 plane in c direction vs. different amount of Pr doping (x): For $x \geq 0.20$, O(2)≡O(3) due to the orthorhombic-tetragonal phase transition. The lines through the points are guide to the eye.

Fig. 4 Site occupation factor percent of Ba atoms at R site vs. different amounts of Pr doping (x). The line through the points is a guide to the eye.

In the Zou's superconducting Pr-123 samples [12], one possibility for the larger c lattice parameter with respect to the typical 123 structures, have been proposed to be the presence of Ba atoms at Pr site (Ba_R). Narozhnyi et al. [13, 14], based on the effective magnetic moment of Pr atom, have concluded that in the Zou's superconducting Pr-123 samples there should be some Ba atoms at Pr site; Ba^{2+} on R site dopes additional mobile holes and compensates the localization of holes by the Pr-O(2,3) hybridization. They have mentioned that the substitution of Ba for Pr could be a natural explanation not only for the superconductivity in Pr-123, but also for the elongation of the distances between the CuO_2 planes observed in Ref. [12]. It is noteworthy that Zou and Nishihara [15] have also confirmed that based on the effective magnetic moment of Pr in Pr-123, Ba_R could be a possible interpretation for Pr-123 superconductivity. Therefore, if in the unit cells with Ba_R superconductivity occurs, the resistivity will decrease at T_p. However, due to the large insulating parts of the sample, which are Pr substituted, the resistivity do not decrease to zero till the Josephson coupling energy between superconducting parts exceeds the thermal energy of the order of k_BT. Then, the superconducting transition occurs. Another evidence in support of our claim is that the temperature at which Ba_R occurs in our samples, T_p, has the same value as the superconducting transition temperature of Pr-123 in Ref. [12].

Table 1 Oxygen content of $Gd(Ba_{2-x}Pr_x)Cu_3O_{7+\delta}$ samples for different Pr-doping (x).

x	0.00	0.05	0.10	0.15	0.20	0.25	0.30	0.35	0.40	0.5
7+δ	6.99	7.03	7.03	7.09	7.01	7.06	7.06	6.97	6.99	6.96

The superconducting parts of the samples with transition temperature T_p cannot be the Gd-123 regions because the T_c of Gd-123 is about 92 K for the optimum value of oxygen [16]. Also, the oxygen content for the oxygen deficient Gd-123 with $T_c \sim T_p \sim 80$ K is less than 6.85 [16], but all of our samples have the oxygen contents of more than 6.96 (Table 1). One scenario for explaining the observed hump anomaly could be the existence of oxygen depleted Gd-123 regions and oxygen rich $Gd(Ba_{2-x}Pr_x)Cu_3O_{7+\delta}$ regions (due to the Pr^{3+} at Ba^{2+} site). The oxygen depleted Gd-123 have T_c varying from 90 to 0 K for different oxygen contents. But, the humps occur at almost constant 80K temperature in all the corresponding samples. Therefore, this possibility should also be ruled out.

In summary, in the controversial case of Pr-123 in HTSC cuprates, the mis-substitution of Ba atoms at R site may be a decisive idea. This scenario seemingly could explain when Pr-123 compound is a superconductor or a non-superconductor. The final test to check the legitimacy of this model requires more investigations including the experiments, which could directly detect the exact atomic positions in the compound. Ba_R in our samples occurs at $T_p \sim 80$–90 K, which could be the origin of the humps on $\rho(T)$ curve.

Acknowledgements This work was supported in part by the Offices of Vice President for Research and Dean of Graduate Studies at Sharif University of Technology.

References

[1] Z. Yamani and M. Akhavan, Phys. Rev. B **56**, 7894 (1997).
[2] H.A. Blackstead and J.D. Dow, Solid State Commun. **115**, 137 (2000), and references therein.
[3] M. Akhavan, Physica B **321** (2002) 265.
[4] H.A. Blackstead, D.B. Chrisey, J.D. Dow, J.S. Horwitz, A.E. Klunzinger, and D.B. Pulling, Phys. Lett. A **207**, 109 (1995).
[5] M.J. Kramer, K.W. Dennis, D. Falzgraf, R.W. McCallum, S.K. Malik, and W.B. Yelon, Phys. Rev. B **56**, 5512 (1997).
[6] M.R. Mohammadizadeh and M. Akhavan, Phys. Rev. B **68**, 104516 (2003).
[7] Z. Yamani and M. Akhavan, Supercond. Sci. Technol. **10**, 427 (1997).
[8] D.B. Wiles and R.A. Young, J. Appl. Cryst. **14**, 149 (1981);
R.A. Young, in: The Rietveld Method, ed. R. A. Young (Oxford University Press, New York, 1993), p. 1.
[9] M. Covington and L.H. Greene, Phys. Rev. B **62**, 12440 (2000).
[10] H. Khosroabadi, V. Daadmehr, and M. Akhavan, Mod. Phys. Lett. B **16**, 943 (2002).
[11] A. Carrington, A.P. Mackenzie, C.T. Lin, and J.R. Cooper, Phys. Rev. Lett. **69**, 2855 (1992).
[12] Z. Zou, J. Ye, K. Oka, and Y. Nishihara, Phys. Rev. Lett. **80**, 1074 (1998).
[13] V.N. Narozhnyi and S.-L. Drechsler, Phys. Rev. Lett. **82**, 461 (1999).
[14] V.N. Narozhnyi, D. Eckert, K.A. Nenkov, G. Fuchs, K.-H. Muller, and T.G. Uvarova, cond-mat/9909107.
[15] Z. Zou and Y. Nishihara, Phys. Rev. Lett. **82**, 462 (1999).
[16] N. Ichikawa, S. Uchida, J.M. Tranquada, T. Niemoller, P.M. Gehring, S.-H. Lee, and J.R. Schneider, Phys. Rev. B **59**, 14712 (1999); cond-mat/9910037.

Effects of Pr doping and magnetic field on vortex pinning in Gd-123 based HTSC

M. R. Mohammadizadeh and **M. Akhavan**[*]

Magnet Research Laboratory (MRL), Department of Physics, Sharif University of Technology, P.O. Box 11365-9161, Tehran, Iran

Received 31 August 2003, accepted 31 December 2003
Published online 1 April 2004

PACS 74.25.Fy, 74.25.Ha, 74.72.Jt

The magnetoresistance of single phase polycrystalline $Gd(Ba_{2-x}Pr_x)Cu_3O_{7+\delta}$ samples have been studied within thermally activated flux creep model. The decrease of pinning energy with applied magnetic field can be scaled to two power law relations for magnetic fields, smaller and also larger than about 1 kOe. The derived pinning energy shows that the Pr-doping, similar to weak links, decreases the vortex flux pinning energy. It is also concluded that the substitution of Pr at Ba site has a more destructive effect on the flux dynamics than Pr at rare earth site.

© 2004 WILEY-VCH Verlag GmbH & Co. KGaA, Weinheim

1 Introduction

The electronic properties of high-T_c superconductors (HTSC) are affected by application of magnetic field [1]. The temperature width of the resistive transitions of different families of HTSC have been investigated in single crystal, oriented thin film, and polycrystalline samples [2]. In the first two cases, strong magnetic fields are required to produce appreciable temperature broadening in the transition, while for the polycrystalline materials quite small fields cause similar broadening [3].

Palstra et al. [4], using single crystal material, have pointed out that the thermally activated flux creep (TAFC) model describes the data quite well for very low levels of the resistivity. They found a temperature independent energy barrier U_0 and obtained its field dependence from the slopes of the Arrhenius resistivity plots.

We have studied the Pr substitution at Ba site in $GdBa_2Cu_3O_{7-\delta}$ (Gd-123) compound since in spite of different models to explain the destructive effect of Pr in $RBa_2Cu_3O_{7-\delta}$ (R-123, R is a rare earth) superconductor compounds, the superconducting Pr-123 has been reported by Zou et al. [5]. One scenario for explaining superconducting Pr-123 samples is the substitution of larger Ba atoms at Pr sites. So, the real positions of Pr/Ba atoms and effect of Pr in R-123 and other families of HTSC is a crucial subject in studying HTSC. In this paper, we will report the magnetoresistance measurements of $Gd(Ba_{2-x}Pr_x)Cu_3O_{7+\delta}$ (GdBaPr-123) system, and compare it with the magnetic properties of $(Gd_{1-x}Pr_x)Ba_2Cu_3O_{7-\delta}$ (GdPr-123) [6]. The TAFC model will be applied to fit the resistivity broadening curve under magnetic field with decreasing temperature.

2 Experimental details

The $Gd(Ba_{2-x}Pr_x)Cu_3O_{7+\delta}$ single phase samples with x = 0.0, 0.05, 0.1, 0.15, 0.2, and 0.25 were synthesized by the standard solid state reaction technique. Appropriate amounts of Gd_2O_3, Pr_6O_{11}, $BaCO_3$, and CuO powders with 99.9% purity were mixed, ground, and calcined at 840 °C for 24 h in an air atmosphere. Calcination was repeated twice with intermediate grinding. Then powders were reground, pressed into pellets, and synthesized at 930 °C for 24 h in an oxygen atmosphere. The samples were cooled to 550 °C and retained under oxygen flow for 16 h. Finally, they were furnace

---

[*] Corresponding author: e-mail: akhavan@sharif.edu, Phone: +98 21 6164510, Fax: +98 21 6012983

cooled to room temperature. The oxygen content of samples has been determined by the iodometric titration technique with ±0.03 accuracy.

An ac four-probe method with f = 33 Hz and 0.3 A/cm^2 current density was used for the conductivity measurements of the samples within the temperature range of 10 to 300 K. The electrical contacts were attached to the long side of the samples by silver paste. A Lake Shore-330 temperature controller with two Pt-100 resistors was used for measuring and controlling temperature to within ±10 mK. For the magnetoresistivity measurements a dc magnetic field of maximum value 20 kOe was applied perpendicular to the current and sample's surface.

3 Results and discussion A typical magnetic field dependence of the resistivity is shown in Fig. 1 for x = 0.05 sample in the range of 0 to 20 kOe. A very small change of the normal state resistivity and a noticeable broadening of superconducting transition is observed in the resistivity curve with the magnetic field. With increasing the applied magnetic field, the T_c^{onset} remains almost constant, while $T_c(\rho = 0)$ decreases considerably. The change of resistivity in the mixed state region, due to the thermally activated flux creep, near T_c^{onset} is less sensitive to the magnetic field; the tail part near T_c^{mid}, associated with the weak links between the grains [7], is more sensitive to the magnetic field.

For the x = 0.00, 0.05, and 0.10 samples, there are two parts in the mixed state of magnetoresistance curves: a steep part, associated with the onset of the superconductivity in the individual grains, and a transition tail due to the weak links coupling the grains. For the mentioned x, the steep parts remain almost unchanged with the applied magnetic field, while the tail parts shift considerably to lower temperatures. In addition, for the x = 0.15, 0.20, and 0.25 samples, the distinction between the corresponding two parts seems less pronounced [2]. This shows that for x < 0.20, due to small amount of Pr-doping, there are intragrain regions, where the magnetic field has no influence on. So, there are two distinct steep and dissipative parts. With doping larger amount of Pr i.e. x ≥ 0.20, there are enough Pr-doped unit cells, which play the role of weak links. Hence, for these values of Pr-doping, the magnetic field influences the weak links, including the porous parts and the Pr-doped unit cells in between the superconducting cells. Therefore, for x = 0.25, application of magnetic field causes a superconductor–insulator transition (SIT). This is an important effect, which is similar to two-dimensional electronic systems [8]. It is interesting that the critical value of Pr-doping for the corresponding distinction is the critical value of Pr-doping for metal–insulator transition (i.e., x_c^{MIT} = 0.20). Therefore, the x_c^{MIT} value for Pr-doping in our samples is enough for the crossover from TAFC-tail part to weak link dominant region.

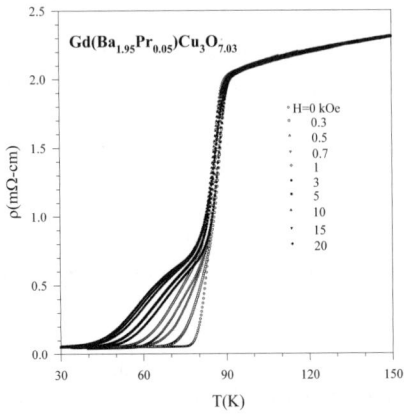

Fig. 1 Resistivity vs. temperature of Gd(Ba$_{1.95}$Pr$_{0.05}$)Cu$_3$O$_{7.03}$ sample under different applied magnetic fields.

The resistivity transition in the polycrystalline HTSC in the applied magnetic field has been found to obey the Arrhenius relation as $\rho(H,T) = \rho_0 \exp(-U_0(H,T)/k_B T)$, where U_0 is the pining energy that depends on temperature and magnetic field [9]. The functional dependence of $\rho(H,T)$ is expected to give information about the mechanism of dissipation. The U_0 value can be directly deduced from the linear part of slope of $\ln\rho$ vs. $1/T$ plot, where the linear part decreases with the increase of H (Fig. 2). From the linear data in the tail part of the plot, we have calculated the slop of each line at different applied magnetic fields. In the following, we will fit our data to this relation.

The derived flux pinning energy, U_0 from the above fitting at different magnetic fields for different amounts of Pr-doping is presented in Fig. 3. For each x, the flux pinning energy decreases with increase of applied magnetic field. This is in agreement with other reports [10,11]. The deduced value of pinning energy is in good agreement with the corresponding values in GdPr-123 samples [11], and it is in the order of magnitude of other approaches for finding U_0, e.g. magnetic relaxation measurements in YPr-123 [10, 12].

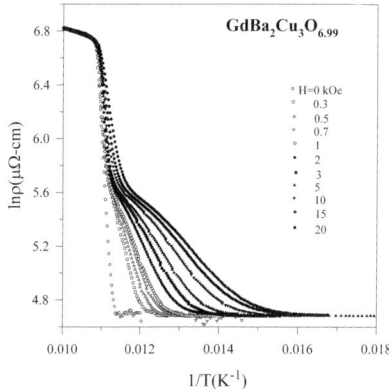
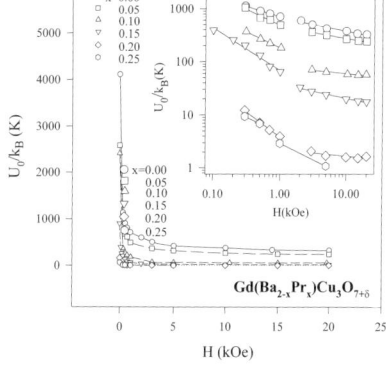

Fig. 2 Arrhenius plot of $GdBa_2Cu_3O_{6.99}$ sample under different applied magnetic fields.

Fig. 3 Pinning energy vs. applied magnetic field for different amounts of Pr-doping (x). The inset is the log–log plot.

We have fitted a power law relation for U_0 vs. H as $U_0 \sim H^{-\beta}$ for each sample. The values of β and quality of power law fit, χ^2 for different amounts of Pr-doping are presented in Table 1; the value of β begins from 0.297 for x = 0.00 and increases to 0.782 for x = 0.25. The more important point is that with increasing x, as is evident from χ^2, the power law relation becomes less valid. Our result is consistent with the Palstra et al. [4] result for $Bi_2Sr_2CaCu_2O_{8-\delta}$ system, where β depends on the range of magnetic field. The nonlinearity and the positive curvature of $\ln U_0$ vs. $\ln\beta$ for GdBaPr-123 show that Pr at Ba site has weaker flux pinning effect than Pr at R site, since the $\ln U_0$–$\ln\beta$ curves of GdPr-123 are linear [6]. Therefore, the destructive effect of Pr at Ba site in comparison with the case, where Pr is substituted at R site requires less energy to move fluxes i.e., suppressing the superconductivity with increasing H. The more destructive effect of Pr at Ba site with respect to Pr at R site has also been concluded through other measurements such as pseudogap [13], and T_c [14]. Based on our data, we conclude that U_0 does not scale with H^{-1}, in agreement with Palstra et al. [9].

If we re-examine the $\ln U_0$ vs. $\ln H$, there is a break about H = 1 kOe as is shown in the inset to Fig. 3. The values of β and χ^2 in different regions of H are presented in Table 2. Now, as is evident from χ^2, the $U \sim H^{-\beta}$ fit is more suitable for H < 1 kOe (completely) and H > 1 kOe (nearly). We think that the origin of this effect is that the linear part in $\ln\rho$–$1/T$ plots become less valid with increasing x. In other words, with increasing Pr-doping, the temperature dependence of U_0 becomes stronger. This shows that the more advanced models such as thermally activated Ambegaokar and Halperin phase slip might be more suitable for this system [15].

With increasing x, the pinning energy decreases. As we concluded earlier, the Pr substitution behaves like weak links. With increasing weak links, the vortices creep easier, whereas the pinning energy decreases. Our result is consistent with the effect of Pr in GdPr-123 system [6], and Ca-doped GdPr-123 [16].

Table 1 The β factor and quality of fit χ^2 for different amounts of Pr doping (x).

x	β	χ^2
0.00	0.297	0.987
0.05	0.323	0.952
0.10	0.477	0.924
0.15	0.623	0.947
0.20	0.467	0.895
0.25	0.782	0.980

Table 2 The β factor and χ^2 for different amounts of Pr doping (x) for two ranges of magnetic field: smaller and larger than 1 kOe.

x	H < 1 kOe		H >1 kOe	
	β	χ^2	β	χ^2
0.00	0.394	0.996	0.261	0.971
0.05	0.611	0.966	0.223	0.985
0.10	0.594	0.996	0.121	0.923
0.15	0.804	0.978	0.269	0.980
0.20	0.942	0.995	0.125	0.683
0.25	0.609	1	0.618	1

In summary, based on our magnetoresistance measurements in GdBaPr-123 samples, it is concluded that within thermally activated flux creep model, the pinning energy decreases with increasing Pr doping. A universal β in power law relation of U_0 vs. H is not accessible, but there is a break in the linear behavior of logU_0–logH curve. This means that corresponding to smaller or larger than 1 kOe magnetic field, β has two different values for each Pr doping x. In comparison with GdPr-123 system, it is concluded that Pr at Ba or R site plays the role of weak links, but the effect of Pr at Ba site is more destructive on superconductivity than Pr at R site. The inherent relation between the normal, superconducting, and mixed state of HTSC could be a guide for understanding the superconducting mechanism(s).

Acknowledgements This work was supported in part by the Offices of Vice President for Research and Dean of Graduate Studies at Sharif University of Technology.

References

[1] M.W. Coffey and J.R. Clem, Phys. Rev. Lett. **67**, 368 (1991).
[2] M.R. Mohammadizadeh and M. Akhavan, Supercod. Sci. Technol. **16**, 538 (2003).
[3] H. Hilgenkamp and J. Mannhart, Rev. Mod. Phys. **74**, 485 (2002).
[4] T.T.M. Palstra, B. Batlogg, L.F. Schneemeyer, and J.V. Waszczak, Phys. Rev. Lett. **61**, 1662 (1988).
[5] Z. Zou, J. Ye, K. Oka, and Y. Nishihara, Phys. Rev. Lett. **80**, 1074 (1998).
[6] H. Khosroabadi, V. Daadmehr, and M. Akhavan, Physica C **384**, 169 (2003).
[7] V. Ambegaokar and B.I. Halperin, Phys. Rev. Lett. **22**, 1364 (1969).
[8] M.R. Mohammadizadeh and M. Akhavan, Supercond. Sci. Technol. **16**, 1216 (2003).
[9] J.J. Kim, H.K. Lee, J. Chung, H.J. Shin, H.J. Lee, and J.K. Ku, Phys. Rev. B **43**, 2962 (1991).
[10] L.M. Paulius, C.C. Almasan, and, M.B. Maple, Phys. Rev. B **47**, 11627 (1993).
[11] V. Daadmehr and M. Akhavan, phys. stat. sol. (a) **193**, 153 (2002).
[12] Y.G. Xiao, B. Yin, J.W. Li, Z.X. Zhao, H.T. Ren, L. Xiao, X.K. Fu, and J.A. Xia, Supercond. Sci. Technol. **7**, 623 (1994).
[13] M.R. Mohammadizadeh and M. Akhavan, Physica B **336**, 410 (2003).
[14] M.R. Mohammadizadeh and M. Akhavan, Phys. Rev. B **68**, 104516 (2003).
[15] M.R. Mohammadizadeh and M. Akhavan, Physica C **390**, 134 (2002).
[16] H. Shakeripour and M. Akhavan, Supercond. Sci. Technol. **14**, 234 (2001).

Magnetic field effects on electrical resistivity of $(Gd_{1-x}Pr_x)Ba_2Cu_3O_{7-\delta}$ and $Gd(Ba_{2-x}Pr_x)Cu_3O_{7+\delta}$

M. R. Mohammadizadeh and M. Akhavan[*]

Magnet Research Laboratory (MRL), Department of Physics, Sharif University of Technology, P.O. Box 11365-9161, Tehran, Iran

Received 31 August 2003, accepted 31 December 2003
Published online 1 April 2004

PACS 74.25.Fy, 74.25.Ha, 74.72.Jt

The magnetoresistance measurements of single phase polycrystalline $Gd(Ba_{2-x}Pr_x)Cu_3O_{7+\delta}$ and $(Gd_{1-x}Pr_x)Ba_2Cu_3O_{7-\delta}$ samples have been compared and analyzed by the Ambegaokar and Halperin phase slip model. The derived power law dependences of pinning energy to applied magnetic field and temperature have been studied for different amounts of Pr doping. The power factors depend on the Pr doping. Moreover, in both systems, the magnetic field dependence of the pinning energy power factor increases for small amounts of Pr doping, while for larger amount of Pr doping, it decreases. The critical current power law dependence to magnetic field decreases with increasing Pr doping. Pr substitution at Ba or rare earth site, decreases the pinning energy and the critical current density, playing the role of weak links.

© 2004 WILEY-VCH Verlag GmbH & Co. KGaA, Weinheim

1 Introduction

There are some different models for interpretation of resistivity broadening under magnetic field. One approach to the problem of flux dynamics in a magnetic field, has been achieved by applying the Ambegaokar and Halperin (AH) phase slip model [1]. This model describes the effects of thermal fluctuation of the phases of the order parameters cross a highly damped, current driven Josephson junction. Within this theory, when k_BT becomes comparable with the Josephson coupling energy, the resistance $\rho(T)$ in the limit of low current, $I \ll I_c(T)$, is given by $\rho(T) = \rho_n[I_0(\gamma/2)]^{-2}$, where, I_0 is the modified Bessel function, ρ_n is the average normal state resistivity of the junction, and γ is the normalized barrier height for thermally activated phase slip, defined as $\gamma = U_0/k_BT = A(1-t)^q/H$ [1], where, U_0 is the activation energy, $t \equiv T/T_c$ is the reduced temperature, H is the applied magnetic field, and A is a constant.

In this paper, we will compare the effects of Pr substitution at Ba site i.e., $Gd(Ba_{2-x}Pr_x)Cu_3O_{7+\delta}$ (GdBaPr-123) [2], and at R site i.e., $(Gd_{1-x}Pr_x)Ba_2Cu_3O_{7-\delta}$ (GdPr-123) [3] on magnetoresistance for different amounts of Pr-doping. Based on the AH theory, we will investigate the temperature and magnetic field dependences of the pinning energy as a function of Pr-doping. The critical current at zero temperature can be derived from the corresponding theory, which will be presented as a function of magnetic field.

2 Experimental details

The $Gd(Ba_{2-x}Pr_x)Cu_3O_{7+\delta}$ single phase samples with x = 0.0, 0.05, 0.1, 0.15, 0.2, and 0.25 were synthesized by the standard solid state reaction technique. Appropriate amounts of Gd_2O_3, Pr_6O_{11}, $BaCO_3$, and CuO powders with 99.9% purity were mixed, ground, and calcined at 840 °C for 24 h in an air atmosphere. Calcination was repeated twice with intermediate grinding. Then powders were reground, pressed into pellets, and synthesized at 930 °C for 24 h in an oxygen atmosphere. The samples were cooled to 550 °C and retained under oxygen flow for 16 h. Finally, they were furnace coo-

---
[*] Corresponding author: e-mail: akhavan@sharif.edu, Phone: +98 21 6164510, Fax: +98 21 6012983

cooled to room temperature. An ac four-probe method was used for the conductivity measurements of the samples within the temperature range of 10 to 300 K. A Lake Shore-330 temperature controller with two Pt-100 resistors was used for measuring and controlling temperature to within ±10 mK. For the magnetoresistivity measurements a dc magnetic field of maximum value 20 kOe was applied perpendicular to current and surface of sample.

3 Results and discussion With increasing H, the broadening tail part of the resistivity curve expands to lower temperatures. The main steep part of the resistivity remains almost unchanged with the application of H. This is due to strong intragrain pinning energy, which does not allow any vortex motion. However, for larger amounts of Pr-doping i.e., $x \geq 0.15$, due to the weak link effect of Pr, the onset part of the resistivity changes too under the same applied magnetic field [4]. Under this range of applied magnetic fields, the normal state resistivity of the samples does not change, considerably [9].

A typical AH fit to the resistivity under magnetic field are shown in Fig. 1 for GdBaPr-123 compound. In order to avoid the ambiguity of the double transition during the fitting process for $x < 0.15$, the higher-temperature transition, which corresponds to the transition of the superconducting grains was excluded in the fitting of resistivities. In our fittings and also other reports [6], the $\rho(T = 0)$ part has some deviations from the AH model, where the thermally activated flux creep model is more preferable; and has been discussed for GdBaPr-123 compounds elsewhere [4].

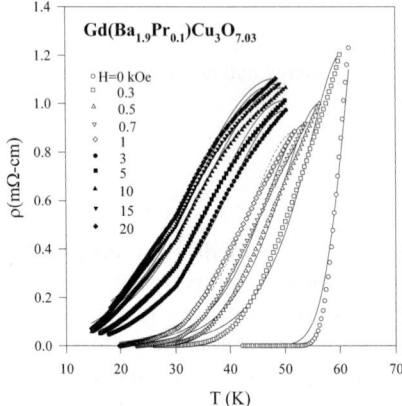

Fig. 1 Resistivity vs. temperature of $Gd(Ba_{1.90}Pr_{0.10})Cu_3O_{7.03}$ sample under different applied magnetic fields.

Figure 2 shows the γ vs. H curves for GdBaPr-123 compound. The insets to the figure shows clearly that $\gamma(H)$ is quite linear in the log–log plot. In GdPr-123 samples $\gamma(H)$ also decreases with increasing H. For $x = 0.05$, the $\gamma(H)$ has a power law dependence, but for larger x, it is not simply a power law relation. The corresponding slopes i.e., power factor in $\gamma \sim H^{-\beta}$, are presented in Table 1 for different amounts of Pr-doping at Ba and R sites. β is the power factor of $C(H) \equiv A/H \sim H^{-\beta}$. The average value of β is about 1.2 for both systems. With increasing magnetic field, the coupling energy between the Josephson vortices decreases. For each magnetic field, with increasing Pr-doping, γ also decreases. The power law decreasing of γ with H is also reported by others [6–8, 9–11].

Tinkham [11], phenomenologically derived the magnetic field dependence of pinning energy as being proportional to H^{-1} in single crystal of Y-123 with fields up to 90 kOe. Kim et al. [10] for Y-123 thin films have derived $\beta = 0.73$ in fields up to 100 kOe. In $Bi_2Sr_2CaCu_2O_{8+\delta}$ single crystals with fields up to 55kOe [12], and in granular Y-123 samples with fields up to 1 kOe [6], and 75 Oe [7], β is reported equal 0.5; in granular Ca-doped GdPr-123 with magnetic fields up to 17 kOe, β is reported about 0.3 [9]. Therefore, we can confirm only the power law dependence of γ to H. Although, our derived power factor is close to the Tinkham's result [11], the dispersion in the generally reported amounts of β is wide: from 0.3 [8] to 1.2 [6]. In addition, it seems that the range of applied magnetic field has an effect on the value of β. As we summarized in the last paragraph, for high fields, β approaches the larger values i.e., 1, and for low fields, it goes to smaller values i.e., 0.5. Moreover, the rate of fall off of $\gamma(H)$ depends on the size and orientation of the Josephson junctions, which are determined by the sample microstructure. That is, single crystal or polycrystalline samples could lead to different magnetic dependence of γ. So, it is difficult to find a universal $\gamma \sim H^{-\beta}$ relation.

Table 1 The α and β factor, in Ba (GdBaPr-123) or R (GdPr-123) site, and oxygen content of GdBaPr-123 samples for different amounts of Pr doping.

	GdPr-123		GdBaPr-123		
x	α	β	α	β	$7+\delta$
0.00	0.20	1.20	0.20	1.20	6.99
0.05	0.02	1.02	0.07	1.07	7.03
0.10	0.24	1.24	0.15	1.15	7.03
0.15	0.19	1.19	0.32	1.32	7.09
0.20	–	–	0.27	1.27	7.01
0.25	–	–	0.27	1.27	7.06

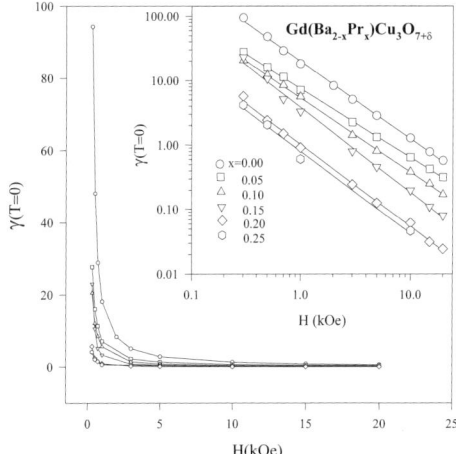

Fig. 2 The parameter γ at zero temperature vs. H of GdBaPr-123 samples for different amounts of Pr-doping. The lines are guides to the eye. Inset shows the $\ln\gamma$–$\ln H$ fit.

The critical x for a crossover from intergrain to intragrain flux dynamics is $x_c = 0.15$ [4]. For x = 0.00, 0.05, and 0.10 samples, there are two unaffected steep and affected tail parts by magnetic field in the magnetoresistance curves, but for x = 0.15, 0.20, and 0.25 samples, there is only one affected part, which dissipates under magnetic fields from onset temperature down. For more details see Ref. [4]. Using the AH model in both limits, confirms the capability of this model to explain not only the intergranular weak links, but also the strongly intragranular interactions [12]. The magnetoresistance of samples with larger amounts of Pr-doping (x > 0.20) shows that with increasing magnetic field there is a superconductor–insulator transition.

From the AH phase slip theory, the parameter A at temperatures close to T_c is given by $A/H = J_{cj}(0)\hbar a^2/ek_BT_c$, where $J_{cj}(0)$ is the critical current density at zero temperature and a is the average grain size [6]. Considering the typical grain size in our samples to be ~1 μm by the SEM measurements, and using the last equation., the estimated values of $J_{cj}(0)$ for the GdBaPr-123 samples are presented in Fig. 3. Due to using the intergranular MR data, this $J_{cj}(0)$ has more intergranular character. However, with increasing magnetic field, the critical current in the grain boundaries decreases significantly as a power law dependence $J_{cj}(0) \sim H^{-n}$, with n ~ 0.2. The $J_{cj}(0)$ decreases with increasing Pr-doping also in GdPr-

123 samples. Decreasing of $J_{cj}(0)$ with Pr-doping in both systems, suggests that Pr ions act as weak links, as has also been concluded through thermally activated flux creep model [4]. Decreasing of critical current with increasing H is an indication of the sensitivity of superconducting grain junction to the applied magnetic field in these compounds. The increase in the magnetic field results in the penetration of more magnetic flux lines into the sample. Hence, the increase in the Lorentz force due to the field increment causes more flux lines to move; the $J_{cj}(0)$ decreases with increasing H.

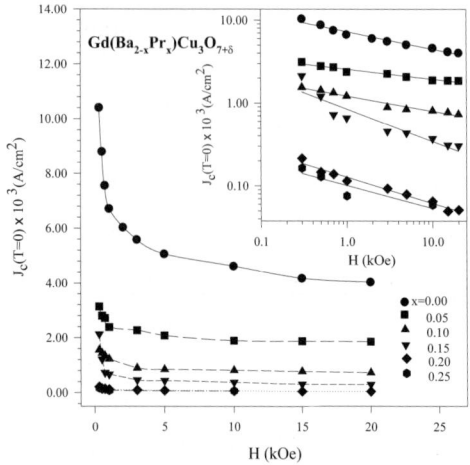

Fig. 3 Critical current density at zero temperature vs. H of GdBaPr-123 samples for different amounts of Pr-doping (x). The lines are guides to the eye. Inset shows the $\ln J_c - \ln H$ fit.

In summary, based on the AH phase slip model applied to the GdBaPr-123 and GdPr-123 samples, the power law dependence of γ to magnetic field and temperature has been obtained for different amounts of Pr-doping. Pr substitution both at Ba and R sites, in each magnetic field, decreases the pinning energy and critical current density. So, the role of Pr in magnetoresistance properties is like weak links. Since, in our MR curves, the intergranular properties are dominant, the effects of Pr at Ba and R sites appear to have similar influences on the mixed-state phase as weak links. The successful interpretation of magnetoresistance by using the AH phase slip model enhances the validity of this model. Our comparison of AH model predictions with experimental results confirms the validity of this model in both weakly coupled and strongly interacting grains.

Acknowledgements This work was supported in part by the Offices of Vice President for Research and Dean of Graduate Studies at Sharif University of Technology.

References

[1] V. Ambegaokar and B.I. Halperin, Phys. Rev. Lett. **22**, 1364 (1969).
[2] M.R. Mohammadizadeh and M. Akhavan, Phys. Rev. B **68**, 104516 (2003).
[3] M.R. Mohammadizadeh, H. Khosroabadi, and M. Akhavan, Physica B **321**, 301(2002).
[4] M.R. Mohammadizadeh and M. Akhavan, Supercond. Sci. Technol. **16**, 538 (2003).
[5] M.R. Mohammadizadeh and M. Akhavan, Eur. Phys. J. B **33**, 381 (2003).
[6] H.S. Gamchi, G.J. Russell, and K.N.R. Taylor, Phys. Rev. B **50**, 12950 (1994).
[7] C. Gaffney, H. Petersen, and R. Bednar, Phys. Rev. B **48**, 3388 (1993).
[8] A.C. Wright, K. Zhang, and A. Erbil, Phys. Rev. B **44**, 863 (1991).
[9] H. Shakeripour and M. Akhavan, Supercond. Sci. Technol. **14**, 234 (2001).
[10] J.J. Kim, H.K. Lee, J. Chung, H.J. Shin, H.J. Lee, and J.K. Ku, Phys. Rev. B **43**, 2962 (1991).
[11] M. Tinkham, Phys. Rev. Lett. **61**, 1658 (1988).
[12] W. Chen, J.P. Franck, and J. Jung, Physica C **341–348**, 1195 (2000).
[13] A.C. Wright, T.K. Xia, and A. Erbil, Phys. Rev. B **45**, 5607 (1992).

Thermally activated phase slip and variable range hopping in Tm(Ba$_{2-x}$Pr$_x$)Cu$_3$O$_{7+\delta}$

Z. Mokhtari, H. Khosroabadi, and M. Akhavan[*]

Magnet Research Laboratory (MRL), Department of Physics, Sharif University of Technology, P.O. Box 11365-9161, Tehran, Iran

Received 31 August 2003, accepted 31 December 2003
Published online 23 April 2004

PACS 74.25.Fy, 74.25.Ha, 74.72.Jt

The magnetoresistance measurement of single phase polycrystalline TmBa$_{2-x}$Pr$_x$Cu$_3$O$_{7+}$ samples have been analyzed by the Ambegaokar and Halperine (AH) phase slip model. The magnetic field dependence of pinning energy power factors increase with the increase of Pr doping. The derived critical current density from the AH theory decreases with increasing magnetic field and Pr-doping for all superconducting samples. It shows that the Pr-doping plays the role of weak link. We have also investigated the normal state resistivity with the hopping model. We have found a CG-VRH cross-over with the increase of Pr content near metal insulator transition. The deacrease of the localization length of the carriers show that Pr-doping strongly localizes the carriers, and finally causes the suppression of superconductivity.

© 2004 WILEY-VCH Verlag GmbH & Co. KGaA, Weinheim

1 Introduction Many research on the rare earth-based HTSC show that Pr atom is the only rare earth with stable orthorhombic 123 structure, and is an insulator in contrast with the other R123 compounds. Different anomalous effects have been observed when Pr atom is substituted in R-123 and other HTSC families. For a recent review on the role of Pr in HTSC see Ref. [1]. Although, many efforts have been made to explain the insulating behaviour of Pr-123, some groups have reported observation of superconductivity in this compound [2, 3]. To explain the insulating behaviour of Pr123, the Pr/Ba mis-substitution effect has been proposed [4]. Upon this explanation, many different experiments have been planned, in which Pr is substituted at Ba site in different R-123 systems. Doping Pr at Ba site has strong effects on the superconducting and normal state properties of these systems.

In contrast to single crystals [5], and oriented thin films [6], in the polycrystalline materials quite small fields show an appreciable broadening in resistive transition curves, even for a magnetic field applied parallel to the transitional current. To explain this unusual observation, models such as flux entanglement [7], flux line melting [8], and Ambegaokar-Halperin thermally activated phase-slip (AH) [9] have been proposed. It has been shown that AH model well describes the broadening resistivity under different magnetic fields for compounds such as Bi–Pb–Sr–Ca–Cu–O [10] and Y123 [11]. Also, the effect of Pr-doping on magnetoresistance has been investigated by the AH model in a systematic order [12, 13], and has shown good agreement with experiment.

Hopping conduction between localized states has been widely used for describing the normal state of different HTSC [14, 15]. Also, the effect of Pr-doping on normal state conduction has been investigated by this model in Tm$_{1-x}$Pr$_x$Ba$_2$Cu$_3$O$_{7-\delta}$ [16] and Gd$_{1-x}$Pr$_x$Ba$_2$Cu$_3$O$_{7-\delta}$ [17] with good agreement.

The purpose of this paper is to report on the experimental results on resistivity and magnetoresistivity of Tm-based 123 system with Pr-doping; we also compare our results with other Pr-doped systems.

---

[*] Corresponding author: e-mail: akhavan@ sharif.edu, Phone: +98 21 6164510, Fax: +98 21 6012983

2 Experimental details A series of TmBa$_{2-x}$Pr$_x$Cu$_3$O$_{7+\delta}$ samples with $0.00 \leq x \leq 0.50$ were synthesized by solid-state reaction technique. We have followed the procedures in accordance with Ref. [12]. Our samples were calcinated and sintered at 840 °C and 930 °C, respectively, and were retained under oxygen flow for 2h at 650 °C before cooling to room temperature. The oxygen content of samples was determined by iodometric titration technique. The XRD measurements were done by Philips PW-3710 powder diffractometer with Cu K$_\alpha$ radiation and $\lambda = 1.5406$ Å in room temperature. The XRD patterns were analyzed by the Rietveld structure refinement method, with a modified version of the DBW3.2 program. Also, the electrical resistivity and magnetoresistivity measurements were performed on rectangular samples by using an ac four-probe method, with 5 mA current according to Ref. [12].

3 Results and discussion The average mass density of the samples is 5.2 g/cm^3. The oxygen content of the samples increase from 7 for x = 0.00 to 7.13 for x = 0.50 with the increase of Pr content, due to the difference between the valance of Pr and Ba ions. The results of XRD analysis by the Rietveld method, shows an orthorhombic to tetragonal (O–T) structure phase transition at x = 0.45, which corresponds to the superconductor–insulator transition (SIT) in our samples. Also, it shows a distinct change in the rate of variation of cell parameter c at x = 0.25, which corresponds to the metal–insulator transition (MIT).

With the increase of x, the superconducting transition temperature (T$_c$) decreases and the width of transition temperature (ΔT$_c$) as well as the normal state resistivity increase. With the increase of number of insulting parts in grains (i.e. Pr substituted unit cells), the homogeneity of grains decreases, which leads to larger ΔT$_c$. The normal state resistivity for x < 0.25 samples is metallic (dρ/dT > 0), and for x > 0.25 is semiconducting-like (dρ/dT < 0). So, the critical doping for MIT is about 0.25–0.30. In the sample with x < 0.45, the superconducting transition occurs, while for x $\geq$ 0.45, there is no transition. Therefore, the critical doping for SIT is x$_c^{SIT}$ = 0.45. It is important to distinguish the difference between the critical x values for SIT and MIT.

Upon the application of magnetic field, the resistive transition shows two distinct parts: (1) A steep part near the onset of superconductivity, and (2) a transition tail part. The experimental data in the tail part were fitted with the AH model. Fig. 1 shows the experimental magnetoresistance data and the fitted curves to the AH model in the tail region for TmBaPr123 (x = 0.15). This figure indicates that the AH model fits the experimental data better under low fields. Also, in our samples the T(ρ = 0) region has some deviation from the AH model. In this region, the thermally activated flux creep model is more suitable. This deviation has also been reported by others [12]. In the AH theory, the resistance $\rho(T)$ is given by [9]:

$$\rho(T,H) = \rho_n \left[I_o \left(\frac{\gamma}{2} \right) \right]^{-2} \qquad (1)$$

where, $\gamma = U_0 / k_B T = C(H)(1-t)^q$, and t = T/T$_c$. For fitting the data of each compound, we have chosen three free fitting parameters: ρ_n, C, q. The value of ρ_n is observed to be close to the resistivity value of the branching point in the magnetoresistance curve. The branching point tends to lower resistance with increasing magnetic field. Therefore, ρ_n decreases slowly with the increase of magnetic field. Also, the ρ_n increases by adding Pr to Tm123, like other similar compounds (Gd123) [13]. The reason is probably due to the intergranular weak links caused by the increase of Pr-doping, which increases the system's resistivity. It is resulted that magnetic field induced dissipation in high temperature superconductors can be explained in terms of phase slip through an effective medium of Josephson weak links.

The value of q (the temperature power law dependence of γ) decreases with the increase of magnetic field until near 200 Oe (~ H$_{c1}$); for magnetic fields larger than it, q increases. Different behaviours of q, lower and higher than 200 Oe, show the importance of penetration of flux lines in the grains. Different values have been proposed for q: 0.85 [11], 1.5 [18], 2 [19] and 1.53–2.8 [20]. Our results of q have been distributed between 1.25–2.15.

Moreover, it is shown that with increasing applied magnetic field, the parameter C that equals to $\gamma(T = 0)$, and is proportional to the pinning energy, decreases. It also decreases with the increase of Pr content. Both of these results are in agreement with other reports [12, 13, 21].

From the AH theory, the parameter $C(H)$ is given by:

$$C(H) = \frac{J_{cj}(O)\hbar a^2}{ek_B T_c}, \qquad (2)$$

Using the obtained $C(H)$ through fitting, and Eq. (2), we have estimated the critical current density at zero temperature, $J_{cj}(O)$ for our samples, which is shown in Fig. 2. To see that the region of fitting is related to the intergrain region, the derived $J_{cj}(O)$ has also the intergrain nature. From Fig. 2, we can see that with increasing magnetic field, the critical current density decreases, in agreement with [12, 13, 21]. This demonstrates the sensitivity of weak links to the magnetic field in these samples.

$J_{cj}(O)$ also depends on H as $J_{cj}(O) \sim H^{-n}$ with n = 0.43 for Tm123 and n ~ 0.2 for Pr-doped samples. This is comparable with the average value of n = 0.2 for GdBaPr123 [12].

For TmBaPr123 samples, the obtained values of $J_{cj}(O)$ are more than the corresponding values for GdBaPr123 [12], which can be due to the better quality of TmBaPr123 samples.

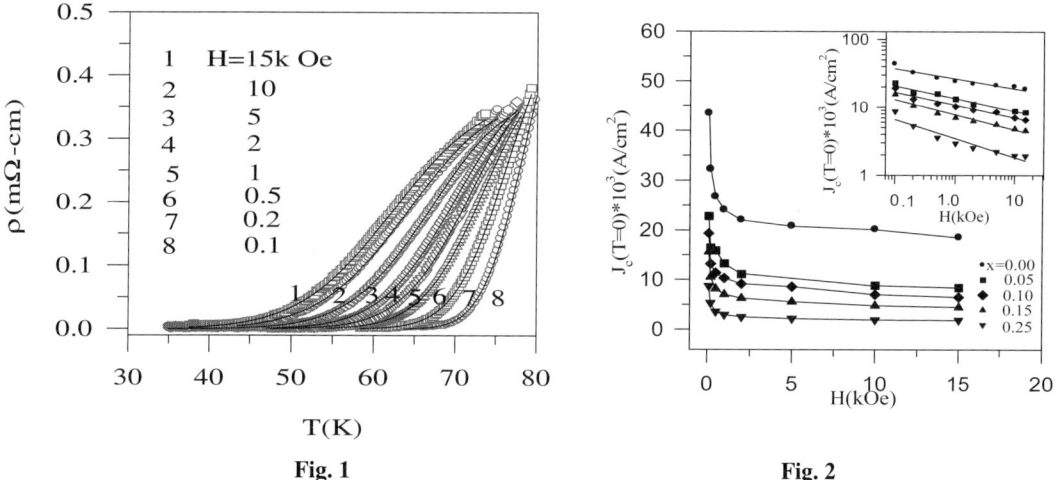

Fig. 1 The AH fit to the resistivity of $Tm(Ba_{0.85}Pr_{0.15})Cu_3O_{7+\delta}$ for different magnetic fields. The dots are the experimental data, and the lines are the fit results.

Fig. 2 Critical current density $J_c(T = 0)$, versus magnetic field for different amounts of Pr-doping (x). The solid lines are guides to the eye.

In order to evaluate if the VRH or CG conducting mechanism corresponds to Pr-doped systems, we have fitted the normal state resistivity of TmBaPr123 samples. It is to be noted that different amount of oxygen content of the samples and different temperature domains used in finding the dominant conducting regime, affects the obtained regime. We have fitted 2D-VRH, 3D-VRH, and CG regimes separately for all the samples with:

$$\rho = \rho_0 (T/T_0)^{2p} \exp(T_0/T)^p. \qquad (3)$$

For the metallic region of the normal state resisitivity, the CG regime is dominant till x = 0.25, but for the non-metallic region ($0.25 \leq x \leq 0.50$), the 2D-VRH conduction is the dominant mechanism. There is a little difference between the 2D & 3D-VRH mechanisms in the samples with insulating behaviour.

We find T_0 (fitting parameter in Eq. (3)) and d (hopping length) for different mechanisms. As expected, the hopping length decreases with the increase of Pr content in each regime. The length of the localized wave functions is higher for Tm123 (x = 0.0). This means that due to the very large d with respect to the distance of the neighbouring atoms, the overlap of the carriers' wave functions is enough for the conduction to perform easily. Further, by Pr-doping, the hopping length decreases.

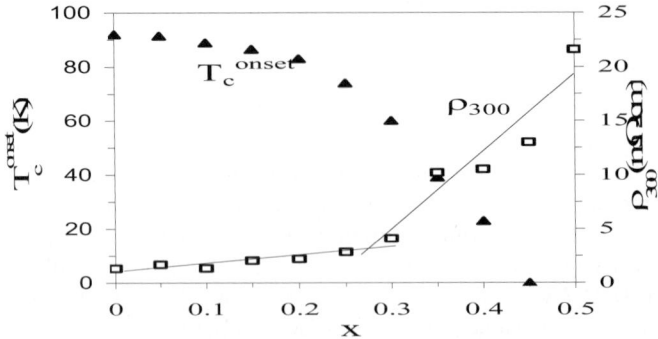

Fig. 3 The superconducting transition temperature (T_c), and resistivity at T = 300 K versus different amounts of Pr doping (x).

4 Conclusion Decreasing the critical current density with increasing magnetic field and Pr-doping shows that the Pr-doping plays the role of weak link. It is interesting that the transition from CG to 2D-VRH conduction mechanism occurs near MIT. There is also a distinct change in the slope of T_c and ρ_{300} curves near x = 0.25 (Fig. 3) corresponding to MIT. These results all imply the importance of MIT in the TmBaPr123 samples, and the correlation of Pr-doping with electronic characteristic of 123 HTSC systems.

Acknowledgments We wish to thank M. Kariminezhad, M. Mirzadeh and P. Maleki for assistance and useful discussions. We also thank M.R. Mohammadizadeh for helpful discussions. This work was supported in part by the Offices of Vice President for Research and Dean of Graduate Studies at Sharif University of Technology.

References

[1] M. Akhavan, Physica B **321**, 265 (2002).
[2] Z. Zou, J. Ye, K. Oka, and Y. Nishihara, Phys. Rev. Lett. **80**, 1074 (1998).
[3] H.A. Blackstead and J.D. Dow, Solid State Commun. **115**, 137 (2000).
[4] H.A. Blackstead and J.D. Dow, Phys. Rev. **51**, 11830 (1995).
[5] T.T.M. Palstra, B. Batlogg, R.B. van Dover, L.F. Schneemeyer, and J.V. Waszczak, Phys. Rev. B **41**, 66 (1990).
[6] R.H. Koch, V. Foglietti, W. J. Gallagher, G. Koren, A. Gupta, and M.P.A. Fisher, Phys. Rev. Lett. **63**, 1511 (1989).
[7] M.P.A. Fisher, Phys. Rev. Lett. **62**, 1414 (1989).
[8] D.R. Nelson, Phys. Rev. Lett. **60**, 1973 (1988).
[9] V. Ambegaokar and B. I. Halperin, Phys. Rev. Lett. **22**, 1364 (1969).
[10] A.C. Wright, K. Zhang, and A. Erbil, Phys. Rev. B **44**, 863 (1991).
[11] G. Gaffney, H. Petersen, and R. Bednar, Phys. Rev. B **48**, 3388 (1993).
[12] M.R. Mohammadizadeh and M. Akhavan, Physica C **390**, 134 (2003).
[13] H. Shakeripour, M. Akhavan, Supercond. Sci. Technol. **14**, 234 (2001).
[14] B. Dabrouski, D.G. Hink, J.D. Jorgensen, R.K. Kalia, P. Vashishta, D.R Richards, D.T. Marx, and A.W. Mitcher, Physica C **156**, 24 (1988).
[15] E.J. Osquiguil, L. Civale, R. Decca, and F. dela Cruz, Phys. Rev. B **38**, 2840 (1988).
[16] X.W. Cao, Z.H. Wang, Y.J. Tang, Physica B **212**, 411 (1995).
[17] Z. Yamani, M. Akhavan, Solid State Commun. **107**, 197 (1998).
[18] M. Tinkham, Phys. Rev. Lett. **61**, 1658 (1988).
[19] G. Deutscher and K.A. Muller, Phys. Rev. Lett. **59**, 1745 (1987).
[20] D.H. Kim, K.E. Gray , R.T. Kampwirth and D.M. Mckay, Phys. Rev. B **42**, 6249 (1990).
[21] H.S. Gamchi, G.J. Russel and K.N.R. Taylor, Phys. Rev. B **50**, 12950 (1994).

Fabrication of BSCCO thin films using sputtering technique

Hadi Salamati[*,1], **Parviz Kameli**[1], and **Mohammad Akhavan**[2]

[1] Solid State Lab., Physics Dept., Isfahan University of Technology, Isfahan 84154, Iran
[2] Magnet Reaserch Lab., Physics Dept., Sharif University of Technology, Tehran, Iran

Received 31 August 2003, accepted 31 December 2003
Published online 8 April 2004

PACS 74.62.Bf, 74.72.Hs, 74.78.Bz, 81.50.Cd

Investigation of BSCCO 2223 phase formation in superconducting thin films of Bi-based cuprates deposited on MgO substrate is reported. A series of films were made by *in situ* dc-magnetron sputtering method and another series were made by *ex situ* rf-magnetron sputtering technique. Same target have been used in both technique. In the case of the ex situ method, the films were annealed for different period of time at 800 °C, and in the case of in situ method, the films were deposited at different temperature substrate. The influence of annealing time and the temperature of the substrate on the quality and the phase formation for ex situ and in situ method have studied, respectively. The results of our studies show, although the samples prepared by in situ method have a better mechanical and superconductivity properties, the post-annealed ex situ prepared samples have a superb structural and superconductivity properties.

© 2004 WILEY-VCH Verlag GmbH & Co. KGaA, Weinheim

1 Introduction

Although, among the high-T_c superconductors, $YBa_2Cu_3O_{7-\delta}$ is an easier materials for making in situ and ex situ thin films [1], the $Bi_2Sr_2Ca_{n-1}Cu_nO_x$ system is more challenging and interesting one.

It is well known that, there are several phases in the BSCCO system. The composition of each phase is expressed by a general formula of $Bi_2Sr_2Ca_{n-1}Cu_nO_x$, (n = 1, 2, 3) and the T'$_c$s of 10, 85,110 K respectively. Here after, we used the abbreviation 2201, 2212, 2223 phases. In these series, 2223 is the most attractive, because it has the highest superconducting transition temperature, T_c, about 110 K. In spite of this potential of BSCCO materials, most applications of high-T_c superconducting thin films are based on YBCO materials, since the in situ and ex situ growth of BSCCO thin films is comparatively difficult. The difficulties are due to the fact that, the superconducting properties are strongly dependent on variations of stoichiometry. At the first stage, it is very difficult to make a single-phase target of 2223 phase. For the deposition of BSCCO thin films additional complexity arises from possible phase mixture and oxygen over doping. Because of the complex crystallographic structure and small coherence length, extreme requirements are imposed on the uniformity and stability of deposition process.

Different techniques such as molecular beam epitaxy (MBE) [2, 3], metal organic chemical vapor deposition (MOCVD) [4, 5], sputtering [6, 7], liquid phase epitaxy (LPE) [8], and pulsed laser deposition (PLD) [9, 10] have been employed.

In this paper, we report on the preparation of target and the deposition of BSCCO (2223) phase thin films, using in situ dc-magnetron sputtering system and ex situ rf-magnetron sputtering technique. We have used a single target method and its effect on the superconductivity and the structure of the films are examined.

---

[*] Corresponding author: e-mail: salamati@cc.iut.ac.ir, Phone: +98 311 391 2375, Fax: +98 311 391 2376

2 Experiment

2.1 Target preparation

Precursor powder with 2(Bi, Pb):2Sr:3Ca:4Cu cation ratios were prepared by solid state reaction of extra pure (better than 99.99%) Bi_2O_3, $SrCO_3$, $CaCO_3$, and CuO powders. Proportioned powders were ball milled for 20 h in ethyl alcohol and then dried at 75 °C for 30 h.

The dried powder was calcined at 840 °C for 24 h in air. The calcinations were repeated three times at 840 °C for 24 h in order to improve the homogeneity by intermittently grinding the powder. The calcined powder was compressed into a disk of 55 mm in diameter and 7 mm in height in a 150 ton press. We have used the polyvinyl alcohol as a binder. The compacted powder disk was heat-treated at 860 °C for 170 h. in air. Same target have been used in ex situ and in situ depositions.

2.2 In situ film preparation

For the in situ technique, we employed a dc-sputtering system equipped with a magnetron source. The total pressure of the sputtering gas mixture O:Ar (2:1) was kept constant at 600 mTorr. The discharge power was 20 W. The MgO (100) substrates were fixed to the homemade heater block by silver paste and set in front of the gun at the distance about 4 cm. The substrate temperature varied between 700 and 800 °C, and controlled by using resistive heater and a thermocouple placed inside the heater block and just below the substrate. After deposition, the films were immediately cooled down to 600 °C within 3 min in the deposition atmosphere, and then kept for 30 min in 760 Torr O_2 before cooling to room temperature at 30 min. The preparation conditions for each sample are given in Table 1.

Table 1

Sample	Substrate temperature (°C)	Total gas pressure (mTorr)	Sputtering time (h)
Thin1	700	600	2
Thin2	750	600	2
Thin3	800	600	2

Superconductivity transitions were determined by four-probe-resistive measurements. The structure of the films were analysed by X-ray diffraction (XRD) with CuK_α radiation. The microstructures of the films were observed by scanning electron microscope (SEM).

2.3 Ex situ film preparation

For the ex situ technique, we have employed an rf-sputtering system equipped with a magnetron source. The thin films were deposited using a planar magnetron sputter gun with a reduced gap between target and substrate. The sputtered gun was mounted in a turbomolecular pumped vacuum chamber; the basic pressure was 2×10^{-6} Torr. The sputtering atmosphere was pure argon with the pressure of 300 mTorr. The sputtering time was 30 min and the rf-power of 200 W generated a cathode with no bias voltage. The MgO (100) substrates were fixed to the substrate holder with silver paste. The substrate temperature was held at 20 °C during film growth. After deposition, the films were taken out and tested. These films showed no sign of conductivity and they have an amorphous structure. These films were post annealed at 800 °C for 15 min (sample thin41). Once again, the same samples were annealed at 800 °C for 3h (sample thin42).

Superconductivity transitions were determined by four-probe-resistive measurements. The structure of the films were analysed by X-ray diffraction (XRD) with CuK_α radiation.

3 Results and discussion

Figure 1 shows the XRD patterns of the films prepared at different substrate temperature, keeping the total gas pressure and the O_2/Ar ratio constant. In this figure, all strong diffraction lines are assigned to (00l) of 2212 and 2223 phase.

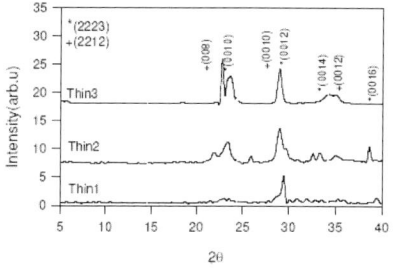

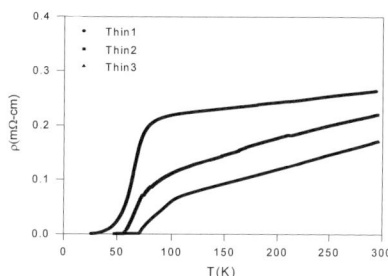

Fig. 1 XRD pattern for *in situ* prepared samples.

Fig. 2 Typical resistive transitions for *in situ* prepared samples.

But, as one can see clearly from the pattern, the proportions of 2223 phase will increase as the substrate temperature increases. Figure 2 shows the results of resistivity measurements for these series of the samples. In this figure, the three samples show superconducting transitions, and they have a metallic behavior in their normal state. But, as the temperature of substrates increase to 800 °C the slop of resistivity at the normal state increases, the transitions temperature of the films are increase, and the normal resistivity of the films (above 110 K) are decreases.

So, as pointed out by Kula et al. [11], by increasing the substrate temperature (up to a few degree from melting point, 860 °C), the ratio of the formation of 2223 phase will increase. This is consistent with the established view that the heat treatment as close as possible to the temperature of melting point is indispensable to obtain well-oriented BSCCO films [12]. The scanning electron micrographs of these three films are presented in Fig. 3.

 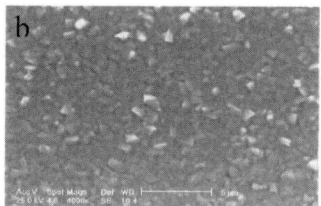 

Fig. 3 Micrographs of the surface of *in situ* prepared samples, (a) Thin1, (b) Thin2, (c) Thin3.

As seen in the picture, the films surface are almost smooth and homogen. As the temperature of the substrate increases, the formation of large grain become more favorable and we have a better link between the grains. This result is consistent with the resistivity measurements too.

Figure 4 shows the results of resistivity measurement for the samples prepared by ex situ method, namely Thin41 and Thin42. The thickness of the films is about 200 nm. The Thin41 sample was post-annealed at 800 °C for 15 min. As seen in the Fig. 4, Thin41, the normal resistivity of sample shows a semiconductor behavior and the indications of superconducting transition have taken place at different temperatures and at low temperature. For this sample, it seems that the period of post-annealing was not enough to change the amorphous structure of the sample and the connections between the grains have not been linked completely.

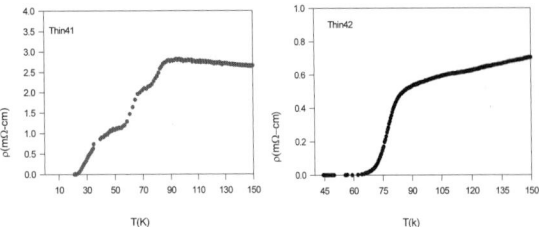 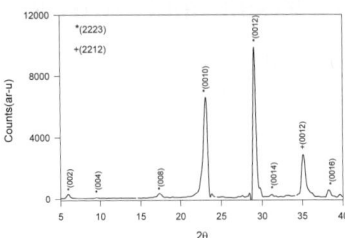

Fig. 4 Typical resistive transition of *ex situ* prepared samples.

Fig. 5 XRD pattern for the Thin42 samples.

For the above reason we have post annealed this sample one more time at 800 °C for 3 h. The result of resistivity measurements for this sample is shown in Fig. 4, Thin42. As it shows from the graph, the normal state of resistivity indicate a metallic behavior and the superconductivity transition has taken place at much higher temperature with much smaller transition width. Figure 5 shows the results of XRD measurements obtained on thin42 film. As can be seen from the peaks in this figure, the thin42 film is c-axis oriented, (00l), with the predomination of 2223 phase.

4 Conclusions

The process formation of the BSCCO, 2223 superconducting phase in ex situ rf-magnetron-sputtering thin films has been studied, and a reproducible fabrication procedure leading to an almost single 2223 phase BSCCO compound was achieved, using MgO (100) as a substrate. The fabricated films were highly oriented with c-axis perpendicular to the surface of substrate. Structural studies, based on the X-ray diffraction measurements, revealed that the predominate phase of the films are BSCCO 2223 phase. The results of our studies show, although the samples prepared by in situ method have a better mechanical and superconductivity properties, the post-annealed ex situ prepared samples have a superb structural and superconductivity properties.

Acknowledgements The authors would like to thank Isfahan University of Technology for supporting this project. We also would like to thank Dr. Solimani, Dept. of Electrical Engineering, Tehran University, and Dr. Dorodian, Material and Energy Research Lab. Karaj, for their help and assistance in experiments. This project was financially supported by the Ministry of Science, Research and Technology under the grant project 503495.

References

[1] C. B. Eom, J. Z. Sun, B. M. Lairson, S. K. Streiffer, A. F. Marshal, K. Yamamoto, S. M. Anlage, J. C. Bravman, T. H. Geballe, S. S. Ladermam, R. C. Tabet, and R. D. Jacowitz, Physica C **171**, 354 (1990).
[2] D. J. Rogers, P. Bove, and F. Hosseini Tehrani, Supercond. Sci. Technol. **12**, R75 (1999).
[3] H. Ota, S. Migita, Y. Kasai, H. Matsuhata, and S. Sakai, Physica C **311**, 42 (1999).
[4] S. J. Golden, F. F. Lange, D. R. Clarke, L. D. Chang, and C. T. Necker, Appl. Phys. Lett. **61**, 351 (1992).
[5] T. Sugimoto, N. Kubota, Y. Shiohara, and S. Tanaka, Appl. Phys. Lett. **63**, 2697 (1993).
[6] Z. Mori, E. Minamizono, S. Koba, T. Doi, S. Higo, and Y. Hakuraku, Physica C **339**, 161 (2000).
[7] S. I. Karimoto, S. Kubo, K. Tsuru, and M. Suzuki, Jpn. J. Appl. Phys. **36**, 84 (1997).
[8] G. Balestrino, M. Marinelli, E. Milani, A. Paoletti, and A. Paroli, J. Appl. Phys. **70**, 6939 (1991).
[9] L. Ranno, D. Martines-garcia, J. Perriere, and P. Barboux, Phys. Rev. B **48**, 13945 (1993).
[10] S. Zhu, D. H. Lowndes, B. C. Chakoumakos, J. D. Budai, D. K. Christen, X. Y. Zheng, E. Jones, and B. Warmack, Appl. Phys. Lett. **63**, 409 (1993).
[11] W. Kula, R. Sobolewski, J. Gorecka, and S. Lewandowski, J. Appl. Phys. **70**, 3171 (1991).
[12] R. Henn, T. Kroener, J. Greek, G. Linker, and O. Meyer, Physica C **221**, 405 (1994).

Power law behavior of $Tl_1Ba_2Ca_2Cu_3O_9$ superconducting tapes

B. A. Albiss[*,1], **A. El-Ali**[2], and **K. A. Azez**[3]

[1] Physics Department, Jordan University of Science and Technology, Irbid, Jordan
[2] Physics Department, Yarmouk University, Irbid, Jordan
[3] Physics Department, King Faisal University, Saudi Arabia

Received 31 August 2003, accepted 31 December 2003
Published online 1 April 2004

PACS 74.25.Fy, 74.72.Jt

Specimens of $Tl_1Ba_2Ca_2Cu_3O_9$ superconducting tapes (Tl-1223) have been prepared using the powder-in-tube (PIT) method. Dissipative mechanisms and power law behavior were investigated using magnetoresistance measurements and I–V characteristics. The temperature and magnetic field dependences of the resistance R(T,B) were fitted to the Arrhenius relation from which the magnetic field dependence of the pinning energy $U_0(B)$ was derived. The I–V data were fitted to a power law expression $V \sim I^{\alpha(T,B)}$ in which the exponent α was found to decrease gradually with increase of temperature and magnetic field. The variations of the critical current densities with applied magnetic field $J_c(B)$ were also studied. The correlation of the pinning energy to the critical current density was attempted. These results are discussed and explained in terms of the thermally activated flux flow (TAFF), grain boundaries, weak links, and Josephson junctions between the grains.

© 2004 WILEY-VCH Verlag GmbH & Co. KGaA, Weinheim

1 Introduction

Since the discovery of high temperature superconductors (HTSCs) [1], many problems concerning both their phenomenology and basic microscopic theories remain controversial. However, a systematic experimental and theoretical investigation of the physical properties of large number of HTSCs with a various preparation and characterization methods remains important. For their potential applications, the fundamental requirement for the HTSCs is the current carrying capacity in applied magnetic fields at relatively high temperatures, which is crucial for nearly every usage of bulk materials. In this regard, HTSCs behave very differently and all of them suffer from the problem of weak links at the grain boundaries and the dissipative behavior in the mixed state.

Dissipative or resistive behavior occurs when the quantized vortices move in the presence of current flow and magnetic field. Many models have been proposed to explain the dissipative mechanisms in HTSCs interims of flux creep [2], flux flow [3], thermal fluctuations [4], vortex glass [5] and Josephson junctions [6], but none of these approaches alone has been successful.

Different phases of the vortex matter have been studied very intensively after the discovery of the high-T_c superconductivity. Vortex solids, such as lattices and glasses can be moved under the influence of the Lorentz force. This leads to dissipation and under these conditions the critical current density is very low. Fluxon confinement ("pinning") prevents this motion and then J_c can be substantially increased. Therefore, for improving J_c, the optimization of the pinning of vortices is the decisive factor. Since the vortex has a normal core with a size comparable to the superconducting coherence length, it is energetically favourable to superimpose the vortex cores spatially with normal state defects in a superconductor. Knowing that, intense efforts to optimize the vortex confinement by various defects have

---

[*] Corresponding author: e-mail: baalbiss@just.edu.jo, Fax: ++962-2-7095014

been made. As a rule in most cases a random spatial distribution of defects (impurity phases, voids, irradiation defects, etc.) with different sizes was used. In this case neither the size nor the distribution of the pinning centres is properly controlled and the resulting pinning phenomena are very complicated and, in this case, a disordered vortex matter is stabilized by these random pinning arrays.

In applied magnetic fields where the repulsive interaction between vortex lines becomes significant, the pinning of vortices to fixed position in the superconductor can deform the vortex lattice from its ideal configuration. At low magnetic fields, the interaction energy between the vortex lines is weak, so that random pinning centres will cause slight increase in the elastic energy of the lattice. However, at high applied magnetic fields, weak pinning centre cannot compete with the increase strength of the vortex–vortex interactions. In this case only strong pinning sites will hold individual vortex lines in place independently of the repulsive interaction with neighbouring vortices. In high temperature superconductors (HTSCs), the vortex lines are particularly susceptible to pinning because the weak coupling between the CO_2 planes which gives way to highly flexible vortices.

At low temperatures, the vortices are essentially frozen into their distorted configuration. As the temperature is raised, thermal fluctuation of the vortex line positions become important. Thermal fluctuations in HTSCs are considerably stronger than in conventional superconductors. This is due to the, small value of the in-plane coherence length, the high critical temperature, the layered and anisotropic nature of these materials.

It is known that at nonzero temperature, motion of flux creep is possible with the help of thermal activation. According to the Anderson theory of flux creep [7], the energy dissipated (the resistance) at which the fluxoids jump over the pinning barrier is given by the usual Arrhenius expression:

$\rho(T) = \rho_0 \exp(-U_0/kT)$. Where U_0 itself is a function of temperature and magnetic field, and $U_0 \to 0$ as $T \to T_c$; so the exponential relationship is by no means pure. For HTSCs, several researchers [8] found experimentally that the pinning energy falls off as $(1-T/T_c)^{3/2}/H$.

Flux creep measurements are very informative in understanding the pinning mechanisms of type-II superconductor, and useful for comparing the pinning of different samples. However, special care must be taken for obtaining pinning energy from flux creep measurements.

One of the HTSCs suitable for power application (transmission of electrical energy, various magnet systems) is the $TlBa_2Ca_2Cu_3O_x$ phase of the Tl–Ba–Ca–Cu–O system (known as Tl-1223) because of its high irreversibility line, good intergrain connectivity and relatively good magnetic flux pinning. In this work we present our results concerning the synthesis and characterization of Tl-1223 silver sheathed tape using the magneto-resistance, I–V characteristics measurement methods.

2 Experimental

The samples were prepared by a solid-state reaction method using Tl_2O_3, BaO and CuO as starting materials. The high purity powders were weighed in the appropriate amounts to form a nominal composition at $Tl_1Ba_2Ca_2Cu_3O_y$. The powders were thoroughly mixed in dry atmosphere using an agate mortar and pressed into disc-shaped pellets under a pressure of ~ 400 kg/cm^2. The pellets were wrapped into gold foil to prevent loss of thallium and then sintered at 860 °C for 5 h in oxygen flow. The pellets were reground, wrapped again with gold foil, and post-annealed at 900 °C for 2 h.

The superconductor made in this experiment is a tape. Although more fragile than a wire, a tape is more flexible than a wire, allowing the superconductor to bend without damaging itself. A superconductor, to be practical, must be able to be installed into pre-existing conduits. A rigid superconductor is only useful when a straight line can be utilized. Since a tape is a more useful form, it was chosen for this experiment.

The tape is prepared using the powder-in-tube (PIT) method using a high TBCCO-1223 ($TlBa_2Ca_2Cu_3O_9$) ceramic powder is packed into a silver tube (a billet) and then transformed into a wire and then rolled into a tape. A detailed study of the tape fabrication was reported by El-Ali et al. [9].

The structural characterizations of the samples were carried by X-ray diffraction methods with CuKα radiation (λ = 1.542 Å). The lattice parameters of the Tl-1223 sample are calculated by the least square method using a computer program as follows: a = 3.8240 Å and c = 15.335 Å. Using a scanning electron microscope (SEM), samples showed well-shaped plate-like grains. The average grain size was estimated to be 1–2 µm in diameter.

The V–I characteristics and R–T curves were measured using the standard dc four-probe method with magnetic field B applied perpendicular to the surface of the tape. The V–I curves were measured over temperature range of 80–120 K in applied magnetic fields up to 0.5 T. The critical current densities were determined using an electric field criterion of 1 µV/cm. The current density used for electrical resistance measurement was 1 mA/cm^2. The temperature was measured with a calibrated platinum resistance sensor with an accuracy of ± 0.1 K. In all measurements, the sample was first zero-field cooled down to $T = 80$ K at which the measurements were to be done. Then, applying magnetic field up to 0.5 T, reducing it to zero and repeating the measurement for higher temperatures very close to the critical temperature $T_c = 120$ K.

3 Results and discussion

The normalized resistance as a function of temperature R(T)/ R(300 K) is shown in Fig. 1. The Tl-2223 tape exhibit a sharp transition with a transition width of a bout 5 K, the zero resistance transition temperature is $T_c \sim 120$ K.

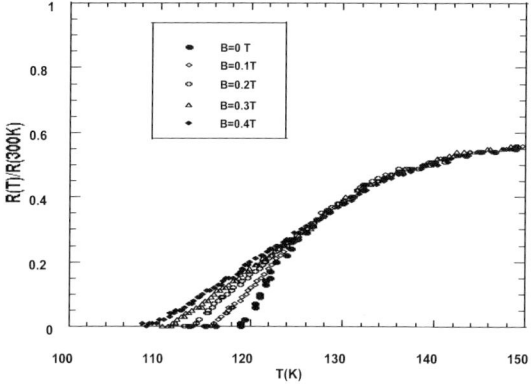
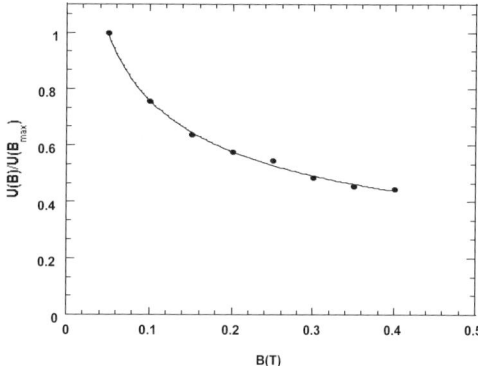

Fig. 1 Temperature dependence of the normalized resistance for the Tl-1223 tape at different applied magnetic fields perpendicular to the tape surface.

Fig. 2 Magnetic field dependence of the normalized pinning energy of Tl-1223 tape.

The resistive transition in the polycrystalline HTSCs in the applied magnetic field has been found to obey the Arrhenius relation $R(T, B) = R_0 \exp(-U(B,T)/kT)$, where U_0 is the pinning energy that depends on temperature and magnetic field. The functional dependence of $R(T,B)$ shown in Fig. 1 is expected to give information about the mechanisms of dissipation. In the simplest case, the temperature dependence of U_0 is neglected. Then $U(B)$ values can be directly deduced from the slopes of the plots of $\log(R)$ vs. $1/T$. From the linearization of the data in the middle part of the plots, and despite the scattering in the data, we have calculated the slopes of each line at different applied magnetic fields. The magnetic field dependence of the pinning energy is shown in Fig. 2. It is found that the field dependence of the normalized pinning energy U/U_{max} follows the form $U(B)/U_{max} \sim B^{-0.5}$. Based on our data we noticed that the value of the power exponent depends on the range of the magnetic field. For wider ranges of magnetic of magnetic fields, we expect a crossover of the magnetic field dependences. This crossover is associated to the polycrystalline nature of the sample and the change in the dissipation mechanism with increase of the magnetic field. These results are quite difference from other results for HTSCs [2, 10] and seem to support that the dissipation related with magnetoresistance transition quite below T is dominated by a thermally activated flux flow TAFF model. In this model, the pinning energy is given by $U \sim (1-T/T_c)^n/B$

where T_c is the transition temperature at zero field. We believe that for the case of polycrystalline sample in the form of wires and tapes, a more sophisticated analysis is needed to deduce the temperature dependence of the pinning energy. A large number of parameters are involved in the relation, which makes the curve fitting very delicate. The observed power law behavior of the pinning energy $U(B)$ over a magnetic field range up to 0.5 T for the Tl-1223 tape depends upon the size and orientation of the grains, the weak links between the grains and the direction of the applied magnetic field. However, improving the grain boundaries, grain alignment and the weak links in polycrystalline tapes will make them more practical in applied magnetic field.

The temperature dependence of the transport critical current density $J(T)$ at zero applied magnetic field perpendicular to the tape surface is shown in Fig. 3. The data were fitted to the relation $J_c(T) \sim (1-T/T_c)^n$ with $n \approx 2$ and $T_c = 120$ K. The observed value $n \approx 2$ is higher than that of the Ginzburg–Landaue value $n \approx 1.5$ for temperatures near T usually observed in thin films and single crystals of HTSCs, but is quite close to $n \approx 2$, and is expected to rise due to different weak links in polycrystalline samples such as wires and tapes. In polycrystalline materials, the superconducting grains can be separated by a non-superconducting material, which strongly determines the temperature dependence of J. The $J(T)$ relation in our Tl-1223 tape was found to follow the Ambegoakar–Baratoff relationship [11], in which the grain boundary regions are treated as superconducting–normal–superconducting SNS junctions or superconducting–insulator–superconducting SIS junctions. For the SNS case, De Gennes [12] showed that $J_c(T) \sim (1-T/T_c)^n$ with $n \approx 2$ for T close to T_c. The nearly square law in our case suggests that mainly SNS junctions determine the electrical transport mechanism.

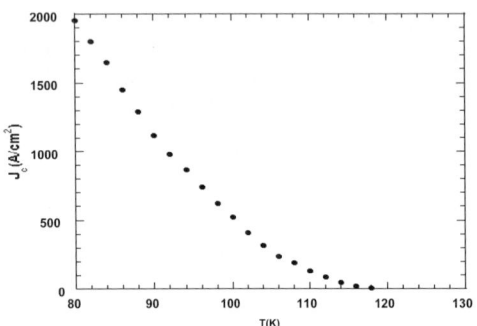

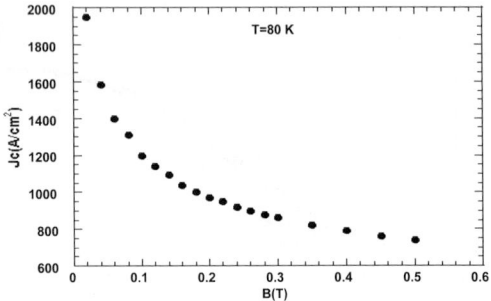

Fig. 3 Temperature dependence of the critical current density of Tl-1223 tape at zero magnetic field.

Fig. 4 Magnetic field dependence of the critical current density of Tl-1223 tape at T = 80 K.

The magnetic field dependence of J_c is shown in Fig. 4. The observed variation is qualitatively the same as that observed for polycrystalline samples [13, 14] and can be fitted rather well with $J_c(B)$. Considerably weaker $J(B)$ dependence has been c observed in BSCCO tapes exhibiting enhanced flux pinning [15].

By taking into account that the intragranular pinning energy is related to the pinning force and thus the critical current density [16], one can conclude that, for polycrystalline samples of Tl-1223 tape $U(T, B)$ is given by: $U(T, B) \propto (1-T/T_c)^2/B^{-0.5}$.

I–V characteristics of the sample have been measured at different magnetic fields in the range from 0 to 0.5 T and at different temperatures in the range from 80 to 120 K. It is evident from the I–V characteristics that these plots exhibit a power law behavior $V \propto I^\alpha$ when plotted on a log I–log V scale, the slope of each line α is calculated at each value of the magnetic field and temperature. It should be noted that the exponent $\alpha(T, B)$ depends both temperature and magnetic field. For example, the temperature dependence of the exponent $\alpha(T, B)$ at zero magnetic field is shown in Fig. 5. At temperature quite below T_c and relatively low magnetic field, a non-ohmic dissipation ($\alpha > 1$) was observed. While at temperatures close to T_c, the sample exhibits an ohmic behavior ($\alpha \sim 1$). In this temperature dependence, no sign of the critical transition at ($\alpha = 3$) expected by the KT theory [17]. The smearing out of this KT transition seems to be strongly sample-dependent. Also we could not observe such a transition for previous work on polycrystalline Bi-2223 [18] may be because of the wide temperature and magnetic field range used in our measurements.

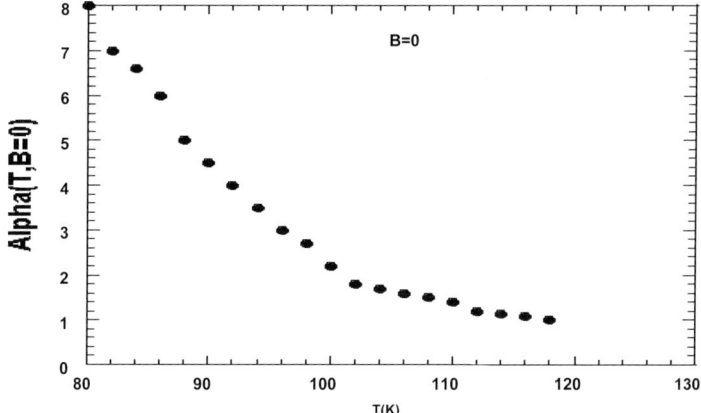

Fig. 5 Temperature dependence of the exponent α(T, B) of Tl-1223 tape at zero magnetic fields.

Though our study aimed at grasping a better insight into the power law behaviors and the dissipation mechanisms in $TlBa_2Ca_2Cu_3O_9$ HTSCs. Furthermore, a much more detailed study on the correlation of the pinning energy to the critical current density is needed for wider ranges of temperatures and magnetic fields.

Acknowledgements This work was supported by Jordan University of Science and Technology grant no: 2000/28 and Yarmouk University grant no:19/2000.

References

[1] J. G. Bednorz and A. Muller, Z. Phys. **64**, 189 (1986).
[2] T. T. M. Plastra, B. Batalogg, L. F. Schneemeyer, and J. V. Waszczak, Phys. Rev. Lett. **61**, 1662 (1988).
[3] M. Tinkham, Phys. Rev. Lett. **61**, 1658 (1988).
[4] B. Oh, K. Char, A. D. Kent, M. Naito, M. R. Beasley, T. H. Geball, R. H. Hammond, and A. Kaptulnik, Phys. Rev. B **37**, 7861 (1988).
[5] D. S. Fisher and P. A. Fischer, Phys. Rev. B **43**, 130 (1991).
[6] D. H. Kim et al., IEEE Trans. Magn. **27**, 1383 (1991).
[7] P. W. Anderson, J. Phys. **9**, 309 (1962).
[8] D. E. Farell et al., Phys. Rev. Lett. **67**, 1165 (1991).
[9] A. El-Ali, B. A. Albiss, and K. Khasawinah, to be published in phys. stat. sol. (2004); this conference.
[10] M. Takasa and H. Itozaki, Appl. Phys. Lett. **59**, 1236 (1991).
[11] V. Ambegaokar and A. Bratoff, Phys. Rev. Lett. **10**, 486 (1963).
[12] P. G. De Gennes, Rev. Mod. Phys. **36**, 225 (1964).
[13] S. X. Dou, H. K. Liu, and Y.C. Guo, Appl. Phys. Lett. **60**, 2929 (1992).
[14] S. Jin and T. H. Tiefel, Appl. Phys. Lett. **58**, 868 (1991).
[15] H. K. Liu et al., Supercon. Sci. Tech. **5**, 591 (1992).
[16] M. Tinkham and J. C. Lobb, Solid State Phys. **42** (1989).
[17] J. M. Kosterlitz and D. J. Thouless, J. Phys. C **5**, L124 (1972).
[18] B. A. Albiss et al. J. Low-Temp. Phys. **105**, 957 (1996).

The noise power spectral density in thin epitaxial YBa$_2$Cu$_3$O$_{7-\delta}$ films

A. Labrag[*,1], **A. Taoufik**[1], **S. Senoussi**[2], and **A. Ramzi**[1]

[1] Laboratoire des Matériaux Supraconducteurs à Haute Température Critique, Département de Physique, Faculté des Sciences, Université Ibn Zohr, B.P. 8106, Agadir, Morocco

[2] Laboratoire de Physique des Solides (associé au CNRS. URA. 0002), Université Paris Sud, Bâtiment 510, 91405 Orsay Cedex, France

Received 31 August 2003, accepted 31 December 2003
Published online 8 April 2004

PACS 74.25.Fy, 74.40.+k

We have studied the noise power spectral density $S(f, T)$ in YBa$_2$Cu$_3$O$_{7-\delta}$ thin films. Measurements were performed for various magnetic field, up to 14 T, and in a large range of temperatures. We found that in thin films deposed by the laser ablation method on the surface (001) of SrTiO$_3$ substrate, the noise power spectral density $S(f)$ exhibits a lorentzian shape behavior in the mixed state and that all the peaks of temperature dependent noise power spectral density $S(T)$ at a determined frequency value were appeared in large game of the temperature. Our results suggested that the noise power spectral density under the applied magnetic field in the mixed state does not come from thermal fluctuation and that the flux motion remains the most probable origin of the noise.

© 2004 WILEY-VCH Verlag GmbH & Co. KGaA, Weinheim

1 Introduction

In recent years significant study was achieved towards the understanding the origin of low frequency noise [1–4] in thin films made from epitaxially grown of YBa$_2$Cu$_3$O$_{7-\delta}$ and SrTiO$_3$ insulating layers. R. Straub et al. [5] have demonstrate that, in a weak applied magnetic field, the single fluctuating vortex is responsible for a large excess low-frequency flux noise measurements in high-T$_C$ superconducting. However, M. J. Ferrari et al. [6] have shown that low-frequency noise is correlated with quality of high-T$_c$ films, which is affected by the presence of a variety of defects in these films.

On the other hand, others results found in the literature [5, 7] indicate that the low frequency noise, under the applied magnetic field, may be caused by the flux motion of the high-T$_C$ superconducting or may come from the resistance fluctuation which is caused by thermal fluctuation in the mixed state superconducting according to the theory of the temperature fluctuation [8] which predicts that the noise power spectral density, at a constant frequency, is proportional to $(d\rho/dT)^2$ obtained from the $\rho(T)$ curve.

In this paper we have examined the effect of resistance fluctuations, produced by the temperature fluctuations under the intense magnetic field on the noise power spectral density $S(f, T)$ in YBa$_2$Cu$_3$O$_{7-\delta}$ thin films. We have also investigate, the origin of the noise in the high-T$_C$ superconducting by studying the noise power spectral density $S(T)$ at frequencies $f = 115.84$ and 607.37 Hz.

---

[*] Corresponding author: e-mail: zizlab@hotmail.com, Phone: +212 48 22 09 57, Fax: +212 48 22 01 00

2 Experimental

C-axis oriented epitaxial YBa$_2$Cu$_3$O$_{7-\delta}$ films with a thickness of 400 nm and a width of 7.53 µm are deposited by the laser ablation method on the surface (100) of SrTiO$_3$ substrate. The resistance vanished, in zero magnetic field, at T_c = 90 K. Electrodes of measurement are in gold and deposited on the surface of the sample in situ by evaporation. The distance between electrodes of power measurement is 135 µm. Contact resistances were less than 1 Ω. A direct current, perpendicular to the magnetic field, is applied on edges of the sample. The sample central region voltage signal goes through a low-noise transformer of report n = 100, then in a preamplifier of gain equal to 100 and finally in a RC filter. The signal is visualized on a programmable oscilloscope then recorded and analyzed by computer [9].

3 Results and discussion

In Fig. 1, we plotted the frequency dependence of the voltage noise in an applied magnetic field of 14 T and at a constant current for two values of the temperature 83.4 K (Fig. 1a) and 87.2 K (Fig. 1b). It was found that, in a large range of frequencies 1–1000 Hz, the noise power spectral density S(f) exhibits tow regimes: the first region is the low frequency noise 1–10 Hz and the second is high frequency noise 10–1000 Hz. It was also found that $S(f)$, in the mixed state, follows a Lorentzian shape, of $A\left[(1+\pi f/f_0^3)\right]^{-1}$, where A is constant and f_0 is a characteristic frequency.

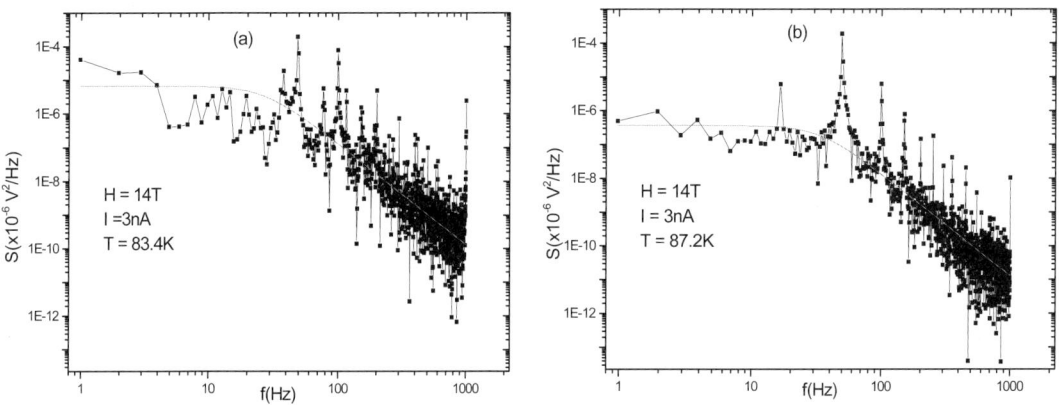

Fig. 1 The frequency dependence of the voltage noise in an applied magnetic field of 14 T and at a constant current for two values of the temperature 83.4 K (Fig. 1a) and 87.2 K (Fig. 1b). The solid line represents the fit of our data by the form A(1+($\pi f/f_0$)^3)$^{-1}$.

In Fig. 2(a), The noise power spectral density S(f) at frequency f = 115.84 Hz was plotted as a function of temperature T. On the even figure the square line presents variations of (dρ/dT)^2 obtained from the curve $\rho(T)$, which is shown in Fig. 2(b). It can be seen that the shape of the noise power spectral density $S(T, f$ = 115.84 Hz) is in remarkable separation with that of (dρ/dT)^2. It can be also seen that the temperatures where the maximum of $S(T)$ and (dρ/dT)^2 occur a different place. Hence no correlation between the noise measurements and the temperature derivative of the resistivity was found.

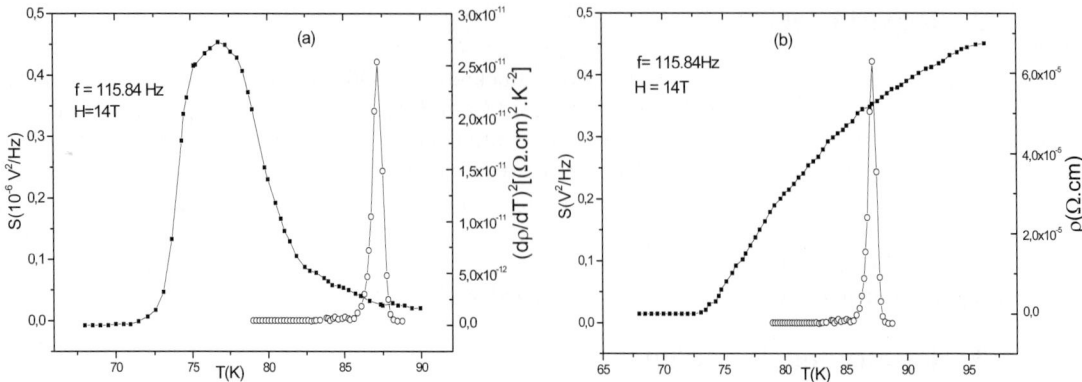

Fig. 2 (a) The noise power spectral density $S(T)$ at frequency $f = 115.84$ Hz and the temperature derivative of the resistivity in an applied magnetic field of 14 T. (b) The noise power spectral density $S(T)$ at frequency $f = 115.84$ Hz and resistivity versus the temperature.

Figure 3 shows the temperature dependency of $S(T)$ for $H = 14$ T and a current of 3 nA at frequencies $f = 115.84$, 607.37 Hz. As it can be seen in this figure, the peak of the noise power spectral density S(T) for $f = 607.37$ Hz appears at 83.4 K but for $f = 115.84$ Hz this peak appears at 87.2 K.

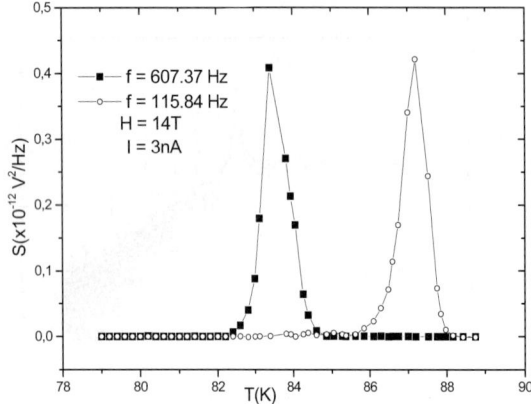

Fig. 3 The noise power spectral density $S(T)$ versus the temperature for low frequency values 115.84 Hz and 607.37 Hz, in an applied magnetic field of 14 T.

In conclusion, we have measured the noise power spectral density in YBa$_2$Cu$_3$O$_{7-\delta}$ thin films in the mixed state. Our results show yhat (a) the $S(f)$ follows a lorentzain shape in the form of $A(1+(\pi f/f_0)^3)^{-1}$ were A is constant and f_0 is a characteristic frequency, (b) the noise origin is not the resistivity fluctuation and consequently is not the thermal fluctuation, (c) the noise under the applied magnetic field effect most likely come from the flux motion thermally activated.

References

[1] S. Keil, R. Straub, R. Gerber, R.P. Huebener, D. Koelle, R. Gross, and K. Barthel, IEEE Trans. Appl. Supercond. **9**, 2961–2966 (1999).
[2] E. Dantsker, S. Tanaka, P.A. Nilsson, R. Kleiner, and J. Clarke, Appl. Phys. Lett. **69**, 29 (1996).
[3] Y. Paltiel, G. Jung, Y. Myasoedov, M. L. Rappaport, E. Zeldov, S. Bhattacharya, and M. J. Higgins, Fluct. Noise Lett. **2**, 1, L31–L36 (2002)
[4] D. Babic, T. Nussbaumer, C. Strunk, C. Schoeneneberger, and C. Suergers, Phys. Rev. B **66**, 014537 (2002).

[5] R. Straub, S. Keil, R. Kleiner, and D. Koelle, Appl. Phys. Lett. **78**, 23, 3645 (2001).
[6] M. J. Ferrari, M. Johnson, F. C. Wellstood, J. J. Kingston, T. J. Shaw, J. Clarke, J. Low, Tem. Phys. **94**, 15 (1994).
[7] H. E. Horng, J. M. Wu, S.Y. Yang, and J. D. Chern, Physica C **282**, 1433 (1997), Physica C **287**, 1434 (1997).
[8] A. Taoufik, S. Senoussi, and A. Tirbiyine, Ann. Chim. Sci. Mat. **24**, 227 (1999).
[9] R. F. Voss and J. Clarke, Phys. Rev. B **13**, 556 (1976).

Glass temperature and critical current density in $YBa_2Cu_3O_{7-\delta}$ thin films

A. Ramzi[*,1], **A. Taoufik**[1], **S. Senoussi**[2], and **A. Labrag**[1]

[1] Laboratoire des Matériaux Supraconducteurs à Haute Température Critique, Département de Physique, Faculté des Sciences, Université Ibn Zohr, B.P. 8106, Agadir, Morocco
[2] Laboratoire de Physique des Solides (associé au CNRS. URA. 0002), Université Paris Sud, Bâtiment 510, 91405 Orsay Cedex, France

Received 31 August 2003, accepted 31 December 2003
Published online 1 April 2004

PACS 74.25.Dw, 74.25.Sv

We have studied the glass temperature T_g and the critical current density J_c in high quality $YBa_2Cu_3O_{7-\delta}$ thin films. We measured the J–E curves at different temperatures near the transition temperature region in applied magnetic field parallel to the c-axis, the J–E support the existence of a continuous phase transition at temperature T_g from vortex-liquid to vortex-glass phase. A straight line in the $log(1-T_g/T_c)$ versus $log(H)$ plot gives a relationship of $H = H_0(1-T_g/T_c)^{2.10}$ where v_0 is zero critical exponent. To determine J_c as a function of magnetic field, we have realized the systematical measure of the J–E characteristic in a large range of temperatures with an increment of 0.2 K for each magnetic field value. The critical current density decreases as the applied magnetic field increases. For each temperature value, we observed two regimes in the critical current density $J_c(H)$ behavior.

© 2004 WILEY-VCH Verlag GmbH & Co. KGaA, Weinheim

1 Introduction

In high T_c-superconductors, the nature of the mixed state at high field value $H > H_{c1}$ was the object of numerous studies. One of the more interests phenomena is the phase transition of the second order of an ordered state of vortex-glass type toward a vortex-liquid state [1]. Experimental evidence has been presented for the existence of a vortex-glass in thin films by Koch et al. [2]. The glass transition is expected to be a second-order phase transition at the temperature $T_g(H)$, characterized by a power law $E \propto J^{(z+1)/(d-1)}$ for $T = T_g$ [3], which is a straight line in the $logE$–$logJ$ plot. From the slop $(z+1)/(d-1)$ of this line the critical exponent z which describes the divergent length of the vortex–vortex interaction [4]. Considering the physical dimension $d = 3$ [5].

In $YBa_2Cu_3O_{7-\delta}$ films, J_c is orders of magnitude larger up (to 10^{12} A/m^2) at small magnetic fields and low temperatures. This high J_c is generally attributed to strong pinning of vortices by extended defects. Dam et al. [6] have shown, that in thin films of $YBa_2Cu_3O_{7-\delta}$ vortices are pinned by linear defects extending throughout the whole film-thickness. They have shown that dislocations are responsible for the high critical current density in $YBa_2Cu_3O_{7-\delta}$ films. In high critical temperature superconductors, the CuO_2 layers are strong pinning centers for vortices which are aligned along these layers. The defects and precipitates cannot act strongly as the pinning centers when the applied magnetic field is adjusted with the c-axis. This is because the coherence length along the c-axis in high T_c-superconductors is very short [7] and thus the vortex core size is very small.

---

[*] Corresponding author: e-mail: ab_ramzi@yahoo.fr, Phone: +212 48 22 09 57, Fax: +212 48 22 01 00

It is so very interesting from both viewpoint of fundamental research and practical application to understand the vortices pinning mechanisms of high T_c-superconductors. On the other hand from a practical point of view the problem is to find the mechanism of strong pinning of vortices and to obtain samples with higher critical current density.

2 Experimental

The sample for this study was a high quality $YBa_2Cu_3O_{7-\delta}$ thin film, which was prepared by laser ablation deposited on $SrTiO_3$ substrate in the plane (001). The c-axis is perpendicular to the film surface. The film has a thickness of 400 nm and a width of 7.53 µm. The distance between electrodes of power measurement was 135 µm. The resistance vanished at $T_c = 90$ K in zero magnetic field. Golden electrodes of measurement were deposited on the sample surface in situ by evaporation. Contact resistances were less than 1 Ω. For these experiments, a DC transport current was perpendicularly to the applied magnetic field direction. The signal was passed through a low-noise transformer with a ratio n = 100, then it was amplified with a preamplifier and finally in the RC filter. The signal was recorded and analyzed by computer [8].

To rule out distortions of the J–E curves by extensive heating that could be induced by the very high dissipation levels employed here, a pulsed current power supply was used with a time duration $\tau \approx 10$ ms, a wave from repeat time ~ 2 s and an average over 64 pulses. The applied magnetic field, the angle between the applied magnetic field and the c-axis and the current still unchanged during the experiment the temperature was incremented by 0.2 K.

3 Results and discussions

Figure 1 represents two series of the J–E curves for two fields magnetic applied 0.6 T and 10 T parallel to the c-axis. These characteristics are gotten while descending in temperature from the normal state. For every field we define a T_g temperature at which happens the transition of the vortex glass. To a certain temperature and a certain current density, the resistance is of type "flux flow" (or Stephen Bardeen). The derivative $dlog(E)/dlog(J)$ changes at a certain temperature denoted T_g. According to Fisher et al. [1], T_g would represent the temperature of the transition toward the vortex-liquid phase. It is clear according to these curves that a linear behavior happen to $T_g = 85$ K for $H = 0.6$ T and $T_g = 71$ K for $H = 10$ T, this linearity continues on the entire studied E interval.

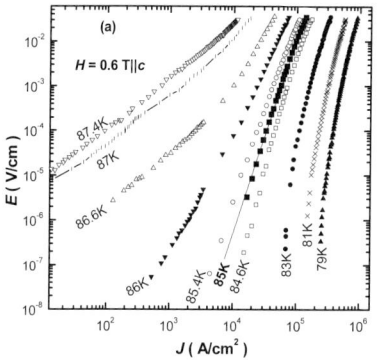
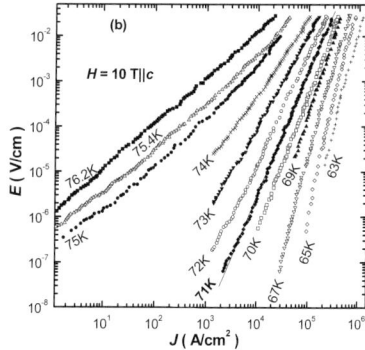

Fig. 1 The J–E characteristics in log–log scales, for different temperatures and H is parallel to the c-axis. (a) $T_g = 85$ K for $H = 0.6$ T and (b) $T_g = 71$ K for $H = 10$ T.

For temperatures higher than T_g, the isotherms change curvature. As one can see it on the Fig. 1, the shape of the isotherms of each series changes gradually with the temperature decrease. These isotherms are strictly linear for temperatures near T_c; they have a positive curvature for ($T_c < T < T_g$) and become again linear at $T = T_g$ that is the signature of the transition toward the vortex-liquid state. This behavior is observed for all fields except for $H = 0$ T (Fig. 2). In fact, it is difficult to define only one T_g temperature to $H = 0$ T because of the anisotropy of these materials and the proper field H_p (0.04 T when we applies the high currents) which is higher to the applied outside field. The problem comes because the direction of H_p is not very definite. this is why we are not able to determine only one linear curve of the log$E$–log$J$ characteristics and, so, only one T_g temperature.

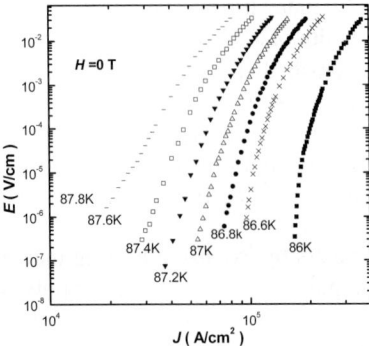

Fig. 2 The J–E characteristics in log–log scale, for $H = 0$ T at different temperatures.

The transition line deducted from the experiments is shown in Fig. 3 (solid squares). A straight line in the $\log(1 - T_g / T_c)$ versus $\log(H)$ plot gives the following relationship

$$H \propto \left(1 - \frac{T_g}{T_c}\right)^{2\nu_0},$$

with zero critical exponent $\nu_0 = 1.5 \pm 0.1$ (solid line in Fig. 3). In fact, we have found evidence of the transition from vortex-liquid to vortex-glass and the existence of the vortex-glass phase in $YBa_2Cu_3O_{7-\delta}$ thin films. The characteristic behaviors in the transition and in the vortex-glass phase are consistent with the vortex-glass theories. This dependence T_g on H is similar to that for $Tl_2Ba_2CaCuO_8$ thin films and $YBa_2Cu_3O_{7-\delta}$ single crystals [9].

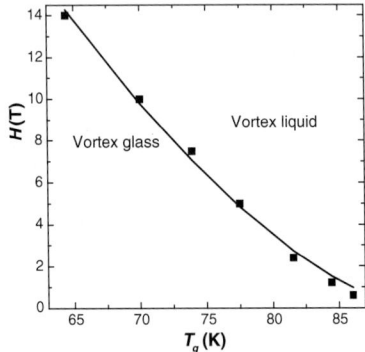

Fig. 3 The H dependence of the T_g, the solid squares represent the experimental values and the solid line is the theoretical curve.

The critical current density is deduced from the systematical measures of J–E characteristic. The adopted field criterion was 1 µV/cm. In Fig. 4, we present the critical current density $J_c(H)$ variations in perpendicular-field orientation.

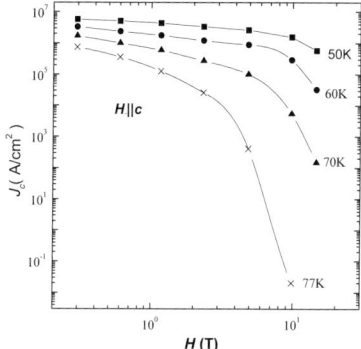

Fig. 4 The critical current density J_c as a function of the H for different temperature values.

Two regimes are presented in $J_c(H)$ variations. $J_c(H)$ exhibits a slight decrease below an applied magnetic field H^*. H^* is the field where the slope of $J_c(H)$ curve changes. A gradual transition to a power low behavior that is $J_c(H) \approx H^{-\alpha}$ is found above this field.

Current densities of about 10^6 A/cm^2 can be carried at 50 K up to 14 T. Typically Nb$_3$Sn filaments carry such current densities only up to 8 T at 4.2 K. The J_c drops a further order of magnitude [10, 11]. Thus YBa$_2$Cu$_3$O$_{7-\delta}$ films can carry more current at high fields than conventional superconductors.

The strong decrease of the current density $J_c(H)$ when $H > H^*$ can be explained by a collective pinning behavior of vortices. The vortices number becomes greater when the applied magnetic field increases. Not all defect sites can accommodate a vortex, as the interaction between vortices dominates over the pinning energy wherever the defects are too close to each other. This suggests the importance of interactions between adjacent vortices in preventing their motion.

Dam et al. [6] shown that dislocations are the main source of pinning in YBa$_2$Cu$_3$O$_{7-\delta}$ thin films and that each dislocation can pin exactly one vortex. They have shown that the field, where the critical current density decreases strongly is proportional to the measured dislocation density.

Another explanation of the J_c decrease is that YBa$_2$Cu$_3$O$_{7-\delta}$ thin films can be considered as a totality of superconductor grains connected between them by weak junctions. These junctions can be: poor region in superconductivity, grains boundaries, insulator barriers, normal metallic layers or semiconductor. These layers block the super-current movement what reduced the critical current density.

References

[1] M. P. A. Fisher, Phys. Rev. Lett. **62**, 1415 (1989).
[2] R. H. Koch, V. Foglietti, W. J. Gallagher, G. Koren, A. Gupta, and M. P. A. Fisher, Phys. Rev. Lett. **63**, 1511 (1989).
[3] A. Taoufik, A. Tirbiyine, El. Assif, and A. Ramzi, Inst. Phys. Conf. Ser. N° **167**, 887 (1999).
[4] A. Tirbiyine, A. Taoufik, S. Senoussi, and A. Ramzi, Physica C **341–348**, 1334 (2000).
[5] J. M. Roberts, B. Brown, B. A. Hermann, and J. Tate, Phys. Rev. B **49**, 6890 (1994).
[6] B. Dam, J. M. Huijbregtse, F. C. Klaassen, R. C. F. van der Geest, G. Doornbos, J. H. Rector, A. M. Testa, S. Freisem, J. Aarts, J. C. Martinez, B. Stäuble-Pümpin, and R. Griessen, Nature **399**, 439 (1999).
[7] T. K. Worthington, W. J. Gallagher, D. L. Kaiser, F. H. Holtzberg, and T. Dinger R., Physica C **153–155**, 32 (1988).
[8] A. Taoufik, S. Senoussi, and A. Tirbiyine, Ann. Chim. Sci. Mat. **24**, 227 (1999).
[9] N.-C. Yeh, W. Jiang, D. S. Reed, U. Kriplani, and F. Holtzberg, Phys. Rev. B **47**, 6146 (1993).
[10] J. W. Ekin, IEEE Trans. Magn. **19**, 900 (1983).
[11] P. A. Hudson, F. C. Yin, and H. Jones, IEEE Trans. Magn. **19**, 903 (1983).

The YBa$_2$Cu$_3$O$_{7-\delta}$ anomalous second peak and irreversible magnetic field in the magnetization hysteresis cycles

A. Taoufik[*,1], A. Ramzi[1], S. Senoussi[2], and A. Labrag[1]

[1] Laboratoire des Matériaux Supraconducteurs à Haute Température Critique, Département de Physique, Faculté des Sciences, Université Ibn Zohr, B.P. 8106, Agadir, Morocco
[2] Laboratoire de Physique des Solides (associé au CNRS. URA. 0002), Université Paris Sud, Bâtiment 510, 91405 Orsay Cedex, France

Received 31 August 2003, accepted 31 December 2003
Published online 1 April 2004

PACS 74. 25.Ha, 74.72.Bk

The flux jumps, the second peak and the irreversible magnetic field in the magnetization hysteresis cycles have been investigated in the high temperature superconductor YBa$_2$Cu$_3$O$_{7-\delta}$ single crystals. These cycles were obtained for different temperature values, the applied magnetic fields up to 6 T and the angle θ between the applied magnetic field and c-axis. The magnetization curves exhibit a remarkable second peak "fishtail", this second peak was not observed for the low temperature, but we observed the flux jumps "saw tooth". The temperature dependence of the irreversible magnetic field, H_{irr}, for the applied magnetic field perpendicular to the ab planes is given by an extended expression, $H_{irr} \alpha (1-T/T_c)^\alpha$, where α is a constant, the Abrikosov flux dynamics can explain this behavior. The H_{irr} as a function of θ has been strongly influenced by the flux pinning and the thermally assisted flux motion.

© 2004 WILEY-VCH Verlag GmbH & Co. KGaA, Weinheim

1 Introduction

Observations of various anomalous peaks on magnetization hystersis cycles in high critique temperature samples have revived the interest to study specific properties of magnetization processes in these novel materials. Up to now, three types of maxima have been observed, each with a distinct temperature behavior: the central or low-field peak at fields close to 0 T, the fishtail or second peak at relatively high fields, and sometimes even an intermediate peak between the central and the fishtail peak, whereas the origin of the central peak and the intermediate peaks are widely discussed in literature [1–4]. All these peaks aren't observed for the low temperatures. However, the cycle shows several flux jumps "saw tooth" at these temperatures [1].

The fishtail [5] in the magnetization curves has been observed in the mixed state of high-T_c oxide superconductors [6, 7]. In YBa$_2$Cu$_3$O$_{7-\delta}$, the magnetization takes a broad maximum in magnitude near the irreversible magnetic field after a pronounced increase of the diamagnetic magnetization and a subsequent minimum due to magnetic flux penetration. This behavior becomes remarkable at high temperatures.

In this paper, we measure the magnetization hysteresis cycles as a function of temperature and the angle, θ, between the applied magnetic field and c-axis. Then, we report a comparative investigation of the flux jumps, the second peak and the irreversible magnetic field.

---

[*] Corresponding author: e-mail: ataoufik@hotmail.com, Phone: +212 48 22 09 57, Fax: +212 48 22 01 00

2 Experimental

YBa$_2$Cu$_3$O$_{7-\delta}$ single crystals structure was characterized using X-ray diffraction. The X-ray analysis showed that the crystal had an orthorhombic structure with space group Pmmm and the lattice parameters of $a = 2$ mm, $b = 1.7$ mm and $c = 0.8$ mm. The superconducting transition temperature is about 91 K and a transition width of 1 K (DC-susceptibility in $H = 10^{-3}$ T). Transmission electron microscopic (T.E.M.) revealed not only the presence of the usual twin boundaries (as the major visible defect) but also dislocations, stacking faults and small angle (1–2°) sub-grain boundaries. In addition, the sample certainly contains also point defects, in particular oxygen vacancies, not visible in the experimental conditions.

The magnetic measurements were made by means of vibration sample magnetometer (V.S.M.) in external magnetic fields up to 6 T. The orientations of the sample on the desired directions were performed at room temperature using a home made rotatable sample holder with a relative error in θ of the order 0.2 ° [8].

3 Results and discussion

Figure 1 shows the magnetization hysteresis cycles at temperature $T = 4$ K for $\theta = 80°$. As it can be seen, at this temperature, the cycle present several flux jumps "saw tooth" [9]. This phenomenon appears for deferent directions at low temperature (T < 10 K). At this temperature the applied magnetic field is not sufficient to saturate the sample. In this case, the field is not penetrated until the sample center, i.e., that the maximum applied magnetic field (here $H = 6$ T) is lower than the complete penetration field H_p of the crystal. H_p is equal to the applied magnetic field at which the first flux lines just arrive at the sample center. The jumps are due to the sample internal heating. The heating depends both on the sweep rate dH/dt of the applied magnetic field and on its volume, when the sweep rate is low the internal heating decreases. In fact, we always observed this kind of cycle for large single crystals of YBa$_2$Cu$_3$O$_{7-\delta}$ at very low temperature but we never see it for fine grain sample with $R < 10$ µm [1]. The phenomenon disappears with higher temperatures.

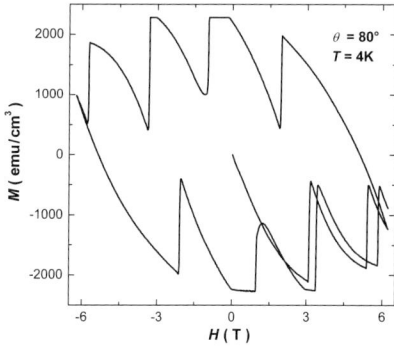

Fig. 1 The M versus H of the YBa$_2$Cu$_3$O$_{7-\delta}$ single crystal at T = 4 K for $\theta = 80°$, this kind of "saw tooth" cycle is very common for large single crystals at low-T.

We present in Fig. 2 the magnetization hysteresis cycles for three directions of the applied magnetic field with the c-axis ($\theta = 0°$, 60 ° and 90°) at temperature $T = 70$ K and 80 K. The magnetization curve exhibits a remarkable second peak "fishtail" and a large hysteresis cycle indicating a strong flux-pinning line. The fishtail decrease with increasing angle and almost disappear for the applied magnetic field perpendicular to the c-axis, this peak appears when the applied magnetic field is not parallel to the superconducting CuO_2 layers. Several interpretations have been advanced to explain the fishtail, certain tell that it is probably due to the interaction between the vortex and structure defect system [10], other have shown that gaps of oxygen can be responsible [11].

At $T = 80$ K, the irreversible magnetization M_{irr} disappears at $H \approx 2.7$ T for $\theta = 0°$ and at $H \approx 3.5$ T for $\theta = 60°$ (Figs. 2(a) and 2(b)), $M_{irr} = (M^+ - M^-)/2$ where M^+ and M^- are magnetization at decreasing and increasing applied magnetic fields, respectively. The hysteresis cycle branches M^+ and M^- meet for the irreversible magnetic field H_{irr}, which depends on the temperature and the angle.

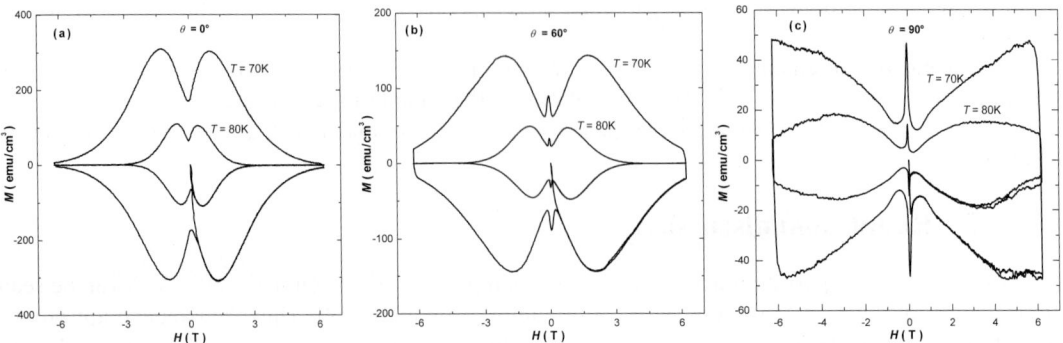

Fig. 2 The magnetization hysteresis cycles for three directions of the applied magnetic field with the c-axis ($\theta = 0°$, 60° and 90°) at temperature $T = 70$ K and 80 K.

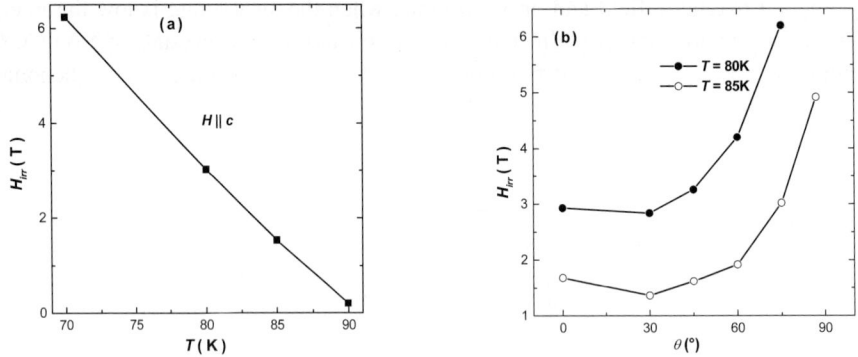

Fig. 3 (a) The variation of H_{irr} with T for applied magnetic field parallel to the c-axis, the dots is experimental data and lines are fitting results with Eq. (1). (b) The field H_{irr} versus θ at $T = 80$ K (solid circles) and $T = 85$ K (open circles).

The Fig. 3(a) illustrate $H_{irr}-T$ for the applied magnetic field parallel to the c-axis. The continuous line is the fit with the power law:

$$H_{irr} = H_0 \left(1 - \frac{T}{T_c}\right)^\alpha, \qquad (1)$$

where $H_0 = 32 \pm 2$ T, $T_c = 91$ K and $\alpha = 1.12 \pm 0.05$. As it can be seen, this law give a good fit for both high and low field studied here, like in [12]. This proves that the nature of the flux dynamics has one regime in the whole region. It is the irreversible limit of the Abrikosov flux dynamics. In the others granular superconductors, several different regimes are clearly apparent [13].

© 2004 WILEY-VCH Verlag GmbH & Co. KGaA, Weinheim

The irreversible magnetic field H_{irr} versus the angle θ for $T = 80$ K (solid circles) and $T = 85$ K (open circles) is plotted in Fig. 3(b), beyond $\theta = 30°$, H_{irr} increases when θ increases. N. Kobayashi et al. [14] and K. Watanabe et al. [15] described H_{irr} measured by the effective-mass model. Since H_{irr} is usually defined by a certain voltage criterion, it seems that the H_{irr} is strongly influenced by the flux pinning and the thermally assisted flux motion.

In conclusion, the temperature dependence of the irreversible magnetic field H_{irr} which exhibits a power law with power larger than unity. The angular dependence of H_{irr} is described by the effective-mass model. This result means that the anisotropy in H_{irr} is dominated by the anisotropic vortex structure intrinsic to the material and the second peak arises from the correlation between flux lines.

References

[1] S. Senoussi, S. Mammond, and M. F. Mosbah, Studies of HTs, edited by A. Narlikar, Vol. 14 (Nova Sciences, Commack, NY, 1994), p. 107.
[2] M. Däumling, E. Walker, and R. Flükiger, Phys. Rev. B **50**, 13024 (1994).
[3] M. Däumling and W. Goldacker, Z. Phys. B **102**, 331 (1997).
[4] S. B. Roy, A. K. Pradhan, and P. Chaddah, Physica C **250**, 191 (1995).
[5] C. A. Cardoso, M. A. Avila, R. A. Ribeiro, and O. F. de Lima, Physica C **341–348**, 1291 (2000).
[6] G. Yang, P. Shang, S. D. Sutton, I. P. Jones, J. S. Abell, and C. E. Gough, Phys. Rev. B **48**, 4054 (1993).
[7] T. Tamegai, Y. Iye, I. Oguro, and K. Kishio, Physica C **213**, 33 (1993).
[8] S. Senoussi, F. Mosbah, O. Sarrhini, and S. Hammond, Physica C **211**, 288 (1993).
[9] V. V. Chabanenko, V. F. Rusakov, A. I. D'yachenko, S. Piechota, A. Nabialek, and H. Szymezak, Physica C **341–348**, 2031 (2000).
[10] M. Werner, F. M. Sauerzopf, H. W. Weber, B. D. Veal, F. Licci, K. Winzer, and M. R. Koblischka, Physica C **235–240**, 2833 (1994).
[11] A. Fert, J. P. Redoulès, Ph. Odier, and N. Pellerin, Physica C **235–240** 2809 (1994).
[12] Y. Yeshurum and A. P. Malozemoff, Phys. Rev. Lett. **47**, 220 (1988).
[13] X. J. Fan, X. F. Sun, J. Zhang, X. Zhao, and X.-G. Li, Physica C **341–348**, 1161 (2000).
[14] K. Kobayashi, K. Hirano, T. Nishizaki, H. Iwasaki, T. Sasaki, S. Awaji, K. Watanaba, H. Asaoka, and H. Takei, Physica C **251**, 255 (1995).
[15] K. watanabe, S. Awaji, N. Kobayashi, H. Yamane, T. Hirai, and Y. Muto, J. Appl. Phys. **69**, 1543 (1991).

Corrosion process of MgB$_2$ superconductor in a moisture atmosphere

M. Annabi, A. M'Chirgui[*], F. Ben Azzouz, M. Zouaoui, and M. Ben Salem

Material Physics Laboratory, Faculty of Sciences, Bizerte, Tunisia

Received 31 August 2003, accepted 31 December 2003
Published online 8 April 2004

PACS 74.70.Ad, 81.65.Rv

The degradation behavior of polycrystalline magnesium diboride (MgB$_2$) superconductor in a highly humid air has been investigated. The MgB$_2$ samples with grain size around 10 µm were exposed to the action of dry carbon dioxide, to water-vapour-saturated air and to water vapour in a closed system. It has been shown that prolonged exposure to highly humid air completely destroys superconductivity. Infrared transmission spectroscopy analyses indicate that water simply acts as a catalyst for the formation of magnesium carbonate (MgCO$_3$) and diboride oxide (B$_2$O$_3$) species, and the rate of degradation strongly depends on the water partial pressure in air. Microstructure analyses reveal that the degradation process was observed to initiate more likely at the surface and also at the grain boundaries of small MgB$_2$ grains. Resistivity versus temperature measurements indicates that on exposure to humid air, the residual resistance ratio decreases and the material becomes insulator with prolonged exposure time.

© 2004 WILEY-VCH Verlag GmbH & Co. KGaA, Weinheim

1 Introduction The recently discovered copperless superconductor MgB$_2$ has generated extensive scientific interest for basic and applied research. The intrinsic characteristics of MgB$_2$ make up this material a very promising candidate for technological uses. Polycrystalline MgB$_2$ samples appear to be electrically isotropic [1] in spite of their layer-like crystalline structure with a negligible small effect of their grain boundary on the super current. Indeed since the coherence length ξ = 50 Å, is larger than the junction length between the grains, weak-link phenomena seem to be less operative in MgB$_2$, compared to high-T$_c$ cuprates [2]. However, it was commonly reported that samples with poorly connected and small size grains (0.1–5 µm) were particularly highly affected when exposed to ambient environment [3]. Most studies of MgB$_2$ degradation are focused on atmospheric air with initial H$_2$O wetting [4, 5]. In this contribution, an investigation of the degradation behavior of polycrystalline magnesium diboride superconductor with grain size around 10 µm in a humid air with different water partial pressure is presented. The samples were exposed to the action of dry carbon dioxide, to water-vapor-saturated air or just to water vapor in a closed system.

2 Experiment process Polycrystalline MgB$_2$ superconductors were synthesized by solid-state route. Details of preparation have been published previously [6]. The resulting MgB$_2$ pellets displayed typical parabolic resistive variations in the normal state and underwent a sharp resistive transition to a superconducting state with zero resistance at 37 K. All samples were nearly single hexagonal MgB$_2$ with an insignificant amount of secondary phases. The size of MgB$_2$ grains is ranging from 5 to 30 µm. MgB$_2$ bar-shaped pellets with typical dimensions of 5 x 2 x 1 mm^3 were exposed to the action of carbon dioxide, water vapor saturated air and just to water vapor in a closed system for different periods. One set of pellets was exposed to an ambient atmosphere during almost one year and other samples were exposed to air containing water vapour at a partial pressure of 4 and 40 kPa for different periods. The superconduct-

---

[*] Corresponding author: e-mail: Ali.mchirgui@fsb.rnu.tn

ing and chemical changes of samples were investigated. The alteration products were studied by infrared transmission spectroscopy in the 1200–4000 cm^{-1} domain with 2 cm^{-2} resolutions, at ambient temperature. The sample morphology was investigated using scanning electron microscopy equipped with a dispersive energy system of emitted X rays.

3 Results The resistivity-versus-temperature plots recorded for the fresh MgB_2 sample and after exposing to air containing water vapor at a partial pressure of 40 kPa for various periods of time are shown in Fig. 1. For clarity, the data in Fig. 1 are normalized to 40 K values. The fresh sample shows onset of superconducting transition temperature (T_c^{onset}) at ~ 38.2 K with a sharp transition width ($\Delta T \sim 1$ K) to superconducting state. The room temperature resistivity was found to increase with increasing humid air exposure time. For the sample exposed for 24 hrs, although the T_c^{onset} remained at 38.2 K, the transition to superconducting state developed a tail and zero resistance is observed only at ~ 33 K. The transition width was seen to increase with humid air exposure time while the T_c^{onset} remained constant at 38.2 K. For the sample exposed for 96 hrs in humid air with 40 kPa partial pressure, the zero resistance state could not be observed till 19 K. The sample become insulating for 40 kPa humid air exposure time greater than 120 hrs.

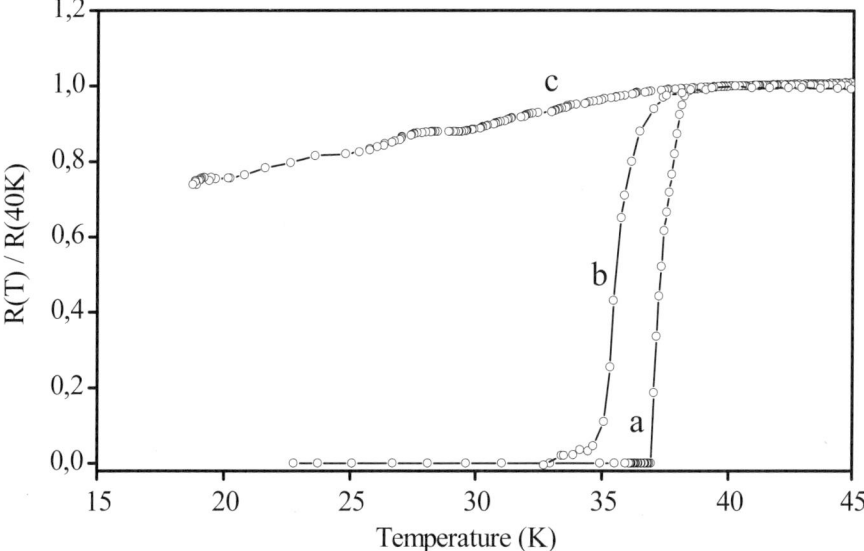

Fig. 1 Temperature dependence of resistivity recorded on polycrystalline MgB_2 samples: (a) as prepared, (b) after exposure to water vapour at a partial pressure of 40 kPa for 24 hrs, (c) as (b) but for 96 hrs.

Figure 2 shows IR transmission spectra in the range 1200–4000 cm^{-1} of freshly prepared polycrystalline MgB_2 pellets after exposure to ambient atmosphere and after exposure to air containing water vapour at a partial pressure of 4 and 40 kPa. The spectrum of freshly prepared MgB_2 (Fig. 2a) shows small shoulder in the range 2300–2400 cm^{-1}. These features are due to fluctuation in the atmospheric CO_2 concentration inside the spectrometer. For the sample exposed during almost one year (Fig. 2b) in ambient atmosphere, the IR transmission spectrum does not present any additional bands. This shows that the rate of degradation of MgB_2 pellets exposed to air in a temperature climate of moderate humidity is rather slow. Figure 2c shows IR transmission of polycrystalline MgB_2 pellets exposed for 744 hrs to air containing water vapour at a partial pressure of 4 kPa. Two shoulders around 2460 and 1470 cm^{-1} were observed and attributed to alteration species. In the case of samples exposed to air containing water vapour at a partial pressure of 40 kPa, the previous bands, associated with alteration products are seen to appear earlier. Indeed, after exposing MgB_2 pellets for 24 hrs (Fig. 2d), a small band appear at 1250 cm^{-1}

along with two weak bands centred at 2460 and 1470 cm^{-1}. The intensity of the bands associated with alteration products increases with exposure duration. After 288 hrs (Fig. 2e) a new band appears at 3692 cm^{-1}. The broad band centred at 1460 cm^{-1} is likely consisting of various bands, not clearly determined. This band broadening may result from the combination of various vibration modes corresponding to several ions in that region. On the other hand, the spectrum of polycrystalline MgB$_2$ pellet immersed in distilled water during 5 hrs (Fig. 2f), clearly shows a band at 3692 cm^{-1} along with two separated bands at 1422 and 1480 cm^{-1}. The two previous bands, observed at 2460 and 1250 cm^{-1} in the IR transmission of MgB$_2$ sample exposed to humid air are still observable herein, but their intensities are significantly reduced. This is indicating that alteration products related to these later bands are soluble in water.

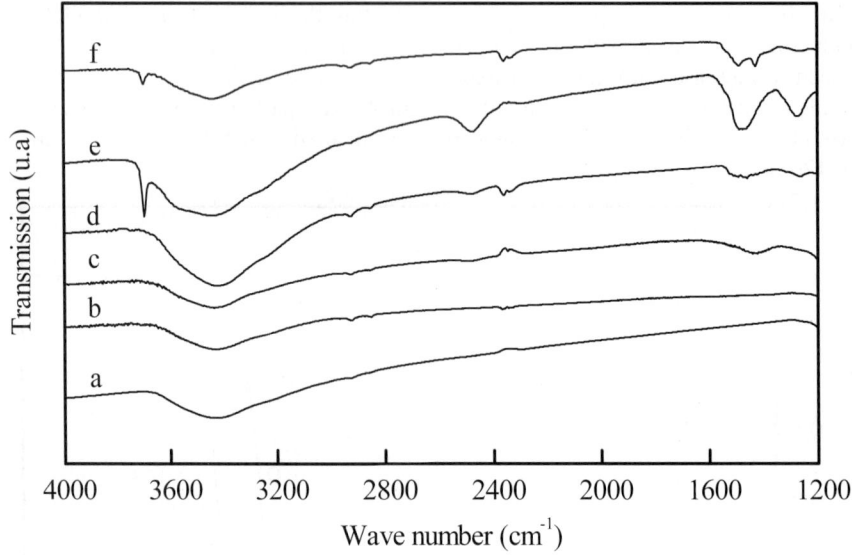

Fig. 2 IR transmission spectra of MgB$_2$ pellets: (a) as prepared; (b) after exposure to ambient environment during one year; (c) after exposure to water vapour at a partial pressure of 4 kPa for 744 hrs; (d) after exposure to water vapour at a partial pressure of 40 kPa for 24 hrs; (e) as (d) but after 288 hrs; (f) after immersion in water for 5 hrs.

All the IR transmission spectra recorded for the degraded MgB$_2$ samples are compared with those of ground Mg(OH)$_2$ MgCO$_3$ and B$_2$O$_3$ powders over the same frequency range. The sharp band at 3692 cm^{-1} is associated with the stretching vibration modes of free OH ion in Mg(OH)$_2$. The bands appearing at 1422 and 1482 cm^{-1} in Fig. 2f are associated with carbonate ions. In contrast, in the spectra for MgB$_2$ exposed to water vapour saturated air, these bands are not clearly separated and a large band is observed. This broad band corresponds to the complex vibrations involving the simultaneous participation of MgCO$_3$ and B$_2$O$_3$, with however a predominant influence of the B$_2$O$_3$. Indeed B$_2$O$_3$ possesses a large band in the same spectral region. Moreover, we should notice that the solubility of magnesium carbonate and hydroxide (S = 0.00125 mol l^{-1} and 0.00014 mol l^{-1}, respectively) are smaller than the solubility of B$_2$O$_3$ (S = 0.15800 mol l^{-1}) in water and that the immersion of MgB$_2$ samples in water leads to the formation of Mg(OH)$_2$ and MgCO$_3$. Consequently, one can conclude that the strong bands observed at 1250 and 2460 cm^{-1} in the IR transmission spectra recorded for pellets exposed to water vapour saturated air could be assigned to B$_2$O$_3$ modes. MgB$_2$ samples decompose when immersed in water or exposed to water vapour saturated air and the rate of degradation strongly depends on partial pressure of water vapour. The effects of pure water and dry carbon dioxide on MgB$_2$ sample were also investigated. This was performed in order to isolate the effect of water vapour and carbon dioxide present in air.

© 2004 WILEY-VCH Verlag GmbH & Co. KGaA, Weinheim

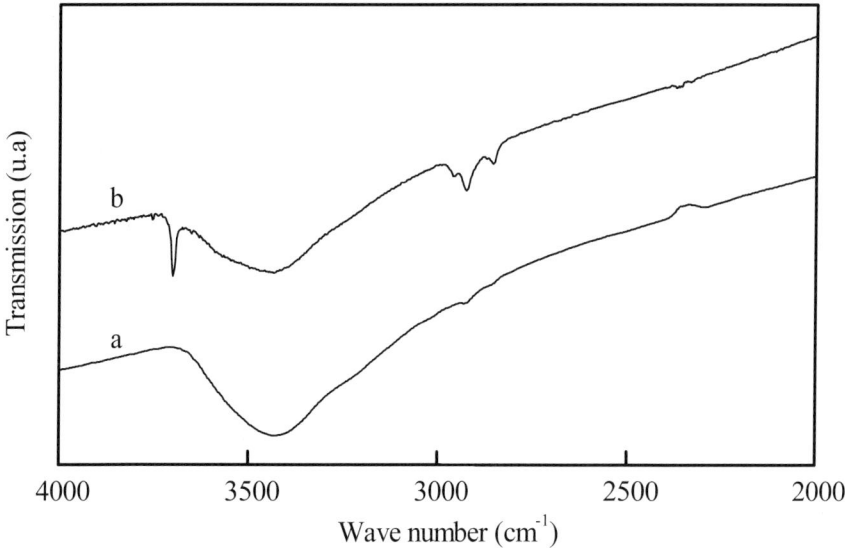

Fig. 3 IR transmission spectra of MgB_2 pellets: (a) as prepared, (b) after exposure to water vapour for 24 hrs.

Figure 3 illustrates the IR transmission spectrum measured for an MgB_2 pellet that was exposed to pure water vapour during 24 hrs. The spectrum (Fig. 3b) shows clearly a sharp band at 3692 cm^{-1} that is attributed to magnesium hydroxide product. Alternatively, during dioxide carbon exposure of MgB_2, no appreciable alteration product is apparent on the IR spectrum even over several hours. The MgB_2 is not sensitive to the action of pure carbon dioxide. Therefore, we conclude that the reaction of the MgB_2 pellets with atmospheric CO_2 is catalyzed by the presence of water vapour.

During degradation, Mg^{2+} ion sites must attract first water molecules from surrounding environment in order to produce magnesium hydroxide and boron. The metal hydroxide and boron formed then react with CO_2 and O_2 molecules respectively to produce magnesium carbonate $MgCO_3$, and B_2O_3. At later stages, more and more water molecules will interact with Mg^{2+} ion sites, so that sufficient amounts of $Mg(OH)_2$ will be produced.

4 Conclusion The degradation behaviour of polycrystalline MgB_2 superconductor has been investigated by analysing samples, before and after exposure to water vapour saturated air or dry carbon dioxide for various periods of time, using resistivity measurements and infrared transmission spectroscopy. The results show that on exposing to humid air, MgB_2 samples get decomposed into B_2O_3, Mg carbonate and hydroxide. The rate of degradation is found to strongly depend on partial pressure of water vapour.

References

[1] D.K. Finnemore, J.E. Ostensen, S.L. Bud'ko, G. Lapertot, and P.C. Canfield, Phys. Rev. Lett. **86**, 4656 (2001).
[2] D.C. Larbalestier, M.O. Rikel et al., Nature **410**, 186 (2001).
[3] A. Serquis, Y.T. Zhu, D. E. Peterson, F. M. Mueller, R. Schulze, V.F. Nesterenko, and S.S. Indrakanti, Appl. Phys. Lett. **80**, 4401 (2002).
[4] B.D.S. Aswal, K. P. Muthe, Ajaya Singh, Shahswati Sen, Kunjal Shah, L.C. Gupta, S.K. Gupta, and V.C. Sahni, Physica C **363**, 208 (2001).
[5] C.H. Cheng, Y. Zhao, Y. Feng, X.T. Zhu, N. Koshizuka, and M. Murakami, Supercond. Sci. Technol. **16**, 125 (2003).
[6] M. Zouaoui, A. M'chirgui, F. Ben Azzouz, B. Yangui, and M. Ben Salem, Physica C **382**, 217 (2002).

Effects of nano-Al$_2$O$_3$ particles on the superconductivity of Pb-doped BSCCO

M. Annabi, A. M'Chirgui[*], F. Ben Azzouz, M. Zouaoui, and M. Ben Salem

Laboratoire de Physique des Matériaux, Faculté des Sciences de Bizerte, Tunisia

Received 31 August 2003, accepted 31 December 2003
Published online 8 April 2004

PACS 74.25.Fy, 74.25.Qt, 74.72.Hs

Nanometer Al$_2$O$_3$ particles were added to (Bi,Pb)$_2$Sr$_2$Ca$_2$Cu$_3$O$_x$ (Bi-2223) precursor powders during the final sintering cycle of a multi_step preparation process. The influence of Al$_2$O$_3$ on the melting temperature, microstructure and transport property of ceramics was studied by means of DTA, XRD, SEM/EDX, electrical and magnetic measurements. We report that the addition of a small amount of Al$_2$O$_3$ (0.2 wt%) increased the pellet's critical current density J$_c$ at 77 K by ~35% and improved J$_c$ behaviour in applied magnetic field (parallel or perpendicular to the sample surface). The results indicate that the introduction of a proper amount of nano-Al$_2$O$_3$ particles during the final processing of BPSCCO samples can effectively improve the flux pinning ability and has a little detrimental effect on the (Bi,Pb)-2223 formation process.

© 2004 WILEY-VCH Verlag GmbH & Co. KGaA, Weinheim

1 Introduction

Among the currently known high-T$_c$ superconductors (HTSCs), (Bi,Pb)-2223 appears to be the most promising candidates for the application of power transmission cables at liquid-nitrogen temperature. However, in spite of their very high transition temperatures, their performance in magnetic fields has several drawbacks. Indeed, due to their low flux pinning energy, the critical current of high T$_c$ superconductors decreases fast with increasing temperature and applied magnetic field. In order to overcome this problem, the introduction of crystal defects that are on the order of the coherence length has shown clear advantages. It provides flux pinning centres and thus enhances the critical current density of polycrystalline samples. The route of artificially introducing nanometer-sized oxide particles as strong flux pinning centres in the processing of large bulk BSCCO or further PIT-textured Ag/BSCCO thick tapes is the most practical and promising technique for industrial-scale application. In this case, nanosize second-phase particles must be trapped within the superconducting grains and become therefore a second-phase defects. There are several report results of improved J$_c$ values through introducing nanometer particles such as SiC, MgO, ZrO$_2$, Al$_2$O$_3$ and SrZrO$_3$ [1–3] into high T$_c$ superconductors. Most of these studies reported that nanometer-scale inclusions were mainly observed to form aggregates between lamellae, act thus as barriers and hinder the grain growth. Few reports deal with nano-size particles associated with defects such as dislocations or stacking faults within the superconducting grains. This feature was mainly ascribed to the adopted processing technique, large doping amount or incomplete detachment of nano-size particles with temperature. It has been suggested that added particles did not serve in this case as pinning centres and the flux pinning enhancement has been attributed to the interfaces between particles and the superconductor matrix. Further investigations and analyses are required to give better insight on these points.

---

[*] Corresponding author: e-mail: Ali.mchirgui@fsb.rnu.tn

In this work, the effects of Al_2O_3 nanophase addition on microstructure and especially transport properties of (Bi,Pb)-2223 phase were investigated. In addition, the present work attempts to shed additional light on the influence of nanosize particles on the intergrain weak links in terms of intergranular critical current.

2 Experimental procedure

A bismuth-based superconducting powder with a nominal cation ratio $Bi_{1.6}Pb_{0.4}Sr_{1.9}Ca_{2.1}Cu_3O_{10-\delta}$ was elaborated by the solid-state synthesis route through a two-cycle annealing process. The experimental details for this multi-step procedure are described elsewhere [4]. Nanometer Al_2O_3 was added during the second thermal cycle to the precursor powder by mixing and hand grinding both powders. The diameter of the Al_2O_3 particles is around 50 nm. Several powder mixtures with Al_2O_3 weight concentration ranging from 0% to 1% were prepared. All powder mixtures were characterized by differential thermal analysis (DTA) to determine their melting temperatures. Since the DTA results showed that the gross melting temperature of the powders decreased slightly with varying Al_2O_3 concentration, preliminary experiments were carried out to determine the optimum sintering temperature for each sample. The temperatures tested ranged from 820 °C to 850 °C with 5 °C intervals. Each powder mixture is thus sintered during the second heat treatment cycle at the particular temperature for 36 h.

The critical current densities were measured by the standard DC four-probe technique using a criterion of 5 µV/cm. For the ceramics after the final sintering cycle, J_c dependence on applied magnetic field was measured. The magnetic field was applied in two directions: parallel and perpendicular to the ceramic's wide surface, but always perpendicular to the current-flow direction. X-ray powder diffraction measurements, microstructural observation and chemical analysis were performed on ceramics after final sintering by using a scanning electron microscope (SEM) equipped with a dispersive energy system of emitted X rays (EDAX).

3 Results and discussion

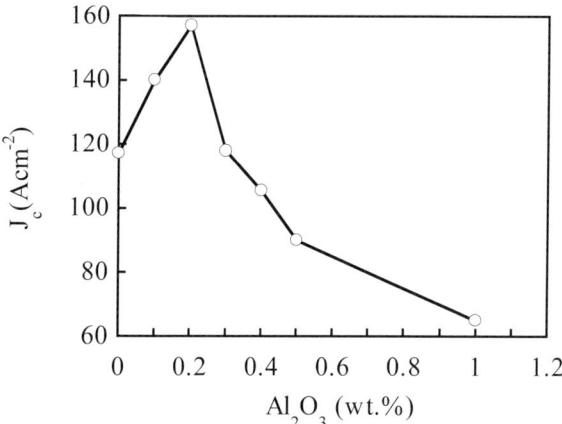

Fig. 1 Highest J_c at 77 K achieved among all sintering temperatures (820 °C to 840 °C) for (Bi,Pb)-2223 pellets added with various amounts of Al_2O_3.

Critical current densities were measured at 77 K and self magnetic field for all samples. Sintering was carried out at four temperatures: 825, 830, 835 and 840 °C, and was 72 h. The highest J_c achieved among all sintering temperatures for each sample was compared in Fig. 1. It is clear that the addition of a small amount of Al_2O_3 (0.2 wt%) enhanced the current-carrying capacity of (Bi,Pb)-2223 ceramics by about 35%. However, the addition of a larger amount of Al_2O_3 decreased the current-carrying capacity compared to the non-added sample. The larger the amount of Al_2O_3 added, the lower the J_c for the samples.

The negative effect of high Al_2O_3 content on the critical current density is probably due to weak links at grain boundaries. Indeed, excessive doping may induce severe agglomerations of Al_2O_3 particles at the boundaries of superconducting grains, which hinder the circulation of inter-granular supercurrents.

Figure 2 shows the dependence of critical current densities on applied magnetic field for undoped and doped samples at 77 K. The external magnetic field was applied either parallel to the sample wide surface or perpendicular to the sample wide surface. The measured J_c values in magnetic fields were normalized to their zero-field values $J_c(0)$. In this figure, one can see that for both H directions at 77 K, $J_c/J_c(0)$ of the ceramic with added 0.2 wt% Al_2O_3 is larger than that of the non-added sample, similar to the case of the ceramic with 0.5 wt% Al_2O_3, though the improvement is less pronounced. It is notable that the improvement of $J_c/J_c(0)$ becomes higher with the increasing doping amount, but comes quickly to a highest value at a relatively small amount of 0.2 wt%, and after the J_c decreases. It is reported that, in low-field regime, J_c is mainly controlled by the weak links between superconducting grains. Whereas, in high fields, J_c behaviour is mostly governed by flux pinning when H is perpendicular, and by both flux pinning and grain alignment, when H is parallel to sample surface [5, 6]. Furthermore, the flux-pinning effect is more pronounced for H perpendicular [6]. Compared to the non-added ceramic, the ceramic with added 0.2 wt% Al_2O_3 showed a slower J_c decrease with increasing magnetic field, particularly for H perpendicular. Moreover the normalized critical current densities of the ceramic with 0.2 wt% Al_2O_3 are four times larger than for the reference ceramic when H is applied parallel to the c direction. All the above results suggest that such increasing of J_c and $J_c/J_c(0)$ in applied magnetic field, is probably from the enhanced flux pinning due to the introduction of a proper amount of nano-Al_2O_3 particles.

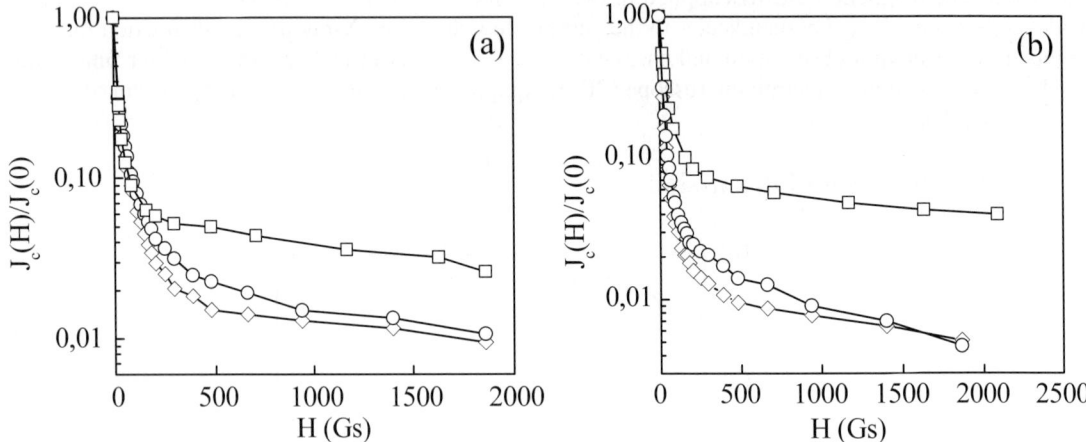

Fig. 2 Comparison of normalized J_c at 77 K dependence on magnetic field for (Bi,Pb)-2223 pellets added with 0.0 (◊), 0.2 (□) and 0.5 (o) wt% Al_2O_3. Magnetic field was applied in two directions: parallel and perpendicular to the sample wide surface, but always perpendicular to the current flow.

The microstructure of the variously added ceramics was checked by using SEM and EDX. Shown in Figs. 3a–c are the SEM micrographs corresponding to the ceramics which were added with 0, 0.2 and 0.5 wt% Al_2O_3, respectively. Each fractured surface of the processed pellets consisted primarily of large (Bi,Pb)-2223 plate-form grains. Al_2O_3 rich zones found in our samples may be roughly divided into two groups. (1) The first group consists of small areas situated between (Bi,Pb)-2223 grains. These areas are not homogenously distributed in the sample. The EDX analysis reveals agglomerates of alumina particles and no Al rich secondary phases were depicted. These areas are rarely observed in the sample with 0.2 wt% Al_2O_3 and are visibly more numerous in the sample with higher Al_2O_3 amount as shown in Fig. 3b. Agglomerated Al_2O_3 particles keep their original nanometer size which is around 50 nm. This result indicates that in highly doped samples a large portion of the added Al_2O_3 remained as discrete particles even after the thermal treatment. (2) The second group is represented in SEM micrographs as nanometer entities with almost regular forms; 50 nm, imbedded in the light-grey matrix of (Bi,Pb)-2223 plates as shown in Fig. 3c. The EDX analysis resolution in the SEM observations is about 1–2 μm, so it is not possible to get the correct composition of these entities under SEM. Nevertheless, EDX analyses on these (Bi,Pb)-2223 plates detect the presence of Al, whereas EDX analysis on similar plates which do not contain such entities does not detect the presence of Al. No such entities were observed in no-added samples. It is highly probable in this case that, during processing of the material, the small Al_2O_3 particles have been captured by the growing Bi-2223 grains forming finely dispersed inclusions in the superconductor matrix. Further study by TEM is needed to investigate these nanoscale entities and the possible associated crystal defects such as dislocations or stacking faults. In contrast to the doped samples, the interior of the Bi-2223 plates looks clean and without light-grey entities. These fine entities might serve as potential pinning centres and contribute to the improvement of J_c and J_c behaviour in magnetic field for the ceramic added with small amount of Al_2O_3.

In summary, we report herein that the introduction of a proper amount of nano-Al_2O_3 particles during the final processing of BPSCCO samples increased J_c by ~ 35% and improved J_c behaviour in applied magnetic field. This improvement is believed to be due to pinning effect of the fine Al_2O_3 particles.

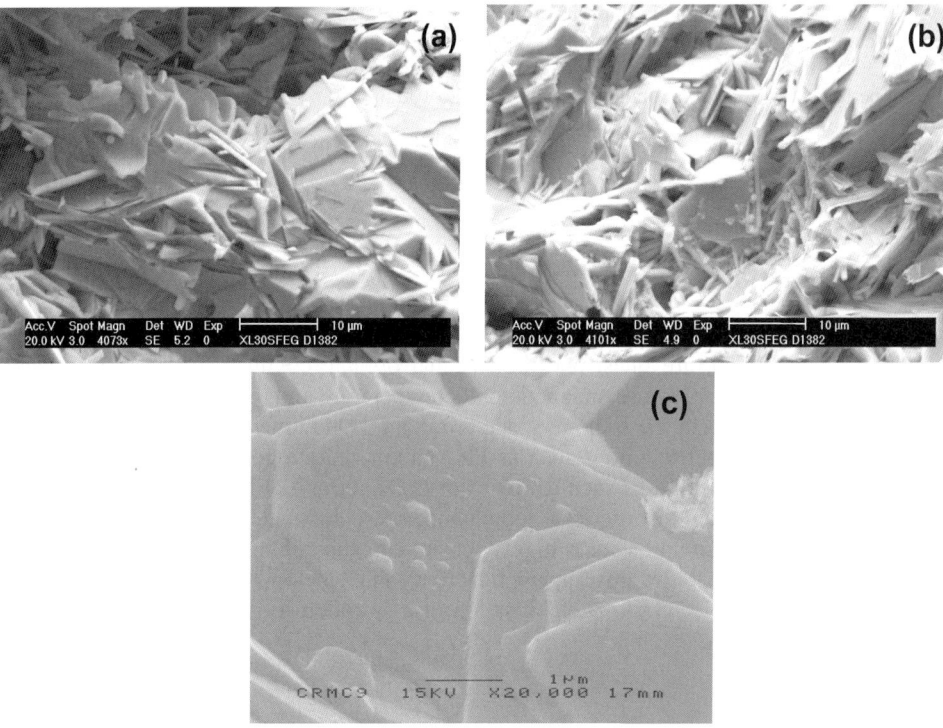

Fig. 3 SEM micrographs for (Bi,Pb)-2223 pellets added with various amounts of wt% Al_2O_3: (a) 0.0, (b) 0.5 and (c) 0.2.

References

[1] Y.C. Guo, Y. Tanaka, T. Kuroda, S.X. Dou, and Z.Q. Yang, Physica C **311**, 65 (1999).
[2] E. Guilmeau, B. Andrzejewski, and J.G. Noudem, Physica C **387**, 382 (2003).
[3] S. Sengupta, V.R. Todt, P. Kostic, Y.L. Chen, M.T. Lanagan, and K.C. Goretta, Physica C **264**, 34 (1966).
[4] F. Ben Azzouz, A. M'chirgui, B. Yangui, C. Boulesteix, and M. Ben Salem, Physica C **356**, 83 (2001).
[5] B. Hensel, J.-C. Grivel, A. Jeremie, A. Perin, A. Pollini, and R. Flukiger, Physica C **205**, 329 (1993).
[6] E. Mezzetti, R. Gerbaldo, G. Ghigo, L. Gozzelino, and L. Gherardi, Phys. Rev. B **59**, 3890 (1999).

Non-magnetic anion substitution in $(TMTSF)_2ClO_4$: consequences on superconductivity

N. Joo[*,1,2], **C. Pasquier**[1], **P. Auban Senzier**[1], **D. Jérome**[1], and **K. Bechgaard**[3]

[1] Laboratoire de Physique des Solides, Bât. 510, Université Paris-Sud, 91405 Orsay, France
[2] Faculté des Sciences de Tunis, LPMC, Campus Universitaire, 1060 Tunis, Tunisie
[3] Department of Solid State Physics, Risø National Laboratory, 4000 Roskilde, Denmark

Received 31 August 2003, accepted 31 December 2003
Published online 8 April 2004

PACS 74.25.Fy, 74.20.Rp, 74.70.Kn, 74.62.Dh

The influence of non magnetic impurities on the superconductivity of the quasi-one dimensional organic conductor $(TMTSF)_2ClO_4$ is investigated. Non magnetic impurities are added through the substitution of up to 3% of the ClO_4 anions by ReO_4 anions. Transverse resistivity and magnetoresistance measurements were performed in the superconducting state down to 100 mK.

© 2004 WILEY-VCH Verlag GmbH & Co. KGaA, Weinheim

1 Introduction

Charge transfer salts of the $(TMTSF)_2X$ family, where TMTSF is the donor molecule and $X = PF_6$, ClO_4, ReO_4... is the acceptor anion, are quasi one dimensional organic conductors. These Bechgaard salts exhibit a rich phase diagram as pressure, magnetic field, temperature and the anion are varied [1–3].

$(TMTSF)_2X$ salts crystallize in the triclinic space group P1⁻. One inversion centre is located on each anion site, so that centrosymmetric anions such as PF_6^- have a unique position in the structure. On the other hand noncentrosymmetric anions such as ReO_4^- or ClO_4^- exhibit at least two equivalent positions. In $(TMTSF)_2ClO_4$, the anions are disordered at room temperature and undergo a disorder-order transition at $T_{AO} = 24$ K. The resulting ground state is highly sensitive to the cooling rate at T_{AO1}: a slow cooling rate drives pure $(TMTSF)_2ClO_4$ towards superconductivity (SC) below 1.2 K (relaxed state). On the opposite, a large cooling rate leads to the stabilisation of a spin density wave (SDW) state below 5 K (quenched state).

2 Results

We report here an experimental study of eight $(TMTSF)_2(ClO_4)_{1-x}(ReO_4)_x$ alloys with a nominal value of x ranging from 0 to 3.1 %. However, because of the very low concentrations of ReO_4^- anions, the actual concentration could not be checked in each sample. Therefore, we used the residual resistivity measured in the relaxed state rather than the nominal value of x to estimate the rate of impurities. The results show a drastic influence of non magnetic impurities on superconductivity.

In order to avoid a mixing of two different effects (alloying and disorder driven by the cooling rate), we first cooled the alloyed samples very slowly at a rate of about –0.04 K/min at T_{AO}.

We observed for this range of concentrations, that alloying ReO_4 into $(TMTSF)_2ClO_4$ does not influence the ordering temperature of the anion lattice ($T_{AO} = 24$ K) (Fig. 1) in keeping with X ray measurements [4].

---

[*] Corresponding author: e-mail: joo@lps.u-psud.fr

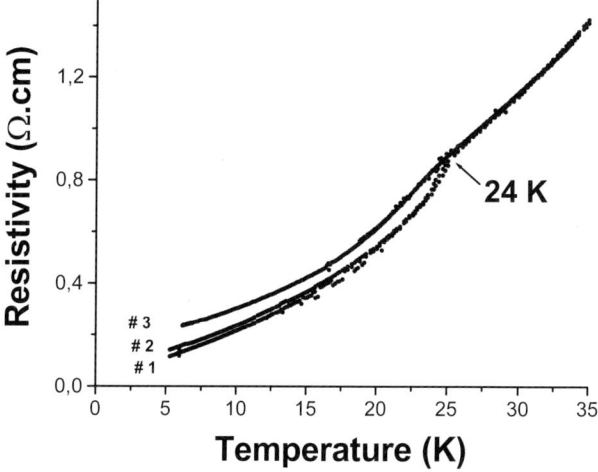

Fig. 1 Anion ordering transition of three different $(TMTSF)_2(ClO_4)_{1-x}(ReO_4)_x$ alloys.

At low temperatures, in the relaxed state, the increase of the residual resistivity (due to the increased concentration of impurities) is correlated with the decrease of the SC critical temperature, T_c (Fig. 2). For the highest concentrations of impurities, the critical temperature is only $T_c \approx 0.5$ K.

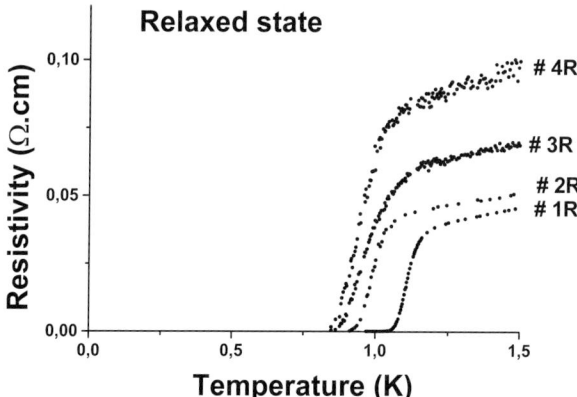

Fig. 2 Superconducting transition in the relaxed state for five different $(TMTSF)_2(ClO_4)_{1-x}(ReO_4)_x$ alloys.

Then we measured the samples with larger cooling rates (–4 K/min and –26 K/min), and observe an enhancement of the residual resistivity and a decrease of the critical temperature (Fig. 3).

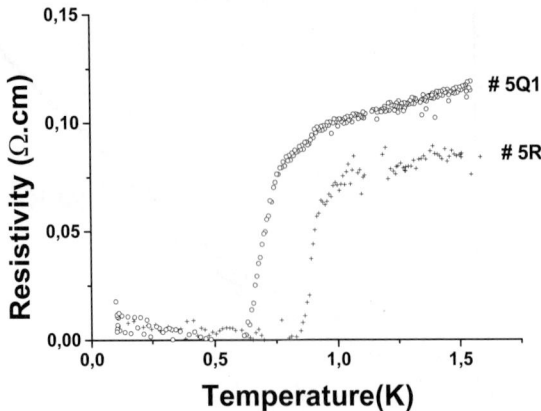

Fig. 3 Effect of the cooling rate on the superconducting transition of a $(TMTSF)_2(ClO_4)_{1-x}(ReO_4)_x$ alloy. Q1 means quenched state with a cooling rate at $T_{AO} = 24$ K of -4 K/min.

We noticed that a large cooling rate is equivalent to the addition of more impurities as it manifests as an enhancement of the residual resistivity, a decrease of the critical temperature and the upper critical field.

We believe that this is related to the anions ordering: in the neighborhood of ReO_4 anions, a rapid cooling keeps the ClO_4^- anions from being ordered and they remain in the high temperature disordered state, so little domains of ClO_4^- anions around ReO_4^- anions behave as if they were impurities. A quenched state is therefore more easily reached in presence of impurities.

We show in Fig. 4 the critical temperature versus residual resistivity for all the samples, in the relaxed state R = -0.04 K/min and some quenched states Q1 = -4 K/min and Q2 = -26 K/min.

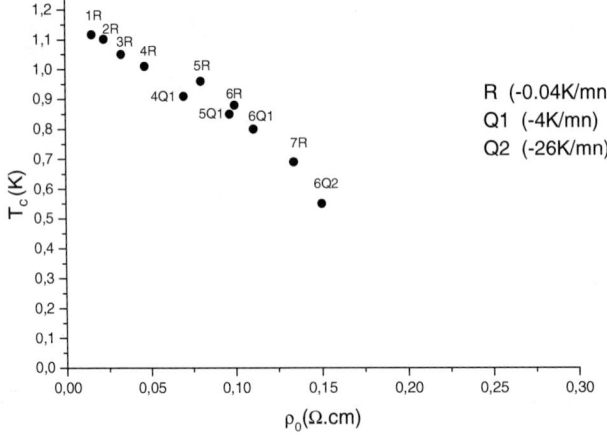

Fig. 4 Superconducting critical temperature versus residual resistivity of the different $(TMTSF)_2(ClO_4)_{1-x}(ReO_4)_x$ alloys.

Finally, we performed magnetoresistance measurements for all the alloys, first in the relaxed state then for large cooling rates. The evolution of the upper critical field, H_{c2}, with alloying and cooling rate is shown in Fig. 5 for some samples.

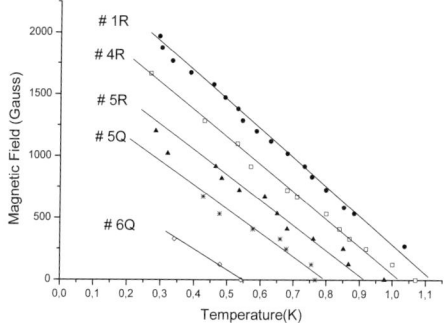

Fig. 5 Variation of the upper critical field with temperature of eight $(TMTSF)_2(ClO_4)_{1-x}(ReO_4)_x$ alloys and for different cooling rates. R = relaxed state, Q=Quenched state with a cooling rate of –4 K/min.

From Fig. 5, it is therefore evident that, at a fixed temperature, H_{c2} is decreasing with increasing the ReO_4^- concentration or the cooling rate, leading to a strong suppression of superconductivity by the introduction of impurities. We also notice that in a wide range of temperature, H_{c2} varies linearly with temperature and that independently of the conditions (alloying and various cooling rates) so that the slope $\frac{dH_{C2}}{dT}$ is nearly constant.

3 Discussion and conclusion

We have shown from resistivity and magnetoresistance measurements that non magnetic impurities have an important influence on the superconducting state of $(TMTSF)_2ClO_4$. We noticed that fast cooling behaves as adding more impurities by the enhancement of the residual resistivity and the drop of the critical temperature and the upper critical field. This is interesting insofar as a conventional superconductor that presents an isotropic gap is not affected by the presence of non magnetic impurities [5, 6]. The extreme sensitivity of the SC state to the presence of non magnetic impurities may now be discussed in the framework of theoretical models depending on whether this ground state is singlet [7] or triplet [8].

Higher impurities concentrations should lead to the disappearance of superconductivity (SC) and its replacement by a spin density wave state (SDW) [9] but it is not yet completely clear whether a quantum critical point or a SDW-SC phase coexistence [10] should appear.

References

[1] D. Jérome et al., J. Phys. Lett. **41**, L95 (1980).
[2] K. Bechgaard et al., Phys. Rev. Lett. **46**, 852 (1981).
[3] D. Jérome and H.J. Schulz, Adv. Phys. **31**, 299 (1982).
[4] V. Ilakovac et al., Phys. Rev. B **56**, 13878 (1997).
[5] B.T. Matthias, H. Suhl, and E. Corenzwit, Phys. Rev. Lett. **1**, 92 (1958).
[6] P. W. Anderson, J. Phys. Chem. Solids **11**, 26 (1959).
[7] A.I. Larkin and V.I. Melnikov, Sov. Phys. JETP **44**, 1159 (1976).
[8] A.A. Abrikosov, J. Low-Temp. Phys. **53**, 359 (1983).
[9] S. Tomic et al., Physica B **143**, 357 (1986).
[10] T. Vuletic et al., Eur. Phys. J. B **25**, 319–331 (2002).

Effects of gamma irradiation on the superconducting properties of $TlBa_2Ca_2Cu_3O_x$ tapes

A. El-Ali[*,1], **B. A. Albiss**[2], and **K. Khasawinah**[1]

[1] Physics Department, Yarmouk University, Irbid, Jordan
[2] Physics Department, Jordan University of Science and Technology, Irbid, Jordan

Received 31 August 2003, accepted 31 December 2003
Published online 1 April 2004

PACS 61.80.Ed, 74.72.Jt

Specimens of Tl-1223 superconducting tapes have been prepared using the powder-in-tube (PIT) method. Effects of gamma irradiation on the structure and transport properties of Tl-1223 tapes have been studied up to a gamma dose of 400 MR. A gradual decrease in the transition temperature and increase in the normal state resistance have been observed. A considerable decrease in the intensities of the major diffraction peaks of the XRD pattern was observed due to gamma irradiation. The I–V data were fitted to the power law $V \sim I^{\alpha(B,T)}$ in which the power exponent α was found to increase gradually with gamma dose. The dependence of α on the temperature was studied and the effect of gamma irradiation on the pinning energy was discussed and explained in terms of the defects introduced on the grain boundaries, weak links, and Josephson junctions.

© 2004 WILEY-VCH Verlag GmbH & Co. KGaA, Weinheim

1 Introduction

High temperature superconducting compounds (HTSCs) have a wide range of practical applications. They can be used in different environments including environments that contain radiation. It is quite important to know the effect of different types of radiation on HTSCs. Researchers studied the effect of electrons [1, 2], protons [3], neutrons [4, 5], heavy ions [6] and γ-irradiation [7–9] on these materials. γ-rays have a particular importance since they can penetrate through the protecting layers and shielding and may penetrate into the superconductor itself.

Polycrystalline Tl-2223 superconductor was investigated under the influence of both transport critical current and applied magnetic field [10]. The I–V characteristics of the sample obey a power law $V \propto I^{\alpha}$ and the H–V curves obey a power law $V \propto H^{\beta}$. The exponent α depends on both temperature and applied field while the exponent β depend on both temperature and applied current. The ratio α/β depends on the microstructure of the sample. The temperature dependence of the vortex flux pinning with γ-irradiation on polycrystalline Tl-2223 samples was studied [11]. γ-irradiated sample showed enhancement in the pinning at the same temperature by introducing defects that reduce the effect of weak links between grains which act as a new pinning agents to prevent motion of vortices.

A highly textured Tl-1223 /Ag superconducting tapes were prepared and studied [12]. Transport properties were found to be dominated by weak links phenomena indicating that Tl-1223 is a promising material for HTSCs applications. Tl-1223 poly crystalline samples were irradiated at room temperature [13], a decrease in T_c and increase in the normal state resistance with increasing gamma dose were observed. The critical current density increase considerably with the radiation dose. The magnetic and thermal properties of Tl-1223 were reported elsewhere [14].

---

[*] Corresponding author: e-mail: abedali@yu.edu.jo

The effect of γ-irradiation on $Tl_2Ba_2Ca_2Cu_3O_{10}$ tapes were studied [15]. The enhancement of the critical current density in the tapes with γ-irradiation was suggested to be due to generation of pinning sites and defects that can improve the grain boundaries and alignments. The effect of γ-irradiation on the silver-sheathed $Bi_{1.6}Pb_{0.4}Ba_2Ca_2Cu_3O_{10}$ tapes have also been investigated [16]. Results indicated increase in the normal state resistance up to a dose of 100 MR with a slight decrease in T_c with increasing γ-dose. The critical current density $J_c(B)$ increased with irradiation showing no effect of irradiation on it at relatively high fields.

HTSCs bulk or polycrystalline are always disordered and inhomogeneous in the regions of grain boundaries. These regions can be not only non-superconductive but insulators as well. In the HTSCs, regions of grain boundaries and nearby regions are strongly depleted with charge carriers and thus should be sensitive to γ-rays and other kind of radiations. For this reason the superconducting properties in HTSCs can be markedly affected by γ-rays at various ranges of radiation doses. For wires and tapes, there is an interest in Tl-1223 compound due to its promising transport properties in technologically useful magnetic fields at 77 K. In the present paper, we report on the effect of γ-irradiation on the structure and transport properties of the $Tl_1Ba_2Ca_2Cu_3O_y$ tapes based on the resistivity and I–V characteristics measurements. These tapes were prepared by the powder-in-tube method usually used to help solve the problems of flux pinning and grain boundaries seen in polycrystalline TBCCO 1223 compounds.

2 Experimental

The samples were prepared using Tl_2O_3, BaO and CuO as starting materials. The high purity powders were weighed in the appropriate amounts to form a nominal composition at $Tl_1Ba_2Ca_2Cu_3O_y$. The powders were thoroughly mixed in dry atmosphere using an agate mortar and pressed into disc-shaped pellets under a pressure of ~ 400 kg/cm^2. The pellets were wrapped into gold foil to prevent loss of thallium and then sintered at 860 °C for 5 h in oxygen flow. The pellets were reground, wrapped again with gold foil, and post-annealed at 900 °C for 2 h.

The tape is prepared using the powder-in-tube (PIT) method using a high TBCCO-1223 ($TlBa_2Ca_2Cu_3O_9$) ceramic powder packed into a silver tube and then transformed into a wire and then rolled into a tape as follows:

The ceramic material (in this case, TBCCO-1223) had to be ground into a fine powder to maximize the amount that will fit inside the silver tube. Also, the finer the powder the more it will flow during the drawing process, keeping the tape from having pockets with no powder, which would prevent the tape from being superconductive (although there would be stretches of tape that would be superconductive, provided that there are no gaps with no powder). A Controlled Atmosphere Glove Box filled with pure oxygen, was used to limit the TBCCO's exposure to contaminants. If the TBCCO is contaminated by carbon, the carbon will react with the oxygen in the TBCCO to become carbon dioxide during the annealing process. The gas would be trapped and would render the tape useless as a superconductor. Inside the box, the powder was ground using a mortar and the pestle. After grinding the powder for an hour and a half, it had turned into a very fine powder.

When the powder was ready, it had to be encased in a pre-manufactured silver tube. The tube was about 10cm long, inner diameter of 4 mm and outer diameter of 6mm with one end plugged and the other open. The powder was then compacted by inserting a silver cylinder whose diameter was approximately the same as the inside diameter of the tube. The powder had to be compacted slowly as the escaping air could spray the powder outside of the tube.

With the tube finished, it could now be drawn into a wire. To do this, one end has to be swaged to a diameter smaller than that of the draw die. The drawing and swaging steps were repeated until the wire reached a diameter of 2 mm. This step-down procedure insures that the powder inside is distributed throughout the wire.

Next, the wire was flattened into a tape, using a rolling mill. To prepare the wire, it was pinched on both ends and a small hole drilled into the flattened ends. Two thin wires, one tied to each end and then connected to different wheels, kept the tension in the wire constant, facilitating the rolling process. This process is repeated several times to get the suitable tape thickness and length (in our case the thickness was about 0.5 mm and the length is 15 cm).

The tape was now finished, but it was not yet superconductive. In order to become a superconductor, the tape had to undergo annealing, a heat treatment to align the grains of the powder inside the tape by melting the powder and letting it re-crystallize into the correct alignment. The tape was cut out into pieces of 5 cm lengths, which was the optimal size for the probe that would be used to obtain the critical current value later on. The length was also small enough to occupy a constant temperature zone inside the furnace. The tape was annealed at several different temperatures, with the goal to maximize its critical current value. In our case, the best annealing temperature is about 850 K.

Any change in sintering time and cooling rate may affect the transition temperature, electrical resistivity, critical current density, oxygen content and phase purity. To avoid such effects, the treatments for the sample were carried out under the same conditions. The structural characterizations of the samples were carried by X-ray diffraction with CuK$_\alpha$ radiation (λ = 1.542 Å). The lattice parameters of the Tl-1223 sample are calculated by the least square method using a computer program as follows: a = 3.8240 Å and c = 15.335 Å. Using a scanning electron microscope (SEM), samples showed well-shaped plate-like grains. The average grain size was estimated to be 1–2 µm in diameter.

The I–V characteristics and R–T curves were measured using the standard dc four-probe method with magnetic field B applied perpendicular to the surface of the tape. The I–V curves were measured over temperature range of 80–120 K in applied magnetic fields up to 0.5 T. The critical current densities were determined using an electric field criterion of 1 µV/cm. This is essential when taking into account the resistivity of Ag, which is a good conductor and any value of J_c below 10^3 A/cm^2 is insignificant. The current density used for electrical resistance measurement was 1 mA/cm^2. The temperature was measured with a calibrated platinum resistance sensor with an accuracy of ± 0.1 K. In all measurements, the sample was first zero-field cooled down to T = 80 K at which the measurements were to be done. Then, applying magnetic field up to 0.5 Tesla, reducing it to zero and repeating the measurement for higher temperatures very close to the critical temperature T_c= 120 K.

Each piece of the tape was divided into two parts of thickness ~0.5 mm and ~5cm in length. One of the samples was γ-irradiated at room temperature using a Co^{60} γ-source with a rate of 0.5 MR/h doses in a direction perpendicular to its surface.

3 Results and discussion

The dependence of the resistance on temperature R(T) for Tl-1223 tapes before and after γ -irradiation is presented in Fig. 1. Above the transition temperature (T_c = 120 K), the normal state resistance increases considerably up to a γ -dose of 100 MR. However, the effect at higher doses is hardly observed. It can be seen from Fig. 2 that T_c of the sample falls off quickly up to a dose of 100 MR with small linear change up to 400 MR.

X-ray diffraction patterns of the tapes with composition $Tl_1Ba_2Ca_2Cu_3O_y$ were indexed before and after irradiation. Figure 3 shows the dependence of the normalized intensities for some peaks on the radiation dose. It is noticed that the major peaks intensities decrease rather rapidly at lower doses but the positions of the peaks did not change. At higher doses the relative decrease was smaller. This could be explained due to considerable amount of lattice defects created by the radiation with no expansion of the lattice itself.

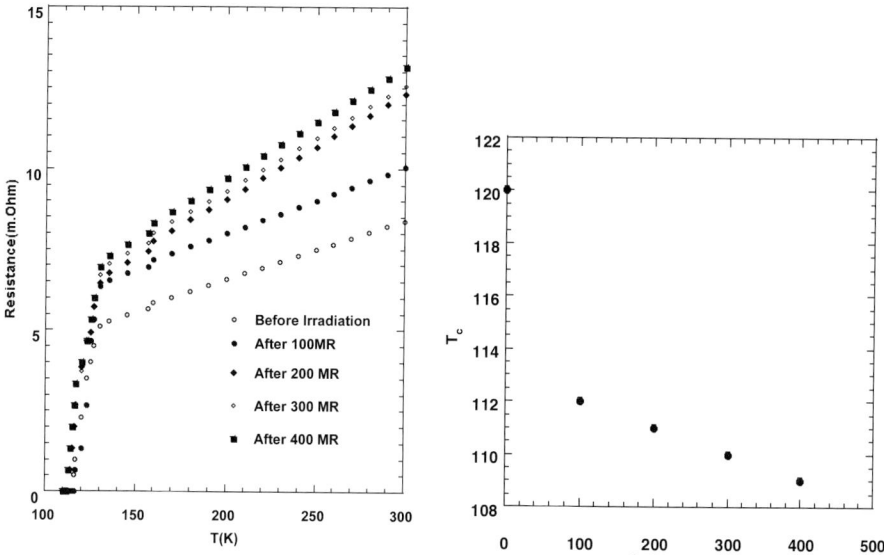

Fig. 1 Resistance versus temperature of $Tl_1Ba_2Ca_2Cu_3O_y$ before and after gamma irradiation.

Fig. 2 Variations of T_c with gamma doses.

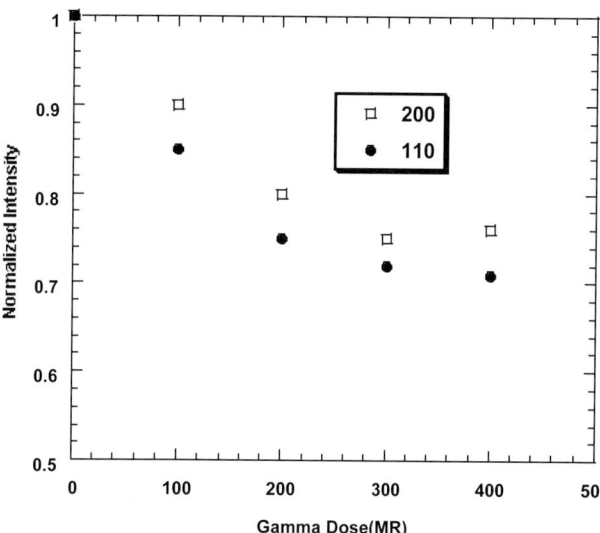

Fig. 3 Dependence of the normalized intensities of the diffraction peaks of Tl-1223 on the radiation dose.

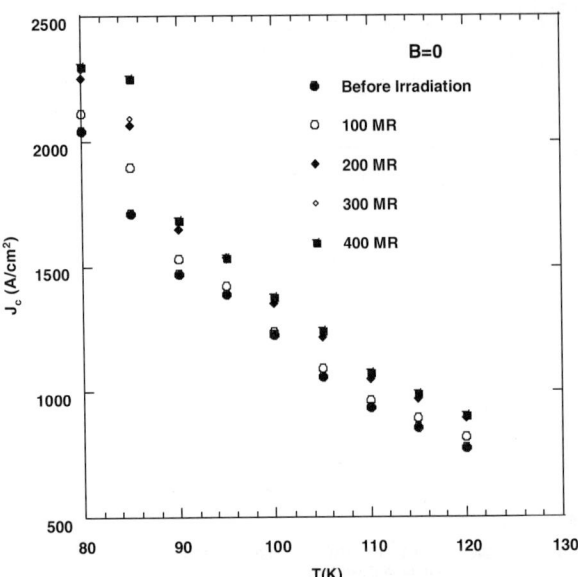

Fig. 4 Temperature dependence of the critical current density at zero applied magnetic field on various γ-doses.

Figure 4 shows the temperature dependence of the critical current density $J_c(T)$ at zero applied magnetic filed. At temperatures quite below T_c, the critical current density increases considerably with irradiation dose, while at temperatures very close to T_c, the change was relatively small. It was found to fit rather well with the relation $J_c(T) \sim (1-T/T_c)^n$ with $n \approx 2$ and $T_c = 120$ K. The observed value of n is higher than the Ginsburg–Landau [17] value for temperatures near T_c, usually observed in thin films and single crystals of HTSCs, but quite close to n = 2 as expected to rise due to different weak links between the grains in polycrystalline sample such as tapes. The superconducting grains may be separated by a non-superconducting material, which strongly determines the temperature dependence of J_c. The $J_c(T)$ dependence in our tape was found to follow the Ambegaokar–Baratoff relationship [12], in which grain boundary regions are treated as superconductor–insulator–superconductor (SIS) junctions or superconductor-normal-superconductor (SNS) junctions. This nearly square law (n ≈ 2) behaviour suggests that the transport mechanism is determined mainly by SNS junctions as reported for polycrystalline samples [13]. However, the temperature dependence of J_c was almost the same as for unirradiated samples. These results indicate that the gamma irradiation improves the grain boundaries and the weak links between the grains. It should be noted that the J_c value is lower in our tape samples compared to polycrystalline bulk samples. This is because in the tapes the Tl-1223 grains are well aligned and rolling helps to orient the grains by moving the anisotropic grains along their slip planes and this results in less pinning centres.

The magnetic field dependences of the critical current density before and after irradiation is shown in Fig. 5. The data is fitted rather well with the relation $J_c(B) \sim B^{-0.5}$ for fields higher than 0.15 T as reported Hasan et al. [15]. At relatively small values of magnetic fields, considerable increase in J_c was observed due to γ-irradiation but the increase is rather small at relatively higher fields. In regions of weak fields and high doses J_c behavior is rather anomalous. This could be connected to the structure and the presence of considerable amount of weak link junctions.

The *I–V* characteristics of our samples exhibit a power law $V \propto I^\alpha$. Figure 6 shows the temperature dependence of α (with B = 0) at various radiation doses. At temperatures quite below T_c a non-ohmic (α > 1) behaviour is seen for the tapes while at temperatures close to T_c, α is independent of the dose and the sample exhibits an ohmic behaviour. It should be noted that α, in fact, depends on magnetic field with similar behaviour as reported in Ref. [10].

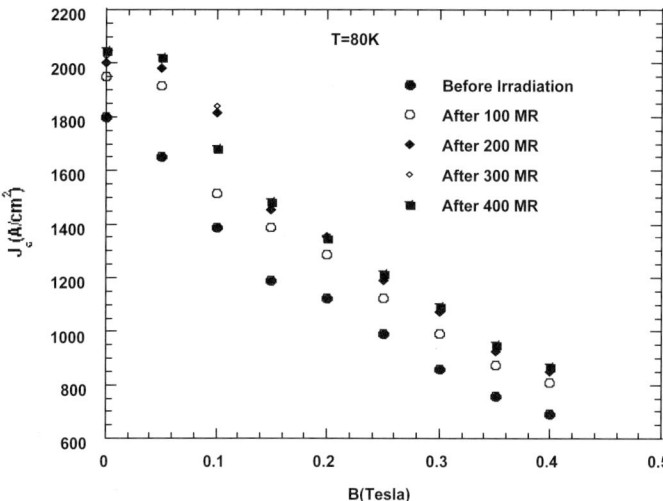

Fig. 5 The magnetic field dependence of J_c for Tl-1223 tapes on γ-irradiation dose at T = 80 K.

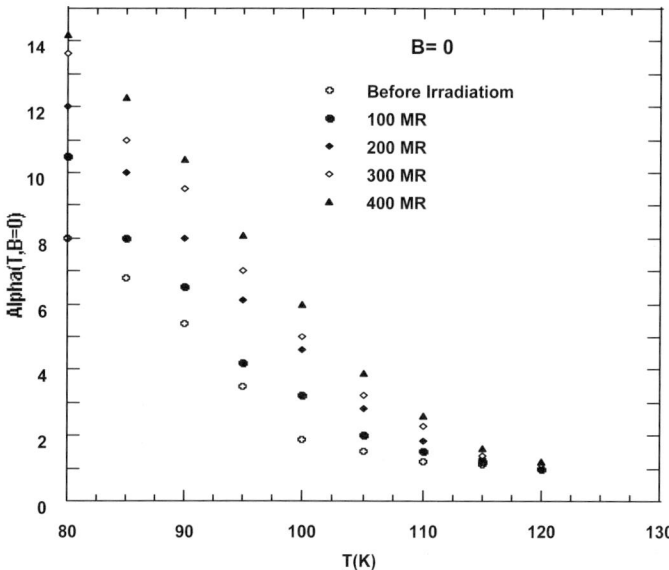

Fig. 6 The temperature dependence of α (with B = 0) at various radiation doses.

4 Conclusion

The observed decrease in T_c and increase in normal state resistance suggests that our tapes behave like polycrystalline granular systems. The broadening of the transition width and the increase in the normal state resistance provides information about the microstructure of the polycrystalline tape and shows that there is a considerable amount of weak link junctions in it. γ-irradiation generates some defects, improves the grain boundaries, and decreases the number of weak link junction between the grains and improves the critical current density $J_c(T, B)$ but no drastic crystal lattice expansion can take place. Samples exhibit a power law $V \alpha I^\alpha$ behaviour, and the exponent α depends on radiation dose. It turns out that introducing different types of structural defects to the polycrystalline HTSCs, especially in the form of tapes, is quite important for studying the pinning and dissipation mechanisms in these materials. The conductivity of granular metals in the normal state is determined by tunnelling of single particle excitation (unpaired electrons) through the boundaries. In superconductors, the coherence between the grains can be established by Josephson junctions. Much more independent (theoretical and experimental) detailed studies are needed to understand the mechanisms of producing defects, atomic displacements and the ionizing influence of gamma rays on HTSCs.

Acknowledgements This work was supported by Jordan University of Science and Technology grant no: 2000/28 and Yarmouk University grant no: 19/2000.

References

[1] M. Konczykowski, Physica A **168**, 291 (1990).
[2] J. Giapintzakis, W. C. Lee, J. P. Rice, D. M. Ginsberg, I. M. Robertson, M. A. Krik, and R. Wheeler, Phys. Rev. B **45**, 10677 (1992).
[3] B. M. Vlce, H. K. Viswanathan, M. C. Friscggerz, S. Flesher, K. Vandervoot, J. Downy, U. Welp, M. K. Akrik and G. W. Crabtree, Phys. Rev. B **48**, 4067 (1993).
[4] A. Umezawa, G. W. Crabtree, J. Z. Liu., H. W. Weber, W. K. Kwok, L. H. Nanez, T. J. Moran, C. H. Sowers, and H. Claus, Phys. Rev. B **36**, 7151 (1987).
[5] R. B. Van Dover, E. M. Gyorgy, L. F. Scheemeyer, J. W. Mitchell, K. V. Rao, R. Puzania, and J. V. Waszczak Nature **342**, 55 (1989).
[6] M. Xu, J. E. Ostenson, D. K. Finnermore, M. J. Kramer, J. D. Fendrich, U. Welp, G. W.Crabtree, B. Dabrowaski, and K. Zhang, Phys. Rev. B **53**, 5815 (1996).
[7] I. M.Obaidat, H. P. Goeckner, and J. S. Kouvel, Physica C **291**, 8 (1997).
[8] M. K. Hasan, J. Shobaki, I. A. Al-Omari, B. A. Albiss, M. A. Al-Akhras, K. A. Azez, A. K. El-Qisari, and J. S. Kouvel, Supercond. Sci. Technol. **22**, 1 (1999).
[9] B. A. Albiss et al., Physica C **331**, 297 (2000).
[10] A. El-Ali et al., Abhath Al-Yarmouk **11 2B**, 877 (2002)
[11] J. Shobaki et al., Supercond. Sci. Technol. **13**, 909 (2000).
[12] Ambegaokar and A. Baratoff, Phys. Rev. Lett. **10**, 1963 (1963).
[13] B. A. Albiss et al., Physica B **321**, 324 (2002).
[14] G. Triscone, A. Junod, and R. E. Gladyshevskii, Physica C **264**, 233 (1996).
[15] M. K. Hasan et al., Supercond. Sci. Technol. **12**, 1106 (1999).
[16] Albiss, A. Borhan, and El-Ali Abdul Raouf, IEEE Trans. Appl. Supercond. **12**, 1121 (2002).
[17] V. L. Ginsburg and L. D. Landau, Zh. Eksp. Teor. Fiz. **20,** 1064 (1950).

Identification of pinning centres in high T_c superconducting thin films by AC susceptibility

D.-G. Crété[*,1], **R. Bernard**[1], **J.-H. Pommereau**[1], **C. Gadois**[1], **S. Berger**[**,1], **J. Briatico**[1], **J.-P. Contour**[1], **O. Durand**[2], **J.-L. Maurice**[1], **J. Grollier**[1], and **K. Bouzehouane**[1]

[1] Unité Mixte de Physique CNRS / THALES, Domaine de Corbeville, 91404 Orsay, France
[2] THALES Research and Technology, Domaine de Corbeville, 91404 Orsay, France

Received 31 August 2003, accepted 31 December 2003
Published online 1 April 2004

PACS 74.25.Qt, 74.62.Bf, 74.78.Bz, 74.78.Fk

We compare vortex pinning properties of high temperature superconductors (HTS) thin films and correlate these to microstructure to identify pinning centres. We have grown $YBa_2Cu_3O_{7-\delta}$ (YBCO), $YBCO/PrBa_2Cu_{2.8}Ga_{0.2}O_{7-\delta}$ (YBCO/PBCGO), $NdBa_2Cu_3O_{7-\delta}$ (NBCO) thin films and superlattices on (100) $SrTiO_3$. We show that pinning by intersections of twin boundaries usually dominates except for strained superlattices [1] and for the NBCO thin films presenting only one [1, 1, 0] twin variant (the [1 $\bar{1}$ 0] being completely eliminated), thereby drastically reducing the density of twin boundary intersections (TBI). As expected, AC susceptibility measurements show low pinning properties of these samples. Using High Resolution AC susceptibility, we have discriminated several mechanisms contributing to vortex pinning in a given sample. Indeed, when we eliminated TBI pinning in NBCO thin films, several peaks were observed in $\chi(T)$ curve, each one presumably associated with a specific pinning mechanism.

© 2004 WILEY-VCH Verlag GmbH & Co. KGaA, Weinheim

1 Introduction Vortex pinning properties of superconducting materials are generally studied for large current applications such as energy transport/storage and high field magnets where strong pinning properties are required. However, most of electronic devices based on vortex dynamics, such as flux-flow DC transformer (Giaever transformer) [2, 3] or vortex-flow transistor (VFT) [4], involve very weak pinning properties. It is thus necessary to identify the dominant contributions to pinning in order to reduce them.

A large variety of defects can pin vortices: vacancies, foreign atoms in interstitial or substitutional position, dislocations, stacking faults, grain and twin boundaries, large inclusions, surface roughness... Oxygen vacancies in CuO_2 planes could depress the superconducting order parameter if they are clustered. However, this hypothesis is controversial because it is difficult to characterise the distribution of this kind of defect. Inclusions of non-superconducting parasitic phase play a similar role when their size is of the order of a few nanometres [5]. According to the literature [6], the surface roughness of the films is also at the origin of vortex pinning because of the thickness modulation. Larger inclusions can pin several vortices. Outgrowths, nucleated in the bulk of the films, can be large non-superconducting zones in the volume of the superconductor and are prone to pin vortices. Meanwhile, their pinning efficiency depends on the potential profile at the interface, which is not well defined.

Linear defects, because of the shape similitude with vortices, can have a strong pinning energy if their transversal extension is of the order of the coherence length ξ. Such are dislocations [7], appearing during island coalescence or during strain relaxation [8], and amorphous tracks created by heavy ion irradiation [9]. Grain boundaries (in cuprates thin films) and twin planes (in monocristals and films, in particu-

---

[*] Corresponding author: e-mail: denis.crete@thalesgroup.com, Phone: +33 1 69 33 91 16, Fax: +33 1 69 33 07 40
[**] Now at CEA, 31-33 rue de la Fédération, 75752 Paris Cedex 15, France.

lar in YBCO) are planar defects likely to pin the vortices very efficiently. Particular attention is given to the dependence of pinning with the relative orientation of twin planes [10]. Their thickness is a few angstroms, close to the value of ξ in these materials.

The studies meet difficulties in interpreting the origin of pinning mainly because of the great variety of the microstructure of the synthesised materials. There is no systematic correlation between sample morphology and experimental pinning data. We have identified several vortex-pinning sources in HTS thin films and heterostructures made by pulsed laser ablation. For each kind of sample, we present the main features of their microstructure. We then compare their pinning properties evaluated from AC susceptibility measurements assuming that not only the nature of the defects, but also their density determine the importance of their contribution to pinning. Finally, we identify the main centres involved in vortex pinning for these samples.

2 Experimental procedure The samples have been grown in a pulsed laser deposition machine that has been described in detail in ref. [11]. Scanning electron microscopy was used to evaluate the outgrowth density. X-ray diffraction (XRD) has been used to characterise the crystallographic orientation, the twinning of the films [12, 13] and to confirm the absence of any inclusion. Depending on the SrTiO$_3$ (STO) substrate orientation, the NBCO samples present preferential twinning along [110] [8], respectively with the following percentage P_{110}: LDM596, 73 % and LDM567, 100%.

AC susceptibility is measured by placing square thin films with typical area of 10×10 mm² between two coils (diameter 5 mm) and cooled in a helium-flow cryostat. A weak alternating current in the primary coil induces a magnetic field H_{ac} perpendicular to the thin film, with calibrated amplitude of 0.01–0.4 mT. A uniform static magnetic field H_{dc} can be superimposed with the same direction. The temperature is measured near the sample by a Si diode calibrated between 4–100 K and a 0.1 K relative accuracy.

3 Results and discussion We first present the effect of outgrowths found in YBCO thin films grown by pulsed laser ablation on STO substrates. For these samples, we use for comparisons the magnetic relaxation rate $S = -d(\ln M)/dt$ (where M is the magnetisation, t the time) which can be derived from AC susceptibility data by the procedure described in [1] and which describes how fast the vortices can relax to the equilibrium. In addition, it is related to the pinning energy U_p by $S = k_B T/U_p$, where k_B is Boltzmann's constant and T, the temperature. Table 1 reports on samples with different microstructures and $t_{0.1}$, the reduced temperature for which $S = 0.1$. We note a significant correlation of $t_{0.1}$ only with the outgrowth density, smaller with the thickness – not completely zero as it is correlated to outgrowth density – and not with the roughness of the samples. We claim that pinning by outgrowths has a major effect in samples with an outgrowth density larger than ~10^7 cm^{-2}, when no other defect has been created intentionally.

Table 1 Samples are YBaCuO thin films, characterised by their morphology and by AC susceptibility with a DC field H_{dc} of 10 mT and $H_{ac} = 0.5$ mT between 1 and 100 kHz. D_{out} is the outgrowth density, R_a is the roughness average, d the thickness and T_c is the critical temperature measured by resistive transition. The correlation of the reduced temperature $t_{0.1}$ is significant only with the outgrowth density.

Sample No.	D_{out} (×10^7 cm^{-2})	R_a (nm)	d (nm)	$t_{0.1} = T_{S=0.1}/T_c$
962	2.1	2.8	50	0.984
979	1.5	3.4	100	0.990
971	4.4	6	300	0.994
973	1.2	6.5	300	0.991
1057	3.8	6	500	0.992
443	0.6	5.2	100	0.972
1155	4.8	–	30	0.988

However, the correlation is not strong indicating that another pinning mechanism is important. We investigated on the effect of twinning by a study of the pinning properties of YBCO/PBCGO superlattices [1]: by introducing PBCGO layers, we can reduce strain relaxation, thereby reducing the amount of twinning and eventually eliminating it completely. Table 2 shows characterisation of the heterostructures, where Λ is the period of the superlattice, M (resp. N) is the number of unit cells in the YBCO (resp. PBCGO) layers, c is the c-axis lattice parameter. Their outgrowth density is in the range $1...5 \times 10^7 \, cm^{-2}$. It shows that samples with a large M/N ratio have the smaller lattice parameter, close to the value for the YBCO thin film. Samples with a large lattice parameter are tetragonally strained, consistent with the observation of no twinning. More extensive characterisation of the superlattices shows that there is no strain in AL1481 and that twinning is important for both of the [110] and [1 $\bar{1}$ 0] twin variants.

Table 2 Summary of the XRD analysis of a YBCO thin film and the superlattices $[(YBCO)_M/(PBCGO)_N]_{10}//STO$.

Sample	Λ (nm)	M	N	c (nm)	T_c (K)
AL1319	11.47	3	6	1.1749	60.5
AL1235	11.0	4	5	1.17135	68.9
AL1318	31.0	9	18	1.1740	77.7
AL1481	25.3	15	6	1.1714	89
AL959	Total = 100	(film)		1.1687	

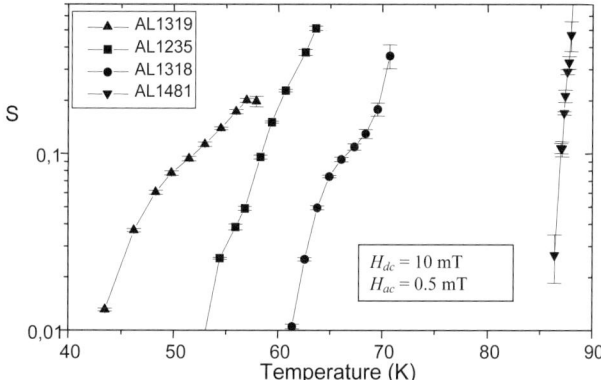

Fig. 1 Derived $S(T)$ curves from AC susceptibility measurements of the superlattices characterised in Table 2.

The $S(T)$ curve for the superlattices are visible Fig. 1. As expected, we observe an increase of the relaxation rate with temperature for all the samples, the pinning energy vanishing at T_c. Although there is a significant difference between the critical temperature of a superlattice and that of a superconducting thin film, this graph shows that the variation of $S(T)$ for the strained films (AL1319 and AL1318) are similar while it looks closer to a single YBCO film [1] for the relaxed films (AL1235 and AL1481), and not correlated to M. It indicates that the main contribution to vortex pinning comes from twin boundaries in spite of the outgrowth density. Note that although there is a large number of interfaces in the superlattices, pinning by surface roughness does not dominate the pinning properties for the relaxed films.

In addition, previous study [2] has shown that it is possible to obtain NBCO films with a single [110] twin variant as shown by the TEM picture Fig. 2a. Whereas sample LDM596 (with a twinning proportion P_{110} of 73%) exhibits one narrow peak in $\chi''(T)$ the dissipative component of the magnetic susceptibility, films with P_{110} of 100 % present a broad transition of ~0.1T_c (see LDM567, Fig. 2b). Thus, we claim that twin boundary intersections (TBI) have the most efficient pinning mechanism for samples with both twin variants, including the relaxed superlattices: TBIs (in sample LDM596) have a pinning energy large enough to hide the contribution of other pinning sources in AC susceptibility measurements and induce a single peak in $\chi''(T)$. As

TBIs are almost absent in NBCO thin films with only one twin variant (e.g. LDM567, Fig. 2a), the multiple peaks in $\chi''(T)$ originates from defects such as dislocations, oxygen deficiency, surface roughness, joined grains and outgrowth ... none of them strongly dominating the pinning behaviour over the full temperature range as for TBIs, but only in the vicinity of their associated peak.

Figure 2b also shows that up to 5 peaks can be observed with LDM567 when the applied magnetic field is very small, typically less than 0.05 mT, otherwise the fine features are washed out: this is the basis for High Resolution AC susceptometry [14], a very promising tool for characterisation of pinning centres.

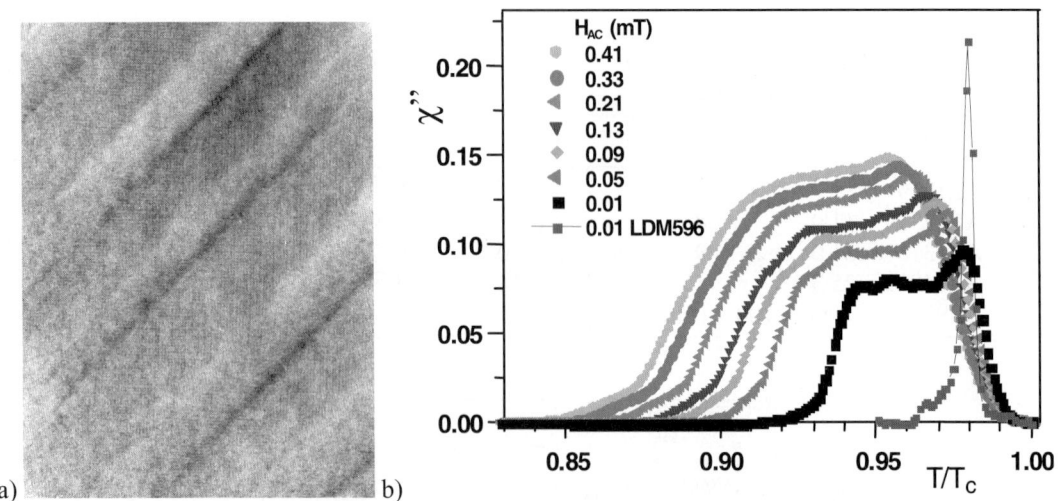

Fig. 2 Characterisation of very low outgrowth density NdBaCuO thin films LDM567 & LDM596 on $SrTiO_3$. a) 100 nm × 140 nm Transmission Electron Microscope image of LDM567: we observe only one direction for the twinning planes, i.e. no twin boundaries intersections. b) $\chi''(T/T_c)$ for different AC field amplitudes H_{AC} at 11.52 kHz: fine features (up to 5 peaks) can be observed with LDM567 for the lower values of H_{AC}.

4 Conclusion We have shown that twin boundary intersections and outgrowth density are the dominant pinning centres for most of the YBCO and NBCO thin films. We could prepare samples where these contributions are strongly reduced, allowing the observation of residual pinning by several other mechanisms which have not yet been identified with microstructural features.

Acknowledgements We thank E. Jacquet (UMP CNRS/THALES) for samples fabrication, K. van der Beek (Ecole Polytechnique, Palaiseau) and B. Mercey (CRISMAT, Caen) for helpful discussions.

References

[1] S. Berger, D.-G. Crété, J.-P. Contour, K. Bouzehouane, J.-L. Maurice, and O. Durand, Phys. Rev. B **63** 144506 (2001).
[2] D.-G. Crété, S. Berger, K. Bouzehouane, J. Briatico, J.-P. Contour, O. Durand, J.-L. Maurice, and K. van der Beek, Physica C **372–376**, 634 (2002).
[3] S. Berger, K. Bouzehouane, D.-G. Crété, and J.-P. Contour, Eur. Phys. J. Appl. Phys. **6**, 111 (1999).
[4] P. Bernstein, C. Picard, M. Pannetier, P. Lecoeur, J.-F. Hamet, T..D. Doan, J.-P. Contour, M. Drouet, and F.-X. Regi, J. Appl. Phys. **82**, 5030 (1997).
[5] P. J. Kung, M. P. Maley, M. E. McHenry, J. O. Willis, M. Murakami, and S. Tanaka, Phys. Rev. B **48**, 13922 (1993).
[6] D. Daldini, P. Martinoli, J. L. Olsen, and G. Berner, Phys. Rev. Lett. **32**, 218 (1974); P. Mathieu and Y. Simon, Europhys. Lett. **5**, 67 (1988).
[7] B. Dam, J. M. Huijbregtse, F. C. Klaassen, R. C. F. van der Geest, G. Doornbos, J. H. Rector, A. M. Testa, S. Freisem, J. C. Martinez, B. Stäuble-Pümpin, and R. Griessen, Nature **399**, 439 (1999).

[8] J.-L. Maurice, J. Briatico, D.-G. Crété, J.-P. Contour, and O. Durand, Phys. Rev. B **68**, 115429 (2003).
[9] S. Hébert, V. Hardy, G. Villard, M. Hervieu, Ch. Simon, and J. Provost, Phys. Rev. B **57**, 649 (1998).
[10] S. Sanfilippo, A. Sulpice, O. Laborde, D. Bourgault, Th. Fournier, and R. Tournier, Phys. Rev. B **58**, 1518 (1998).
[11] J.-P. Contour, C. Sant, D. Ravelosona, B. Fischer, and L. Patlagan, Jpn. J. Appl. Phys. **32**, L1134 (1993).
[12] J.-L. Maurice, O. Durand, K. Bouzehouane, and J.-P. Contour, Physica C **351**, 5 (2001).
[13] Y. Li and K. Tanabe, J. Appl. Phys. **83**, 7744 (1998).
[14] R. Bernard, J.-H. Pommereau, C. Gadois, S. Berger, D.-G. Crété, J. Briatico, J.-P. Contour, O. Durand, and J.-L. Maurice, ICM 2003 (Rome, 27th July-1st August 2003).

Magnetic properties of superconducting ceramics $Bi_{1.6}Pb_{0.4}Sr_2Ca_2Cu_3O_{10+d}$ prepared by different methods

A. Aït-Kaki[1], **O. Belkhen**[2], **A. Amira**[2,3], and **M.-F. Mosbah**[*,2]

[1] Université d'Oum-el-Bouaghi, Institut d'électrotechnique, Oum-el-Bouaghi, Algérie
[2] Université Mentouri, Laboratoire des Couches Minces et Interfaces, Campus de Chaabet-Erssas, 25000 – Constantine, Algérie
[3] Université de Jijel, Institut de Technologie, B.P. 98, Cité Ouled Aïssa, 18000 Jijel, Algérie

Received 31 August 2003, accepted 31 December 2003
Published online 5 April 2004

PACS 74.25.Ha, 74.62.Bf, 74.72.Hs, 81.20.Ev

Samples of $Bi_{2-x}Pb_xSr_2Ca_2Cu_3O_{10+d}$ have been elaborated by use of three different methods. The first method uses a precursor obtained from $CaCO_3$, $SrCO_3$ and CuO. In the second method the CuO is not used to obtain the precursor while the third method is without precursor. At each step of the elaboration, the samples are analyzed by means of X-ray powder diffraction. Final samples are characterized by Scanning Electron Microscopy. The three methods allows to obtain a pure $Bi_{2-x}Pb_xSr_2Ca_2Cu_3O_{10+d}$ phase. AC and DC Susceptibility measurements show the same critical temperature, T_c, but different dependance temperature when the constant applied field is changed. Differences are also observed in magnetization measurements at 5 and 80 K. The results are correlated with the different granular structures obtained by the three methods.

© 2004 WILEY-VCH Verlag GmbH & Co. KGaA, Weinheim

1 Introduction

Commercial power application of long length wires based on $Bi_{2-x}Pb_xSr_2Ca_2Cu_3O_{10+d}$ (Bi(Pb)-2223) are available and under test in public electricity supply grid. This material is, among the great number of high T_c superconducting materials, the most promising for practical applications. For applications such as magnets, fault current limiters and transformers, its critical current properties under applied magnetic fields have a great importance.

The phase diagram of the system $Bi_{1.6}Pb_{0.4}Sr_2CuO_6$–$CaCuO_2$ suggests that the Bi(Pb)-2223 phase is stable in a narrow range of temperature just below the melting temperature [1]. On the other hand, the influence of the Bi:Pb ratio in the starting composition on the formation of the Bi-2223 phase showed that the pure phase may be obtained in few cases [2]. Generally the phases were obtained with an excess of Bi and in one case with Ca in excess and Sr deficient. Critical current densities of Bi(Pb)-2223 silver sheathed tapes have been correlated by Müller et al. [3] to conditions of preparation in the powder in tube method. Their results show that higher values of the critical current density are obtained with higher partial pressure of oxygen and higher sintering time leading in the same time to higher Bi(Pb)-2223 fraction. In Bi(Pb)-2223 tapes made from Bi(Pb)-2212 precursors, the powder calcination temperature has an influence on the sintering temperature leading to the maximum critical current density [4]. Enhanced critical current densities have been obtained by induced Cr-ion defects [5] on Bi(Pb)-2223 tapes and MgO addition on Bi(Pb)-2223 powder [6]. The critical current density may be obtained by transport or magnetic measurements. Although the existence and the definition of the critical state seem not clearly defined for high T_c superconductors [7], the magnetic measurements give a qualitative and, until some

---

[*] Corresponding author: e-mail: faycalmos@yahoo.fr, Phone: 213 31 614342, Fax: 213 31 614342.

extent, a quantitative idea about the pinning of vortices. Pinning force of vortices may have, depending on the macroscopic nature of the material (bulk or film), many origins [8] but is always proportional to the critical current density. In the same time, magnetic measurements show directly the intergrain weak links which are the major problem encountered in these materials due to their great anisotropy. Starting from the same composition and varying the method of preparation, not necessarily the same final phase composition is obtained. The magnetic properties should depend on the phase composition and the structural quality of the samples. In the present work, using the solid state reaction, we study the effect of the method of preparation on the magnetic properties of three samples of Bi(Pb)-2223 ceramics.

2 Experimental

Starting from powders of Bi_2O_3, PbO, $SrCO_3$, $CaCO_3$, and CuO (purity over 99.9 %), samples of superconducting ceramics Bi(Pb)-2223 have been prepared in air. The starting compounds were weighted in the molar ratio 0.8/0.4/2/2/3 corresponding to the formula $Bi_{1.6}Pb_{0.4}Sr_2Ca_2Cu_3O_{10+d}$. The calcination is made in three steps for two methods and in one step for the third method. An excess of PbO corresponding, in weight, to a percentage of the PbO used firstly was added at the beginning for the third method and at the final calcination step for the other methods. The precursor was made from a mixture of $SrCO_3$, $CaCO_3$, and CuO in the first method, while in the second method only $SrCO_3$ and $CaCO_3$ were used. For the first and the second method respectively, the first calcination was at 920 °C during 48 hours and at 1250 °C during 5 to 10 hours, the second calcination at 835 °C during 144 hours and at 845 °C during 5 to 10 hours, the final calcination at 835 °C during 60 hours and at 845 °C during 60 hours. For the third method, the mixture of all the powders (PbO weighted with 10 % in excess) was calcined in one step at 820 °C during 22 hours. After forming in pellets shape, under a pressure varying from 1 to 6 t/cm^2, the sintering was made during 80 to 192 hours at 850 °C to 860 °C for the first method, during 116 to 144 hours at 850 to 855 °C for the second method and during 80 to 192 hours at 850 to 860 °C for the third method.

The samples have been characterized by X-ray diffraction (XRD) analysis at each step of the preparation using a Siemens D8-Advance powder diffractometer using a Bragg-Brentano geometry and copper Kα radiation ($\lambda_{CuK\alpha}$ = 1.54060 Å). Diffraction spectra peaks were identified using ICDD data base [9]. Without preparation, samples have been photographed using a Scanning Electron Microscope (SEM) Philips XL 30 with a tilt of 0°. DC and AC susceptibility versus temperature and magnetization versus field measurements have been made on a commercial Quantum Design SQUID.

3 Results and discussion

In the presented results sample 1 has been prepared by the first method, sample 2 by the second method and sample 3 by the third method. The analysis of the XRD patterns recorded after the preparation of the precursors, the calcination and the sintering have been presented in a previous work[10]. These analysis have shown that after the final sintering, sample 1 presents, besides the Bi(Pb)-2223 ($Bi_{1.6}Pb_{0.4}Sr_2Ca_2Cu_3O_x$) phase, traces of Ca_2PbO_4, $Ca_{0.4}Sr_{0.6}CuO_2$ and $SrCO_3$. In sample 2, besides the Bi(Pb)-2223 ($Bi_{1.5}Pb_{0.5}Sr_2Ca_2Cu_3O_x$) phase, the detected phases are traces of Bi(Pb)-2212 and Bi-2234. In sample 3, the Bi(Pb)-2223 phase is the same as in sample 2. Traces of Bi(Pb)-2212, Ca_2PbO_4 and $CaPbO_3$ phases are also detected. SEM photographs have shown that grain size varies from 3 to 10 µm in the three samples.

Figure 1 shows the AC susceptibility versus temperature $\chi'(T)$ measured with zero applied magnetic field. The three samples have the same temperature of transition $T_c \approx 107$ K. When for sample 2 and sample 3, χ' show a second transition at lower temperature indicating traces of Bi(Pb)-2212 phase [6], for sample 1 the second apparent transition indicate the more pronounced presence of weak links or coupling transition [11]. The temperature of the second transition in that sample is too low to be correlated with the Bi(Pb)-2212 phase. Sample 2 and sample 3 show the same feature in $\chi''(T)$ indicating the same apparent mean grain radius. For these samples the peak in $\chi''(T)$ is more broadened confirming the presence of the Bi(Pb)-2212 phase. This presence is more pronounced in sample 3 where an unusual second weak peak may be observed at lower temperature in $\chi''(T)$.

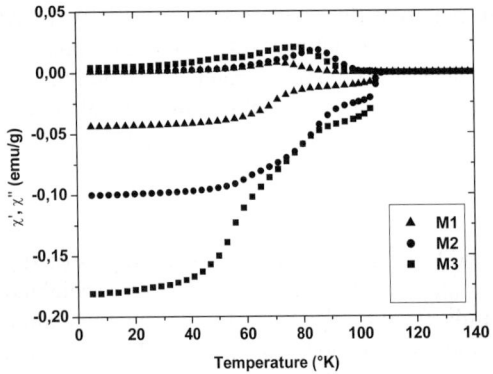

Fig. 1 AC susceptibility versus temperature of sample 1 (M1), sample 2 (M2) and sample 3 (M3).

The ZFC DC susceptibility curves (Fig. 2) present features different from those in the real part of the AC susceptibility curves. For the 5 G applied magnetic field, the low temperature transition is at about 70 K for samples 1 and 2, and at about 80 K for sample 3. For the 100 G applied magnetic field curves, the low temperature transition is hardly detected by a weak change of the slope at about 80 K for samples 1 and 2 and at about 85 K for sample 3. On the other hand, for the two values of the applied magnetic field, in the FC curves the second transition is at nearly at the same temperature.

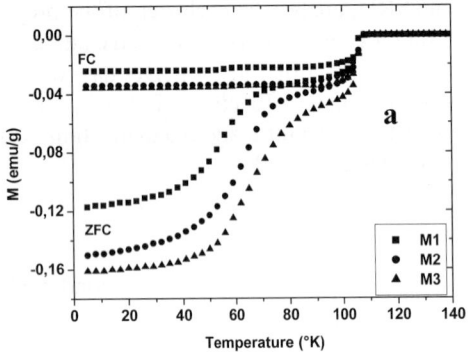
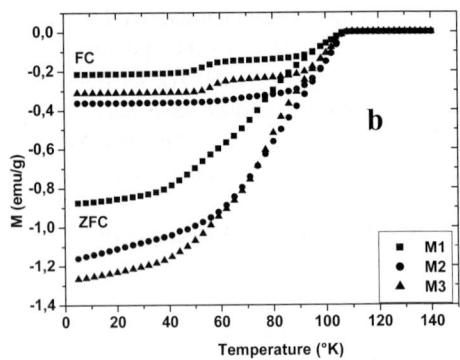

Fig. 2 ZFC and FC DC susceptibility versus temperature of the three samples with an applied magnetic field of 5 G (a) and 100 G (b).

The low field magnetization curves M(H) at 5 K and 80 K (Fig. 3a) present the same appearance. These curves show that initial slope is not the same for all the samples. This slope is enhanced by the demagnetizing factor [12]. The demagnetizing factor is enhanced when the grain is plate like, extended and perpendicular at the direction of the field. Thus, when the field is applied perpendicular to the pellet, the sample having the maximum slope has the best texture. In our case it is sample 3. The low temperature curves show a portion of the low field cycle induced by weak links (A–A', B–B', C–C'). The weak link cycle of sample 1 seems to have less area than those of samples 2 and 3.

The M(H) cycles have been recorded at 5 K with a maximum applied magnetic field of 4.5 T. From these cycles $\Delta M(H) = (M_+ - M_-)/2$ have been extracted where M_+ is upper part of the cycle and M_- is the lower part (Fig. 3b). These curves show a peak at nearly the same field. Assuming nearly the same density for the samples, the pinning force and the critical current are much higher in sample 3.

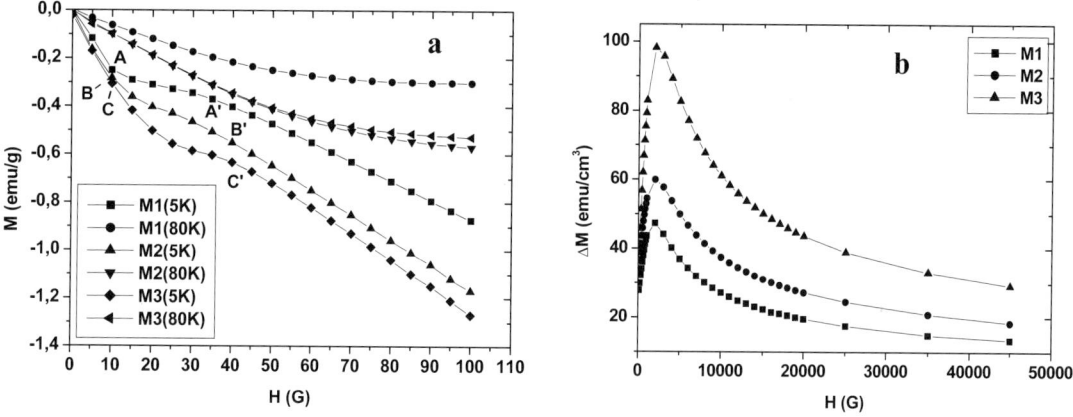

Fig. 3 Low field magnetization M(H) curves at 5 K and 80 K (a) and ΔM(H) extracted from the high field cycles at 5 K (b).

4 Conclusion

Three methods have been used to prepare superconducting ceramics samples of Bi(Pb)2223 with the use of solid state reaction route and an excess of PbO. The methods give the same temperature of transition T_c. The magnetic measurements show a presence of traces of Bi(Pb)-2212 phase in the three methods. The AC and DC susceptibility measurements, the magnetization measurements under low and high applied field show that the simplest way for calcination, the third method, gives the best results.

Acknowledgments We are grateful to Drs P. Molinié and A. Leblanc-Soreau for magnetic measurements. This work was supported by CNEPRU D2501/04/02 project and ANDRU CU 19933 project.

References

[1] P. Strobel, J. C. Toledano, D. Morin, J. Schneck, G. Vacquier, O. Monnereau, J. Primot, and T. Fournier, Physica C **201**, 27 (1992).
[2] S. Bernik, Supercond. Sci. Technol. **10**, 671 (1997).
[3] J. Müller, O. Eibl, B. Fischer, and P. Herzog, Supercond. Sci. Technol. **11**, 238 (1998).
[4] C. S. Li, P. X. Zhang, Z. M. Yu, H. L. Zheng, X. M. Xiong, Y. S. Liu, Q. Y. Wang, F. S. Liu, Y. F. Wu, P. Ji, and L. Zhou, Physica C **386**, 127 (2003).
[5] M. H. Pu, Y. Feng, P. X. Zhang, L. Zhou, J. X. Wang, Y. P. Sun, and J. J. Du, Physica C **386**, 41 (2003).
[6] E. Guilmeau, B. Andrzejewski, and J. G. Noudem, Physica C **387**, 382 (2003).
[7] S. J. Park, M. K. Hasan, and J. S. Kouvel, Physica C **325**, 13 (1999).
[8] M. H. Pu, Y. Feng, P. X. Zhang, L. Zhou, and J. X. Wang, Physica C **386**, 47 (2003).
[9] PDF-2 DATABASE 47 6A JUN 97 JCPD-ICDD (1997) USA.
[10] O. Belkhen, M.-F. Mosbah, and A. Amira, CIMTEC 2002, Proceedings Part I, 3rd Forum on New Materials – Vol. III, 4th Int. Conference "Science and Engineering of HTC Superconductivity", Techna Srl Faenza – RA Italy (2002), pp. 69–76.
[11] S. Senoussi, J. Phys. III France **2**, 1041 (1992).
[12] S. Senoussi, S. Hammond, and M.-F. Mosbah, Studies in High Temperature Superconductors, Vol. 14, Ed. Anant Narlikar (Nova Science Pub., Inc., Commack, New York, 1995), pp. 107–184.

Structure and transport properties of (calcium, fluorine) co-doped Y-based superconducting ceramics, effect of heat treatment

A. Amira[*,1], **M.-F. Mosbah**[2], **A. Leblanc**[3], **P. Molinié**[3], and **B. Corraze**[3]

[1] Institut de Technologie, Centre Universitaire de Jijel, B.P. 98 Jijel 18000, Algeria
[2] Laboratoire de Couches Minces et Interfaces, Route Ain El Bey, Constantine 25000, Algeria
[3] Institut des Matériaux de Nantes, 2 rue de la Houssinière, Nantes 44322, Cedex 3, France

Received 31 August 2003, accepted 31 December 2003
Published online 1 April 2004

PACS 74.25.Fy, 74.25.Ha, 74.72.Bk, 81.20.Ev

The oxygen content and the orthorhombicity of $Y_{0.7}Ca_{0.3}Ba_2Cu_3O_yF_{0.2}$ are reduced compared to those of $YBa_2Cu_3O_y$ free samples. Further addition of fluorine decreases them more and more in $Y_{0.7}Ca_{0.3}Ba_2Cu_3O_yF_{0.4}$ and reduces the superconducting volume fraction. As a consequence, $Y_{0.7}Ca_{0.3}Ba_2Cu_3O_yF_{0.2}$ is optimally doped with charge carriers. On the contrary, $Y_{0.7}Ca_{0.3}Ba_2Cu_3O_yF_{0.4}$ is underdoped. After an air-annealing at 450 °C and air quench, the normal state resistivity variation with temperature indicates the co-existence of weak metallic and semiconducting characters in both samples. An additional annealing in flowing oxygen at 920 °C makes $Y_{0.7}Ca_{0.3}Ba_2Cu_3O_yF_{0.2}$ optimally doped with charge carriers while $Y_{0.7}Ca_{0.3}Ba_2Cu_3O_yF_{0.4}$, that becomes pseudo-tetragonal, is more underdoped than after preparation.

© 2004 WILEY-VCH Verlag GmbH & Co. KGaA, Weinheim

1 Introduction The dependence of superconducting properties on the hole concentration in the CuO_2 planes of cuprates is still one of the intensively studied phenomenon. The varaiation of charge carriers density can be made through chemical doping and heat treatment. The substitution of Y by Ca in $YBa_2Cu_3O_y$ decreases oxygen content, changes charge carriers density and promotes the othorhombic-tetragonal transition [1–4].

Fluorine substitution for oxygen in $Tl_{0.5}Pb_{0.5}Sr_{1.6}Ba_{0.4}Ca_2Cu_3O_y$ increases the onset transition temperature, critical current density and modifies charge carriers density [5]. On the contrary, T_c and superconducting volume fraction were observed to decrease with fluorine doping in $Bi_2Sr_2Ca_2Cu_3O_y$ samples [6].

The mechanism of charge doping in CuO_2 planes could be correlated with normal state resistivity versus temperature variation [7–11]. Underdoped samples are characterized by a temperature T* at which the slope of resistivity changes. This temperature is related to opening of a normal state pseudo-gap and decreases with increasing charge carriers density [12, 13]. No change in the resistivity slope is revealed for optimally doped samples (metallic like character). In their work, Arulsamy et al. [14, 15] have used Fermi–Dirac statistics by introducing an average ionization energy to derive a phenomenological resistivity model. When temperature decreases, this model allows explanation of the transition of resistivity from metallic to semiconducting like behaviors in the normal state for overdoped samples.

In this paper, we present the effects of (calcium, fluorine) co-doping and heat treatment on structural and transport properties of Y-based ceramics.

---

[*] Corresponding author: e-mail: amira_abderrezak@hotmail.com, Phone: +213 34 49 55 78, Fax: +213 34 49 55 78

2 Experimental Two bulk ceramics of $Y_{0.7}Ca_{0.3}Ba_2Cu_3O_yF_x$ (x = 0.2, 0.4) were prepared by conventional solid state reaction. Starting materials were pure Y_2O_3, $BaCO_3$, $CaCO_3$, CuO and CaF_2. The materials were mixed and pressed into pellets under a pressure of 5 tons, calcined at 920 °C for 24 hours, sintered in air at 920 °C for 60 hours with two intermediate grindings and finally oxygenated in flowing O_2 at 500 °C for 24 hours. The samples were submitted to two additional heat treatments in order to see the effect on transport properties. The first one was in air at 450 °C during 14 hours with an air quench. The second was in flowing oxygen at 920 °C during 16 hours with a slow cooling to room temperature at the rate of 50 °C/hour.

The XRD patterns were registered by a Siemens D-5000 diffractometer with Cu Kα radiation. Cell parameters were refined by use of DICVOL 91 software [16]. EDX and SEM analysis were performed on a Jeol 5800 microscope. The four probe resistivity measurements were carried out by use of an Oxford cryostat. Magnetization measurements were made on a commercial SQUID magnetometer.

3 Results and discussion XRD analysis has shown that samples are of high purity. A part of these patterns showing the main lines, are displayed on Fig. 1. The pattern of $YBa_2Cu_3O_y$ prepared in the same conditions as the substituted ones, is also shown for comparison. The dedoubling of the main lines is less significant in substituted samples. The doping promotes then the orthorhombic-tetragonal transition.

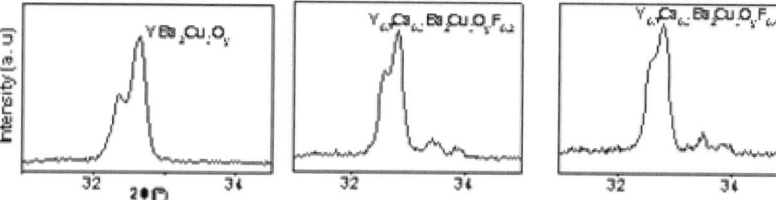

Fig. 1 XRD patterns of samples.

The degree of orthorhombicity (b–a)/b and oxygen content obtained from iodometric titration are given on Table 1. As it is clear, Ca and F co-doping reduces orthorhombicity and oxygen content in

Table 1 Orthorhombicity and oxygen content of samples.

Sample	(b–a)/b	y
$YBa_2Cu_3O_y$	1.57 %	6.91
$Y_{0.7}Ca_{0.3}Ba_2Cu_3O_yF_{0.2}$	1.17 %	6.64
$Y_{0.7}Ca_{0.3}Ba_2Cu_3O_yF_{0.4}$	0.93%	6.57

$Y_{0.7}Ca_{0.3}Ba_2Cu_3O_yF_{0.2}$. Fluorine addition decreases them more and more in $Y_{0.7}Ca_{0.3}Ba_2Cu_3O_yF_{0.4}$. The incorporation of calcium and fluorine into the grains was confirmed by EDX qualitative analysis [17]. The average size of the grains is about 10 µm in both samples.

ZFC magnetizations measured under a field of 100 G are shown in Fig. 2. The Meissner signal and

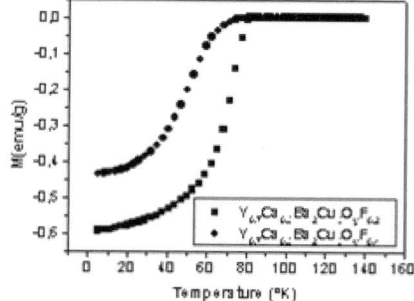

Fig. 2 ZFC magnetizations of samples.

thus the real superconducting volume of $Y_{0.7}Ca_{0.3}Ba_2Cu_3O_yF_{0.4}$ is lower than that of $Y_{0.7}Ca_{0.3}Ba_2Cu_3O_yF_{0.2}$. This may be related to oxygen content values [18].

The temperature dependence of resistivities of samples just after preparation are given on Fig. 3. The resistivity of $Y_{0.7}Ca_{0.3}Ba_2Cu_3O_yF_{0.2}$ has the lowest value and exhibits a metallic-like character suggesting that this sample is optimally doped with charge carriers [8–11]. On the contrary, The resistivity behavior of $Y_{0.7}Ca_{0.3}Ba_2Cu_3O_yF_{0.4}$ indicates that this one is in the underdoped region as it is clear from the slope $d\rho/dT$ variation. The crossover temperature T* is about 187 K. The oxygen content reduction by fluorine addition in $Y_{0.7}Ca_{0.3}Ba_2Cu_3O_yF_{0.4}$ leads then to opening of the normal state pseudo-gap.

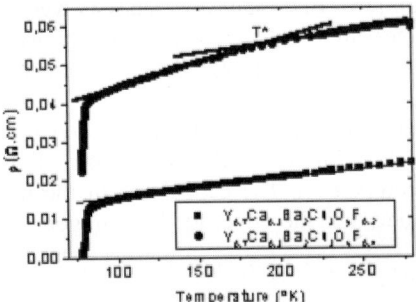

Fig. 3 Resistivity of samples just after preparation.

After an annealing in air at 450 °C for 14 hours with an air-quench, the resistivity variations are plotted in Fig. 4 where we see a co-existence of weak metallic and semiconducting characters in both samples. Resistivity increases with decreasing temperature until reaching a maximum and will again decrease until the transition to superconducting state. For our knowledge, this behavior of the normal stateresistivity was not observed yet in polycrystals and was not predicted by any model. The model proposed by A. D. Arulsamy et al. [14, 15] assumes that samples are purely polycrystalline with negligible impurity phases and defect-free with no concentration fluctuations. Other quenched pure polycrystalline samples of $Y_{1-x}Ca_xBa_2(Cu_{3-y}Zn_y)O_{7-}$ do not exhibit this scenario too [11]. We have also to note that resistivity of $Y_{0.7}Ca_{0.3}Ba_2Cu_3O_yF_{0.4}$ is lower than for $Y_{0.7}Ca_{0.3}Ba_2Cu_3O_yF_{0.2}$. This behavior of resistivity may be then related to defects caused by the quench such as artefacts, vacancies or impurity phases. The maximums in these resistivites may not be related to T*, i.e, to opening of a pseudo-gap.

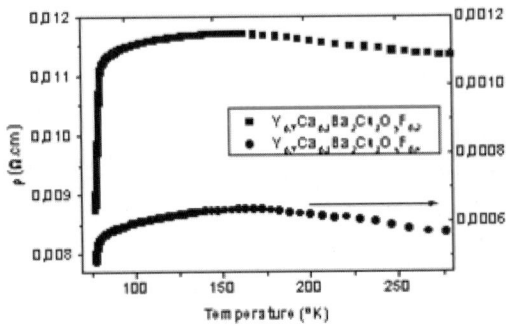

Fig. 4 Resistivity of samples after annealing in air at 450 °C with air quench.

After annealing in flowing oxygen during 16 hours at 920 °C with a slow cooling to room temperature at a rate of 50 °C/hour, the resistivity behaviors of the samples become like after prepartion. It is clearly shown in Fig. 5 that $Y_{0.7}Ca_{0.3}Ba_2Cu_3O_yF_{0.2}$ is optimally doped with charge carriers. The observed right shift of the crossover temperature T* (from 187 K to 222 K) for $Y_{0.7}Ca_{0.3}Ba_2Cu_3O_yF_{0.4}$ indicates that this sample is more underdoped than in case of Fig. 4. A refinement of the XRD spectrum of this sample gave a degree of orthorhombicity of about 0.24%. This sample is then pseudo-tetragonal.

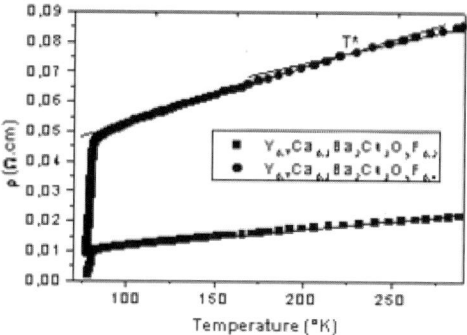

Fig. 5 Resistivity of samples after annealing in flowing oxygen at 920 °C with slow cooling.

4 Conclusion

The oxygen content and orthorhombicity of $YBa_2Cu_3O_y$ by co-doping with calcium and fluorine. Adding more fluorine reduces them more and more and decreases superconducting volume fraction. $Y_{0.7}Ca_{0.3}Ba_2Cu_3O_yF_{0.2}$ is then optimally doped with charge carriers and $Y_{0.7}Ca_{0.3}Ba_2Cu_3O_yF_{0.4}$ is underdoped. After annealing in air at 450 °C with air quench, the normal state resistivity of both samples indicates the co-existence of metallic and semiconducting characters. Additional annealing in flowing oxygen at 920 °C redrives $Y_{0.7}Ca_{0.3}Ba_2Cu_3O_yF_{0.2}$ sample to optimally doped region. $Y_{0.7}Ca_{0.3}Ba_2Cu_3O_yF_{0.4}$ with a pseudo-tetragonal structure, is more underdoped than after preparation.

Acknowledgements We are pleased to aknowledge Prof. A. D. Arulsamy from National University of Singapore for helpful discussions.

References

[1] S. I. Schlachter, W. H. Fietz, K. Grube, Th. Wolf, B. Obst, P. Schweiss, and M. Klasser, Physica C **328**, 1 (1999).
[2] A. Ulug and B. Ulug, Mater. Chem. Phys. **9262**, 1 (2001).
[3] C. W. Chang, J. G. Lin, C. Y. Chang, R. S. Liu, and C. Y. Huang, Chin. J. Phys. **36**, 360 (1998).
[4] A. Schmehl, B. Goetz, R. R. Schulz, C. W. Schneider, H. Bielefeldt, H. Hilgenkamp, and J. Mannhart, Europhys. Lett. **47**, 110 (1999).
[5] N. M. Hamdan, Kh. A. Ziq, and A. S. Al-Harthi, Physica C **314**, 125 (1999).
[6] E. Bellingeri, G. Grasso, R. E. Gladychevskii, M. Dhallé, and R. Flükiger, Physica C **329**, 267 (2000).
[7] H. Alloul, T. Ohno, and P. Mendels, Phys. Rev. Lett. **63**, 1700 (1989).
[8] M. B. Maple, C. Almasan, C. Seaman, S. H. Han, K. Yoshiara, M. Buchgeister, L. M. Paulius, B. W. Lee, D. A. Gajewski, R. F. Jarnim, C. R. Fincher, Jr. G. Blanchet, and R. Guertin, J. Supercond. **7**, 97 (1994).
[9] K. Q. Ruan, Y. Feng, C. Y. Wang, X. H. Chen, and L. Z. Cao, Physica C **282–287**, 1165 (1997).
[10] T. Watanabe, T. Fujii, and A. Matsuda, Phys. Rev. Lett. **79**, 2113 (1997).
[11] S. H. Naqib, J. R. Cooper, J. L. Tallon, and C. Panagopoulos, Physica C **387**, 365 (2003).
[12] G. A. Levin and K. F. Quader, Phys. Rev. B **62**, 11879 (2000).
[13] J. L. Tallon and J. W. Loram, Physica C **349**, 53 (2003).
[14] A. D. Arulsamy, Phys. Lett. A **300**, 691 (2002).
[15] A. D. Arulsamy, P. C. Ong, and M. T. Ong, Physica B **325**, 164 (2003).
[16] A. Boultif and D. Louer, J. Appl. Cryst. **24**, 987 (1991).
[17] A. Amira, M.-F. Mosbah, A. Leblanc, P. Molinié, and B. Corraze, Proceedings of the International Conference on Magnetism, Roma, Italy, July 27 August 2003, to appear in J. Magn. Magn. Mater.
[18] A. J. Jacobson, J. M. Newsam, D. C. Johnson, D. P. Gorshom, J. T. Lewandowski, and M. S. Alvarez, Phys. Rev. B **39**, 254 (1989).

Praseodymium and oxygen role on magnetic properties of $PrBa_2Cu_3O_{6+x}$

A. Harat[1], **G. Fillion**[*,2], **P. Haen**[3], **J. Hejtmanek**[4], **M. F. Mosbah**[5], and **M. Guerioune**[1]

[1] Laboratoire LEREC, Université Badji Mokhtar, BP 12 El Hadjar, Annaba, Algérie
[2] Laboratoire Louis Néel, CNRS-UJF, BP166, 38042 Grenoble Cedex 9, France
[3] CRTBT, CNRS-UJF, BP 166, 38042 Grenoble cedex 9, France
[4] Institute of Physics, ASCR, Cukrovarnicka 10, 16200 Praha 6, Czech Republic
[5] Laboratoire des Couches Minces, Université Mentouri, Constantine, Algérie

Received 31 August 2003, accepted 31 December 2003
Published online 1 April 2004

PACS 74.25.Ha, 74.72.Jt

We report on magnetic investigation on $PrBa_2Cu_3O_{6+x}$ compounds ($0.4 < x < 1.0$) where the Pr_{3+} ions exhibit an antiferromagnetic order (AF) at $T_N \sim 11$ K for $x \sim 0.5$ and 17 K for $x \sim 1$. For better understanding of the role of the praseodymium and oxygen content on magnetic and transport properties, we measured temperature and field dependence of susceptibility and magnetization of $PrBa_2Cu_3O_{6+x}$ in two oxygenation states $x = 0.4$ and 0.9. We show that a careful analysis of these data can bring to evidence the existence of several tiny transitions, especially a sharp one around 13 K which may be related to a change of direction of copper moments.

© 2004 WILEY-VCH Verlag GmbH & Co. KGaA, Weinheim

1 Introduction

In the superconducting RE123 system (RE = rare earth), the Pr substituted compounds present many intriguing properties, mainly suppression of superconductivity and high antiferromagnetic (AF) Pr ordering temperatures T_N. This is why they are considered as good candidates to get a deeper insight on the transport and magnetic properties of these systems. Previous investigations of AF ordered Pr123 compounds have shown that T_N can range from 9 to 11 K and from 14 to 20 K for the under- and fully oxygenated systems, respectively [1–3]. According to recent studies [4], which showed bulk superconductivity in $PrBa_2Cu_3O_7$, a proposed explanation of the Pr123 insulating behavior is the partial substitution of Pr in Ba sites. It seems that the value of T_N not only depends on the rate of oxygen, but also on the Pr and Ba content, and that the respective roles of these elements might be intimely intricated. A study of the $Pr_{1+x}Ba_{2-x}Cu_3O_{7+y}$ system showed a decrease of T_N with increasing x, attributed to an increase of Pr–O bond lengths in the CuO2 bilayer, which, assuming a strong Pr 4f–O 2p orbital hybridization, lead to weaker superexchange for Pr–O–Pr coupling [5, 6]. However, an investigation of $Pr_xY_{1-x}Ba_2Cu_3O_{7-\delta}$ single crystals [7] showed, in contrast to studies on polycrystals, that T_N is unchanged for the entire insulating range of x ($0.4 < x < 1$), and that only the effective magnetic moment increases with x in the range 2–3 μ_B, meaning that the exchange interactions seem not affected by x. These few examples show how it is actually difficult to compare the physical properties of samples of different preparations, leading often to contradictory results, and hampering any reliable interpretation. In order to only underline the role of oxygen content, we have started a study of a $PrBa_2Cu_3O_{6+x}$ ceramic for two oxygenation states ($x = 0.4$

---

[*] Corresponding author: e-mail: fillion@grenoble.cnrs.fr, Phone: +33 476 887 904, Fax: +33 476 881 191

and 0.9), which should keep constant the Pr–Ba substitution, and rule out the influence of other parameters such as impurities.

2 Results

After preparation, our $PrBa_2Cu_3O_{6+x}$ ceramic was underdoped and showed the tetragonal structure. One part of it was cut and submitted to controlled oxygenation, absorbing half a mol of oxygen. It then showed the orthorhombic structure, and its composition was estimated to $x = 0.9$, leading thus to the initial one, $x = 0.4$. Details on preparation and thermogravimetric analysis (TGA) are given elsewhere [8]. The two (non- and oxygenated) samples were then studied separately. To check for the Pr/Ba substitution, SEM and electron probe microanalysis (EPMA) quantitative analysis have been made on both samples. This later EPMA analysis, though overestimating the oxygen content, confirms the difference of ~ 0.5 on x and give a cation composition of $Pr_{1.13}Ba_{1.92}Cu_{2.95}$ for both samples. Some tiny traces of $BaCuO_2$ secondary phase have been found in the $O_{6.4}$ sample but not in the $O_{6.9}$ one.

The magnetic measurements were performed in a very sensitive home made vibrating magnetometer. The magnetic susceptibility, $\chi(T)$, was taken as M/H in a magnetic induction of 10 mT. The magnetization M has been measured versus magnetic field H up to $\mu_0 H = 2.1$ T for different temperatures, and as well as at constant field in function of the temperature. The susceptibilities of both samples and their derivatives have been partly reported elsewhere [8, 9]. For both samples, $\chi(T)$ increases continuously on cooling whereas T_N appears only as a small kink, as shown in Fig. 1a for $x = 0.4$, and is rather determined from the derivative $d\chi(T)/dT$ [8].

For $T > 40$ K, $\chi(T)$ is well fitted by a modified Curie–Weiss law (MCW) $\chi_0 + C/(T-\theta_p)$ and the results are summarized in Table 1. Although T_N decreases with x, the Curie constant C remains practically the same as well as the offset χ_0, but θ_p differs in about the same proportions than T_N, suggesting that only the magnetic interactions are affected by a change in x but not the moments.

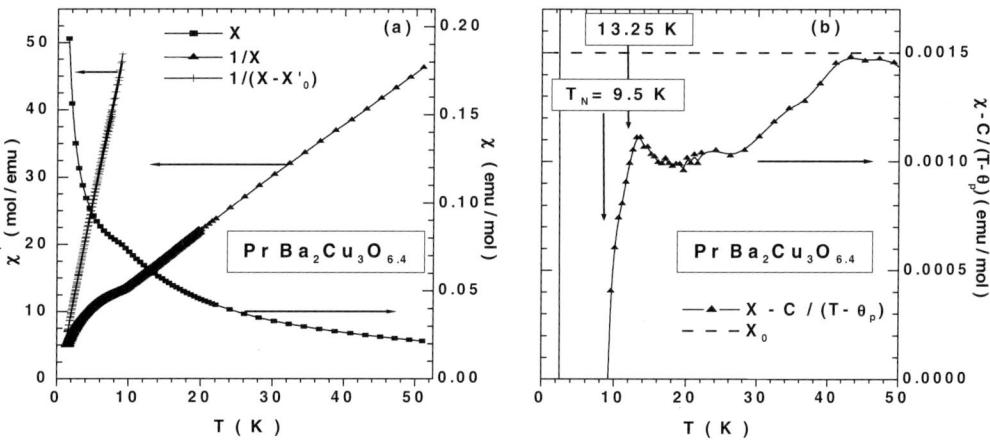

Fig. 1 a) Magnetic susceptibility $\chi(T)$, its reciprocal $\chi^{-1}(T)$ and Curie–Weis law fit for $T < T_N$ for $PrBa_2Cu_3O_{6.4}$, b) plot vs. temperature of the difference with the fit, $\chi - C (T-\theta_p)$, for $T > 40$ K.

Table 1 Pr AF ordering temperatures T_N and $\chi(T)$ fitting parameters to $\chi_0 + C/(T-\theta_p)$ for $T > 40$ K.

$PrBa_2Cu_3O_{6+x}$	T_N (K)	χ_0 (emu/mole)	C (emu.K/mol)	μ_{eff} (μ_B)	θ_p (K)
$x = 0.4$	9.5	0.0015	1.15	3.1	–6
$x = 0.9$	14.5	0.0015	1.15	3.1	–8

As shown in Fig. 1b, the plot of the remaining difference obtained by subtracting the fitted Curie–Weiss law from $\chi(T)$ in the whole temperature range, brings to evidence, aside a change of slope at 40 K, a sharp anomaly located at 13.25 K, as well in field cooled (FC) or in zero field cooled (ZFC) experiments.

For $T < T_N$, the large increase of $\chi(T)$, which is different from common AF ordering, follows another MCW law $\chi'_0 + C'/(T-\theta'_p)$ (Fig. 1a). The parameters of these laws are given in Table 2. One can see that they depend strongly on oxygen content x but also on ageing in air for the oxygenated sample.

Table 2 Results of susceptibility data fitting to modified Curie–Weiss law $\chi'_0 + C'/(T-\theta'_p)$ for $T < T_N$.

$PrBa_2Cu_3O_{6+x}$	χ'_0 (emu/mole)	C' (emu.K/mol)	θ'_p (K)
x = 0.4	0.0550	0.180	+0.25
x = 0.4 after 18 months	0.0550	0.180	+0.25
x = 0.9	0.0345	0.315	−2.00
x = 0.9 after 9 months	0.0400	0.155	−1.40

Such an increase was also observed by other authors [1, 2]. It can become much less pronounced in the variation of $M(T)$ measured in fields larger than those applied in the present $\chi(T)$ measurements (see for instance $M(T)$ in 2 T from Ref. [2, 10]). In accordance, our $M(H)$ data show a curvature at low temperature (Fig. 2a). This means that the low temperature increase of $\chi(T)$ is not completely intrinsic and that it is partially due to some amount of remaining paramagnetic ions. However, this increase can never be completely saturated, even in large fields. Moreover, it does not disappear above T_N, but only for $T > \sim 40$ K. The subtraction of the linear term gives a partially saturated moment M_S defined as in Fig. 2a for $H = 2$ T and which is represented as function of temperature in Fig. 2b. As one can see, the increase of M_S for $T < 40$ K from zero value for high temperatures to a large one for $T < T_N$ indicates the existence of a weak ferromagnetic component and can be associated with the decrease of χ at 40 K (Fig. 1b), which is well above the Pr ordering temperature. To explain this feature, one can invoke either some progressive change in the ordering of Cu(1) on chains [11] or a change of the magnetic structure of the Cu(2) in planes from collinear to non collinear, induced by the growing of the Pr moments. For this last explanation, a magnetic coupling between Pr and Cu(2) spins is needed, as argued by many authors in order to account for the high AF ordering of Pr [12, 13].

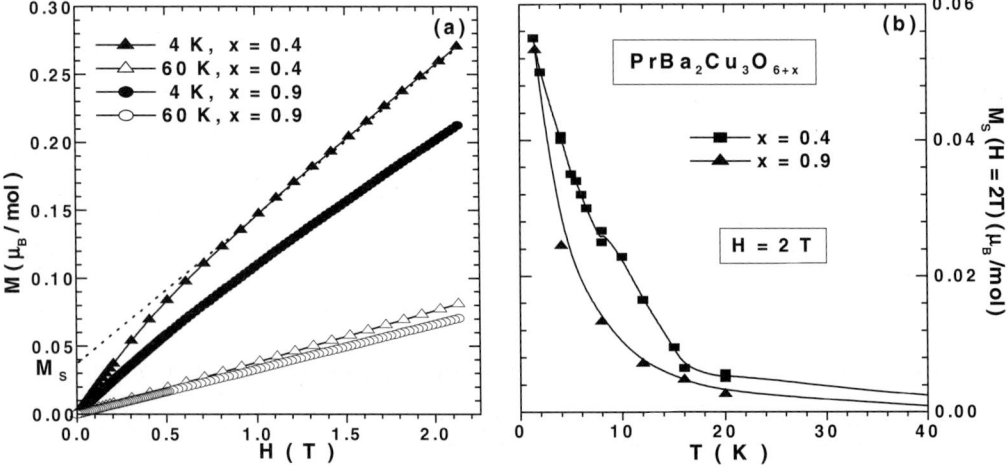

Fig. 2 a) Magnetization vs. field at $T = 4$ K and $T = 60$ K, b) temperature evolution of M_s at $H = 2$ T.

3 Discussion

The fact that T_N is lowered by only decrease in oxygen content x means that the exchange interactions between Pr sites depend on the concentration of holes in the CuO2 planes and hybridization of Pr–4f orbitals with the O–2p orbitals.So these exchange interaction could be mediated by the charge carrier's band in the CuO2 plane rather than a classical superexchange.

Below T_N(Pr), the large increase of χ, well fitted by a MCW law, show that there is a large number of remaining paramagnetic centers which strongly increases with oxygenation. If a full occupation of the Pr site by Pr and complete AF order is assumed, the Pr on Ba site cannot account alone for this increase of the Curie constant with x (75%, see Table 2), and the Cu(1) in the chains, much closer to the oxygen site which vary in occupation, must be involved. Even a possible substitution of Ba on the Pr site would be x independent. This fact is in favour of an increase of disorder or frustration in the Cu chains with oxygenation, as assumed in several related studies [14].

An anomaly around 13 K should be detected for $x = 0.9$, as it has been previously observed for $x = 1$ in polycrystals [3] and by neutron diffraction experiment on single crystals [15, 16],but in our sample, this temperature is too close to T_N (14.5 K) to be well distinguished by this procedure. The fact that it is clearly seen on our $x = 0.4$ (Fig. 1b), well above the corresponding T_N, means that this transition is oxygen independent and that the change in the magnetic structure is not driven by the Pr ordering but is an intrinsic spin reorientation of the Cu(2) moments.

In conclusion, a simple analysis of the magnetization and magnetic susceptibility data is able to highlight the transitions occurring in these compounds. So, it can be used for further systematic study of these transitions for different well defined composition and oxygen content.

References

[1] G. Hilscher, E. Holland-Moritz, T. Holubar, H.-D. Jostarndt, V. Nekvasil, G. Shaudy, U. Walter, and G. Fillion, Phys. Rev. B **49**, 535 (1994).
[2] S. Uma, T. Sarkar, M. Seshasayee, G. Rangarajan, Changkang Chen, Yongle Hu, B. M. Wanklyn, and J. W. Hodby, Phys. Rev. B **53**, 6829 (1996).
[3] G. Levy, B. Maiorov, M. S. Corvalan, A. Fainstein, and G. Nieva, Physica B **320**, 333 (2002).
[4] Z. Zou, Jinhua Ye, K. Oka, and Y. Nishihara, Phys. Rev. Lett. **80**, 1074–1077 (1998).
[5] S. R. Hwang, W.-H. Li, K. C. Lee, J. W. Lynn, H. M. Luo, and H. C. Ku, Phys. Rev. B **63**, 172401 (2001).
[6] H. M. Luo, B. N. Lin, Y. H. Lin, H. C. Chiang, Y. Y. Hsu, T. I. Hsu, T. J. Lee, H. C. Ku, C. H. Lin, H.-C. I. Kao, J. B. Shi, J. C. Ho, C. H. Chang, S. R. Hwang, and W.-H. Li, Phys. Rev. B **61**, 14825 (2000).
[7] B. Jayaram, H. Srikanth, B. M. Wanklyn, C. Changkang, E. Holzinger-Schweiger, and G. Leising, Phys. Rev. B **52**, 89 (1995).
[8] M. Lahoubi, A. Harat, G. Fillion, D. Schmitt, J. Marcus, P. Haen, and J. Hejtmánek, Lebanese Sci. Res. Rep. **3**, 205 (1998).
[9] P. Haen, M. Lahoubi, A. Harat, G. Fillion, D. Schmitt, H. Bioud, and J. Hejtmánek, Physica B **284–288**, 1035 (2000).
[10] V. N. Narozhnyi, D. Eckert, K. A. Nenkov, G. Fuchs, T. G. Uvarova, and K.-H. Müller, Physica C **312**, 233 (1999).
[11] H. Kadowaki, M. Nisha, Y. Yamada, H. Takeya, H. Takei, S. M. Shapiro, and G. Shirane, Phys. Rev. B **37** 7932 (1988).
[12] M. W. Pieper, F. Wiekhorst, and T. Wolf, Phys. Rev. B **62**, 1392 (2000).
[13] S. Skanthakumar, J. W. Lynn, N. Rosov, G. Cao, and J. E. Crow, Phys. Rev. B **55**, R3406 (1997).
[14] V. M. Browning, M. S. Osofsky, J. M. Byers, A. C. Ehrlich, and F. Dosseul, Phys. Rev. B **54**, 13058 (1996).
[15] S. Uma, W. Schnelle, E. Gmelin, G. Rangarajan, S. Skanthakumar, J. W. Lynn, R. Walter, T. Lorentz, B. Büchner, E. Walker, and A. Erb, J. Phys.: Condensed Matter **10**, L33 (1998).
[16] A. T. Boothroyd, A. Longmore, N. H. Andersen, E. Brecht, and Th. Wolf, Phys. Rev. Lett. **78**, 130 (1997).

Effect of doping on properties of Bi-based superconductors

L. D. Sýkorová[*], O. Smrčková, and V. Jakeš

Institute of Chemical Technology Prague, Technická 5, 166 28 Prague 6, Czech Republic

Received 31 August 2003, accepted 31 December 2003
Published online 8 April 2004

PACS 61.72.Ww, 74.25.Sv, 74.62.Dh, 74.72.Hs

The influence of alkaline metals on the properties of Bi-based high temperature superconductors has been studied. The samples $Bi_{3.2}Pb_{0.8}Sr_4Ca_5Cu_7A_xO_y$ (A = Li or Cs, x = 0, 0.2, 0.4, 0.6, 0.8) were prepared and characterized by XRD analysis, resistivity and susceptibility measurements. The results showed that all the doped samples had T_c in the range 104–110.5 K. The transition width ΔT of the Bi-2223 phase in the lithium doped samples increased with higher concentration of dopant. In the cesium doped samples the transition width decreased with increasing Cs level up to x = 0.6 then slightly increased. Values of J_c of lithium containing samples decreased with increasing amount of doped lithium. Samples doped with lower amount of cesium (x = 0.2, 0.4) had notably higher critical current densities than the undoped standard. Nevertheless, J_c's of samples with higher amount of cesium were worse than that of the standard. Lithium had a positive influence on the formation of Bi-2223 phase in all concentration except x = 0.6. In samples doped with cesium the volume of Bi-2223 phase was higher than in the undoped sample as well, the values were not as high as in lithium doped samples.

© 2004 WILEY-VCH Verlag GmbH & Co. KGaA, Weinheim

1 Introduction

The properties of BiSrCaCuO superconductor can be controlled by the addition or substitution of the elements having a different ionic radius and a different bonding character. We have investigated [1] the influence of the transition metals with variable valence (V, Nb, Ta, Mn) on the properties of Bi-based superconductors. The study reveals that for all substituents an addition of small amount of them enhances the content of the Bi-2223 phase and improves the transport properties. On the other hand greater amount of substituents deteriorates weak links in the samples. Furthermore, the influence of column 3A elements (B, Al, Ga, In) on the properties of Bi-system has been studied [2]. All samples were mixed phases of Bi-2212 and Bi-2223. The lattice parameters of the superconducting phases were obtained, and no systematic variations with the dopant level were found. Boron enhanced the formation of the Bi-2223 phase, the samples had a narrow transition with the T_c = 108 K. The amount of Bi-2223 phase was increased with small concentration of Al and In but Ga deteriorated the Bi-2223 phase and the Bi-2212 phase became dominant.

Alkaline metals (Li, Na, K, Rb, Cs) could be the candidate for the substitution, since their ionic radii (73–181 pm) overlap those of Bi, Pb, Sr, Ca and Cu. Furthermore, alkaline metals have a +1 valence state, so that their addition is attractive from the point of changing carrier concentrations.

Kawai et al. [3] first studied the effects of substituting alkaline metals for Bi, Sr, Ca and Cu in $Bi_2Sr_2CaCu_2O_x$. They found that alkaline metals drastically decrease the formation temperature of the Bi-2212 phase. The T_c was increased by Li and Na doping, but was decreased by K and Rb doping. Although, there has been considerable work on alkaline metal substitution/addition in the Bi-based system, it has mainly focused on the 2212 phase. Only a few authors studied the influence of alkaline metal

---

[*] Corresponding author: e-mail: Dagmar.Sykorova@vscht.cz

dopants on the structure and superconducting properties of the Bi-2223 system [4]. Doping with K, Rb and Cs to level of 30 mole % leads to an increase in T_c, critical current density J_c and decrease in the normal state resistance.

Zhigadlo et al. [5] reported results of a study of the effect of Cs addition and heat treatment on the phase formation, crystal structure and the superconducting properties of $Bi_{1.7}Pb_{0.3}Sr_2Ca_2Cu_3Cs_xO_y$, x = 0.0–1.0. The best results have been observed with x > 0.4. Addition of Cs to the Bi(Pb)SrCaCuO system was found to be effective in improving the superconducting characteristic of the system.

We have been studied the influence of Li and Cs addition on the superconducting properties and the phase composition in the Bi(Pb)–Sr–Ca–Cu–O system.

2 Experimental

Samples with nominal composition $Bi_{3.2}Pb_{0.8}Sr_4Ca_5Cu_7A_xO_y$ (A = Li or Cs, x = 0, 0.2, 0.4, 0.6, 0.8) were prepared by solid state reaction in air, using high purity powders of oxides and carbonates. The starting composition with excess of Ca, Cu and Pb doping promotes the growth of Bi-2223 phase [6]. Mixed powder was twice calcined at 800 °C for 24 hours in air. The calcined powders were reground and pressed into pellets. Pellets were then sintered at 840 °C for 168 hours in air.

Prepared samples were characterized by measuring of temperature dependence of their resistivity using a standard four-probe method and critical current density. Critical current density (J_c) was determined as current (per section area) whose flow through the sample caused voltage of $1 \cdot 10^{-5}$ V on it. Obtained temperature dependence of J_c was extrapolated to the temperature of 77 K. AC susceptibility was measured using an SQUID susceptometer to distinguish intergrain and intragrain critical temperatures. Phase composition was determined by XRD analysis using DRON 3 with CuK_α radiation.

3 Results and discussion

In all X-ray diffractograms of the substituted samples $Bi_{3.2}Pb_{0.8}Sr_4Ca_5Cu_7A_xO_y$ (A = Li or Cs, x = 0, 0.2, 0.4, 0.6, 0.8) all major peaks were identified (Fig. 1) as phases Bi-2212 and Bi-2223.

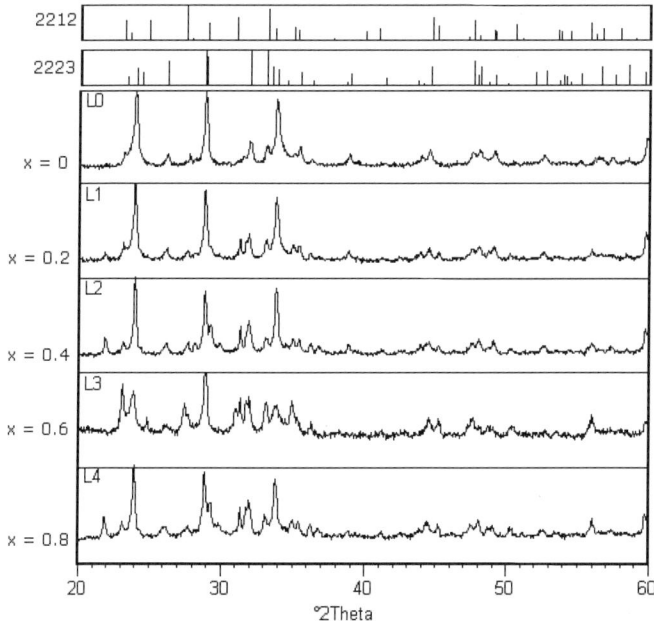

Fig. 1 X-ray diffraction patterns of Li doped samples.

Other identified phases in lower amounts were various mixed oxides such as $Bi_{12}PbO_{19}$, Ca_2PbO_4, $CaCuO_2$ and $Ca_{0.7}Sr_{0.3}O$. Ratio of superconducting phases Bi-2212 and Bi-2223 was then calculated omitting presence of other phases (i.e. assuming that samples contain only two above mentioned phases) with the help the database of standard diffractograms PDF-2 from of intensity of reflexes (008) Bi-2212 and (0010) Bi-2223. Lithium had a positive influence on the formation of Bi-2223 phase in all concentration except x = 0.6. In the lithium doped samples the volume of Bi-2223 phase increased above 90 % (95 % when x = 0.4) when compared with sample without additives (Table 1). In samples doped with cesium the volume of Bi-2223 phase was higher than in the undoped sample as well, the values were not as high as in lithium doped samples (88 % when x = 0.6).

Table 1 The volume content of Bi-2223 and Bi-2212 in lithium doped samples.

x	0	0.2	0.4	0.6	0.8
Phase Bi-2223 [%]	81	93	95	63	91
Phase Bi-2212 [%]	19	7	5	37	9

All the samples doped with lithium and cesium exhibited reproducible superconducting behaviour with T_c (determined from the inflection point of the resistance curve) in range 104.0–110.5 K. The samples with x = 0.2 had T_c higher than the undoped sample (Table 2). The transition width ΔT of the Bi-2223 phase in the lithium doped samples increased with higher concentration of dopant. In the cesium doped samples the transition width decreased with increasing Cs level up to x = 0.6 then slightly increased. The increase of the transition width reflects the weakness of the intergrain connection.

Table 2 The critical temperatures T_c and ΔT_c of Li- and Cs-doped samples.

x	Li-doped samples		Cs-doped samples	
	T_c [K]	ΔT_c [K]	T_c [K]	ΔT_c [K]
0	108	4.8	109	6.1
0.2	110.5	5.7	110.2	4.3
0.4	107	7.0	109.4	4.3
0.6	104	8.9	107	4.6
0.8	107	5.9	106.8	8.3

Values of J_c of lithium containing samples decreased with increasing amount of doped lithium. Samples doped with lower amount of cesium (x = 0.2, 0.4) had notably higher critical current densities than the undoped standard. Nevertheless, J_c's of samples with higher amount of cesium were worse than that of the standard (Table 3).

Table 3 The critical current densities of Li- and Cs-doped samples.

x	J_c (77 K) [A cm^{-2}]	
	Li-doped samples	Cs-doped samples
0	16.41	12.91
0.2	11.86	20.42
0.4	11.91	20.30
0.6	6.91	11.93
0.8	3.70	4.53

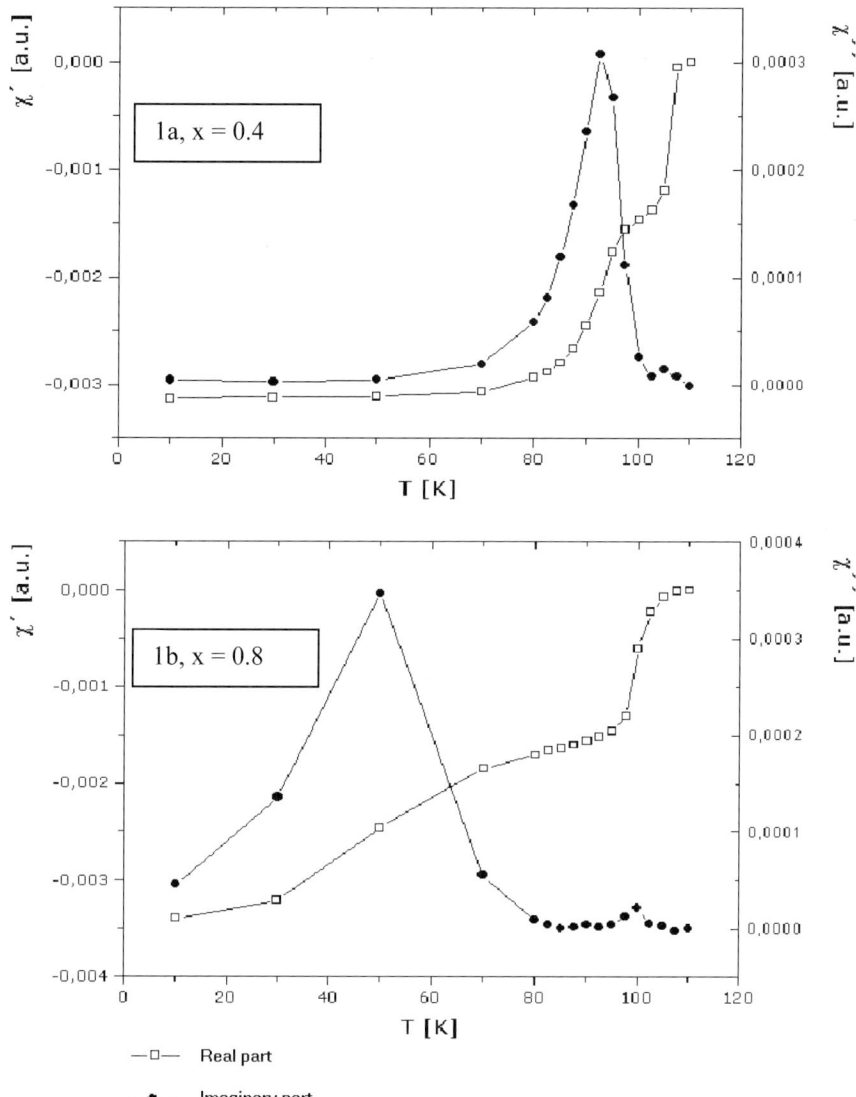

Fig. 2 a), b) Temperature dependence of the real and imaginary parts of the dynamic susceptibility cesium doped samples

The temperature dependence of AC susceptibility for the undoped and cesium doped samples ($x = 0.4$, 0.8) have shown two-step transition, corresponding to the diamagnetic signal of the grains and of the whole ceramic sample. Measurement of AC susceptibility showed (Fig. 2) increase of critical intergrain temperature T_{c1}^{χ} when $x = 0.4$ and decrease with higher amount of cesium ($x = 0.8$). The results confirmed the deterioration of intergrain connection with increasing concentration of cesium.

Acknowledgements This work was supported by Ministry of Education under project MSM 223100002.

References

[1] D. Sýkorová, O. Smrčková, K. Nováková, and P. Vašek, Phys. C **161**, 851 (1997).
[2] D. Sýkorová, O. Smrčková, K. Rubešová, and K. Knížek, Phys. B **321**, 297 (2002).
[3] T. Kawai, T. Horiuchi, K. Mitsui, K. Ogura, S. Takagi, and S. Kawai, Phys. C **161**, 561 (1989).
[4] V.V. Petrashko, N.D. Yhigadlo, and Z.A. Semenenko, Technol. Phys. Lett. **22**, 48 (1996)
[5] N.D. Zhigadlo, V.V. Petrashko, Y.A. Semenenko, C. Panagopoulos, J.R. Cooper, and E.K.H. Salje, Phys. C **299**, 327 (1998).
[6] B.W. Statt, Y. Wang, M.J.G. Lee, J.V. Zakhmi, and P.C. Camargo, Phys. C **156**, 251 (1988).

Electron backscattered Kikuchi diffraction technique: for a better understanding of epitaxial superconducting film growth on buffered Ni (RABiTS) tapes

S. Donet[*,1,3], **P. Chaudouet**[1], **F. Weiss**[1], **C. Jimenez**[1], **H. P. Ng**[1] **C. E. Bruzek**[2], and **J. M. Saugrain**[3]

[1] LMGP-ENSPG, Grenoble, France
[2] NEXANS, Jeumont, France
[3] NEXANS, Clichy, France

Received 31 August 2003, accepted 31 December 2003
Published online 1 April 2004

PACS 61.14.–x, 68.55.–a, 74.78.Bz, 81.15.Gh

A reel-to-reel MOCVD system has been designed to synthesize high quality YBCO coated tapes on Ni RABiTs (Rolling Assisted Biaxially Textured Substrates). Special buffer layer stackings have been deposited to improve the quality of the superconductor. Several buffer layers architectures rose in dense YBCO films with a similar texture, but fluctuations in critical current densities (Jc) were measured. Therefore, a more accurate and detailed measurement method such as Electron Backscattered Kikuchi Diffraction (BKD) has been used to assess the crystalline quality of the stackings. The studied sequence here was NiW /NiO /YSZ /CeO_2 /YBCO. The grain and the subgrain structure as well as their size and disorientation have been analysed for each layer. First on Ni tape, secondly on the protecting NiO (200) buffer layer, then on the following CeO_2 (200) film. Finally, unpublished BKD diagrams of the subsequent YBCO (00l) films have been reported here. The grain evolution has been successfully studied showing the buffer layer effect on the grain growth. In this work we highlight the correlation between the misorientation of the grains (inducing NiO (111) growth and cracks) and the grain boundary morphology (size, grooves).

© 2004 WILEY-VCH Verlag GmbH & Co. KGaA, Weinheim

1 Introduction

Coated conductors based on YBCO films remain the most promising approach for using superconductors in motors, transformers and such relevant applications [1]. The Ni RABiTS tapes remain widely used as a substrate material because a very strong cube texture can be induced by a rolling and recrystallisation process [2]. On the other hand, the nickel have disadvantageous properties such as: ferromagnetism, low tensile strength and natural tendency to form misoriented grains. Actually the performances of the coated tapes are severely limited by the presence of large grain boundaries [3]. So the Ni tape is caped with adequate buffer layers in order to avoid Ni diffusion into the superconducting film and also to reduce the weak-link behaviour of the substrate. Therefore an improvement of the grain-to-grain alignment of the YBCO layer can be performed thanks to adequate buffer layers stacking.

2 Experimental method

High quality YBCO tapes were deposited on Ni buffered substrate. An on-line reel-to-reel Metal Organic Chemical Vapour Deposition (MOCVD) reactor has been developed for continuous deposition of the

---

[*] Corresponding author: e-mail: Sebastien.Donet@inpg.fr, Patrick.Chaudouet@inpg.fr

buffer layers and the YBCO film. The solution of precursors is introduced into an evaporator through an injection system as described by Senateur *et al.* [4]. The precursors used for the buffer layers and YBCO are metalorganic ones: $Zr(thd)_4$, $Ce(thd)_4$, $Y(thd)_3$, $Ba(thd)_2$, and $Cu(thd)_2$ dissolved in 1,2-dimethoxyethan. Both, Ni substrate, buffer layers and the final YBCO film were characterized using SEM, AC susceptibility and XRD. Moreover, AFM snapshots were correlated with BKD analysis allowing us a more precise understanding of the weak-link behaviour of the Ni tapes. Performances of the BKD techniques may be highlighted as: the large-scale observation window and the access to the grain and sub-grain morphological and crystallographic informations (boundaries, orientation, size, repartition). Also, BKD gave us a quantitative characterization of the defects that weaken the current density. In this work, BKD data have been correlated with X-ray diffraction results, AFM, SEM, TEM and AC susceptibility measurements for a better understanding of the relationships between microstructure, crystallography and superconducting properties.

3 Results and discussion

3.1 NiW/NiO//YSZ/CeO$_2$/YBCO deposited by MOCVD

The NiW (micro-alloy at 0.1%WT) tape undergoes a rolling and annealing process to confer a ccf structure [5]. The annealed NiW tape showed an extremely weak thermal etching effect and rare Ni (111) spots were observed (BKD in Fig. 1a). A consecutive treatment under oxygen at 1150 °C lead to a 2µm-thick NiO film with a columnar type structure.

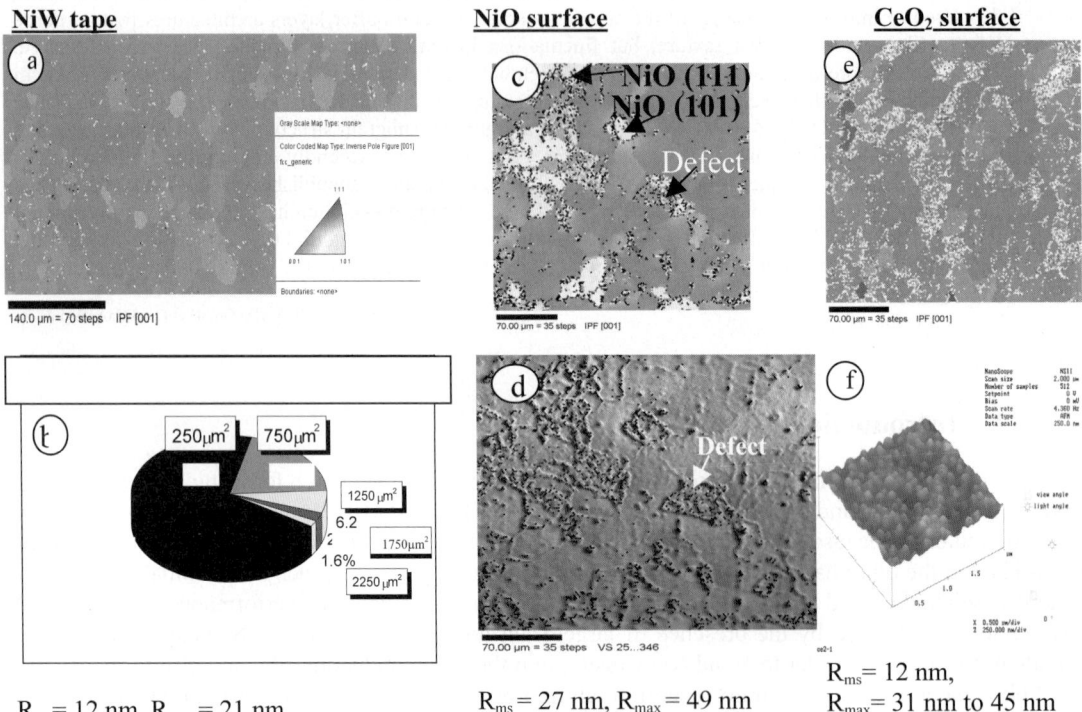

Fig. 1 a) BKD: Ni inverse pole figure, b) BKD: Ni grain size repartition, c) BKD: NiO inverse pole figure, d) SEM: NiO view of the corresponding zone, e) BKD: CeO$_2$ inverse pole figure, f) AFM of the CeO$_2$ surface.

The thermal growth of the biaxially textured NiO (100) film shows a pure cube on cube structure (XRD and TEM). However, during this rapid growth (2 µm in 5 min), dislocations inside the Ni tape have been amplified throughout the Ni-NiO interface (Fig. 4). These defects also visible on SEM picture (Fig. 1d)) have been identified as NiO (111) and NiO (101) grains as shown on Fig. 1c. Morover, the rms roughness of the annealed NiW substrate was 12 nm (obtained by AFM) whereas for NiO it increased to

27 nm. On the other hand, a widening grooving effect appears: from mean grooving depth of 21 nm for Ni to 49 nm for NiO. In view of these results, a 200 nm thick (200)-textured YSZ film was deposited to serve as a chemical and mechanical diffusion barrier. The long dislocations were then stopped at the NiO-YSZ interface and did not propagate across the boundaries (Fig. 2). At the NiO-YSZ interface a large 40 nm thick, well crystallised layer nucleated beneath the interface (Fig. 3) whereas the dislocations pile-up stoppped at the interface. Then, when depositing YBCO directly on the YSZ buffer layer low critical current densities (J_c) were measured mainly due to YSZ-YBCO mismatch (5.4%) as well as a high porosity [6]. In fact, TEM observation showed that though the YBCO-YSZ interface remained sharp the interface look irregular (Fig. 3) but no sign of inter duffusion was detected. YBCO on YSZ was polycrystalline and the crystals were separated by almost vertical boundaries.

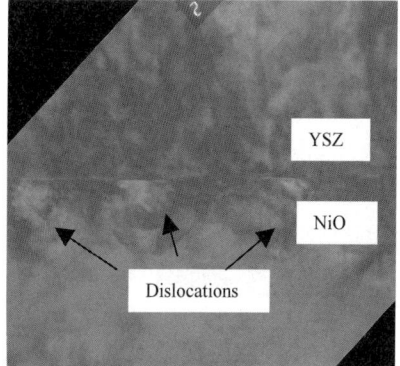
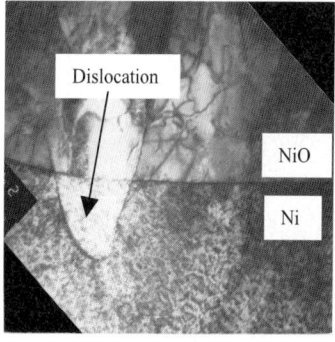

Fig. 2 TEM: YSZ-NiO interface.

Fig. 3 TEM: YBCO-YSZ-NiO cross section view.

Fig. 4 TEM of the Ni/NiO interface

Bright lattice fringes (~14 Å) correspond to YBCO (001) orientation. The planes are not perfectly aligned across the boundaries, with a misorientation as much as 10° (comparable with XRD in-plane texture of 9.3°). The YBCO (001) planes are not straight with a grain but they are especially curled when crossing the grain boundaries (Fig. 3).

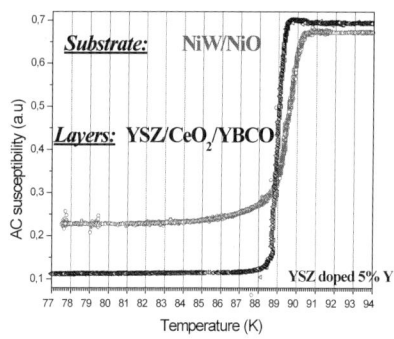
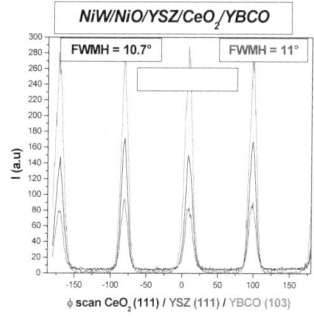
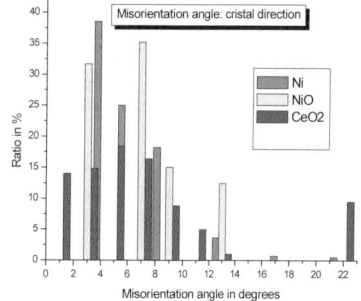

Fig. 5 a) Ac susceptibility measurement of the heterostructure, $J_c = 0.45$ MA/cm^2.

b) XRD: epitaxial growth of the Ni, NiO, YSZ and CeO$_2$ materials.

c) BKD: Grain repartition against misorientation angle.

Thus we added a 150 nm CeO$_2$ film which capped the grooves, and diminished the lattice mismatch with the YBCO, giving denser and smoother YBCO coated Ni tapes as reported elsewhere [6]. Small grains size of the CeO$_2$ films lead to a flatter surface (rms roughness of the grain 12 nm and grooving from 31 nm to 45 nm deep) with CeO$_2$ (111) rare misoriented spots (Fig. 1e). The orientation of the grains layer after layer has been followed by BKD showing a final mean grain misorientation of 7° (Fig. 5 c). Finally BKD diagrams of YBCO were obtained on a second kind of buffered Ni substrate:

NiW/Ni/Y$_2$O$_3$/YSZ/CeO$_2$. BKD overview of the top CeO$_2$ film is repoted on Fig. 6a. Small tapes (20 mm long) were covered with 550 nm YBCO film at a deposition temperature of 800 °C followed by a post annealing oxygenation. Ac suceptibility measurements showed critical temperature onset at 91 K ($\Delta T_c = 0.5$ K) and $J_c = 0.8$ MA/cm^2. Scanning electron microscope (SEM) of the film display some pores on top of the YBCO but no evidence of any [100]/[010] micro-cracking is detected in the YBCO film. YBCO films were then characterised by BKD. The image quality was measured for each individual Kikuchi pattern during the indexation. In Fig. 6c, black spots denotes a bad IQ (Quality Index). Bad IQ implies that the pattern recognition software did not obtain a solution. Thus defects (cracks, pores, amorphous grains) will trigger a larger scatter of the electrons and will clearly appear on the diffraction pattern. Furthemore we find that almost all the grains indexed are (001) oriented and no (110) orientation of the grain was detected.

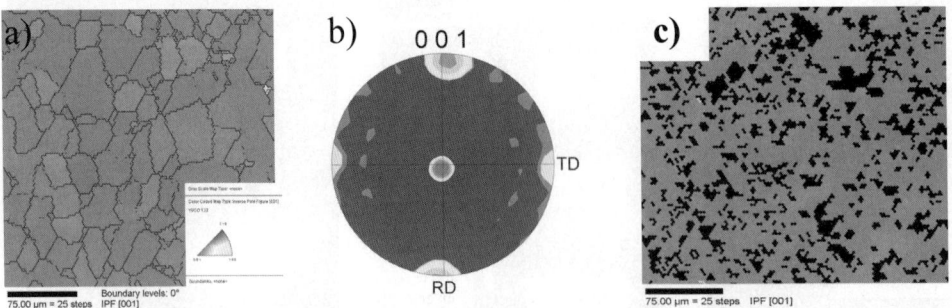

Fig. 6 BKD of the CeO$_2$ buffered tapes (NiW/Ni/Y$_2$O$_3$/YSZ/CeO$_2$) and YBCO films: **a)** Inverse pole figure of CeO$_2$, **b)** Pole figure of YBCO, **c)** Inverse pole figure of YBCO.

4 Conclusions

YBCO films with critical current J_c up to 0.8 MA/cm^2 were performed on Ni pre-buffered tapes whereas on NiO tapes buffer layers and YBCO were obtained by MOCVD showing $J_c = 0.45$ MA/cm^2. Samples were characterised by SEM, TEM, XRD and BKD technique was used to identify and to quantify the weakness of the heterostructure. BKD bridges the XRD and TEM methods and adds the benefits of high spacial resolution. The study of the intergranular properties of the different layers allowed us to establish a statistical evolution of the grain (size, orientation, connection) and revealed that YBCO films, despite high Jc values still have a high proportion of non c-axis oriented grains (and/or zones) due to porosity and dislocations. By using BKD, technique we evaluate that a large J_c improvement is possible by removing these defects from the YBCO films.

Acknowledgements: We would like to thank Dr D.Selbmann and J.Eickmeyer (from IFW, Dresden, Germany) for the NiO tapes and Dr Thieme (from American Superconductor) for the NiW/Y$_2$O$_3$/YSZ/CeO$_2$ tapes. The European Community as part of the Brite Euram READY project, BE97-4572, financially supported this work.

References

[1] P. Tixador, HTS applications, present and future prospects; to be published.
[2] K. Matsumoto, T. Watanabe, T. Tanigawa, and T. Maeda, Long length Y-Ba-Cu-O coated conductors produced by surface-oxidation epitaxy method, IEEE, Trans. Appl. Supercond. **11**, 1 (2001).
[3] D. Dimos, P. Chaudhari, J. Mannhart, and F. K. LeGoues, Phys. Rev. Lett. **61**, 219 (1988).
[4] J.-P. Senateur, C. Dubourdieu, F. Weiss, M. Rosina, and A. Abrutis, Pulsed injection MOCVD of functionnal electronic oxides, Adv. Mater. Opt. Electron. **10**, 155–161 (2000).
[5] J. Eickmemeyer, D. Selbmann, R. Opitz, B. de Oer, B. Holzapfel, L. Schultz, and U. Miller, "Nickel-refractory metal substrate tapes with high cube texture stability°, Supercond. Sci. Technol. **14**, 152–159 (2001).
[6] S. Donet, F. Weiss, J-P. Senateur, P. Chaudouet, A. Abrutis, A. Teiserkis, Z. Saltyte, D. Selbmann, J. Eickmeyer, O. Stadel, G. Wahl, C. Jimenez, and U. Miller, YBCO coated nickel-based tapes with various buffer layers, J. Phys. IV France Pr **11**, 319–323 (2001).

© 2004 WILEY-VCH Verlag GmbH & Co. KGaA, Weinheim

Superconductivity in high-quality $(Hg_{0.9}Re_{0.1})Ba_2CaCu_2O_{6+\delta}$ HTSC thin films

A. Salem[*,1,2], **G. Jakob**[1], and **H. Adrian**[1]

[1] Institute of Physics, Johannes Gutenberg–University, 55099 Mainz, Germany
[2] Physics Department, Faculty of Science, South Valley –University, Qena, Egypt

Received 31 August 2003, accepted 31 December 2003
Published online 8 April 2004

PACS 74.25.Fy, 74.25.Qt, 74.72.Jt, 74.78.Bz, 81.15.Fg

High-quality epitaxial $(Hg_{0.9}Re_{0.1})Ba_2CaCu_2O_{6+\delta}$ (HgRe-1212) HTS thin films were successfully prepared using pulsed laser deposition (PLD) of the $Re_{0.1}Ba_2CaCu_2O_{6+\delta}$ precursor and subsequent Hg vapor annealing. The thin films exhibit a sharp superconducting transition at $T_c \approx 120$ K. The resistive transitions have been investigated in magnetic fields up to 6 T parallel and perpendicular to the c-axis. We have determined the activation energy of thermally activated flux-motion for both magnetic field orientations.

© 2004 WILEY-VCH Verlag GmbH & Co. KGaA, Weinheim

Mercury based high-temperature superconductors, $HgBa_2Ca_{n-1}Cu_nO_{2n+2+\delta}$ [Hg-12(n-1)n, n=1-4], [1, 2] have attracted much attention due to their high superconducting critical T_c. Thin Hg-12(n-1)n films are of great interest for basic research and superconducting devices operating above 100 K. Hg-1212 films [3, 4] have been successfully used for SQUIDs [5]. The high volatile nature, the toxicity of Hg, and the the complexity of processing, have limited the number of studies performed on these thin films. In addition, due to the high hygroscopicity of the precursor materials, extra care should be taken during the preparation to minimize the exposure to air. Stabilization of the Ba-Ca-Cu-O precursors in air is possible by adding a small portion of rhenium [6, 7, 8]. Here we focus on the synthesis and characterization of HgRe-1212 thin films. Our study demonstrates that maintaining the phase equilibrium during the Hg-vapor annealing is important for preparation of high-quality films.

The preparation of HgRe-1212 thin films involves three main steps, (1) preparation of the the precursor target, (2) ablation of the precursor by pulsed laser deposition on (100) oriented $SrTiO_3$ substrates, (3) formation of HgRe-1212 by gas/solid diffusion. The precursor target with the nominal composition $Re_{0.1}Ba_2CaCu_2O_{6+\delta}$ was prepared by solid state reaction from a well ground stoichiometric mixture of ReO_2, $BaCO_3$ and CuO. A calcination at 870-915 °C for 72 h with three intermediate grinding processes was carried out to ensure the homogeneity of composition. The calcinated mixtures were reground and compressed into 23 x 5 mm^2 disk shape using a force of 50 kN. The target was obtained by sintering the compacted disk in air, using a rate of 5 °C/min to 915 °C and maintaining it at this temperature for 72 h, then cooling at a rate of 5 °C/min to room temperature. The slow cooling and heating were crucial to avoid target failure. All steps have been carried out under ordinary laboratory conditions without using special atmosphere. Precursor thin films with a very smooth surface were deposited onto (100) $SrTiO_3$ substrates at room temperature in 0.3 mbar oxygen atmosphere using 28800 pulses (650 mJ/pulse at 8 Hz).

The formation of HgRe-1212 thin films could not be obtained by direct reaction of the thin film precursor with HgO vapor in a sealed quartz tube. The Hg released through the decomposition of HgO at ≈ 500 °C reacted with the CaO component readily to form $CaHgO_2$ at low temperatures (550-700 °C), and fast heating rates have been thought to inhibit the formation of $CaHgO_2$[9]. Use of mixtures of HgO and

---

[*] Corresponding author: e-mail: asalem@mail.uni-mainz.de, Phone: +49 6131 83 88 165, Fax: +49 6131 39 24 076

© 2004 WILEY-VCH Verlag GmbH & Co. KGaA, Weinheim

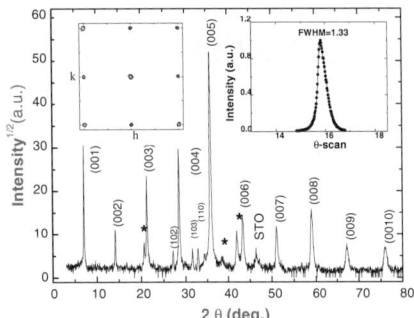

Fig. 1 XRD pattern for a HgRe-1212 thin film. The right inset shows the rocking curve of the (005) peak, the left an intensity mapping of the (hk4) plane of the reciprocal space. Impurity phases are marked by (*).

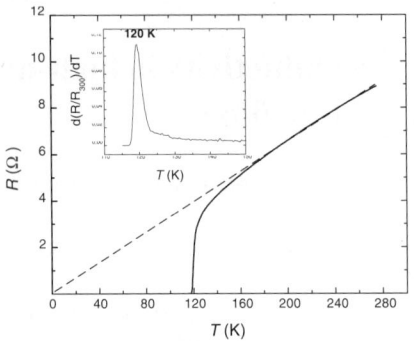

Fig. 2 Temperature dependence of the electrical resistance for a HgRe1212 thin film measured in zero field. The inset is a relation between the temperature derivative of the resistive transition and temperature.

BaCaCuO was shown to control the Hg vapor pressure during the reaction [8]. Therefore the precursor films were sealed into quartz tubes together with pressed Re-212 and HR-212 precursors, which provide the Hg atmosphere. HR-212 was prepared by mixing HgO and the sintered precursor powder according to the formula $(Hg_{0.9}Re_{0.1})Ba_2CaCu_2O_{6+\delta}$ (HgRe-1212). The ratio between Re-212 and HR-212 was 1:3. The quartz tubes were sealed under vacuum (1 mbar), then the ampoule was placed in a steel cylinder as a safety precaution. The evacuated tubes were fast heated to 750 °C in 1 h, and then to 850 °C in 30 min and held at this temperature for 1 h. Finally the samples were cooled to room temperature in 6 hours. The X-ray diffraction (XRD) measurements reveal the structure of the films and the crystallographic orientation. The strong peaks in Fig. 1 are identified as Hg-1212 (00l), which indicates that the *c-axis* of the film is perpendicular to the surface. A *c-axis* = 12.52 Å was thus obtained, which is slightly smaller than that of the bulk Hg-1212 material (*c-axis* = 12.7 Å). The full width at half maximum (FWHM) of the rocking curve of the (005) peak is 1.33 ° (right inset of Fig. 1). Additionally, the crystalline in-plane order was verified by 4-circle X-ray scans (left inset of Fig. 1). The central intensity spot is the (004) reflection. The surrounding spots are the (104), (114), and symmetry equivalent reflections. The electrical resistance of the films was measured by a standard dc four probe method, using a current density of about 10 A/cm^2. Figure 2 shows the electrical resistance as a function of temperature for a HgRe-1212 film measured in zero field. The value of T_c (midpoint) is 120 K. The measurements of the magnetoresistance versus the temperature shown in Fig. 3 were performed with dc magnetic fields up to 6 T parallel and perpendicular to the *c-axis*. The current direction was perpendicular to the magnetic field in both measurements. In the presence of a magnetic field the resistive transition shows a remarkable broadening, that is generally discussed within the framework of thermally activated flux-flow [10], superconducting fluctuations [11], and the vortex-glass transition [12]. For magnetic fields perpendicular to the c-axis, i.e. parallel to the CuO_2 planes, the broadening is much smaller than in the parallel orientation. The anisotropy is stronger than that of YBCO but less pronounced than in Bi-2212 and Tl-2212. The reduced broadening when the field is aligned perpendicular to the *c-axis* is due to the intrinsic pinning provided by the non-superconducting layers in the structure.

Figure 4 shows Arrhenius plots of the resistance for the same data. These Arrhenius plots of the resistive transitions of all films are close to linear in the middle part of these plots as indicated by two dashed lines. At the lowest resistivity levels deviations from linearity occur due to a residual weak link behavior of the sample. For thermally activated flux motion [13] with

$$R(T, B) = R_0 \exp\{-U(B,T)/k_B T\} \quad (1)$$

© 2004 WILEY-VCH Verlag GmbH & Co. KGaA, Weinheim

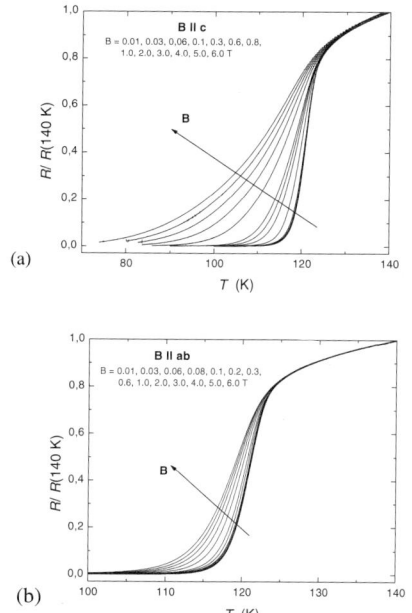
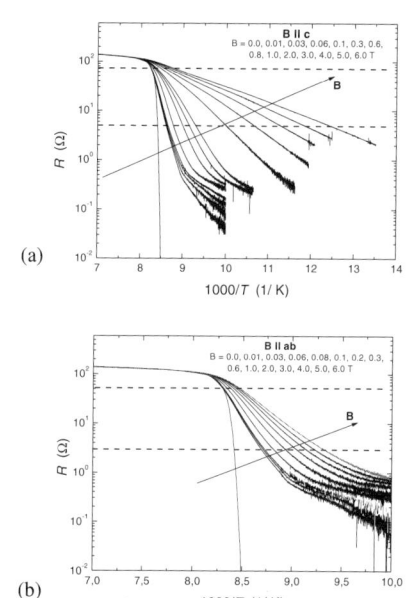

Fig. 3 Temperature dependence of the electrical resistance for a HgRe-1212 film as a function of magnetic field up to 6 T parallel (a) and perpendicular to the c-axis (b).

Fig. 4 Arrhenius plots of the resistive transitions in magnetic fields parallel (a) and perpendicular to the c-axis (b).

the activation energy $U(B,T)$ is determined from the slope of the Arrhenius curves. Also, the activation energy can be obtained by inverting expression Eq. 1. This gives

$$U(B,T) = -k_B T \ln\left(\frac{R(T,B)}{R_0}\right) \qquad (2)$$

The activation energy as a function of temperature $U(T)$ in the presence of a constant magnetic field parallel and perpendicular to the c-axis of HgRe-1212 epitaxial thin film is shown in Figs. 5-a and 5-b, respectively.

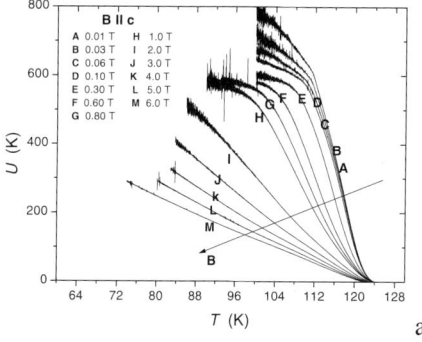
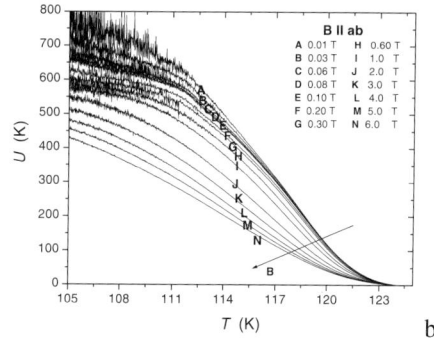

Fig. 5 Temperature dependence of the activation energy $U(B,T)$ determined from Eq. 2 for a HgRe1212 film as a function of magnetic field up to 6 T parallel (a) and perpendicular to the c-axis (b).

© 2004 WILEY-VCH Verlag GmbH & Co. KGaA, Weinheim

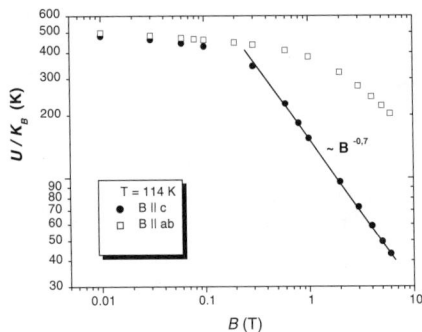

Fig. 6 Magnetic field dependence of the activation energy $U(B,T)$ determined from Eq. 2 for a HgRe1212 film at a temperature of 114 K shown in double logarithmic scale.

For evaluation of Eq. 2 R_0 has been taken as the normal-state resistance at $T=$ 124 K in zero field, i.e. just above the superconducting transition. This choice is physically reasonable and it enforces $U(B,T)$ to go to zero at $T \approx T_c$. The activation energy $U(B,T)$ is plotted as a function of applied magnetic fields in Fig. 6 for a fixed temperature T = 114 K. For fields parallel to the *c-axis* $U(B,T)$ is observed to decrease slightly for fields $B \leq 0.6$ T and a stronger decrease is observed for higher field where $U(B) \propto 1/B^{0.7}$. This indicates a change in dissipation mechanism at higher fields. A similar result has been demonstrated in detail in Bi-2212 thin films in **B** $\parallel$ c orientation [14]. In the field orientation perpendicular to the *c-axis* $U(B,T)$ exhibits the same shape in the low field range and shows a weaker decrease at high fields. Evaluating the U at different temperatures of 111 K and 117 K yields the same behavior.

Preparation and scaling behavior of activation energy of epitaxial superconducting films of HgRe-1212 are reported. [Measurements of the magnetic field anisotropy of flux-motion of HgRe-1212 thin films ($\gamma \simeq$ 7.7) will be presented in a separated paper]. Rhenium stabilized Ba-Ca-Cu-O precursor films are deposited on (001) $SrTiO_3$ substrate by PLD and post-annealed in mercury atmosphere. XRD investigations reveal an oriented film structure with the c-axis parallel to the c-axis of the substrate and in-plane texture. The films show sharp superconducting transitions at $T_c = 120$ K with $\Delta T_c \approx 2$ K. The electrical resistance has been measured in magnetic fields up to 6 T for parallel and perpendicular field orientation. The activation energy for thermally activated flux-motion has been determined as a function of temperature in various magnetic fields.

We acknowledge support by the government of Egypt, the Deutsche Forschungsgemeinschaft (grant AD87-2), the Materials Science Research Center of the University of Mainz and the third international conference on magnetic and superconducting materials.

References

[1] S. N. Putilin, E. V. Antipov, O. Chmaissem, and M. Marezio, Nature **362**, 226 (1993).
[2] A. Schilling, M. Cantoni. J. D. Guo, and H. R. Ott, Nature **363**, 56 (1993).
[3] Y. Q. Wang, R. L. Meng, Y. Y. Sun, Z. J. Huang, K. Ross, and C. W. Chu, Appl. Phys. Lett. **63**, 3084 (1993).
[4] C. C. Tsuei, A. Gupta, G. Trafas, and D. Mitzi, Science **263**, 1259 (1994).
[5] A. Tsukamoto, T. Sugano, S. Adachi, and K. Tanabe, Appl. Phys. Lett. **73**, 990 (1998).
[6] F. Foong, B. Bedard, Q. L. Xu, and S. H. Liou, Appl. Phys. Lett. **68**, 1153 (1996).
[7] Y. Tsabba and S. Reich, Physica C **254**, 21 (1995).
[8] W. N. Kang, R. L. Meng, and C. W. Chu, Appl. Phys. Lett. **73**, 381 (1998).
[9] S. H. Yun, J. Z. Wu, S. C. Tidrow, and D. W. Eckart, Appl. Phys. Lett. **68**, 2565 (1996).
[10] M. Tinkham, Phys. Rev. Lett. **61**, 1658 (1988).
[11] R. Ikeda, T. Ohmi, and T. Tsuneto, J. Phys. Soc. Jpn. **60**, 1051 (1991).
[12] D. S. Fisher, M. P. A. Fisher, and David A. Huse, Phys. Rev. B **43**, 130 (1991).
[13] P. H. Kes, J. Aartz, J. Van den Berg, C. J. Van den Beek, and J. A. Mydosh, Supercond. Sci. Technol. **1**, 242 (1990).
[14] J. T. Kucera, T. P. Orlando, G. Virshup, and J. N. Eckstein, Phys. Rev. B **46**, 11004 (1992).

Neutron powder thermo-diffraction: a very useful tool for the study of crystallisation kinetics and phase segregation in metastable materials

P. Gorria[*,1], **D. Martínez-Blanco**[1], **J. A. Blanco**[1], **J. S. Garitaonandia**[2], **J. Campo**[3], and **R. I. Smith**[4]

[1] Departamento de Física, Universidad de Oviedo, Calvo Sotelo, s/n, 33007 Oviedo, Spain
[2] Departamento de Física Aplicada II, UPV/EHU, P.O. Box 644, 48080 Bilbao, Spain
[3] Instituto de Ciencia de Materiales de Aragón, CSIC-Universidad de Zaragoza, 50009 Zaragoza, Spain
[4] ISIS Facility, Rutherford Appleton Laboratory, Chilton, Didcot, Oxon OX11 0QX, UK

Received 31 August 2003, accepted 31 December 2003
Published online 5 April 2004

PACS 61.12.Ld, 75.50.Bb, 75.50.Tt, 81.20.Ev

This paper gives an overview of what kind of information and how deep can we go into the structure determination of materials, in metastable states, using neutron powder thermo-diffraction, whether temperature induced crystallisation and phase segregation processes have to be completely understood. We have obtained, by means of high energy ball milling technique, four different Fe-based compounds, showing unique characteristics (nanostructured Fe, FeNi alloys, FeCu solid solutions and FeZr amorphous alloys). In situ *neutron diffraction experiments have been carried out in the temperature range between 300 and 1220 K in order to study the great variety of structural changes that takes place in these compounds.*

© 2004 WILEY-VCH Verlag GmbH & Co. KGaA, Weinheim

1 Introduction

It is well established that the physical and chemical properties of a material (electric, magnetic, ...) are intimately related to its microstructure, being this knowledge of primary importance when a complete understanding of these properties is pursued [1]. Neutron diffraction is considered as a fundamental tool when the nuclear and/or magnetic structure of materials has to be determined [2]. Nowadays, Large Facilities as ILL in France or ISIS in UK offer the opportunity, via high flux diffractometers, to collect high resolution diffraction patterns in periods of time of the order of 10^2 seconds. This fact allows making thermo-diffraction experiments in which the kinetics of the structural phase transformations and segregations can be studied in a long range of heating rates [3].

On the other hand, high energy ball milling fabrication technique gives the possibility to obtain a great variety of single phase alloys, in amorphous or crystalline state, and in systems which are immiscible by other conventional synthesis procedures [4]. In concrete, a set of such type of materials are the binary Fe-TM (with TM = Cu, Zr, Ag, ...). The structural and magnetic phase diagram of these metal alloys is still an open question, and constitutes an important task for the benefit of future information in technology developments. For example the phase diagram of the iron–nickel alloy in the bulk is characterised by a structural phase transition from body centred cubic (BCC) to a face centred cubic (FCC) structure, around the compositional range 30 at% of Ni and 70 at% of Fe [5]. This transition is accompanied by a severe loss of the Fe magnetic moment, while the situation in the FeCu case is quite different [6]. These alloys have a number of interesting aspects from both fundamental and applied points of view, especially

---
[*] Corresponding author: e-mail: pgk@pinon.ccu.uniovi.es, Phone: +00 34 98 5102899, Fax: +00 34 98 5103324

regarding their magnetic behaviour (coexistence of ferro and antiferromagnetic character, spin-glass, etc). Moreover, ball milling technique allows to obtain nanostructured metastable materials, with striking magnetic behaviour which is still a matter of discussion [7]. In this contribution a series of *in situ* neutron thermo-diffraction experiments in Fe, Fe–Ni, Fe–Cu and Fe–Zr ball milled samples are presented and the different scenarios that appear (nanostructured BCC-Fe and BCC-FeNi, FCC-FeCu solid solution of "a priori" immiscible elements, and FeZr with amorphous structure) will be discussed in terms of the structural changes that occur during the heating processes.

2 Experimental details

Four samples with nominal compositions, Fe, $Fe_{80}Ni_{20}$, $Fe_{50}Cu_{50}$ and $Fe_{75}Zr_{25}$, in powder geometry have been produced using a Retsch PM 4 high-energy planetary ball mill, from starting powder elements Fe, Ni, Cu and Zr (99.9 % purity and around 10 microns of average diameter size), and using a ball to powder weight ratio between 8:1 and 10:1. The milling time ranges from 40 to 100 hours depending on sample composition. A series of neutron diffraction patterns were collected each 5 min, with $\lambda \approx 1.28$ Å, in D1B diffractometer at ILL (Grenoble, France) and in a temperature range from 300 to 1220 K, for the Fe, FeCu and FeZr samples. The heating rate was 0.5 or 1 K/min. Also, a similar experiment was done for the FeNi sample, in POLARIS time of flight diffractometer at ISIS (RAL, Didcot, UK). FULLPROF software [8] was used for the Rietveld refinement of the patterns.

3 Results and discussion

The BCC-Fe, BCC-FeNi and FCC-FeCu as-milled samples posseses common characteristics such as average grain sizes in the nanometer scale and a high degree of microdeformation due to the mechanical fabrication procedure [6, 9, 10]. These figures are clearly evidenced, as will be shown in the following, by large widening of the diffraction peaks. With the aim to differentiate the behaviour of the four compositions, a separate sub-section will be devoted to the results obtained in each one.

3.1 Nanostructured BCC-Fe

In this case, the effect of high energy ball milling on polycrystalline BCC-Fe is, as mentioned above, the progressive decrease of the average grain size of the crystallites as well as the introduction of strong microstresses during the milling process. The average grain size and the internal deformation values tend to stabilize after about 60–70 hours milling with values below 20 nm and around 0.5–0.6 %, respectively [10]. This nanostructured BCC-Fe shows a decrease of about 10–15 % in the value of the magnetic moment at 5 K respect to that of conventional BCC-Fe, which has been attributed to the large induced microstresses and the existence of non-magnetic Fe atoms in the intergranular region [9]. Moreover, anomalous magnetic behaviour associated with a spin-glass-like transition at low temperatures has been reported [9]. In order to show the structural changes during the thermal process and the possible formation of unknown phases containing Fe, that could be the ultimate responsible of this decrease in the Fe magnetic moment, μ_{Fe}, a sample milled for 80 hours was selected to the study the kinetics between 300 and 1220 K, above the BCC↔FCC reversible martensite↔austenite transformation for pure Fe (1187 K). In Fig. 1, three neutron diffraction patterns at three selected temperatures (as-milled sample at 340 K; above the BCC–FCC transition for Fe, at 1220 K; and at 380 K when the temperature is lowered from 1220 K). Three features have to be noted:

(i) The as-milled sample present wide peaks corresponding to a BCC-Fe phase. Furthermore, a small amount (below 4 wt%) of a metastable Fe oxide (probably γ-Fe_2O_3) is present. The lattice parameter of the BCC-Fe phase is slightly larger (≈ 0.2 %) than that of conventional BCC-Fe. During the heating procedure the progressive sharpening of the peaks is observed, this evidences two thermal induced processes, the grain growth of the nanocrystallites and the disappearance of internal micro-stresses.

(ii) Above 850 K a structural transformation takes place in the sample, γ-Fe_2O_3 transforms into FeO (a paramagnetic iron oxide phase with FCC structure, and called wustite), while BCC-Fe phase shows the well known reversible BCC–FCC transition above 1187 K (see Fig. 1b).

(iii) When cooling the sample only peaks associated with reflections belonging to BCC-Fe and FCC-FeO crystal structures are present (see Fig. 1c). The value for the lattice parameter of BCC-Fe corresponds to the conventional one.

From these results we can say that ball milling technique allows us to obtain massive Fe samples with nanometer structure which can be controlled by milling conditions.

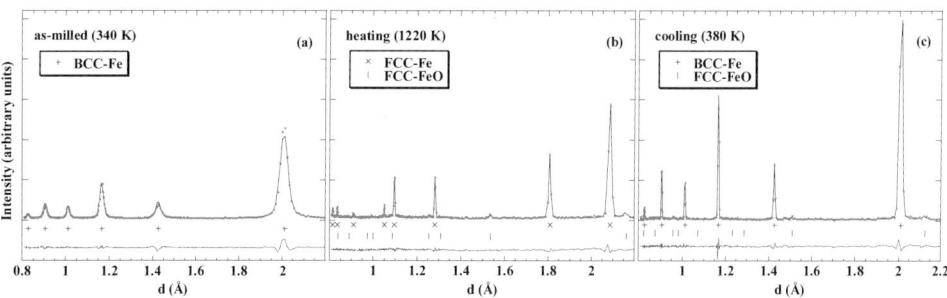

Fig. 1 Neutron diffraction patterns of the Fe powder, milled for 80 hours, at three selected temperatures, a) as-milled sample at 340 K, b) at 1220 K during heating, c) at 380 K during cooling down from 1220 K.

3.2 BCC-$Fe_{80}Ni_{20}$

Fe–Ni binary alloys have been studied for decades due to their technological applications [11]. It is well known that single phase alloys can be obtained by conventional alloying procedures in nearly the whole compositional range, with BCC structure in the Fe-rich side (above 75 at%), and with FCC structure for Fe content below 65 at%. Besides that, a reversible martensite↔austenite transformation is observed in the BCC alloys [5], and an increase up to 50 K is required to complete such transformation. The transformation temperature, T_{MA}, depends on composition, it decreases when Ni content is increased, and shows hysteresis, i.e. $T_{MA}^{heating} > T_{MA}^{cooling}$. We have obtained BCC-$Fe_{80}Ni_{20}$ powder alloy in order to study the possible structural changes induced by ball milling, and we can point out the following noticeable aspect. The martensite↔austenite transformation is observed (see Figs. 2b, 2c), but the beginning of such transformation occurs 200 K below and ends slightly above ($\approx$ 30 K) than expected (around 860 K and 885 K respectively [5]). This fact is due to the atomic diffusion between the interphase regions and the nanograins promoted by thermally induced grain growth processes. Finally, when the sample is cooled down to room temperature (Fig. 2d), the initial martensite structure is recovered.

3.3 FCC-$Fe_{50}Cu_{50}$

The third type of material that we present is a FCC-FeCu solid solution. As we have mentioned in the introduction, Fe and Cu are inmiscible by conventional alloying procedures, but new fabrication techniques permits the obtention of single-phase FeCu materials in nearly the whole compositional range, with BCC structure in the Fe-rich side (above 80 at%), FCC structure below 70 at% of Fe and a mixture of both in between [6, 12, 13]. The origin of the interest of these compounds lies in the fact that even FCC-Cu and FCC-Fe are non-magnetic, FCC-FeCu solid solutions with more than 15 at% of Fe present ferromagnetism [12–15]. However, in this work only the structural features that takes place during the heating process will be presented, the explanation of the magnetic behaviour can be found elsewhere [15]. As can be seen in Fig. 3a, the pattern corresponding to the as-milled sample at 300 K presents peaks associated with reflections corresponding to a phase with a FCC crystal lattice, being a = 3.635(2) Å. This crystal structure is assigned to a FeCu supersaturated solid solution.

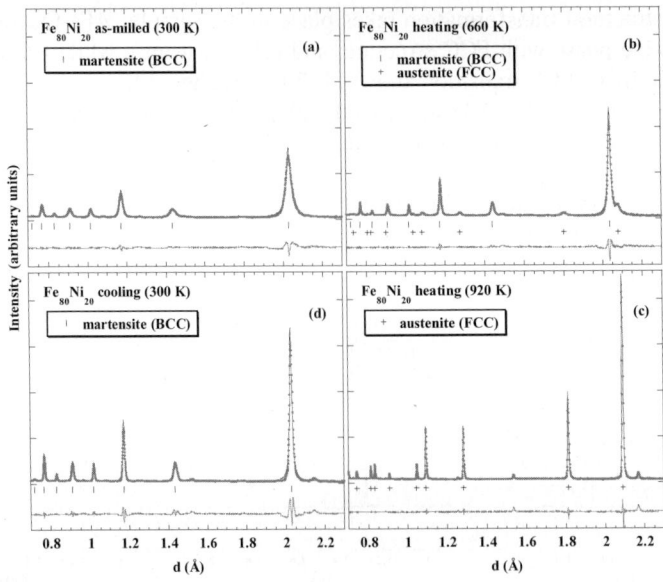

Fig. 2 Neutron diffraction patterns of the Fe$_{80}$Ni$_{20}$ sample at four selected temperatures, a) as-milled sample at 300 K, b) at 660 K and c) at 920 K during heating, and d) at 300 K after cooling down from 1073 K.

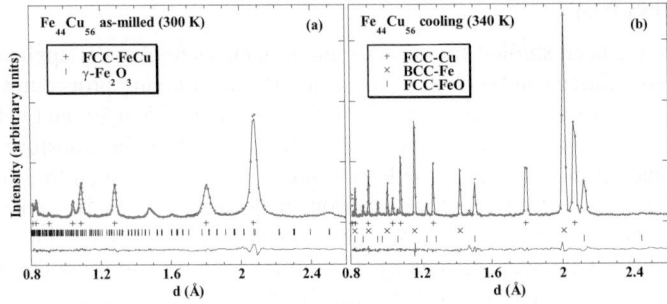

Fig. 3 Neutron diffraction patterns of the Fe$_{44}$Cu$_{56}$ sample, a) as-milled sample at 300 K, b) at 340 K after cooling down from 1200 K.

Small peaks, not detected by X-ray diffraction, and coming from other phase are also observed, and they have been identified as belonging to an iron oxide with tetragonal structure, γ-Fe$_2$O$_3$ (maghemite). Taking into account the relative wt. % of each phase obtained from the analysis of the neutron measurements the real composition of the FCC-FeCu phase is Fe$_{44}$Cu$_{56}$. The most noticeable characteristic of this FeCu system is the thermal decomposition of the initial FCC-FeCu phase into FCC-Cu and BCC-Fe [6, 14]. The segregation of Fe atoms, that will form the BCC phase, begins at around 500 K and ends above 850 K [9]. Together with this segregation process, a transformation from γ-Fe$_2$O$_3$ to FCC-FeO takes place, finishing at the same temperature (850 K). No further structural changes are observed when heating up to 1200 K, but the expected reversible BCC$\leftrightarrow$FCC transformation for BCC-Fe. After cooling down to 340 K (see Fig. 3b), only peaks characteristic of the three mentioned phases (FCC-Cu, BCC-Fe and FCC-FeO) are observed. These results evidences that even it is possible to mix Fe and Cu at the atomic level using ball milling technique, the crystalline FCC phase is metastable in nature, and when thermal energy is supplied to the system a segregation process is induced. From the detailed analysis of the neutron diffraction patterns, we can assure that the final products are pure FCC-Cu and pure BCC-Fe.

© 2004 WILEY-VCH Verlag GmbH & Co. KGaA, Weinheim

3.4 Amorphous-Fe$_{75}$Zr$_{25}$

As we have mentioned in the introduction, single-phase amorphous binary Fe metallic compounds can be obtained by means of ball milling technique. One of such systems is Fe–Zr [4, 16, 17]. Apart of that, ball milling largely extends the compositional range for amorphisation of this system (between 30 and 78 at. % of Fe) compared with other fabrication techniques [4]. Fe–Zr amorphous alloys have been largely studied due to their striking magnetic behaviour including magneto-volume effects [18], however, the study of the magnetic properties is out of the scope of this work.

In order to follow the crystallisation process when it is submitted to a heating process, one Fe-rich FeZr amorphous alloy, Fe$_{75}$Zr$_{25}$, has been chosen for the study. An old controversy exists concerning the Fe-rich intermetallic crystalline phase. It is accepted the existence of two of such phases, Fe$_2$Zr and a second phase that some authors assign two different compositions Fe$_3$Zr or Fe$_{23}$Zr$_6$ [19]. We have selected an Fe$_{75}$Zr$_{25}$ alloy (its stoichiometry corresponds to Fe$_3$Zr) because we expected a single crystallisation process that could lead to pure Fe$_3$Zr or to a mixture of Fe$_2$Zr and Fe$_{23}$Zr$_6$. However, as we can observe in Fig. 4, the crystallisation of the Fe$_{75}$Zr$_{25}$ amorphous alloy is far from be a simple process. From room temperature to around 870 K only haloes characteristic of solid materials without long range atomic order are present At this temperature reflections belonging to two cubic crystalline phases appear superimposed to the amorphous haloes. Both phases have been identified as FCC-Fe$_2$Zr and BCC-Fe. If the temperature is increased, the amount of these two phases increases at expenses of the remaining amorphous matrix, but above 950 K, peaks associated with other crystalline phase with FCC structure begin to appear, while those from Fe$_2$Zr and BCC-Fe begin to decrease in intensity and disappear above 1000 K. At higher temperatures as well as room temperature when cooling the sample, only the diffraction peaks corresponding to the third mentioned phase are present, and taking into account the initial composition of the sample we propose Fe$_3$Zr as the composition of such phase [3, 20].

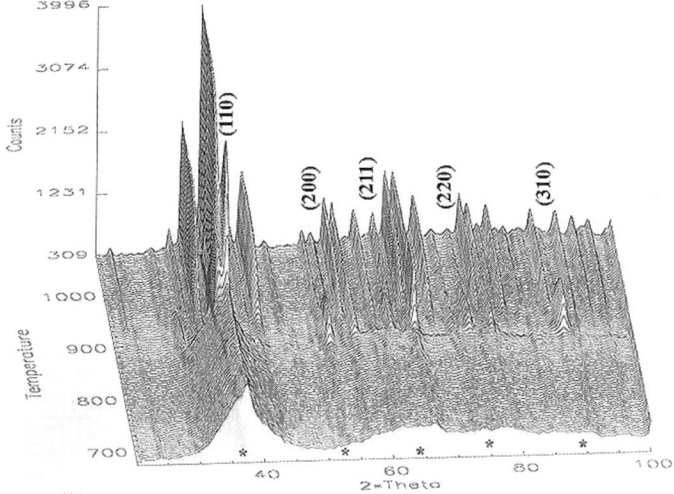

Fig. 4 3D-Neutron thermo-diffractogram of the Fe$_{75}$Zr$_{25}$ between 700 and 1073 K. The reflections corresponding to the metastable BCC-Fe phase are marked with (*).

Besides that, it is worth to note the appearance and disappearance of a metastable BCC-Fe phase as a crystallisation product of an FeZr metallic glass, not yet observed in these kind of materials.

4 Summary and concluding remarks

Ball milling technique allows to obtain a great variety of structural phases for iron, in single or binary metallic compounds, outside thermodynamic equilibrium conditions, i.e. from amorphous to FCC supersaturated and nanostructured solid solutions. This fact offers a unique opportunity to study the magnetism of Fe in different structural environments, previously unknown. However, the understanding of any physical or chemical property, and in particular the magnetic behaviour, of these Fe metastable phases, passes through the good knowledge of the microstructure. Taking into account that these kind of compounds are bulk materials, the most appropriate technique to achieve a good structure determination is neutron powder diffraction, in contrast with X-rays, because neutrons gives information of the whole sample. In this way, intergranular phases as iron oxides, in nanostructured Fe or FeCu samples, can be easily recognised and their relative amount estimated with quite good accuracy. On the other hand, high resolution neutron diffraction patterns can be measured in short periods of time ($\approx 10^2-10^3$ s) with good statistics. Hence, we have shown that *in situ* neutron thermo-diffraction is a very powerful probe when the study of the transformation kinetics at very low heating rates, in metastable materials, is pursued.

Acknowledgements This work was partially supported by the Research Grants MAT2002-04178-C04, MAT2000-1047 and by EU under IHP programme. We also thank the ILL and the CRG-D1B, together with ISIS for the allocation of neutron beam time. One of us, D.M.-B. thanks Spanish CICyT for PhD FPI grant.

References

[1] M. T. Dove, Structure and Dynamics (Oxford Univ. Press, Oxford, 2003).
[2] E. Balcar and S. W. Lovesey, Theory of Magnetic Neutron and Photon Scattering (Oxford Univ. Press, Oxford, 1989).
[3] P. Gorria, J. S. Garitaonandia, R. Pizarro, and J. Campo, in: ILL 2001 Ann. Report, G. Cicognani and C. Vettier (eds.) 48 (2002).
[4] L. Schultz, and J. Eckert, in: Glassy Metals III, H. J. Güntherodt and H. Beck (eds.) (Springer, Berlin, 1994), pp. 69–120.
[5] Magnetic Properties of Metals, d-Elements, Alloys and Compounds, H. P. J. Wijn (ed.) (Springer-Verlag, Berlin/Heidelberg, 1991).
[6] J. Eckert, J. C. Holzer, and W. L. Johnson, J. Appl. Phys. **73**, 131 (1993).
[7] L. Del Bianco, C. Ballesteros, J. M. Rojo, and A. Hernando, Phys. Rev. Lett. **81**, 4500 (1998).
[8] J. Rodríguez Carvajal, Physica B **192**, 55 (1993).
[9] D. Martínez-Blanco, P. Gorria, J. A. Blanco, and J. Campo, Physica B (submitted).
[10] D. Martínez-Blanco, P. Gorria, J. A. Blanco, L. Fernández Barquín, J. S. Garitaonandia, and J. Campo (in preparation).
[11] R. M. Bozorth, Ferromagnetism (Wiley-IEEE, New York, 1993).
[12] C. L. Chien, S. H. Liou, D. Kofalt, Wu Yu, T. Egami, and T. R. McGuire, Phys. Rev. B **33**, 3247 (1986).
[13] E. Ma, M. Atzmon, and F. E. Pinkerton, J. Appl. Phys. **74**, 955 (1993).
[14] P. Crespo, A. Hernando, R. Yavari, O. Drbohlav, A. García Escorial, J. M. Barandiarán, and I. Orue, Phys. Rev. B **48**, 7134 (1993).
[15] P. Gorria, D. Martínez-Blanco, J. A. Blanco, A. Hernando, J. S. Garitaonandia, L. Fernández Barquín, J. Campo, and R. I. Smith (in preparation).
[16] E. Hellstern and L. Schultz, J. Appl. Phys. **63**, 1408 (1993).
[17] R. Pizarro, J. S. Garitaonandia, F. Plazaola, J. M. Barandiarán, and J. M. Grenèche, J. Phys.: Condens. Matter **12**, 3101 (2000).
[18] K. Fukamichi, in: Amorphous Metallic Alloys, F. E. Luborsky (ed.) (Butterworth, London, 1983), Chap. 17.
[19] F. Aubertin, U. Gonser, S. J. Campbell, and H.-G. Wagner, Z. Metallkunde **76**, 237 (1985).
[20] P. Gorria, J. S. Garitaonandia, R. Pizarro, D. Martínez-Blanco, J. Campo, and F. Plazaola, Physica B (submitted).

Theoretical study of diluted magnetic semiconductor trilayers

H. Dakhlaoui and **S. Jaziri**[*]

Département de Physique, Faculté des Sciences de Bizerte, France

Received 31 August 2003, accepted 31 December 2003
Published online 25 March 2004

PACS 73.40.Kp, 75.50.Pp, 85.75.–d

We focus on transport of electron spins for which spin currents can be controlled and manipulated via the electron energy and momentum. In our work we analyze the properties of the hole gas formed in $Ga_{(1-x)}Mn_xAs/GaAs/ Ga_{(1-x)}Mn_xAs$ heterostructure: the electronic structures, polarization and coupling energy E_C are obtained using an efficient self-consistent procedure to solve simultaneously the Schrödinger and Poisson equations tacking account the interaction with Mn magnetic moments and for different parameters of the magnetic $Ga_{(1-x)}Mn_xAs$ layer and the nonmagnetic GaAs spacer.

© 2004 WILEY-VCH Verlag GmbH & Co. KGaA, Weinheim

Diluted magnetic semiconductors have recently attracted a great deal of attention, for their possibility in combining ferromagnetic and semiconductor properties in a single material. The recent discoveries of ferromagnetism with high Curie temperature T_c, in a number of conventional semiconductors doped with magnetic impurities is promising for the implementation of spintronics in semiconductors. Spin injection in low dimensional semiconductors have a great potentiality in the field of magnetoelectronics and spintronics; spin valves and spin injectors are the first practical applications of spintronics [1–3]. Therefore a big effort is being done in two directions: first the study of the origin of ferromagnetism and the high Curie temperature T_c, and second the design and growth of devices. Experimentally the optimal Mn concentration for obtaining high T_c's is near $x \approx 0.06$; thus the magnetic impurities are rather diluted, hence the name "diluted magnetic semiconductors" (DMS). The prototype DMS material is $Ga_{(1-x)}Mn_xAs$ with x = 1–10%. In the ideal material, the Mn ions replace Ga at the cation sites, and their incorporation into GaAs matrix play two roles: they act both as S = 5/2 local moments and as acceptors generating holes in the material.

In this work we study the properties of heterostructure, formed by two magnetic layers GaMnAs of thickness (d_m) separated by a nonmagnetic semiconductor layer GaAs of thickness (d_p) as shown in Fig. 1. DMS's are described by the following Hamiltonian:

$$H = H_h + J\sum_{I,i} S_I \; s_i \; \delta(R_I - r_i) + W\sum_{I,i} n_i \delta(r_i - R_I) \tag{1}$$

The first term describes the carriers, it is the sum of the kinetic energy of the holes and the hole–hole interaction energy. For the range of carrier density of interest in DMS's the carrier–carrier interaction is not relevant and we neglect it. We treat the kinetic energy within the envelope function approximation. In this approach we describe the hole electronic states of the host semiconductor by a parabolic band. The second term proportional to J is the antiferromagnetic exchange interaction between the spin of the Mn^{2+} ions located at R_I and the spins s_i of the itinerant carriers, the last term is an interaction between the carrier charge n_i and the potential arising from the magnetic dopants. We solve Hamiltonian (1) in the mean field approximation, similar to the Jellium model. The local magnetic interaction of the spin carriers with

---

[*] Corresponding author: e-mail: Sihem.Jaziri@fsb.rnu.tn

the Mn spins is substituted by the interaction with an effective magnetic potential $V_c = \frac{S_I J x}{a^3}$ [4], In this expression a^3 is the unit cell volume of the host semiconductor, S_I is the ion spin. J is the antiferromagnetic local coupling. In the same model the electronegativity difference between the carriers and the Mn ions is described by an effective potential $V_w = \frac{Wx}{a^3}$; since there is no reliable experimental information on the value of W, we consider it as a parameter with value 0<W<J. Within this approach, in our heterostructure the holes are free to move in the x–y plane and the one particle wave functions and eingenvalues have the following forms:

$$\psi_{i,\vec{k},\sigma} = \frac{\exp(i\vec{k}\cdot\vec{r})}{\sqrt{s}} \phi_{i,\sigma}(z); \quad \varepsilon_{i,\vec{k},\sigma}(z) = \frac{\hbar^2 \vec{k}^2}{2m} + \varepsilon_{i,\sigma}(z)$$

Here s is the areal dimension of the sample, $\vec{r}$ and $\vec{k}$ are the position and the momentum of the carriers in the plane perpendicular to the growth direction, i is a subband index, $\phi_{i,\sigma}(z)$ and $\varepsilon_{i,\sigma}$ are obtained from the one-dimensional Schrödinger equation in the z-direction:

$$-\frac{\hbar^2}{2m}\frac{\partial^2 \Phi_{i,\sigma}(z)}{\partial z^2} + V_{tot}(z) \Phi_{i,\sigma}(z) = \varepsilon_i \Phi_{i,\sigma}(z) \qquad (2)$$

with $V_{tot}(z) = V_h(z) + V_\sigma^\alpha(z)$. $V_\sigma^\alpha(z)$ is an effective potential and has the following form [5]:

$$V_\pm^\alpha(z) = \begin{cases} \mp V_c + V_w & \text{for} \quad 0 < z < d_m \\ 0 & \text{for} \quad d_m < z < d_m + d_p \\ \mp C_\alpha V_c + V_w & \text{for} \quad d_m + d_p < z < 2d_m + d_p \\ \infty & \text{otherwise} \end{cases} \qquad (3)$$

where C_α is +1(–1) in the ferromagnetic (antiferromagnetic) coupling case, α stands for the solution with ferromagnetic or antiferromagnetic coupling and σ is the carriers spin index up (+) or down (–). $V_c = \frac{SJx}{2a^3}$ and $V_w = \frac{wx}{a^3}$, the effective potential can be designed by the following profiles.

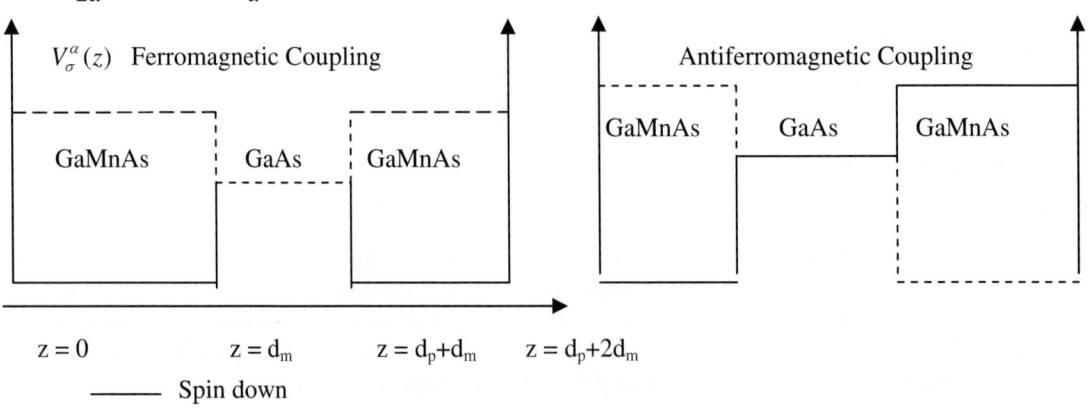

Fig. 1 Potential profiles for both spin carriers directions in the ferromagnetic and antiferromagnetic coupling.

In the Schrödinger equation the Hartree potential $V_h(z)$ is obtained by solving the one dimensional Poisson equation:

$$\frac{\partial^2 V_h(z)}{\partial z^2} = \frac{4\pi e^2}{\varepsilon_0 \kappa}[-C_M(z) + n_+(z) + n_-(z)] ; \quad (4)$$

$C_M(z)$ is the density of Mn ions, $n_\sigma(z)$ is the spin dependent density of the holes, κ is the dielectric constant.

$$n_\sigma(z) = \sum_i \int \frac{d^2\vec{k}_\parallel}{(2\pi)^2} f(\varepsilon_i, \sigma) |\Phi_{i,\sigma}(z)|^2 , \quad (5)$$

where $f(\varepsilon_i, \sigma)$ is the Fermi distribution function defined as :

$$f(\varepsilon_i, \sigma) = \frac{1}{1+\exp(\frac{\varepsilon_{i,\sigma}+\mu-\varepsilon_K}{K_B T})} . \quad \text{where } \varepsilon_k = \frac{\hbar^2 k_\parallel^2}{2m_\parallel}, \; \mu \text{ is the chemical potential, } K_B \text{ the Boltz-}$$

mann constant, T is the temperature. We solve equations (2)–(5) self-consistently [6, 7], using the finite difference technique. For convenience we express the density of holes and Mn ions with $n_0 = (a_B)^{-3}$, where $a_B = \frac{4\pi\varepsilon_0 \kappa\hbar^2}{m e^2}$ is the Bohr radius of hole; $m = 0.5\, m_0$ is the effective hole mass and $k = 12.7$ is the dielectric constant. The parameters of the heterostructure are $d_p = 2.5\, nm$ and $d_m = 7\, nm$, T=50° K, μ = 25 meV. We have plotted in Fig. 2 the density of carriers in the ferromagnetic coupling case as function of V_W; according to our previous results, the spatial charge distribution n(z) should be influenced by the band offset V_W. We notice that the hole concentration increases in both magnetic layers GaMnAs, however there is a penetration in the nonmagnetic layer GaAs which separates them. This penetration depends on V_W. Our results indicate that through the parameter W which describes the interaction between magnetic Mn ions and holes we can control the density of carriers along the heterostructure.

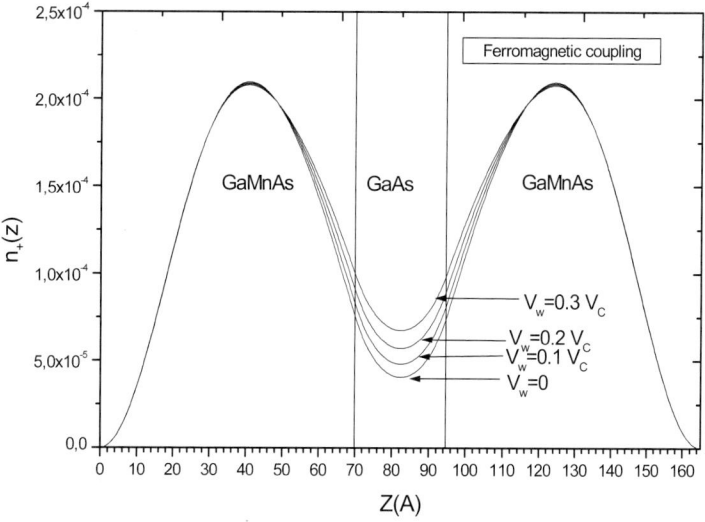

Fig. 2 Profiles of the carriers density for different values of V_W.

We define the magnetic coupling energy as $E_C = E_{AF} - E_F$, where E_{AF} and E_F are respectively the energy of the antiferromagnetic and the ferromagnetic configuration. In Fig. 3, we plot E_C as a function of the band offset V_w, for two values of J. The coupling energy E_C decreases when V_w increases and it changes sign as a function of the band offset V_w. This coupling is non zero because there is a paramagnetic holes gas in the central layer which mediates the interaction between the two magnetic layers. Note that the value of the exchange coupling energy is smaller than in metallic systems. However in DMS heterostructures the magnetic field necessary to overcome the antiferromagnetic coupling is

$$B \approx \frac{E_{AF} - E_F}{g\ \mu B\ S} \approx 100 Gauss$$, i.e. it is large enough for magnetoresistance applications.

In order to study the spin distribution in the structure, we have plotted in Fig. 4 the polarization

$$p(z) = \frac{n_+(z) - n_-(z)}{n_+(z) + n_-(z)},$$ in both ferromagnetic and antiferromagnetic configurations, as a function of

the position z. The polarization is constant in the two magnetic layers, and presents some variation in the non magnetic layer GaAs, due to the mixing of spin up and spin down in this central layer.

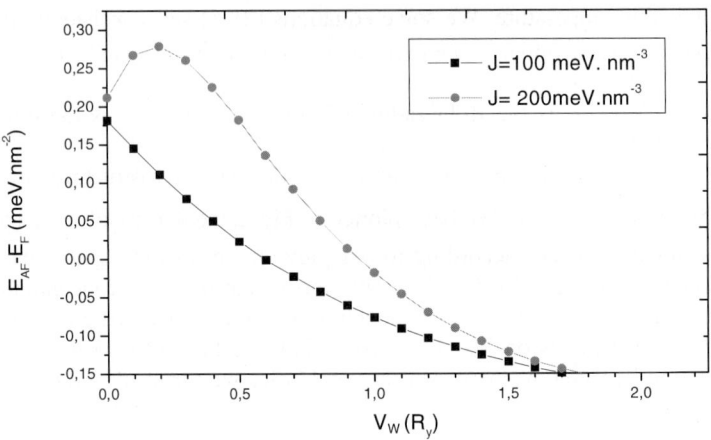

Fig. 3 Energy difference between the antiferromagnetic and ferromagnetic coupled magnetic layers as function of the band offset V_w / V_c.

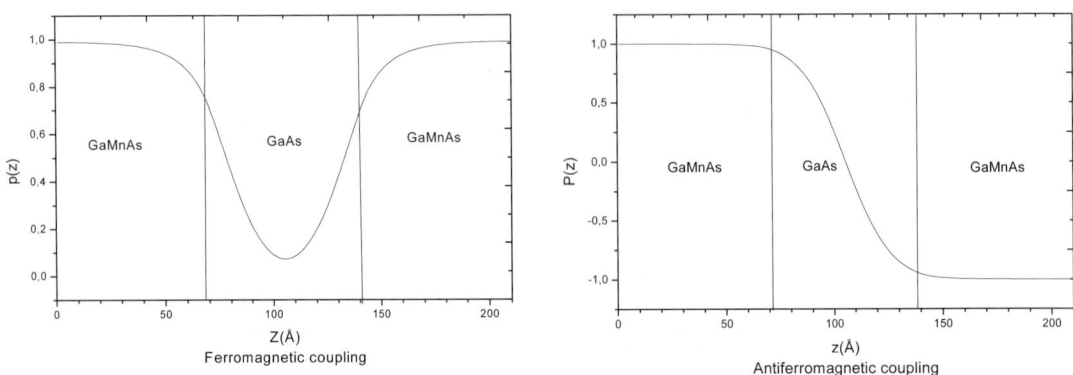

Fig. 4 Polarization in the two magnetic couplings (d_m=7 nm, d_p = 7 nm).

© 2004 WILEY-VCH Verlag GmbH & Co. KGaA, Weinheim

In summary, we have studied an heterostructure formed by two DMS layers separated by a nonmagnetic semiconductor using self consistent procedure; we have investigated the properties of this strucure (charge distribution and the polarization P(z)) and we have evaluated the coupling energy as function of the band offset V_w. Our resultsshow that it is possible to control the sign of coupling between the magnetic layers by mean of the band offset V_w which describes the interaction between the Mn ions and the charges carriers. Our results may be useful to perform new devices based on these materials.

References

[1] Semiconductor Spintronics and Quantum Computation, D. D. Awschalom, D. Loss, and N. Smarth (eds.), series Nanosciences and Technology (Springer, Berlin, 2002).
[2] S. A. Wolf et al., Science **294**, 1488 (2001).
[3] E. I. Rashba, Fiz. Tverd. Tela (Leningrad) **2**, 1224 (1960), [Sov. Phys. – Solid State **2**, 1109 (1960)].
[4] J. Fernandez-Rossier and L. J. Sham, cond-matt/0201326, (2003).
[5] M. P. Lopez-Sancho. M. C. Munoz, and L. Brey, cond-matt/0207287, (2002).
[6] José A. Cuesta, Angel Sanchez, and Francisco Dominguez-Adame, J. Semicond. Sci. Technol. (1995).
[7] J. Konig, J. Schliemann, T. Jungwirth, and A. H. MacDonald, cond-matt/0111314, (2001).

Zeeman coupling and Swap action in spin-based diluted magnetic semiconductor quantum dot quantum computer

A. Hichri[1] **and S. Jaziri**[*,2]

[1] Laboratoire de Physique de la Matière Condensée, Faculté des Sciences de Tunis, Tunisie
[2] Département de Physique, Faculté des Sciences de Bizerte, 7021 Zarzouna, Bizerte, Tunisie

Received 31 August 2003, accepted 31 December 2003
Published online 25 March 2004

PACS 71.70.–d, 75.50.Pp, 75.70.–i, 85.75.–d

We present theoretically the Zeeman coupling and exchange induced swap action in spin-based quantum dot quantum computer models in the presence of magnetic field. We study the valence and conduction band states in a double quantum dots made in diluted magnetic semiconductor. The later have been proven to be very useful in building an all-semiconductor platform for spintronics. Due to a strong p–d exchange interaction in diluted magnetic semiconductor ($Cd_{0.57}Mn_{0.43}Te$), the relative contribution of this components is strongly affected by an external magnetic field, a feature that is absent in non magnetic double quantum dots. We determine the energy spectrum and the swap time as a function of magnetic field and the inter dot distance within the Hund–Mulliken molecular-orbit approach and by including the Coulomb interaction.

© 2004 WILEY-VCH Verlag GmbH & Co. KGaA, Weinheim

Modern information technology utilizes the charge degree of freedom of electrons in semiconductors to process the information and the spin degree of freedom in magnetic materials to store the information. From the physical points of view, the enhanced spin related phenomena due to the coexistence of the magnetism and semiconductor properties have been recognized in magnetic semiconductors and diluted magnetic semiconductors (DMSs). DMSs are based on non-magnetic semiconductors, and are obtained by allowing them with a sizable amount (a few percents or more) of magnetic elements, such as Mn. The family of magnetic semiconductors have been extensively studied, because of their peculiar properties resulting from the coupling between itinerant electrons and localized magnetic spins (s–d exchange interactions). In this paper we focus the strong carrier's magnetic ion Zeeman splitting which leads, e.g., to the effect of giant spin splitting of energy bands [1]. The discovery of new principles of computation based on quantum mechanics has led to the idea of using coupled quantum dots (QDs) for quantum computation [2]. The requirement for the quantum bit of information (qubit), which is at the heart of the quantum computer, is that it can exist in any state of a quantum two two-level system, i.e. $|\psi\rangle = \alpha|0\rangle + \beta|1\rangle$, where $|0\rangle$ and $|1\rangle$ are the states of the "classical" bit, and $|\alpha|^2 + |\beta|^2 = 1$. It is crucial to explore the challenges facing coherent control of qubits in solid state structures, particularly, the issue of possible error corrections in realistic systems. Among various microscopic degrees of freedom that have been considered for the role of qubits in solid state quantum computer, spins of electrons are natural candidates because of their well-defined Hilbert spaces and their relatively long decoherence time compared to the orbital degrees of freedom. In the proposed spin-based quantum computer, the exchange energy J (the energy difference between the two lowest states) and Zeeman coupling to an external magnetic field plays a fundamental role of establishing two qubit entanglements [3]. Swaps are used to move spin states around so that an arbitrary pair of spins can be brought into controlled entanglement, which is

---

[*] Corresponding author: e-mail: Sihem.Jaziri@fsb.rnu.tn, Phone : +216 72 591 906, Fax: +216 72 591 566

© 2004 WILEY-VCH Verlag GmbH & Co. KGaA, Weinheim

essential to quantum computer [1]. In this paper, we consider the theoretical issue of controlling the swap operation in the proposed solid state quantum computer involving QD spin entanglement, with externally controlled magnetic field. Our analysis is based on an adaptation of a Hund–Mulliken variational technique to parabolically confined coupled QDs. We calculate the energy of pairs of QDs in which two carriers are vertically coupled via quantum tunneling and are subject to the Coulomb interaction.

The main difference between calculations for DMS based and non-magnetic structures arises from the presence of the carrier-magnetic ion (sp–d) exchange interaction. The s–d exchange interaction can be written in terms of Zeeman-like Hamiltonian $H_{s-d} = N_0\, \alpha\, x\, \langle S \rangle\, s^e$ in the case of the conduction electrons, and the p–d interaction [4], $H_{p-d} = N_0\, \beta\, x\, \langle S \rangle\, s^h/3$ in the case of the valence band holes. Here s^e and s^h are operators of electrons spin and the total angular momentum of hole, respectively, and x is the molar fraction of magnetic ions. To describe the average value of the component of the localized spins along the applied magnetic field direction $\langle S_j^z \rangle$, a modified Brillouin function with two phenomenological parameters $S_0 = S_0(x)$ and $T_0 = T_0(x)$ is commonly used [5]

$$\langle S_j^z \rangle = S_0 B_s\left(\frac{g\mu_B BS}{k_B(T+T_0)}\right) \qquad (1)$$

here $g=2$ is the Landé factor of manganese d-electrons, $S = 5/2$ is a magnitude of the spin of the magnetic ions, and μ_B is the Bohr magneton. The remaining two components of average spin vanish in the materials in question $\langle S_j^x \rangle = \langle S_j^y \rangle = 0$. The corresponding exchange constants $N_0\alpha$ for the conduction band states and $N_0\beta$ for the valence band states in majority of II–VI DMS, are usually of opposite signs [6]. Usually also, the absolute value of $N_0\beta$ is considerably greater than $N_0\alpha$.

We consider a system of two vertically coupled QDs containing one carrier each. It is essential that the carriers are allowed to tunnel between the dots, and the total wav function of the coupled system must be anti–symmetric. It is this fact which introduces correlations between the spins via the charge (orbital) degrees of freedom. The Hamiltonian which we use for the description of two electrons or holes confined in vertically coupled QDs is

$$H_{orb} = \sum_{i=1,2} h_{\alpha i} + C + H_z \qquad (2)$$

where $h_{\alpha i} = \frac{1}{2m_{\alpha i}}(p_{\alpha i} - \frac{q_{\alpha i}}{c}A(r_{\alpha i}))^2 + V_{\alpha i, l}(r_{\alpha i}) + V_{\alpha i, v}(r_{\alpha i})$ is the single particle Hamiltonian for the i-th electron $(\alpha = e, q_e = -1)$ or hole $(\alpha = h, q_h = +1)$ in three dimensions with coordinate r_α. Electrons or holes have effective masses m_α and confinement energies $\hbar w_\alpha$. The coupling of the dots V_v along the inter-dot axis is modelled by a double well potential [7], whereas for the lateral confinement we choose the parabolic potential. The Coulomb interaction is included by $C = q_\alpha^2/\kappa|r_{\alpha 1} - r_{\alpha 2}|$ with the dielectric constant κ. We allow for a magnetic field $B = (0,0,B)$ applied along the z-axis and which couples to the carrier charge via the vector potential $A(r) = B(-y,x,0)/2$, and to the spin via the Zeeman coupling term H_z.

We consider a Hund–Mulliken method [8] of molecular orbits which include the states with double occupation $|\psi_\pm^d\rangle = (|0\rangle \pm |1\rangle)(|0\rangle \pm |1\rangle)/(2(1\pm S))$ and the states with simple occupation $|\psi_\pm^s\rangle = \{(|0\rangle + |1\rangle)(|0\rangle - |1\rangle) \pm (|0\rangle - |1\rangle)(|0\rangle + |1\rangle)\}/2\sqrt{1-S^2}$, here $|0\rangle$ is the ground state wave function of the one-particle confined in upper of the double dot system, whereas $|1\rangle$ corresponding to the lower dot. A non-vanishing overlap $S = \langle 0|1 \rangle$ implies that the carrier can tunnel between the dots. At low temperature where $k_B T \ll \hbar w_z$, we can restrict our selves to the two lowest orbital eigenstates of H_{orb} [9], one of which is symmetric (singlet state) and the other one is anti-symmetric (triplet state). The exchange energy $J = E_T - E_s$ originates from the Coulomb interaction (due to Pauli principle), whereas Zeeman splitting is purely a spin effect, thus we can safely ignore the Zeeman effect when we discuss the orbital degrees of freedom.

The magnetic field B also couples to the carrier spin via the Zeeman term, hence to take it into account means to introduce an additional parameter into the equations, describing splitting of spin levels in the magnetic field. According to this notation, the Hamiltonian describing the electrons and holes in DMS QD takes the form

$$H_c = \sum_i h_i + C + N_0 \alpha \, x \, \langle S \rangle \, s_i^e - g_c \mu_B B \, s_i^e \,, \tag{3}$$

$$H_v = \sum_i h_i + C + N_0 \beta \, x \, \langle S \rangle \, s_i^h / 3 - g_v \mu_B B \, s_i^h \tag{4}$$

where g_c, g_v are the Landé factor correspondingly for $\pm 1/2$ electrons in the conduction band and $\pm 3/2$ heavy holes in the valence band.

For the sake of simplicity, to describe s–d and p–d exchange interactions we use in this paper the values of these constants equal to their values in the bulk. This simplification becomes invalid only in very small QDs (with radius smaller than about 30 Å). Admittedly, for small quantum dots one should replace bulk by the values of p–d exchange constants, calculated in [10]. This does not change, of course, the qualitative results obtained in this work. Throughout this paper, we have evaluated our results for two vertically equal large $Cd_{0.57}Mn_{0.43}Te$ QDs with 125 Å in diameter and 70 Å high. Calculations are performed using the following parameters $\kappa = 8.5$, $T = 4$ K, for holes, $m_{//} = 0.14$, $m_\perp = 0.48$, $g_v = 0.15$, for conduction electrons, $m_c = 0.095$, $g_c = -5$ [6]. In $Cd_{1-x}Mn_xTe$ $N_0\alpha = 220$ meV and $N_0\beta = -880$ meV [11]. We plot in Fig.1 the energy spectrum as a function of magnetic field. The inter-dot distance $a = 225$ Å for holes and $a=85$ Å for electrons system. We see immediately that the character of the hole ground state via the Zeeman splitting differ strongly by an applied magnetic field than the electron spectrum. Since the crossing between the two lowest ground states is depicted at a field of few T for holes.

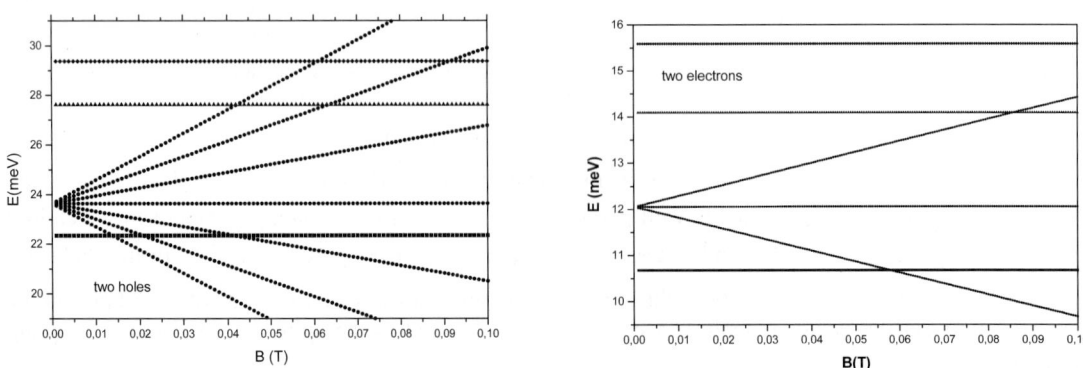

Fig. 1 Field dependence of the lowest carrier levels for two vertically coupled $Cd_{0.57}Mn_{0.43}Te$ dots, with including Zeeman coupling effect.

When the overlap of the electronic wave function is small, we can assume that the electron orbital degrees of freedom are frozen, so that an effective spin Hamiltonian quite faithfully describes the two-electron spin system

$$H_s = Js_1^e s_2^e + \left(\langle 0|N_0\alpha \times \langle S \rangle - g_c\mu_B B|1\rangle\right) s_1^e + \left(\langle 1|N_0\alpha \times \langle S \rangle - g_c\mu_B B|0\rangle\right) s_2^e$$
$$= Js_1^e s_2^e + \gamma_1 s_1^e + \gamma_2 s_2^e \ . \tag{5}$$

Here, we have implicitly assumed, based on the small inter-dot wave function overlap, that the two spins are distinguishable, with spin 1 on the left dot and spin 2 on the right dot. We have also assumed that the field is entirely along the z direction.

Hamiltonian (5) can be expressed in the basis of four two-spin states $|\uparrow\uparrow\rangle = |00\rangle$, $|\downarrow\downarrow\rangle = |11\rangle$, $|\uparrow\downarrow\rangle = |01\rangle$ and $|\downarrow\uparrow\rangle = |10\rangle$ and its eigenstates can be easily obtained. The two polarized states are decouples from the other two, which are mixtures of singlet and triplet states : $|\psi_1\rangle = |\uparrow\uparrow\rangle$, $|\psi_2\rangle = |\downarrow\downarrow\rangle$, $|\psi_3\rangle = (|\uparrow\downarrow\rangle + |\downarrow\uparrow\rangle)/\sqrt{2}$ and $|\psi_4\rangle = (|\downarrow\uparrow\rangle - |\uparrow\downarrow\rangle)/\sqrt{2}$ with the energies $E_1 = J + \Delta$, $E_2 = J - \Delta$, $E_3 = J$ and $E_4 = -3J$, where $\Delta = \gamma_1 + \gamma_2$ represents the overage magnetic field. An important question here is whether theses mixtures and shifts will cause any error in quantum computation in the schemes based on the exchange energy. After all, the swap action in theses models depends on the perfect phase matching in the evolution of singlet and triplet states, as we will show later. Since swap operation is an essential component of the spin based QD quantum computer and several other architectures, we need to precisely quantify the effects of mixtures in singlet and triplet states on the swap action.

To determine whether swap is affected, we explore whether a product state of spin 1 and 2 with evolve into a product state again. Our strategy here is to calculate the concurrence C, which varies from C=0 for an unentangled state (in the usual sense that they are factorized into single particle states) to C=1 for a maximally entangled state. The most general pure states in the case of the two-qubit model can be written as $|\psi\rangle = a_1|\uparrow\uparrow\rangle + a_2|\uparrow\downarrow\rangle + a_3|\downarrow\uparrow\rangle + a_4|\downarrow\downarrow\rangle$ where $\sum_{i=1}^{4}|a_i|^2 = 1$. The concurrence of the pure state is simply given by [12] $C(|\psi\rangle) = 2|a_2 a_3 - a_1 a_4|$, thus, the pure state is entangled if and only if $a_1 a_4 \neq a_2 a_3$. We can then find out whether this pure state corresponds to a swapped state. We consider that the initial state is an unentangled state given by $|\phi(0)\rangle = (\alpha_1|\uparrow\rangle + \alpha_2|\downarrow\rangle)(\beta_1|\uparrow\rangle + \beta_2|\downarrow\rangle)$. If the two electrons are located in two well-separated QDs in the beginning, the above product state does not violate the anti-symmetry requirement of a two fermions state. This state can be expanded in the basis of the eigenstates of Hamiltonian (5). The two spin state at time t takes the form

$$|\phi(t)\rangle = \alpha_1\beta_1 e^{-iE_1 t/\hbar}|\uparrow\uparrow\rangle + \alpha_2\beta_2 e^{-iE_2 t/\hbar}|\downarrow\downarrow\rangle + \frac{1}{\sqrt{2}}(\alpha_1\beta_2 + \alpha_2\beta_1) e^{-iE_3 t/\hbar}|\uparrow\downarrow\rangle + \frac{1}{\sqrt{2}}(\alpha_1\beta_2 - \alpha_2\beta_1) e^{-iE_4 t/\hbar}|\downarrow\uparrow\rangle$$

to have an unentangled state means $C = \left|2\alpha_1\beta_1\alpha_2\beta_2 - (\alpha_1^2\beta_2^2 - \alpha_2^2\beta_1^2) e^{i4Jt/\hbar}\right| = 0$. The solution here are $e^{i\theta} = \pm 1$, where $\theta = 4Jt/\hbar$. When $e^{i\theta} = 1$, $|\phi_1(t)\rangle = (\alpha_1|\uparrow\rangle + \alpha_2 e^{i\Delta t}|\downarrow\rangle)$, the state of the first spin returns to its initial state with a phase shift between the two coefficients, the swap time $t_{swap} = \pi\hbar/2J$. When $e^{i\theta} = -1$, $|\phi_1(t)\rangle = (\beta_1 e^{i\Delta t}|\uparrow\rangle + \beta_2|\downarrow\rangle)$, the swap is achieved with the exception of an additional phase that can be corrected easily with a single qubit operation.

In Fig. 2, we plot the swap time as a function of magnetic field and the QD separation for the $Cd_{0.57}Mn_{0.43}Te$ QD.

Physically, a uniform field means that the Zeeman coupling couples to the total spin (including both electron spins), so that the Zeeman term commutes with the exchange term in the Hamiltonian (5), and therefore does not change the eigenstates. The shifts in the energy levels of the polarized states cause additional phase shift, but can be corrected by applying an opposite magnetic field with the same pulse shape, magnitude, and length.

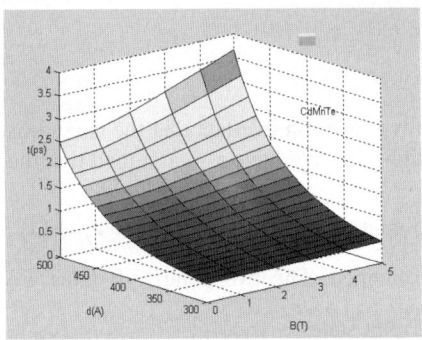

Fig. 2 Swap time to perform a quantum logic gate operation between two spins as a function of the magnetic field and a QD separation. The swap time depend only on J but do not depend to the spin.

In summary, using the Hund–Mulliken approach, we have calculated the energy as a function of magnetic field for carriers confined in a pair of vertically coupled QDs, and have compared the two holes spectra to the two-electron spectra. For two-hole filling in the presence of a magnetic field, the ground-state crossing occurs at fields of a few T, depending on the strength of the confinement, the distance coupling, and the dots size. The swap time given by $t_{swap} = \pi\hbar/2J$ takes a bit to flip from a state $|\uparrow\rangle$ to an orthogonal state $|\downarrow\rangle$. In summery, an external uniform magnetic field does not qualitatively change the proposed quantum computer algorithm logistically it makes the quantum computer operation more difficult because of the necessary correction pulses.

References

[1] S. M. Ryabchenko, Izv. Akad. Nauk SSSR, Se. Fiz. **46**, 440 (1982), Diluted Magnetic Semiconductors, J. K. Furdyna and J. Kossut, Semiconductors and Semimetals, Vol. 25 (Academic Press, New York, 1987).
[2] V. N. Golovach and D. Loss, cond-mat/0201437.
[3] F. Yamaguchi and Y. Yamamoto, Appl. Phys. A **68**, 1 (1999);
D. Loss and D. P. Di Vincenzo, Phys. Rev. A **57**, 120 (1998);
B. E. Kane, Nature (London) **393**, 133 (1998).
[4] I. A. Merkulov, D. R. Yakovlev, A. Keller, W. Ossau, J. Geurts, A. Waag, G. Landwehr, G. Karczewski, T. Wojtowicz, and J. Kossut, Phys. Rev. Lett. **83**, 1431 (1999).
[5] J. A. Gaj, R. Planel, and G. Fishman, Solid State Commun. **29**, 435 (1997).
[6] V. V. Platonov, O. M. Tatsenko, and A. I. Bykov, Physica B **246**, 319 (1998).
[7] A. Hichri and S. Jaziri, J. Nanosci. Nanotech. **3**, 162 (2003).
[8] D. C. Mattis, in: The Theory of Magnetism I, Vol. 17, Springer Series in Solid State Sciences (Springer, New York, 1988), Chap. 4.5.
[9] G. Burkard, D. Loss, and D. P. Di Vincenzo, Phys. Rev. B **59**, 2070 (1999).
[10] A. K. Bhattachcharjee, Phys. Rev. B **58**, 15660 (1998).
[11] F. V. Kyrychenko and J. Kossut, Phys. Rev. B **61**, 4449 (2000).
[12] W. K. Wooters, ibid. **80**, 2245 (1998);
W. K. Wooters, e-print quant-phys/0001114; M. C. Arnesen, S. Bose, and V. Vedral, Phys. Rev. Lett. **87**, 017901 (2001).

Microstructure evolution after thermal treatments of nanocrystalline Ni₃Al and Ni₃Al+B produced by filling

M. Khitouni* and **N. Njah**

Laboratoire de Métallurgie Appliquée, FSS, B. P. 802–3018, Sfax, Tunisie

Received 31 August 2003, accepted 31 December 2003
Published online 1 April 2004

PACS 61.46.+w, 61.72.Ww, 81.16.–c

Changes in the microstructure of nanocrystalline Ni_3Al have been investigated by comparing a high-purity material with boron-doped (0.1 and 0.2 wt%) compounds. The nanocrystalline microstructure was obtained by filing. X-ray diffraction was used to characterize the material properties after different thermal treatments. After filing the materials were completely disordered, with small crystallite size of about 10 nm, similar in all materials. The influence of boron on the structural evolution was also investigated. For increasing annealing temperatures, the following phenomena took place successively: an increase in the LRO with crystallites; then a general improvement of the structure, with a decrease in the concentration of lattice defects. By comparison with pure Ni_3Al, the presence of boron results in an increase of the temperature at which microstructure improvements take place. This latter should be related to the interactions of the boron atoms with point defects as well as dislocations.

© 2004 WILEY-VCH Verlag GmbH & Co. KGaA, Weinheim

1 Introduction

The intermetallic compound Ni_3Al takes the $L1_2$-type ordered crystal structure based on f.c.c. lattice. The minor element Al is located at the corner of a unit cell and the major element Ni at the face center. Plastic deformation of such a material results in the introduction of high concentrations of lattice defects and also in some disordering of the structure [1–5]. Its application potentialities are much extended if the intrinsic grain-boundary brittleness is removed by doping with small amounts of boron [6]. However, boron might influence the kinetics of the processes occurring during heat treatments. The disorder and nanocrystalline structure introduced during deformation are subsequently removed by recovery and recrystallization, when the temperature of the material is increased. Zhou et al. [7] have performed a comparative study of polycrystalline undoped and boron-doped Ni_3Al after a compression by 16 %; they found a very weak retarding effect of the boron addition on recrystallization. Baker et al. [8] found that boron retards recrystallization but does not affect the grain growth rate for extruded rods of Ni_3Al.

The aim of the present work was to determine the effect of boron additions on the evolution of the microstructure during annealing of the deformed Ni_3Al powders produced by filing and to gain more insight into the physical mechanisms that are operative. For this purpose, X-ray diffraction measurements was performed.

2 Experimental

The intermetallics chosen for this investigation were binary $Ni_{76}Al_{24}$ (atomic percentages) and boron-doped $Ni_{76}Al_{24}$(0.1 wt%B, referred to (Ni+Al) and $Ni_{76}Al_{24}$(0.2 wt%B) (Table 1). They were prepared by melting together weighed quantities of high purity Ni, Al and a (Ni-18 wt % B) master alloy in an induc-

---

* Corresponding author: e-mail: khitouni@yahoo.fr, Phone: +00 216 276400, Fax: +0021674274437

tive plasma furnace. No significant change of mass occurred during melting. The starting materials and the alloy synthesis in the plasma furnace were kindly provided by CECM-CNRS, Vitry, France. The ingots were homogenized by annealing for 40 hours at 1323 K under a vacuum of 6.10^{-4} Pa and filed to powders. X-ray diffraction (XRD) was performed on a wide angle diffractometer in the θ-2θ step scan mode by using CuKα radiation. The long-range order (LRO) parameter S was calculated from the integrated intensities of the (100) or (110) superlattice reflections normalized to (200) or (220) fundamental reflections, respectively [4, 5]. The integral breadth $β_e$, defined as the ratio between the integrated intensity I_{int} and its height I_{max}, was used to calculate the crystallite sizes D and the rms-lattice distortions $(ε^2)^{1/2}$ by using the Halder-Wagner approach [9]. I_{int} and I_{max} were determined by using a WinPLOTR software [4]. The experimental measured breadth $β_e$ of a sample reflection was corrected from the instrument effects $(β_i)$ [4] and $β_i$ was taken as the breadth of the same reflection for the annealed powder at 1173 K/3h.

The calculation of intrinsic stacking fault probability α was carried out from (111) and (200) peaks shifts using the following formula [10]:

$$α = 8.3 × \left(\frac{2}{\sqrt{3}} - \left(\frac{\sin θ_{200}}{\sin θ_{111}} \right) \right) \quad (1)$$

where $θ_{111}$ and $θ_{200}$ are peak positions in the deformed in the studied powder containing stacking faults.

Table 1 Alloy compositions (*referred to (Ni +Al)).

Alloy	Ni (at%)	Al (at%)	B (wt%)
Ni₃Al	76.04±0.01	23.96±0.01	–
Ni₃Al(0.1 wt%B)	(76.04±0.01)*	(23.96±0.01)*	0.1
Ni₃Al(0.2 wt%B)	(76.04±0.01)*	(23.96±0.01)*	0.2

3 Results and discussions

Figure 1 shows the XRD patterns of deformed powders over a 2θ range of 20°–80°. They contains the (111), (200), and (220) reflections and as shown, no superlattice reflection is detectable for all alloys. The deformed powder, exhibits then a fully disordered structure. The profiles of all Bragg reflections of deformed powders are significantly broadened, this is related to the reduction of crystallite size and to the important lattice strain introduced by filing. The mean size of the crystallite <D>, the rms-strain $(ε^2)^{1/2}$, the lattice parameter a and the intrinsic stacking fault probability α of the deformed powders are given in Table 2. Figure 2 shows the morphology of the as filed powder of the investigated alloy. A clear refinement is observed.

Table 2 Structural parameters of the deformed powders.

Alloy	<D>(nm)	$(ε^2)^{1/2}$ (%)	α (%)	a(Å)
Ni₃Al	10.020	1.090	4.850	3.591
Ni₃Al(0.1 wt%B)	11.775	0.632	2.401	3.583
Ni₃Al(0.2 wt%B)	9.040	0.489	2.000	3.582

During the annealing of the deformed powders, the transition from a disordered structure consisting of high concentration of defects, to an ordered structure of technical importance, was observed. Figure 3 shows the changes in the long range ordering (LRO) on isochronal annealing. It was observed that the

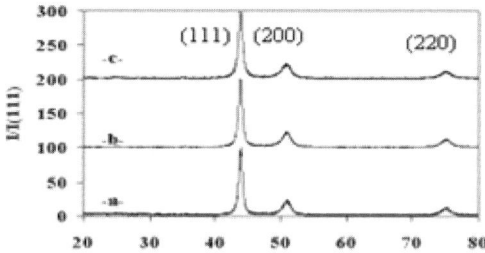 

Fig. 1 X-ray diffraction patterns of filed materials: (a) Ni$_3$Al, (b) Ni$_3$Al (0.1 wt%B) and (c) Ni$_3$Al(0.2 wt%B).

Fig. 2 Ni$_3$Al powder observed by SEM.

doped alloys ordering start at a temperature between 573 and 673 K. By contrast, the undoped alloy ordering start at temperature between 473 and 573 K. Figure 4 represents the evolution of the deformation (intrinsic) stacking fault probability calculated during isochronal annealing. It is clear that a subsequent anneal reduces the lattice defects.

The isochronal microstructure changes in terms crystallite size and lattice strains were presented in Fig. 5. For increasing annealing temperatures, the following phenomena took place successively: a decrease in the density of defects with an increase in the crystallite size. Thus an anneal at 573 K of the boron-free compound causes some increase in the average crystallite size and decrease of the internal stress. The boron-doped materials exhibited the same processes of structure evolution, but shifted to higher temperatures.

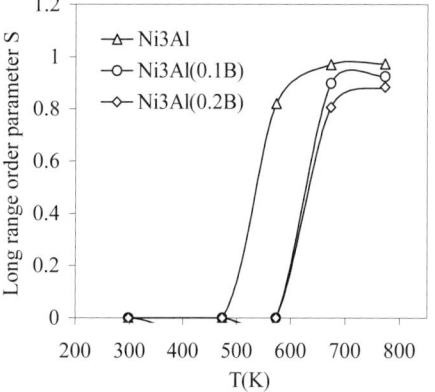 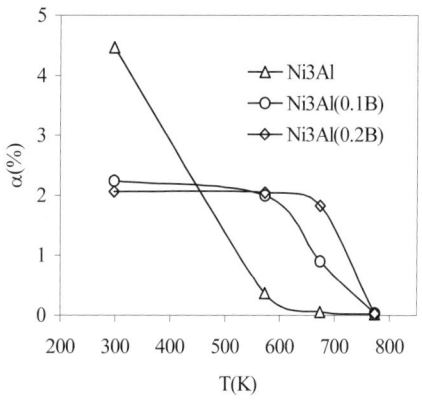

Fig. 3 Evolution of the long range order parameter S during isochronal annealing.

Fig. 4 Evolution of the intrinsic stacking fault probability during isochronal annealing.

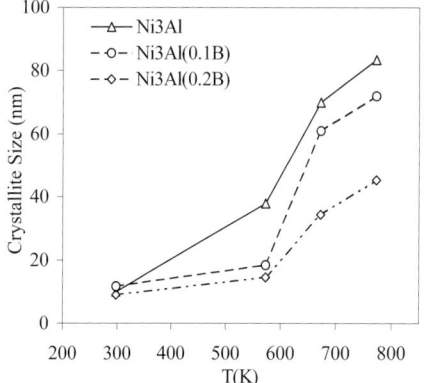 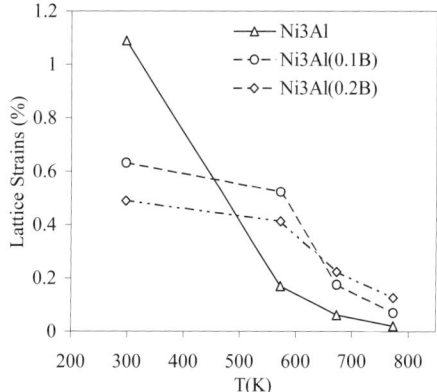

Fig. 5 Evolution of the crystallite size and the lattice strains during isochronal annealing.

A very interesting question concerns the relation between the obtained results and the boron additions. Although boron segregates on grain boundaries, we cannot a priori exclude its influence on the long-range kinetics, because the segregation cannot significantly change the bulk B-concentration as the active areas of grain boundaries are only several atomic layers thick [11, 12]. Boron additions were shown above to delay the long-range ordering of deformed Ni$_3$Al, resulting in a shift of the corresponding process to higher temperatures by about 373 K (Fig. 3). So, the boron addition could affect the ordering kinetics by the subsequent mechanisms; as reported by Yang et al. [6], recrystallization in Ni$_3$Al was retarded by boron, since the boron atoms presumably from atmospheres around dislocations, reducing their mobility, as well as reducing grain boundary mobility. On the other hand it reduces the effective vacancy jump rate with its direct interaction in solid solution with the migrating vacancies, released from the above defects [13]. Then, a decrease of the atomic mobility necessary for ordering.

4 Conclusion

Filing allowed both pure and boron-doped Ni$_3$Al to be highly strained, and resulted in completely disordered materials with a mainly nanocrystalline structure, about 10 nm in size. For increasing annealing temperatures of the deformed powder, the following phenomena took place successively: an increase in the LRO with crystallites; then a general improvement of the structure, with a decrease in the concentration of lattice defects. By comparison with pure Ni$_3$Al, the presence of boron results in an increase of the temperatures at which long-range ordering and microstructure improvement take place. The delayed microstructure recovery of the doped material should be related to the interactions of the boron atoms with point defects as well as dislocations.

Acknowledgements We would like to thank Dr. O. Dimitrov (C.E.C.M, Vitry) for kindly providing the materials and for the possibility to elaborate the alloys. We are grateful to Dr. A. Srasra (I.N.R.S.T, Tunis) for the possibility to use the X-ray diffractometer.

References

[1] R. Z. Valiev, Metas. Mechan. Alloyed and Nanocryst. Mater. **343**, 773 (2000).
[2] A. V. Korznikov, O. Dimitrov, G. F. Korznikova, J. P. Dallas, B. S. R. Idrisova, R. Z. Valiev, and F. Faudot, Acta Mater. **47**, 3301 (1999).
[3] J. S. C. Jang and C. Koch, J. Mater. Res. **5**, 3 (1990).
[4] M. Khitouni, A. W. Kolsi, and N. Njah, Ann. Ch. Sci. Mater. **28**, 17 (2003).
[5] M. Khitouni, A. W. Kolsi, and N. Njah, Phys. Chem. News **9**, 131 (2003).
[6] Y. Yang and I. Baker, Scripta Metall. **34**, 803 (1996).
[7] B. Zhou, S. L. Yu, Y. T. Chou, and C. T. Lu, Intermetallic Compounds, Ed. O. Izumi (The Japan Institute of Metals, Sendai), p. 779 (1991).
[8] I. Baker, D. V. Viens, and E. M. Schulson, J. Mater. Sci. **19**, 1799 (1984).
[9] N. C. Halder, C. N. Wagner, and C. N. J., Adv. X-ray Anal. **9**, 91 (1966).
[10] A. I. Salimon, A. M. Korsunsky, and A. N. Ivanov, Mater. Sci. Eng. **A271**, 196 (1999).
[11] C. T. Liu, C. L. White, and J. A. Horton, Acta Metall. **33**, 213 (1985).
[12] I. Baker, B. Huang, and E. M. Schulson, Acta Metall. **36**, 493 (1988).
[13] N. Njah, D. Gilbon, and O. Dimitrov, Scripta Metall. **33**, 1379 (1995).

Synthesis of potassium chloroapatites, IR, X-ray and Raman studies

H. El Feki[*1], **M. Amami**[1], **A. Ben Salah**[1], and **M. Jemal**[2]

[1] Laboratory of Material Sciences and Environment, University of Sfax, 3038 Sfax, Tunisia
[2] Faculté des Sciences de Tunis, 1060 Belvedre Tunis, Tunisie

Received 31 August 2003, accepted 31 December 2003
Published online 25 March 2004

PACS 61.10.Nz, 61.66.Dk, 78.30.Hv, 81.20.–n

New conditions of synthesis Ca-chlorapatite and K-Ca-chlorapatite by using the melt salts method were determined. The results of X-ray diffraction show that the compound Ca-chlorapatite crystallizes in the hexagonal system (space group P63/m) with the crystallo-graphic parameters (a = b = 9.6228(3) Å and c = 6.7610(3) Å) that are closely equal to those of literature. Two potassium calcium chlorapatites were prepared by this method. Samples were identified by X.R diffraction, IR spectroscopy, chemical analysis and density measurements. Results allow to propose a global general formula $(Ca_{10-x}K_x(PO_4)_6(Cl)_{2-x}$ with $0 < x < 2$). IR absorption and Raman diffusion studies allow to attribute different modes of vibration and to show the structural disorder.

© 2004 WILEY-VCH Verlag GmbH & Co. KGaA, Weinheim

1 Introduction

Apatites constitute a large family of components, which are isomorphous to the fluorapatite. One of the essential characteristics of this structure is its aptitude to form solid solutions and to accept a large number of substitutents [1, 2, 3, 4, 5]. Some works exist concerning the synthesis of substituted chlorapatites [6, 7]. However, attempt to synthesize Ca- chlorapatites and Sr-chlorapatites in aqueous solution leads always to the hydroxyapatites because of the competition between hydroxyl and chloride ions. Some authors affirmed that the calcium chlorapatite could be obtained only by reaction in the solid state [8, 9]. Klement [10] prepared chlorapatite in a melted salt medium.
On the other hand, IR absorption and Raman diffusion experiments performed on apatites, show that the replacement of calcium by other cations causes perturbations of the vibration modes of the PO_4 groups [11]. In order to seek the influence of potassium on the spectroscopic properties of the apatites, samples of chlorapatite have been synthesized then characterized using IR absorption, X-ray diffraction and Raman scattering.

2 Preparation and characterization of the samples

Potassium chlorapatite samples, ClApK1 and ClApK2, have been prepared according to a melt salts method which has been used by Klement et al. [10]. The method is modified by putting the re-agents in a KCl melt. So a succession of $CaCl_2$, $2H_2O$ and $CaHPO_4$, $2H_2O$ layers, alternated with the KCl layer is introduced in a platinum crucible. The mixture is then heated at 900 °C during 12 hours. The sample exempt of potassium ions ClAp has been prepared by substituting $CaCl_2$ to KCl. The amounts of reagents

---

[*] Corresponding author: e-mail: Hafed.ElFeki@fss.rnu.tn, Phone: +00216 74274923, Fax: +00216 742 7437

are given in Table 1. The IR study has been carried out on Shimadzu IR470 spectrometer. X-ray measurements were determined by using a XRD 3000 Seifert powder diffractometer. The cell parameters have been deter-mined by the Dicvol91 program [12]. Raman spectra were recorded at room temperature using com-puterized Dilor RTI 30 triple monochromator. The samples have been submitted to an elementary chemical analysis. Calcium has been determined by complexometry [13]. Analysis of potassium is carried out by atomic absorption spectrometry with a Shimadzu spectrometer. Phosphorus ions have been determined colorimetrically [14]. The titration of the chloride ions has been realized by chromatography in liquid phase. Finally measurements of density have been carried out according to the flotation method [15]. The molar weight of a sample is determined from its density and crystallographic parameters.

Table 1 Operative preparation conditions of the samples.

Samples	ClAp	ClApK1	ClApK1
Reagents	0.735 g of $CaCl_2$, $2H_2O$	1.5 g of KCl	3 g of KCl
	1.29 g of $CaHPO_4$, $2H_2O$	0.162g of $CaCl_2$, $2H_2O$	0.265 g of $CaCl_2$, $2H_2O$
		0.344 g of $CaHPO_4$, $2H_2O$	0.691 g of $CaHPO_4$, $2H_2O$

3 Results and discussion

The X-ray diffraction (Fig. 1) shows that all the synthesized samples contain a unique crystallized apatitic phase. All the peaks have been indexed in the space group P63/m. Crystallographic parameters of the ClAp sample (Table 2) are in agreement with those obtained by Young from the data recorded on single crystal (a = 9.628(5)Å, c = 6.764(5)Å) [16]. The examination of crystallographic parameters of the prepared samples, shows that a parameter decreases and c increases with the potassium content. The decrease of a can be explained by the coexistence of two antagonist effects: the substitution of calcium ($r_{ca_2^+}$ = 0.99Å) by a more voluminous cation (r_{k+} = 1.33Å) inducing the a extension, but, the creation of vacancies in the channel, due to the substitution of a bivalent cation by a monovalent one, causes a decrease of the channel diameter and consequently a decrease of the a parameter. The second effect seems to be prevalent, since a diminution of the a parameter occurs. On the basis of chemical analysis results (Table 3), there is a simultaneous decrease of the mole number of calcium and chloride when potassium content increases. Therefore, we can attribute to the synthesized samples the following general formula: $Ca_{10-x}K_x(PO_4)_6(Cl)_{2-x}$ with $0 < x < 2$, x = 0.7 and 1.3 for the compounds ClApK1 and ClApK2 respectively.

Table 2 Crystallographic parameters and chemical analysis of the synthesized samples.

Samples	a (0.001Å)	c (0.001Å)	% Ca	%K	% P	% Cl
ClAp	9.620	6.760	38.420	0.000	17.867	6.820
ClApK1	9.580	6.850	36.635	2.690	18.317	4.545
ClApK2	9.534	6.931	35.026	5.106	18.727	2.501

Infrared absorption spectra (Fig. 2) show that all the studied samples have an apatitic structure and are pure chlorapatites. We can notice particularly the absence of bands at 3560 and 630 cm^{-1}, due to hydroxyl stretch (ν_S) and hydroxyl libration (ν_L) respectively [2]. They only contain four groups of bands due to phosphate ions that are all infrared active. We can note that all these bands shifted to lower wavelengths, when the potassium content increases (Table 4). Those displacements are probably due to a

Table 3 Number of ions per unit cell contents of ClAp and ClApK.

Samples	Density	Mexp(g/mol)	n_{Ca} (0.10)	n_K (0.03)	n_P (0.05)	n_{Cl}(0.02)
ClAp	3.191 (0.008)	1041 (3)	10.00	0.000	6.00	2.00
ClApK1	3.095 (0.008)	1015 (3)	9.30	0.700	6.00	1.30
ClApK2	3.022 (0.007)	993 (2)	8.70	1.300	6.00	0.70

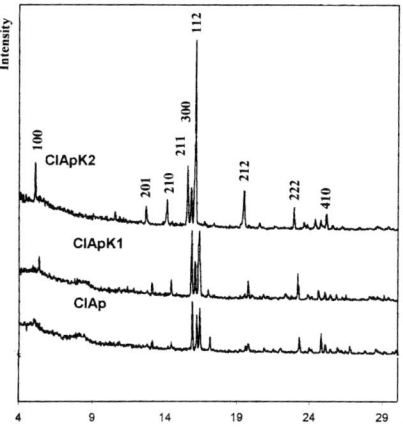

Fig. 1 XRD spectra of the samples ClAp, ClApK1 and ClApK2.

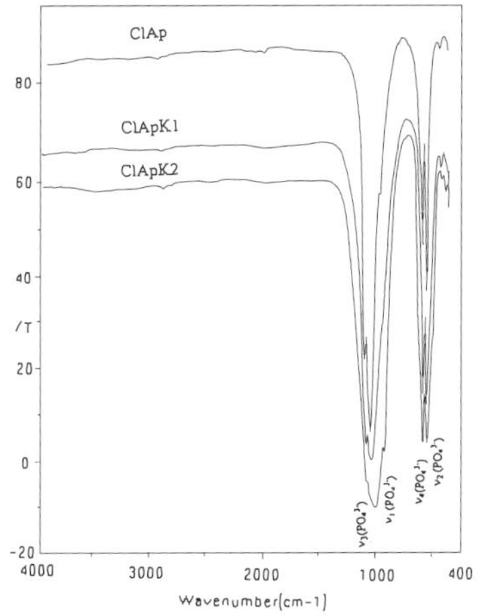

Fig. 2 Evolution of potassium chloroapatites IR spectra with the potassium content.

change of the PO_4 group environment caused by the potassium introduction. In fact, the P-O bond decrease of consecutive to the increase of the O-Metal bond subsequently to the substitution of Ca by K or a vacancy.

Table 4 Attribution of infrared absorption bands of chlorinated apatites.

Samples	ν_1 (cm$_1$)	ν_2 (cm$_1$)	ν_3 (cm$_1$)	ν_4 (cm$_1$)
ClAp	961	474, 486	1035, 1053, 1081	564, 590, 603
ClApK1	954	469	1022, 1030, 1075	558, 586, 593
ClApK2	949	460, 465	1015, 1050, 1070	550, 581, 590

The results of the Raman spectroscopy (Figure 3) are in agreement with those of IR ones, that show an enlargement of the bands and a less resolution when potassium substitutes to calcium ions. In the frequency domain between 1200 and 300 cm^{-1}, the Raman spectrum contains only the bands due to each of the four phosphate internal vibrational modes. One $\nu_1(PO_4)$ band at 960 cm^{-1}, two $\nu_2(PO_4)$ bands at 425 and 440 cm^{-1}, four $\nu_3(PO_4)$ bands at 1020, 1048, 1059 and 1080 cm^{-1} and three $\nu_4(PO_4)$ bands at 575, 590 and 610 cm^{-1} were resolved. It is noticeable that the $\nu_1(PO_4)$ band was extremely intense in the Raman spectrum. Moreover, the Raman spectrum of external vibration modes, that are localized in the frequency domain inferior to 300 cm^{-1}, shows two types of vibration: a band at about 62 cm^{-1} relative to the vibration of translation T'(PO_4) and a band at about 83 cm^{-1} corresponding to the libration motion R'(PO_4). In addition, we notice the existence of weak bands at about 120 and 26 cm^{-1} that can be attributed to the vibration of the K$^+$ or Ca^{2+} cations. The enlargement of bands and the appearance of a shoulder in ClApK2 spectrum, at about 962 cm^{-1} is probably due to a variation of the P-O bond length [17]

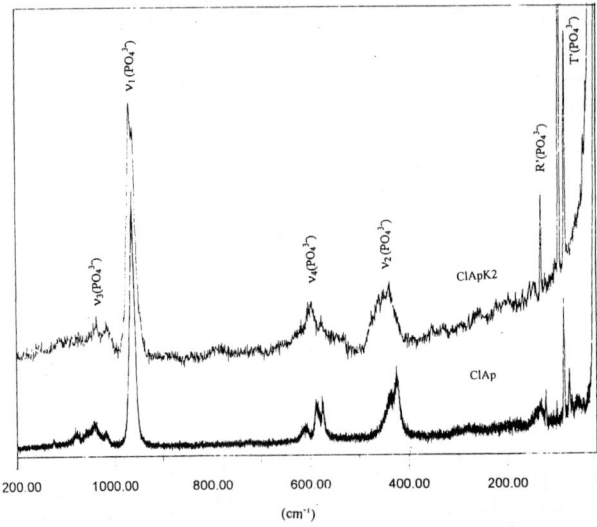

Fig. 3 Raman Spectra of the samples ClAp and ClApK2.

and consequently to a deformation of the PO_4 tetrahedron when the potassium content increases, provides evidence that the potassium insertion (element more voluminous than calcium) results in the creation of a structural disorder.

4 Conclusion

In This work we determine a new method of synthesis of K-Ca-Chlorapatite using a melt salts method. The X-ray investigations shows that the K-Ca-Chlorapatite is isomorphous to the hydroxylapatites. A complimentary study by Chemical analysis, IR, and Raman Spectroscopy, allows to point out the structural disorder in the K-Ca-Chlorapatite.

References

[1] J. C. Trombe and G. Montel, C. R. Acad. Sci. **280**, 567 (1975).
[2] G. Montel, G. Bonel, J. C. Heughebaert, J. C Trombe, and C. Rey. J. Cryst. Growth **53**, 74 (1981).
[3] A. Nounah, J. L. Lacout, and J. M. Savariault, J. Alloys Compd. **188**, 141 (1992).
[4] H. El Feki, I. Khattech, M. Jemal, and C. Rey, Thermochim. Acta **237**, 99 (1994).
[5] E. A. P. De Maeyer, R. M. H. Verbeeck, and I. Y. Pieters, Inorg. Chem. **35** (1996).
[6] E. C. Moreno, T. M. Gregory, and W.E. Brown, J. Res. Nat. Bur. Stand. A **72**, 733 (1968).
[7] H. Mcdowell, T. M. Gregory, and W. E. Brown, J. Res. Nat. Bur. Stand. A **81**, 273 (1977).
[8] U. S. Rai, K. K. Rao, and T. S. B. Narasaraju, Ind. J. Chem. **18A**, 168 (1979).
[9] T. S. B. Narasaraju and D. E. Phebe, J. Mater. Sci. **31**, 1 (1996).
[10] R. Klement and S. Harth, Chem. Ber. **94**, 1452 (1961).
[11] H. El Feki, C. Rey, and M. Vignoles, Calcif. Tissue Int. **49**, 269 (1991).
[12] A. Boultif and D. Lour, J. Appl. Cryst. **24**, 987 (1991).
[13] G. Charlot, Les méthodes de la Chimie Analytique, Masson, (1961).
[14] A. Gee and V. R. Deitz, Ann. Chim. **25**, 1320 (1953).
[15] M. Marraha, Thèse d'Etat, I.N. P. Toulouse (1989).
[16] R. A. Young and J. C. Elliot, J. Arch. Oral. Biol. **11**, 699 (1966).
[17] W. E. Klee and G. Engel, J. Inorg. Nucl. Chem. **32**, 1837 (1970).

Author Index

Abakumov, A. M. . . . . (a) 1403
Adachi, H. . . . . (c) 1752
Adrian, H. . . . . (c) 1961
Aeppli, G. . . . . (b) 1223
Agrestini, S.. . . . . (c) 1832
Ahmed, Z. . . . . (c) 1732
Aichele, T. . . . . (a) 1398
Aït-Kaki, A.. . . . . (c) 1940
Akhavan, M. . (b) 1242, (c) 1828, (c) 1851, (c) 1855,
. . . (c) 1859, (c) 1863, (c) 1867, (c) 1871, (c) 1875,
. . . (c) 1879, (c) 1883, (c) 1887, (c) 1891, (c) 1895
Al Omari, I. . . . . (c) 1796
Al Rawas, A. . . . . (c) 1796
Al Shalabi, B. A. . . . . (c) 1840
Albiss, B. A. . . . . (c) 1899, (c) 1928
Albrecht, J. . . . . (c) 1836
Al-Busaidi, M.. . . . . (c) 1740
Alekseeva, A. M. . . . . (a) 1403
Alinejad, M. R. (c) 1719, (c) 1724, (c) 1728, (c) 1788
Al-Omari, I. A. (c) 1711, (c) 1805, (c) 1809, (c) 1821
Al-Rawas, A. . . . . (c) 1740
Amami, M. . . . . (c) 1985
Ambrosch-Draxl, C. . . . . (b) 1199
Amighian, J. . . . . (c) 1769
Amira, A.. . . . . (c) 1940, (c) 1944
Amirabadizadeh, A. . . (c) 1719, (c) 1724, (c) 1728,
. . . . (c) 1788
Ammar, A. . . . . (c) 1645
Anderson, H. U. . . . . (a) 1428
Annabi, M. . . . . (c) 1916, (c) 1920
Antipov, E. V.. . . . . (a) 1403
Arnal, T. . . . . (a) 1392
Auban Senzier, P. . . . . (c) 1924
Auer, H. . . . . (b) 1199
Azez, K. A. . . . . (c) 1821, (c) 1899

Barbara, B. . . . . (b) 1167, (c) 1595
Barone, Antonio . . . . (b) 1192
Bartenlian, B. . . . . (c) 1620
Bass, J. (a) 1379
Beauvillain, P.. . . . . (c) 1620
Bechgaard, K. . . . . (c) 1924
Belkhen, O. . . . . (c) 1940
Ben Ali, A. . . . . (c) 1813
Ben Azzouz, F. . . . . (c) 1916, (c) 1920
Ben Kraiem, M. S. . . . (c) 1691, (c) 1697, (c) 1701,
. . . . (c) 1706
Ben Salah, A. . . . . (c) 1985
Ben Salem, M.. . . . . (c) 1916, (c) 1920
BenAli, A. . . . . (b) 1211

Bennaceur, R.. (b) 1211, (b) 1216, (b) 1229, (b) 1236,
. . . . (c) 1813, (c) 1817
Berger, S. . . . . (c) 1935
Bernard, R.. . . . . (c) 1935
Bernas, H. . . . . (a) 1386
Bessais, L. . . . . (c) 1716
Bianconi, A. . . . . (c) 1832
Birge, Norman O. . . . . (a) 1379
Blackstead, H. A. . . . . (a) 1428
Blanco, J. A. . . . . (c) 1965
Bouguerra, A.. . . . . (c) 1683
Boujelben, W. . (a) 1410, (a) 1416, (a) 1421, (c) 1631
Bouregba, R. . . . . (c) 1687
Bourges, P. . . . . (b) 1204
Bouzehouane, K. . . . . (c) 1935
Bouziane, K. . . . . (c) 1740
Bramnik, K. G. . . . . (c) 1669
Briatico, J. . . . . (c) 1935
Brun, A. . . . . (c) 1620
Bruzek, C. E. . . . . (c) 1957

Cai, Q. (a) 1428
Campi, G. . . . . (c) 1832
Campo, J. . . . . (c) 1965
Chang, Chengkang . . . . (c) 1784
Chappert, C. . . . . (a) 1386
Charfi-Kaddour, S. . . . (b) 1211, (b) 1216, (b) 1229,
. . . . (b) 1236, (c) 1813, (c) 1817
Charguia, R. . . . . (c) 1813
Chaudouet, P.. . . . . (c) 1957
Cheikh-Rouhou Koubaa, W. . . . . (c) 1620
Cheikh-Rouhou, A. . . . (a) 1410, (a) 1416, (a) 1421,
. . . (c) 1625, (c) 1631, (c) 1645, (c) 1655, (c) 1660,
. . . (c) 1664, (c) 1669, (c) 1675, (c) 1691, (c) 1697,
. . . . (c) 1701, (c) 1706
Cheikh-Rouhou, W. . . . . (c) 1625
Chen, Shipu . . . . (c) 1776, (c) 1780
Chérif, S.-M. . . . . (c) 1587, (c) 1591
Chikazawa, S.. . . . . (c) 1752
Chniba, N. . . . . (c) 1675
Christensen, N. B. . . . . (b) 1223
Clausen, K. N. . . . . (b) 1223
Contour, J.-P. . . . . (c) 1935
Corraze, B. . . . . (c) 1944
Crété, D.-G. . . . . (c) 1935
Cristiani, G. . . . . (a) 1436, (c) 1836

D'Orazio, F. . . . . (c) 1748
Dakhlaoui, H.. . . . . (c) 1971
De Francisco, C. . . . . (c) 1610, (c) 1792

De Negri, S. . . . . (c) 1832
De Raedt, H. . . . . (b) 1180
Dhahri, J. . . . . (c) 1679
Djéga-Mariadassou, C. . . . . (c) 1716
Dmitriev, V. M. . . . . (c) 1824
Donet, S. . . . . (c) 1957
Dong, Chengyuan . . . . (c) 1776, (c) 1780
Dorolti, E. . . . . (c) 1716
Durand, O. . . . . (c) 1935

Ehrenberg, H. . . . . (c) 1669
El Feki, H. . . . . (c) 1985
El-Ali, A. . . . . (c) 1899, (c) 1928
El-Ghanem, H. M. . . . . (c) 1840
Elkestawy, M. A. . . . . (c) 1664
Ellouze, M. . . . (a) 1410, (c) 1655, (c) 1660, (c) 1691,
. . . . (c) 1697, (c) 1701, (c) 1706
Elzain, M. . . . . (c) 1796

Ferré, J. . . . . (a) 1386
Ferreira, A. R. . . . . (c) 1792
Filippi, M. . . . . (c) 1832
Fillion, G. . . . . (c) 1683, (c) 1948
Fröhlich, K. . . . . (c) 1649
Fuess, H. . . . . (a) 1410, (c) 1669

Gadois, C. . . . . (c) 1935
García, D. . . . . (c) 1610
Garg, Anupam . . . . (c) 1581
Garitaonandia, J. S. . . . . (c) 1965
Georges, P. . . . . (c) 1620
Ghazi, Mohammad E. . . . . (c) 1637, (c) 1641
Ghedira, M. . . . . (c) 1649
Giovannini, M. . . . . (c) 1832
Giraud, R. . . . . (b) 1167
Gismelseed, A. . . . . (c) 1740, (c) 1796
Görnert, P. . . . . (a) 1398
Gorria, P. . . . . (c) 1965
Grollier, J. . . . . (c) 1935
Guerioune, M. . . . . (c) 1948
Guo, Zhenghong . . . . (c) 1736

Habermeier, H.-U. . . . . . (a) 1436, (c) 1614, (c) 1836
Haddad, S. . . . . (b) 1216
Hadermann, J. . . . . (a) 1403
Haen, P. . . . . (c) 1948
Haghiri-Gosnet, A. M. . . . (a) 1392, (a) 1416, (c) 1687
Hamdeh, H. H. . . . . (c) 1809
Harat, A. . . . . (c) 1948
Hasan (Qaseer), M. K. . . . . (c) 1821
Hasanpour, A. . . . . (c) 1769
Hatton, P. D. . . . . (c) 1637, (c) 1641
Hejtmanek, J. . . . . (c) 1948
Hergt, R. . . . . (a) 1398
Héritier, M. . . . (b) 1211, (b) 1216, (b) 1229, (b) 1236,
. . . . (c) 1813, (c) 1817
Hernández-Gómez, P. . . . . (c) 1610, (c) 1792
Hichri, A. . . . . (c) 1976

Hong, Chin-Yih . . . . (c) 1604
Hori, Y. . . . . (c) 1756
Horng, H. E. . . . . (c) 1604
Hsu (Xu Zuyao), T. Y. . . (c) 1736, (c) 1772, (c) 1776,
. . . . (c) 1780
Hussey, N. E. . . . . (b) 1223

Ibarra-Palos, A. . . . . (c) 1625
Ihzaz, N. . . . . (c) 1679
Íñiguez, J. . . . . (c) 1610, (c) 1792
Ishida, S. . . . . (c) 1756

Jakeš, V. . . . . (c) 1952
Jakob, G. . . . . (c) 1961
James, W. J. . . . . (a) 1428
Jamet, J.-P. . . . . (a) 1386
Jaziri, S. . . . . (c) 1971, (c) 1976
Jemal, M. . . . . (c) 1985
Jérome, D. . . . . (c) 1924
Jiang, Ling . . . . (c) 1784
Jimenez, C. . . . . (c) 1957
Joo, N. . . . . (b) 1236, (c) 1924
Jumas, J. C. . . . . (c) 1846

Kallel, N. . . . . (c) 1649
Kameli, Parviz . . . . (c) 1895
Kamoun, I. . . . . (c) 1631
Kanekiyo, H. . . . . (c) 1719
Kariminezhad, M. . . . . (c) 1855
Keimer, B. . . . . (b) 1204
Khalifeh, Jamil M. . . . . (c) 1760, (c) 1765
Khasawinah, K. . . . . (c) 1928
Khène, S. . . . . (c) 1683
Khitouni, M. . . . . (c) 1981
Khosroabadi, H. . (c) 1855, (c) 1859, (c) 1863, (c) 1867,
. . . . (c) 1871, (c) 1875, (c) 1891
Kostylev, M. . . . . (c) 1587
Kostyuchenko, V. V. . . . . (c) 1595
Koubaa, M. . . . . . . (a) 1392, (a) 1416, (c) 1687
Kreisel, J. . . . . (c) 1679

L'Héritier, Ph. . . (c) 1691, (c) 1697, (c) 1701, (c) 1706
Labrag, A. . . . . (c) 1904, (c) 1908, (c) 1912
Lake, B. . . . . (b) 1223
Lataifeh, M. S. . . . . (c) 1711
Latini, A. . . . . (c) 1832
Lebedev, O. I. . . . . (a) 1403
Leblanc, A. . . . . (c) 1944
Leblanc-Soreau, A. . . . . (c) 1846
Lecoeur, Ph. . . . . (c) 1687
Lefmann, K. . . . . (b) 1223
Lehlooh, A.-F. . . . . (b) 1186
Lombardi, Filomena . . . . (b) 1192
Lorenz, A. . . . . (a) 1398
Lu, Ping . . . . (c) 1772
Lucari, F. . . . . (c) 1748

Author Index

M'Chirgui, A. . . . . (c) 1916, (c) 1920
Madar, R. . . . (a) 1410, (c) 1631, (c) 1655, (c) 1660
Mahmood, S. H. . . . . (b) 1186
Maleki, P. . . . . (c) 1871
Mankorntong, M. . . . . (b) 1223
Mao, Dali . . . . (c) 1784
Marcus, J. . . . . (c) 1679
Maroutian, T. . . . . (c) 1620
Martínez-Blanco, D. . . . . (c) 1965
Mathet, V. . . . . (a) 1386, (c) 1620
Matsuda, M. . . . . (c) 1756
Maurice, J.-L. . . . . (c) 1935
McMorrow, D. F. . . . . (b) 1223
Meddeb, D. . . . . (b) 1236, (c) 1817
Megdiche, S. . . . . . (c) 1655, (c) 1660, (c) 1664
Mercey, B. . . . . (c) 1687
Metallo, C. . . . . (c) 1832
Michielsen, K. . . . . (b) 1180
Mirzadeh, M. . . . . (c) 1875
Mischenko, A. S. . . . . (c) 1595
Miyashita, S. . . . . (b) 1180
Moch, P. . . . . (c) 1591
Modarreszadeh, M. . . . . (c) 1867
Mohammadizadeh, M. R. (c) 1828, (c) 1851, (c) 1879,
. . . . (c) 1883, (c) 1887
Mokhtari, Z. . . . . (c) 1891
Molinié, P. . . . . (c) 1846, (c) 1944
Monaco, Antonia . . . . (b) 1192
Montero, O. . . . . (c) 1610
Morresi, L. . . . . (c) 1748
Mosbah, M.-F. . . . . . (c) 1940, (c) 1944, (c) 1948
Moshfegh, A. Z. . . . . (c) 1744
Mossalla, B. . . . . (c) 1859
Mougin, A. . . . . (a) 1386
Mozaffari, M. . . . . (c) 1769
Murri, R. . . . . (c) 1748

Nandra, A. . . . . (c) 1716
Ng, H. P. . . . . (c) 1957
Nickel, C. . . . . (b) 1216
Njah, N. . . . . (c) 1981
Nohara, M. . . . . (b) 1223

Oumezzine, M. . . . . (c) 1649, (c) 1679

Pailhès, S. . . . . (b) 1204
Pasquier, C. . . . . (b) 1211, (c) 1924
Passacantando, M. . . . . (c) 1748
Peng, Wenyi . . . . (c) 1772
Perdigao, J. M. . . . . (c) 1792
Pernet, M. . . . . (c) 1625
Picozzi, P. . . . . (c) 1748
Pignard, S. . . . . (c) 1649, (c) 1679
Pinto, N. . . . . (c) 1748
Pommereau, J.-H. . . . . (c) 1935
Poullain, G. . . . . (c) 1687
Pourarian, F. . . (c) 1719, (c) 1724, (c) 1728, (c) 1788
Pratt, W. P. . . . . (a) 1379
Prellier, W. . . . . (c) 1687
Prentslau, N. N. . . . . (c) 1824
Pullini, D. . . . . (c) 1600

Rais, A. . . . . (c) 1711, (c) 1796, (c) 1821
Rammeh, N. . . . . (c) 1669
Ramzi, A. . . . . . . (c) 1904, (c) 1908, (c) 1912
Raposo, V. . . . . (c) 1610, (c) 1792
Regnault, L. P. . . . . (b) 1204
Renard, J. P. . . . . (a) 1392
Repain, V. . . . . (a) 1386
Ritter, C. . . . . (c) 1669
Rizal, C. L. S. . . . . (c) 1752, (c) 1756
Rong, Yonghua . . . . (c) 1736
Rønnow, H. M. . . . . (b) 1223
Roussel, H. . . . . (c) 1631
Roussigné, S.-Y. . . . . (c) 1591
Roussigné, Y. . . . . (c) 1587
Rozova, M. G. . . . . (a) 1403

Saafan, S. A. . . . . (c) 1664
Sab, S. . . . . (c) 1716
Saccone, A. . . . . (c) 1832
Salamati, Hadi . . . . (c) 1895
Saleh, A. S. . . . . (b) 1186
Salem, A. . . . . (c) 1961
Sangpour, P. . . . . (c) 1744
Sanipoli, C. . . . . (c) 1832
Sarnelli, Ettore . . . . (b) 1192
Sasagawa, T. . . . . (b) 1223
Saugrain, J. M. . . . . (c) 1957
Senoussi, S. . . . . . . (c) 1904, (c) 1908, (c) 1912
Sergeeva, N. . . . . (c) 1587
Sfar, I. (b) 1229
Shawagfeh, Naseem T. . . . . . (c) 1760, (c) 1765
Sherman, E. Ya. . . . . (b) 1199
Shimizu, K. . . . . (a) 1421
Sidis, Y. . . . . (b) 1204
Smeibidl, P. . . . . (b) 1223
Smith, R. I. . . . . (c) 1965
Smrčková, O. . . . . (c) 1952
Soltan, S. . . . . (c) 1836
Soulimane, R. . . . . (a) 1392, (c) 1687
Stachkevitch, A. . . . . (c) 1587
Stępień-Damm, J. . . . . (c) 1824
Strobel, P. . . . . (c) 1625
Suski, W. . . . . (c) 1824
Sýkorová, L. D. . . . . (c) 1952

Tafuri, Francesco . . . . (b) 1192
Taheri, P. . . . . (c) 1867
Tajabor, N. . . . (c) 1719, (c) 1724, (c) 1728, (c) 1788
Takagi, H. . . . . (b) 1223
Takakura, W. . . . . (c) 1752
Talik, E. . . . . (c) 1824

Taoufik, A. . . . . (c) 1904, (c) 1908, (c) 1912
Tatara, G. . . . . (b) 1174
Taubert, J. . . . . (a) 1398
Testa, Gianluca . . . . (b) 1192
Thonhauser, T. . . . . (b) 1199
Tkachuk, A. M. . . . . (b) 1167
Tomáš, I. . . . . (c) 1800
Torres, C. . . . . (c) 1792
Triki, M. . . . . (c) 1675

Ueda, Y. . . . . (c) 1752, (c) 1756
ul Haq, A. . . . . (c) 1732
Ulrich, C. . . . . (b) 1204
Urazhdin, S. . . . . (a) 1379

Van Tendeloo, G. . . . . (a) 1403
Vértesy, G. . . . . (c) 1800
Vincent, H. . . . . (c) 1649, (c) 1679
Vordewisch, P. . . . . (b) 1223

Wagner, F. E. . . . . (b) 1186
Walha, I. . . . . (a) 1416
Wang, Changzheng . . . . (c) 1736
Weiss, F. . . . . (c) 1957
Widatallah, H. . . . . (c) 1796

Yamada, A. . . . . (c) 1756
Yang, J. B. . . . . (a) 1428
Yang, S. Y. . . . . (c) 1604
Yelon, W. B. . . . . (a) 1428
Yousif, A. A. . . . . (c) 1711
Yousif, A. . . . . (c) 1796

Zhang, Jihua . . . . (c) 1772
Zheng, Jianchao . . . . (c) 1776
Zhou, X. D. . . . . (a) 1428
Zouaoui, M. . . . . (c) 1916, (c) 1920
Zouari, S. . . . . . . (c) 1625, (c) 1645, (c) 1675
Zvezdin, A. K. . . . . (c) 1595, (c) 1600
Zvezdin, K. A. . . . . (c) 1600

NEW JOURNAL

Editor-in-Chief:
Pavel P. Pashinin,
Moscow, RUS

Deputy Editors-in-Chief:
Valery M. Yermachenko,
Moscow, RUS

Igor V. Yevseyev,
Moscow, RUS

Free print or online subscription in 2004!*
Register at:
www.lphys.org

Editorial Board:
W. Becker, Berlin, GER
D. Chorvat, Bratislava, SK
M. V. Fedorov, Moscow, RUS
S. A. Gonchukov, Moscow, RUS
M. Jelinek, Prague, CZ
J. Lademann, Berlin, GER
J. T. Manassah, New York, USA
P. Meystre, Tuscon, USA
R. B. Miles, Princeton, USA
P. P. Pashinin, Moscow, RUS
G. Petite, Saclay, F
L. P. Pitaevskii, Trento, I
K. A. Prokhorov, Moscow, RUS
V. M. Shalaev, West Lafayette, USA
J. E. Sipe, Toronto, CAN
K.-I. Ueda, Tokyo, JP
I. A. Walmsley, Oxford, GB
H. P. Weber, Bern, CH
E. Wintner, Vienna, A
E. Yablonovitch, Los Angeles, USA
V. M. Yermachenko, Moscow, RUS
I. V. Yevseyev, Moscow, RUS
V. I. Yukalov, Dubna, RUS
A. M. Zheltikov, Moscow, RUS

2004, Volume 1, 12 Issues
ISSN print: 1612-2011 · ISSN online: 1612-202X

Laser Physics Letters is an international journal with rapid publication. It offers you a comprehensive overview of:

- Theoretical and experimental laser research and applications.
- Every aspect of modern laser physics and quantum electronics – emphasizing physical effects in various media (solid, gaseous, liquid).
- Peculiarities of propagation of laser radiation.
- Impact of laser radiation on various substances.
- Use of lasers and laser spectroscopy.

* Free subscription available to personal registrants only.

www.lasphys.com

Wiley-VCH · Customer Service Department · P.O. Box 101161 · D-69451 Weinheim · Germany
Tel.: (+49) 6201 606-400 · Fax: (+49) 6201 606-184 · E-Mail: service@wiley-vch.de · www.wiley-vch.de

NEW JOURNAL

Editors-in-Chief:

Riccardo d'Agostino, Bari, I

Pietro Favia, Bari, I

Christian Oehr, Stuttgart, GER

Michael R. Wertheimer, Montréal, CAN

Free print or online subscription in 2004!* Register at:
www.plasma-polymers.org

International Advisory Board:

F. Arefi-Khonsari, Paris, F

E. Fischer, Fort Collins, USA

F. Fracassi, Bari, I

K. K. Gleason, Cambridge, USA

A. Granier, Nantes, F

H. J. Griesser, Mawson Lakes, AUS

A. Haljaste, Tartu, EST

A. Holländer, Golm/Potsdam, GER

U. Kogelschatz, Hausen, CH

M. Kogoma, Tokyo, JPN

M. Kushner, Urbana, USA

Y. Segui, Toulouse, F

M. Strobel, St. Paul, USA

A. M. Wrobel, Lodz, PL

2004, Volume 1 · 2 Issues in 2004 · 9 Issues in 2005
ISSN print: 1612-8850 · ISSN online: 1612-8869

Aims and Scope:

Plasma Processes and Polymers publishes articles on **low-temperature plasma sources** and **processes** operating at pressures ranging from **partial vacuum to atmospheric.**

Processes include **plasma deposition, etching** and **surface modification** of materials as well as processing with photons, radicals and ions. Also of interest are studies on **materials characterization, plasma diagnostics** and **modeling,** and other related subjects.

Readership

Materials scientists, physicists, chemists, and engineers – both in academia and industry.

* Free subscription available to personal registrants only.

Wiley-VCH · Customer Service Department · P.O. Box 101161 · D-69451 Weinheim · Germany
Tel.: (+49) 6201 606-400 · Fax: (+49) 6201 606-184 · E-Mail: service@wiley-vch.de · www.wiley-vch.de

www.optics-encyclopedia.com

The Optics Encyclopedia
Basic Foundations and Practical Applications

Edited by Thomas G. Brown, Univ. of Rochester, USA; Katherine Creath, Creath Optineering Services, USA; Herwig Kogelnik, Lucent Technologies, USA; Michael E. Kriss, Sharp Laboratories of America, Inc., USA; Joanna Schmit, Veeco Instruments, Inc., USA; Marvin J. Weber, LBNL, USA

3527-40320-5   2004   3530 pp with 1798 figs, 39 in color   Hbk
€ 1099.- / £ 590.- / US$ 985.-

THE Reference Work in Optics!

5 Volume Set

With 94 expert articles in 5 volumes, this is both a comprehensive review as well as an introduction to the entire field. The contributions range from classical optics right up to the latest applications, including:

- IT and telecommunications
- Optical sensing and metrology
- Material processing
- Biomedicine
- Optical components and systems
- Laser design and technology.

The international editor team paid great attention to ensuring fast access to the information, and each carefully reviewed article features:

- an abstract
- a detailed table of contents
- continuous cross-referencing
- references to the most relevant publications in the field, and
- suggestions for further reading, both introductory as well as highly specialized.

In addition, a comprehensive index provides easy access to the enormous number of key words beyond the 94 headlines.

The result is a rapid reference for skilled professionals on all topics of modern photonics, while newcomers from physics and engineering will appreciate the readily comprehensible style and structure.

Read more in a free sample chapter at: www.pro-physik.de

Highlights

Laser Cooling and Trapping of Neutral Atoms, Harold Metcalf, Stony Brook, State Univ. NY, USA, and Peter van der Straten, Debye Inst., Utrecht Univ., The Netherlands

Electrodynamics, J. David Jackson, Univ. of California at Berkeley, California, USA

Optical Design, Pantazis Mouroulis, California Inst. of Technology, Pasadena, USA

Geometric Optics, Roland Shack, Univ. of Arizona, Tucson, USA

Photography, Digital, Michael E. Kriss, Sharp Lab. of America, Inc., Camas, WA, USA

Solid State Lasers, Tso Yee Fan, Lincoln Lab., Massachusetts Inst. of Technology, Lexington, USA

Optical Metrology, Peter de Groot, Zygo Corporation, Connecticut, USA

Optoelectronics, Safa O. Kasap, Univ. of Saskatchewan, Saskatoon, CAN

Physiological Optics, Martin Jüttner, Neuroscience Research Inst., Aston Univ., Birmingham, UK

X-ray Optics, Alan Michette, Dep. of Physics, King's College London, UK

John Wiley & Sons, Ltd. • Customer Services Department • 1 Oldlands Way • Bognor Regis • West Sussex • PO22 9SA England
Tel.: +44 (0) 1243-843-294
Fax: +44 (0) 1243-843-296
www.wileyeurope.com

Wiley-VCH • Customer Service Department
P.O. Box 101161• D-69451 Weinheim, Germany • Tel.: +49 (0) 6201 606-400
Fax: +49 (0) 6201 606-184
e-Mail: service@wiley-vch.de
www.wiley-vch.de

WILEY   **WILEY-VCH**

Wiley makes an Impact...

... dominating the Field of Materials Science!

ISI Impact Factor: 6.801

Please visit
www.advmat.de
for further details

Advanced Materials – THE top journal in the materials sciences, with the highest independently assessed Impact Factor of any professional, peer-reviewed primary materials science journal.

ISI Impact Factor: 4.656

Visit
www.afm-journal.de
for further details

Advanced Functional Materials has become the leading journal that publishes full papers on primary materials science research.

Access these journals from your desktop.
The journals are available through Wiley InterScience

Wiley-VCH Customer Service Department
e-Mail: service@wiley-vch.de
John Wiley & Sons, Inc., Subscription Dept
e-mail: subinfo@wiley.com

Information for conference organizers and guest editors

The third journal section *physica status solidi (c) – conferences and critical reviews* is devoted to the publication of proceedings, ranging from large international meetings to specialized workshops, as well as collections of topical reviews on various areas of current solid state physics research. The new series has been launched in December 2002 with volume **0** (2002/03). It is available both as a regular journal both online and in hardcover print volumes, to be delivered to subscribers, conference contributors and participants (upon arrangement with the organizers). Single copies of pss (c) may be ordered as a book using its ISBN number. Regular subscriptions to pss (c) are offered in combination with pss (a) and/or pss (b).

Essential details concerning layout and organization of the new journal series are:

- pss (c) is published as a full hardcover-bound series, carrying a standard green-coloured cover design, individually adapted according to the organizers' request which includes conference designation, logo, names of Guest Editors etc.

- Proceedings issues contain all conference contributions which have been peer-reviewed and accepted by the Guest Editors. Upon special agreement between the pss journal editors and the Guest Editors, part of the conference papers may also be published simultaneously in an issue of pss (a) or (b). For all papers, strict criteria for journal publications, i.e. positive peer-review by independent referees, are obligatory. All papers are unambiguously citable as phys. stat. sol. (a), (b), or (c) journal articles and will be covered by standard reference databases.

- All articles are published online in PDF format at Wiley InterScience. Access for registered users (e. g. conference participants with special password) may be installed. The online version contains colour figures at no additional cost, regardless of their colour or black/white representation in print.

- The Editorial Office provides document templates and style files for Word and LaTeX, respectively, to be used by all authors, allowing an easy manuscript preparation and length estimate of their paper with respect to the page limits given by the organizers.

- The issue is completed by a table of contents in topical order, an author index, a preface, listings of conference committee members, organizers and sponsors, and any additional material, if desired.

- The usual service of the Editorial Office is available and includes support in the refereeing process, acceptance messages, PDF proofs (for typesetted papers), free PDF reprints (hardcopy reprints may be ordered) as well as individual communication with authors and organizers. The use of a Web-based software system for online submission and refereeing of papers is offered to Guest Editors.

- The editors of pss (c) aim at a timely, professional, and high-quality print and online publication of proceedings, typically within only four to six months after a conference.

- Various service packages for production are available, including either full typesetting of papers using electronic manuscript data or publication-ready delivery of manuscript files (prepared using the template/style files) by the organizers.

For further details as well as an individual offer for the publication of the proceedings of your forthcoming conference or of a special issue containing topical reviews, please contact the Editorial Office at pss@wiley-vch.de (for other contact information see the title page).